Multiple Objective Decision Making for Land, Water, and Environmental Management

Proceedings of the First International Conference on Multiple Objective Decision Support Systems (MODSS) for Land, Water, and Environmental Management: Concepts, Approaches, and Applications

Edited by

S.A. El-Swaify
Department of Agronomy and Soil Science
College of Tropical Agriculture and Human Resources
University of Hawaii
Honolulu

and

D.S. Yakowitz
U.S. Department of Agriculture
Agricultural Research Service
Southwest Watershed Research Center
Tucson, Arizona

(WASWC)
World Association of Soil and Water Conservation

LEWIS PUBLISHERS
Boca Raton Boston London New York Washington, D.C.

Library of Congress Cataloging-in-Publication Data

International Conference on Multiple Objective Decision Support
    Systems for Land, Water and Environmental Management: Concepts,
    Approaches, and Applications (1st : 1996 : Honolulu, Hawaii)
        Multiple objective decision making for land, water, and
    environmental management : proceedings of the First International
    Conference on Multiple Objective Decision Support Systems (MODSS)
    for Land, Water and Environmental Management: Concepts, Approaches,
    and Applications / edited by S.A. El-Swaify and D.S. Yakowitz.
            p.   cm.
        "Honolulu, Hawaii, September, 1996"--P. xvi.
        Includes bibliographical references and index.
        ISBN 1-57444-091-8
            1. Natural resources--Management--Congresses.   2. Multiple
    criteria decision making--Congresses.   I. El-Swaify, S. A. (Samir
    Aly)   II. Yakowitz, D. S. (Diana S.)   III. Title.
    HC13.I544   1996
    333.7--dc21   97-48330
                                                                                   CIP

This book contains information obtained from authentic and highly regarded sources. Reprinted material is quoted with permission, and sources are indicated. A wide variety of references are listed. Reasonable efforts have been made to publish reliable data and information, but the author and the publisher cannot assume responsibility for the validity of all materials or for the consequences of their use.

Neither this book nor any part may be reproduced or transmitted in any form or by any means, electronic or mechanical, including photocopying, microfilming, and recording, or by any information storage or retrieval system, without prior permission in writing from the publisher.

All rights reserved. Authorization to photocopy items for internal or personal use, or the personal or internal use of specific clients, may be granted by CRC Press LLC, provided that $.50 per page photocopied is paid directly to Copyright Clearance Center, 27 Congress Street, Salem, MA 01970 USA. The fee code for users of the Transactional Reporting Service is ISBN 1-57444-091-8/98/$0.00+$.50. The fee is subject to change without notice. For organizations that have been granted a photocopy license by the CCC, a separate system of payment has been arranged.

The consent of CRC Press LLC does not extend to copying for general distribution, for promotion, for creating new works, or for resale. Specific permission must be obtained from CRC Press LLC for such copying.

Direct all inquiries to CRC Press LLC, 2000 Corporate Blvd., N.W., Boca Raton, Florida 33431.

© 1998 by CRC Press LLC.
Lewis Publishers is an imprint of CRC Press LLC

No claim to original U.S. Government works
International Standard Book Number 1-57444-091-8
Library of Congress Card Number 97-48330
Printed in the United States of America   1 2 3 4 5 6 7 8 9 0
Printed on acid-free paper

Foreword

Wise and sustainable use of natural resources is essential for meeting the needs of both present and future generations. The alarming magnitude of ongoing and persistent natural resource and environmental degradation and its detrimental impacts worldwide suggest that "fresh" alternative strategies are needed for addressing natural resources management and environmental problems. A critical step in any rational strategy is to gather, manage, and disseminate appropriate conservation planning information to decision makers in all society sectors, including land users, extension advisors, planners, administrators, policy makers, and researchers.

The present worldwide "revolution" in innovative and efficient technologies for information management and "knowledge engineering" allows the systematic and timely sharing of natural resource databases and other necessary information nationally, regionally, and internationally. Transforming available information into *quantitative* aids that facilitate decision making is a logical outcome of this revolution. Challenges in meeting this need include the complexity of individual land use and management issues, the potential conflicts among resource use objectives and expectations of various stakeholders, the incompleteness of certain data or nonuniversal applicability of natural resource models, and the difficulty in setting management goals because of diverse value judgments.

This volume is the product of the first international conference held to address these issues and share available state-of-the-art developments on multiple objective decision support systems for land, water, and environmental management. While it is not now clear when more future conferences will be held or whether the always desirable "networking" will be an outcome of the conference, our feedback indicates that the conference was a success. We believe that this book will be a valuable and timely publication for the field of natural and renewable resource management and that many scientists and professionals around the world will be more active in addressing some of the many still open questions and difficulties discussed herein and in using and contributing to the tools, models, and concepts presented here and at the conference. This would bode well for addressing the increasing nonsustainability concerns and for reconciling the potentially conflicting objectives of profitable land use with minimized detriment to the environment.

We thank all conference participants and book contributors for their valuable contributions to this effort.

The Editors
Honolulu, Hawaii
September, 1996.

The Editors

Samir A. El-Swaify is Vice President for the Pacific Region, World Association of Soil and Water Conservation (WASWC) and Professor of Soil and Water Conservation, Department of Agronomy and Soil Science, College of Tropical Agriculture and Human Resources, University of Hawaii, Honolulu. He is a graduate of the University of Alexandria (Egypt) and the University of California (Davis). He has been with the University of Hawaii since 1965, serving as Department Chairman for 9 years. He has served international aid and scientific organizations and published widely on soil erosion problems in the tropics, including the first book dedicated to the subject. He hosted and co-organized several international gatherings on soil conservation and natural resources management (including Malama Aina 83), chaired the Subcommission on Soil Conservation and the Environment of the International Society of Soil Science, served as a member of Study Group for World Soil Conservation (Commission on Ecology, IUCN), and, since its founding, has served as a member of the Board of Directors of the International Soil Conservation Organization (ISCO). He has published and edited several books in the arena of natural resource conservation and authored many technical and journal papers on these subjects.

Diana S. Yakowitz is a systems engineer and research hydrologist since 1991 with the U.S. Department of Agriculture's Agricultural Research Service. She is stationed at the Southwest Watershed Research Center in Tucson, Arizona. She is also an adjunct assistant professor at the University of Arizona, Department of Agricultural and Biosystems Engineering. Dr. Yakowitz's educational background includes a Bachelor of Science degree in physics, and a Masters and Ph.D. in systems and industrial engineering from the University of Arizona. She has been active in developing multiple objective decision tools and decision support systems to aid in land management decisions. She is an author of many papers on these subjects as well as several on the topic of stochastic programming.

Acknowledgments

The Editors are indebted to the following technical reviewers who served as members of our Editorial Board:

Khaled Abu-Zeid	Safwat Abdel Dayem	Robert M. Caldwell
Elizabeth Dunn	Carl Evensen	Richard E. Green
Philip Heilman	Mariano Hernandez	Keith W. Hipel
Dana Hoag	Arthur Hornsby	Mohammed A. Kishk
Paul Lawrence	PingSun Leung	John Li
John Masterson II	Ted L. Napier	Ginger Paige
Tony Prato	James A. Silva	Jeffry Stone
D. P. Stonehouse	Russell Yost	

The MODSS Conference Chair (S. A. El-Swaify) was fortunate to receive the support of Cochairs D.S. Yakowitz and L. J. Lane and the following members of the Organizing Committee:

C. R. Amerman	Robert Caldwell	Carl Evensen
Jeff Fox	T.E. Hakonson	C. Hernandez
Patrice Janiga	Paul Lawrence	PingSun Leung
Victor Phillips	Mark Ridgley	Roger Shaw
Gordon Tsuji	R. S. Yost	

The World Association of Soil and Water Conservation (WASWC, Pacific Region) served as the primary Society sponsor of the Conference. The Soil and Water Conservation Society (International Chapter), International Society of Soil Science (Subcommission on Conservation and the Environment), and the International Society on Multiple Criteria Decision Making were cosponsors.

Institutional cosponsors were the University of Hawaii, Southwest Watershed Research Center (ARS/USDA), East-West Center (EWC, Program on Environment), Oak Ridge National Laboratory (ORNL, Center for Global Environmental Studies), USDA Forest Service (USFS), Queensland Department of Primary Industries (QDPI, Australia), Instituto Nacional de Investigaciones Forestales y Agropecuarias (INIFAP, Mexico), and the University of Arizona. The financial support received from EWC, ARS, ORNL, and USFS towards meeting certain conference and publication expenses is gratefully acknowledged.

Contributors

Abdel-Dayem, Mohamed Safwat, Director, Drainage Research Institute, National Water Research Center, El-Kanater, Cairo, Egypt

Abdel-Gawad, Shaden T., Deputy Director, Drainage Research Institute, National Water Research Center, El-Kanater, Cairo, Egypt

Abu-Zeid, Khaled M., Senior Researcher, National Water Research Center, P.O. Box 6, El Qanater, Qalubia, 13621 Egypt

Allen, W.J., Manaaki Whenua — Landcare Research, PO Box 282, Alexandra, New Zealand

Alocilja, E., Dept. of Ag. Engineering, Michigan State University, East Lansing, MI USA

Amien, Istiqlal, Center for Soil and Agroclimate Research, Jalan Ir. H. Juanda 98, Bogor, Indonesia

Arias-Rojo, Hector M., Centro de Investigaciones y Desarrollo de los Recursos Naturales de Sonora, Hermosillo, Sonora, Mexico

Ascough II, James C., USDA-ARS-NPA, Great Plains Systems Research Unit, 301 S. Howes St., P.O. Box E, Fort Collins, CO 80522 USA

Ashwood, Tom L., Environmental Sciences Division, Oak Ridge National Laboratory, Oak Ridge, Tennessee, USA

Averill, James, Department of Psychology, University of Massachusetts, Amherst, MA 01003 USA

Babu, Ram, Central Soil and Water Conservation Research and Training Institute, 218, Kaulagarh Road, Dehradun-248 195 India

Bentham, Murray J., Agriculture & Agri-Food Canada, Research Branch, PARI DSS, 5C 26 Agriculture Building, University of Saskatchewan, 51 Campus Drive, Saskatoon, SK, Canada 57N 5A8

Bergstrom, John C., Department of Agricultural Economics, 208 Connor Hall, University of Georgia, Athens, GA 30602 USA

Bessembinder, J.J.E., Department of Theoretical Production Ecology, Wageningen Agricultural University, P.O. Box 430, 6700 AK Wageningen, The Netherlands

Bosch, O.J.H., Manaaki Whenua — Landcare Research, PO Box 282, Alexandra, New Zealand

Brebber, Lindsay, Queensland Department of Primary Industries, Resources Management Institute, Meiers Road, Indooroopilly, Queensland 4068, Australia

Camboni, Silvana M., Ohio Agricultural Research and Development Center, 2120 Fyffe Road, Columbus, OH 43210 USA

Carnes, S.A., Energy Division, Oak Ridge National Laboratory, P.O. Box 2008, Oak Ridge, TN 37831-6206 USA.

Chen, Shi, Institute of Mountain Hazard & Environment, The Chinese Academy of Sciences, Chengdu, 610041, P.R. China

Cikánová, S., Research Institute for Soil and Water Conservation, Žabovřeská 250, 156 27 Praha 5, Czech Republic

Coffey, S.W., North Carolina State University, Department of Biological and Agricultural Engineering, NCSU Water Quality Group, Box 7637, Raleigh, NC 27695-7637 USA

Cogle, Lex, Queensland Department of Primary Industries, P.O. Box 4880, Mareeba, Queensland 4088, Australia

Connolly, R. D., Agricultural Production Systems Research Unit, PO Box 102, Toowoomba, Queensland 4350, Australia

da Silva, L. Mira, Department of Agriculture, University of Reading, Earley Gate, Reading, RG6 2AT, UK

Deer-Ascough, Lois Ann, Department of Soil and Crop Science, Colorado State University, Fort Collins, CO 80523 USA

Dennis, Donald F., USDA Forest Service, Northeastern Forest Experiment Station, P.O. Box 968, Burlington, VT 05402 USA

Doherty, John, Queensland Department of Primary Industries, Resources Management Institute, Meiers Road, Indooroopilly, Queensland 4068, Australia

Dhyani, B.L., Central Soil and Water Conservation Research and Training Institute, 218, Kaulagarh Road, Dehradun-248 195 India

Doležal, F. Research Institute for Soil and Water Conservation, Žabovřeská 250, 156 27 Praha 5, Czech Republic

Dumanski, J., Eastern Cereal and Oilseed Research Centre, Agriculture and Agri-Food Canada, Ottawa, ON, Canada K1A 0C6

Dunn, Elizabeth G., Department of Agricultural Economics, University of Missouri-Columbia, 200 Mumford Hall, Columbia, MO 65211 USA

Ehman, J.L., Geographic Information System Laboratory, Indiana University — School of Public and Environmental Affairs Bloomington, IN 47405 USA

El-Swaify, S.A., Department of Agronomy and Soil Science, University of Hawaii, Honolulu, HI 96822 USA

Fan, W., Geographic Information System Laboratory, Indiana University — School of Public and Environmental Affairs Bloomington, IN 47405 USA

Faulkner, David, Virginia Natural Resources Conservation Service, Blacksburg, VA 24061 USA

Fang, Liping, Department of Mechanical Engineering, Ryerson Polytechnic University, 350 Victoria Street, Toronto, ON, Canada M5B 2K3

Flašarová, V., Agricultural Research Institute, Ltd., Havlíčkova 2787, 767 01 Kroměříž, Czech Republic

Foster, M.A., The Pennsylvania State University, Department of Entomology and Center for AI Applications in Water Quality, 505 ASI Building, University Park, PA 16802 USA

Freebairn, D.M., Agricultural Production Systems Research Unit, PO Box 102, Toowoomba, Queensland 4350, Australia

Fresco, L.O., Department of Agronomy, Wageningen Agricultural University, P.O. Box 431, 6700 AK Wageningen, The Netherlands

Fulcher, Chris, Center for Agricultural, Resource and Environmental Systems, University of Missouri-Columbia, 200 Mumford Hall, Columbia, MO 65211 USA

Gale, J.A., North Carolina State University, Department of Biological and Agricultural Engineering, NCSU Water Quality Group, Box 7637, Raleigh, NC 27695-7637 USA

Gameda, S., Eastern Cereal and Oilseed Research Centre, Agriculture and Agri-Food Canada, Ottawa, ON, Canada K1A 0C6

Gannon, R.W., North Carolina Department of Environment, Health, and Natural Resources, P.O. Box 29535, Raleigh, NC 27626 USA

Gibson, R.G., Manaaki Whenua — Landcare Research, PO Box 282, Alexandra, New Zealand

Glass, Ronald J., USDA Forest Service, Northeastern Forest Experiment Station, So. Burlington, VT 05403 USA

Goodrich, Philip R., Department of Biosystems and Agricultural Engineering, University of Minnesota, 1390 Eckles Ave., St. Paul, MN 55108 USA

Graham, R., Environmental Science Division, Oak Ridge National Laboratory, P.O. Box 2008, Bldg. 1000, Oak Ridge, TN 37831-6335 USA

Greer, J.E., ARIES Laboratory, Department of Computer Science, University of Saskatchewan, Saskatoon, SK, Canada

Hakonson, T.E., Center for Ecological Risk Assessment and Management, Colorado State University, Ft. Collins, CO 80521 USA

Hannam, I., NSW Department of Land and Water Conservation, 23-33 Bridge Street, Sydney 2000, Australia

Hanson, Jon D., USDA-ARS-NPA, Great Plains Systems Research Unit, 301 S. Howes St., P.O. Box E, Fort Collins, CO 80522 USA

Heilman, Philip, USDA-Agricultural Research Service, Southwest Watershed Research Center, 2000 E. Allen Rd., Tucson, AZ 85719 USA

Hernandez, Mariano, USDA-ARS-Southwest Watershed Research Center, 2000 E. Allen Rd., Tucson, AZ 85719 USA

Hipel, Keith W., Departments of Systems Design Engineering and Statistics and Actuarial Science, University of Waterloo, Waterloo, ON, Canada N2L 3G1

Hoag, D.L., Department of Agricultural and Resource Economics, Colorado State University, Fort Collins, CO 80523-5210 USA

Holtfrerich, David R., Texas A&M, STARR Lab, Department of Rangeland Ecology and Management, College Station, TX USA

Hope, Bruce K., Oregon Department of Environmental Quality, Waste Management and Cleanup Division, P.O. Box 8099, Portland, OR 97207-8099 USA

Hornsby, A.G., Soil and Water Science Department, University of Florida, Gainesville, FL 32611-0290 USA

Huszar, Paul C., Department of Agricultural and Resource Economics, Colorado State University, Ft. Collins, CO U.S.A.

Imam, Bisher, Department of Hydrology, University of Arizona, Tucson, AZ 85721 USA

Jeck, S.C., Pacific Agri-Food Research Centre, Agriculture and Agri-Food Canada

Johnston, J.J., Geographic Information System Laboratory, Indiana University — School of Public and Environmental Affairs Bloomington, IN 47405 USA

Jones, Alice J., Agronomy Department, University of Nebraska, Lincoln, NE 68583-0910 USA

Jopp, A.J., Farmer, Moutere Station, RD 3 Alexandra, New Zealand

Keller, James M., Department of Computer Engineering and Computer Science, University of Missouri-Columbia, 217 Engineering Bldg. West, Columbia, MO 65211 USA

Kilgour, D. Marc, Department of Mathematics, Wilfrid Laurier University, Waterloo, ON, Canada N2L 3C5

Kishk, Mohammed Atif, Soil Science Department, Faculty of Agriculture, Minia University, Minia, Egypt

Klímová, P., Research Institute for Soil and Water Conservation, Žabovřeská 250, 156 27 Praha 5, Czech Republic

Knott, C.B., Luucent Technologies, 2000 Regency Park, Suite 500, Cary, NC 27511 USA

Köbrich, Claus, Department of Livestock Production, Universidad de Chile, Casilla 2 Correo 15, La Granja, Chile

Kovács, G.J. Research Institute for Soil Science and Agricultural Chemistry, 1022 Budapest Herman Otto u. 15. Hungary

Křen, J., Agricultural Research Institute, Ltd., Havlíčkova 2787, 767 01 Kroměříž, Czech Republic

Kubát, J., Research Institute of Crop Production, Drnovská 507, 161 06 Praha 6, Czech Republic

Lait, Rob, Queensland Department of Primary Industries, Center for Wet Tropics Agriculture, P.O. Box 20, South Johnstone, Queensland 4859, Australia

Lane, L.J., USDA-ARS-Agricultural Research Service, Southwest Watershed Research Center, 2000 E. Allen Rd., Tucson, AZ 85719 USA

Lawrence, Paul A., Department of Natural Resources, PO Box 6014, Rockhampton, Queensland 4702, Australia

Lehning, D.W., The Pennsylvania State University, Agricultural and Biological Engineering Department, 226 Ag Engineering Building, University Park, PA 16802 USA

Li, John Z.C., Department of Agronomy and Soil Science, University of Hawaii at Manoa Honolulu, HI 96822 USA

Li, Mengbo, Department of Agronomy and Soil Science, University of Hawaii at Manoa. Honolulu, HI 96822 USA

Li, Tongyang, Institute of Mountain Hazard & Environment, The Chinese Academy of Sciences, Chengdu, 610041, P.R. China

Line, D.E., North Carolina State University, Department of Biological and Agricultural Engineering, NCSU Water Quality Group, Box 7637, Raleigh, NC 27695-7637 USA

Lipavský, J., Research Institute of Crop Production, Drnovská 507, 161 06 Praha 6, Czech Republic

Littlefield, B.Z., Geographic Information System Laboratory, Indiana University — School of Public and Environmental Affairs Bloomington, IN 47405 USA

Loftis, David L., USDA Forest Service, Bent Creek Exp. Forest, 1577 Brevard Road, Asheville, NC 28806 USA

Madden, J.C., CSIRO Division of Water Resources, Griffith, NSW 2680, Australia

Marks, Leonie A., Department of Agricultural Economics, University of Missouri-Columbia, 200 Mumford Hall, Columbia, MO 65211 USA

Matthew, P.L. Department of Plant Production, University of Queensland, Gatton College 4343, Australia

Meyer, W.S., CSIRO Division of Water Resources, Griffith, NSW 2680, Australia

Mielke, Lloyd N., Agronomy Department, University of Nebraska, Lincoln, NE 68583-0910 USA

Moon, D.E., Pacific Agri-Food Research Centre, Agriculture and Agri-Food Canada

More, Thomas A., USDA Forest Service, Northeastern Forest Experiment Station, So. Burlington, VT 05403 USA

Napier, Ted L., Ohio Agricultural Research and Development Center, 2120 Fyffe Road, Columbus, OH 43210 USA

Németh, T. Research Institute for Soil Science and Agricultural Chemistry, 1022 Budapest Herman Otto u. 15. Hungary

Nissapa, Ayut, Department of Agricultural Development, Prince of Songkhla University, Hat Yai, 90112 Thailand

Nofziger, D.L., Department of Agronomy, Oklahoma State University, Stillwater, OK 74078, USA

Novák, P., Research Institute for Soil and Water Conservation, Žabovřeská 250, 156 27 Praha 5, Czech Republic

Offerle, B., Geographic Information System Laboratory, Indiana University — School of Public and Environmental Affairs, Bloomington, IN 47405 USA

Oropeza-Mota, Jose L., Instituto de Recursos Naturales, Departamento de Fisica de Suelos, Colegio de Postgraduados, Montecillo, Edo.de Mex., Mexico

Osmond, D.L., North Carolina State University, Department of Biological and Agricultural Engineering, NCSU Water Quality Group, Box 7637, Raleigh, NC 27695-7637 USA

Paige, Ginger B., USDA-Agricultural Research Service, Southwest Watershed Research Center, 2000 E. Allen Rd., Tucson, AZ 85719 USA

Park, J., Department of Agriculture, University of Reading, Earley Gate, Reading, RG6 2AT, UK

Parton, Kevin A., School of Health, University of New England, Armidale, NSW 2351, Australia

Pease, James W., Department of Agricultural and Applied Economics, Virginia Polytechnical Institute, Blacksburg, VA 24061 USA

Peasley, B.A., Department of Conservation and Land Management, Inverell Research Service Centre, Inverell, NSW 2360, Australia

Peng, Xiaoyong (John), Department of Systems Design Engineering, University of Waterloo, Waterloo, ON, Canada N2L 3G1

Phillips, K.A., Tetra Tech, 10306 Eaton Place, Suite 340, Fairfax, VA 22030 USA

Pinto, P.A., Departamento de Agricultura, Instituto Superior de Agronomia, 1399 Lisboa Codex, Portugal

Pokorný, E., Agricultural Research Institute, Ltd., Havlíčkova 2787, 767 01 Kroměříž, Czech Republic

Prathapar, S.A., CSIRO Division of Water Resources, Griffith, NSW 2680, Australia

Prato, Tony, Center for Agricultural, Resource and Environmental Systems, University of Missouri-Columbia, 200 Mumford Hall, Columbia, MO 65211 USA

Pýcha, M., Research Institute for Soil and Water Conservation, Žabovřeská 250, 156 27 Praha 5, Czech Republic

Rabbinge, R., Department of Theoretical Production Ecology, Wageningen Agricultural University, P.O. Box 430, 6700 AK Wageningen, The Netherlands

Rafea, Ahmed, Central Laboratory for Agricultural Expert System, P.O. Box 100 Dokki, Giza, Egypt

Randolph, J.C., Geographic Information System Laboratory, Indiana University — School of Public and Environmental Affairs Bloomington, IN 47405 USA

Reed, R.M., Energy Division, Oak Ridge National Laboratory, P.O. Box 2008, Oak Ridge, TN 37831-6206 USA

Rehman, Tahir, Department of Agriculture, The University of Reading, PO Box 236, Earley Gate, Reading RG6 2AT, UK

Reiling, Stephen D., Department of Resource Economics and Policy, 5782 Winslow Hall, University of Maine, Orono, ME 04469-5782 USA

Ritchie, J.T., Department of Crop and Soil, Michigan State University, East-Lansing, MI 48824-1325 USA

Roach, John, Department of Computer Sciences, Virginia Polytechnical Institute, Blacksburg, VA 24061 USA

Robillard, P.D., The Pennsylvania State University, Agricultural and Biological Engineering Department, 226 Ag Engineering Building, University Park, PA 16802 USA

Robotham, M.P., Department of Agronomy and Soil Science, University of Hawaii at Manoa, 1910 East-West Rd., Honolulu, HI 96822 USA

Ross, John, NSW Department of Land and Water Conservation, 23-33 Bridge Street, Sydney 2000, Australia

Sample, Bradley E., Environmental Sciences Division, Oak Ridge National Laboratory, Oak Ridge, TN USA

Scheckler, Rebecca K., Department of Entomology, Virginia Polytechnical Institute, Blacksburg, VA 24061 USA

Selby, C.J., Pacific Agri-Food Research Centre, Agriculture and Agri-Food Canada

Selley, Roger, South Central Research and Extension Center, Box 66, University of Nebraska, Clay Center, NE 68933 USA

Shaffer, Marvin J., USDA-ARS-NPA, Great Plains Systems Research Unit, 301 S. Howes St., P.O. Box E, Fort Collins, CO 80522 USA

Shaw, Roger, Queensland Department of Primary Industries, Resources Management Institute, Meiers Road, Indooroopilly, Queensland 4068, Australia

Šimon, J., Research Institute of Crop Production, Drnovská 507, 161 06 Praha 6, Czech Republic

Spooner, J., North Carolina State University, Department of Biological and Agricultural Engineering, NCSU Water Quality Group, Box 7637, Raleigh, NC 27695-7637 USA

Stone, Jeffry J., USDA-Agricultural Research Service, Southwest Watershed Research Center, 2000 E. Allen Rd., Tucson, AZ 85719 USA

Stone, Nicholas D., Department of Entomology, Virginia Polytechnical Institute, Blacksburg, VA 24061 USA

Stonehouse, D. Peter, Department of Agricultural Economics and Business, University of Guelph, Guelph, ON, Canada N1G 2W1

Stout, Susan L., USDA Forest Service, Forestry Sciences Lab, P.O. Box 928, Warren, PA 16365 USA

Suter, II, Glenn W., Environmental Sciences Division, Oak Ridge National Laboratory, Oak Ridge, TN USA

Turner, M.H., The Waste Policy Institute, 470 L'Enfant Plaza East, Suite 7105, Washington, D.C. 20025 USA

Twery, Mark J., USDA Forest Service, 705 Spear Street, Burlington, VT 05403 USA

van Ittersum, M.K., Department of Theoretical Production Ecology, Wageningen Agricultural University, P.O. Box 430, 6700 AK Wageningen, The Netherlands

Van Huylenbroeck, G., Department of Agricultural Economics, University of Gent, Coupure Links 653 B-9000 Gent, Belgium

Vickner, Steven S., Department of Agricultural and Resource Economics, Colorado State University, Ft. Collins, CO 80523 USA

Washington-Allen, Robert A., Environmental Sciences Division, Oak Ridge National Laboratory, Oak Ridge, TN USA

Whitehouse, Ian, Land Management, Landcare Research, Lincoln, New Zealand

Williams, Stephen B., USDA Forest Service, 3825 E. Mulberry St., Fort Collins, Colorado 80524 USA

Yakowitz, Diana S., USDA-Agricultural Research Service, Southwest Watershed Research Center, 2000 E. Allen Rd., Tucson, AZ 85719 USA

Yost, Russell, Department of Agronomy and Soil Science, University of Hawaii at Manoa, Honolulu, HI 96822 USA

Zhang, Xianwan, Institute of Mountain Hazard & Environment, The Chinese Academy of Sciences, Chengdu, 610041, P.R. China

Contents

SECTION I:
INTRODUCTION, RATIONALE, AND RELEVANCE OF MODSS

1 **Introduction to Multiple Objective Decision Making for Land, Water, and Environmental Management** 3
   S.A. El-Swaify and Diana S. Yakowitz

2 **Land Use Planning for Sustainable Agriculture: Are Environmental Conflicts Inevitable?** 9
   S.A. El-Swaify

3 **Using the Decision Support System GMCR for Resolving Conflict in Resource Management** 23
   Keith W. Hipel, D. Marc Kilgour, Liping Fang, and Xiaoyong (John) Peng

SECTION II:
ADDRESSING DIVERSE CLIENT NEEDS

4 **Co-Learning Our Way to Sustainability: An Integrated and Community-Based Research Approach to Support Natural Resource Management Decision Making** 51
   W.J. Allen, O.J.H. Bosch, R.G. Gibson, and A.J. Jopp

5 **Consequences of Short-term, One-Objective Decision Making in Land and Water Use Planning in Egypt** 61
   Mohammed A. Kishk

6 **Pesticide Economic and Environmental Tradeoffs: User's Perspective** 75
   A.G. Hornsby, D.L. Hoag, and D.L. Nofziger

7   **Pesticide Economic and Environmental Tradeoffs: Developer's Perspective** 83
    D.L. Nofziger, Arthur G. Hornsby and D.L. Hoag

8   **Multi-Objective Decision Support Strategies for Cropping Steep Land** 93
    Alice J. Jones, Roger Selley, and Lloyd N. Mielke

9   **A Multicriteria Decision Support System for Evaluating Cropping Pattern Strategies in Egypt** 105
    Khaled M. Abu-Zeid

10 **An Integrated Approach to Ecological Risk Assessment of a Large Forested Ecosystem** 121
    Tom L. Ashwood, Bradley E. Sample, Robert A. Washington-Allen, and Glenn W. Suter, II

11 **Managing Environmental Cleanup Risks as a Multiobjective Decision Problem** 133
    T.E. Hakonson and L.J. Lane

12 **A Multiple Criteria Decision-Making Model for Comparative Analysis of Remedial Action Alternatives** 143
    Bruce K. Hope

13 **Overview of a Decision Support System for the Evaluation of Landfill Cover Designs** 153
    Ginger B. Paige, Jeffry J. Stone, Leonard J. Lane, and Diana S. Yakowitz

14 **AGFADOPT: A Decision Support System for Agroforestry Project Planning and Implementation** 167
    M.P. Robotham

15 **Developing an Integrated Nutrient Management Decision Aid** 177
    Russell Yost and John Z.C. Li

SECTION III:
MODSS METHODOLOGIES, TOOLS, COMPONENTS, AND INTEGRATION

16 **Smart Pitchfork** 189
    Philip R. Goodrich

17 **Using Desired Future Conditions to Integrate Multiple Resource Prescriptions: The Northeast Decision Model** 197
    Mark J. Twery, Susan L. Stout, and David L. Loftis

18 **A Multiattribute Tool for Decision Support: Ranking a Finite Number of Alternatives** 205
    D.S. Yakowitz

19 Effects of Optional Averaging Schemes on the Ranking of Alternatives by the Multiple Objective Component of a U.S. Department of Agriculture Decision Support System 217
Bisher Imam, Diana S. Yakowitz, and Leonard J. Lane

20 Integrating Agricultural Expert Systems with Databases and Multimedia 233
Ahmed Rafea

21 WATERSHEDSS©: A Decision Support System for Nonpoint Source Pollution Control in Predominantly Agricultural Watersheds 241
D.L. Osmond, D.E. Line, J.A. Gale, R.W. Gannon, C.B. Knott, K.A. Phillips, M.H. Turner, S.W. Coffey, J. Spooner, P.D. Robillard, M.A. Foster, and D.W. Lehning

22 A Knowledge-Based Reasoning Toolkit for Forest Resource Management 251
Stephen B. Williams and David R. Holtfrerich

23 Subjective Evaluation of Decision Support Systems Using Multiattribute Decision Making (MADM) 269
James C. Ascough, II, Lois Ann Deer-Ascough, Marvin J. Shaffer, and Jon D. Hanson

24 CERES Models in Multiple Objective Decision-Making Process 281
G.J. Kovács, J.T. Ritchie, and T. Németh

25 Advances in Ration Formulation for Beef Cattle Through Multiple Objective Decision Support Systems 291
Steven S. Vickner and Dana L. Hoag

26 Conflict Resolution Among Heterogeneous Reusable Agents in the PARI DSS 299
Murray J. Bentham and Jim E. Greer

27 Integrated Decision Making for Sustainability: A Fuzzy MADM Model for Agriculture 313
Elizabeth G. Dunn, James M. Keller, and Leonie A. Marks

28 Elements of a Decision Support System: Information, Model, and User Management 323
D.E. Moon, S.C. Jeck, and C.J. Selby

SECTION IV:
APPLICABILITY TO ECONOMIC, SOCIAL, POLICY, RISK, AND SUSTAINABILITY ISSUES

29 Adoption of Environmental Protection Practices in the Scioto River Watershed: Implications for MODSS 337
Ted L. Napier and Silvana M. Camboni

30 SOILCROP — A Prototype Decision Support System for Soil
Degradation-Crop Productivity Relationships 349
S. Gameda and J. Dumanski

31 Evaluation of a Prototype Decision Support System for
Rangelands in the Southwest United States 363
Paul A. Lawrence, Leonard J. Lane, Jeffry J. Stone,
and Diana S. Yakowitz

32 Managing Soil Structure, Particularly with Respect to
Infiltration, for Long-Term Cropping System Productivity 379
R.D. Connolly and D.M. Freebairn

33 Protecting Soil and Water Resources Through Multiobjective
Decision Making 385
Tony Prato and Chris Fulcher

34 Socioeconomics of Upland Soil Conservation in Indonesia 395
Paul C. Huszar

35 Issues and Concerns in Multiple Objective Management
of Natural Resources in the Agri-Food Industry 407
D. Peter Stonehouse

36 A Multicriteria Approach for Trade-off Analysis Between
Economic and Environmental Objectives in Rural Planning 419
G. Van Huylenbroeck

37 Long-Term Explorations of Sustainable Land Use 437
J.J.E. Bessembinder, M.K. van Ittersum, R. Rabbinge, and L.O. Fresco

38 A Multicriteria Framework to Identify Land Uses Which
Maximize Farm Profitability and Minimize Net Recharge 447
S.A. Prathapar, W.S. Meyer, E. Alocilja, and J.C. Madden

39 The Use of Crop Models for Land Valuation 455
F. Doležal, J. Lipavský, J. Křen, P. Novák, J. Šimon, J. Kubát, M. Pýcha,
S. Cikánová, P. Klímová, V. Flašarová and E. Pokorný

40 An Agroecological Approach to Sustainable Agriculture 465
Istiqlal Amien

41 Does Existence Value Exist? 481
John C. Bergstrom and Stephen D. Reiling

42 Analyzing Public Inputs to Multiple Objective Decisions
Using Conjoint Analysis 493
Donald F. Dennis

43 Noneconomic Values in Multiple Objective Decision Making 503
Thomas A. More, James Averill, and Ronald J. Glass

44 Quantifying Economic Incentives Needed for Control
of Nonpoint Source Pollution in Agriculture 513
Philip Heilman, Leonard J. Lane, and Diana S. Yakowitz

45 CROPS: A Constraint-Satisfaction System for Whole-Farm Planning 527
Nicholas D. Stone, David Faulkner, Rebecca K. Scheckler, James W. Pease, and John Roach

SECTION V:
APPLICABILITY TO NATIONAL, REGIONAL, AND GLOBAL ISSUES

46 Lessons Learned and New Challenges for Integrated Assessment Under the National Environmental Policy Act 541
S.A. Carnes and R.M. Reed

47 Sustainable Agriculture and the MCDM Paradigm: The Development of Compromise Programming Models with Special Reference to Small-Scale Farmers in Chile's VIth Region ... 557
Claus Köbrich and Tahir Rehman

48 Use of a DSS for Evaluating Land Management System Effects on Tepetate Lands in Central Mexico 571
Mariano Hernandez, Philip Heilman, Leonard J. Lane, Jose L. Oropeza-Mota, and Hector M. Arias-Rojo

49 Climate Change Effects on Agricultural Productivity in the Midwestern Great Lakes Region 585
B.Z. Littlefield, J.L. Ehman, W. Fan, J.J. Johnston, B. Offerle, and J.C. Randolph

50 Decision Support Systems for Irrigated Zones: An Integrated Approach to Land Use Planning and Management in Southern Europe 599
L. Mira da Silva, J. Park, and P.A. Pinto

51 Management of Natural and Renewable Resources on Watershed Basis in India 615
Ram Babu and B.L. Dhyani

52 Water Management Scenario Simulation for Decision Support in Multiobjective Planning 629
Safwat Abdel-Dayem, Shaden Abdel-Gawad, and Khaled Abu-Zeid

53 Use of Goal Programming in Aquaculture Policy Decision Making in Southern Thailand 641
Kevin A. Parton and Ayut Nissapa

54 A Case Study in the Use of an Expert System as a Multiobjective Decision Support System (MODSS) — Boobera Lagoon Environmental Management Plan 655
P.L. Matthew and B.A. Peasley

55 Multiple Objective Decision Support Systems Used in Management of Temperate Forest Ecosystems in Southeast Australia 667
John Ross and Ian Hannam

56 Seasonal No-Tillage Ridge Cropping System: A Multiple Objective Tillage System for Hilly Land Management in South China 685
Xianwan Zhang, Shi Chen, Tongyang Li, and Mengbo Li

57 The Use of Multiobjective Decision Making for Resolution of Resource Use and Environmental Management Conflicts at a Catchment Scale 697
Roger Shaw, John Doherty, Lindsay Brebber, Lex Cogle, and Rob Lait

SECTION VI:
SUMMARY AND CONCLUSIONS

58 A Synthesis of the State-of-the-Art on Multiple Objective Decision Making for Managing Land, Water, and the Environment 719
A.C. Jones, S.A. El-Swaify, R. Graham, D.P. Stonehouse, and I. Whitehouse

Index 733

INTRODUCTION, RATIONALE, AND RELEVANCE OF MODSS

Chapter 1

Introduction to Multiple Objective Decision Making for Land, Water, and Environmental Management

S.A. EL-SWAIFY AND DIANA S. YAKOWITZ

This book is the postconference Proceedings of MALAMA 'AINA (Preserve the Land) 1995, the "First International Conference on Multiple Objective Decision Support Systems (MODSS) for Land, Water, and Environmental Management: Concepts, Approaches, and Applications." The conference was convened in Honolulu, Hawaii, in July 1995 in recognition of the fact that considering multiple objectives and the development of computer-based decision aids are requisite to rational environmental decision making.

The selection of this rather specific conference topic is in direct contrast to MALAMA 'AINA 83, the first (truly) open global conference on soil erosion and conservation also held in Honolulu in January 1983, and indeed most recent international forums on managing natural resources. MALAMA 'AINA 95 was, in a way, a sequel but dealt with a new, specific, timely, and cutting-edge topic which reflects the emerging innovations for addressing natural resource management and environmental problems. This provided a higher dimension of multidisciplinary participation than at any of the previous natural resource conservation conferences.

Why Multiple Objective Decision Making and Decision Support Systems?

Over the past half century or so, agricultural research and technological innovations that depend on intensive external inputs have succeeded in assuring impressive increases in global agricultural productivity. The increases averaged about 3%/year and have kept up with the needs of increased populations on every continent except Africa (El-Swaify, 1994). However, this success story is nearing its limits for many reasons, primarily: (a) the decreasing availability of new productive lands and usable water resources (Buringh, 1982; El-Swaify, 1991), (b) the alarming rates of ongoing human-induced land, soil, and water resource degradation (ISRIC/UNEP, 1990), (c) the increasing encroachment on and over-stressing marginal, fragile, and degradation-vulnerable lands (see Chapter 3), (d) the inability to enhance the adoption of conservation-effective land use systems at rates sufficient to overtake ongoing degradation (Napier et al., 1994), and last but not least (e) the continuing expansion of earth populations at alarming rates.

Natural resource degradation, therefore, represents an important element of total "global change". In the broad sense, such a change encompasses not only changes in climate or atmospheric/stratospheric composition, but also other changes in the global life-support systems of human societies. Much of the international debate and research funding, however, focuses on global climate change, although other dynamic changes may be equally important, if not more so. Natural resource degradation is also a major force that drives global climate change. It is no surprise, therefore, that "Natural Resource Management" (NRM) has emerged globally as a key issue of the highest priority for assuring sustainable agricultural development, and has attracted renewed commitment from such mission-oriented institutions as the International Agricultural Research Institutes (IARIs). The conference's central topic and selected themes were propelled, to a large extent, by the recent paradigm shifts to sustainable development and environmental quality issues.

In view of the emerging paradigm shifts, managing natural resources, mitigating degradation threats, and conforming to proliferating environmental regulations is a complex task that must address the various expectations of many community segments and society sectors. Those expectations, when translated into specific planning objectives, are often in conflict with one another. A prominent example is the likelihood of conflict between "conventional, production-driven" and "environmentally-driven" objectives of natural resource use and management. One approach to reconciling these conflicts lies in "holistic, ecosystem-based" land use planning and development.

This approach requires the selection of an "ecosystem unit" that is most appropriate for planning (e.g., a well-defined hydrologic catchment or watershed), and applying quantitative criteria to plan land use and judge the many impacts of alternative management practices, preferably prior to project implementation. Land use consequences and impacts are often not in harmony and tradeoffs or "optimization" in decision making may be necessary. A specific example of conflict is the urgency in enhancing crop productivity, through intensified management and applied external inputs, while also controlling nonpoint source pollution and protecting both surface and groundwater qualities.

Societies and communities are, therefore, undergoing increased interest in the subject of conflict resolution. Conflict resolution always involves multiple stakeholders and the multiple decision criteria of the interested parties (Hipel et al., 1993). This reality and the complexity of most decision-making situations require that many researchers and decision makers apply multiple-objective tools of decision making rather than those that are merely driven by single objectives.

The considerations discussed above, the complexity of many of the models and large databases required for predicting various scenarios, and the need to examine many contingency plans or points of view with the goal of arriving at optimal choices, underpin the need for MODSS. We envisage an increased adoption of this and other operations research (OR) techniques by developers for aiding decision making in environmental management toward the same goals.

Why a Conference?

Much has been accomplished in the MODSS arena in recent years. Current scientific literature is enriched with considerable knowledge, experience, and a variety of approaches that lend themselves to developing quantitative multiobjective decision support tools to allow making recommendations on "simultaneously" productive, sustainable, and environmentally sound land use. This first international conference was so timed as to take advantage of current progress and to share these accomplishments among the global scientific community. As will be noted in the final chapter of this book, our expectations were by and large realized. A global scope was justified because the impacts of many natural resource decisions indeed extend beyond what previously may have been considered to be of only national or local concern.

Ideal decision tools for valid recommendations on land, water, and environmental management must include quantitative and analytical components; must span and integrate the physical, biological, socioeconomic, and policy elements of decision making; and must be user-friendly and directly relevant to client needs. Fortunately, participants included scientists representing both the "hard" and "soft" scientific disciplines and interdisciplines relevant to land resource utilization and management and who are actively contributing the knowledge required for the development of decision aids. Over 30 countries were represented at the conference by speakers and authors from all continents with the exception of Antarctica. Real world applications from all these regions were presented at the meeting.

One goal of the conference was to provide a forum for the demonstration of DSS currently in use or under development for natural resource planning. Many of the papers presented were exploratory in nature and involve research plans and/or the investigation of appropriate decision factors and techniques to be considered. The topics covered by the authors of such papers underscore the need for continued activity and should spur interaction between researchers and the development of many alternative systems which provide clientele with a set of choices to meet their respective needs. Presented case studies demonstrate how the application of MODSS has addressed, or can address, critical needs in land use planning and management in various countries or regions. As shown in the coming chapters, and in the summation chapter, applications included a spectrum ranging from compliance with environmental regulations to identifying opportunities for contributing to the planning of sustainable land use in developing, especially tropical, countries.

An Overview of Emerging Emphases

Despite the wide diversity among conference contributions, it was possible to logically fit individual presentations and sessions, and indeed the contributions to this book, into one of the following four major "themes":

1. Relevance for addressing user needs, various society sectors, and different levels of clientele
2. MODSS methodologies, tools, components, and integration
3. Economic, social, cultural, and policy issues and applications, including risk assessment, conflict resolution, and sustainable development
4. Applicability to global and regional issues, especially in Asia and the Pacific, including degradation processes in small island ecosystems and major river basins, and competitive uses of water for agriculture and aquaculture

These themes are presented following the introductory Part I, as Parts II, III, IV, and V of this book, respectively. Many of the papers in these sections could have logically been included in other sections as well. Therefore, it should not be assumed that papers in the last section, for example, do not cover topics in the other three.

Several topics were given special emphasis with multiple follow-up workgroup sessions. These included two panels on the role of MODSS in integrated assessment, relevance for and applications of MODSS to sustainability issues, explanations and demonstrations of available MODSS software, and incorporating socioeconomic considerations in multiple objective decision making. A mid-week field excursion was designed to demonstrate the relevance of conference topics to the dynamics of rapidly changing natural resource use in the "highly compressed" whole-island ecosystems of the state of Hawaii, and to also highlight the wisdom of certain indigenous historic concepts of land allocation and use as symbolized by the "holistic watershed-plus" of Ahupua`a addressed briefly in the next chapter. The final day was dedicated to presentations of session synthesis by assigned theme spokespersons and open discussions among all participants of the week's contributions. A synthesis of those is presented in the closing chapter of the book.

As planned, environmental issues received considerable emphasis at the conference and are addressed by several authors in this volume. These have been addressed in the context of:

- Water management, with a focus on water supply and quality
- Environmentally compatible agricultural land use and management, including crop selection, irrigation management, tillage systems, pesticide and nutrient management, erosion control, combating soil degradation, assessing market demand, and enhancing farm profitability

- Land use planning, conservation planning, and reclamation (restoration), with emphasis on economic, social, and environmental implications
- Waste management and remediation of contaminated sites, with concern for human, wildlife, water, and vegetation health
- Natural areas and steep land management, with concern for deforestation, wildlife, erosion control, recreation, and other economic issues
- Aquaculture, with concern for food supply, competition, and water quality
- Livestock management, with concern for economics, food supply, and fertilizer

Further details are presented in the respective chapters and a follow-up synthesis is provided in Part VI.

The conference and material included in this volume illustrate the diversity of perspectives and points of view held by those involved in environmental management in our global community. We believe that awareness of and sensitivity to such diverse issues are healthy steps in reconciling conflicts that may arise in striving for technically sound, economically viable, socially acceptable, and environmentally compatible natural resource management systems.

References

Buringh, P. 1982. Potentials of world soils for agricultural production. Transactions of the 12th International Congress of Soil Science. Indian Society of Soil Science, New Delhi.

El-Swaify, S.A. 1991. Land-based limitations and threats to world food production. *Outlook Agric.* 20(4):235–242. C.A.B. International.

El Swaify, S.A. 1994. State of the art for assessing soil and water conservation needs and technologies: a global perspective. Pages 13–27 in T. Napier, S. Camboni, and S.A. El-Swaify, Eds. *Adopting Conservation on the Farm — An International Perspective on the Socioeconomics of Soil and Water Conservation*. Soil and Water Conservation Society, Ankeny, Iowa.

Hipel, K.W., K.J. Radford, and L. Fang. 1993. Multiple participant-multiple criteria decision making. *IEEE Trans. Syst. Man Cybern.* 23(4):1184–1189.

International Soil Reference and Information Center (ISRIC) and United Nations Environment Program (UNEP). 1990. Global assessment of human-made land degradation. UNEP, Nairobi.

Napier, T., S. Camboni, and S.A. El-Swaify, Eds. 1994. *Adopting Conservation on the Farm — An International Perspective on the Socioeconomics of Soil and Water Conservation*. Soil and Water Conservation Society, Ankeny, Iowa.

Chapter 2

Land Use Planning for Sustainable Agriculture: Are Environmental Conflicts Inevitable?

S.A. EL-SWAIFY

Abstract

The productivity of agroecosystems and their vulnerability to environmental degradation are both determined by the very same fundamental factors, processes, and constraints. These may be climatic, topographic, soil-based, or a consequence of selected land use and applied management. This commonality strongly argues for compatibility, rather than conflict, between production-driven and environmental objectives of agricultural land use. However, the increasingly intensive use of "productive lands" and ever-increasing encroachment upon "marginal lands" to meet the needs of exploding human populations have led to increasing conflicts between these sets of objectives. Even so, it is our hypothesis that the objectives are less likely to clash if certain conditions are met in the planning and management of land and natural resources for agriculture. These conditions include assuring comprehensive *ex ante* accounting of the likely changes in the ecosystem as a consequence of development; conducting the planning, implementation, and evaluation of land use systems at meaningful spatial and temporal scales; maximizing the match between site attributes, the planned land use, and applied management; emphasizing the increased use efficiency of externally added, production-boosting inputs; defining meaningful objective indicators and realistic thresholds for judging changes in the agroecosystem's productive capacity and impact on environmental quality; monitoring these indicators for regular *ex post* evaluations of project performance upon implementation; and making judicious use of available data, models, geographic information systems, and decision aids for selecting sustainable land uses and designing environmentally sound management. Multicriteria analyses allow definition of the most critical bottlenecks to sustainable land use and can lead to designing quantitative decision aids for optimizing management. Such decision aids need not be complex. Rather, systems of varying complexity are necessary for meeting the needs of various clientele and the many diverse levels of decision making.

Introduction

There is general agreement that agricultural activities and enterprises have had a commendable historic track record for meeting global human needs. Statistics show that agricultural production continues to match or exceed the pace of human and animal demand for food on all continents except Africa. The criterion of sustained food production, however, is no longer acceptable by itself for assuring the welfare of future generations. History shows that many early civilizations collapsed as a result of excessive exploitation of natural resources and degradation of lands and soils. The emergence of the "sustainable agriculture paradigm" in recent years has been prompted by the failing or deficient performance of many "modern" agricultural enterprises because of physical, biological, ecological, economic, social, and/or political factors. Technological successes have allowed farmers to practice "improved" and productive land use, but were often short lived, particularly in areas with marginal land quality. Desirable changes in production but without deliberate "land husbandry" can be easily offset by long-term instability and degradation of lands and declines in environmental quality. Degradation concerns apply to both small and large enterprises, as well as to developing and developed countries. Much of the degradation and many, perhaps most, of the associated failures are predictable, and are triggered by large mismatches between the attributes and impopsed use and management of land. "Correcting" such mismatches to meet expected (often optimistic) production goals or to restore the land's productive capacity is often unsuccessful and expensive. A very timely question facing decision makers at every level is: "Can agriculture continue to cope with the expanding demand for food, fodder, fuel, and fiber by the earth's ever-increasing human populations (mostly in the less-developed countries)?" The answer lies in the rigor of land use planning and implementation actions. Steps include recognizing, quantifying, predicting, prioritizing, and arresting or reversing the causes and symptoms of agricultural "unsustainability" at an early stage. Identifying sustainability determinants, early degradation indicators, and judgment criteria represents a monumental challenge which must be met to assure the welfare of future generations. In this paper, we will address obstacles to achieving sustainability goals and suggest management strategies to facilitate achieving them through holistic planning and enhanced harmony between desirable land use consequences.

Natural Resources-Based Constraints to Sustainable Agriculture

Intrinsic Constraints

The quality of land for productive use is determined collectively by a host of intrinsic attributes. These attributes, whether physical, chemical, or biological, can favor, limit, or completely inhibit agricultural activity; they include climatic, pedologic, and topographic characteristics. For instance, climatic constraints determine not only the adequacy of a site for productive and economic use but also its vulnerability to degradation (e.g., rainfall and wind erosivity). Slope characteristics may favor or restrict land use; as they determine accessibility, trafficability, stability against surface erosion and mass movement, potential runoff and flooding, and exposure to climatic influences, particularly wind and solar radiation. Despite such obvious and predictable degradation threats, human encroachment on marginal lands is increasing due to the scarcity of productive, high-quality resources which were placed first in agriculture. Statistics show that less than 25% of the total global land area is considered potentially productive, with only 9% possessing high or medium productivity. About half of the potentially productive area is already cultivated; a grand total of 1,700 million ha remain undeveloped in the form of grasslands or forests. These comprise the theoretical sum of land globally "available" for future expansion in agriculture (Buringh, 1982). Much of this land is marginal and there are mounting social, economic, political, and ecological reasons why it may not be available for development. Thus, preserving the quality and high productive capacity of lands now in use is a must so that degradation will not further exacerbate the land scarcity problem.

Natural Resource Degradation

Recognizing, predicting, and avoiding degradation are critical for implementing productive soil and land use with a long-term sustainability perspective. It is important to note, however, that while human intervention is often an exacerbating force, it is not always responsible for degradation. Natural forces and processes can degrade land, soil, and water. Both natural and human-induced land degradation impose productivity constraints (El-Swaify, 1991; 1994). The most common processes of degradation are:

- Accelerated runoff and erosion
- Breakdown of soil structure
- Reduced water infiltration, storage, and availability to crops
- Increased waterlogging; reduced aeration
- Increased compaction and impedance to root development
- Increased surface soil sealing and crusting
- Excessive accumulation of chemicals (salinity, sodicity, alkalinity, acidity, toxicity by pollutants)
- Nutrient and organic matter depletion
- Desertification arising from changes in the hydrologic regime towards aridity
- Urban encroachment

The most recent, and somewhat quantitative, information on the extent of human-induced degradation is contained in a generalized report titled "Global Assessment of Land Degradation" or "GLASOD" (ISRIC/UNEP, 1991). It categorized the various degradation classes by severity and provided a geographic breakdown by continent. The area of land affected by degradation was estimated as 1,200 million ha for Asia, 400 for Africa, 245 for South America, 215 for Europe, 150 for Central and North America, and 112 for Australia. Among the various causative processes, soil erosion (by water and wind) stands out as the most extensive form of degradation facing mankind today, accounting for over 60% of the total degraded land. Soil nutrient depletion and salinization are next in importance; both reflect insufficient safeguards against chemical degradation under high intensity, particularly irrigated, farming. While "modern" forms of degradation (especially soil pollution) are considerably less extensive globally than those discussed above, they are far more important in industrialized than in primarily agrarian regions.

Certain other factors and human activities (such as deforestation or irrigation) may bring about degradation and have, on occasion, been included among the important degradation actors. Additional considerations concern other forms of "depletion" of natural, renewable, or nonrenewable resources. These include the declining availability and quality of water resources, increasing scarcity of affordable sources of energy, insufficient soil recycling of organic material and animal waste, and dwindling biological diversity.

The shortage of water available for irrigation has reduced the global rate of expansion of irrigated land area to about 2.3 millian ha/year, nearly 50% of the peak rate achieved during the 1970s; scarcity in water quantity is further compounded by increasing degradation in the quality of water supplied (Postel, 1989). Historically, irrigation has been responsible for abundant and secure crop production under the "Green Revolution." The decline in irrigation development requires that strong emphasis be placed on sustainable land use in rainfed areas (El-Swaify, 1988).

Excessive exploitation and alteration of natural ecosystems, including sensitive tropical rainforests, is seriously impacting global biological diversity. The sincere and noble efforts on genetic "improvement" of economic crops which was perhaps the essence of the "Green Revolution" generally gave rise to increasing dependence on productive, stress-tolerant, but "genetically narrow" varieties for meeting population needs. Lack of diversity in agroecosystems often results in aggravated pest problems, reduced resilience and tolerance to environmental stresses, and increased dependency on high external and management inputs.

For some of these discussed forms of degradation, the causes, potential for occurrence in different environments, and the impacts on soil productivity and environmental quality are rather well understood and largely predictable. These include soil erosion and salinization. For other forms, such as structural breakdown, surface sealing and crusting, compaction, and restricted aeration, quantitative information on extent and impacts, and management options to reverse the problems may be scarce of lacking altogether. For still other forms, such as excessive exhaustion of soils, water supplies, and energy sources, the basic underlying causes may be neither physical nor biological, but rather economic, social, cultural, or policy-driven.

It may be concluded that strategies for avoiding the potential conflict between productive agricultural land use and detrimental environmental impacts should be based, to a great extent, on understanding and combating natural resource degradation. Actions based on this understanding will likely assure the maintenance of ecosystem "health" and maximize harmony between land users and their surrounding interlinked ecosystems. It is the author's contention that this task is easier to accomplish if points of commonality and departure between various management goals are well understood.

Potential for Harmony Between Production and Environmental Goals

From the "technical" (or "biological-physical") perspectives, there appears to be no good justification for perceived or actually encountered conflicts between productive and environmentally sustainable goals of land use. This somewhat intuitive vision is conveyed by some classical definitions of agriculture as: "humans utilizing and managing their land in harmony with their natural environment." History shows that such vision of "sustainable agriculture" was realized by many an ancient civilization; a thorough analysis of the situation in Egypt before the onset of perennial irrigation was provided by Hughes (1992). That society had an excellent track record for meeting human needs as well as setting high ecological standards. Historians agree that the healthy attitudes and practices of Egyptians were rooted in a world view that affirmed the sacred values of all nature, and of land in particular. To be sure, there were difficult periods in that civilization when pharaonic governments collapsed and these have been correlated mainly with anomalies in Nile floods, whether as excesses or shortages. However, the traditional patterns of culture reasserted themselves after such periods with remarkable tenacity. The reliable natural regime, faith derived from it, and the perceptive efforts of the people, provided the environmental and human ingredients necessary for a sustainable society. So did the "sciences" of geometry, astronomy, and records (hieroglyphics), and the "technology" of irrigation, all of which were marshalled to assure the dependability of relationships to the environment. In spite of some environmental problems induced by periods of overpopulation which led to deforestation, shrinkage of wetland habitats, and salinization, Egypt remained productive, "full of life", and self-sufficient in most respects at the end of the ancient period. The experience of that society, with maintained sustainability over so many centuries, offers many lessons to the modern world; most have yet to be learned.

The argument that harmony is a realistic expectation can be lent scientific credibility because the productivity of agroecosystems and their vulnerability to environmental degradation are both controlled by the very same fundamental factors and processes, i.e., climatic, soil-based, topographic, and applied management. Examples abound where each of these land attributes can contribute to productivity or degradation. These common features are illustrated in Figures 2.1 and 2.2.

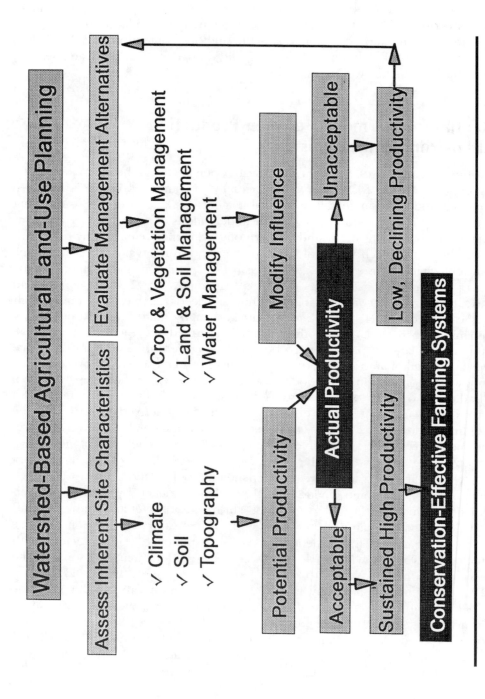

Figure 2.1 Factors determining the productivity and sustainability of agricultural land use and management practices.

Land Use Planning for Sustainable Agriculture: Are Environmental Conflicts Inevitable? 15

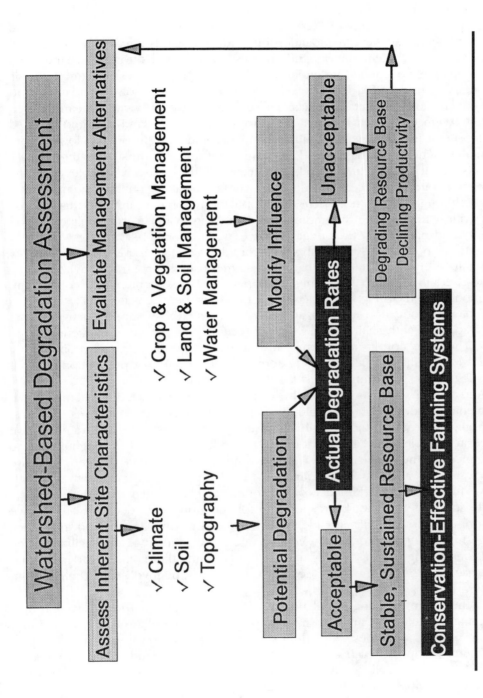

Figure 2.2 Factors determining the vulnerability of natural resources to degradation.

Points of Departure Between Production and Environmental Goals

The ever-increasing number of earth occupants, mostly in developing and less-developed countries, has exerted great pressures on the natural resource base and forced the excessive exploitation of productive lands already in use as well as increasing encroachment on, and unrealistic expectations from, new lands with low or marginal quality. More often than not, demographic pressures are the basic or original cause of food shortages, accelerated land and soil degradation, and subsequent environmental impacts. These events move in vicious, nearly closed cycles. When the match between site attributes, planned land use, and selected management technologies is minimal or nil, degradation becomes a logical consequence. When degradation and environmental impacts are severe, they pose a serious threat to human survival, particularly if they are irreversible. Soil, the primary natural resource, is realistically "nonrenewable" on the "human" time scale. Although we will not discuss the population problem further in this chapter, it is critical to note the abounding and mounting evidence relating population density and natural resource degradation (e.g., Dazhong, 1993).

National policy and leadership vision contribute directly to a society's sensitivity and will to ensure sustainable natural resource use and confront environmental issues. Certain authors (e.g., Soule and Piper, 1992) also contend that a large gap in emphasis and conflicts in expectations may exist between "farming" (the actual farm-based production activity) and "agribusiness" (those support activities that are associated with selling farm inputs, processing of products, marketing them, etc.). The latter's profit motives may lead to exerting undue pressures on farming enterprises to intensify the use of external inputs and follow production-maximization strategies; with an increased risk to environmental quality.

Agricultural scientists do not always play the role expected of them in environmental advocacy. They should avoid isolation from the root causes of unsustainability over which they may have no direct control as they have an obligation to be actively involved in providing factual information and in heightening the awareness of the "powers to be" as to sound natural resource planning and use policies. They also have an obligation to sell the "holistic perspective" which allows full accounting of impacts of environmental change in managed ecosystems. While scientists may not be responsible for rendering final decisions, they have the primary responsibility for identifying and quantifying degradation indicators, suggesting acceptable or nonacceptable rates of degradation, and facilitating the decision-making process. Policy makers must assume the ultimate leadership and land users, who take the major economic risks, must be involved in the decision-making process.

In the face of meeting population needs, developing countries are increasingly adopting agricultural management technologies which earlier characterized those of "developed" countries. Capital-intensive dependence on farm machinery and agrichemical inputs which can cause "modern" forms of land and environmental degradation (e.g., soil pollution and structural deterioration) no longer separate so-called developed and developing nations.

An essential element of truly sustainable land use is to ensure the environmental integrity of "whole ecosystems." Emphasis on the productivity or environmental quality of a single farm alone, or even of an arbitrary collection of farms, is insufficient. The definition of a "whole ecosystem" requires careful delineation of its boundaries, formulation of meaningful indicators of its "health," and ensuring that the delineated spatial scale is meaningful for quantifying these indicators and addressing the needs of all ecosystem occupants. The appropriate scale may be a catchment, watershed, hydrologic unit area, whole river basin, or some other unit which allows a holistic expression of interdependencies and continuities between landscape elements. Whatever the scale, an integrated quantitative accounting of development impacts is necessary (e.g., Pimentel et al., 1994).

A similar argument may be made for temporal scale. Production-driven, profit-minded goals prevail over environmental concerns particularly in short-term planning and in developing, heavily populated and/or rapidly industrializing countries. The urgency of meeting society's present material needs often overwhelms leaders and planners to such an extent that the long-term degradation in "ecosystem health," which threatens the welfare of future generations, are considered a luxury. Plans, measurements, or observations which do not monitor changes in all meaningful physical, biological, and socioeconomic attributes for a sufficiently long time may provide false or incomplete impressions of sustainability.

Contrasting Examples and Case Studies

Examples from the fields of soil erosion and conservation and nonpoint source pollution will be used to illustrate the influence of selected scale on achieving harmony between competing objectives for conservation-effective land use and management.

Small Tropical Volcanic Island Ecosystems

Small tropical islands of volcanic origin, such as the Hawaii Archipelago, are characterized by steep topography, aggressive rainfall, and a small proximity between sediment sources on land and ultimate sediment destinations in the surrounding ocean. The potential and actual water erosion are high and both the temporal spans and spatial scales between on-site production and off-site impacts of sediment are highly compressed, resulting in high sediment yields. The high frequency of rapid sediment movement to shoreline areas following significant rainstorms poses strong challenges to contaminant control and coastal zone management efforts (El-Swaify, 1992). In addition, opportunities for conflict between communities which pursue different means of livelihood (e.g., farming vs. fishing), are many and frequent. Holistic conservation planning must incorporate the full continuum of landscape segments including contaminant sources, pathways, and destination areas. Considerable community and agency efforts have been dedicated to addressing these problems within the Hawaii Interagency Water Quality Action Program. These efforts benefited from what historians have documented about the "Ahupua'a" concept, which was embraced by early Hawaiians as a framework for policy on land use and was applied as a tool for land allocation and management. Land access and right of use for deriving food needs were granted by ruling monarchs as continuous land strips extending from the uplands into the sea. It therefore represents an expansion of "watershed" use as a management unit by integrating landscapes, emphasizing good "husbandry" of all land segments, and sustaining benefits to all resource users and occupants.

Large River Basins

While awareness of upland-lowland interactions is so readily attained for small landscapes, it can escape attention in areas such as complex large catchments and river basins, etc. In addition, such large hydrologic areas, while intimately linked and unified by "bonds" of flowing water, are nonetheless often divided by political boundaries. An instructive case study is that of the Nile River, where sediments borne by seasonal floods were historically considered favorable, some believe essential, for sustaining Egypt's early agriculture (Hughes, 1992). Seasonal floods from the Ethiopian and Eritrean highlands and plateaus contribute approximately 86% of the total annual water supplies to the lower Nile and the vast majority of erosional sediments (Said, 1991). The Atbara and Blue Nile rivers are the primary conveyers of these seasonal waters. Watershed catchments of the two rivers belong to four countries, namely Eritrea, Ethiopia, Sudan, and Egypt.

Sediments no longer have the same value to agricultural productivity in the Nile's lower basin and, instead, both upland erosion and downstream sedimentation have severe impacts on co-basin states (El-Swaify and Hurni, 1996). Upstream, a 1 to 2% annual reduction in soil productivity was estimated for the Ethiopian highlands; actually observed declines reached up to 10%/year in the western parts of Ethiopia. Downstream, sediment deposition threatens the usefulness and longevity of several water storage structures, dams, and reservoirs designed to stabilize water resources and availability, and to generate hydroelectric power for use by the respective communities. For example, Egypt's population of over 60 million inhabitants is almost entirely dependent on the capacity and function of the Aswan High Dam. Sediment accumulated in the lake south of the High Dam during the period 1987 to 1992 is estimated at 594 mm^3, an average of nearly 120 mm^3/year (El-Dardir cited by El-Swaify and Hurni, 1996). The potential effect of siltation on the course of the Nile river also causes serious concern, should sediment build-up continue as at present.

It can be shown that all sister countries comprising the Nile basin are detrimentally impacted by erosion, sedimentation, or associated phenomena and thus can benefit from integrated measures of soil and water conservation and restored land quality and productivity in sediment source areas. Existing information shows high availability of well-suited soil and water conservation technologies. Implementing appropriate measures for control of erosion in the upper catchments will arrest land degradation and reduce the need for further exploitation of remaining natural areas, enhance watershed performance, base flows, and reliability of water supplies into the eastern tributaries of the Nile, reduce the frequency of catastrophic flood events, maintain hydroelectric power capacity, enrich habitats for fish and wildlife, reduce the sediment loads of the region's rivers, enhance the economic uses of water bodies, reduce siltation of waterways and storage dams, and preserve the present, natural course of the Nile. Technical actions and necessary steps can be readily prescribed. However, it is recognized that implementing technical actions must follow political agreements endorsing the need for basin-wide cooperation, and the blessing of communities and leaders of co-basin states, together with especially targeted funds to allow such implementation.

Conclusions and Recommendations

The preceding discussion and examples indicate that planners, policy makers, and scientists may be able to achieve maximum harmony between production-driven and environmentally sensitive objectives of land use by:

1. Keeping a holistic perspective of sustainability during the planning process. Where possible, comprehensive, *ex ante* accounting of land use and management impacts on the ecosystem as a consequence of development should be performed.
2. Identifying physical, biological, economic, and sociocultural-political determinants of "success" and quantifying specific measurable criteria for exercising "holistic" judgment. Research and education targeted to quantifying sustainability should be enhanced; inference is inadequate for rigorous decision making.
3. Conducting the planning, implementation, and evaluation of land use systems at meaningful temporal and spatial scales. The first should provide a sufficient time horizon for planning, implementation, and evaluation and allow detecting "steady state" consequences of development. The latter should be landscape-based, even if the required continuity crosses political boundaries.
4. Maximizing the match between site attributes, planned land use, and selected management technologies. Substantial mismatches often necessitate massive landscape modifications and the use of large quantities of external inputs to achieve unrealistic (optimistic) production targets, generally with negative environmental consequences.
5. Emphasizing incorporation of concepts and lessons learned from natural systems, including maximization of biological diversity and recycling, and minimization of soil disturbance.
6. Employing protective rather than corrective strategies against natural resource degradation. Even if the degradation is reversible, the economics and length of time required for reclaiming or rehabilitating degraded lands are such that "prevention is better than cure."
7. Planning of production ventures for maximum use-efficiency of added inputs rather than for maximum yield. Over-prescribed agrichemicals pose both productivity and pollution hazards. Tools include integrated pest management, intercropping, crop residue management, and conservation tillage.
8. Defining objective indicators and realistic thresholds for the agroecosystem's productive capacity and ecological integrity, and monitoring of these indicators for regular *ex post* evaluations of project performance upon implementation.

9. Using sound site-specific information (databases, geographic information systems, models, decision aids, etc.) for quantitative optimization of land use plans and management from a balanced (sustainability) perspective.
10. Easing barriers to long-term research to gain confidence in predictive models and decision tools.
11. Avoiding scientists' isolation from the root causes of the unsustainability over which they have no direct control (e.g., population expansion, land use planning, and trade policies) and urging their active involvement in heightening the awareness of the "powers to be" of merits and pitfalls of evolving policy.
12. Benefiting from, rather than emotionally dismissing, emerging innovations such as biotechnology and irradiation which often displace agrichemical inputs.

These considerations may not guarantee full harmony but will likely reduce the conflicts often encountered between the production and environmental emphases of agricultural land use. One healthy perspective of integrating these objectives allocates productivity values with commensurate economic accounting to all environmental changes in managed ecosystems. Multicriteria analyses in these situations will assist in defining the most critical bottlenecks determining the sustainability of land use. When conflicts are unavoidable, quantitative MODS systems will allow optimization among competing management objectives. Such decision aids need not be complex. Rather, systems of varying complexity are necessary for meeting the needs of various clientele and decision makers.

References

Buringh, P. 1982. Potentials of world soils for agricultural production. *Transactions of the 12th International Congress of Soil Science*. Indian Society of Soil Science, New Delhi.

Dazhong, Wen. 1993. Soil erosion and conservation in China. Pages 63–85 D. Pimentel, Ed. in *World Soil Erosion and Conservation*. Cambridge University Press.

El-Swaify, S.A. 1988. Conservation-effective farming systems for the tropics. Pages 134–136 P.W. Unger, T.V. Sneed, P.W. Jordan, and R. Jensen, Eds. in *Challenges to Dryland Agriculture, A Global Perspective*. Proc. International Conference on Dryland Farming. Texas Agricultural Experiment Station, Temple.

El-Swaify, S.A. 1991. Land-based limitations and threats to world food production. *Outlook Agric.* 20(4):235–242.

El-Swaify, S.A. 1992. Land degradation hazards and protection elements in small, tropical, volcanic island ecosystems. In *Proceedings of the 7th International Conference on Soil Conservation*, Soil Conservation Service of New South Wales, Sydney.

El-Swaify, S.A. 1994. State of the art for assessing soil and water conservation needs and technologies: a global perspective. Pages 13–27 in *Adopting Conservation on the Farm — An International Perspective on the Socioeconomics of Soil and Water Conservation*. Soil and Water Conservation Society, Ankeny, Iowa.

El-Swaify, S.A. and Hans Hurni. 1996. Transboundary effects of soil erosion and conservation in the Nile Basin. *Land Husbandry*. 1(1&2):5–21.

Hughes, J.D. 1992. Sustainable agriculture in ancient Egypt. *Agric. Hist.* 66(2):12–22.

International Soil Reference and Information Center (ISRIC) and United Nations Environment Program (UNEP). 1991. Global assessment of human-made land degradation. UNEP, Nairobi.

Pimentel, D., C. Harvey, P. Resosudarmo, K. Sinclair, D. Kurz, M. McNair, S. Crist, L. Shpritz, L. Fitton, R. Saffouri, and R. Blair, 1994. *Science.* 267:1117–1123.

Postel, S. 1989. Water for Agriculture: Facing the Limits. Worldwatch Paper 93, Worldwatch Institute, Washington, D.C., 54 pages.

Said, Rushdi. 1992. *The River Nile.* Dar El-Hilal Press, Cairo.

Soule, J.D. and J.K. Piper. 1992. *Farming in Nature's Image — An Ecological Approach to Agriculture.* Island Press, Washington, D.C., 286 pages.

Chapter 3

Using the Decision Support System GMCR for Resolving Conflict in Resources Management

KEITH W. HIPEL, D. MARC KILGOUR, LIPING FANG, AND XIAOYONG (JOHN) PENG

Abstract

The graph model for conflict resolution is presented as a novel decision technology for systematically studying disputes that can arise in resource management and elsewhere. This comprehensive conflict resolution methodology possesses solid and realistic mathematical foundations which allow it to accurately model a strategic situation, forecast compromise solutions, and help assess the political and social viability of a resource management project. To permit the graph model for conflict resolution to be conveniently applied by practitioners to practical real-world conflict problems, a basic design is put forward for a new decision support system (DSS) called GMCR II. In order to illustrate how GMCR II works in practice, it is used to formally model and analyze an international environmental management dispute involving governments in both Canada and the U.S.

Introduction

The development of natural resources and preservation of the natural environment are two social objectives that are often in conflict. Within the sustainable development paradigm, decision makers attempt to reach a suitable balance between

Table 3.1 Classifying Decision Making Techniques

		Objectives	
		One	Two or more
Decision makers	One	Most OR methods	Multiiple criteria decision making (MCDM)
	Two or more	Team theory	Graph model for conflict resolution

these two goals. However, resource development projects usually affect a wide range of interest groups and often give rise to serious conflicts. To find resolutions that are viable both socially and politically, yet sustainable from environmental and economic perspectives, managers and other decision makers require techniques to assist them in understanding strategic decision making. The major objective of this paper is to use an international resource management conflict as an illustrative example to show how the graph model for conflict resolution and the accompanying DSS GMCR II can be used to understand environmental disputes, advise decision makers, and forecast compromise solutions. Based upon the results of a conflict analysis study, a decision maker can identify courses of action that fall within the social and political constraints of the problem, and are most likely to lead to a favorable resolution that is both justifiable and sustainable.

Formal modeling approaches for studying a wide variety of conflict situations that can arise in the real world are the metagame analysis method of Howard (1971) and the conflict analysis technique of Fraser and Hipel (1979; 1984). The recently developed graph form of the conflict model extends and refines these methodologies to describe more accurately the behavior of participants in a strategic conflict (Fang et al., 1989; 1993; Kilgour et al., 1987). The graph form takes states, rather than individual decisions, as the basic units for describing a conflict. Possible types of social behavior in a conflict are then represented by appropriate solution concepts.

Many types of decision techniques were developed within a field called Operations Research or simply OR. Table 3.1 portrays how OR methods can be categorized according to the criteria of the number of decision makers and the number of objectives. As is shown, most OR techniques reflect the viewpoint of one decision maker having one objective. For instance, linear programming can be employed as an optimization tool by an organization to minimize its costs. Multiple criteria decision making (MCDM) techniques (see, for example, Goicoechea, 1982; Hipel, 1992; Radford, 1989; Roy, 1985; Vincke, 1992) are designed for finding the more preferred alternative solutions to a problem when the discrete alternatives are evaluated against criteria ranging from cost (a quantitative criterion) to aesthetics (a qualitative criterion). The evaluations of the criteria for each alternative reflect the objectives or preferences of the decision maker.

In team theory, each team has the single objective of winning. Hence, the approach is categorized as shown in Table 3.1. The graph model for conflict resolution, in addition to some other game theory techniques, constitute methods that are designed for use in multiple participant-multiple objective decision making. As a matter of fact, this is the category of OR for which there is high demand for the development of decision aids but in which the least effort has been devoted in the past.

The second column in Table 3.1 classifies techniques that can be used with multiple objectives. Useful relationships between MCDM and conflict resolution are put forward by Hipel et al. (1993). When a decision maker is involved in a dispute, he or she can be envisioned as having a MCDM problem from his/her own perspective and hence the problem is located in the top right cell in Table 3.1. This is because the decision maker must somehow use his/her personal objectives and values to rank the possible alternative scenarios or the states in the dispute. When the decision maker strategically interacts with the other decision makers in order to arrive at an eventual resolution, the situation becomes a conflict or game, which is the lower right cell in Table 3.1.

In the next section, the theory underlying the graph model for conflict resolution is outlined. Subsequently, a general design for the new DSS called GMCR II is presented for allowing practitioners to conveniently and expeditiously apply the graph model approach to the actual conflict. To demonstrate the efficacy of the graph modeling approach to conflict resolution in resource development, GMCR II is applied here to an important international resource dispute. This conflict was occasioned by a proposed open-pit coal mining development in the Canadian province of British Columbia along a tributary of the Flathead River, which flows into the U.S. state of Montana. Because effluent from the mining operation would have adverse effects downstream, especially on recreational water use in the U.S., the development faced international opposition from the governments of the U.S. and the State of Montana. In accordance, the Boundary Waters Treaty of 1909, the governments of the U.S. and Canada requested the International Joint Commission (IJC) to examine the entire situation. GMCR II is used to model and analyze this intriguing environmental conflict at a crucial point in its evolution, just before the IJC made its recommendation.

As the reader can appreciate, this paper focuses upon conflict resolution methodologies developed by the authors as well as other researchers with whom they had the pleasure of working. Nonetheless, overviews of game theoretic model and how they can be used in decision making are provided by Fang et al. (1993) in Chapter 1 of their book as well as by Hipel et al. (1993). These authors also provide key references for various kinds of decision-making models.

The Graph Model for Conflict Resolution

A game or conflict model is a systematic structure for describing the main characteristics of a conflict. After developing the game model, one can employ the model as a basic structure upon which one can extensively analyze the possible strategic interactions among the *decision makers* (DMs) in order to determine the possible compromise resolutions or equilibria. The output from this stability analysis as well as related sensitivity analyses can be useful, for example, to help support decisions made by people who can exercise real power in a conflict. The next two subsections outline some of the key ideas behind modeling and stability analysis, respectively. Subsequently, the computer implementation of these ideas is described in the section for the DSS called GMCR II.

Modeling

The graph model for conflict resolution represents a conflict as moving from state to state (the vertices of a graph) via transitions (the arcs of the graph) controlled by the DMs. One inherent advantage of the graph model is that it can incorporate *irreversible moves*, whereby a DM can unilaterally move from state k to state q but not from q to k. Examples of irreversible moves are represented later in the case study. Another key advantage of the graph model is its ability to describe easily *common moves*, in which more than one DM can cause the conflict to move from one particular state to the another common state. For instance, the greenhouse effect can be caused by greenhouse gases emitted from any one of many countries around the world.

A graph model for a conflict consists of a directed graph and a payoff function for each DM taking part in the dispute. Let $\mathbf{N} = \{1, 2, \ldots, n\}$ denote the set of DMs and $\mathbf{U} = \{1, 2, \ldots, u\}$ the set of *states* or possible scenarios of the conflict. In the application section, it is shown how the option form can be used in practice to generate the states in a dispute. A collection of finite *directed graphs* $D_i = (\mathbf{U}, \mathbf{A}_i)$, $i \in \mathbf{N}$, can be used to model the course of the conflict. The vertices of each graph are the possible states of the conflict and therefore the vertex set, $\mathbf{U}$, is common to all graphs. If DM i can unilaterally move (in one step) from state k to state q, there is an arc with orientation from k to q in $\mathbf{A}_i$. For each DM $i \in \mathbf{N}$, a *payoff function* $P_i : \mathbf{U} \to \mathbf{R}$, where $\mathbf{R}$ is the set of real numbers, is defined on the set of states. A payoff function measures the worth of states to a DM. It is assumed that values of the payoff function represent only the DM's ordinal rankings of the states.

DM i's graph can be represented by i's *reachability matrix*, $\mathbf{R}_i$, which displays the unilateral moves available to DM i from each state. For $i \in \mathbf{N}$, $\mathbf{R}_i$ is the $u \times u$ matrix defined by

$$R_i(k,q) = \begin{cases} 1 & \text{if DM } i \text{ can move (in one step)} \\ & \text{from state } k \text{ to state } q \\ 0 & \text{otherwise} \end{cases}$$

where $k \neq q$, and by convention

$$R_i(k,k) = 0.$$

A more economical expression of DM i's decision possibilities is his or her *reachable list*. For $i \in \mathbf{N}$, DM i's reachable list for state $k \in \mathbf{U}$ is the set $\mathbf{S}_i(k)$ of all states to which DM i can move (in one step) from state k. Therefore,

$$\mathbf{S}_i(k) = \{q \in \mathbf{U} : R_i(k,q) = 1\}.$$

The payoff function for DM i, P_i, measures how preferred a state is for i. Thus, if $k, q \in \mathbf{U}$, then $P_i(k) \geq P_i(q)$ iff i prefers k to q, or is indifferent between k and q. When this inequality is strict for all pairs of distinct states for every DM, the conflict is called strict ordinal. Beyond the ordinal information of preference or indifference, nothing can be inferred from the value of P_i. For instance, $P_i(k) > P_i(q)$ indicates that i prefers k to q, but the value of $P_i(k) - P_i(q)$ gives no meaningful information about the strength of this preference. For convenience, small positive integers are used as the values of $P_i(\bullet)$.

A unilateral improvement from a particular state for a specific DM is any preferred state to which the DM can unilaterally move. To represent *unilateral improvements*, each DM's reachability matrix can be used to define a matrix $\mathbf{R}_i^+$, according to

$$R_i^+(k,q) = \begin{cases} 1 & \text{if } R_i(k,q) = 1 \text{ and } P_i(q) > P_i(k) \\ 0 & \text{otherwise} \end{cases}$$

Similarly, DM i's reachable list, $\mathbf{S}_i(k)$, can be replaced by $\mathbf{S}_i^+(k)$, defined by

$$S_i^+(k) = \{q \in \mathbf{S}_i(k) : R_i^+(k,q) = 1\}.$$

Thus, $\mathbf{S}_i^+(k)$ is called the unilateral improvement list of DM i from state k.

Stability Analysis

The stability analysis of a conflict is carried out by determining the stability of each state for every DM. A state is *stable* for a DM iff that DM has no incentive to deviate from it unilaterally, under a particular stability definition or solution concept. A state is an *equilibrium* or possible resolution under a particular solution concept iff all DMs find it stable under the stability definition.

Overview of Solution Concepts

A *solution concept* constitutes a mathematical description of a behavior pattern. Because DMs can react to conflict situations in many ways, there are many different solution concepts. In their book, Fang et al. (1993) define (Chapter 3) and mathematically compare (Chapter 5), using the graph form of the model, the solution concepts listed in Table 3.2. Furthermore, they demonstrate how the graph model and an associated solution concept can be equivalently expressed in extensive form, which is much more complicated and hence not as well suited for practical applications (Chapter 4).

The solution concepts in Table 3.2 are developed for application to conflicts having two DMs or more than two DMs. The first column gives the names of the solution concepts and their acronyms, while the second one provides original references. The last three columns furnish ways of characterizing the solution concepts in a qualitative sense according to the three criteria of foresight, disimprovement, and knowledge of preferences. *Foresight* refers to the ability of a DM to think about possible moves that could take place in the future. If the DM has high or long foresight, he or she can imagine many moves and countermoves into the future when evaluating where the conflict could end up because of an initial unilateral move on his part. Notice, for example, that in Nash stability the

Table 3.2 Solution Concepts and Human Behavior

Solution concepts	References	Foresight	Disimprovements	Knowledge of Preferences
Nash stability (R)	Nash, 1950, 1951	Low	Never	Own
General metarationality (GMR)	Howard, 1971	Medium	By opponents	Own
Symmetric metarationality (SMR)	Howard, 1971	Medium	By opponents	Own
Sequential stability (SEQ)	Fraser and Hipel, 1979, 1984	Medium	Never	All
Limit-move stability (L_h)	Kilgour, 1985; Kilgour et al., 1987; Zagare, 1984	Variable	Strategic	All
Nonmyopic stability (NM)	Brams and Wittman, 1981; Kilgour, 1984, 1985; Kilgour et al., 1987	High	Strategic	All

foresight is low, whereas it is very high for the solution concept of nonmyopic stability. The criterion of *strategic disimprovements* in the fourth column means that a particular DM may temporarily move to a worse state in order to reach a more preferred state eventually. *Disimprovements by opponents* indicate the other DMs may put themselves in worse positions in order to block unilateral improvements by the given DM. The last column entitled *knowledge of preferences* refers to the preference information used in a stability analysis. For example, when carrying out a stability analysis for a given DM under R, GMR, or SMR the preferences of all other DMs are not used in the calculations, although their ability to move unilaterally to other states is taken into account. These solution concepts can be quite useful in situations where there is high uncertainty by the particular DM over the preferences of his or her competitors.

Unilateral Moves

For a conflict involving more than two DMs, one must define movements that can involve more than one DM. Let $\mathbf{H} \subseteq \mathbf{N}$ be any subset of the DMs, and let $\mathbf{S}_\mathbf{H}(k)$ denote the set of all states that can result from any sequence of unilateral moves, by some or all of the DMs in $\mathbf{H}$, starting at state k. In this sequence, the same DM may move more than once, but not twice consecutively. If all of the DMs' graphs are *transitive*, for any sequence of moves in which one DM moves twice consecutively, there is always an equivalent single move available by transitivity. The simplified definition of a unilateral move by $\mathbf{H} \subseteq \mathbf{N}$ is:

Unilateral Move: Let $k \in \mathbf{U}$ and $\mathbf{H} \subseteq \mathbf{N}$, $\mathbf{H} \neq \varnothing$. *A unilateral move by* $\mathbf{H}$ *is a member of* $\mathbf{S}_\mathbf{H}(k) \subseteq \mathbf{U}$, *defined inductively by*

1. if $j \in \mathbf{H}$ and $k_1 \in \mathbf{S}_j(k)$, then $k_1 \in \mathbf{S}_H(k)$,
2. if $k_1 \in \mathbf{S}_H(k)$, $j \in \mathbf{H}$, and $k_2 \in \mathbf{S}_j(k_1)$, then $k_2 \in \mathbf{S}_H(k)$.

For computational implementation, the induction stops as soon as no new state (k_2) is added to $\mathbf{S}_H(k)$.

By replacing $\mathbf{S}_H(k)$, $\mathbf{S}_j(k)$, and $\mathbf{S}_j(k_1)$, by $\mathbf{S}_H^+(k)$, $\mathbf{S}_j^+(k)$, and $\mathbf{S}_j^+(k_1)$, respectively, in the above definition one obtains the definition of a *unilateral improvement* by **H** when all DMs' graphs are transitive. More general definitions for a unilateral move and unilateral improvement are given by Fang et al. (1993, Section 3.4) for the case when the graphs can be intransitive.

$\mathbf{S}_H(k)$ and $\mathbf{S}_H^+(k)$ can be thought of as **H**'s reachable list and unilateral improvement list, respectively. Specifically, the sets $\mathbf{S}_{N-i}(k)$ and $\mathbf{S}_{N-i}^+(k)$ represent the possible states of "response sequences" of i's opponents against a move by i to k.

Illustrative Definitions of Solution Concepts

To provide the reader with an appreciation of possible types of strategic interactions that can take place among DMs involved in a dispute, the first four solution concepts listed in the left column of Table 3.2 are defined. Definitions for all of the solution concepts along with illustrative examples are provided by Fang et al. (1993, Chapter 3).

Nash Stability: *Let $i \in \mathbf{N}$. A state $k \in \mathbf{U}$ is Nash Stable (R) for DM i iff $\mathbf{S}_i^+(k) = \emptyset$.*

General Metarationality: *For $i \in \mathbf{N}$, a state $k \in \mathbf{U}$ is general metarational (GMR) for DM i iff for every $k_1 \in \mathbf{S}_i^+(k)$ there is at least one state $k_x \in \mathbf{S}_{N-i}(k_1)$ with $P_i(k_x) \leq P_i(k)$.*

Under general metarationality, DM i expects that the other DMs ($\mathbf{N} - i$) will respond to hurt i, if it is possible for them to do so, by a sequence of unilateral moves. DM i anticipates that the conflict will end after the DMs of $\mathbf{N} - i$ have responded. Additionally, DM i's opponents are assumed to ignore their own payoffs when making their sanctioning moves.

Symmetric Metarationality: *For $i \in \mathbf{N}$, a state $k \in \mathbf{U}$ is symmetric metarational (SMR) for DM i iff for all $k_1 \in \mathbf{S}_i^+(k)$, there exists $k_x \in \mathbf{S}_{N-i}(k_1)$, such that $P_i(k_x) \leq P_i(k)$ and $P_i(k_3) \leq P_i(k)$ for all $k_3 \in \mathbf{S}_i(k_x)$.*

The SMR solution concept postulates that a DM, i, expects that he or she will have a chance to counterrespond (k_3) to the other DMs' response (k_x) to his original move (k_1). DM i anticipates that the conflict will end after his counterresponse.

Sequential Stability: *For $i \in \mathbf{N}$, a state $k \in \mathbf{U}$ is sequentially stable (SEQ) for DM i iff for every $k_1 \in \mathbf{S}_i^+(k)$ there is at least one state $k_x \in \mathbf{S}_{N-i}^+(k_1)$ with $P_i(k_x) \leq P_i(k)$.*

The main difference between the GMR and SEQ stability is the requirement that sanctions be credible by being unilateral improvements for the sanctioning DMs.

Table 3.3 Existing DSSs for Conflict Analysis

Acronyms	Purposes	Ref.
GMCR I	The graph model for conflict resolution is used along with the solution concepts in Table 3.2 to model and analyze disputes having two or more DMs.	Fang et al., 1993; Kilgour et al., 1995
DecisionMaker	The option form and solution concept of sequential stability (SEQ) are used for analyzing both small and large conflicts.	Fraser and Hipel, 1988, 1989
Conan	Metagame Analysis is used to interactively model and analyze conflicts.	Howard, 1990
INTERACT	The option form and graphical displays are used to analyze situations under the control of several interested parties.	Bennett et al., 1994
SPANNS	Designed for use in negotiation support, SPANNS has a rule-based expert system for the tactical component and a conflict analysis model for the strategic component.	Meister and Fraser, 1994

The Decision Support System GMCR II

Existing Systems for Multiple Participant Decision Making

In order to conveniently apply a given formal decision model to practical problems, a flexible DSS is required (Sage, 1991). A wide variety of decision-making models and associated DSSs have been developed in fields which include OR, management sciences, systems engineering, and statistics. The purpose of this section is to present the design of a new DSS called GMCR II which implements the graph model for conflict resolution approach outlined in the previous section for systematically studying real-world conflict that can arise in resource management, international trade, and elsewhere.

In the past, some DSSs have been developed for modeling decision situations involving more than one DM. When designed for employment in negotiations, a DSS is commonly referred to as a negotiation support system. Kilgour et al. (1995), Thiessen and Loucks (1992), and Jelassi and Foroughi (1989) provide overviews and comparisons of existing negotiation support systems. Papers describing the theory and application of existing negotiation support systems include contributions by Angus (1990), Anson and Jelassi (1990), Gauvin et al. (1990), Jarke et al. (1987), Nagel and Mills (1989), Nunamaker (1989), Singh et al. (1985), and Winter (1985).

Table 3.3 presents a list of DSSs that can be utilized for aiding decision making under conditions of conflict. This table constitutes a modified version of the table originally presented by Radford et al. (1994) which provides contact addresses for the developers of the DSSs given in Table 3.3.

As can be seen in Table 3.3, GMCR I is the only DSS that is based on the graph model for conflict resolution. GMCR I was developed by Fang et al. (1993) and included on diskette with their book *Interactive Decision Making: The Graph*

Model for Conflict Resolution. It has been successfully used to analyze a variety of real-world conflicts. This DSS is basically a stability analysis program that implements all of the solution concepts given in Table 3.2. When used on a microcomputer, GMCR I can handle medium-sized conflicts that have no more than about 200 states. It requires an input file which contains information on the number of DMs, number of states, reachable lists for each DM, and a preference ranking of the states for each DM. Subsequent to carrying out a stability analysis, GMCR I produces a huge output file containing detailed stability results for every solution concept, every state and every DM. The algorithms for GMCR I are described in Chapter 3 of Fang et al. (1993) while a tutorial and user's manual are given in the appendices.

Although GMCR I possesses a powerful analysis engine, most of the modeling work has to be done outside of the system. For large or medium-sized conflicts, it is not easy to determine feasible states, reachable lists, and ranking of states by hand, and, therefore, input errors can easily arise. Moreover, the analysis is only based on state numbers and the interpretation of what each state represents has to be done outside of the system. Finally, GMCR I does not currently possess the capability to carry out extensive sensitivity analyses, hypergame analyses (Wang et al., 1988), or coalition analyses.

GMCR II — The Next Generation

The objective of GMCR II is to provide the next generation of a DSS for the graph model for conflict resolution that significantly expands the capability of the previous generation, GMCR I. Figure 3.1 depicts the overall design of GMCR II. The model subsystem in Figure 3.1 uses the GMCR I engine for calculating stability results for all of the solution concepts listed in Table 3.2. However, the memory management for the engine is being improved to allow larger conflict to be expeditiously analyzed. The User Interface for both input and output and Data Subsystem are completely new designs that are presently being incorporated into GMCR II.

Figure 3.2 portrays how stability results in the data subsystem in Figure 3.1 can be organized and thereby be easily retrieved for purposes such as displaying appropriate findings, updating the model, and executing different kinds of sensitivity analyses. Figure 3.2, which originally appeared in Fang et al. (1993, page 195), can be interpreted in a variety of ways which include:

1. For each DM, the *decision maker's plane* (parallel to the STATE/STABILITY TYPE plane) indicates the stability types or solution concepts (if any) under which each state is stable for that specific DM.
2. For each stability type the *stability-type plane* (parallel to the DECISION MAKER/STATE plane) provides a complete analysis of the model according to that stability type.
3. For each state, the *state plane* (parallel to the DECISION MAKER/STABILITY TYPE plane) identifies the DMs for whom the particular state is stable (s), under each possible stability type, or unstable (u).
4. The STABILITY TYPE/STATE plane itself, referred to as the *equilibrium plane*, contains the projection of stability results for each DM (E if s for all DMs, blank otherwise), indicating all equilibria for each stability type.

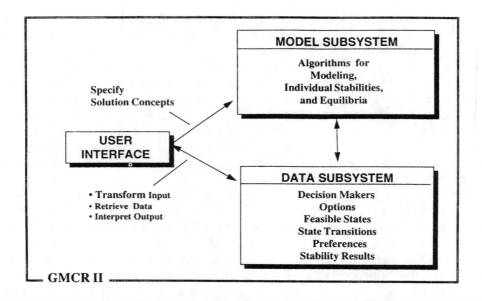

Figure 3.1 Design of GMCR II.

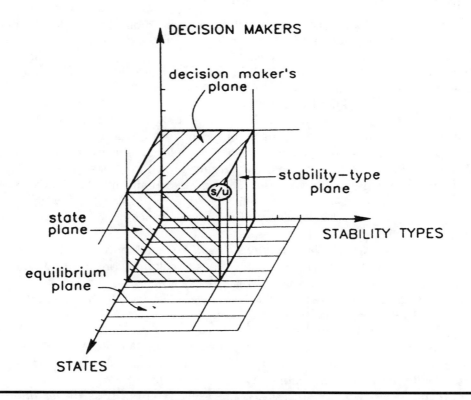

Figure 3.2 Organization of stability results in the data subsystem.

GMCR II is being developed using Borland C^{++} within the Microsoft Windows 3.1 environment. In this way, the system will be widely accessible, user-friendly, and have powerful graphical capabilities. Table 3.4 furnishes the menu for the User Interface for GMCR II. As can be seen, GMCR II can employ the option form for defining states, which is used with the application in this paper. The option form is especially useful when one is aware of the specific courses of actions or options that are available to each DM. Because a large number of states can be automatically calculated by GMCR II using a relatively much smaller number of options, the option form can be used with very large conflicts.

In some situations, one may prefer to discuss a conflict in terms of overall states and not worry about the details of how each state is defined. For example, high-level executives in a corporation may be interested in final possible results or states to a conflict and generally how more preferred situations can be achieved. The graphical input of states mentioned in Table 3.4 may be very helpful in these cases, especially for smaller conflicts at early stages of development. A circle or node can be drawn on a screen to represent a state while movement among states can be depicted using arrows. This graphical information can then be converted by GMCR II into a form which can be processed by the analysis engine. For output, the user will be able to select different aspects of the analysis to be illustrated graphically. It is also desirable to show possible paths from a status quo state to an equilibrium. For both the input and output interface, it is required to have aesthetically pleasing drawings of graphs (Di Battista et al., 1993).

The option form can present problems in memory and execution time. For example, the DSS INTERACT mentioned in Table 3.3 has difficulty obtaining a list of all feasible scenarios, even in models with only 10 options (Bennett et al., 1992). Fraser and Hipel (1988) have devised efficient algorithms based on the preference tree concept which create significant savings in memory and execution time during a stability analysis using the solution concept of sequential stability. In GMCR II, a 32-bit double word is used to represent a state in option form. This approach is efficient and can be used in analyses employing all of the solution concepts given in Table 3.2. For a given digit or bit, a "0" or "1" indicates if the option is taken by the DM controlling it or not selected, respectively. Because there are 32 bits, this format can handle up to 32 options, which is more than sufficient for use in real-world applications. If there are n options in the game, GMCR II can automatically calculate the resulting 2^n states.

An example of a state for a conflict having 10 options is represented in binary form as

$$\overbrace{\ldots\ldots\underbrace{000}_{DM_4}\ \underbrace{1}_{DM_3}\ \underbrace{001}_{DM_2}\underbrace{001}_{DM_1}}^{32\,bits}\,(binary)$$

which is the number 73 in decimal form. This state is actually the state depicted in Table 3.5 in the application section. There are four DMs in the conflict and the brackets indicate which options a given DM controls.

In practice, there is usually a large number of states that cannot occur and therefore should be eliminated from the conflict model in option form. Accordingly, one of the menu items under the Model Input in Table 3.4 is State Elimination.

Table 3.4 Menu Tree for GMCR II

1st Level	2nd Level	3rd Level
File	New	
	Open	
	Save	
	Save As ...	
	Exit	
Edit	Undo	
	Cut	
	Copy	
	Paste	
Model Input	Option Form	Input DMs/Options
		Irreversible Moves
		Common Moves
		State Elimination
		State Combination
	Graphical	
Preview	Conflict Description	
	Feasible State List	
	Reachable Lists	
Preferences	Preference Tree	Basic/Conditional
		Interactive
	Option Weighting	
	Further Ranking	
Stability Output	Option Form	
	Graphical Output	
Interpretation	Status Quo Analysis	
	Graphical Evolution Path	
Further Analysis	Sensitivity Analysis	
	Hypergame Analysis	
	Coalition Analysis	
Help	Contents	
	About ...	

In the State Elimination dialogue box, for example, a user may specify information such as "DM_1 can select option 1 only if DM_2 chooses option 4." A logic interpretation program can then transform this information and put a "(option 1) and (not option 4)" into the "Infeasible Situation" list, which in turn requires all states with the pattern • • • • • • — — — 0--1 to be eliminated. The dash given as "-" indicates that the entry can be either a 0 or 1. The following two "mask"s are used to execute this elimination: m_0 = • • • • • •1111110111; m_1 = • • • • • •0000000001. If a state k satisfies: (1) $k|m_0 = m_0$, and, (2) $k\&m_1 = m_1$, it can be proven that k must belong to the above pattern which should be eliminated.

With the aforementioned efficient date structure and related modeling algorithms, the generation of a list of all feasible states can always be done very quickly. The other modeling and analysis processes also greatly benefit from this data structure and bitwise operations.

In an irreversible move, a DM can, for instance, cause a conflict to go from state k to q by unilaterally changing his or her option selection, but cannot make the transition from q to k. Accordingly, within the Irreversible Move dialogue box, a user can highlight the option or options which cause the irreversible move.

After the Model Input stage in Table 3.4, GMCR II can automatically generate the Feasible State List and Reachable State Lists, which are available to the user under the Preview menu. Subsequently, the user must input the preference information.

The preference tree approach was first suggested by Fraser and Hipel (1988) and is also described by Hipel and Meister (1994). The preference tree ranks the states for a given DM based upon lexicographical preference statements about the options. Because the preference information for a particular DM appears as a tree when portrayed graphically, the preference information is called a preference tree.

The Preference Tree or Option Weighting method can be used to obtain a ranking of the states for each DM from most to least preferred where there may be some blocks containing equally preferred states. The Further Ranking dialogue box can be used to refine the ranking list for a given DM by allowing the user to move a state from one location to another.

As can be seen in Table 3.4, a variety of output can be generated by the engine for GMCR II after the user specifies which solution concepts to employ for each DM. The hypergame analysis refers to analyzing a conflict in which there are misperceptions by one or more of the DMs (Bennett, 1977; Wang et al., 1988). Finally, examples of sensitivity analyses are presented with the resources allocation conflict in the next section.

The Flathead River Resource Development Dispute

Background of the Flathead River Resource Development Dispute

A brief summary of the background of the Flathead River resource development dispute is given here, based on Carroll (1983) and IJC (1988). The Flathead River Region is shown in Figure 3.3. The Flathead River flows from its source in the southeastern corner of British Columbia south across the international boundary into Montana, then into Flathead Lake, and eventually into the Columbia River. The discovery of coal in the British Columbia portion of the Flathead valley was first reported in 1910. In 1970, a new company called Sage Creek Coal Limited was formed to explore and develop the site. Its Stage I proposal for developing the mine was approved by the province of British Columbia in October 1976. In February 1984, the government of British Columbia granted Sage Creek Coal Limited approval-in-principle for Stage II of the proposal — a 2.2 million tonnes/year thermal coal mine located on Howell and Cabin Creeks, tributaries of the Flathead River, 10 km upstream from the international boundary. The mine plan was based on 21 years of mining at the proposed rate, but coal reserves would likely justify a further 20 years of mining at this same rate.

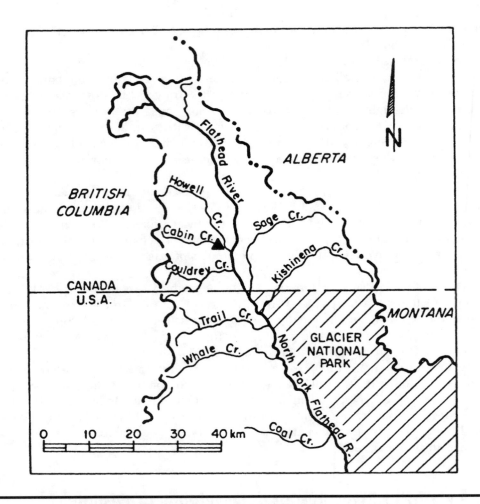

Figure 3.3 Development location in the Flathead River drainage basin.

The governments of the U.S. and the State of Montana were concerned about the possible effects of this proposed mine on the Flathead River system, Glacier National Park, and Flathead Lake. In response to these concerns, the U.S. and Canadian governments requested that the IJC examine the possible impacts of the proposed mine on water quality and quantity, fisheries, and other water uses associated with the Flathead River, and make recommendations.

Modeling and Analysis of the Flathead River Resource Development Dispute

The Flathead River Resource Development Dispute is modeled at a time just prior to the IJC decision in December 1988. The modeling and analyses of the conflict are based upon the conference paper by Kilgour et al. (1991). The main DMs were:

Table 3.5 The DMs and Options of the Flathead River Resource Development Dispute

DMs and options		Status quo	
Sage Creek	1. Continue: *Continue* original development	Y	Strategy for Sage Creek
	2. Modify: *Modify* to reduce impact	N	
	3. Stop: *Stop* project	N	
British Columbia	4. Original: Support *original* project	Y	Strategy for British Columbia
	5. Modification: Require *modification*	N	
	6. Deny: *Deny* license	N	
Montana	7. Oppose: *Oppose* any development	Y	Montana strategy
IJC	8. Original: Recommend *original* project	N	Strategy for IJC
	9. Modification: Recommend *modification*	N	
	10. No: Recommend *no* project	N	

1. **Sage Creek Coal Limited (Sage Creek)** — Sage Creek Coal Limited was owned by Rio Algom Mines (60%) and Pan Ocean Oil Limited (40%). As of 1988, Sage Creek already had substantial financial and other commitments to the Flathead River Development project. Sage Creek hoped for an IJC recommendation that would favor the continuation of the Cabin Creek coal development, which would pave the way for the province of British Columbia to issue a full Stage II license.
2. **Province of British Columbia (British Columbia)** — The provincial government had the authority to grant the license for mining, but had to be concerned with the environmental impact assessment and would be subject to pressure from the federal government of Canada.
3. **State of Montana (Montana)** — The Montana government feared that the proposed mining development would result in transboundary pollution and environmental degradation. Environmental groups and the U.S. Department of Interior strongly agreed with the Montana government on this issue.
4. **International Joint Commission (IJC)** — The IJC is defined under Section IX of the Boundary Waters Treaty of 1909 and is composed of three members from Canada and three from the U.S. Its mandate includes making recommendations and judgments on conflicts falling under this treaty.

Other DMs, such as the federal governments of Canada and the U.S., were considered not to be directly involved in the dispute at this stage. Environmental groups, including some from Canada, were so closely identified with the Montana state government that they could be included with it.

The DMs and their options are listed in Table 3.5. As of December 1988, Sage Creek could continue the original project, modify it to reduce environmental impact, or stop it all together. The government of British Columbia could license the original development, or a modified project, or it could deny a license for the project. Montana could continue to oppose any development or drop its active opposition. The IJC could support either the original, or a modification, or no project at all.

Figure 3.4 displays the DM and option input screen using GMCR II for the Flathead River conflict. Notice that in the upper left portion of the screen the user can add, modify, and delete DMs involved in the conflict. Because the DM British Columbia is highlighted, the lower left portion shows its options entered by the user. In the table, two options have already been entered and the third option call deny is about to be added to the list.

The option form for representing a state is shown as a column of Y's and N's in Table 3.5. A "Y" indicates "Yes" the option opposite the Y is taken by the DM controlling it, whereas an "N" means "No" the option is not selected. A *strategy* is a situation where a given DM decides which options to invoke. For example, in Table 3.5, Sage Creek is selecting the option of "Continue" and rejecting its other two options. The strategies for the other three DMs are also shown in Table 3.5.

A state is formed after each DM selects a strategy. Written horizontally in text, the state (YNN YNN Y NNN) given in Table 3.5 is formed by Sage Creek, British Columbia, Montana, and the IJC following strategies (YNN), (YNN), (Y), and (NNN), respectively, in order to create the overall state. This state is actually the status quo situation for when the Flathead River dispute is studied.

In the Flathead dispute, there is a total of 10 options. Because each option can be either selected or rejected, there is a total of $2^{10} = 1,024$ states. Nonetheless, many of these states cannot take place in the actual conflict since they are infeasible for a variety of reasons. For example, because Sage Creek cannot build a project that exceeds its license, any such strategy combinations is infeasible. When infeasible combinations are removed, there remain 37 states, numbered 0, 1, ..., 36, which appear as columns in Table 3.6.

The reachable lists for this dispute are given in Table 3.7 while some of the possible state transitions are indicated graphically in Figure 3.5. As shown by the horizontal lines in Table 3.7 and the three groupings to the right in Figure 3.5, states 1 to 36 can be divided into three sets, each containing 12 states, according to the recommendation of the IJC. State 0 is the status quo position. Notice that because Sage Creek cannot build a project that exceeds its license, a move by British Columbia from the original project to the modified project or no project forces Sage Creek to do likewise. For example, British Columbia can move unilaterally from state 25 to states 28 and 30. However, as shown by the arrows in Figure 3.5 these movements are irreversible — British Columbia cannot move from state 28 to 25 or from 30 to 25. Finally, to produce the detailed graph in Figure 3.5 for states 1 to 12 as well as states 13 to 24, simply subtract the number 24 and 12, respectively, from each of the state numbers in the lower detailed graph.

Recall that a DM's payoff function indicates his preferences — a state with a greater payoff value is more preferred. In this conflict, payoff functions were determined by assessing the relative preferences of the DMs. For example, Sage Creek most prefers the maximum possible project consistent with British Columbia's license, with IJC's support and no opposition from Montana. British Columbia's preference can be characterized by its priorities of matching with IJC's

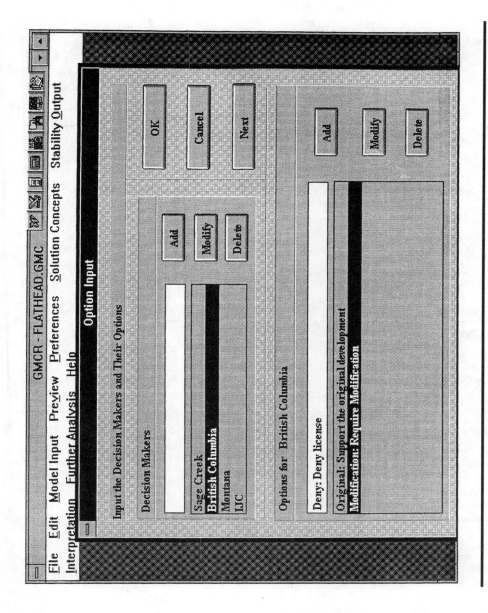

Figure 3.4 DM and Option Input Dialogue Box of GMCR II.

Table 3.6 States of the Flathead River Resource Development Dispute

		0	1	2	3	4	5	6	7	8	9	10	11	12	13
Sage Creek	1. Continue	Y	Y	N	N	N	N	N	Y	N	N	N	N	N	Y
	2. Modify	N	N	Y	N	Y	N	N	N	Y	N	Y	N	N	N
	3. Stop	N	N	N	Y	N	Y	Y	N	N	Y	N	Y	Y	N
British Columbia	4. Original	Y	Y	Y	Y	N	N	N	Y	Y	Y	N	N	N	Y
	5. Modification	N	N	N	N	Y	Y	N	N	N	N	Y	Y	N	N
	6. Deny	N	N	N	N	N	N	Y	N	N	N	N	N	Y	N
Montana	7. Oppose	Y	N	N	N	N	N	N	Y	Y	Y	Y	Y	Y	N
IJC	8. Original	N	Y	Y	Y	Y	Y	Y	Y	Y	Y	Y	Y	Y	N
	9. Modification	N	N	N	N	N	N	N	N	N	N	N	N	N	Y
	10. No	N	N	N	N	N	N	N	N	N	N	N	N	N	N
State numbers		0	1	2	3	4	5	6	7	8	9	10	11	12	13

		14	15	16	17	18	19	20	21	22	23	24	25	26	27
Sage Creek	1. Continue	N	N	N	N	N	Y	N	N	N	N	N	Y	N	N
	2. Modify	Y	N	Y	N	N	N	Y	N	Y	N	N	N	Y	N
	3. Stop	N	Y	N	Y	Y	N	N	Y	N	Y	Y	N	N	Y
British Columbia	4. Original	Y	Y	N	N	N	Y	Y	Y	N	N	N	Y	Y	Y
	5. Modification	N	N	Y	Y	N	N	N	N	Y	Y	N	N	N	N
	6. Deny	N	N	N	N	Y	N	N	N	N	N	Y	N	N	N
Montana	7. Oppose	N	N	N	N	N	Y	Y	Y	Y	Y	Y	N	N	N
IJC	8. Original	N	N	N	N	N	N	N	N	N	N	N	N	N	N
	9. Modification	Y	Y	Y	Y	Y	Y	Y	Y	Y	Y	Y	N	N	N
	10. No	N	N	N	N	N	N	N	N	N	N	N	Y	Y	Y
State numbers		14	15	16	17	18	19	20	21	22	23	24	25	26	27

		28	29	30	31	32	33	34	35	36
Sage Creek	1. Continue	N	N	N	Y	N	N	N	N	
	2. Modify	Y	N	N	N	Y	N	Y	N	
	3. Stop	N	Y	Y	N	N	Y	N	Y	Y
British Columbia	4. Original	N	N	N	Y	Y	Y	N	N	
	5. Modification	Y	Y	N	N	N	N	Y	N	
	6. Deny	N	N	Y	N	N	N	N	Y	
Montana	7. Oppose	N	N	N	Y	Y	Y	Y	Y	
IJC	8. Original	N	N	N	N	N	N	N	N	
	9. Modification	N	N	N	N	N	N	N	N	
	10. No	Y	Y	Y	Y	Y	Y	Y	Y	
State numbers		28	29	30	31	32	33	34	35	36

recommendation, for Sage Creek not building a project larger than IJC's recommendation, and for no opposition from Montana. Montana's preference is determined by the size of Sage Creek's project, IJC's decision, and British Columbia's license. Because the IJC is required by treaty to make a recommendation on the project using only technical (rather than strategic) criteria, its only preference is to make a recommendation. Hence, all recommendations are equally preferred for IJC. The four payoff functions for the four DMs are shown under column $P_i(k)$ in Table 3.7.

Table 3.7 Reachable Lists [$S_i(\bullet)$] and Payoffs [$P_i(\bullet)$] for the Flathead River Resource Development Dispute

k	Sage Creek $S_1(k)$	$P_1(k)$	British Columbia $S_2(k)$	$P_2(k)$	Montana $S_3(k)$	$P_3(k)$	IJC $S_4(k)$	$P_4(k)$
0		1		1		1	7,19,31	1
1	2,3	37	4,6	37	7	2		37
2	1,3	31	4	31	8	8		37
3	1,2	19	5,6	30	9	20		37
4	5	25	2,6	25	10	10		37
5	4	13	3,6	23	11	22		37
6		7	3,5	21	12	24		37
7	8,9	36	10,12	36	1	3		37
8	7,9	30	10	29	2	9		37
9	7,8	18	11,12	28	3	21		37
10	11	24	8,12	24	4	11		37
11	10	12	9,12	22	5	23		37
12		6	9,11	20	6	25		37
13	14,15	35	16,18	5	19	4		37
14	13,15	29	16	17	20	12		37
15	13,14	17	17,18	11	21	26		37
16	17	23	14,18	35	22	14		37
17	16	11	15,18	27	23	28		37
18		5	15,17	19	24	30		37
19	20,21	34	22,24	4	13	5		37
20	19,21	28	22	16	14	13		37
21	19,20	16	23,24	10	15	27		37
22	23	22	20,24	34	16	15		37
23	22	10	21,24	26	17	29		37
24		4	21,23	18	18	31		37
25	26,27	33	28,30	3	31	6		37
26	25,27	27	28	7	32	16		37
27	25,26	15	29,30	13	33	32		37
28	29	21	26,30	9	34	18		37
29	28	9	27,30	15	35	34		37
30		3	27,29	33	36	37		37
31	32,33	32	34,36	2	25	7		37
32	31,33	26	34	6	26	17		37
33	31,32	14	35,36	12	27	33		37
34	35	20	32,36	8	28	19		37
35	34	8	33,36	14	29	35		37
36		2	33,35	32	30	36		37

State 0 = Status quo.
States 1–12 = IJC recommends original project.
States 13–24 = IJC recommends modified project.
States 25–36 = IJC recommends no project.

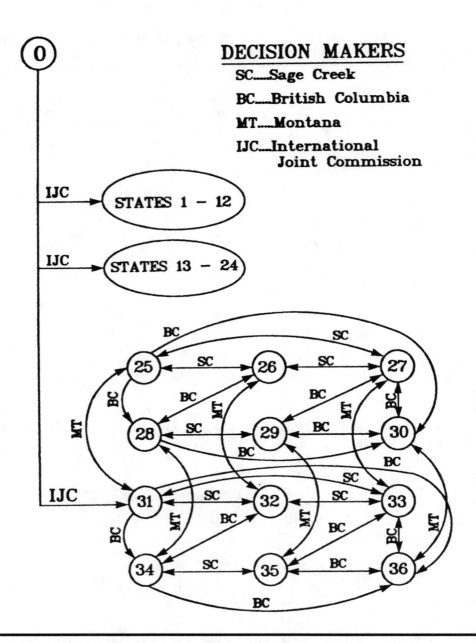

Figure 3.5 The graph model for the Flathead River resource development dispute.

GMCR II was used to carry out a complete stability analysis on the model described in Table 3.7. The equilibria under all solution concepts are summarized in Table 3.8. This table indicates that states 7, 22, and 30 are equilibria for all solution concepts, and possess a much greater degree of stability than any other states.

The status quo at the time of analysis is state 0, in which Sage Creek is proceeding with the original project, the government of British Columbia has approved-in-principle Stage II of the project, Montana is opposing the project,

Table 3.8 Equilibria for the Flathead River Resource Development Dispute

k	Equilibrium
2	GMR, SMR
3	GMR, SMR
7	R, GMR, SMR, SEQ, L_1,L_2,L_3,NM
8	GMR, SMR
9	GMR, SMR
16	GMR, SMR
17	GMR, SMR
22	R, GMR, SMR, SEQ, L_1,L_2,L_3,NM
23	GMR, SMR
30	R, GMR, SMR, SEQ, L_1,L_2,L_3,NM
36	GMR, SMR

Table 3.9 State Transitions from Status Quo to Final Outcome

Sage Creek	1. Continue	Y		Y	→	N		N
	2. Modify	N		N		N		N
	3. Stop	N		N	→	Y		Y
British Columbia	4. Original	Y		Y	→	N		N
	5. Modification	N		N		N		N
	6. Deny	N		N	→	Y		Y
Montana	7. Oppose	Y		Y		Y	→	N
IJC	8. Original	N		N		N		N
	9. Modification	N		N		N		N
	10. No	N	→	Y		Y		Y
State numbers		0		31		36		30

and the IJC has not yet made a recommendation. Equilibria 7, 22, and 30 correspond to each of the IJC's possible recommendations. If it recommends stopping the project, the likely resolution is state 30. The actual outcome was that IJC recommended stopping the project and the state 30 resulted. The sequence of state transitions from the status quo state 0 through the transition states 31 and 36 to the equilibrium 30 is shown in Table 3.9, where arrows indicate which strategy selections change when moving from one state to another.

A sensitivity analysis was conducted by using GMCR II to analyze alternative models to the one defined in Table 3.7. These models differ only in certain aspects of the DM's preference. The objective was to assess the effects of these preference variations on the results of the analysis.

First, British Columbia's preferences were changed to reflect the hypothesis, that even if the IJC recommends stopping the project, British Columbia would prefer to grant a license for a modified, as long as there is no opposition from Montana, i.e., $[P_2(27) = 9, P_2(28) = 33, P_2(29) = 15, P_2(30) = 13)$ instead of $(P_2(27) = 13,$

Table 3.10 Equilibria for the Sensitivity Analysis

k	Equilibrium
2	GMR, SMR
3	GMR, SMR
7	R, GMR, SMR, SEQ, L_1, L_2, L_3, L_4, L_5, NM
8	GMR, SMR
9	GMR, SMR
16	GMR, SMR
17	GMR, SMR
22	R, GMR, SMR, SEQ, L_1, L_2, L_3, L_4, L_5, NM
23	GMR, SMR
28	GMR, SMR
29	GMR, SMR, SEQ
30	GMR, SEQ, L_3, L_4, L_5, NM
35	GMR, SEQ, $L_2, L_3^{II}, L_4^{II}, L_5$, NM
36	GMR, SMR, SEQ, $L_2, L_3, L_4^{II}, L_5^{II}$, NM^{II}

Case I has a fixed point at length 5.
Case II has a fixed point at length 3.
See Fang et al., 1993; Kilgour et al., 1987.

$P_2(28) = 9$, $P_2(29) = 15$, $P_2(30) = 33$]. The equilibria for this case are summarized in Table 3.10.

There is no change in equilibria for the first two sets of states (states 1 to 24), but the changes in last group (states 25 to 36) are substantial. First there is no Nash equilibrium from states 24 to 36. Second, the two forms of limited-move and nonmyopic equilibria defined in Fang et al. (1993) and Kilgour et al. (1987) made different predictions. In Case I, the original DM takes part in the sanctioning: here Montana can move from state 36 to 30 after 36 is reached, explaining why state 36 is a nonmyopic (high foresight) equilibrium only in Case II. As well, there are always two other nonmyopic equilibria 30 and 35. But Sage Creek can stop the project in all of these high foresight equilibria. This means that, even if British Columbia wants some kind of project to proceed, no project will be built, if the DMs behave nonmyopically.

Another sensitivity analysis involves changing British Columbia's preference to reflect that even if the IJC recommend stopping the project and opposition from Montana continued, British Columbia would prefer to grant a license for a modified project. The resulting equilibrium for the last group of states is state 34, for which Sage Creek will build a modified project. If this case occurs, the federal governments of Canada and the U.S. would likely become directly involved in this dispute.

As a consequence of the original analysis and then two sensitivity analyses, it can be concluded that the pressure from Montana (and elsewhere) on British Columbia was as crucial to the final outcome as the IJC's recommendation. If British Columbia had been willing to face national and international opposition in order to achieve at least a modified project, then the analysis implies that such a project would likely have been built. The analysis indicates that British Columbia's sensitivity on Montana's continuing pressure (or the threat of pressure) is decisive.

Conclusions

GMCR II constitutes the next generation of a DSS that can efficiently and effectively examine strategic conflicts that can take place in land, water, and environmental management, as well as many other areas of human endeavor. As demonstrated by the real-world resource management conflict in the previous section, GMCR II can provide decision makers and analysts with decision advice, structural insights, and answers to *what-if* questions. With this enhanced understanding, analysts can better explain strategic relationships and assist DMs, who may have the opportunity to direct the evolution of the conflict toward more favorable results. Certainly, in order to approach ideal solutions to resource management problems that are both economically and environmentally sustainable, the viewpoints of the competing interest groups must be taken into account through a decision technology such as GMCR II.

References

Angus, J. 1990. Negotiate! This software teaches you the fine art of negotiation. *Portable Computing.* 4(3):56.

Anson, R. and M.T. Jelassi. 1990. A development framework for computer-support conflict resolution. *Eur. J. Operat. Res.* 46:181–199.

Bennett, P.G. 1977. Toward a theory of hypergames. *OMEGA.* 5:749–751.

Bennett, P.G., A. Tait, and K. Macdonagh. 1992. INTERACT: developing software for interactive decisions. Working Paper 35. Department of Management Science, University of Strathclyde, U.K.

Bennett, P.G., A. Tait, and K. Macdonagh. 1994. INTERACT: developing software for interactive decisions. *Group Decision Negotiat.* 3(4):351–372.

Brams, S.J. and D. Wittman. 1981. Nonmyopic equilibria in 2 × 2 games. *Confl. Manage. Peace Sci.* 6(1):39–42.

Carroll, J.E. 1983. *Environmental Diplomacy: An Examination and a Prospective of Canadian-U.S. Transboundary Environmental Relations.* The University of Michigan Press, Ann Arbor, Michigan.

Di Battista, G., P. Eades, R. Tamassia, and I.C. Tollis. 1993. Algorithms for drawing graphs: an annotated bibliography. Available via anonymous ftp from wilma.cs.brown.edu, file/pub/gdbiblio.tex.Z.

Fang, L., K.W. Hipel, and D.M. Kilgour, 1989. Conflict models in graph form: solution concepts and their interrelationships. *Eur. J. Operat. Res.* 41(1):86–100.

Fang, L., K.W. Hipel, and D.M. Kilgour. 1993. *Interactive Decision Making: The Graph Model for Conflict Resolution.* Wiley, New York.

Fraser, N.M. and K.W. Hipel. 1979. Solving complex conflicts. *IEEE Trans. Syst., Man Cybern.* SMC-9:805–817.

Fraser, N.M. and K.W. Hipel. 1984. *Conflict Analysis: Models and Resolutions.* North-Holland, New York.

Fraser, N.M. and K.W. Hipel. 1988. Decision support systems for conflict analysis. Pages 13–21 in M.G. Singh, K. Hindi, and D. Salassa, Eds. Managerial Decision Support Systems, Proceedings of the IMACS/IFORS 1st International Colloquium on Managerial Decision Support Systems. North Holland, Amsterdam.

Fraser, N.M. and K.W. Hipel. 1989. Decision making using conflict analysis. *OR/MS Today.* 16(5):22–24.

Gauvin, S., G.L. Lilien, and K. Chatterjee. 1990. The impact of information and computer based training on negotiator's performance. *Theor. Decision.* 28:331–354.

Goicoechea, A., D.R. Hansen, and L. Duckstein. 1982. *Multiobjective Decision Analysis with Engineering and Business Applications*. Wiley, New York.

Hipel, K.W., Ed. 1992. *Multiple Objective Decision Making in Water Resources*. American Water Resources Association (AWRA) Monograph Series No. 18. Herndon, Virginia.

Hipel, K.W., L. Fang, and D.M. Kilgour. 1993. Game theoretic models in engineering decision making. *J. Infrastruct. Planning Manage. Japan Soc. Civil Eng.* 470(IV-20):1–16.

Hipel, K.W. and D.B. Meister. 1994. Conflict analysis methodology for modelling coalitions in multilateral negotiations. *Inf. Decision Technol.* 19:85–103.

Hipel, K.W., K.J. Radford, and L. Fang. 1993. Multiple participant-multiple criteria decision making. *IEEE Trans. Syst., Man. Cybern.* 23(4):1184–1189.

Howard, N. 1971. *Paradoxes of Rationality*. MIT Press, Cambridge, Massachusetts.

Howard, N. 1990. Soft game theory. *Inf. Decision Technol.* 16(3):215–227.

International Joint Commission. 1988. *Flathead River International Study Board Report*. Ottawa, Ontario — Washington, D.C.

International Joint Commission. 1988. *Impacts of a Proposed Coal Mine in the Flathead River Basin*. Ottawa, Ontario — Washington, D.C.

Jarke, M., M.T. Jelasi, and M.F. Shakun. 1987. MEDIATOR: towards a negotiation support system. *Eur. J. Operat. Res.* 31(3):314–334.

Jelassi, M.T. and A. Foroughi. 1989. Neogotiation support systems: an overview of design issues and existing software. *Decision Support Syst.* 5:167–181.

Kilgour, D.M. 1984. Equilibria for far-sighted players. *Theor. Decision*. 16:135–157.

Kilgour, D.M. 1985. Anticipation and stability in two-person noncooperative games. Pages 26–51 in M.D. Ward and U. Luterbacher, Eds. *Dynamic Models of International Conflict*. Lynne Rienner Press, Boulder, Colorado.

Kilgour, D.M., L. Fang, and K.W. Hipel. 1991. Analyzing the Flathead River resource development dispute using the graph model for conflicts. Pages 101–110 in M.G. Singh and L. Travé-Massuyès, Eds. *Decision Support Systems and Qualitative Reasoning, Proceedings of the IMACS International Workshop on Decision Support Systems and Qualitative Reasoning, Toulouse, France.*

Kilgour, D.M., L. Fang, and K.W. Hipel. 1995. GMCR in negotiations. *Negotiat. J.* 11(2):151–156.

Kilgour, D.M., K.W. Hipel, and L. Fang. 1987. The graph model for conflicts, *Automatica*. 23(1):41–55.

Meister, D.B. and N.M. Fraser. 1994. Conflict analysis technologies for negotiation support. *Group Decision Negotiat.* 3(3):333–345.

Nagel, S.S. and M.K. Mills. 1989. Multicriteria dispute resolution through computer aided mediation software. *Mediat. Q.* 7(2):175–189.

Nash, J.F. 1950. Equilibrium points in n-person games. *Proc. Natl. Acad. Sci. U.S.A.* 36:48–49.

Nash, J.F. 1951. Noncooperative games. *Ann. Math.* 54(2):286–295.

Nunamaker, Jr., J.F. 1989. Experience with and future challenges in GDSS group decision support systems. *Decision Support Syst.* 5:115–118.

Radford, K.J. 1989. *Individual and Small Group Decisions*. Springer-Verlag, New York.

Radford, K.J., K.W. Hipel, and L. Fang. 1994. Decision making under conditions of conflict. *Group Decision Negotiat.* 3:169–185.

Roy, B. 1985. *Méthodologie Multicritère d'Aide à la Décision*. Economica, Paris.

Sage, A.P. 1991. *Decision Support Systems Engineering*. Wiley, New York.

Singh, M.G., J.B. Singh, and M. Corstjens. 1985. MARK-OPT — a negotiating tool for manufaturers and retailers. *IEEE Trans. Syst., Man Cybern.* 15(4):483–495.

Thiessen, E.M. and D.P. Loucks. 1992. Computer-assisted negotiation of multiobjective water resources conflicts. *Water Resour. Bull.* 28(1):163–177.

Vincke, P. 1992. *Multicriteria Decision-Aid*. Wiley, Chichester, U.K.

Wang, M., K.W. Hipel, and N.M. Fraser. 1988. Modeling misperceptions in games. *Behav. Sci.* 33(3):207–223.
Winter, F.W. 1985. An application of computer decision tree models in management-union bargaining. *Interfaces*. 15(2):74–80.
Zagare, F.C. 1984. Limited-move equilibria in 2 × 2 games. *Theor. Decision*. 22.

ADDRESSING DIVERSE CLIENT NEEDS

II

Chapter 4

Co-Learning Our Way to Sustainability: An Integrated and Community-Based Research Approach to Support Natural Resource Management Decision Making

W.J. ALLEN, O.J.H. BOSCH, R.G. GIBSON, AND A.J. JOPP

Introduction

Sustainability is an elusive goal. The notorious vagueness of the term, and its scope for varied and seemingly legitimate interpretation by different parties, appear to make it all but useless as an operational guide (O'Riordan, 1988). One has only to consider simple questions — sustain what? how? for whom? over what time period? measured by what criteria? — to appreciate that sustainability can never be precisely defined. Regardless of the ambiguity of the term, however, there appears to be a general consensus that achieving sustainability will place new demands on individuals, society, and science.

Getting serious about sustainability means acknowledging the intricate interdependency of environmental, economic, and social issues on a finite planet.

Within this broader context, science and technology are seen as providing means to achieve ends that are continually redefined by major social concerns. The challenge facing science is how best to structure and undertake research to meet the diverse — and often apparently conflicting — needs of society, local communities, and individual natural resource users. In an uncertain and constantly changing environment, science must strive to develop the understanding, knowledge, forums, and learning environments to better inform and support more sustainable decision making.

From this perspective, successful research and development (R&D) efforts will be participatory in nature, and be based on a process of active adaptive management. In turn, finding out about complex and dynamic situations, then taking action to improve them, forms the basis for developing the necessary learning environment.

Methodological Challenges

The theoretical foundations on which natural resource R&D policies and practices are based are undergoing a paradigm shift. Conceptually, traditional approaches are generally based on reductionist scientific methodologies and often on the expertise within single disciplines (Dahlberg, 1991). Despite a growing recognition of the increasing complexity and social construction of resource management issues, there have been few recent innovations in research methodology other than the development of quantitative modelling and an increased focus on the development of expert systems (Ison and Ampt, 1992). In addition, the research systems in which these DSS (decision support systems) have been developed have been, and still are, largely characterized by the linear transfer of technology (TOT) model of research and development (Russell et al., 1989). The dominant metaphors are those of "information transfer," "channels of communication," and "teaching," most of which arise from mistakenly seeing human communication in the same way as data transferred between computers (Ison, 1991).

Despite the vast amounts of time and money that have been and are being spent on natural resource management R&D, the results, as Hadley (1993) points out, are often illusory or counterproductive. For example, in African countries conservation attempts have largely been ineffective (Bosch, 1989), and few range management projects have had a discernible, positive, and permanent impact on the way communal rangeland is used (Behnke and Scoones, 1991). In New Zealand, the slowness with which land use and land administration have responded to changes in ecological conditions has been noted by O'Connor (1986). Perhaps most telling, the majority of these development initiatives have failed to enlist the active cooperation of the communities they were supposed to serve.

Recognition of such failures has triggered attempts to rethink approaches for linking research with management and policy. Increasingly, alternative approaches are based on concepts of open and evolving systems. They are participatory in nature. There is a growing acceptance of the need to build on principles of experiential learning and systems thinking (Bawden et al., 1985). Research, technology (extension), education, and users are therefore recognized as forming elements of one agricultural information system (Roling, 1988). Such a system must be seen as going beyond the TOT paradigm. An information system, in this sense, cannot be usefully regarded only in terms of its transfer. Rather, it is a "social system," within which people interact to create new knowledge, and broaden their perspective of the world (Land and Hirschheim, 1983; Ison, 1991). Given the diverse set of decision environments inherent in the resource management arena, such a system will, to an increasing extent, rely on information technology for its function.

Because there can never be perfect knowledge of ecological processes within nonequilibrium systems, the ideas underpinning our perceptions of sustainable resource management will change as knowledge expands (Burnside and Chamala, 1994). As evolving economic, technical, and social systems impact on management they also contribute to changing definitions of sustainability. Accordingly, successful resource management must be based on a process of active adaptive management, or "learning by doing" (Walters and Hilborn, 1978; Westoby et al., 1989).

Integrated Systems for Knowledge Management

The Integrated Systems for Knowledge Management (ISKM) approach is designed to support such an ongoing process of constructive community dialogue and to provide practical resource management decision support for land managers and policy makers. This framework has been developed in the South Island high country of New Zealand to help communities (policy makers, land managers, and other interest groups) share their experiences and observations to develop the knowledge needed to support sound resource management decision making.

The focus of the ISKM framework (Figure 4.1) is to provide an organized set of principles and methodologies which will guide our actions as we go about "managing" real-world problem situations. It builds on principles of experiential learning and systems thinking, and is applicable to developing the knowledge and action needed to change real situations constructively. In practice, the process is cyclical and highly iterative with many steps likely to be carried out simultaneously. There are also numerous entry points. The process comprises two phases, which together serve to create an effective learning environment for those involved. The ISKM framework can, however, be usefully viewed as having four main steps as illustrated in Figure 4.1.

Step 1: Scoping Goals and Objectives

The first phase of the approach emphasizes developing a common understanding of any perceived issue or problem. This entails an initial scoping process to help those involved to clearly define the nature of the system under consideration. This serves as a basis for determining the needs of the different interest groups involved, and the specific goals and outcomes they wish to achieve. Because this provides an opportunity to involve all interested parties in the research process from the outset, it is more likely to lead to the development of opportunities and outcomes relevant to community needs.

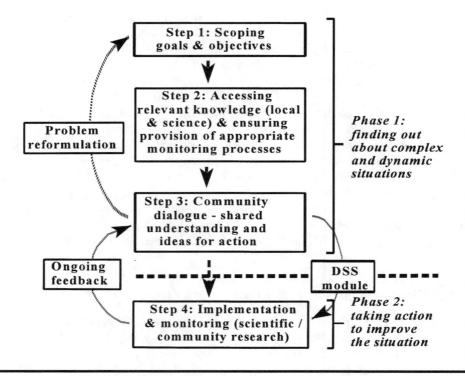

Figure 4.1 ISKM — a participatory research framework to facilitate the identification and introduction of more sustainable resource management practices. The two phases interact to create an effective learning environment.

Step 2: Accessing Relevant Knowledge

This emphasis on problem formulation ensures a focus on the collation and development of "relevant" knowledge. This clarity also ensures the provision of relevant monitoring tools by which the success of measures to achieve the stated goals can be assessed.

Years of experience have provided land managers with a wealth of knowledge on their local systems. This information, unfortunately, is rarely documented and not readily available to land managers on a collective basis. Similarly, much of the valuable knowledge that scientists have accumulated is fragmented, held in different databases, and not always readily available. Accordingly, the second step of the ISKM process focuses on bringing local and scientific knowledge systems together. In this regard, the initial scoping activities also provide a basis for the design of appropriate processes (interviews, focus groups, questionnaires, etc.) to unlock and access the relevant existing data and information from both local and research communities. In turn, the process supports the successful implementation of relevant monitoring programs as land managers become involved in the interlinked processes of monitoring and adaptive management.

Step 3: Community Dialogue

Given the complexity and different social perceptions of many agricultural and environmental situations, the process actively supports improved communication flows among all those involved to develop the "useful knowledge" needed to provide practical decision support.

Facilitated workshops provide a learning environment within which participants develop a shared understanding of how others see the world and how that shapes the way they act in it (e.g., manage their land, carry out their research, develop policy). Importantly, the process recognizes the contextual nature of information. A strategy or goal suggested by a farmer, policy maker, or environmental group will always have been derived from within a particular social, economic, and ecological setting. Scientific results are similarly derived from a particular context, which will include factors such as scale, site, and the researcher's personal world view. Accordingly, the community dialogue process is designed to seek the active cooperation of participants in developing a common understanding of the context in which any individual piece of information becomes relevant.

Generating Useful Knowledge

The ongoing community dialogue is designed to produce useful knowledge to help all those involved in the process. It provides those land managers and policy makers who participated in the process with immediate access to new ideas and perspectives which may help them reevaluate their current management or policy strategies. Because many sustainability issues need to be addressed simultaneously at a number of different levels of decision making, the workshops emphasize the generation of appropriate strategies for different system hierarchies (from block/field goals through individual enterprise objectives to catchment community goals). At the same time, it helps the different interest groups involved to develop a shared understanding which can reduce the level of conflict that currently surrounds many resource management issues.

The process automatically aids the identification of new and relevant research initiatives as knowledge gaps are identified. Importantly, these forums also provide farmers, conservators, policy makers, and others with the opportunity to provide researchers with a greater appreciation of their information and technical needs.

Capturing Knowledge for Decision Support

As demonstrated above, this ongoing community dialogue provides all those directly involved with a learning environment in which "useful knowledge" is developed. There is a need to capture this knowledge to benefit all of those who have not had the opportunity to be directly involved. Where appropriate this can be done through a range of presentation techniques such as manuals, posters, or decision trees on paper. However, in many cases computer-based DSS are not only appropriate, but essential, to help deal with the complexity inherent in environmental decision making.

The prototyping approach that is inherent in the ISKM framework encourages an interactive process where DSS developers and users collaboratively discover new requirements and refinements, which are then incorporated in succeeding versions. This is especially useful when DSS development is seen as a process that can be enhanced by the use of iterative 'soft' systems methodologies involving processes of feedback and learning among all the different participants in the situation under inquiry (Miles, 1988). In this way, the development process allows the user to learn and experience the system at an early stage. This process is important because it encourages user confidence in subsequent working versions (Brittan, 1980).

In the long term, such a computer-based DSS is designed to integrate a diverse array of information sources and provide users with a more holistic perspective of a complex situation. It is perhaps best viewed as a library incorporating a wide range of experiential knowledge in the form of expert systems, DSS modules, software packages, and databases.

The design encourages the user to define and then select a management goal. By answering simple questions and being prompted to provide further information with the help of associated models and specialist packages, the user can create new information (allowing for ecological diversity, etc.) relevant to the issue under investigation. Prompts act to provide a pathway towards the provision of management advice. Through the use of hypertext the user can obtain further explanation and clarification of the assumptions behind selected answers, along with the ability to access associated subject areas. In some cases this will simply require access to another part of the system, but it is envisaged that in the future access will also be provided to external information sources through links such as Internet. These related abilities are important as they allow users to assess the reliability of the decision support on their own terms (Stafford Smith and Foran, 1990), and to create a personal learning environment (Jonassen, 1992).

Step 4: Monitoring and Adaptive Management

Importantly, the ISKM framework allows the substance and context of the required information flows to be updated as more knowledge becomes available, and different goals are set.

An Ongoing Role for Land Managers

In normal practice, land managers manipulate ecosystems primarily to achieve a management objective, rather than to find out how the system works. However, as a number of researchers observe, a management action can also be regarded as an experiment (Walters and Hilborn, 1978; Dankwerts et al., 1992). As land managers measure the outcomes of their management actions they continually gain new "experimental results." These results provide new information whereby the community-derived knowledge base developed through the third step of the ISKM process is reevaluated and expanded in collaboration with scientists and other stakeholders (Figure 4.1).

Involvement in the participatory processes of monitoring and adaptive management in this way, means that individual land managers acquire greater technical expertise — building on both collective local knowledge and an associated scientific awareness of their physical environment. At the same time, by achieving specific objectives for improving their resource position through a collective effort, land managers develop greater confidence, which, in turn ensures the uptake needed for the process to continue.

An Ongoing Role for Science

At any given time the research process can play an important role in helping the community and scientists to determine new research priorities jointly. It helps identify knowledge gaps and assists in prioritizing new research initiatives. This is a continual process as evolving knowledge, technologies, and value systems inevitably change our perceptions and provide new areas and issues for research (Stuth et al., 1991). It also acts to provide automatic feedback of research results to end users.

The addition of new local knowledge through monitoring and adaptive management will add to the range of strategies to be evaluated, and strategies and options will continually change in response to social, economic, and ecological pressures. This creates a role for ongoing research to determine the wider applicability and environmental implications of management options and strategies.

Concluding Remarks

The process is thus iterative, with each repetition serving to maximize the knowledge available at any time to support decision making by those in the community. As those involved cooperate to develop the necessary knowledge and knowledge-based tools, new issues will be raised and the process expanded. Ultimately this concept of "learning by doing" could also be broadened to include policy initiatives. As Rondinelli (1983) argues, all social development activities must be seen primarily as experiments and dealt with as complex and uncertain ventures in which the participation of those who are expected to benefit is essential.

In the South Island high country of New Zealand the ISKM framework was initially used to help the community find practical land management strategies to address the problem of an invasive weed, *Hieracium* spp. But using this approach to look at the problem from the point of view of management also highlights how ecological, social, and economic issues are inexorably linked. No one manages for *Hieracium* alone. For example, farmers are primarily concerned with managing for goals such as increased stock production or available forage supply, while conservators will place an emphasis on management to protect a particular species or threatened ecosystem. Both these groups will also be concerned with other issues such as watershed and landscape management. Accordingly, the ISKM process is now being used in the high country to address a number of related issues such as conservation, grazing management, burning, and water quality.

As we bring different knowledge systems together through this process, it becomes clear that what you look for is what you get. As Argyris and colleagues (1985) point out, depending on the community in which they operate, each different interest group will look for different facts and solutions in accord with their own set of norms for inquiry. In the example above, we find scientists concentrating on determining the effects of grazing on *Hieracium* (describing and accounting for some phenomenon). In contrast, farmers ask more focused questions such as the effects of different grazing regimes (rotational grazing vs. set stocking, different grazing intensities and frequencies, etc.), and are concerned with applying the answers to real-life contexts "amidst all the complexity and multiple dilemmas of values they pose" (Argyris et al., 1985).

Much of the apparent conflict surrounding many resource management issues relates to the fact that different interest groups fail to appreciate the perspectives and values inherent in the actions of others. If these groups can be encouraged to share their experiences and viewpoints there will be a greater understanding of why these differences exist. Equally important, the involvement of different groups may well provide useful ideas and strategies that lie outside the normal perspective of those with the primary responsibility for managing any particular resource.

Collaboratively developing new management options and strategies through the ISKM process provides all interested parties with the opportunity to learn from local experiences gained within enterprise and catchment-level systems. This provides all those involved with an appreciation of management concerns and issues, and allows groups such as scientists and policy makers a better feeling of how their contributions fit into the total system. The result of such cooperation automatically leads to the design of relevant research that will directly benefit both land managers and policy makers.

Although cooperative ventures such as those described here may not yet offer definitive solutions to such elusive issues as sustainability, they can begin to offer a variety of knowledge-based tools and possible courses of action to enable the community to make better informed decisions. In turn, as communication flows between different sectors of the community are expanded and improved, the level of needless conflict surrounding a number of land management issues should be minimized. Accordingly, this participatory approach represents a framework through which different segments of society can cooperate to develop and work towards a more coordinated set of environmental goals.

References

Argyris, C., R. Putnam, and D.M. Smith. 1985. *Action Science*. Jossey-Bass, San Francisco.
Bawden, R.J., R.J. Ison, R.D. Macadam, R.J. Packham, and I. Valentine. 1985. A research paradigm for systems agriculture. *Agricultural Systems Research for Developing Countries, Proceedings of an International Workshop*. J.V. Remenyi, Ed. ACIAR.
Behnke, R.H. and I. Scoones. 1991. Rethinking range ecology: implications for rangeland management in Africa. Overview of paper presentations and discussions at the technical meeting on savanna development and pasture production. Woburn (UK), 19–21 November 1990. London, Commonwealth Secretariat, Overseas Development Institute and International Institute for Environment and Development.
Bosch, O.J.H. 1989. Rangeland inventory and monitoring: the African experience. *Proceedings of the International Conference and Workshop on Global Natural Resource Monitoring and Assessments: Preparing for the 21st Century*. Volume 1. Pages 221–231.
Brittan, J.N.G. 1980. Design for a changing environment. *Comput. J.* 23(1):36–42.

Burnside, D.G. and S. Chamala. 1994. Ground-based monitoring: a process of learning by doing. *Rangeland J.* 16(2):221–237.
Dahlberg, K.A. 1991. Sustainable agriculture — fad or harbinger? *BioScience.* 41(5):337–340.
Danckwerts, J.E., P.J. O'Reagain, and T.G. O'Connor. 1992. Range management in a changing environment: a South African Perspective. *Rangeland J.* 15(1):133–144.
Hadley, M. 1993. Grasslands for sustainable ecosystems. Pages 12–18 in M.J. Baker, Ed. *Grasslands for Our World.* SIR Publishing, Wellington.
Ison, R.L. 1993. Changing community attitudes. *Rangeland J.* 15(1):154–166.
Ison, R.L. and P.R. Ampt. 1992. Rapid Rural Appraisal: a participatory problem formulation method relevant to Australian agriculture. *Agric. Syst.* 38:363–386.
Jonassen, D.H. 1992. What are cognitive tools? In P.A.M. Kommers, D.H. Jonassen, and J.T. Mayes, Eds. *Cognitive Tools for Learning.* Springer-Verlag, New York.
Land, R. and R. Hirschheim. 1983. Participative systems design: rationale, tools and techniques. *J. Appl. Syst. Anal.* 10:91–107.
Miles, R.K. 1988. Combining 'soft' and 'hard' systems practice: grafting or embedding? *J. Appl. Syst. Anal.* 15:55–60.
Nores, G.A. and R.R. Vera. 1993. Science and information for our grasslands. Pages 23–27 in M.J. Baker, Ed. *Grasslands for Our World.* SIR Publishing, Wellington.
O'Riordan, T. 1988. The politics of sustainability. In R.K. Turner, Ed. *Sustainable Environmental Management: Principles and Practice.* Westview Press, Boulder, Colorado.
O'Connor, K.F. 1986. The influence of science on the use of tussock grasslands. *Rev. J. Tussock Grasslands and Mountain Lands Inst.* 43:15–78.
Roling, N. 1988. *Extension Science: Information Systems in Agricultural Development.* University Press, Cambridge.
Rondinelli D.A. 1983. Projects as instruments of development administration: a qualified defence and suggestions for improvement. *Publ. Admin. Dev.* 3(3):307–327.
Russell, D.B., R.L. Ison, D.R. Gamble, and R.K. Williams. 1989. A Critical Review of Extension Theory and Practice. AWC/UWS, Richmond, Virginia.
Russell, D.B. and R.L. Ison. 1991. The research-development relationship in rangelands: an opportunity for contextual science. In Proceedings IVth International Rangelands Congress, Montpelier, France.
Stafford Smith, D.M. and B.D. Foran. 1990. RANGEPACK: the philosophy underlying the development of a microcomputer-based decision support system for pastoral land management. *J. Biogeogr.* 17:541–546.
Stuth, J.W., C.J. Scifres, W.T. Hamilton, and J.R. Connor. 1991. Management systems analysis as guidance for effective interdisciplinary grazingland research. *Agric. Syst.* 36:43–63.
Walters, C.J. and R. Hilborn. 1978. Ecological optimization and adaptive management. *Annu. Rev. Ecol. Systemat.* 9:157–188.
Westoby, M., B. Walker, and I. Noy-Meir. 1989. Opportunistic management for rangelands not at equilibrium. *J. Range Manage.* 42(4):266–274.

Chapter 5

Consequences of Short-Term, One-Objective Decision Making in Land and Water Use Planning in Egypt

MOHAMMED ATIF KISHK

Abstract

Contemporary Egypt faces many challenges. To face them, decisions are being made, at all levels, under stress situations. On one hand, Egyptian farms are extremely small (more than 95% of the holdings are smaller than 2 ha) and the majority are very poor. The decisions in using land and water are governed by the day to day need to feed large families. The options they have are very limited. On the other hand, the public sector, unable to invest enough to face pressing needs of the current population, very often fails to invest for the sake of future generations.

This paper gives examples of conflicts between current and future needs of small-holder irrigated agriculture in Egypt. It shows how short-term, one-objective decision making in the use and management of both land and water contributes to the degradation of these resources. It will cite examples related to the conversion of arable land to non-agricultural uses, the allocation of land and water for different crops, and the pricing policies and some legal and institutional aspects related to soil and water conservation.

Introduction

Egypt suffers a serious scarcity of both arable land and water. The per capita share of both land and water is declining all the time. The result is that in spite of the obvious developments in the agricultural sector, the level of self-sufficiency of major food commodities is worsening. (By 1973, Egypt had become a net importer of agricultural products.) Overall production should increase to turn the trend of deteriorating degree of self-sufficiency. This is just one challenge Egypt is facing. The other challenge is that competition for land, and particularly for water, is bound to increase. It will increase between sectors and between various uses within each sector, and more importantly between current and future generations (Kishk and Lundquvist, 1994).

To face these challenges, decisions are being made, at all levels, under stress situations. On one hand, Egyptian farm holdings are extremely small (more than 95% of the holdings are smaller than 2 ha) and the majority are very poor. The decisions in using land and water are governed by the day to day need to feed large families. Their options are very limited On the other hand, the public sector, unable to invest enough to face pressing needs of the current population, very often fails to invest for the sake of future generations.

This paper gives examples of conflicts between current and future needs of small-holder irrigated agriculture in Egypt. It shows how the short-term, one-objective decision making in the use and management of both land and water contributes to the degradation of these resources. It will cite examples related to the conversion of arable land to non-agricultural uses, the allocation of land and water for different crops, and the pricing policies and some legal and institutional aspects related to soil and water conservation.

The examples given show that planners and decision makers in Egypt, as well as in many developing countries, are sometimes facing serious difficulties in making the right decisions. However, in some cases costly mistakes can be avoided and many losses can be saved.

Land, Water, People, and Food in Egypt

Primarily because of the scarcity of water, arable land in Egypt has always been limited. It was enough, nevertheless, to feed the Egyptian population and allow some exports until the 1940s. Increasingly from the 1950s to the 1970s, Egypt became a net food importer and lost the self-sufficiency that had lasted for centuries. The figures in Table 5.1 summarize the situation. As can be seen, the land/man and water/man ratios steadily decline and the food gap widens. The situation reached crisis proportions and had several consequences. Apart from economic and political dependence and the alarming social unrest that may arise as a result of poverty and hunger, the human pressures on the limited resource base were expected to be tremendous. Agricultural production practices became more intensive (two crops a year, the second crop is usually sown before the first one is harvested, the haphazard use of huge amounts of chemicals, the promotion of high yielding varieties...etc.).

Table 5.1 Population, Land, Water, and Food Gap in Egypt (1800-1990)

Year	Population (millions)	Land/man (fed[a]/person)	Water/man (m³/person)	Food gap Million ton	Food gap Million $
1800	2.0	1.00	—	—	—
1850	4.6	0.87	—	—	—
1897	9.7	0.51	5,084	—	—
1907	11.2	0.48	4,414	—	—
1927	14.2	0.39	3,484	—	—
1937	15.9	0.33	3,484	—	—
1947	19.0	0.31	2,604	—	—
1960	26.1	0.23	1,893	1.3	150
1970	33.2	0.18	1,713	1.7	984
1980	42.1	0.14	1,351	7.4	10,090
1990	55.0	0.13	1,034	?	?

[a] One feddan = 0.42 ha.
Hamdan (1983) and CAPMAS (1989).

What makes the present situation even worse is not only the problem of overpopulation. The population density on the limited arable land area is among the highest in the world. The majority of the rural population are very small, poor farmers, living and operating on a very small resource base. More than 95% are holding less than 2 ha, often depleted and fragmented. They have a very low level of human capital in terms of education, knowledge, and health. They suffer chronic indebtedness and lack accessibility to institutional credit and inputs. They face unstable markets and prices and receive no or very little extension support. They have no control on rural institutions and lack access to public services. The socioeconomic-political conditions of those small farmers are in the heart of land degradation/conservation potentials and possibilities.

Under the pressures of this challenging situation, decisions are being made and it is only natural that the decisions made at all levels are strongly affected by those pressures. Let us now examine a few examples.

Decisions Related to Land Degradation Problems

There is near consensus that land degradation problems in Egypt are serious and far-reaching in terms of areas affected and the millions of poor farmers who suffer the consequences to the extent that the whole economy is actually retarded by the occurrence of such problems. Those problems are related to the climate, but they have also an intimate link with the conditions under which the small Egyptian farmers have to live and work. The decisions being made to deal with those problems often fail and in some cases worsen the problems rather than solving them. There are many reasons behind this decision failure. Among these reasons the following are just a few examples.

Inadequate Definitions and Very General Assessments

In many cases of desertification debate and discussions, there was the assumption that desertification is primarily a technical land management issue. This shortcoming has been stressed by Olsson (1993) when he stated that "...the main reason for the failure in dealing with land degradation problems was that these problems have mainly been considered as an environmental issue, dealt with primarily by natural scientists using natural science based methodologies and data." He then continues "...If we see dry land degradation as a symptom of a socioeconomic political disease rather than an environmental one, we may be more successful." The same argument was also used by Mortimore (1993) who stated that "the vast majority of the disruptive changes fostering desertification stem from social causes. The deterioration of the physical landscape is the symptom of the problem rather than its root cause." Bakhit (1993), using the analogy between medicine and desertification, concluded that "emphasis on treating the symptoms of desertification, such as the restoration of depleted vegetative cover, is an endeavor which does not contribute substantially to control the problem, as long as the root causes which drive people to deplete natural resources remain unchallenged."

Because of loose and inadequate definitions there have been many general assessments and rough estimates of land degradation problems. In Egypt, for instance, El Gabaly (1972) estimated that one-third of the irrigated land in Egypt was more or less salinized or threatened with salinity. Kishk (1986) stated that in 1982, almost all irrigated area in Egypt was potentially salt-affected and at least one-half of this area is more or less affected. Biswas (1991), while discussing these estimates, said that these are really meaningless statements because they did not "even indicate what is meant by salinized land, for example does it mean decline in agricultural productivity due to salinity by 1%, 5%, 10%, 50% or 80% or land completely withdrawn from agricultural activities."

Dregne and Chou (1992) gave estimates of the areas affected by different levels of desertification worldwide. Their estimates of areas affected in Egypt were as follows:

- 1,735,000 ha weakly affected
- 700,000 ha moderately affected
- 50,000 ha strongly affected
- 1,000 ha very strongly affected

In their definition, the very strongly affected areas are those in which more than 75% of their productivity is lost due to desertification. From just very preliminary observations of a frequent traveler throughout Egypt, it seems that the very strongly affected areas are much more than 1,000 ha. In this and other general assessments, both underestimates and overestimates are obvious, and many important aspects related to the problems are usually missing.

This is not the case in Egypt only. Worldwide, general assessments like these have been criticized by many scientists who even deny the very existence of desertification in some Sahelian landscapes which were generally considered seriously desertified (Forse, 18989; Hellden, 1991; Olsson, 1993; Mortimore, 1993).

The general statements are perhaps useful in popularizing the severity of land degradation problems. However, they are not at all useful in giving practical orientation to deal with those problems effectively on a reliable scientific basis. They may draw the attention of the public but on the level of the decision-making process, they are meaningless or misleading. They are based on an abstract ecological ideal rather than the reality of contemporary resource use and needs (Olsson, 1993).

One of the major reasons for such general statements on desertification is the inadequacy of definitions. The lack of proper definitions is one of the most crucial reasons for what is recognized now as the desertification credibility gap (Warren and Agnew, 1988) and for the fact that the whole concept of desertification is still confusing and misused. Needless to say, there is still widespread confusion about the meaning of the term even in the scientific community, not to mention the media, the policy makers, and the public in general.

Olsson (1993) rightly criticized the global estimates of areas affected by desertification given by UNEP in 1977 and 1984 (reviewed by Mabbutt, 1985) and Dregne et al, (1991) as being "extremely rough figures that were quoted and manipulated by other authors and contributed to the false picture of an advancing desert."

Even the World Atlas of Desertification (UNEP, 1992) is not based on primary research into the regions most affected, but again, as in 1977 and 1984, based on estimates and guesses. The Global Assessment of Soil Degradation (Oldeman, 1991) which was the main source of information for the World Atlas of Desertification, is just a compilation of existing information and of the knowledge of over 250 different experts "each one with potentially different perceptions of the problem, its definitions and probably also with different goals" (Olsson, 1993).

One example of false general statements was cited recently about Sudan (Olsson, 1993). Olsson first quoted some of the statements previously made:

- "It has been estimated that 650,000 square kilometers of the Sudan had been desertified over the last 50 years and that the front-line has been advancing at a rate of 90-100 kilometers annually during the last 19 years (Suliman, 1988).
- "Vice-President Bush has been urged to give aid to the Sudan because desertification was advancing at 9 km per annum (Warren and Agnew, 1988).

He then summarized the results of 30 years of research in the Western Sudan by scientists from Sweden and the University of Khartoum, reviewed by Hellden (1991). The main conclusions were

- Creation of long-lasting desert-like conditions could not be found.
- Degraded areas surrounding settlements did not expand.
- The northern cultivation limit did not change significantly.
- No changes in vegetation cover took place that could not be explained by climatic variations.
- Crop yield variations could mainly be explained by climatic variations, rather than a secular trend.

What has been said so far does not mean that there is no such problem called desertification. It only means that many general statements made in the last 20 years are not accurate and not based on real field work, and therefore they may be misleading and confusing. If there is enough evidence that will convince us to accept this as a fact, the conclusion then is there is a serous need for real improved assessments of land degradation problems both on global, national, and local levels.

Assessment and monitoring should be concerned not only with answering the questions of how much, where, and at what rate, but also the why question, and what is relevant or appropriate and what is not. Why are the problems worsening in spite of the fact that the technologies for solving them are available? What are we really lacking? Is it the money, the facilities, the will or what? What kind of institutions are able to help the small poor farmers conserve their limited land and water resources? Are those small poor farmers able to do so after all? What are the relationships between the socioeconomic-political status of the majority of farmers and the options they might have in using and managing their resources in a sustainable way? Usually, there are no adequate answers for these questions and similar others mainly because the problems are not very well defined.

Being aware of the many limitations in the current approaches and methodologies in land degradation assessment, scientists have recently proposed alternatives (see for example Ibrahim, 1993; Olsson, 1993; Bakhit, 1993). The proposed alternatives mainly widen the scope of land degradation problems to include all relevant human dimensions in the assessment and monitoring of such problems. While these alternatives will certainly be useful, they might not be the final solution. The world community, including the scientists, will have to continue improving its understanding of desertification. The first step in this endeavor is better assessments and monitoring. As Bakhit (1993) nicely puts it, "failure to 'win the battle' was not because the 'enemy' desertification was stronger than it was first assumed, but mainly because the 'enemy' was not clear."

The International Convention to Combat Desertification signed in 1994 has recognized the complexity of the challenge of combating desertification and proposed to meet it with a holistic approach. Furthermore, the Convention identifies the primacy of the fight against poverty to restore degraded land. It anticipated the people-centered approach to development that was the focus of the recent World Summit for Social Development in Copenhagen.

During the Convention it was stressed that "desertification is about the degradation of arid lands and related natural resources potential, a complex process which still needs to be much better understood but which certainly includes a broad range of multidisciplinary measures to fight land degradation and poverty." It was then recommended that "priority must be given to participatory eco-development programs that may remove the pressure of vulnerable groups on the fragile environment by diversifying their economic options" (UN, 1994).

Lack of Complete and Consistent Data

Data on land degradation are produced by many different institutions including different departments, agencies, authorities, councils, and committees belonging to several ministries and universities. It is very difficult, and always has been impossible, to achieve enough coordination in the planning and execution of interrelated programs. This is just one aspect of the problem. Another is that each department or institute has its own objectives, definitions of the problems, approach and methodologies. It is often difficult, therefore, to compile data from different sources in a meaningful context.

It is recognized that sometimes there are more data than is actually required, but we still may face the difficulty of collecting them in a way that permits comparisons between various places during various periods of time. Data available will remain meaningless unless adequate analysis is undertaken to transform such data into information. Even then, much work remains to be done to give this information the form and content that will enable decision makers to use it safely. In Egypt, as may be the case in many other developing countries, we may have much data on land degradation problems, but it seems we do not have enough consistent information to find possible or easy solutions.

Even for some problems that are easy to assess like the areas of farm land lost for urbanization in Egypt, there are contradictions in the published data. The estimates range between 20,000 fed (Parker and Coyle, 1981) to 100,000 fed (World Bank, 1990). The most reliable figure perhaps is from 30,000 to 50,000 fed annually (Hamdan, 1983; Ghabbour and Ayyad, 1990).

The global attempts are no better, the world Map of Desertification (UNCOD, 1977) shows that in Egypt:

- The majority of the country is actual desert.
- The risk of desertification is high to very high.
- Most areas of Egypt are subject to sand movement, soil stripping, salinity, and human pressure.

In this map and again in the World Map of Soil Degradation published by FAO, UNEP, UNESCO in 1980, many details and facts were missing. For example, while salinity was spreading in many areas in the Nile Valley, on both maps it was confined only to the Northern Delta and Fayoum depression.

Lack of a Multidisciplinary Approach

Land degradation problems are usually multidimensional. Their solutions require contributions from the fields of natural science, resource use and management technology, economics, sociology and political sciences. The multidimensional nature of land degradation has been made clear in the definition of "land" so nicely stated in the "Framework for Land Evaluation" (FAO, 1976): "Land comprises the physical environment, including climate, soils, hydrology and vegetation to the extent that these influence potential for land use. It also includes the result of past and present human activity." In Egypt, almost all projects make sure to mention their intention to follow a multidisciplinary approach. However, as in many other cases in science and society, good intentions are not enough. In practice, even if there are different disciplines, they are not working together as one mind. When the results of each discipline are combined with the results of others in the final report, it is very obvious then that most of the important issues and relations are missing.

Unfair Competition and Biased Decisions

Everywhere, there are often many (individuals, groups, classes, regions, generations, enterprises, etc.) competing for few opportunities. This is particularly the case in poor countries. The competition between rich and poor countries or classes within countries is even worse, and the adverse effects of this competition always apply to the poor.

The decisions, made at all levels, regarding the use of resources, will normally favor some of the competing parties at the cost of others, depending on the power structure directing the decision-making process. Even if the decisions take care of the needs of all parties, they will directly affect people in other places/countries or at other times (future generations). The conflicts are unavoidable mainly because those people are not represented, and therefore their needs are not fully considered in our decision making. Sometimes, either by design or by circumstance, their needs are completely neglected.

Chambers and Conway (1992), in their very interesting discussion about sustainable rural livelihoods, stated that in practice, future generations and their livelihoods are undervalued in decision making for four reasons:

1. Inumeracy (the failure to recognize the numbers involved),
2. Undemocratic democracy (future generations have no votes),
3. Discounting (devaluing the future),
4. Uncertainty (inability to predict the future)

I would say that these practical difficulties are only valid in a few cases and they apply particularly to the rich. They are valid only when the decision making can afford to consider the needs of the future generations. In many cases, we are obliged to neglect future needs completely simply because we cannot afford to do otherwise. The opportunities are too limited and the only possible choice is either we or them. Considering the future needs or what is called "sustainable development" is like simply asking the poor people "to die today in order to live well tomorrow". The poor cannot exercise this option as Robert Mugabe put it in his address to the UN General Assembly in 1987. The only option the poor might have is to repeat the famous Egyptian saying "let me live today and you may kill me tomorrow." After all, why should anyone work or look for the future, if he can see no future?

Let us have a closer look at some examples. The international financial institutions (i.e., IMF, IBRD, and USAID) pose pressures on Egypt and many other countries as well as to implement what is called "Structural Adjustment" policies. Those countries are requested to radically increase the share of the market within the Egyptian political economy by privatization of public regulation. The main argument is that competition is the judge of efficiency (Brombley and Bush, 1994). This is also the main feature of USAID strategy for Egypt (USAID, 1992) which stresses that Egyptian agriculture must capitalize on its comparative advantage. This entails a shift towards the production of high-value, low-nutrition foodstuffs for the Egyptian market. This only favors large commercial farmers at the expense of most of the rural producers.

It seems that the government of Egypt is not in a position to refuse such policies. After a thorough analysis of USAID role in Egypt's policies, Mitchell

(1991) concluded that Egypt's dependence on the U.S. and the level of debts that goes with it, "has given the United States a powerful position of influence within the Egyptian state." The Egyptian state, in fact, is only concerned with slowing down the process and trying to accommodate its effects. The government of Egypt has heard many times, both from inside and outside Egypt, warnings that say (see for example Brombley and Bush, 1994):

- "This policy will further empower the strong and weaken the poor."
- "There has been no example of successful development in the history of late industrializing capitalism that did not involve significant levels of state support for particular sector, selective protection of parts of the economy, targeted investment and the (intended or unintended) getting of relative prices wrong."
- "Given the scope of the reform program, its failure to deliver high and sustained levels of growth, and hence to cut into unemployment and to raise living standards, might produce levels of conflict and instability that more resemble recent events in Algeria than the traditional picture of Egypt as a bulwark of regional stability."

The government of Egypt has been lately worried watching the increased terrorist activity of "Islamic fundamentalists". However, this activity was mainly treated by police intervention like normal crimes. The impact of such adjustment on the socioeconomic inequalities, unemployment and hence, the political stability is denied, or at best, minimized in spite of the fact that this movement derives its elements from the very poor, unemployed people.

The English saying "where there is a will, there is a way" is encouraging perhaps, but certainly not true. In the real everyday life we often meet many situations where there is a will but all ways are closed. The Egyptians met such situations throughout their long history and therefore their saying is more realistic "the sight is long but the hand is short". When the hand is short and the focus is on problems and crises, many opportunities are missed.

The growing disparity in income between rich and poor, supported by the government bias, has enabled the better off to divert the country's resources from the production of staples to the production of luxury items. As a result, Egypt now grows more food for animals than for humans and even the smallest farmers have been forced to shift from self-provisioning to the production of animal products and to rely increasingly on subsidized imported flour for their stable diet (Mitchell, 1991).

Another example of unfair competition and biased decisions is quite obvious in the allocation of water resources. As is known, water resources in Egypt are seriously limited. No country in the world is more dependent on irrigated agriculture than Egypt. Currently, the annual renewable water supplies are less than 1000 m^3 per person (the limit for what is called water famine). Egypt is totally dependent on the Nile for its food production. While the rate of population growth is still high, the flow of the Nile is constant or even diminishing by recurrent droughts or by water-diverting projects in one or more of the reparian countries.

These limited water resources are being wasted in several ways particularly in agriculture, the largest consumer of water in Egypt. Kishk and Lundquvist (1994) examined water productivity in Egyptian agriculture. They found that the water used to irrigate 116,000 ha of sugarcane (with a net revenue of $234 million) could be otherwise used to irrigate 488,000 ha of wheat (with a net revenue of $349 million). However, the current policies reflect a certain bias against the poor farmers and consumers as well. The water-wasting crops (i.e., rice, sugarcane, and fruit trees) are mainly grown and intensively consumed by the rich. On the other hand, cereals are mainly grown by the small, poor farmers and are the main food for the majority poor. Because water is almost free and the growers of water-wasting crops are politically powerful, they can get higher prices for their products and pay nothing for the extra water they use. It is also difficult to ask the poor farmers to pay for water either. However, it was proposed that water may be given free up to a certain limit, and this will be a good motivation for farmers to grow water saving crops.

But again the small farmers have no power to affect the decision making, and therefore, the most precious resource of the country is wasted in favor of a few rich powerful farmers. Fortunately, in allocating water for different cropping patterns, there is no conflict between being fair and being efficient.

Lack of Sound Alternatives

Sometimes, there is no real decision because sound alternatives are lacking. One obvious example in Egypt is the conversion of farm land to nonagricultural uses. The bulk of farm land in Egypt is located in the Nile Valley and Delta. More than 99% (58 million) of the total population live in less than 3% (3 million ha) of the total area of the country. Almost all housing, industry, and infrastructure are located also in the same area. The result is that considerable cropland is being converted to nonagricultural uses of an irreversible nature. There are no reliable data about the acres lost. As mentioned earlier, the minimum estimate for the areas lost is 10,000 ha annually. The land lost is among the richest and most productive land in the world. To compensate for these losses, marginal land in the desert is being reclaimed at high cost and doubtful returns.

There are strategies and policies to reduce urban encroachment on prime farm land. By law, building on agricultural land is a crime. There is a fine of about $3,000 and imprisonment liability. However, if you move around any town or village in Egypt, you will see buildings coming up every day, many are government buildings.

Strategies, policies, and legislation to stop or reduce the irreversible loss of farm land very often do not work. Everywhere, this problem feeds on itself. Each plot lost not only breeds new houses that demand and consume more resources and services, but also creates speculation among new farmers who will be happy to sell the land and make more money than they can make by keeping it as a producing farm (Halcrow et al, 1982). In this way, land becomes an asset to be converted to money. In Egypt, the average net return from using the land to produce food was $439/ha/year in 1987 (Ministry of Agriculture, 1991). If the land is used for building, the normal price may range from $0.2 million to more than $5 million/ha depending on the location. Is there any farmer, rich or poor, who can resist this? In such a situation, the absurdity of asking the farmers to conserve the land for their children, does not make any practical sense.

It was just for mental exercise that 300 farmers from 5 villages around Minia were asked: Would you be willing to sell your land for building? On every face there was a look saying something like: "Are you fool or what?" or "What a silly question!" Then the sole answer was "Of course, I'll sell it, who wouldn't?", and some of them will add, "it will be very good luck for my children and I can buy a larger plot somewhere else, build a house and do so many things I was not able to do before."

Removal of top soil for brick making is another similar instance. The brick factories around cities and towns used to offer to farmers what was to them muck money as a price for 50 or 100 cm of top soil of their fields. The average price used to be around $10,000 for the top 100 cm of a hectare. By law, this is forbidden since 1985. However, the practice still goes on in many places.

Our interview with 300 farmers in 5 villages near Minia indicated that all the farmers were aware of the harmful effects of removing the top soil on both the land and the crop yields. All, however, said that the price offered is more than enough to compensate for the losses. They also said they usually remove parts from their own fields to make bricks for their own use because it is cheaper and they have no other alternatives. In such a situation, there is no real decision making because there are no alternatives.

Decisions Related to Investments for Land Conservation

What is given above is just one example. There are many more examples. In Egypt, as is the case everywhere, land conservation needs heavy investments. The public sector, being unable to invest enough to face the pressing needs of the current population, will often forget to invest for the sake of the future generations. On the other hand, even when farmers are rich, they cannot make conservation investment on which there is no short-term return. For poor farmers, even if the short-term return is obvious, they cannot afford investing money simply because they do not have any savings. Borrowing money is perhaps the answer in this case, but is this possible for the small farmers?

A recent survey showed that out of 150 farmers in the reclaimed land west of Minia, 120 said their income is not enough to cover their needs. About what they usually do to cope with their low income situation, 94 said they borrow money from relatives, 33 said they sell "something", 17 said they sell their work, 18 said they sell their children's work, and 15 said they do not know what to do. Only 2 out of 150 farmers said they can borrow money from the Agricultural Credit Bank (those two were relatively rich farmers owning more than 15 ha each). (Kishk, 1993).

In a similar study in a village near Minia, 98 farmers were interviewed (Ibrahim et al, 1993). Only two said they have very small savings for emergencies and none had received any investment loan from the Agricultural Credit Bank. Of those farmers is there one who will be able to invest money in soil conservation?

Inadequate Legislation

In Egypt, and in many other developing countries, the seriousness and urgency of land degradation problems have been generally recognized. Yet legislative action to meet such problems, which is very important response, remains very poor, hesitating and inconsistent. According to Rauschkolb (1971), adequate legislation should necessarily include several aspects:

- Clear statement of purpose
- Identification of the various aspects of the problem
- Establishment of priorities
- Government financial contribution
- Creation of institutions and authorities responsible for land conservation
- Coordination and interconnection between different legal enhancements affecting land
- Education of the public to appreciate land and soil conservation
- Implementation and enforcement through police powers, penalties, and similar measures

I may add to this list that the availability of sound alternatives to restricted activities is highly important.

Conclusion

In conclusion, it can be said that decision making for more sustainable use and management of resources is not an easy task. The saying: "for every problem there is a solution that is easy, straight and wrong" is applicable to the issue and is very wise indeed.

References

Bakhit, A.A. 1993. Desertification: reconciling intellectual conceptualization and intervention effort. *GeoJournal*. 31(1):33-40.
Biswas, A.K. 1991. Land and water management for sustainable agricultural development in Egypt: opportunities and constraints. Report to FAO Project TCP/EGY/0052.
Brombley, S. and R. Bush. 1994. Adjustment in Egypt? The political economy of reform. Review of African Political Economy No. 60:201-213.
CAPMAS (Central Agency for Public Mobilization and Statistics).1989. Statistical Year Book, Arab Republic of Egypt 1952–1988. Cairo.
Chambers, R. and G.R. Conway. 1992. Sustainable rural livelihoods: practical concepts for the 21st century. Discussion paper No. 296, Institute of Development Studies, Sussex, UK.
Dregne, H.E. and Tan-Ting Chou. 1992. Global desertification dimensions and costs. H.E. Dregne, Ed. Degradation and Restoration of Arid Lands. Int. Center for Arid and Semi-Arid Land Studies, Texas Tech. Univ., USA.
Dregne, H.E., M. Kassas, and B. Rosanov. 1991. A new assessment of the world status of desertification. Desertification Control Bulletin 20, UNEP, Nairobi.
El-Gabaly, M. 1992. Opening address. Proceedings of the International Symposium on New Development in the Field of Salt Affected Soils. Cairo.
FAO and UNESCO. 1977. World map of desertification. UNCOD, A/CONF. 74/2. UNEP, Nairobi.
FAO. 1976. A Framework for Land Evaluation. FAO Soil Bulletin No. 32, FAO, Rome.
Forse, B. 1989. The myth of the marching desert, *New Scientist*. 31-32.
Ghabbour, S.I. and M.A. Ayyad. 1990. The state of rural environment in developing countries. Academy of Scientific Research and Technology. Cairo. 538 pages.

Halcrow, H.G.. EO. Heady, and M.L. Contener, Eds. 1982. Soil Conservation Policies, Institutions and Incentives. Soil Conservation Society of America. USA. 330 pages.
Hamdan, G. 1983. *Personality of Egypt.* Volume 3. Book World, Cairo (in Arabic).
Hellden, U. 1991. Desertification-time for an assessment? *AMBIO* 20(8):372-383.
Ibrahim, F. 1993. A reassessment of the human dimension of desertification. *GeoJournal.* 31(1):5-10.
Ibrahim, F.N., M.A. Kishk, and M.S. El-Zanaty. 1993. Rural and urban poverty: comparative case studies in Egypt. In: Bohle, H.G.; T.E. Downing; J.O. Field and F.N. Ibrahim (Eds.) Coping with Vulnerability and Criticality: Case Studies on Food-Insecure People and Places. Verlag breitenbach Publishers Saarbrucken; Fort Lauderdale: Breitenbach.
Kishk, M.A. 1986. Land degradation in the Nile Valley. *AMBIO.* 15(4):226-230.
Kishk, M.A. 1993. Agro-sociological survey of different farmer groups in Samalout and El-Fashn. Consultant report to the Dutch-Egyptian Project "Control of Waterlogging and Salinization in the Fringes of the Nile Valley. 56 pages (unpublished).
Kishk, M.A., M. El-Zanati, and J.B. Fitch. 1988. Comparison of different settler groups in the West Samalout and West El-Fashn Region of Middle Egypt. Research Project Report submitted to the Ford Foundation. 24 pages (unpublished).
Kishk, M.A. 1990. Conceptual issues in dealing with land degradation/conservation problems in developing countries. *GeoJournal.* 20(3):187-190.
Kishk, M.A. 1994. Poverty, Hunger and Land Degradation, Which comes First? Agro-Sociological case Study of the Small Egyptian Farmers. Presented in the 9th ISCO Conference, New Delhi, India, December, 1994.
Kishk, M.A. and J. Lundquvist. 1994. Water productivity in Egyptian agriculture: Scenarios for more sustainable use. Presented in the 8th IWRA Water Resources Conference, Cairo, Nov. 21-25, 1994.
Mabbutt, J.A. 1985. Desertification of the worlds range lands. *Desertification Control Bulletin* 12:1-11.
Ministry of Agriculture, Egypt. 1991. Agriculture Economics Bulletin 1985-1987 Cairo.
Mitchell, T. 1991. Americas' Egypt: discourse of the development industry. Middle East Report, March, April:18-34.
Mortimore, M. 1993. Population growth and land degradation. *GeoJournal.* 31(1):15-22.
Oldeman, L.R. 1992. Global assessment of soil degradation. Draft report for the UNEP State the Environment Report, Int. Soil Reference and Information Center, Wageningen. The Netherlands.
Olsson, L. 1993. Desertification in Africa — a critique and an alternative approach. *GeoJournal.* 31(1):23-32.
Parker, J.B. and J.R. Coyle. 1981. Urbanization and agricultural policy in Egypt. FAER No. 196. Economic Research Services, USDA, Washington, D.C.
Rauschkolb, R.S. 1971. Land degradation. FAO Soil Bulletin No. 13, FAO, Rome.
Suliman, M.N. 1988. Dynamics of range plants and desertification monitoring in the Sudan. UNEP Desertification Control Bulletin 16:27-31.
UN. 1994. Convention on Desertification. A/AC.241/27.
UNCOD. 1977. World Map of Desertification, A/CONF. 74(UN, New York).
UNEP. 1984. Activities of the United Nations Environment Programme in the Combat Against Desertification. A report prepared by the Desertification Branch of UNEP.
UNEP. 1992. World Atlas of Desertification. Arnold Publisher, London.
UNESCO. 1977. Map of the World Distribution of Arid Regions, MAB Technical Notes.
USAID. 1992. Country Program Strategy FY 1992-1996: Agriculture. USAID, Cairo.
Warren, A. and C.T. Agnew. 1988. An assessment of desertification and land degradation in arid and semi-arid areas. A report for Greenpeace International. 90 pages.
Warren, A. 1993. Desertification as a Global Environmental Issue. *GeoJournal.* 31(1):11-14.
World Bank. 1990. Egypt: Environmental Issues. Draft discussion paper, World Bank, Washington, D.C. 45 pages.

Chapter 6

Pesticide Economic and Environmental Tradeoffs: User's Perspective

A.G. HORNSBY, D. L. HOAG, AND D. L. NOFZIGER

Abstract

A multiple-objective decision support system (MODSS) has been developed that permits simultaneous evaluation of profitability and groundwater hazard while selecting herbicides for weed control in peanut production. This interactive software allows the user to quickly evaluate economic loss and groundwater hazard of alternative herbicide treatment strategies. Ease of use and understandability were major design criteria for this DSS. The system uses two graphics screens to present decision consequences. Context-specific help is available for all input and output screens using the "F1" key. Results of a session can be printed for a permanent record using the "F2" key. A companion chapter describes the technical development of this tool (see Chapter 7).

Introduction

With the increasing national concern about the quality of drinking water, farmers are expected to manage pesticide applications in such a way that water contamination is avoided or at least minimized. However, farmers are given little information on the product label about how to accomplish this expectation. They do not have ready access to environmental fate and toxicity data or to decision aids that will guide them to better resource management decisions. In choosing alternative pesticides, they must consider not only water quality goals but also economic viability of farming. This paper introduces a MODSS entitled "Pesticide

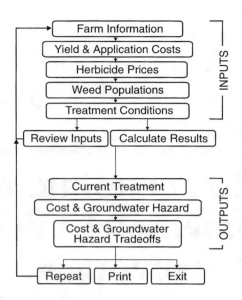

Figure 6.1 Flow chart for PEET.

Economic and Environmental Tradeoffs (PEET)" that simultaneously addresses these issues for peanut growers.

PEET was developed to provide simultaneous evaluation of profitability and groundwater hazard while selecting herbicides for weed control. PEET has been implemented in North Carolina, Florida, and Oklahoma. This interactive software allows the user to quickly evaluate economic loss and groundwater hazard of each different herbicide treatment strategy.

The PEET user interface contains two major parts: user inputs and program outputs (see Figure 6.1). Most of the inputs needed are already in the program's default databases. The user supplies the remaining information about their location which accesses the appropriate preprocessed information needed for output results. Location-sensitive defaults can be used or replaced by the user. The program uses component models to estimate leaching and economic impacts which can be replaced by others if desired. However, data from these models must be consistent in format with the existing database structure to assure that PEET will function properly. The objective of this paper is to present a user's perspective of PEET.

User Inputs

Five interactive user input screens elicit information about the user's operation and other relevant factors. The first screen elicits farm information including user's name, field name and acreage, county and soil survey map unit ID, tillage and irrigation practices, and herbicide application type. The latter five are chosen from "pick lists" activated by pressing the "F1" HELP key. The cursor is moved to the appropriate choice using the up or down arrow key. Input formats for irrigation, tillage practices, and herbicide application type can be customized to match the prevailing practices in the state or region where PEET is used. For example, herbicide application types choices for North Carolina and Florida are preplant incorporated, preemergent, at-crack, and postemergent, whereas for Oklahoma the choices are preplant incorporated, preemergent, spot treatment, and postemergent. This flexibility allows the software to be easily implemented in other states and to be modified for a particular state where a new application type is recommended. The application types mirror the pesticide recommendations made by state cooperative extension pest control manuals.

The second screen elicits information on expected low, typical, and high yields that might be achieved for the selected soil. It also asks for information on the costs per acre for pesticide application (one pass across the field) and scouting costs if postemergent treatment was chosen as application type. Finally, the expected selling price for the crop is requested. All of these entry fields contain default values for each county prepared by commodity specialists.

The third screen lists default herbicide prices for all products recommended for weed control in peanuts. The user may change the prices to reflect the true cost of purchase which may vary due to amount purchased, time of year, product promotion, or other local factors. The default prices are average prices for herbicide products marketed in the state or region within the state. It can be customized to a localized area by making the revisions in the appropriate support database.

The fourth screen contains a table of weeds that are commonly encountered in peanut production and elicits an assessment of weed densities. If the application types "preplant incorporated" or "preemergent" were selected on the first screen, this screen requests that the expected density of weed species to be controlled be identified by the letters N, L, H indicating no weeds, low infestation, or high infestation levels. If the application types "spot treatment," "at-crack" or "postemergent" were selected, then the screen would request results of a scouting report for weeds to be controlled. In this latter case, the user must enter the number of weeds found per 100 ft^2 for each weed specie.

If "spot treatment," "at-crack," or "postemergent" was selected the fifth screen will elicit information about soil moisture, average weed height, and treatment date. Soil moisture and weed height determine the efficacy of control. Treatment date is used to select the groundwater hazard result. This screen does not appear when preplant incorporated or preemergence herbicide treatments are selected.

Program Outputs

The output screens are designed to inform the user of the economic and groundwater hazard consequences of available herbicide treatment strategies. Rather than presenting only the "best choice" (least cost least hazard) strategy, all alternative strategies are presented. Pressing the "F10" key will initiate calculations and move to the first of three output screens.

The first output screen lists all recommended herbicide treatment strategies and requests that one strategy be selected by the user. The choice made in this screen will be used later to compare with all other alternative strategies.

The second screen shows the costs and groundwater hazard of various treatment strategies (Figure 6.2). This screen contains treatment strategies, expected

COST AND POTENTIAL GROUNDWATER HAZARD BY HERBICIDE

Herbicide	Potential Loss to Weeds by Yield (lb/ac)			Rank	Potential Groundwater Hazard Index (% of HAL or MCL)
	3600	4000	4400		Preferred <—— 1% 10% 100% 1000%
	($/acre)				
NO TREATMENT	421	468	515	NA	
Pre-Plant Incorporate					
BALAN EC 4.00QT	350	387	423	1	<1
BALAN DF 2.50LB	352	388	425	1	<1
TREFLAN 5 0.80PT	340	376	412	1	<1
TREFLAN EC 1.00PT	332	368	405	1	<1
SONALAN EC 2.50PT	367	406	445	17	<1
SONALAN EC 3.00PT	346	383	420	18	<1
PROWL 3.3EC 1.80PT	369	409	449	21	<1
PROWL 1.50PT	370	410	450	22	<1
PROWL 3.3EC 2.40PT	339	375	412	23	<1
PROWL 2.00PT	339	376	412	24	<1
BALAN EC 3.00QT	257	283	309	27	<1
+VERNAM 7E 2.33PT					
BALAN DF 2.00LB	260	286	312	27	<1
+VERNAM 7E 2.33PT					
VERNAM 7E 3.50PT	308	340	372	30	<1
PURSUIT 4.00FL OZ	140	153	167	34	<1
DUAL 1.50PT	387	428	469	41	29
DUAL 2.00PT	352	389	426	43	38

Figure 6.2 Example of "Cost and Groundwater Hazard" output screen for a preplant incorporated, periodically irrigated, Cobb SiL soil in Caddo County, Oklahoma.

monetary losses due to uncontrolled weeds, rank of treatment strategies based on groundwater hazard or stochastic dominance, and groundwater hazard expressed as a percent of HAL (USEPA Health Advisory Level [USEPA, 1989]) or MCL (maximum contaminant level) displayed both graphically and numerically. By examining the potential loss and potential groundwater hazard information, the user can discover herbicide treatment strategies that are most cost effective and least hazard to groundwater. In those cases where the most profitable herbicide presents some groundwater hazard, the user can make a judgment about the severity of the hazard and choose to use or not use that chemical. An additional feature is the ability to examine what is behind the loss estimates. By using the right cursor arrow, the reduction in weed load is shown for each weed specie selected and herbicide treatment. This allows the user to learn how efficacious each selected chemical is on the target weeds.

The final screen facilitates comparisons of economic losses and groundwater hazards for the treatment choice made in the first output screen with all other alternative treatment strategies. In the upper left quadrant of the screen two vertical bar graphs display the groundwater hazard for the treatment selected in output screen one and another treatment strategy. In the lower left quadrant of this screen two vertical bar graphs display the economic loss for the selected treatment compared to another treatment strategy. On the right-hand side of the screen is a graph of economic loss vs. groundwater hazard with asterisks representing the treatment strategies. Where several strategies give nearly equivalent values, a single asterisk (*) may represent several strategies on the screen. This screen is divided vertically at 10 and 100% HAL and color coded for emphasis. Asterisks on this graph are highlighted and values shown on the bar graphs are changed using the arrow keys. The strategies being compared are listed at the bottom of the screen.

Program Setup Menu

A setup menu allows site-specific aquifer information and risk probability level to be set by the user. Colors for the cost/groundwater hazard plane can also be selected. The user can also choose the method of ranking the chemicals for groundwater hazard. The two ranking methods used are (1) the ratio of the concentration in groundwater to the USEPA health advisory level, HAL, or MCL (HoHo index as published by Hoag and Hornsby, 1992) at the user-specified probability level and (2) by stochastic dominance (Hoag and Nieuwoudt, 1995).

Execution Time

Within 3 to 5 min the user can enter inputs and have the economic and groundwater hazard tradeoffs displayed for his/her situation. The groundwater hazard for each county, soil type, irrigation and tillage practice, and herbicide product is precalculated using an environmental fate model (Nofziger and Hornsby, 1994) and long-term weather records and is stored in a database. The executable program gathers data from various support databases and computes profitability (using the method of Coble and Mortensen, 1992) and groundwater hazard rank on the fly.

Computation time ranges from about 6 to 30 sec (on a 486, 25 Mhz PC) depending on the number of weed species chosen.

Help Features

PEET contains extensive context relevant help messages and "pick lists" for most input items. The pop-up help menu for the output screens included "purpose of screen," "herbicide information," "potential economic loss," "potential groundwater hazard," "treatment rank," "hazard index," and "special keys." The "herbicide information" help message presents information on application rate, application depth, application window, irrigation application, weed names and populations before and after treatment, and information (sorption coefficient, degradation half-life, and HAL/MCL values) about the active ingredient in the selected herbicide.

Ease of Use

PEET requires minimal effort to use. The input information is usually readily available and easily entered or chosen from a "pick list." Default values are provided in most cases to aid the user. The output screens provide easily interpreted tabular and graphical results. Options chosen or data entered become default until the user exits the program. The help options clarify concepts and facilitate ease of use.

The setup options menu permits use of site-specific aquifer porosity and the choice of the level of risk the user feels comfortable using. This feature is important if the support system is to be widely used in varying geological settings and by individuals with widely different risk acceptance philosophies.

Session Printout

A printed record of input and output screen results can be obtained using the "PRINT" key. This can be very useful for record-keeping purposes and can provide a take-home record for those who do not have a home computer.

References

Coble, H.D. and D.A. Mortensen. 1992. The threshold concept and its application to weed science. *Weed Technol.* 6:191–195.

Hoag, D.L. and A.G. Hornsby. 1992. Coupling groundwater contamination with economic returns when applying farm pesticides. *J. Environ. Qual.* 21:579–586.

Hoag, D.L. and L. Nieuwoudt. 1995. Ranking leaching using stochastic dominance (unpublished manuscript), Department of Agricultural and Resource Economics, Colorado State University, Ft. Collins, Colorado.

Nofziger, D.L. and A.G. Hornsby. 1994. CMLS94b: Chemical movement in layered soils model for batch processing. Division of Agriculture, Oklahoma State University, Stillwater, Oklahoma. 74 pages.

Nofziger, D.L., A.G. Hornsby, and D.L. Hoag. 1998. Pesticide economic and environmental tradeoffs: developer's perspective. Pages 83–92 in S.A. El-Swaify and D. Yakowitz, Eds. *Multiple Objective Decision Making for Land, Water and Environmental Management*. St. Lucie Press, Boca Raton, Florida.

USEPA (U.S. Environmental Protection Agency). 1989. Drinking Water Health Advisory. Lewis Publishers, Chelsea, Michigan. 819 pages.

Chapter 7

Pesticide Economic and Environmental Tradeoffs: Developer's Perspective*

D.L. NOFZIGER, A.G. HORNSBY, AND D.L. HOAG

Abstract

A decision-support system (DSS) was developed to evaluate both economic and environmental impacts of different pest management practices. Chapter 6 describes the use of this tool. A weed competition model which is capable of estimating crop yield for any mixture of weed species in combination with treatment efficacy data form the basis of the economic module. Pesticide movement to groundwater is simulated for each potential treatment using site-specific soil properties, weather, and management practices. The groundwater hazard incorporates the amount of pesticide leached and its toxicity. PEET (the Pesticide Economic and Environmental Tradeoff) was implemented initially for weed control in peanuts in North Carolina, Florida, and Oklahoma. The interactive software is written so it can be used at new locations and with new crops with little or no additional programming. Instead, information for the area and crop are stored in databases created by the local team of experts.

Introduction

The PEET DSS was designed to aid farmers in selecting pesticides. Most farmers are interested in earning money without harming the natural resources. When

---

* The work reported here was supported by the Oklahoma Agric. Exp. Stn. and is published with the approval of the Director.

selecting pesticides, they often have several options available. PEET is a tool which allows them to see which pesticides, if any, will decrease their economic loss from the pest. PEET also evaluates the potential impact of each pesticide upon groundwater quality. In some cases, a pesticide can be used which minimizes economic loss as well as groundwater quality degradation. In other cases, a tradeoff between economics and groundwater quality must be made. PEET is designed to make decisionmakers aware of these tradeoffs so they can make informed decisions. This paper describes how PEET was designed and how it can be used for new locations and crops. Chapter 6 presents PEET from the perspective of the end user.

System Design

The following design requirements were established for PEET:

1. It must provide economic and water quality information for specific sites and management practices specified by the user.
2. It must incorporate uncertainties in the information and convey that to the user.
3. It must operate on standard MS-DOS computers with a hard disk.
4. It must be convenient to enter information and to interpret results.
5. Its speed must be satisfactory for use in interactive mode.
6. It must be usable at different locations by changing data used, without reprogramming.

To meet these requirements, the PEET program must carry out four major tasks. Those include getting inputs from the user, calculating economic impacts for each treatment, evaluating the groundwater hazard posed by each treatment, and presenting results of the analysis in different forms. Details of the economic and groundwater hazard calculations are given on the following pages.

Task 1: Obtaining User Inputs

Table 7.1 contains a list of inputs required for the DSS. The user interface in PEET was designed to be easy to use and to assist the user in making these entries. It enables the user to select answers from menus in many cases. The options are often reduced by other selections already made. For example, the soils displayed in the soil menu are only those for the county already selected. When numerical values are expected, default values are provided. The user can replace these values with more appropriate ones when known. Although five input screens are used, the user can page forward or backward as desired to change current values. Help messages are provided for each entry. The user can get these help messages by pressing the <F1> function key or by configuring the system to automatically display all help messages. Since we expect users to explore tradeoffs for different conditions, responses entered for one situation become the default values for the next one. That means the user must only make the changes of interest rather than redefine all the values.

Table 7.1 Parameters Entered or Selected by the PEET User

Name of farmer or grower
Name and size of field of interest
County in which field is located
Soil type
Tillage practices used
Irrigation practices used
Type of herbicide treatment
Low, normal, and high weed free yields for this field
Cost of applying a pesticide per unit area
Cost of scouting for weeds per unit area
Expected market price of crop per unit harvested
Purchase price and purchase units of each potential pesticide
Density of each weed species in the field
Soil moisture conditions at time of treatment
Weed size at time of treatment
Approximate date of treatment

To facilitate use in different locations, all options which can be selected, default values displayed, help messages written to the screen, and credits for local developers are stored in databases and external files. These items can be changed by the team of local experts implementing PEET for a particular area. No programming skills are required.

Task 2: Calculating Economic Impact

Yield Loss due to Weeds

Yield loss can be computed for any combination and density of weeds. In turn, yield savings can be calculated for any pesticide based on its efficacy on each weed type. The yield reduction, $Y_{Loss\text{-}Abs}$ due to weed population and infestation is given by

$$Y_{Loss\text{-}Abs} = Y_{Loss\text{-}Rel} Y \quad (1)$$

where $Y_{Loss\text{-}Rel}$ is the relative loss in yield and Y is the projected weed-free yield. PEET displays economic loss for low, normal, and high weed free yield values to provide the user with insight into the impact of yield upon the economics of herbicide use. The relative yield loss is given by

$$Y_{Loss\text{-}Rel} = \frac{AID}{A + ID} \quad (2)$$

where D is the total competitive load, A is the maximum relative yield loss, and I is the yield loss per unit competitive load for low loads (Coble and Mortensen, 1992). The total competitive load D is the sum of the competitive loads for the different weed species. It is calculated using the equation

$$D = \sum_{j=1}^{n} c_j d_j \qquad (3)$$

where n is the number of weed species infesting the field, c_j is the competitive index for weed j and d_j is the density of weed j. The competitive index is a weighting factor for each weed species reflecting the impact of that weed species upon the yield of the crop relative to the impact of a selected standard weed species. This index is determined by the local weed scientist when PEET is implemented at a site.

Economic Loss Due to Weeds

The economic loss, E_{Loss}, due to weeds is given by

$$E_{Loss} = Y_{Loss-Abs} V + C_{Appl} + C_{Scout} + \sum_{Herb=1}^{m} C_{Herb} R_{Herb} \qquad (4)$$

where V is the expected value of the crop per unit harvested, C_{Appl} is the cost of herbicide application, C_{Scout} is the cost of scouting for weeds, C_{Herb} is the cost of herbicide Herb per unit applied, R_{Herb} is the rate of application of herbicide Herb, and m is the number of herbicides used in the treatment. If no herbicide is applied, the last three terms in the equation are zero.

Economic Loss With Treatments

The density of each weed species after treatment with a herbicide is calculated using the equation

$$d_{jk} = d_j \left(1 - e_{jk}\right) \qquad (5)$$

where d_{jk} is the density of weed j after treatment k, d_j is the density of weed j before treatment, and e_{jk} is the efficacy of treatment k for weed j. Equations 1 to 4 are used to calculate economic loss for each treatment by replacing d_j with d_{jk} in Equation 3.

Computing Time Requirements for Economics

Computing times for economic analysis is on the order of 1 sec. This component of the project provides no problem for interactive use.

Task 3: Evaluating Groundwater Hazard

Definition of Groundwater Hazard

In PEET, the groundwater hazard (GWH) associated with a particular pesticide is defined as the ratio of the estimated concentration (C) of the active ingredient in groundwater to the USEPA lifetime health advisory level (HAL) or to the maximum contaminant load (MCL) for the active ingredient. That is

$$GWH = \frac{C}{C_{critical}} \qquad (6)$$

where $C_{critical}$ is the HAL or MCL for the pesticide. If a treatment contains more than one active ingredient, the *GWH* for the treatment is taken as the sum of the GWH values for all active ingredients. This index, introduced by Hoag and Hornsby (1992), is useful in that it incorporates both the predicted concentration of the pesticide in groundwater and the toxicity of the pesticide. As defined here, values of GWH less than 1 correspond to active ingredient concentrations below the HAL or MCL of the chemical. Thus, those products in those conditions would not be thought to pose a significant risk to human beings.

Calculating Groundwater Hazard

With this definition, calculating the GWH requires that we calculate the concentration of each active ingredient in the groundwater below the soil of interest. In general, the amount of chemical leaching to groundwater depends upon soil properties, chemical properties, management practices such as irrigation management and tillage, application dates, application amounts, application depths, weather, water infiltration and runoff, and plant use of water. Various models exist which could be used to estimate the leaching and degradation of pesticides. *Any model capable of estimating the pesticide concentration in groundwater can be used with PEET.*

Since PEET is intended to be a decision-making tool, we are interested in predicting the concentration of chemicals in groundwater. However, we do not know what the future weather at a site will be. Large differences in leaching due to weather are common (Haan et al., 1994). Therefore, PEET incorporates this uncertainty into GWH estimates. That is done by simulating movement for many weather records at a site and obtaining distributions of expected concentrations. Figure 7.1 shows probability distributions for bentazon sodium salt and acifluorfen sodium salt concentrations in the Cobb fine sandy loam soil under irrigated conditions in Caddo county, Oklahoma. The HAL for acifluorfen is 1 µg L^{-1} and that for bentazon is 20 µg L^{-1} as shown. Figure 7.2 shows the GWH probability distribution for each chemical. The cumulative distributions in Figure 7.2 were obtained by dividing the concentration distribution for each chemical (shown in Figure 7.1) by the HAL for that chemical. The solid circles show the calculated GWH for the two pesticides at a probability of 0.10. PEET displays these GWH for all of the potential pesticides treatments at the probability level selected by the user.

Ranking Treatments

One of the types of output presented to the user of PEET is a ranking of treatments by GWH. This ranking is more complex because GWH is a probability distribution,

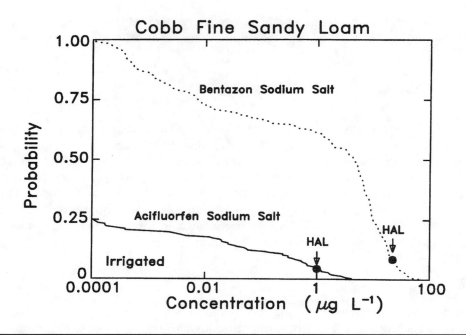

Figure 7.1 Probability of exceeding different concentrations for two pesticides applied at a rate of 0.56 kg ha^{-1} on an irrigated soil in Caddo county, Oklahoma. The arrows indicate the HAL for each pesticide.

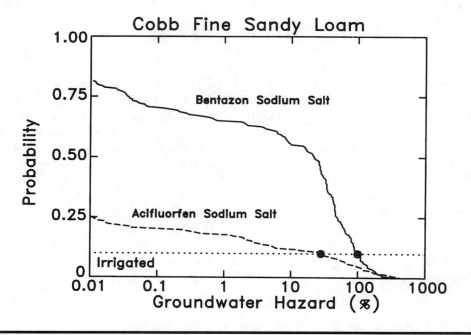

Figure 7.2 Probability of exceeding different GWH values for the two pesticides shown in Figure 7.1.

not a simple number. Moreover, the probability distributions frequently intersect and cross-over one another. Two forms of ranking are supported in PEET. The user can select the method preferred. In the first case, treatments are ranked on the basis of their GWH values *at the probability level specified by the user*. That is, the treatments are ranked on the basis of the GWH values at a single probability level without regard for the behavior of the curves at other probabilities. This ranking is easily understood but may be misleading. We observed that the probability distributions for different treatments sometimes intersect and cross-over one another. Therefore, one herbicide may have a lower GWH value than another at one probability level, but a higher GWH value at another probability level. A second ranking scheme based on principles of stochastic dominance is also supported in PEET. This ranking process considers the probability distributions for their entire range in making the ranking (Hoag and Nieuwoudt, 1995).

Pesticide Transport Model Used in Florida, North Carolina, and Oklahoma

The CMLS94 model of Nofziger and Hornsby (1994) was used to estimate the amount of each active ingredient leaching below a depth of 1 m. We assumed degradation below that depth was negligible and hence that amount represents the amount entering the groundwater. The concentration (C) in Equation 1 was estimated by mixing the calculated amount of pesticide leached in an aquifer with porosity f and mixing depth t. Values of f and t can be specified by the user.

CMLS94 is an update to the original CMLS model of Nofziger and Hornsby (1986, 1988) written to serve as a management tool and a decision aid for the application of organic chemicals, especially pesticides, to soils. CMLS can be used to estimate the location of the center of mass of nonpolar organic chemicals in soils as a function of time. Downward movement of water is the transport process considered in CMLS94. The model also estimates the degradation of the chemical and the amount remaining in the soil profile. All of the parameters required by the model are relatively easily obtained.

CMLS94 improves upon CMLS by incorporating new understandings we have gained since CMLS was introduced. These new capabilities can enhance our understanding of the transport and fate of chemicals through soils. The original CMLS required the user to supply daily infiltration and evapotranspiration values for the simulation period. CMLS94 includes routines to estimate these values from daily weather records so those data can be more easily obtained. CMLS94 enables users to assess uncertainty in predictions due to unknown future weather at a site by simulating chemical movement for many different weather sequences for that site. Probability distributions can then be obtained for any outputs of interest.

Computing Time Requirements

Obtaining probability distributions for GWH provides a great deal of important information, but it requires a large amount of computing resources. A typical analysis for a user could require more than 10,000 simulations. This would require more than 120 min on personal computers. Clearly, real-time simulation on these machines will not be satisfactory for interactive use.

To overcome this time limitation, GWH were calculated for each soil and chemical using sets of management practices which span the practices used by farmers. Those values along with treatment ranks by stochastic dominance were stored in a GWH database. That database is queried by the interactive PEET program. This reduces the time for a response to only a few seconds. The database for peanuts in Oklahoma occupies approximately 4 MB of disk space and contains approximately 40,000 records. This represents more than 25 million simulations. Several Sun SparcStations were used to make the simulations in approximately 1 week.

Task 4: Displaying Results

Information on the economic and environmental consequences of using each weed control practice are presented three ways as illustrated by Hornsby et al. (1996). Once again, the user can move between the three screens as desired. Help messages are available here also to explain the meaning of different parts of the output.

Special software was developed for horizontal and vertical scrolling of selected portions of the output screen. This enables the user to scroll vertically through all of the treatment options while keeping header information in place. Horizontal scrolling is used to replace the economics window with information on the weed density and competitive load for each weed species and each treatment. In this way, the user can see tradeoffs between the GWH and weed density for any particular weed species.

Implementing Peet at a New Location

Although implementing PEET for a new geographical area or crop is not trivial, most of the effort is spent gathering needed information. No programming will be required in PEET. People interested in implementing PEET for a particular geographical area must select a pesticide transport model and assemble the following information:

1. Soil names on which the crop is grown; soil properties required by the chemical transport model
2. Irrigation types used and their characteristics
3. Tillage types used and their impact upon the transport model parameters
4. Weather data for the weather stations needed to span the area
5. Chemical transport properties and HAL or MCL values for each active ingredient
6. Weed species expected and the competitive index of each
7. Potential pest management practices including the products used, application rates, application types, and times of application
8. Efficacy of each treatment on each weed species at different weed sizes for different growing conditions
9. High, low and typical yields for each irrigation practice and soil (these become default values which the PEET user can change)

10. Costs of each product in units purchased by farmer (these can be modified by PEET user)
11. Maximum yield loss and yield loss per competitive index required in Equation 2

Once these data are assembled, the GWH index and rankings among treatments must be calculated. The databases must also be entered into the form expected by PEET. If CMLS94b is the pesticide transport model selected, that software is available along with tools to generate the required input files and to process model outputs to obtain GWH values for each soil-chemical-management system. Software for ranking treatments by stochastic dominance is also available. All databases are stored in Paradox format. Software is available from the authors for use in entering all data into databases of the structure required by PEET. Tools are also provided for customizing help messages, credits, and contact persons for local needs.

Concluding Remarks

The PEET DSS was designed for use at different locations and for many crops. It has been implemented for weeds in peanuts in Florida, North Carolina, and Oklahoma. Work is underway to implement it in other crops. If you would like to implement PEET for a location or crop of interest, feel free to contact the authors to obtain the software and more detailed instructions.

Additional work is underway in the development of PEET. This work includes incorporating risks due to preferential flow of chemicals through large cracks and pores, calculating risks for management units consisting of more than one soil type, and implementing PEET for use on the world-wide web. Additional work is needed to include risks to surface water.

References

Coble, H.D. and D.A. Mortensen. 1992. The threshold concept and its application to weed science. *Weed Technol.* 6:191–195.

Haan, C.T., D.L. Nofziger, and F.K. Ahmed. 1994. Characterizing chemical transport variability due to natural weather sequences. *J. Environ. Qual.* 23:349–354.

Hoag, D. and A.G. Hornsby. 1992. Coupling groundwater contamination with economic returns when applying farm chemicals. *J. Environ. Qual.* 21:579–586.

Hoag, D. and L. Nieuwoudt. 1995. Ranking leaching using stochastic dominance. Unpublished manuscript, Department of Agricultural and Resource Economics, Colorado State University, Ft. Collins, Colorado.

Hornsby, A.G., D.L. Hoag, and D.L. Nofziger. 1998. Pesticide economic and environmental tradeoffs: user's perspective. Pages 75–81 in S.A. El-Swaify and D.S. Yakowitz, Eds. *Multiple Objective Decision Making for Land, Water, and Environmental Management.* St. Lucie Press, Boca Raton, Florida.

Nofziger, D.L. and A.G. Hornsby. 1986. A microcomputer-based management tool for chemical movement in soil. *Appl. Agric. Res.* 1:50–56.

Nofziger, D.L. and A.G. Hornsby. 1988. Chemical movement in layered soils: user's manual. Computer Software Series CSS-30. Agricultural Experiment Station, Oklahoma State University, Stillwater, and University of Florida, Gainesville. IFAS. Cir. 780. 44 pages.

Nofziger, D.L. and A.G. Hornsby. 1994. CMLS94 chemical movement in layered soils. Computer Software Series. Oklahoma Agricultural Experiment Station, Oklahoma State University, Stillwater. 76 pages.

Chapter 8
Multi-Objective Decision Support Strategies for Cropping Steep Land

ALICE J. JONES, ROGER SELLEY, AND LLOYD N. MIELKE

Abstract

A multi-objective decision support strategy (MODSS) is presented for cropping steep lands. Profitability and soil erosion control are considered. Different crops are evaluated on six landscape positions. Gross margin is used as an indicator of profitability. In developing the strategy, a series of five questions are answered regarding crop yield, profitability, and soil loss by landscape position; and the cost of erosion control using crop substitution. Linear programming is used to identify a variety of cropping options that result in different gross margins for the field. These cropping options are then evaluated for erosion control using the revised universal soil loss equation. Any crop substituted on one or more landscape positions must provide greater erosion control than the crop being replaced. A number of tillage systems are also considered as a means of reducing erosion at little or no additional cost. The result is a matrix of cropping and tillage options that each correspond to a specific level of profitability and soil erosion control. The matrix can be used to prioritize choices for profitability, erosion control, and/or cropping and tillage systems. This strategy can be useful in assessing real field situations, and in developing land use planning alternatives and policy.

Introduction

Agricultural producers have and will likely continue to farm steep land. Among the greatest concerns are water-induced soil erosion and its control, profitability, and land use policies. Decision management strategies are needed that may allow farming to be continued on steep lands with minimal environmental impact.

Crop production and profitability are usually evaluated on a field or farm basis. However, within a given field, many landscape positions exist. Crop yields on different landscape positions in eastern Nebraska can vary by as much as 17% for maize, 44% for soybean, and 32% for sorghum (Jones et al., 1989). This variability may contribute to lower economic returns if production inputs are applied uniformly across all landscape positions.

Farming by landscape position encompasses the concepts of "precision farming" and "site-specific farming," and can be applied to crop and tillage selection as well as production inputs. On steep land, "farming by landscape position" will allow for practices that optimize the productivity of the natural resource base and maximize the efficiency of production inputs. As a result maximum profitability can be obtained subject to environmental constraints.

Decision Criteria

Many considerations must be taken into account when developing systems approaches to decision making. The criteria used in our strategy include soil erosion control, profitability, and land use policies. Factors influencing soil erosion control options include quantitative measures of the natural resource base, production and management practices that can be used to enhance erosion control, and the ability of a producer to make and carry out sound management decisions in a timely manner.

Profitability will be influenced by factors that are and are not controlled by a producer. Selection of purchased vs. "on-farm" inputs, crop selection, and management practices that maximize production are under the control of a producer; potential productivity of the natural resource base, (i.e., soil type, climate, and topography), and market price are not.

A loose interpretation of land use policies may include governmental laws and ordinances as well as the availability of technical assistance and the farming experience and traditions of a producer and local peer group. Other social, cultural, political, and economic priorities may also be intertwined in formal or informal land use policy.

DSS for cropping steep land can easily incorporate quantitative data on soil erosion control and profitability, thus, producing a multitude of cropping and tillage options. The third criterion, land use policy, is that portion of the strategy that is subjectively applied to the resultant database. The evaluation process then, can provide "best "options for farming steep land for specific soil erosion control that are also profitable and can guide policy makers as they establish goals/laws for erosion control that also reflect a producer's need for economic survival.

Objective

Our objective was to develop a DSS that will identify cropping and tillage options that are appropriate to maximize profits for a wide range of soil erosion control levels. Our DSS is based on results obtained from sequentially answering the following research questions:

- What is the variability of crop production as a function of landscape position?
- What is the profitability of different crops as a function of landscape position?
- What is the potential soil loss for different crops and landscape positions?
- What is the cost associated with erosion control using crop substitution?
- What are the profits and soil losses associated with cropping systems options?

Approach

Modeling

Profitability and soil loss calculations are made using a linear programming model (Pfeiffer, 1984) and the Revised Universal Soil Loss Equation (RUSLE) (SWCS, 1993), respectively. Inputs to the linear programming model are crop yield by landscape position, proposed cropping and tillage options, and costs and returns. Inputs to RUSLE are field data on slope length, slope gradient, soil type, rainfall zone/location, conservation practices, crop selection, and farming operations. Other data needed for RUSLE are provided as part of the RUSLE model or may be added from localized databases.

Field Data

Fields under consideration are divided into six landscape positions to more accurately calculate soil loss and crop yield, and to provide a basis for assigning cropping options to different positions (Figure 8.1). Landscape positions considered include upper (UI) and lower (LI) interfluve, shoulder (SH), upper (UL) and lower (LL) linear, and footslope (FT). The interfluve positions are the uppermost portion of the hillslope and receive little overland flow. The interfluve can contribute runoff to downslope positions. The shoulder is a convex transitional zone at the base of the interfluve and above the linear slope. The linear positions receive overland flow from upslope positions and contribute runoff to the footslope. The footslope represents the base of the hill. Water and sediment running off of the footslope may enter a waterway or other surface water system.

For each landscape position, slope length, slope gradient, and area as a proportion of the entire field, are measured in the field. Grain and forage samples are collected from each landscape position to determine yield.

Economics

Profitability is expressed as gross margin, the difference between revenue and operating costs. Operating costs include inputs such as seed, fertilizer, insecticide,

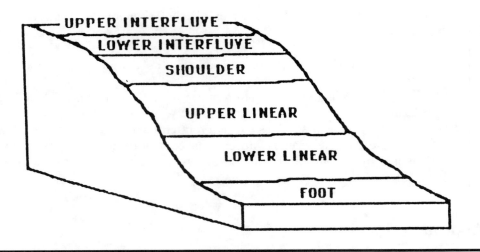

Figure 8.1 Schematic diagram of landscape positions.

labor, fuel, and repairs (AEES, 1989). These costs can vary from one crop to another. Any effect crops can have on depreciation, interest on investment, property taxes, and insurance were not considered. Ownership costs were ignored because it was assumed that the total annual use of equipment was unaffected by cropping and tillage options. Market prices can be obtained from current data or historic records.

Soil Erosion

Parameters for estimating soil loss are based on field measurements, published data, and existing databases. The RUSLE factor K is based on soil type, R on geographic location, LS on slope length and gradient, C on cropping and management practices, and P on conservation support practices.

Case Study

In our evaluation, crop yields were obtained for a 2-year period on five farms in eastern Nebraska. Maize, sorghum, and soybean were evaluated. All field data used in this analysis represent average values across crop-location years. Alfalfa yields were estimated based on the average level of production for the other crops. Market prices were based on long-term price ratios among crops; topographic factors for the landscape positions were measured in the field (Jones et al., 1990).

Several tillage systems were considered in the analysis because high residue tillage systems that greatly improve erosion control were assumed to be employed at little or no additional cost. The tillage systems evaluated were conventional tillage, reduced tillage, and no-till. The cropping and management factor, C, for RUSLE was based on crop and tillage combinations (Jones et al., 1990). Values for R, K, LS, and P in RUSLE were either calculated from field data or obtained from published bulletins.

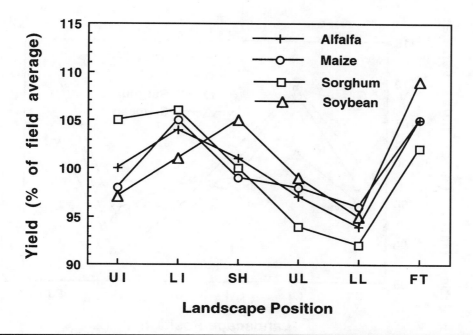

Figure 8.2 Relative crop yields for landscape positions.

Analysis

What is the Variability in Crop Production as a Function of Landscape Position?

Answers to this question help establish the inherent productivity of different landscape positions and increases associated with management inputs. Average field yields for maize, alfalfa, sorghum, and soybean were 3.70, 6.05, 4.05, and 1.86 mg/ha, respectively. Relative yields for each landscape position ranged from 92 to 109% of field average (Figure 8.2). All crop yields were below the field average on the upper and lower linear position. The upper interfluve had above-average yield for sorghum while maize and soybeans had below-average yields. Yields were above average for all crops on the lower linear and footslopes. Based on this first step of our approach we would grow sorghum on the UI and LI, soybean on the SH, UL and FT, and maize on the LL.

What is the Profitability of Different Crops as a Function of Landscape Position?

Increasing production efficiency by growing different crops on various landscape positions does not guarantee maximum profits because of different operating costs and market prices. Maximizing profits is a more important measure of production efficiency than is yield if the economic survival of a producer is considered.

Production on our field using a single crop with uniform management inputs would result in gross margins of $259, $272, $262, and $128/ha for maize, sorghum,

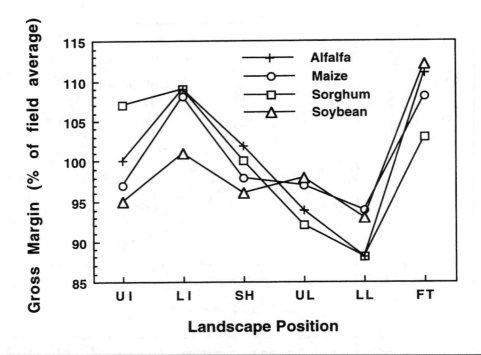

Figure 8.3 Relative gross margin of crops for landscape positions.

soybean, and alfalfa, respectively. Relative gross margin indicates that profits from different landscape positions can be obtained that are up to 12% greater than the field average (Figure 8.3). The UI, LI, and FT provide the greatest return on investments while UL and LL provide the least. The absolute gross margin for alfalfa is no more than 50% of the other crops for any landscape position. The maximum difference in gross margin between maize, sorghum and soybean was $52/ha for UI, $32/ha for LI, $20/ha for SH, $8/ha for UL, $4/ha for LL, and $12/ha for FT.

In eastern Nebraska, continuous crops are not usually grown and crop selection depends upon climatic conditions. In the northeastern part of the state, maize would be rotated with soybean while in southeastern Nebraska, sorghum would be rotated with soybean. Alfalfa is grown throughout the area. Different crops, as well as different crop rotations, then could be considered for different landscape positions based on profitability.

What is the Potential Soil Loss for Different Crops and Landscape Positions?

Controlling soil erosion is essential to minimize agriculture's impact on the environment and to keep productive topsoil on the land. Because of topographic differences among landscape positions, soil does not erode at an equal rate across the field. Calculation of erosion rates by landscape positions aid in the identification of areas where erosion reduction management practices are needed most.

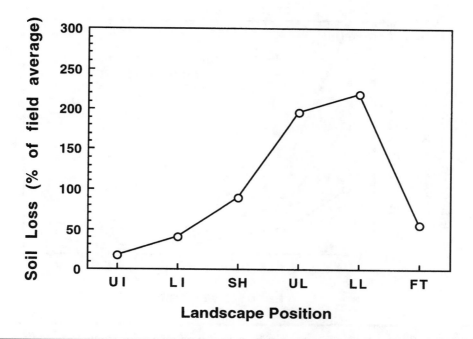

Figure 8.4 Potential soil loss for landscape positions.

Field average soil loss among crops was 20, 2.5, and 33 mg/ha/year for sorghum (or maize), alfalfa, and a sorghum (or maize)/soybean rotation, respectively. Differences among crops depend largely on crop and/or residue cover of the soil surface and growth habits. Alfalfa is a perennial and provides substantial soil protection year round whereas, sorghum (or maize) and soybean grow for approximately 5 months/year. The fragile soybean residue is subject to rapid decomposition and breakage. This results in increased erosion potential whenever soybean is grown. Sorghum and maize have similar growth patterns and residue production levels so that they are analyzed the same in erosion control calculations.

Soil loss variability across landscape positions indicates that erosion potential is highly variable (Figure 8.4). Nearly three-fourths of the soil loss originated on the UL and LL; almost one-fifth came from the SH and FT.

What is the Cost Associated with Erosion Control using Crop Substitution?

Soil loss can be decreased by substituting crops with greater erosion control for crops with less erosion control. This will likely be accomplished at the expense of profits because, as in our situation, crops providing greater erosion control are less profitable. Profit loss can be minimized by substituting crops on landscape positions where the cost of erosion control is least. The cost of erosion control is calculated as the difference in gross margin between a crop being grown and the crop being substituted, divided by the difference in the erosion rate for the two crops.

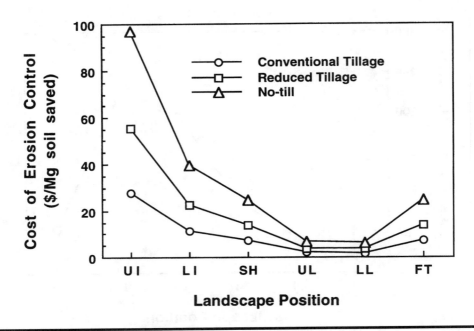

Figure 8.5 Cost of controlling erosion by substituting alfalfa for sorghum on different positions using three tillage systems.

When alfalfa is substituted for sorghum, the cost of controlling erosion can range from $2 to $97/mg of soil saved depending on tillage system used and landscape position where the substitution occurs (Figure 8.5). The profit loss associated with crop substitution is slightly influenced by tillage system; however, the amount of soil saved is greatly influenced by tillage system. The cost of erosion control using crop substitution is greatest with no-till because most erosion control has already been accomplished at little or no cost through residue management. Therefore, the cost of achieving the next increment of soil savings in no-till is quite high as compared to a conventional tillage system.

Independent of tillage system used, the lowest cost of erosion control occurred on the LL; the greatest cost occurred on the UI. The general ranking for alfalfa substitution, and other crops that can be substituted to provide greater erosion control, is LL>UL>FT>SH>LI>UI.

What are the Profits and Soil Losses Associated with Cropping Systems Options?

A number of cropping options can now be selected for the landscapes that will allow different soil erosion control and profitability levels to be considered (Table 8.1).

Cropping options are identified using the linear programming model . Options are considered that result in most profitable to least profitable scenarios. The first cropping option includes continuous sorghum on the UI and LI and a sorghum/soybean rotation on the other four positions. Continuous sorghum is then sequentially substituted by landscape position for the sorghum/soybean rotation

Table 8.1 Cropping Options for Landscape Positions that Result in Greatest to Least Profitability

Cropping option	Crop planted					
	Upper interfluve	Lower interfluve	Shoulder	Upper linear	Lower linear	Foot
1	S	S	S/B	S/B	S/B	S/B
2	S	S	S/B	S/B	S	S/B
3	S	S	S/B	S	S	S/B
4	S	S	S	S	S	S/B
5	S	S	S	S	S	S
6	S	S	S	S	A	S
7	S	S	S	A	A	S
8	S	S	S	A	A	A
9	S	S	A	A	A	A
10	S	A	A	A	A	A
11	A	A	A	A	A	A

S = sorghum; B = soybean; A = alfalfa.
After Jones et al., 1989.

in options 2 through 5. The sequence of substitution is based on results from Figure 8.5 and consideration for land area of each landscape position. Once sorghum has been substituted on all positions, then alfalfa is substituted for sorghum using the same process. This is accomplished in options 6 through 11. Sorghum is substituted before alfalfa because it is a more profitable crop.

Option 1 provides the greatest profit (Figure 8.6) and greatest soil loss (Figure 8.7). As options 2 through 11 are considered profits are gradually diminished and erosion control is increased.

Decision Making

Selection of a specific cropping option depends on goals for profitability and/or soil erosion control. These goals may be self-imposed or imposed by law/ordinance or even by peers. Trade-offs between profits and erosion control can be considered as in Figure 8.8.

Many cropping options allow profits to be maintained ≥ $250/ha with soil loss levels from 10 to 42 mg/ha/year. At soil loss levels below 10 mg/ha/year profits decline rapidly. Use of no-till allows for the greatest profit at very low soil loss levels. Choosing to continue with conventional tillage will limit cropping options, if a profit or erosion control goal has been established.

Conclusions

With site-specific field data, this approach to MODSS for erosion control and profitability can be adapted to any field and farming operation with steep land. Any combination of crops can be considered in this analysis so long as they are feasible for a geographic area. Slight modifications in the procedures may be

102 *Multiple Objective Decision Making for Land, Water, and Environmental Management*

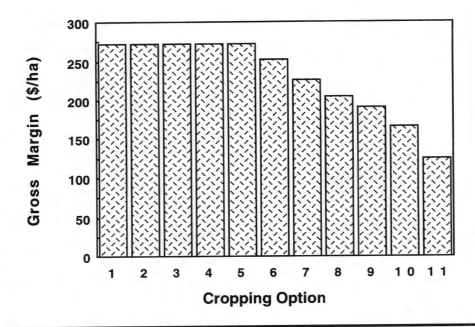

Figure 8.6 Profitability for cropping options.

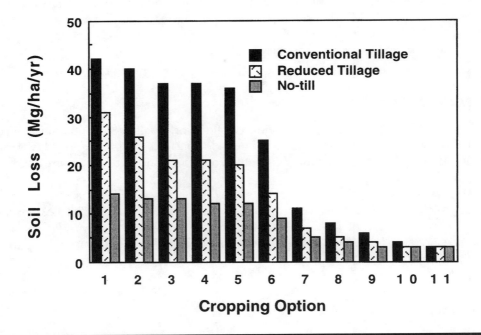

Figure 8.7 Soil loss for cropping options and tillage systems.

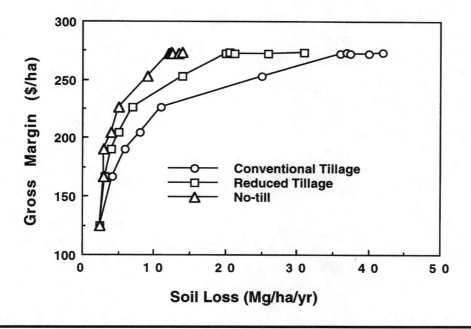

Figure 8.8 Gross margin and soil loss for cropping options and tillage systems.

required depending upon criteria specific to the economics, natural resource base, and growing conditions under consideration. For example, some crops or residue may be harvested for fodder, fuel, or building material. In other instances, environmental costs may also be applied to crop selection and management.

Real and theoretical considerations can be incorporated into the MODSS as well. Field by field evaluation can be conducted. Also, land use planning and policies, and cropping systems can be improved by simulating "what — if" scenarios.

References

AEES (Agricultural Economics Extension Staff). 1989. Nebraska crop and livestock production costs. EC-89-872. Cooperative Extension Service, University of Nebraska, Lincoln.

Jones, A.J., L.N. Mielke, C.A. Bartles, and C.A. Miller. 1989. Relationship of landscape position and properties to crop production. *J. Soil Water Conserv.* 44:328–332.

Jones, A.J., R.A. Selley, and L.N. Mielke. 1990. Cropping and tillage options to achieve erosion control goals and maximum profit on irregular slopes. *J. Soil Water Conserv.* 45:648–653.

Pfeiffer, G. 1984. MPS-PC users manual. Version 2.5. Linear programming system for the IBM personal computer. Research Corp./Research Software. 6840 E. Broadway Blvd., Tucson, Arizona, 85710.

SWCS (Soil and Water Conservation Society). 1993. RUSLE, Revised universal soil loss equation. Version 1.03. Ankeny, Iowa.

Chapter 9

A Multicriteria Decision Support System for Evaluating Cropping Pattern Strategies in Egypt

KHALED M. ABU-ZEID

Abstract

This paper presents a decision support system (DSS) developed to evaluate different cropping pattern and water use strategies in the Eastern Nile Delta of Egypt. The computer-based DSS uses a geographical information system to support the decision making through the two techniques of graphical visualization and multicriteria evaluation. This paper focuses more on the second technique where the detailed logic for an Expert System was developed to perform the multicriteria evaluation and decision-making process. With water being the most essential element for cultivating any cropping pattern, the knowledge base for the Expert System was based upon information obtained from numerous experts involved in the decision-making process for water resources management issues in the study area. An unbiased procedure was implemented to give environmental, ecological, economical, political, and social weights to the experts and to their multicriteria procedures for evaluating water management strategies. A powerful user friendly interface was developed to provide the communication tool linking the decision maker's interaction with the database, the Geographical Information System (GIS), and the Expert System. The DSS enables the user to view the impacts on different parameters, such as soil salinities and evapo-transpiration, due to implementing a certain water use strategy. The parameters are obtained from a simulation model for irrigation and drainage in Egypt and are stored in a database used by the DSS. A graphical user interface is provided for the user to select between strategies and compare impacts of different strategies on water usage and crop production. The user can select to simultaneously view more than one strategy on the computer screen for comparison purposes. The user can also view the short- and long-term effects of a particular strategy on a parameter by showing its distribution at successive time steps. The DSS is developed using PDC PROLOG as a programming language, and it is coupled with IDRISI as a Geographical Information System GIS tool. Every effort has been made to make it a user friendly tool for decision makers.

Introduction

The problem of satisfying water demands in irrigation has been the main concern for many countries all over the world. With the increase in world population and the conflicts over scarce sources of water, this problem has become even more crucial. The international world in its search for nonconventional sources of water considered sources such as the nonrenewable groundwater, and the reuse of drainage water for agriculture. Today, engineers and politicians around the world are very concerned about the environmental and social impacts of the use of these nonconventional sources of water. They are often faced with numerous alternatives for satisfying water, food, and economic demands. It is very important to be able to effectively evaluate these alternatives and to assess the different impacts on crop production and water usage. Living in the new era of computer technology, it is expedient to use computer-based GIS, Expert Systems, and DSS to assist decision makers in their impact evaluation processes. This is demonstrated in the DSS (WRMDSS) presented in this paper.

Decision Support System

Figure 9.1 shows a flow chart of the computerized DSS, WRMDSS (*W*ater *R*esources *M*anagement *D*ecision *S*upport *S*ystem), developed for evaluating the impacts of different cropping pattern strategies using different water quantities and qualities. With rice being the most water consuming crop in the Eastern Nile Delta in Egypt, the Egyptians realized the importance of considering rice as a key in their water management policies. An irrigation and drainage model (SIWARE) was developed for the Eastern Nile Delta (DRI, 1991). The SIWARE model simulates irrigation and drainage water and salt flow in the area for different water management strategies. One of the questions that faces water managers is what to do in a water shortage situation. One might suggest reducing the area cultivated by high water consuming crops. Another could suggest reducing the quantity of water delivered to high water consuming crops while keeping the area cultivated by that crop the same. Both strategies will have adverse effects on crop yield, and different effects on soil salinization and different water uses. In order to evaluate such alternatives, different rice strategies have been simulated in SIWARE to obtain data such as water resources uses, associated water salinities, water and salt balance for each crop, crop evapotranspiration, and soil salinities. The results of these simulations form the database for the DSS developed in this research. The

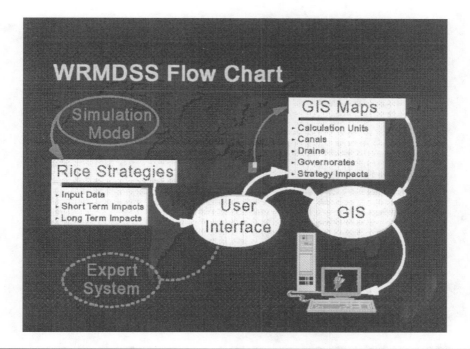

Figure 9.1 WRMDSS flow chart.

DSS utilizes the GIS capabilities to put the large amount of data in a manageable format for decision makers. Often times, decision makers prefer to see information in a graphical form, especially information that varies spatially and temporally (Abu-Zeid, K., 1994). The DSS mainly depend on two techniques to aid decision makers in evaluating the alternative rice strategies for water resources management. The first technique (Abu-Zeid and Sunada, 1994) considers the effective graphical visualization of the different impacts of applying different strategies such as the impact on evapotranspiration shown in Figure 9.2. The second technique which is the focus of this paper is the Expert System for the multicriteria evaluation of different strategies. Both techniques use GIS to perform the strategy comparison.

Multicriteria Evaluation Expert System

The Expert System in this DSS has different aspects in irrigation, agriculture, and economics. It provides criteria for the evaluation of different variables such as the water quantity, water quality, crop yield, crop revenue, water cost, and water resources according to several implemented strategies. Since these variables are spatially and temporally varying, GIS is applied to execute these evaluations. Once the experts' criteria are translated to specific rules, the Expert System can utilize GIS modules to evaluate the strategies and assist one to arrive at a decision (Abu-Zeid and Sunada, 1994). The strategies differ according to the cropping pattern, water duty, and the water resources used (surface water, groundwater, or drainage water reuse) and the associated levels of salinities. The Expert System logic for arriving at a decision in evaluating different cropping pattern and water management

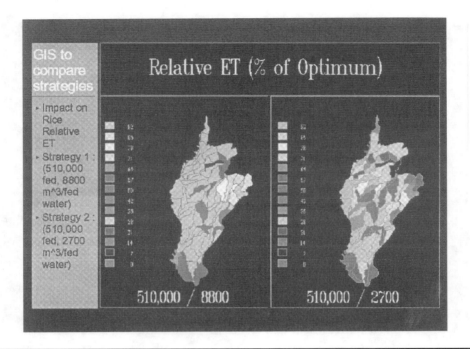

Figure 9.2 **Two strategies comparison.**

strategies was developed. The Expert System evaluates the strategies according to different criteria. One of the criteria was programmed into the DSS using PDC PROLOG language as an example of how this Expert System can be automated. Water resources and irrigation experts in Egypt were interviewed. Their expertise in dealing with water shortage problems, their familiarity with the irrigation system in Egypt and their criteria in evaluating different strategies built the knowledge database for the Expert System.

The DSS was developed by combining the huge numerical database of information in the GIS part with the Expert System which contained the appreciable knowledge database obtained from the experts. A questionnaire was the main source for eliciting the evaluation criteria for the Expert System needed by the DSS for analyzing and evaluating the simulated water management strategies and alternatives. The experts were asked to weigh different category of strategy impacts according to their priority of consideration. The weights were on a scale from "0" to "10" reflecting the importance of the impacts to the experts, with 10 being the weight given to the most important impact to the expert. "0" would be the weight given to the least important (or not important) impact to the expert. These weights were normalized to give the weights shown in the Table 9.1. These weights show the applicable form of weights for the Multicriteria Evaluation Expert System that can be used by the DSS to evaluate the strategy impacts. Figure 9.3 shows the accumulated weights of the impact weights given by the experts which reflects the great concern to economical impacts followed by the environmental, social, political, and then to the ecological impacts.

Different cropping pattern and water management strategies have to be evaluated according to their environmental, ecological, economical, political, and

Table 9.1 Normalized Experts' Strategy Impacts Weights

Experts	Environmental impacts	Ecological impacts	Economical impacts	Political impacts	Social impacts
Expert 1	0.21	0.18	0.26	0.16	0.18
Expert 2	0.22	0.22	0.20	0.20	0.17
Expert 3	0.18	0.18	0.21	0.18	0.24
Expert 4	0.14	0.14	0.27	0.19	0.27
Expert 5	0.25	0.17	0.33	0.13	0.13
Expert 6	0.19	0.17	0.22	0.14	0.28
Expert 7	0.21	0.21	0.21	0.18	0.18
Expert 8	0.27	0.14	0.24	0.19	0.16
Expert 9	0.20	0.17	0.24	0.20	0.20
Expert 10	0.21	0.18	0.24	0.16	0.21
Expert 11	0.22	0.19	0.25	0.16	0.19
Expert 12	0.23	0.18	0.23	0.18	0.18
Expert 13	0.15	0.18	0.23	0.25	0.20
Expert 14	0.25	0.18	0.23	0.15	0.20
Expert 15	0.19	0.14	0.22	0.22	0.24
Expert 16	0.21	0.21	0.24	0.17	0.17

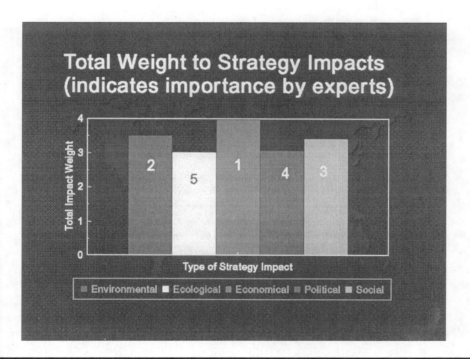

Figure 9.3 Accumulated weights of impact weights given by the experts.

social impacts. Experts have different criteria for evaluating different kinds of impacts. The experts were asked to state their criteria for evaluating each category of impacts for any cropping pattern or water management strategy. Their response

Table 9.2 Environmental Impacts Criteria

Category	Repetition by experts	Environmental impacts criteria
1	11	Decreased water quality (increased water salinity)
2	10	Soil degradation (increased soil salinity)
3	6	Health hazard (air and land pollution)
4	6	Water logging (increased water table)
5	4	Total cultivated area
6	4	Crop yield
7	3	Aquifer depletion (long-term impacts)
8	3	Soil erosion/sedimentation
9	3	Sea water intrusion
10	3	Groundwater pollution
11	3	Increase of aquatic weeds
12	2	Surface water savings
13	2	Effect on wldlife population (birds, trees,..)
14	2	Effect on fish population

Table 9.3 Ecological Impacts Criteria

Category	Repetition by experts	Ecological impacts criteria
1	9	Effect on wildlife
2	5	Effect on fish population
3	5	Water pollution
4	3	Wildlife land/vegetation
5	2	Soil degradation
6	2	Effect on weather patterns
7	2	Effect on weed growth
8	2	Water logging
9	1	Fresh water released to sea or stored in north lakes
10	1	Drainage water released to sea

showed a variety of criteria for each category of impacts ranked according to the experts' priority. These criteria were grouped as shown in the tables below (Tables 9.2 to 9.6) giving the number of repetition of these criteria by the experts. The tables show the environmental, ecological, economical, political, and social impacts criteria identified by a category number. Another weight called the "Expert Weight" was calculated for each expert. Each expert is assigned an environmental, ecological, economical, political, and social "Expert Weight" depending on the number of the experts' repetitions to his suggested criteria (the category of criteria are shown in the previous tables but the itemized criteria for each expert such as Table 9.9 are not shown here). The "Expert Weight" is considered a measure for any expert's expertise since the repetition of the expert's suggested criteria by other experts reflects commonality among experts and some awareness to strategy impacts and their evaluation criteria. The "Expert Weights" were normalized

Table 9.4 Economical Impacts Criteria

Category	Repetition by experts	Economical impacts criteria
1	12	Economical analysis (country)
2	9	Financial analysis (individual)
3	9	Crop yield
4	8	Costs (investment, operation, and maintenance)
5	7	Water savings
6	3	Agricultural expansion
7	3	Equity (supply/demand)
8	2	Drainage water reuse (+ve) (saving water)
9	2	Drainage water reuse (-ve) (pumping costs)
10	1	Tourism (navigation), power generation, and fishing
11	1	Water pricing

Table 9.5 Political Impacts Criteria

Category	Repetition by experts	Political impacts criteria
1	10	Food self-sufficiency and strategic crops; import (wheat and rice), export (cotton)
2	5	Government internal policies (prices)
3	4	Water users associations involvement
4	4	Reclamation projects
5	3	Surface and groundwater use regulations
6	3	Surface water saving
7	3	Coordination between agricultural and irrigational policies
8	3	Impact on external policies (neighboring countries)
9	2	Impact on employment

Table 9.6 Social Impacts Criteria

Category	Repetition	Social impacts criteria
1	8	Farmers adaptability to new strategies
2	7	Effect on individuals' income
3	6	Land reclamation and expansion
4	5	Water users cooperation
5	4	Equity in water distribution
6	2	Considering farmer's ideas
7	2	Imposed prices by government

to give the weights shown in Table 9.7. These normalized weights represent the "Expert Weight" X(i,1) in Table 9.8 that will be presented later. Figure 9.4 shows the distribution of the expert weights among the experts and the type of impacts.

Table 9.7 Normalized Experts' Weights

Experts	Environmental weight	Ecological weight	Economical weight	Political weight	Social weight
Expert 1	4.95	2.53	6.26	10.71	7.58
Expert 2	4.17	0.00	2.01	0.00	0.00
Expert 3	9.38	11.39	4.92	9.18	7.58
Expert 4	8.07	14.56	10.07	5.10	9.60
Expert 5	4.95	8.86	5.82	9.69	0.00
Expert 6	2.34	8.86	5.37	6.63	3.54
Expert 7	6.77	12.03	8.50	13.27	13.64
Expert 8	7.55	8.86	6.26	1.53	4.55
Expert 9	4.95	8.23	6.49	3.06	6.57
Expert 10	8.59	11.39	4.70	5.10	5.05
Expert 11	5.21	3.16	6.71	4.08	6.57
Expert 12	10.16	0.00	7.16	1.53	13.13
Expert 13	8.83	3.16	3.36	7.14	5.56
Expert 14	4.43	0.00	7.16	8.16	9.60
Expert 15	0.00	0.00	8.50	9.69	0.00
Expert 16	9.64	6.96	6.71	5.10	7.07

We need to look at the final form of the logic that the DSS will use to rank the strategies according to the experts' knowledge base of criteria. The following table (Table 9.8) shows how the DSS will rank the strategies according to the combined effect of weighing and ranking which was provided by the experts. Table 9.8 illustrates the evaluation of two strategies by more than one expert to arrive at the final score (or rank) for the strategies according to the combined effect of the experts' preferences.

Definitions of the various weights and scores are also shown. As some of the weights were calculated before an explanation will follow to show how we can arrive at the values to be filled in Table 9.8 for the different weights and variables such as, the "Expert Weight," "Reversed Rank," "Impact Weight," "Combined Weight," "Total Weight," "Boolean Score," "Total Score," and "Final Score."

As the experts were asked to suggest criteria for the evaluation of the water and cropping pattern strategies and also rank them according to their priority of evaluation, these ranks were reversed and then normalized to give the weights such as these shown in Table 9.9 below. The weights shown in Table 9.9 represent the "Reversed Rank" W(i,j,k) for the environmental criteria used in Table 9.8. For the previous weights to be representative, either all 14 categories of the environmental criteria in Table 9.2 should be evaluated, or if only some of the criteria will be evaluated then only these criteria should be used in calculating the previous weights. This is because if an expert had numerous criteria then his normalized weights (W(i,j,k)) will be smaller than another expert who had fewer criteria. These small weights will be compensated for if we evaluated all criteria. If some criteria are not evaluated, the expert who mentioned these criteria will not be given the fair chance to participate in making the decision due to the elimination of some of his criteria and the associated weights, and also due to assigning smaller weights to the rest of the criteria that he mentioned since they should all

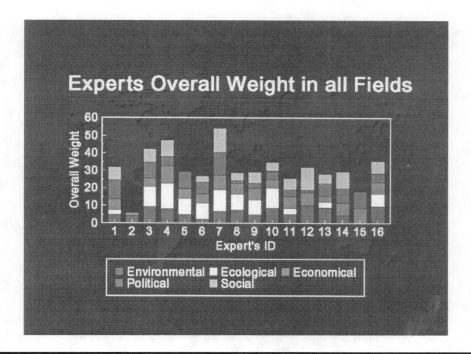

Figure 9.4 Expert weights in different type of impacts.

add up to 1.00 for each expert. This problem needs some more analysis to study the effect of neglecting the evaluation of one of the criteria on the fair participation of all experts, and if there is any conflicting or complementary effects on the Expert Weight (X(i,j)) and the Normalized Criteria Weight (W(i,j,k)) for the Reversed Rank. In other words, if any of the criteria is going to be neglected, it would be worthwhile to first study the effect of this neglection on calculating the Expert Weight (X(i,j)), and the Normalized Criteria Weight (W(i,j,k)), and its influence on the fair participation of all experts in the decision. (e.g., a criteria may be used in calculating the Expert Weight (X(i,j)) (increasing it), and the Normalized Criteria Weight (W(i,j,k)) (decreasing other criteria weights for that expert), but it may not be evaluated resulting in an unfair participation of this expert in the decision making process). Weights were calculated for the ecological, economical, political, and social impact evaluation criteria to develop tables similar to Table 9.9.

Once all the weights are available to form the knowledge base for the Expert System, they are used as demonstrated in Table 9.8. The "Expert Weight", "Reversed Rank" (Criteria Weight), and the "Impact Weight" can be multiplied together for each expert to give a "Combined Weight" for a certain criteria. The "Combined Weights" for each expert will be summed up to give a "Total Weight" reflecting a weight given by all chosen experts for that criteria. This procedure is repeated for all other criteria of interest. In applying this Expert System, usually two strategies are to be evaluated, according to a number of criteria, taking into consideration the experts preferences and weights. We take one criteria at a time and evaluate the strategy according to this criteria. A "Boolean Score" (0 or 1) is assigned to the strategy depending on whether it's better (1) or worse (0) in satisfying this specific criteria. The "Boolean Score" for each strategy will then be

Table 9.8 Two Straegies Evaluation by More than One Expert

Criteria (k) C(j,k)	Expert 1(l)			Expert 2(l)	Total weight Z(k)	Boolean score B(n,k)	Total score ZB(n,k)	Total weight Z(k)	Boolean score B(n,k)	Total score ZB(n,k)
	Expert weight X(i,j)	Reversed rank W(i,j,k)	Impact weight Y(i,j)	Combined weight XWY(i,k)						

(j) (1) Environmental Impacts

						Strategy I (n)			Strategy II (n)	
C(1,1)	X(1,1)	W(1,1,1)	Y(1,1)	XWY(1,1)	Z(1)	B(I,1)	ZB(I,1)	Z(1)	B(II,1)	ZB(II,1)
C(1,2)	X(1,1)	W(1,1,2)	Y(1,1)	XWY(1,2)	Z(2)	B(I,2)	ZB(I,2)	Z(2)	B(II,2)	ZB(II,2)
C(1,3)	X(1,1)	W(1,1,3)	Y(1,1)	XWY(1,3)	Z(3)	B(I,3)	ZB(I,3)	Z(3)	B(II,3)	ZB(II,3)
C(1,4)	X(1,1)	W(1,1,4)	Y(1,1)	XWY(1,4)	Z(3)	B(I,4)	ZB(I,4)	Z(3)	B(II,4)	ZB(II,4)
(1) Environmental impacts total score							TS(n,j) = TS(I,1)			TS(n,j) = TS(II,1)

(j) (2) Ecological Impacts

						Strategy I (n)			Strategy II (n)	
C(2,5)	X(1,2)	W(1,2,5)	Y(1,2)	XWY(1,5)	Z(5)	B(I,5)	ZB(I,5)	Z(5)	B(II,5)	ZB(II,5)
C(2,6)	X(1,2)	W(1,2,6)	Y(1,2)	XWY(1,6)	Z(6)	B(I,6)	ZB(I,6)	Z(6)	B(II,6)	ZB(II,6)
C(2,7)	X(1,2)	W(1,2,7)	Y(1,2)	XWY(1,7)	Z(7)	B(I,7)	ZB(I,7)	Z(7)	B(II,7)	ZB(II,7)
C(2,8)	X(1,2)	W(1,2,8)	Y(1,2)	XWY(1,8)	Z(8)	B(I,8)	ZB(I,8)	Z(8)	B(II,8)	ZB(II,8)
(2) Ecological impacts total score							TS(n,j) = TS(I,2)			TS(n,j) = TS(II,2)

(j) (3) Economical impacts

						Strategy I (n)			Strategy II (n)	
C(3,9)	X(1,3)	W(1,3,9)	Y(1,3)	XWY(1,9)	Z(9)	B(I,9)	ZB(I,9)	Z(9)	B(II,9)	ZB(II,9)
C(3,10)	X(1,3)	W(1,3,10)	Y(1,3)	XWY(1,10)	Z(10)	B(I,10)	ZB(I,10)	Z(10)	B(II,10)	ZB(II,10)
(3) Economical impacts total score							TS(n,j) = TS(I,3)			TS(n,j) = TS(II,3)

(i) (4) Political impacts					Strategy I (n)			Strategy II (n)	
C(4,11)	X(1,4)	W(1,4,11)	Y(1,4)	XWY(1,11)	Z(11)	B(I,11)	ZB(I,11)	Z(11)	B(II,11) ZB(II,11)
C(4,12)	X(1,4)	W(1,4,12)	Y(1,4)	XWY(1,12)	Z(12)	B(I,12)	ZB(I,12)	Z(12)	B(II,12) ZB(II,12)
(4) Political impacts total score							**TS(n,j) = TS(I,4)**		**TS(n,j) = TS(II,4)**

(i) (5) Social impacts					Strategy I (n)			Strategy II (n)	
C(5,13)	X(1,5)	W(1,5,13)	Y(1,5)	XWY(1,13)	Z(13)	B(I,13)	ZB(I,13)	Z(13)	B(II,13) ZB(II,13)
C(5,14)	X(1,5)	W(1,5,14)	Y(1,5)	XWY(1,14)	Z(14)	B(I,14)	ZB(I,14)	Z(14)	B(II,14) ZB(II,14)
(5) Social impacts total score							**TS(n,j) = TS(I,5)**		**TS(n,j) = TS(II,5)**
				Final score FS(n)			**FS(I)**	**Final score**	**FS(II)**

Expert Weight $X(i,j)$: An unbiased normalized weight calculated for each expert, reflecting his awareness about each category of impact.
Reversed Rank Weight $W(i,j,k)$: A weight given to each criteria according to the expert's ranking to the criteria (calculated by reversing the rank order and normalizing).
Impact Weight $Y(i,j)$: A weight given by each expert reflecting the importance of each category of impacts in his strategy evaluation.
Combined Weight $XWY(i,k)$: The multiplication of the Expert Weight, the Reversed Rank, and the Impact Weight

$$[XWY(i,k) = X(i,j) * W(i,j,k) * Y(i,j)].$$

Total Weight $Z(k)$: The summation of all experts' Combined Weights [Sum of $XWY(i,k)$].
Boolean Score $B(n,k)$: It's a value of "Zero" or "1", with "Zero" meaning that the strategy is worse in satisfying the criteria, and "1" meaning that the strategy is better in satisfying the criteria.
Total Score $ZB(n,k)$: It is the multiplication of the Total Weight by the Boolean Score for each strategy

$$[ZB(n,k) = Z(k) * B(n,k)].$$

Final Score $FS(n)$: It is the summation of the Total score for each category of impacts

$$[FS(n) = \text{Sum of } TS(n,j)].$$

Table 9.9 Experts' Reversed Ranks for Environmental Criteria

Experts	Rank	Environmental criteria	Rearranged rank	Reversed rank weight W(i,j,k)	W(i,j,k) X 10
Expert 1	1	Category (5)	4	4/10	4.00
	2	Category (2)	3	3/10	3.00
	3	Category (11)	2	2/10	2.00
	4	Category (12)	1	1/10	1.00
Expert 2	1	Category (6)	3	3/6	5.00
	2	Category (2)	2	2/6	3.33
	3	Category (12)	1	1/6	1.67
Expert 3	1	Category (4)	6	6/21	2.86
	2	Category (2)	5	5/21	2.38
	3	Category (1)	4	4/21	1.90
	4	Category (6)	3	3/21	1.43
	5	Category (14)	2	2/21	0.95
	6	Category (8)	1	1/21	0.48
Expert 4	1	Category (2)	5	5/15	3.33
	2	Category (9)	4	4/15	2.67
	3	Category (11)	3	3/15	2.00
	4	Category (1)	2	2/15	1.33
	5	Category (5)	1	1/15	0.67
Expert 5	1	Category (4)	3	3/6	5.00
	2	Category (1)	2	2/6	3.33
	3	Category (13)	1	1/6	1.67
Expert 6	1	Category (3)	2	2/3	6.67
	2	Category (8)	1	1/3	3.33
Expert 7	1	Category (1)	3	3/6	5.00
	2	Category (2)	2	2/6	3.33
	3	Category (10)	1	1/6	1.67
Expert 8	1	Category (3)	6	6/21	2.86
	2	Category (1)	5	5/21	2.38
	3	Category (11)	4	4/21	1.90
	4	Category (8)	3	3/21	1.43
	5	Category (7)	2	2/21	0.95
	6	Category (9)	1	1/21	0.48
Expert 9	1	Category (2)	3	3/6	5.00
	2	Category (4)	2	2/6	3.33
	3	Category (10)	1	1/6	1.67
Expert 10	1	Category (1)	5	5/15	3.33
	2	Category (2)	4	4/15	2.67
	3	Category (7)	3	3/15	2.00
	4	Category (4)	2	2/15	1.33
	5	Category (9)	1	1/15	0.67
Expert 11	1	Category (3)	3	3/6	5.00
	2	Category (1)	2	2/6	3.33
	3	Category (10)	1	1/6	1.67

Table 9.9 Experts' Reversed Ranks for Environmental Criteria (Continued)

Experts	Rank	Environmental criteria	Rearranged rank	Reversed rank weight W(i,j,k)	W(i,j,k) X 10
Expert 12	1	Category (1)	7	7/28	2.50
	2	Category (3)	6	6/28	2.14
	3	Category (2)	5	5/28	1.79
	4	Category (6)	4	4/28	1.43
	5	Category (13)	3	3/28	1.07
	6	Category (14)	2	2/28	0.71
	7	Category (5)	1	1/28	0.36
Expert 13	1	Category (1)	5	5/15	3.33
	2	Category (4)	4	4/15	2.67
	3	Category (5)	3	3/15	2.00
	4	Category (7)	2	2/15	1.33
	5	Category (2)	1	1/15	0.67
Expert 14	1	Category (3)	2	2/3	6.67
	2	Category (1)	1	1/3	3.33
Expert 15	—	—	—	—	—
Expert 16	1	Category (3)	5	5/15	3.33
	2	Category (1)	4	4/15	2.67
	3	Category (2)	3	3/15	2.00
	4	Category (4)	2	2/15	1.33
	5	Category (6)	1	1/15	0.67

multiplied by the "Total Weight" to give a "Total Score" for each strategy. This is repeated for all other criteria. The sum of all "Total Scores" will give a "Final Score" for each strategy which will decide which strategy is preferable. The Expert System at this stage has simulated the process that the selected experts would have gone through to arrive at a decision regarding evaluating different strategies and their impacts.

The question that arises is how can we arrive at the "Boolean Score" for the strategy according to a certain criteria and within the available data, and if it is possible to examine the applicability of GIS to do such an evaluation and arrive at this "Boolean Score".

GIS in Multicriteria Evaluation

As an example of how GIS can be used to arrive at these Boolean Scores, one criteria from each type of impact will be evaluated to compare two rice water management strategies for the Eastern Nile Delta. The idea is to see the impact of each strategy on the selected parameters using GIS. Simply, when comparing the two strategies, if one strategy had a negative impact on the selected parameter on a larger spatial scale than the other strategy, the first one will be given a Boolean Score of "0". The other strategy which had a negative impact on a smaller spatial scale will be given a Boolean Score of "1" since it negatively affected a smaller area than the first Strategy. On the other hand, a similar procedure will be used if the impact is a positive one.

The first one of the two rice water strategies to be compared is a strategy where the area of cultivated rice is 510,000 feddans and the water requirements given to the rice crop is the optimum 8,800 m^3/feddan. The other strategy is having the same cultivated area of rice (510,000 feddans), but with less water of 2,700 m^3/feddan given to the rice crop.

Using GIS is very useful in calculating the areas affected by applying these strategies. The general idea is to form the two maps for each of the two strategies in the GIS format showing the values of the parameter for the criteria which is evaluated. The two maps in the GIS format have the area divided into cells and each cell is assigned the value for the parameter of interest. The two maps are then subtracted to yield a resultant map. According to the different strategy impacts on different locations of the area, the resultant map will show positive, negative, and zero values assigned to the cells. If we subtract the map for strategy two from that for strategy one, then the cells that have positive values in the resultant map will reflect strategy one having higher values for the impact than strategy two at the locations where these cells are. The opposite will be true if there is negative values associated with cells in the resultant map. Depending on the impact being a positive impact or a negative impact, we can accordingly say if the strategy is better when the value for the impact parameter is higher or when it is lower. The next step would be to calculate the number of cells having positive values and the number of cells having negative values in the resultant map. Depending on which number of cells are more we can tell which strategy is better in satisfying that particular criteria.

Let's take one criteria from the environmental impacts and evaluate it by all experts. This will result in the Table 9.10 where values may be extracted from the Tables 9.1, 9.7, and 9.9.

GIS maps were used to show the soil salinity impact maps for each strategy and the resultant subtraction map was used to calculate the areas. The areas, in terms of number of cells, were then used to determine the Boolean score.

As in the previous example where two strategies were compared and one criteria was evaluated for the environmental impacts, similarly one criteria for ecological, and the same for the economical, social, and political impacts were evaluated. For the environmental impacts the soil salinity criteria was evaluated, the effect on fish population criteria for ecological impacts, supply/demand ratio for economical impacts, water savings for political impacts, and farmer's income for social impacts. The criteria were evaluated using the same procedure mentioned above. After the two strategies were compared according to these criteria and the Boolean scores were assigned to each strategy depending on which strategy is better in satisfying the criteria, the Boolean scores were then multiplied by the weights obtained from the tables that were formed earlier. The final score for each strategy after evaluating a criteria for each type of impact is shown in Table 9.11. It shows the type of impact, the criteria being evaluated, its category number, the areas being affected positively or negatively, reflected in number of cells, the Boolean score, the total weights, and the final scores. The strategy that scores a higher score reflects its preference according to a combined decision based on the suggestions and weighting system by all the experts.

Table 9.10 Weights for Soil Degradation Criteria (Environmental Impact)

Experts	Expert weight	Criteria weight	Impact weight	Combined weight	Strategy I	Strategy II
Expert 1	4.95	3.00	0.21	3.12		
Expert 2	4.17	3.33	0.22	3.05		
Expert 3	9.38	2.38	0.18	4.02		
Expert 4	8.07	3.33	0.14	3.76		
Expert 5	4.95	0.00	0.25	0.00		
Expert 6	2.34	0.00	0.19	0.00		
Expert 7	6.77	3.33	0.21	4.73		
Expert 8	7.55	0.00	0.27	0.00		
Expert 9	4.95	5.00	0.20	4.95		
Expert 10	8.59	2.67	0.21	4.82		
Expert 11	3.21	0.00	0.22	0.00		
Expert 12	10.16	1.79	0.23	4.18		
Expert 13	8.83	0.67	0.15	0.89		
Expert 14	4.43	0.00	0.25	0.00		
Expert 15	0.00	0.00	0.19	0.00		
Expert 16	9.64	2.00	0.21	4.05		
	Area		Cells		0	−42980
	Boolean		Score		1	0
			Total weight	37.75		
			Final	Score	37.75	0.00

Table 9.11 Final Scoring for Strategy Comparison

			Strategy I	Strategy II	
			Area in cells		
Impact	Criteria	Category	Boolean	Score	Total score
Environmental	Soil degradation	2	0	−42980	37.78
			1	0	
Ecological	Fish population	2	−3961	−41040	23.99
			1	0	
Economical	Supply/demand	7	41856	372	10.34
			1	0	
Political	Water saving	6	−44399	−852	8.78
			0	1	
Social	Farmers' income	2	36485	0	50.68
			1	0	
	Final	Score	122.76	8.78	

Strategy one with area cultivated with rice of 510,000 feddans, and rice water application of 8,800 m³/feddan scored a higher score of 122.76 than strategy two of 510,000 feddans and rice water application of 2,700 m³/feddan which scored 8.78. Although Strategy II scored higher in the political impact of water saving, strategy I scored higher in the selected environmental, ecological, economical, and social impacts evaluation.

Conclusions

A computerized DSS that aids in the process of decision making for agriculture and water resources problems was presented. A DSS, that incorporates GIS, provides decision makers with the tools to better view and analyze different alternatives and their impacts. Effective visualization of alternatives and impacts is an important support tool for the decision-making process and is essential in the development of every DSS.

An effective user interface for a DSS should enhance the use of GIS by decision makers who may not have to be familiar with operating GIS. Modeling human experience for supporting decision making is as important as developing simulation models. If the available simulation models are verified for their applicability to the problem of interest, it would be worth it to develop a DSS enhanced by an Expert System to evaluate the output of these simulation models.

GIS may be an effective tool to execute Expert System multirules for multicriteria evaluation of water management and cropping pattern strategies. The procedure, developed for the Expert System to arrive at a decision in evaluating different strategies, arranges the thoughts, criteria, and preferences of experts in a way that may not have been implemented in their previous decisions, but in fact, may be very much desired once such a DSS is available. This was actually stated in many of the experts responses to one of the questions in the questionnaire that they were asked to fill.

The weighting procedure that was used to give the experts some kind of an environmental, ecological, economical, political, and social weight reflecting their awareness about these fields may be very well considered an unbiased and fair way of giving weights to the experts.

The results of the experts' answers to the questionnaires show that the economical impacts of water management and cropping pattern strategies in Egypt are of the most important concern, followed by the environmental impacts, social, political, and then ecological impacts.

References

Abu-Zeid, K. 1994. A GIS Multi-Criteria Expert Decision Support System for Water Resources Management. Dissertation. Colorado State University, Fort Collins, Colorado.

Abu-Zeid, K. and D. Sunada. 1994. GIS: A Tool for Decision Making in Water Resources Management. GIS'94 Symposium Proceedings. Volume 2. Vancouver, British Columbia, Canada. Pages 689–695.

DRI (Drainage Research Institute). 1991. Reuse of Drainage Water Project. Reuse Report 30. DRI, WRC, MPWWR, Egypt.

Eastman, J.R. 1993. IDRISI, Version (4.0–4.1). Technical Reference. Clark University, Graduate School of Geography, Worcester, Massachusetts.

PDC PROLOG. 1992. User's Guide, Version 3.3. Prolog Development Center, Copenhagen, Denmark/Atlanta, Georgia.

Chapter 10

An Integrated Approach to Ecological Risk Assessment of a Large Forested Ecosystem*

Tom L. Ashwood, Bradley E. Sample,
Robert A. Washington-Allen, and Glenn W. Suter, II

Introduction

The purpose of this chapter is to describe an approach for developing an ecological assessment program for the U.S. Department of Energy's (DOE's) Oak Ridge Reservation (ORR). Such a program is required to assess existing ecological risks, to predict changes in those ecological risks from proposed remedial actions, and to monitor the effectiveness of remedial actions in reducing ecological risks. This chapter sets out the basic framework for the ecological assessment program and defines the specific tasks to be completed in the first year.

The principal regulatory impetus for a Reservation-wide ecological assessment program derives from the Federal Facilities Agreement (FFA) that governs implementation of environmental restoration activities on the Reservation. This agreement integrates the provisions of the Resource Conservation and Recovery Act (RCRA) with those of the Comprehensive Environmental Response, Compensation, and Liability Act (CERCLA) and requires that environmental restoration activities be based

---

* Research was sponsored by the U.S. Department of Energy, Office of Waste Management under contract DE-AC05-OR21400 with Lockheed Martin Energy Systems, Inc.

upon human health and ecological risks. Ecological risk assessments must be based on Reservation-level data as well as data for individual contaminated sites.

Given the regulatory impetus and the need for information to support decisions to be made by the FFA parties [i.e., DOE, EPA Region IV, and the Tennessee Department of Environment and Conservation (TDEC)], the program described in this plan was developed based on a series of discussions among those parties. This process, collectively called the Data Quality Objectives (DQO) process, is described in a later section. Results of the DQO process were a set of problem statements, decisions, decision rules, and inputs. This plan has been developed to provide those inputs.

Description of the Oak Ridge Reservation

The ORR covers ~14,000 ha in Anderson and Roane counties of east Tennessee. The ORR is bounded on the south and west by the Clinch River. The residential/commercial portion of the city of Oak Ridge borders the ORR to the north and east.

Three major DOE facilities are located on the ORR. Oak Ridge National Laboratory is a multiple-mission research facility. The K-25 Site, formerly the Oak Ridge Gaseous Diffusion Plant, now houses Environmental Restoration and Waste Management programs, laboratory, engineering, and administrative personnel. The Y-12 Plant is part of the DOE's nuclear weapons complex and also houses the Center for Manufacturing Technology. Most of the ORR outside these major facilities is part of the Oak Ridge National Environmental Research Park.

Operation of the DOE facilities over the past 50 years has resulted in intentional and inadvertent release of contaminants to all ambient media. The nature of these facilities has been such that a wide variety of contaminants have been released. Results of biological monitoring studies over the past 30 years have shown that many of these contaminants — especially radioisotopes — have accumulated in terrestrial and aquatic biota. Areas where contaminants have been released or have accumulated following release have been identified, and operable units have been established for remedial actions.

Conceptual Framework

The conceptual strategy for ecological risk assessment on the ORR, which was developed by the FFA parties through a DQO process, is described by Suter et al. (1994). The purpose of the strategy is to permit the completion of adequate ecological risk assessment on individual operable units while ensuring that all ecological resources on the ORR that are exposed to contaminants are assessed. The strategy distinguishes four types of potentially contaminated areas: (1) source operable units, which may contain waste disposal areas, (2) groundwater aquifers that are potentially contaminated by source operable units, (3) aquatic integrator operable units, which are streams and associated floodplains that drain source operable units, and (4) the terrestrial integrator, which encompasses the Reservation.

Generic conceptual models for these operable were developed. Figure 10.1 shows the flows of contaminants among the four classes of operable units. Figure 10.2 shows the contaminant flows into and within the overall terrestrial

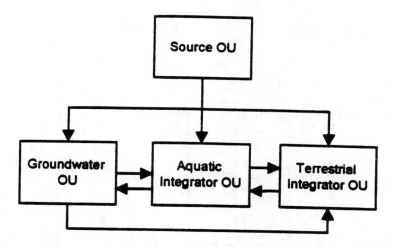

Figure 10.1 Conceptual model of the transport of contaminants among the different classes of operable units (OUs).

ecosystem. Each compartment of these models has been described in terms of its content, inputs, and outputs by Suter et al. (1994).

Implementation of the ORR strategy for ecological risk assessment depends on the exchange of information among the operable units. As part of the remedial investigation, each operable unit is responsible for characterizing: risks to the ecological endpoints that are associated with the operable unit (i.e., occur on the operable unit and have a scale appropriate to the operable unit), contributions to contaminant inputs to "downstream" integrator operable units, and risks resulting from contaminant inputs from "upstream" operable units. In order to perform those characterizations, each operable unit needs estimates of: fluxes of contaminants from "upstream" operable units, and ecological risks resulting from its contributions to downstream operable units. The needed data would be generated if each operable unit characterized sources that occurred within the operable unit, exposures to all endpoint biota on the operable unit (whether associated with the operable unit or with a larger scale integrator operable unit), responses of endpoint biota associated with the operable unit, and fluxes of contaminants off the operable unit and into integrator operable units.

All of these operable unit types have been previously recognized except the terrestrial integrator. The creation of the terrestrial integrator is a result of the recognition by the FFA parties that some ecological endpoints were not being appropriately assessed. Those were the highly mobile animals that move among the operable units and the rest of the Reservation, feeding, drinking, and resting in various locations. Effects on these wide-ranging animals can not be attributed to one operable unit, but without a terrestrial integrator, there would be no mechanism to assess the combined effects of the multiple operable units. In addition, the terrestrial integrator provides a means of assessing the ecological risks associated with habitat destruction during remediation by identifying the habitat types destroyed, the importance of the habitat loss to endpoint species and communities, and the significance of the particular area destroyed (e.g., as

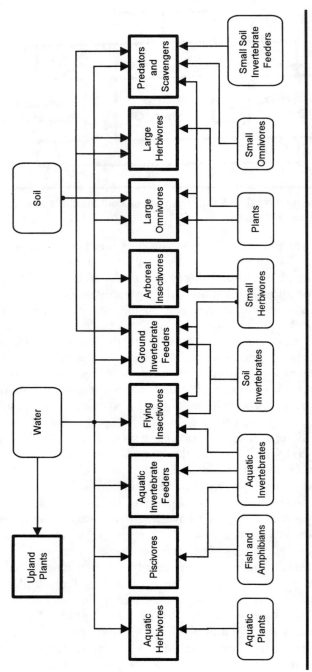

Figure 10.2 Generic conceptual model for contaminant transport and exposure in the terrestrial integrator operable unit.

a habitat corridor). Finally, the terrestrial integrator provides a means of interpreting the significance of effects on individual operable units in an ecosystem context.

Review of Existing Information

The FFA parties agreed during the DQO process that existing monitoring programs for aquatic systems were adequate to provide input for their future decisions. Thus, this program focuses only on the terrestrial ecosystem of the ORR. The remainder of this section provides a summary of information on the terrestrial habitat of the ORR.

The original forests on the ORR were extensively cleared, and the land was cultivated or partially cleared and used for rough pasture by settlers in the area. Except on very steep slopes, most of the forest had been cut for timber or cleared for agriculture by the time the federal government acquired the land in 1942. With the end of cultivation in 1942, fields have developed into forest either through natural succession or through planting of pines.

The ORR is home to many plant species. Plant communities on the ORR are characteristic of those found in the intermountain regions of Appalachia. The dominant association on the ORR is oak-hickory forest. Small cedar barrens are common on the ORR. These drought-tolerant plant communities, which occur on shallow, limestone soils include several species of rare plants.

The ORR also supports a wide variety of wildlife species. A list of amphibians, reptiles, birds, and mammals on the ORR has been compiled (Suter et al., 1994) that includes animals that have been identified on the ORR as well as some animals present in the Ridge and Valley region and for which adequate habitat exists on the ORR. This list includes more than 30 species that are listed by either the Tennessee Wildlife Resources Commission or the U.S. Fish and Wildlife Service as threatened, endangered, or in need of management.

Since 1960 several ecological researchers have measured contaminants in biota on the ORR. One conclusion that can be drawn from these studies is that every investigation of contamination in wildlife has found at least some individuals who were contaminated. This conclusion should be tempered with the realization that, in most cases, researchers were looking for contaminants that were expected to bioaccumulate in areas that were known to be contaminated. Nevertheless, the important point is that biota in contaminated areas are accumulating that contamination.

A second noteworthy feature of the historical data is that mobile animals with large home ranges (e.g., Canada geese, deer, kingfishers, and wild turkeys) are contaminated, and contaminated individuals of those species have been collected at locations outside the boundaries of existing operable units.

Whereas the ecological significance of this widespread contamination is unclear at this time, there is evidence that some wildlife species in some areas of the ORR are suffering adverse effects from existing contamination.

Although numerous contaminants have been observed at source operable units on the ORR, many of these contaminants possess characteristics that make it unlikely that they will present a risk beyond the scale of the source operable unit. Some of these characteristics include high volatility, rapid environmental degradation, low persistence, low environmental mobility, low bioavailability, and low bioaccumulation potential. These characteristics act to limit the area potentially affected by these contaminants to the source operable unit where they occur and its immediate surroundings. Because our emphasis is on wide-ranging species, contaminants that are persistent, mobile, and that may enter food webs and bioaccumulate are the primary concern.

The initial list of inorganic and organic contaminants was obtained from the list of chemicals of potential concern identified in Ross et al. (1992). Additional contaminants identified through current work at source operable units or as residues in previous biota sampling were added to the list. A total of 134 contaminants were identified. The list of chemicals of potential concern is not intended to be static. It is expected that the list will be updated, with contaminants added or deleted, as new information becomes available.

The Process

With the preceding information as background, the process of developing an ecological assessment program was simply a matter of identifying the needs of the various parties and determining an approach that would satisfy those needs. The DQO approach used here provides a methodology for accomplishing these goals.

The DQO process usually involves seven steps: (1) problem statement, (2) decision identification, (3) development of decision rules, (4) establishment of boundaries, (5) identification of inputs to the decision, (6) determination of uncertainty constraints, and (7) optimization of the sampling design. Because of the nature of this project, the FFA parties agreed that determining uncertainty constraints and optimizing sampling design could not be accomplished until further information is available. Therefore, the process proceeded only through the first five steps. For this project, the DQO process was implemented in three meetings in 1994.

Problem statements and decisions are presented in Table 10.1. Tables 10.2 and 10.3 identify the assessment and measurement endpoints that are considered to be inputs in the decision process.

The primary boundary for this project is the Reservation boundary. Each operable unit will address the ecological system within its boundaries. Data from each operable unit will be used to assess the impacts on wide-ranging species. Boundaries may be habitat or species specific. After mapping the habitat, boundaries may need to be reassessed.

Three decision rules were agreed upon:

1. A significant impact has occurred if there is a >20% reduction in abundance or production of an endpoint species (Tables 10.1 and 10.2).
2. A significant impact has occurred if there is a >20% reduction in species richness or abundance in an endpoint community (Tables 10.1 and 10.2).
3. A significant impact has occurred if there is a >20% reduction in survival, growth, or reproduction of organisms in a toxicity test that is representative of an endpoint species or community.

Table 10.1 Problem Statements and Decisions for the ORR Ecological Monitoring and Assessment Program

Problem 1:	Given the previous data, it is important to determine if contamination has resulted in ecological impacts on terrestrial and aquatic species and to determine the extent of the impacts across the ORR.
Problem 2:	Assess the ecological system to the appropriate level required to allow evaluation of impacts of potential activities.
Problem 3:	Once data has been gathered for problems 1 and 2, the effectiveness of remedial action (or no action in selected cases) must be evaluated.
Decision 1:	Determine whether there is an unacceptable ecological impact.
Decision 2:	Determine whether interim actions are necessary due to the ecological impact.
Decision 3:	If an ecological impact exists, the "major" contributing sources to the problem and appropriate remedies must be identified, evaluated, and properly implemented.

Table 10.2 Assessment Endpoints for an Ecological Risk Assessment of the ORR

Group or trophic position	Species or community[a]
Upland plants	Distribution and abundance of plant community types and threatened or endangered plant species
Aquatic herbivores	**Cumberland slider** and Mallard duck
Piscivores	**River otter, bald eagle, osprey, double-crested cormorant, black-crowned night heron,** mink, great blue heron, and belted kingfisher
Aquatic invertebrate feeders	**Hellbender,** leopard frog, and pied-billed grebe
Flying insectivores	**Gray bat, Indiana bat, eastern small-footed bat, Rafinesque's big-eared bat,** and rough-winged swallow
Ground invertebrate feeders	**Long-tailed shrew, masked shrew, smokey shrew, southeastern shrew, six-line racerunner, slender glass snake, Tennessee cave salamander, green salamander,** American woodcock, European starling, and American toad
Arboreal insectivores	Quality and availability of habitat
Large omnivores	Muskrat, raccoon, and wood duck
Large herbivores	**Grasshopper sparrow, Henslow's sparrow, lark sparrow, vesper sparrow,** white-tailed deer, wild turkey, Canada goose, and groundhog
Predators and scavengers	**Golden eagle, northern harrier, Cooper's hawk, red-shouldered hawk, sharp-shinned hawk, barn owl, black vulture, eastern cougar, northern pine snake,** red fox, snapping turtle, and black rat snake

[a] Species in boldface are listed as threatened, endangered, or in need of management by the Tennessee Wildlife Resources Agency or the U.S. Fish and Wildlife Service. Scientific names of all animals are listed in Appendix C.

Table 10.3 Measurement Endpoints for an Ecological Risk Assessment of the ORR

Group or trophic position	Species or community
Upland plants	Distribution and abundance of plant community types and threatened or endangered plant species
Aquatic herbivores	Pond slider and Mallard duck
Piscivores	Great blue heron, belted kingfisher, and northern water snake
Aquatic invertebrate feeders	Leopard frog, and pied-billed grebe
Flying insectivores	Rough-winged swallow and common bats
Ground invertebrate feeders	Short-tailed shrew, American woodcock, European starling, and American toad
Arboreal insectivores	Quality and availability of habitat
Large omnivores	Muskrat, raccoon, and wood duck
Large herbivores	White-tailed deer, wild turkey, Canada goose, and groundhog
Predators and scavengers	Red fox, snapping turtle, and black rat snake

For species listed as threatened, endangered, or in need of management by the State of Tennessee or the U.S. Fish and Wildlife Service, the 20% degradation criteria is not acceptable. Any toxic effect on a listed species is considered a significant impact. Wetlands are also considered separately from the above criteria. No net loss of wetland functions is permitted.

Data Gaps

Data needed to evaluate ecological risk at the Reservation scale may be broken into two categories: population-related parameters and exposure-related parameters.

- Abundance (numbers of individuals of an endpoint species on the ORR)
- Demographics (age and sex distributions and measures of reproduction of an endpoint species on the ORR
- Distribution (spatial arrangement of individuals of endpoint species on the ORR)
- Home range (area where individual resides [in hectares])
- Feeding range (area where individual forages [in hectares])
- Local habitat preference (type and frequency of use of habitats available on the ORR)
- Local food habits (type and frequency of use of foods available on the ORR)
- Contaminant concentrations in food, water, and soil
- Contaminant body burdens
- Bioindicators of contaminant exposure
- Proportion and magnitude of contamination within preferred habitats
- Proportion of time spent in contaminated habitats

Comparison of these data needs to the existing data defines the data gaps that need to be filled so that a Reservation-wide ecological risk assessment can be performed.

Although some data exists with which to assess ecological risks to Reservation-wide endpoints, much information is lacking. For example, ORR-specific population data are only available for three assessment endpoints — upland plants, mallard duck, and white-tailed deer. Other population-related data gaps, such as home and feeding ranges and food and habitat preferences of endpoint species, will be filled using published observations or values from other locations within the range of the species. If published data are not available or if the habits of the endpoint species are so variable that observations from other locations may not be representative of those on the ORR, specific studies may be undertaken to collect the needed data.

The most abundant data are contaminant concentrations in food, water, and soil. These data are available for many endpoint species. However, because data are available only for those operable units for which environmental restoration programs have been initiated and such programs have not been initiated at all operable units, existing data are not sufficient to definitively evaluate risk to ORR-wide populations.

Abundance, demographics, and distribution of wildlife are related to the abundance, diversity, and distribution of habitat. By identifying the habitats available on the Reservation and the habitat requirements of endpoint species, models to estimate the distribution and abundance of populations of endpoint species may be developed. Assessment of available habitat on the ORR will use existing maps and new analyses to be performed as part of this program. Extensive literature reviews will be performed for each endpoint species to identify critical habitat parameters that may be used to predict the presence and abundance of endpoint species in each particular habitat.

To evaluate the proportion of habitat contaminated, habitat maps will be compared to the areas of known contamination. The proportions of various habitat types observed within these contaminated areas will serve as estimates of the proportion of that habitat that is contaminated. The proportion of an endpoint species habitat that is contaminated will be determined by comparing this information to the habitat use information for each endpoint species.

Finally, while efforts will be made to address and fill all data gaps, this may not be possible due to practical limitations. Because a weight-of-evidence approach will be used in this assessment, not all data listed above are necessary to adequately assess risk. The strategy is to use existing data (ORR-specific and literature) to perform screening assessments which identify two types of data needs. The first type is data gaps that prevent performance of a complete screening assessment. The second type is data that are needed to resolve uncertainties revealed by the screening assessment, e.g., actual dose to endpoints vs. literature-derived estimates.

Specific Tasks

In order to provide the information identified in the DQO process and in the subsequent review of data gaps, two specific tasks were identified initially — development of habitat maps for the ORR as a whole and each individual operable unit, and preliminary assessment of ecological risk. Both tasks will require multiple years to complete, and it is likely that additional tasks will be identified as the program proceeds. The following two subsections provide summaries of the status of both tasks at the end of the first year of program implementation.

Habitat Mapping and Analysis

An updated land use/land cover map for the ORR was developed using remotely sensed data which provides information over a larger geographic extent, and at less expense, than data collected on the ground. Landsat Thematic Mapper 5 satellite imagery was used to create the land use/land cover map for the ORR and surrounding areas. The image consists of seven spectral bands spanning the visible, infrared, and thermal-infrared wavelengths; these bands are particularly useful for environmental applications such as vegetation type and health determination, soil moisture, snow and cloud differentiation, and rock type discrimination. Data are stored in digital format as a grid of pixel values for each band. Each image pixel represents the average reflectance over a ground area of 30×30 m.

The image used to create the ORR land use/land cover map was taken from satellite path 19/row 35 on April 13, 1994 at approximately 9:00 am Eastern Standard Time. A spring image was selected to maximize the physiological differences between evergreen and deciduous vegetation types; the April 13 date was chosen for its low (<10%) cloud cover.

The 100×100-km image is georeferenced to U.S. Geological Survey (USGS) ground control points and terrain corrected using a USGS 1:24,000 digital elevation model. The image is resampled to the standard ORR map projection [i.e., Tennessee State Plane zone 5301, North American Datum 83] with 25×25-m pixels. The study area subscene is nearly 189,000 ha in size. The ORR boundary encompasses an area of 14,252 ha.

A detailed description of the classification strategy and the accuracy assessment is presented in Washington-Allen et al. (1995). The land use/land cover map for the entire ORR is presented in Figure 10.3. Individual land use/land cover maps for each operable unit were also prepared. For each operable unit and for the ORR as a whole, the area of each land cover type has been calculated.

In addition to the land use/land cover map, literature reviews are being conducted to identify habitat requirements for each of the endpoint species identified in Table 10. In succeeding years, the map will be updated with layers identifying the habitat distribution of each of these endpoint species.

Ecological Risk Assessment

This program serves two risk assessment purposes. First, it provides the means to assess the ecological implications of contaminant exposures on individual source and aquatic integrator operable units. Second, this program assesses the risks to the terrestrial biota due to the combined effects of multiple contaminant sources and habitat disturbances. This activity will serve to indicate the risks to the Reservation-scale ecological endpoints and, by integrating ecological risks across the Reservation, will provide a basis for ultimately determining that remediation

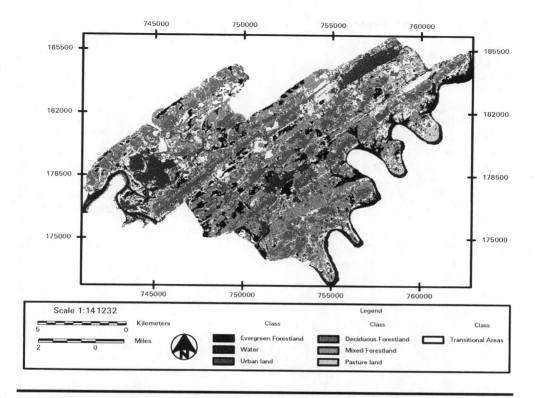

Figure 10.3 April 13, 1994 classified LANDSAT Thematic Mapper image of the ORR. Image projection is Tennessee State Plane (meters), Zone 5301, NAD 83. Adapted from Washington-Allen et al. (1995).

has been successful. Such a determination must occur before the ORR can be removed from the National Priority List.

The preliminary ecological risk assessment accomplished three important objectives. First, we determined which source operable units were most likely to contribute to ecological risks. This permits future investigations to focus on those contaminant sources most likely to pose the greatest risks. In addition, we prioritized the assessment endpoints in terms of their relative potential exposure to contamination. Thus, we can now eliminate some endpoints from further consideration and we can focus our data collection on those species most likely to be impacted.

Second, we assessed the ecological risks to four piscivorous species. The vast majority of contaminant transport on the ORR is via surface water, and all major ORR streams contain measurable contamination (primarily mercury and polychlorinated biphenyls) in aquatic biota. Thus, piscivorous wildlife are potentially at substantial risk. Preliminary results suggest that there are potentially significant population-level risks to belted kingfishers (*Ceryle alcyon*), great blue herons (*Ardea herodias*), mink (*Mustela vison*), and river otter (*Lutra canadensis*) from consuming fish and aquatic organisms from ORR streams.

Finally, risks to herbivores and vermivores were assessed using soil contamination measurements from 12 source operable units on the ORR. This assessment was aimed at demonstrating the feasibility of the approach and identifying data needs for similar assessments. Preliminary results suggest that soil contamination from these 12 operable units do not pose significant population-level risks to the wildlife studied.

Lessons Learned

Probably the most important lesson to be learned from this effort is that it is possible to design a cost-effective ecological assessment program for a highly complex site with multiple contaminants and multiple contaminant sources. Our program considers ecological risks at all appropriate scales.

We also discovered that, when using historical data and data from multiple programs, you need to plan for substantial data management efforts. Finally, agreements between regulators and responsible parties cannot be assumed to hold up through the course of an assessment program. Therefore, it is necessary to maintain communication with the regulators.

References

Ross, R.H., A. Redfern, R.K. White, and R.A. Shaw. 1992. Approach and strategy for developing human health toxicity information for contaminants of concern at sites administered by the U.S. Department of Energy Oak Ridge Field Office Environmental Restoration Program. ES/ER/TM-38. Martin Marietta Energy Systems, Inc.

Suter, G.W., II, B. Sample, D. Jones, and T. Ashwood. 1994. Approach and Strategy for Performing Ecological Risk Assessments for the Department of Energy's Oak Ridge Reservation: 1994 Revision. ES/ER/TM-33/R1. Oak Ridge National Laboratory.

Washington-Allen, R.A., T.L. Ashwood, S.W. Christensen, H. Offerman, and P. Scarbrough-Luther. 1995. Terrestrial Mapping of the Oak Ridge Reservation: Phase 1. ES/ER/TM-152. Lockheed Martin Energy Systems, Inc.

Chapter 11

Managing Environmental Cleanup Risks as a Multiobjective Decision Problem

T.E. HAKONSON AND L.J. LANE

Abstract

This paper describes Federal Facility cleanup programs, such as those being conducted by the Departments of Energy (DOE) and Defense (DOD), from the perspective of a risk management decision problem. Our intent is to describe the risk management decision process and the technical risk assessment procedures to help define the design and operational requirements of multiobjective decision tools for risk management decision making. We also present the case for using human health and ecological risk assessments as a mechanism for integrating a strong technical component into the decision-making process.

The role of science and risk assessments in the environmental decision-making process will likely be governed by the ease with which complex technical relationships, data, and statistical uncertainties are interpreted and used by the decision maker. This interface between the scientist or risk assessor and the risk manager is ideally suited for the use of decision analysis tools that provide a common basis for integrating, synthesizing, and valuing the scientific and policy information. Two limitations to implementing decision support systems for risk management applications are dynamic risk assessment models, particularly for ecological risk assessments, and supporting databases, particularly on the fate and effects of hazardous contaminants.

Introduction

Large, federally funded environmental cleanup programs are projected to cost several hundred billion dollars over the next several decades (Abelson, 1990; 1992; 1993; Breshears et al., 1993; U.S. Department of Energy, 1991; Congress of the United States, 1991). The primary purpose of these programs is to manage risks to human health and the environment from contaminants that have been released as a part of normal operations and accidents. It is widely recognized (Abelson, 1990; 1992; 1993; Breshears et al., 1993; Levin, 1992; Cowling, 1992; Loucks, 1992; Russell, 1992; Schindler, 1992) that risk-based decision making must be used in these programs to set priorities for addressing problems and to select cost-effective solutions that ensure protection of health and the environment.

Lessons learned from the National Acid Precipitation Assessment Program (NAPAP) suggest that the role of science in the environmental decision-making process is largely governed by the ease with which complex technical relationships and data can be interpreted and used by the decision maker (Levin, 1992; Cowling, 1992; Loucks, 1992; Russell, 1992; Schindler, 1992; Breshears et al., 1993). If the results of the research are difficult to interpret and use as a part of the decision process, decision makers will often heavily rely on other, mostly nontechnical, criteria in making decisions. Examples of nontechnical criteria would include some environmental guidance and regulations (arbitrary safety factors are often built into regulatory standards), social-political factors, and costs. Complete reliance on nontechnical criteria for decision making can lead to overly conservative and costly decisions that may or may not reflect the real problems and risks. Moreover, management actions that have a weak technical basis, can enhance the risks to receptors (i.e., the cure is worse than the disease).

This paper describes the large Federal Facility cleanup programs, such as those being conducted by the Departments of Energy (DOE) and Defense (DOD), from the perspective of a risk management decision problem. Our intent is to describe the risk management decision process and the technical risk assessment procedures to help define the design and operational requirements of multiobjective risk management decision tools (Lane et al., 1991; Ascough, 1992). We also present the case for using human health and ecological risk assessments as a mechanism for integrating a strong technical component into the decision-making process.

Cleanup Programs, Risk Assessments, and Decision Making

Calls for ensuring a sound scientific basis for decision making (Abelson, 1990; 1992; 1993; National Research Council, 1989) have come to the forefront over the last few years because of the projected several hundred billion dollar life cycle costs of the environmental restoration programs currently being conducted by DOE and DOD (McGuire, 1989; Pasternak and Cary, 1992). These programs consist of three phases including an initial phase to characterize the types and concentrations of contaminants in biotic and abiotic receptors at a site and the transport processes that mobilize the contaminants. Existing data and additional field sampling serve as the basis for completing this phase. In the second phase, all relevant data are used to conduct a technical assessment of the risks to both humans and ecosystems. State and federal regulations mandate these assessments to ensure

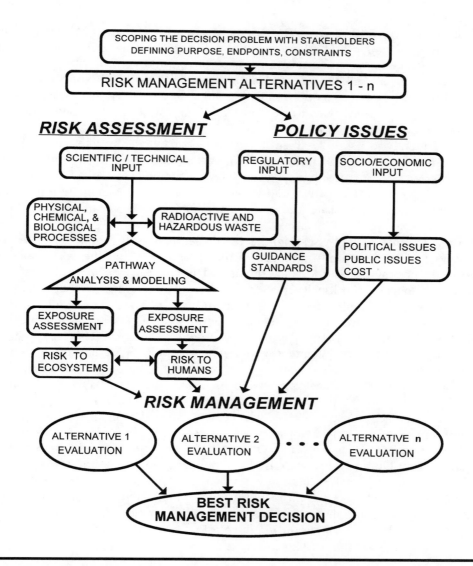

Figure 11.1 Decision process to manage risks.

that human health and the environment are protected from site contaminants (Harwell, 1989; Bartell et al., 1992; Suter, 1993). If the calculated risks are acceptable, then the risk manager has technical justification for no further action at the site. If the risks are unacceptable, then a third phase is implemented to remediate the site until potential risks are reduced to acceptable levels.

The Decision Process to Manage Risks

The general types of information that comprise the decision process to manage risks are presented in Figure 11.1. We find this perspective generally consistent with comments of Levin (1992), Russell (1992), Cowling (1992), and Loucks (1992), based on their experience with NAPAP. Figure 11.1 illustrates the three major factors to be considered in evaluating risks and risk management alternatives. They are (l) scientific and technical input, (2) regulatory requirements, and (3) political, economic, and social input. These factors must initially be considered in concert with all stakeholders to define the overall problem, specify the time and economic constraints associated with political and public policy issues, specify the management alternatives to be considered, and any other constraints to the decision process.

Subsequently, a risk assessment for each risk management alternative should be driven by scientific and technical input and should be relatively independent of regulatory, political, economic, and social considerations except when necessary to clarify a scenario (Figure 11.1). Risk management alternatives might range from doing nothing to remediate the site to eliminating the contaminants via soil removal.

After completing the assessment, scientific input is still required by the risk manager to help interpret the results of the assessments and the corresponding statistical uncertainties in the risk estimates. This interface between the scientist and risk manager is an ideal venue for the use of decision tools that provide a common basis for integrating, synthesizing, and valuing the scientific information.

As shown in Figure 11.1, the scientific information is just one component of the overall decision problem. The risks associated with a particular risk management alternative must be integrated with regulatory, political, social, and economic factors to derive an overall valuation, or "score", for the particular alternative. The "best" risk management decision is obtained by comparing the "scores" from all of the alternatives and then selecting that option (Figure 11.1). The technical risk assessment should represent our best science, while the risk management component should represent our best judgment based on both scientific and policy issues.

The multiobjective nature of the risk management decision process is readily apparent from Figure 11.1. For example, minimizing costs may not be compatible with the best technical solution or applicable regulatory standard. Likewise, management actions taken to protect humans from exposure to contaminants may enhance risks to components of ecosystems. For example, physical disturbances to remove contaminant sources may reduce risks to humans but completely destroy associated ecosystems, including rare and endangered species.

Components of the Technical Risk Assessment

Conducting a technical risk assessment requires some knowledge about the physical and biological makeup of the contaminated site and the environmental processes that are important in the cycling of energy and materials (Figure 11.1). Examples of the latter would include herbivory, carnivory, natality, mortality, erosion, precipitation, water balance, etc.

Depending on the contaminant(s) that are present and the current and future land use practices, information is also needed on the distribution and transport of the contaminants of interest, including concentrations in soils, water, air, biota, foodstuffs, and on the key ecological processes that mediate contaminant transport to receptors. Ideally, lack of key information would be identified very early in the characterization phase of the program so that sampling could be designed to fill knowledge gaps.

Sources of information on the structure and function of the environment and on contaminant distribution and transport are then used in risk assessment models, of varying complexity, to predict the distribution of the contaminant(s) and to estimate exposures and risks to human and ecosystem receptors (Figure 11.1).

Models play a central role in risk assessments due to the need to make projections about the fate of contaminants in the environment and potential changes in exposure of the organisms of interest with time. Seldom are enough data available to answer questions about contaminant fate and effects, particularly over long timeframes. The risk may increase with time if the chance of exposure is greater in the future than it is at present such as might happen when containment strategies for storing toxic wastes fail or restrictions on access to contaminated sites are removed. In contrast, the risk may decline in the future if the contaminant undergoes biological or radioactive decay, if it becomes sequestered, or if transport pathways change through natural processes.

Generally, two levels of modeling are conducted to support risk assessments. Screening level models incorporate conservative assumptions that can be used to rapidly and, relatively, inexpensively evaluate contaminated areas to determine the need for further action. If the screening level assessment identifies potential risks, another, less conservative procedure is used to better represent the dynamic processes and pathways leading to exposure of receptors. This graded approach to risk assessments reduces the costs by avoiding more intensive data collection and analysis for sites that pose little risks to human health and the environment.

Screening models often use analytical solutions based on the assumption of equilibrium conditions and constant coefficient equations. For example, the concentration of contaminant in vegetation (y) is a function of concentration of contaminant in soil (x), or **y = f(x)**. The advantages of using screening level models are that they are simple to use because they do not require extensive knowledge of transport processes, input parameters, or driving variables. Moreover, the conservative assumptions typically eliminate the need for an uncertainty analysis. The disadvantages are that they do not identify transport processes, cannot easily account for changes with time, or accommodate nonequilibrium conditions. They provide worst case estimates rather than realistic estimates of exposure with no estimate of how likely the worst case may be.

The dynamic, or simulation, model can be used if the screening level assessment demonstrates that the site presents a nonnegligible risk to the environment. Dynamic models provide time-dependent simulation of environmental processes and contaminant behavior as a function of multiple inputs and losses. The advantages of dynamic models are that they can more realistically represent transport processes, time-dependent events, and long-term dynamics of the system. They are also more amenable to evaluating consequences of remediation alternatives and provide realistic rather than conservative estimates. Their disadvantages are that they require a better understanding of the system, including rate constants, time-dependent processes and events, and supporting databases. Moreover, risk estimates have uncertainties associated with them that can complicate the interpretation of results. Perhaps one of the biggest disadvantages of dynamic risk assessment models is that they have been developed for very specific purposes. This means that models may have to be developed on a site by site basis, a time consuming and costly endeavor.

138 *Multiple Objective Decision Making for Land, Water, and Environmental Management*

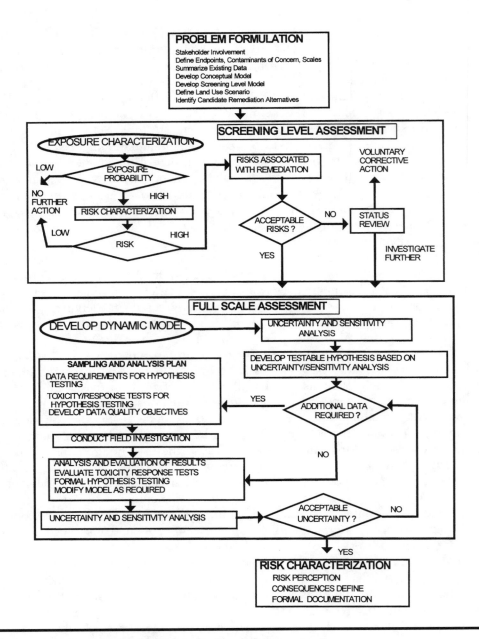

Figure 11.2 Decision logic for conducting risk assessments.

Decision Logic for Conducting a Technical Risk Assessment

The decision logic presented in Figure 11.2 presents a generic approach to conducting a risk assessment at any contaminated site. The approach is consistent with the goals of the U.S. Environmental Protection Agency's Framework for Ecological Risk Assessment (U.S. EPA, 1992) and the National Council on Radiation Protection and Measurement's (1987) guidance for human health assessments. Typically, risk assessments are conducted in three phases that include a problem formulation phase, a technical risk assessment phase (including screening and full scale assessment), and a risk characterization phase.

Problem Formulation

During the problem formulation phase, a conceptual model is developed that reflects all relevant information and data needed to define and scope the decision problem. For example, the study area and site contaminants are identified and described, contaminant toxicity profiles developed from the literature or special studies, existing data are summarized (including contaminant concentrations in soils and biota and environmental relationships that influence the fate and effects of the contaminants), and measurement and assessment endpoints are established. Perhaps the biggest weakness of the risk assessment process is the lack of good quality data to support risk assessment modeling

The selection of assessment and measurement endpoints for ecosystem components are based on several criteria including (1) social relevance, (2) biological relevance, and (3) susceptibility to the stressor (U.S. EPA, 1992; Suter, 1993). Species that are economically or culturally important, such as mule deer, elk, pronghorn, squirrels, nut crops, etc., that are harvested by humans or have other social or cultural significance are possible candidates for an ERA. There is opportunity at this phase to collect supplementary on-site data to fill in major information gaps. The supplementary information required will be largely driven by the data needed to conduct the screening assessment.

Screening Level Assessment

In this phase, predictions of exposure to receptors are made with the screening level model, taking background concentrations into consideration. A decision is made as to the probability of exposure of receptors to above background concentrations of site contaminants. If the probability is low, the site is targeted for no further action. If the exposure is above background levels, then the predicted concentrations are compared to the toxicity profiles of the contaminants (developed in the problem formulation phase) to determine the probability of an effect. If the probability of effect is low, the site is again targeted for no further action. If the probability of effect is not low, then the full scale assessment is triggered to use a dynamic pathway model to evaluate exposures and risk.

Full Scale Assessment

The full scale assessment is designed to allow us to better quantify that risk and its associated uncertainty. This is accomplished through the development and application of a dynamic pathway model that better reflects our understanding of fate and effects of the contaminants at the site. The working hypotheses from the conceptual model provide a framework for developing the pathway model. The pathway model, after uncertainty and sensitivity analysis drives the development of a set of testable hypotheses about the site and its potential impact on human health and the environment.

The results from the pathway analysis provide the means for making risk estimates, usually with large uncertainties in those estimates. Uncertainty and sensitivity analyses of the model quantifies that uncertainty and provides information about where the effort would be most profitably applied to reduce it. Depending on the sources and levels of uncertainty, the procedure may be repeated until the risk estimates are of an acceptable quality. The assessment then proceeds to the risk characterization phase.

Risk Characterization

This is the final phase of the technical risk assessment and includes evaluation of the final risk estimate and its probable consequences. Probable consequences of risk are determined by scientific personnel and then presented to the risk manager and other stakeholders. Formal documentation describes the risk estimates, the process used to arrive at them, their associated uncertainty, and the potential human health and ecological consequences.

Summary

The primary purpose of several on-going, government facility cleanup programs is to protect public health and the environment from harmful effects of contaminants that were intentionally or accidentally released as a part of operations over the last several decades. Risk assessments are legally mandated for these programs and provide a framework for integrating science into the decision-making process. These assessments are structured around prediction models that use knowledge about environmental processes and the fate and effects of contaminants to predict immediate and long-term harm to humans and ecosystems.

Generally, two levels of modeling are conducted to support risk assessments. Screening level models incorporate conservative assumptions that can be used to rapidly and, relatively, inexpensively evaluate contaminated areas on the need for further action. If the screening level assessment identifies potential risks, another, less conservative procedure is used to better represent the dynamic processes and pathways leading to exposure of receptors. In either case, the models should have a firm scientific basis, have parameter values available, and take into account the interactions of subprocesses that influence contaminant fate and effects.

The scientific information is then integrated with regulatory, political, social, and economic factors to derive an overall valuation for the particular cleanup alternative. Iterative use of the process provides evaluations of multiple risk management alternatives from which the best alternative can be selected. Two limitations to implementing decision support systems for risk management applications are dynamic risk assessment models, particularly for ecological risk assessments, and supporting databases, particularly on the fate and effects of hazardous contaminants.

References

Abelson, P.H. 1990. Incorporation of new science into risk assessment. *Science*, 250:1497.
Abelson, P.H. 1992. Remediation of hazardous waste sites. *Science*, 255:901.
Abelson, P.H. 1993. Regulatory costs. *Science*, 259:159.
Ascough, J.C. 1992. A Knowledge-Based Numerical Modeling Approach for Design and Evaluation of Shallow Landfill Burial Systems. Ph.D. dissertation. Purdue University, West Lafayette, Indiana.
Bartell, S.M., R.H. Gardner, and R.V. O'Neill. 1992. *Ecological Risk Estimation*. Lewis Publishers, Chelsea, Michigan.
Breshears, D.D., F.W. Whicker, and T.E. Hakonson. 1993. Orchestrating environmental research and assessment for remediation. *Ecol. Appl.* 3(4):590–594.
Congress of the United States, Office of Technology Assessment. 1991. Complex cleanup: the environmental legacy of nuclear weapons production. OTA-0-484. U.S. Government Printing Office, Washington, D.C.
Cowling, E.B. 1992. The performance and legacy of NAPAP. *Ecol. Appl.* 2:111–116.
Harwell, C.C. 1989. Regulatory framework for ecotoxicology. Pages 497–516 in S.A. Levin, M.A. Harwell, J.R. Kelly, and K.D. Kimball, Eds. *Ecotoxicology: Problems and Approaches*. Springer-Verlag, New York.
Lane, L.J., J.C. Ascough, and T.E. Hakonson. 1991. Multiobjective decision theory-decision support systems with embedded simulation models. *Irrigation and Drainage*. Proc. Honolulu, Hawaii. July 22–26, 1991.
Levin, S.A. 1992. Orchestrating environmental research and assessment. *Ecol. Appl.* 2:103–106.
Loucks, O.L. 1992. Forest response research in NAPAP: potentially successful linkage of policy and science. *Ecol. Appl.* 2:117–123.
McGuire, S.A. 1989. Cleanup or relocation: $128 billion to clean up DOE's wastes? And they want to use my money? *Health Phys. Soc. Newsl.* XVII, (4):9–10.
National Council on Radiation Protection and Measurements. 1987. Ionizing radiation exposure of the population of the United States. NCRP Report No. 93. Bethesda, Maryland.
National Research Council. 1989. The nuclear weapons complex: management for health, safety and the environment. National Academy Press, Washington, D.C.
Pasternak, D. and P. Cary. 1992. A $200 billion scandal. *U.S. News & World Rep.* December 14,1992:34–47.
Russell, M. 1992. Lessons from NAPAP. *Ecol. Appl.* 2:107–110.
Schindler, D.W. 1992. A view of NAPAP from north of the border. *Ecol. Appl.* 2:124–130.
Suter, G.W., II, Ed. 1993. *Ecological Risk Assessment*. Lewis Publishers, Boca Raton, Florida.
U.S. Department of Energy. 1991. Environmental Restoration and Waste Management: Five-Year Plan. Fiscal Years 1993–1997. National Technical Information Service, Springfield, Virginia.
U.S. Environmental Protection Agency. 1992. Framework for ecological risk assessment. U.S. EPA Report EPA/630/R-92/001 (February, 1992).

Chapter 12

A Multiple Criteria Decision-Making Model for Comparative Analysis of Remedial Action Alternatives

Bruce K. Hope

This paper demonstrates that management decisions used in the selection of remedial alternatives for hazardous waste sites are amenable to an analytical methodology based on a fuzzy set model. A systematic technique for comparing a set of remedial alternatives and for identifying the more preferable alternative based on multiple decision criteria is presented. The starting point for this technique is definition of a set of decision criteria applicable to all alternatives. This suite of action- and chemical-specific decision criteria values associated with each alternative are then specified. These values reflect the degree to which stressors in each alternative contribute to that alternative's overall impact as measured by the decision criteria. Once these data are available, mathematical analyses based on concepts from fuzzy set theory are performed to obtain a ranking of alternatives based on the decision criteria. This approach, while relatively simple and direct, is capable of differentiating alternatives in ways that are not possible if data are viewed singly. Ideally, the methodology proposed in this paper will allow risk managers to consider numerous factors in the remedial action selection process, even those previously deemed "too imprecise" for actual recognition and inclusion.

Introduction

There are numerous problems in the environmental management field that cannot be easily or appropriately analyzed on the basis of a single criterion. Usually, both qualitative and quantitative data associated with several criteria need to be systematically considered when evaluating several decision alternatives. In addition, many environmental management problems are characterized by imprecision and the absence of sharply defined criteria of class membership (Juang et al., 1995). What are required are decision procedures that recognize the inherently "fuzzy" nature of environmental problems and do not assume the existence of precise class membership or unambiguous criteria (Kung et al., 1993; Wenger and Rong, 1987). Such decision procedures fall within a large family of multiple criteria methods that have been applied to environmental management problems, but not previously applied to selection of remedial action alternatives (Juang et al., 1995; Shopley and Fuggle, 1984).

At hazardous waste sites, a baseline risk assessment is used to characterize impacts that currently exist and impacts that could be expected in the future if no remedial actions are taken. Once a baseline risk assessment determines that a hazardous waste site presents unacceptable current and/or future risks to the environment, decisions must be made regarding the choice of remedial alternatives. Within the Remedial Investigation/Feasibility Study (RI/FS) and Engineering Evaluation/Cost Analysis (EE/CA) processes, potential remedial alternatives are initially screened for effectiveness, implementability, and cost. For those that pass this screen, a comparative risk analysis is conducted to characterize risks associated with implementation of each remedial alternative and to compare risks among alternatives, including a no-action alternative. Comparative risk analysis is generally a multistep process for comparing: (a) residual chemical-specific and action-specific risks (the potential risks associated with the implementation of an alternative are termed "action-specific" risks) for each alternative, (b) alternatives on the basis of action- and chemical-specific risks, and (c) alternatives on the basis of total (action- plus chemical-specific) risk. It was conjectured that this type of comparative analysis, involving both qualitative and quantitative data with varying degrees of ambiguity, could be made more efficient and useful through application of multiple criteria methods.

This paper presents a systematic technique for comparing a set of remedial alternatives and for identifying the more preferable alternative based on multiple decision criteria. The technique involves: (a) defining a suite of site-specific remedial alternatives, (b) defining a set of action- and/or chemical-specific decision criteria that may be associated with each alternative, (c) assigning estimated or measured decision factor values (values that reflect the degree to which elements comprising each alternative contribute to that alternative's overall impact as measured by the established criteria), and (d) performing mathematical analyses based on concepts from fuzzy set theory to obtain a ranking of remedial action alternatives based on the decision criteria. This approach, while relatively simple and direct, is capable of differentiating alternatives in ways that are not possible if data are viewed singly (Harris et al., 1994).

Methods

The choice of analytical algorithms was strongly influenced by the need to produce a working tool accessible to remedial designers and risk managers, that could be implemented in a spreadsheet environment without specialized software, and whose fundamental concepts were relatively intuitive (Maguire, 1991). Two levels of analysis are provided: a first level ranks alternatives based on pairwise comparisons; a second level identifies the degree of similarity between alternatives. These two levels of analysis are based, respectively, on fuzzy dominance and fuzzy resemblance relations described by Kaufmann (1975). Level 1 ranks alternatives, but those with adjacent positions in the ranking list may or may not be similar. Level 2 provides insights into this issue of similarity. However, Harris et al. (1994) found that, although this second level of analysis could provide additional insights in some cases, the first level of analysis was often sufficient for decision making.

Level 1 Analysis

The decision methodology begins with definition of a set of remedial alternatives, $R_1, R_2, R_j, \ldots, R_m$ and a set of criteria or decision factors, $C_1, C_2, C_k, \ldots, C_n$. The suite of remedial alternatives (R_m) appropriate for a given hazardous waste site are determined through collaboration between the remedial design and risk assessment teams and the client. One possible set of specific criteria (C_n) that can be used to screen each remedial alternative are those described in EPA (1986). In the case of criteria for which the information is qualitative, data on a semantic scale can be transformed to a scale from 0 to 1 (e.g., 0, none; 0.25, minor; 0.5, moderate; 1, major). Because large decision matrices can be cumbersome to manipulate, it is suggested that attempts to differentiate between alternatives be done using only a subset of these criteria, so that $R_m \times C_n \leq 100$.

The remedial alternatives and criteria determine the rows and columns, respectively, of a data matrix $\mathbf{X} = (x_{ik})$, for $x_{ik} > 0$. An entry (x_{ik}) in the matrix is a number which measures the level or value of a decision factor for a given alternative. In those applications where the only interest is to compare alternatives on the basis of equal weights assigned to the decision factors, a matrix, $\mathbf{Y} = (y_{ik})$, is created that is simply a standardized version of $\mathbf{X}$, where entries in each column are standardized with respect to the range of entries in the column as shown in Equation 1.

$$y_{ik} = \frac{x_{ik} - \min(x_{ik})}{\max(x_{ik}) - \min(x_{ik})}, k = 1, 2, \ldots, n. \qquad (1)$$

Although no standardization technique is inherently superior, standardization relative to the range is especially appropriate where a technique is used which performs a pairwise comparison; the disadvantage of this technique is that it produces only an interval scale (Voogd, 1983).

The data matrix **X** can also be transformed in more significant ways by the assignment of weights to the decision factors. The decision factor values (x_{ik}) associated with issues of either particular concern or little concern at a particular site, can be altered by a weighting factor ($\omega_{ik} > 0$) for the *nth* criterion to give their transformed values greater or lesser influence in the analysis. Such weighting can be used to bring factors such as time duration of impacts, prevention management, or restoration management into the decision process without making them actual decision criteria, i.e., C_n (Harris et al., 1994). Regardless of the manner in which data matrix **X** is transformed, the entries in **Y** are assumed to be normalized by columns as described in Equation 1, so that the entries in data matrix **Y** are real numbers between 0 and 1.

Once the data matrix has been determined, several approaches are available to rank the selected alternatives (Harris et al., 1994; Wenger and Rong, 1987). One of the most direct is the dominance method as described by Kaufmann (1975). Its purpose is to describe dominance relationships between pairs of alternatives based on the data matrix Y. For a pair of alternatives *i* and *j*, define:

$$D_k(i,j) = \begin{cases} 1, & \text{if } y_{ik} - y_{jk} > 0 \\ 0, & \text{if } y_{ik} - y_{jk} < 0, k = 1,2,\ldots,n \\ 0.5, & \text{if } y_{ik} - y_{jk} = 0 \end{cases} \quad (2)$$

and then construct the matrix $\mathbf{Z}^1 = (z_{ij})$, where:

$$Z_{ij} = \begin{cases} \sum_{k=1}^{n} D_k(i,j), & \text{if } i \neq j \\ 0, & \text{if } i = j \end{cases} \quad i,j = 1,2,\ldots,m \quad (3)$$

Let s_i and c_j denote the *ith* row sum and the *jth* column sum, respectively, of $\mathbf{Z}^1$. The row sum, s_i, is a measure of the degree to which alternative R_i dominates the other alternatives in terms of its contribution to achieving remedial goals and objectives; that is, the higher the row sum, the less its contribution to an optimal remediation outcome. The column sum, c_j, measures the degree to which alternative R_j is dominated by the other alternatives. The alternatives can be ranked either in order of ascending or descending row sums, the choice being dependent on the manner in which the decision factors are scored.

Level 2 Analysis

The purpose of this additional layer of analysis is to identify the degree of similarity among remedial alternatives based on information in the **Y** matrix. Because two alternatives with adjacent positions in the ranking list may or may not be similar, Level 2 analysis can be applied to discern whether we are dealing with truly distinct alternatives or essentially slight variations of the same theme. Wenger and Rong (1987), based on the work of Kaufmann (1975), provide a full discussion of the theory behind this level of analysis. The following discussion deals only with the implementation of the analysis process. A matrix $Z^2 = (z^*_{ij})$ is defined as:

$$z^*_{ij} = 1 - c \left| \sum_{k=1}^{n} d(y_{ik}, y_{jk}) \right| \tag{4}$$

Here $d(y_{ik}, y_{jk}) = y_{ik} - y_{jk}$ and c is a constant chosen so that $0 \leq z^*_{ij} \leq 1$. The particular value of c is not important, as long as the chosen value yields $0 \leq z^*_{ij} \leq 1$ for all i and j. A "power" matrix, $\mathbf{Z}^P$, which is needed to perform the clustering analysis based on the concept of similarity (Kaufmann, 1975; Kung et al., 1993; Zimmerman, 1985) is then constructed as the "product" of $\mathbf{Z}^2$. With the "$\otimes$" operator indicating a row-column operation similar to that of ordinary matrix multiplication, $\mathbf{Z}^P = \mathbf{Z}^2 \otimes \mathbf{Z}^2 = (z^P_{ij})$, or:

$$(z^P_{ij}) = (z^*_{i1} \wedge z^*_{1j}) \vee (z^*_{i2} \wedge z^*_{2j}) \vee \ldots (z^*_{im} \wedge z^*_{mj}) \tag{5}$$

For real numbers a and b, the operators "$\wedge$" and "$\vee$" are defined as: $a \wedge b$ = maximum(a,b) and $a \vee b$ = minimum(a,b), respectively (after Wenger and Rong, 1987). The $\mathbf{Z}^P$ power matrix can be used to identify clusters of similar alternatives by defining $\mathbf{Z}^P(a) = [z^P_{ij}(a)]$ where:

$$z^P_{ij}(a) = \begin{cases} 1, & \text{if } z^P_{ij} \geq a \\ 0, & \text{if } z^P_{ij} < a \end{cases} \tag{6}$$

Here z^P_{ij} refers to the *ijth* entry in the power matrix $\mathbf{Z}^P$. The variable a provides a "similarity measure." By varying a from 0 to 1, it is possible to determine how clusters are formed and to identify similar alternatives. The important feature of the outcome of this level of analysis is the dynamic clustering process that occurs as the value of the variable a is varied (Wenger and Rong, 1987).

Application

To illustrate the techniques described above, we consider the hypothetical case of a 5-ha marshland contaminated, in what appears to be several hot spots, with a known carcinogen exhibiting low to moderate bioaccumulation and no biomagnification potential. The marshland borders a stream and both are used by recreational fishermen and hunters who walk and wade throughout the area. Slightly above-detection concentration levels of the contaminant have been detected in the tissues of fish collected at the marshland, but none in the tissues of ducks who utilize the marsh. Benthic communities in the stream bordering the marsh do not appear to be impaired when compared to those in an upstream "reference" area. Marshland vegetation shows no signs of stress. Conservative human health and ecological risk assessments suggest the possibility of moderate risk to human health and minimal risk to ecological receptors due to the presence of the contaminant in the marshland.

As a result of these findings, five remedial action alternatives ($R_1, ..., R_5$) are proposed as follows:

- (R_1) Do nothing to alter baseline conditions at the site (the "no-action" alternative).
- (R_2) Install institutional controls to restrict human access; take an observational approach to ecological conditions; institute a long-term monitoring program.
- (R_3) Identify, excavate, and dispose of only the hot spots that appear to be the primary sources of contamination within the marshland; institute long-term monitoring for any residual contamination.
- (R_4) Identify, excavate, and dispose of both the hot spots and all surrounding areas with contamination above a *regulatory criteria threshold*; mitigate the excavated areas to avoid a natural resource damage (NRD) claim from the State.
- (R_5) Identify, excavate, and dispose of both the hot spots and all surrounding areas with contamination above a *risk-based criteria threshold*; mitigate the excavated areas to avoid a NRD claim from the State.

These five remedial action alternatives were arrayed against a subset of criteria, as shown in Table 12.1, derived from those listed in EPA (1986). In selecting the total set of criteria, emphasis was placed on choosing those that could be most easily quantified, either through measurement or estimation. An additional criteria selection factor was a desire to force a greater consideration of ecological risk and impact factors in the decision process; this lead to the inclusion of criteria such as hectares destroyed (C_{24}) and site restoration costs (C_{45}). This is by no means a definitive list and definition of other criteria is certainly possible. For example, only those criteria that have been shown by experience or a sensitivity analysis to "drive" the decision-making process could be selected for analysis. A team composed of risk assessors, remediation engineers, and risk managers would assign decision factor values to these criteria to generate a raw data matrix **X**, as shown in Table 12.2. These values would then be normalized, using Equation 1, to produce data matrix Y, as shown in Table 12.3.

Alternatives could then be ranked using dominance matrix analysis (Equations 2 and 3) as shown in data matrix $\mathbf{Z}^1$ (Table 12.4). The row sum indicates the degree to which a given alternative dominates the others; the column sum indicates the degree to which the alternative is dominated by the others. This analysis shows, assuming all criteria are of equal weight, that alternative R_1, the no-action alternative, is preferred. Alternative R_4, a massive clean-up of the site based on regulatory criteria, is the least preferable alternative.

The rankings would suggest that R_1 and R_2 form one cluster, while the remaining alternatives form another. This supposition can be checked using similarity analysis. Equation 4, with $c = 0.15$, is used to convert data in the **Y** matrix into a $\mathbf{Z}^2$ data matrix, as shown in Table 12.5. Two identical $\mathbf{Z}^2$ matrices are "multiplied" using Equation 5 to produce the power matrix, $\mathbf{Z}^p$, as shown in Table 12.6, that indicates the degree of similarity between pairs of alternatives. This level 2 similarity analysis mirrors the rankings produced by the level 1 analysis. Alternatives R_4 and

Table 12.1 The Subset of Remedial Action Alternative Selection Criteria used for the Hypothetical Case

Criteria category	Specific criteria	Specific criteria description	C_k	Units
Long-term effectiveness and permanence	Magnitude of residual risk	What is the residual human health risk from untreated waste?	1	No. of fatal cancers
		What is the residual ecological risk from untreated waste?	2	Toxicity quotient
	Adequacy and reliability of controls	How long is this alternative expected to afford protection to human health?	8	Years
		How long is this alternative expected to afford protection to ecological receptors?	9	Years
Reduction of toxicity, mobility, and volume	Reduction in toxicity, mobility, or volume	Percentage of reduction in total contaminant mass?	14	%
Short-term effectiveness	Environmental impacts	How much habitat will be removed, destroyed, or otherwise adversely impacted by this alternative?	24	No. of hectares
		How long will wildlife species be exposed to disturbance (noise, human presence, etc.) associated with this activity?	26	Days
	Time until remedial response objectives are achieved	How long until remedial response objectives are achieved?	30	Days
Implementability	Administrative feasibility	How many permits are required to implement this alternative?	36	No. of permits
Cost	Direct capital costs	Expenditure for the equipment, labor, and materials necessary to implement the remedial action?	41	K$
	Site restoration	What are the site restoration costs associated with this alternative? (cost/hectare × total hectares)	45	K$

Table 12.2 Hypothetical Marshland Remediation X Data Matrix

Remedial alternative	\multicolumn{10}{c}{Remedial action selection criteria (from Table 12.1)}
	C_1	C_2	C_8	C_9	C_{14}	C_{24}	C_{26}	C_{30}	C_{36}	C_{41}	C_{45}
R_1	2.0	1.20	0.0	0.0	0.0	0.00	0	0.0	0.0	$0	$0
R_2	1.5	1.00	5.0	5.0	0.0	0.14	14	16.5	2.1	$10	$1
R_3	1.0	0.50	14.1	21.0	48.9	0.50	29	28.1	5.2	$50	$3
R_4	0.5	0.27	13.5	16.0	94.0	2.05	30	30.0	10.8	$350	$12
R_5	0.5	0.22	14.9	12.0	76.7	1.78	35	32.5	9.8	$185	$11

Table 12.3 Hypothetical Marshland Remediation Y Data Matrix

Remedial alternative	\multicolumn{11}{c}{Remedial action selection criteria (from Table 12.1)}
	C_1	C_2	C_8	C_9	C_{14}	C_{24}	C_{26}	C_{30}	C_{36}	C_{41}	C_{45}
R_1	1.00	1.00	0.00	0.00	0.00	0.00	0.00	0.00	0.00	0.00	0.00
R_2	0.68	0.80	0.34	0.24	0.00	0.07	0.40	0.51	0.19	0.03	0.07
R_3	0.35	0.29	0.95	1.00	0.52	0.24	0.83	0.86	0.48	0.14	0.24
R_4	0.03	0.05	0.91	0.76	1.00	1.00	0.86	0.92	1.00	1.00	1.00
R_5	0.00	0.00	1.00	0.57	0.82	0.87	1.00	1.00	0.91	0.53	0.87

Table 12.4 Hypothetical Marshland Remediation Z^1 Data Matrix

Remedial alternative	Remedial alternative					Row sum
	R_1	R_2	R_3	R_4	R_5	
R_1	0	2.5	2	2	2	8.5
R_2	8.5	0	2	2	2	14.5
R_3	9	9	0	4	3	25.0
R_4	9	9	7	0	8	33.0
R_5	9	9	8	3	0	29.0
Column sum	35.5	29.5	19	11	15	

Table 12.5 Hypothetical Marshland Remediation Z^2 Data Matrix

Remedial alternative	Remedial alternative				
	R_1	R_2	R_3	R_4	R_5
R_1	1.00	0.80	0.41	0.02	0.17
R_2	0.80	1.00	0.61	0.22	0.36
R_3	0.41	0.61	1.00	0.61	0.75
R_4	0.02	0.22	0.61	1.00	0.85
R_5	0.17	0.36	0.75	0.85	1.00

Table 12.6 Hypothetical Marshland Remediation Z^P Data Matrix

Remedial alternative	Remedial alternative				
	R_1	R_2	R_3	R_4	R_5
R_1	1.00	0.80	0.61	0.61	0.61
R_2		1.00	0.61	0.61	0.61
R_3			1.00	0.75	0.75
R_4				1.00	0.85
R_5					1.00

R_5 form a cluster at 0.85, as do alternatives R_1 and R_2 at 0.80. Alternative R_3 is somewhat separate, joining R_4 and R_5 at 0.75 and R_1 and R_2 at 0.61. This suggests that alternatives involving some form of active remediation and those favoring little or no direct action are mutually distinct extremes, while the compromise alternative using hot-spot removal occupies the center.

Summary

Fuzzy set models represent a set of decision procedures that can be applied to a range of environmental management problems where alternatives are to be compared on the basis of several criteria (Kaufmann, 1975). As was shown above, the management decisions involved in the selection of remedial action alternatives are amenable to this analytical methodology. It is not necessary for the units used to measure the criteria to be the same or for the information to be in quantitative form. These models are particularly useful when the decision to be made is characterized, as most environmental management problems are, by imprecision and the absence of sharply defined criteria. It should be noted that such "imprecise" factors are elements of the ultimate decision regardless of whether the risk manager chooses to consciously recognize them. Ideally, the methodology outlined above will allow risk managers to consider numerous factors in the remedial action selection process, even those previously deemed "too imprecise" for actual recognition and inclusion.

References

EPA. 1986. Guidance for Conducting Remedial Investigations and Feasibility Studies Under CERCLA [Interim Final]. EPA 540/O-89/004, OSWER Directive 9355.3-0. U.S. Environmental Protection Agency, Washington, D.C.

Harris, H.J., R.B. Wenger, V.A. Harris, and D.S. Devault. 1994. A method for assessing environmental risk: a case study of Green Bay, Lake Michigan, U.S.A. *Environ. Manage.* 18:295–306.

Juang, C.H., S. Wu, and H. Sheu. 1995. A group decisionmaking model for siting LULUs. *Environ. Profess.* 17:43–50.

Kaufmann, A. 1975. *Introduction to the Theory of Fuzzy Subsets.* Volume 1. Academic Press, New York.

Kung, H., L.G. Ying, and Y. Liu. 1993. Fuzzy clustering analysis in environmental impact assessment — a complement tool to environmental quality index. *Environ. Monit. Assess.* 28:1–14.

Maguire, L.A. 1991. Risk analysis for conservation biologists. *Conservat. Biol.* 5:123–125.

Shopley, J.B. and R.F. Fuggle. 1984. A comprehensive review of current environmental impact assessment methods and techniques. *J. Environ. Manage.* 18:25–47.

Voogd, H. 1983. *Multicriteria Evaluation for Urban and Regional Planning.* Pion, London. Pages 77–78, 87–89.

Wenger, R.B. and Y. Rong. 1987. Two fuzzy set models for comprehensive environmental decision-making. *J. Environ. Manage.* 25:167–180.

Zimmerman, H.J. 1985. *Fuzzy Set Theory and Its Applications.* Kluwer Nijhoff Publishing, Norwell, Massachusetts.

Chapter 13

Overview of a Decision Support System for the Evaluation of Landfill Cover Designs

GINGER B. PAIGE, JEFFRY J. STONE, LEONARD J. LANE, AND DIANA S. YAKOWITZ

Abstract

A computer-based prototype decision support system (PDSS) is being developed to assist the risk manager in selecting an appropriate landfill cover design for mixed waste disposal sites. The selection of the "best" design among feasible alternatives requires consideration of multiple and often conflicting objectives. The methodology used in the selection process consists of selecting and parameterizing decision variables or criteria, selecting feasible cover design alternatives, ordering the decision variables, and ranking the design alternatives. Decision variables can be parameterized by the use of data, expert opinion, or simulation models. The simulation models incorporated in the PDSS are the HELP (Hydrologic Evaluation of Landfill Performance) model which is used to simulate the trench cap water balance and the CREAMS (Chemicals, Runoff, and Erosion from Agricultural Management Systems) erosion component which is used to simulate erosion of the cover. The decision model is based on multiobjective decision theory and uses a unique approach to order the decision variables and rank the design alternatives. The decision variables, which are of different magnitudes and dimensions, are normalized to a common 0–1 scale through the use of scoring functions. The scoring functions are parameterized based on a conventional design or existing conditions at the disposal site. Each decision variable for each alternative is compared to the conventional design or existing condition using the scoring functions. The decision variables are ordered and a simple linear program is used to compute the best and worst aggregate scores of each alternative for all possible weights of the decision variables. This approach significantly reduces user subjectivity and bias inherent in most decision-making methodologies and provides the risk manager with a powerful, repeatable method to evaluate landfill covers.

Introduction

The primary purpose of the DOE's Environmental Restoration Program is to manage the health and ecological risks associated with intentional and accidental releases of radioactive and hazardous contaminants to the environment. Several thousand landfills, covering thousands of hectares of real estate, were operated and then decommissioned prior to current regulations. Most of those landfills are currently being evaluated under the DOE's Environmental Restoration Program to determine whether they pose unacceptable health and environmental risks and, if so, the landfills will be remediated to reduce risks to acceptable levels. Most DOE sites that have low to intermediate levels of residual contaminants and pose low risks to humans or ecosystems, can control contaminant migration with natural or synthetic barriers. Migration barriers can be permeable or impermeable depending on the type of contaminant and its mode of transport. The cover (trench cap) is a central feature of most containment strategies and can range from a very simple soil cover to a very complex engineered design that mitigates both the vertical and lateral flow of water and gases (Figure 13.1).

The primary functions of the landfill cover are to isolate the buried waste from the surface environment and to control hydrologic processes, including erosion, that can cause contaminant migration from the site (Hakonson et al., 1982; 1994). Water that infiltrates into the soil cap can lead to enhanced percolation of water and solutes out of the burial environment. Excessive erosion of cap soil can expose buried waste and lead to off-site transport of contaminants. Biological processes, including plant root and burrowing animal intrusion into the waste, can also contribute to off-site migration of contaminants. However, the relative importance of biological processes is strongly related to the hydrologic characteristics of the site (Hakonson et al., 1982).

There are strong interactions among the various hydrologic processes which the cover controls. A reduction or elimination of runoff increases infiltration of water into the soil, resulting in increased soil moisture storage and increased evapotranspiration and/or percolation. Likewise, reducing percolation necessitates that more of the precipitation be partitioned between soil moisture storage, evapotranspiration, and/or runoff. The coupled nature of the processes can be used advantageously in designing landfill caps that minimize or eliminate processes that contribute to contaminant migration (i.e., percolation), while enhancing other terms (i.e., evapotranspiration) that do not.

EPA guidance to permit applicants (USEPA, 1989) recommends that an analysis of the final cover design be presented in the closure plan. The EPA's technical guidance for final covers describes a recommended cover design, often called

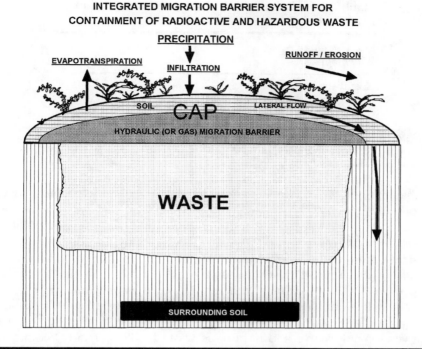

Figure 13.1 Integrated migration barrier system for containment of radioactive and hazardous waste.

EPA's RCRA cap, that will meet the final cover performance standards: minimizes liquid migration, promotes runoff while controlling erosion, and minimizes maintenance. The EPA offers the RCRA cap design as guidance and does not require its use if another design can be shown to meet the technical performance standards. Research in trench cap designs (Hakonson et al., 1994; Lane and Nyhan, 1984; Nyhan et al., 1984; Nyhan et al., 1990) have demonstrated that there may be alternatives to the EPA RCRA recommended design which offer certain technical and economic advantages. The basic problem is to evaluate and compare these alternative designs with the EPA RCRA design for specific waste sites while taking into account the technical, regulatory, and economic issues.

Cost will always be an important criteria for selecting options for remediating contaminated sites. The objective is to reduce costs to a minimum while satisfying technical, regulatory, and political/social constraints. Minimizing cost is often in direct conflict with the best technologic solution derived from the health and environmental risk assessments. Estimated unit costs for construction of several capping alternatives at Los Alamos are compared to the cost of excavating the waste in Table 13.1. The most costly capping alternative, the RCRA cap, is still a factor of 15 less expensive than removal of the waste. Although capping costs are relatively inexpensive compared to other remediation options, they still represent substantial outlays of capital when considering that thousands of hectares of landfill area exist in the U.S.

The process of selecting containment cover technologies requires that alternative cover designs be evaluated in a repeatable, objective, and scientifically

Table 13.1 Estimated 1992 Costs of Remediation Alternatives for Landfills at Los Alamos

Alternative	Cost/unit (million $/ha)
Excavation	80.0
EPA RCRA cover	4.90
Capillary barrier cover	3.70
Bioengineered soil/vegetation cover	0.24
Surface management with erosion control	0.12

defensible manner while taking into account all the necessary technical, regulatory, and economic factors. The complexity of the technical and nontechnical information, and how the information varies in importance across sites, points to the need for decision analysis tools that provide a common basis for integrating, synthesizing, and valuing the decision input. Because the cost of remediating thousands of contaminated DOE sites is projected to be in the 10s or 100s of billions of dollars, methods will be needed to establish cleanup priorities and to help in the selection and evaluation of cost effective remediation alternatives.

This paper presents a PDSS to assist risk managers in evaluating capping alternatives for radioactive and hazardous waste landfills. The PDSS incorporates methods for calculating, integrating, and valuing technical, regulatory, and economic criteria. The goal in developing the PDSS is to improve the quality of technical information used by the risk manager to select landfill capping designs that are cost effective in meeting regulatory performance standards. The specific objectives in designing and testing of the PDSS are to: (1) design and build a computer-based DSS, incorporating multiobjective decision theory, to evaluate the performance of various capping alternatives with respect to applicable regulations and cost; (2) compare the PDSS predictions of cap performance against field data from a study on the hydrologic performance of four capping alternatives; and (3) provide a framework for an operational DSS.

The use of a DSS to design and evaluate barrier cover remediation technology will reduce the likelihood of selecting a barrier cover technology that does not meet performance objectives and imposes the attendant costs of fixing mistakes. Candidate remediation technologies can be evaluated beforehand with the DSS to identify technical and regulatory problems inherent in the technologies, evaluate the projected long-term performance and the practicality of the designs from a construction and economic viewpoint.

Methods and Materials

The multiobjective approach was chosen because of the complex and interrelated technical and regulatory factors that need to be considered in the evaluation of a final cover design (i.e., minimize liquid migration, promote drainage while controlling erosion, minimize maintenance, etc.). In addition, the multiobjective approach allows the evaluation of designs when the objectives of the design factors are in conflict. For example, minimizing percolation into the waste containment area has a potential of increasing the amount of surface runoff and thus increasing erosion of the trench cap surface.

The PDSS presented in this paper follows the approach described by Lane et al. (1991). The methodology used in the selection process consists of selecting and parameterizing decision variables or criteria, selecting feasible trench cap design alternatives, ordering the decision variables and ranking the design alternatives. The components of the PDSS include: simulation models, a decision model, and a graphical user interface. Decision variables can be parameterized using the simulation models, data, or expert opinion. The general category of decision variables for the evaluation of trench cap designs is specified by the EPA guidelines. These include such factors as the elements of the water balance, erosion, and subsidence. The specific decision variables will depend on the current state of the science (i.e., unsaturated flow dynamics, contaminant pathway analysis, erosion mechanics), regulations (minimize percolation and erosion), and socio-economics (cost, site location). Further details of the simulation models, the decision model, and the graphical user interface are given below.

Simulation Model

Lane et al. (1991) recommended using simulation models to parameterize the decision variables when data is not available as a means of incorporating the "best science" in the decision-making process. Due to both the large expense and the extended period of time needed to develop comprehensive data sets of alternative cover designs, not very many exist. Simulation modeling has the advantages of being inexpensive and obtaining long-term evaluations of many design alternatives quickly. The disadvantages are those associated with most simulation models; uncertainty and errors in model structure, parameter estimation, and model output. However, simulation models used in combination with experience offer a feasible method of designing trench cap systems on a regular basis.

In order to demonstrate that a proposed final cover design complies with the regulatory performance standard, the DOE (U.S. DOE, 1990) indicates that it will be necessary to model the hydrologic performance and erosion potential of the proposed cover. It is also recommends that data or model predictions demonstrate that the proposed design will not result in erosion in excess of 5 tons/ha/year. EPA suggests that the HELP model (Hydrologic Evaluation of Landfill Performance, Schroeder et al., 1984; 1994) be used for demonstrating the hydrologic performance.

The simulation models which are embedded in the PDSS are Version 2 HELP landfill water balance model (Schroeder et al., 1984) and the overland flow erosion component of the CREAMS model (Chemicals, Runoff, and Erosion from Agricultural Management Systems, Knisel, 1980). Version 3 of the HELP model (Schroeder et al., 1994), recently released, will replace Version 2 in the PDSS. The HELP model computes the water balance, soil water movement within the trench cap system. The erosion component of the CREAMS model is incorporated as an alternative to the EPA recommended USLE (Universal Soil Loss Equation, Wischmeier and Smith, 1978). CREAMS estimates of water-induced erosion of the cover account more directly for the temporal variation of the erosion process.

HELP is a quasi-two-dimensional model that uses weather, soil, and design data and calculates the infiltration, surface runoff, percolation, evapotranspiration, soil water storage, and lateral drainage in a shallow landfill system with up to 12 different layers. The model simulates water flow within three different soil layer types: vertical percolation, lateral drainage, and barrier soil layers with or without a geomembrane. However, the Help model is unable to simulate flow through a capillary barrier. Both default (USDA soil classes) or user-specified soil characteristics can be used in designing the trench cap system and include the following soil properties: total porosity, field capacity, permanent wilting point, saturated hydraulic conductivity, and initial soil water content. The model also accepts both user-specified and default weather data and includes WGEN, the synthetic weather generator developed by USDA -ARS (Richardson and Wright, 1984), which produces daily precipitation, temperature, and solar radiation values. These last two values are used in HELP to determine snow melt and evapotranspiration. HELP also includes the vegetative growth model from SWRRB (Arnold et al., 1990) to calculate daily leaf area indices. Runoff is estimated using a modified SCS curve number method.

The CREAMS overland erosion component has been added to the HELP model to simulate trench cap erosion. It can be used to predict sediment yield and particle composition of the sediment on an annual or a storm event basis. The erosion component requires the input of hydrologic parameters for each runoff event simulated by the HELP model and an erosion parameter file. The principal outputs from the overland flow component are the soil loss per unit area and the concentration of each particle type for each storm.

Decision Model

The decision model, based on multiobjective decision theory, uses scoring functions as a means of scaling the decision variables to a common scale between 0 and 1. The conventional and viable alternatives are scored on the same set of decision criteria (i.e., percolation of leachate, runoff, evapotranspiration, sediment loss, and cost).

The decision model combines the dimensionless scoring functions of Wymore (1988) with the decision tools presented in Yakowitz et al. (1992; 1993). The scoring were previously used to evaluate shallow land burial systems (Lane et al., 1991; Ascough, 1992). The four generic shapes of scoring functions are: (1) more is worse, (2) more is better, (3) a desirable range, and (4) an undesirable range. These functions can be modified for each criterion by implicitly setting threshold values or allowing the model to set these values by default based upon the data or simulation results. The baseline values can be determined by a standard or conventional practice, federal regulation, or by expert opinion. The scoring functions are set up so that the conventional design or baseline scores 0.5 for each decision variable. All of the other alternative designs are scored relative to the conventional design for each criterion. A design which performs better than the conventional design with regard to a specific criterion will score more than 0.5 for that criterion and one that performs worse will score less than 0.5. A default importance order for the decision criteria is established based on the normalized slopes of the scoring functions. The user is able to change the importance order of the decision criteria interactively.

Once all of the alternatives have been scored, a score matrix is available to complete the analysis. Best and worst composite scores assuming an additive value function are determined by maximizing and minimizing a simple linear program for each alternative and these two composite scores are aggregated to determine the preference ranking of the alternatives. The alternatives are not ranked on a single vector of weights associated with the criteria. The method considers all possible weight vectors consistent with an importance order of the decision criteria discerned by the decision model from the simulation results and scoring functions. The trench cap design with the highest aggregated score, for a given importance order of the criteria, is considered to be the "best" design among the conventional and feasible alternatives.

Algorithm for Ranking Alternatives

Based on the established importance order of the decision criteria, best and worst composite scores for each of the alternatives are determined by the PDSS by maximizing and minimizing Equation (1), respectively (Yakowitz et al., 1992). The solutions to these linear programs are the most optimistic and most pessimistic composite scores (weighted averages) consistent with the importance order.

$$\sum_{i=1}^{m} w(i) Sc(i,j)$$
$$\text{subject to} \quad \sum_{i=1}^{m} w(i) \tag{1}$$
$$w(1) \geq w(2) \geq \ldots \geq w(m) \geq 0$$

Suppose there are m criteria which are ordered in importance as determined above. Let $Sc\ (i,j)$ be the score of the alternative j evaluated with respect to criterion i in the importance order. If $w(i)$ indicates the unknown weight factor associated with criterion i, the highest or best additive composite score for alternative j consistent with the importance order is found by maximizing Equation (1) with respect to the weights $w(i),\ i = 1,.m$. The lowest or worst additive composite score for alternative j consistent with the importance order is found by minimizing Equation (1). In both cases the first constraint normalizes the sum of the weights to 1, the second requires that the solution be consistent with the importance order and restricts the weights to positive values. Thus, the decision maker is not asked to determine an exact weight factor for each criteria. Maximizing and minimizing Equation (1) yields the full range of possible composite scores consistent with a given importance order. Any weight vector that is consistent with the importance order will produce a composite score that falls between the best and the worst composite scores. The designs are then ranked in descending order by the average of the best and worst composite scores. Yakowitz et al. (1993) provide the theoretical justification for this method of ranking the alternatives.

The results of the decision model are presented graphically. Because this system has the potential of recommending design alternatives which may replace the EPA-sanctioned RCRA design, all the steps in the design and evaluation process must be documented. An automatic input/output documentation procedure is implemented along with inclusion of user-supplied notes. All input and output information is listed in an easily read format. Any changes to the default values are noted and included in the output documentation. Output from the simulation models includes both decision variables to be used by the decision model and other variables needed to evaluate the performance of the HELP and CREAMS models. These results can also be presented in text and graphical formats. The text format includes all of the input parameters used and output simulated for a given simulation run.

Testing and Calibration

A preliminary evaluation of the PDSS was conducted using a landfill cover demonstration study at Hill Air Force Base cover in Layton, Utah (Paige et al., 1996b). Four shallow landfill cover design test plots were installed at Hill Air Force Base in Layton, Utah and their performance monitored for a 4-year period from 1990 to 1993 (Hakonson et al., 1994). There are three basic cover designs including a control soil cover, a modified EPA RCRA cover, and two versions of a Los Alamos design (LANL) that contain erosion control measures, an improved vegetation cover to enhance evapotranspiration, and a capillary barrier to divert downward flow of water. The designs were constructed in large 5 by 10 m contained lysimeters. The control soil cover consists of 90 cm of soil over 60 cm of a gravel drainage layer. The EPA RCRA design consists of 120 cm of soil, 30 cm of sand (lateral drainage layer), 60 cm of compacted clay (hydraulic barrier), and 30 cm of a gravel drainage layer. The Los Alamos designs consist of a thin gravel mulch over 150 cm of soil, 30 cm of gravel (capillary break), and 30 cm of a gravel drainage layer. One of the Los Alamos designs was seeded with native perennial grasses and the other with both native perennial grasses and two species of shrubs to enhance evapotranspiration. The surface and all of the underlying layers of the covers were built with a 4% slope (Figure 13.2).

The plots were instrumented to measure the performance of the covers with respect to controlling the hydrology and erosion of the landfill cover. Precipitation, surface runoff and sediment yield, lateral flow, and percolation out of the gravel drainage layer were measured on a daily basis. Soil moisture was monitored approximately biweekly using a neutron probe moisture meter. Evapotranspiration was estimated by solving the water balance equation. The decision criteria considered for evaluating these four designs were: runoff (including lateral flow), evapotranspiration, percolation (leachate production), sediment yield, and construction cost. The annual average and range for each decision variable from 4 years of data collection are presented in Table 13.2.

The ability of the PDSS to evaluate cover designs was assessed using both field water balance and erosion data as well as simulated data from the Hill Air Force Base study. The simulation models embedded in the PDSS, the HELP and

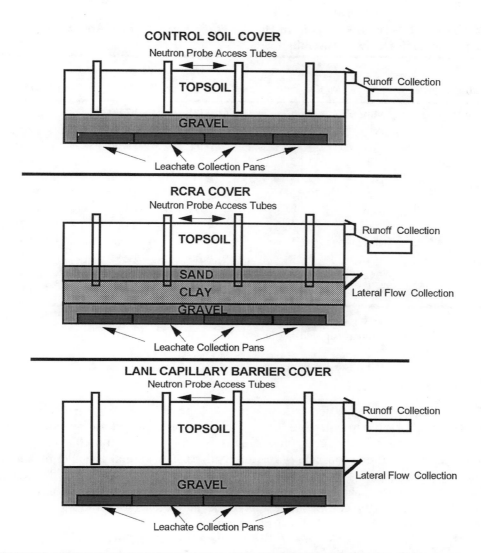

Figure 13.2 Side view of the cover designs at Hill Air Force Base, Layton, Utah (not drawn to scale).

CREAMS models, were calibrated using data from the Hill Air Force Base cover demonstration study. Long-term simulations (200 years) were run using the calibrated model parameters to evaluate the long-term stability of the models. The simulation models were calibrated for two of the Hill Air Force Base cover designs, the Control soil cover and the modified EPA RCRA design (Paige et al., 1996a). Because the HELP model is unable to simulate capillary barrier designs at this point, the PDSS could only be used to evaluate two of the four designs when using the embedded simulation models. The daily precipitation collected at the site, as well as the soil properties and cover characteristics, were used to parameterize and calibrate the simulation models.

Table 13.2 Observed Results from Hill Air Force Base: Average Annual Values and Range for Each Decision Variable

Cap designs		Runoff (cm)+	Sed. yield (kg/ha)	ET (cm)	Percolation (cm)*
			Design criteria		
Control Cap	Min:	0.04	0.00	17.53	2.60
	Ave:	1.40	118.70	27.37	14.74
	Max:	3.98	390.31	35.66	29.43
EPA RCRA	Min:	3.63	0.00	23.34	0.00
	Ave:	12.05	76.70	28.80	0.13
	Max:	20.90	187.82	35.33	0.51
Los Alamos 1	Min:	0.34	0.00	18.00	0.34
	Ave:	5.18	4.50	24.25	6.82
	Max:	8.21	19.98	28.00	13.15
Los Alamos 2	Min:	1.39	0.00	22.92	1.25
	Ave:	3.50	4.80	33.99	7.28
	Max:	4.28	18.81	44.58	17.45

+ Includes lateral flow where applicable.
* Percolation out of trench cap and into waste storage layer.

Discussion

Risk managers are interested in assessing the long-term performance of a landfill cover design for a particular site. Landfill covers are designed and installed to last hundreds of years. The decision model in the PDSS uses the annual average value of the decision criteria for each of the alternatives. The annual average, maximum, and minimum of the conventional design were used to parameterize the scoring functions. The user is able to change the importance order of the decision variables and then compare the composite results of the alternatives for different importance orders side by side in a graphical presentation.

The PDSS was able to differentiate between the alternatives when using both the data and the simulation models to parameterize the decision variables. Detailed results of the preliminary evaluation of the PDSS using the Hill Air Force Base cover demonstration study are presented in Paige et al. (1996a; 1996b). For the Hill Air Force Base cover designs, changing the importance order of the decision variables has a significant affect on the composite scores of the alternative designs and thus their relative ranking (Paige et al., 1996b). When cost is the most important criteria, the plain cover ranks higher than the other designs based on the average of the best and worst possible composite scores. The risk manager may consider minimizing erosion of the trench cap or percolation into the waste layer more important than minimizing cost for a given situation, and therefore adjust the importance order. When percolation is selected as the most important criteria, the EPA RCRA design ranks higher than the alternatives. Changing the importance order so sediment yield is the most important, the Los Alamos designs score higher than the Soil cap and EPA RCRA design. The specific benefits of each of the cover designs were evident in the results of the decision model. The most appropriate design for a particular location also depends on the specific needs and characteristics of the site.

Current Status of the PDSS

Two of the three primary objectives for the development of the PDSS have been accomplished. The two embedded simulation models have been linked and alterations made in the code to provide the decision variables needed in the decision model. The simulation models and the decision model in the PDSS have been evaluated and tested using field data from a cover demonstration study. The inability of the HELP model to simulate flow in a capillary barrier is a major limitation. Currently, capillary barrier designs can only be evaluated in the PDSS using data; however, a capillary barrier simulation component may be integrated into the PDSS depending on progress in capillary barrier research.

In order to provide a complete framework for an operational DSS, the graphical user interface needs to be completed and a complete validation and sensitivity analysis of all the components needs to be conducted. Full validation and extensive testing of the simulation models were not possible with the Hill Air Force Base data because of the short duration of the study and the limited number of covers which could be evaluated by the HELP model. To fully validate and test the PDSS and its embedded simulation models, other data sets of landfill covers for waste disposal sites in a variety of climates are needed.

Summary

The PDSS is being developed for the evaluation of landfill cover designs. In order to evaluate a complete landfill site design, the risk manager would have to consider multiple external factors including a complete risk analysis. The most appropriate or "best" alternative cover design also depends upon the specific needs and characteristics of the site in question, the type of waste and how it is stored, and the potential long-term risks and costs. The ultimate decision would have to be made by the risk manager taking many of these factors as well as local and federal regulations into consideration. The goal of the PDSS is to improve the quality of the technical information used by the risk manager to select cover designs that are cost effective and meet regulatory performance standards. The HELP model is the only model currently sanctioned by the EPA for design and evaluation of landfill covers. With the addition of the CREAMS erosion component and the decision model, the PDSS is a powerful tool for agencies who are concerned with the design and evaluation of landfills. The risk manager will be able to evaluate potential landfill cover technologies with the PDSS in order to identify technical and regulatory problems inherent in the designs and evaluate long-term projected performance.

Acknowledgments

We gratefully acknowledge the source of funding to USDA-ARS-SWRC and Los Alamos National Laboratory for the development of the PDSS from the Department of Energy through the Mixed Waste Landfill Integrated Demonstration (MWLID) at Sandia National Laboratories. The Hill Air Force Base landfill capping demonstration was funded by the U.S. Air Force through the Engineering Services Center at Tyndall Air Force Base and the Department of Energy MWLID.

References

Arnold, J.G., J.R. Williams, and A.D. Nicks. 1990. SWRRB: a basin scale simulation model for soil and water resources management. Texas A & M University Press, College Station. Page 241.

Ascough, J.C. 1992. A Knowledge-Based/Numerical Modeling Approach for Design and Evaluation of Shallow Landfill Burial Systems. Ph.D. dissertation. Department of Agricultural Engineering, Purdue University, W. Lafayette, Indiana. Page 371.

Hakonson, T.E., L.J. Lane, J.G. Steger, and G.L. DePoorter. 1982. Some interactive factors affecting trench cover integrity on low-level waste sites. Proc. Nuclear Regulatory Commission on Low-Level Waste Disposal: Site Characterization and Monitoring, Arlington, Virginia. Pages 377–399.

Hakonson, T.E., K.V. Bostick, G. Truijilo, K.L. Manies, R.W. Warren, L.J. Lane, J.S. Kent, and W. Wilson. 1994. Hydrologic evaluation of four landfill cover designs at Hill Air Force Base, Utah. LA-UR- 93-4469. Los Alamos National Laboratory, Los Alamos, New Mexico. Page 41.

Knisel, W.G. 1980. CREAMS: A Field-Scale Model for Chemicals, Runoff, and Erosion from Agricultural Management Systems. USDA Conservation Res. Rep. No. 26. Page 640.

Lane, L.J. and J.W. Nyhan. 1984. Water and contaminant movement: migration barriers. LA-10242-MS. Los Alamos National Laboratory, Los Alamos, New Mexico. Page 67.

Lane, L.J., J.C. Ascough, and T.E. Hakonson. 1991. Multi-objective decision theory-decision support systems with embedded simulation models. *Irrigation and Drainage Proceedings*. July 1991. ASCE, Honolulu. Pages 445–451.

Nyhan, J.W., G.L. DePoorter, B.J. Drennon, J.R. Simanton, and G.R. Foster. 1984. Erosion on earth covers used in shallow land burial at Los Alamos, New Mexico. *J. Environ. Qual.* 13:361–366.

Nyhan, J.W., T.E. Hakonson, and B.J. Drenon. 1990. A water balance study of two landfill cover designs for semiarid regions. *J. Environ. Qual.* 19:281–288.

Paige, G.B., J.J. Stone, L.J. Lane, and T.E. Hakonson. 1996a. Calibration and testing of simulation models for evaluation of trench cap designs. *J. Environ. Qual.* 25:136–144.

Paige, G.B., J.J. Stone, L.J. Lane, D.S. Yakowitz, and T.E. Hakonson. 1996b. Evaluation of a prototype decision support system for selecting trench cap designs. *J. Environ. Qual.* 25:127–135.

Richardson, C.W. and D.A. Wright. 1984. WGEN: A model for generating daily weather variables. ARS-8. USDA-ARS. Page 83.

Schroeder, P.R., J.M. Morgan, T.M. Walski, and A.C. Gibson. 1984. The hydrologic evaluation of landfill performance (HELP) model. Vol. I and II. EPA/530-SW-84-010. U.S. Environmental Protection Agency, Washington, D.C. Page 143.

Schroeder, P.R., T.S. Dozier, P.A. Zappi, B.M. McEnroe, J.W. Sjostrom, and R.L. Peyton. 1994. The Hydrologic Evaluation of Landfill Performance (HELP) model: engineering documentation for version 3, EPA/600/9-94/168a(b). U.S. Environmental Protection Agency, Cincinnati. Page 87.

U.S. Department of Energy. 1990. Closure of Hazardous and Mixed Radioactive Waste Management Units at DOE Facilities. DOE/EGD-RCRA — 002/0690 (DES91009494). U.S. Department of Energy, Washington, D.C. Page 189.

U.S. EPA. 1989. Technical Guidance Document: Final covers on hazardous waste landfills and surface impoundments. EPA/530-SW-89-047. U.S. Environmental Protection Agency, Washington, D.C.

Wischmeier, W.H. and D.D. Smith. 1978. Predicting rainfall erosion losses — a guide to conservation planning. *USDA Agricultural Handbook 537*. Page 58.

Wymore, W.A. 1988. Structuring system design decisions. *Proceedings of International Conference on Systems Science and Engineering 88*. International Academic Publishers. Pergamon Press. Pages 704–709.

Yakowitz, D.S., L.J. Lane, J.J. Stone, P. Heilman, R.K. Reddy, and B. Imam. 1992. Evaluating land management effects on water quality using multi-objective analysis within a decision support system. *First International Conference on Ground Water Ecology*. USEPA, AWRA. Pages 365–374.

Yakowitz, D.S., L.J. Lane, and Szidarovszky. 1993. Multi-attribute decision making: dominance with respect to an importance order of the attributes. *Appl. Math. Computat.* 54:167–181.

Chapter 14

AGFADOPT: A Decision Support System for Agroforestry Project Planning and Implementation

M.P. ROBOTHAM

Abstract

This paper describes AGFADOPT, a rule-based decision support system (DSS) designed to model some of the social and economic factors that influence small farmer adoption of tree-based innovations. The system is designed to help technically trained personnel incorporate social and economic criteria into project planning and implementation. To create AGFADOPT, information obtained from the literature was used to develop a set of general rules that simulate farmer decision making regarding the adoption of two tree-based technologies: fruit tree orchards and contour hedgerows. These rules were combined in the expert system using a yes/no decision-tree framework. Preliminary verification of the system prototype using data from the Dominican Republic and evaluation of the system by potential users suggest that it is logically consistent, reasonably accurate (83% for tree orchard; 68% for contour hedgerows), and potentially useful.

Introduction

Management problems of uplands and marginal areas in Southeast Asia have been the focus of recent analyses (Allen, 1993; Blair and Lefoy, 1991; Garrity, 1993). This research and other related studies indicate that incorporating trees into existing land management systems can be beneficial. Tree uses can include alley cropping and hedgerow cropping using nitrogen-fixing trees, expanded cultivation of trees for marketable fruit or nuts, and fuelwood and small timber plantations (Garrity, 1993; Juo, 1989; MacDicken and Vergara, 1990; Watson, 1990). However, all of these tree-based technologies, like any introduced technology, can provide benefits only if farmers are willing to adopt them into existing farming systems.

Farmer adoption of introduced technologies involves the interaction of a number of physical-environmental and socioeconomic factors (Feder et al., 1985; Rogers, 1983). However, government personnel at all levels from field technician to chief planning officer are usually trained in technical fields such as agriculture or forestry (Dove, 1992). They often have limited knowledge of the social and economic factors which strongly influence project success. Most government agencies also lack the resources necessary to hire additional staff with social science expertise, or to extensively retrain current staff. Therefore, a computer-based DSS that would provide easy access to knowledge concerning the social and economic factors that affect technology adoption could prove useful.

Although a large number of factors both within and outside the farming system are believed to influence farmer decision making regarding introduced technologies, a relatively limited set of important factors appears repeatedly in a wide variety of different studies (Feder et al., 1985; Rogers, 1983). AGFADOPT, the rule-based expert system presented in this analysis, is based on a simplified model of the decision-making process using general factors drawn from the existing scientific knowledge base. As such, it does not presume to include all the factors that can affect farmer decision making or to exactly replicate the decision-making process. Instead, this model relies on general principles to provide a first approximation of probable farmer behavior in a wide variety of different circumstances.

Existing Knowledge Base

There is a small but growing amount of literature examining the factors which influence the adoption of tree-based innovations such as alley cropping and contour hedgerows (Fujisaka, 1994; Raintree, 1983; Repollo and Castillo, 1989), fuelwood plantings (D. Kummer, personal communication) and tree crops such as fruits, coffee, cacao, and rubber (Watson, 1990). Inferences can also be made from the studies of adoption behavior for soil conservation practices (Napier, 1991; Napier et al., 1991) and from reviews of adoption behavior in annual-crop-based agricultural systems (Feder et al., 1985; Rogers, 1983).

A set of socioeconomic factors that are believed to influence the adoption of tree-based innovations was derived from this knowledge base for use in AGFA-DOPT (Robotham, 1996). Some of the socioeconomic factors are related to local perceptions about the innovation itself. These include: the innovation must be designed to address a perceived problem and local residents must believe that the innovation will help to ameliorate the problem (Fujisaka, 1994). These two factors are often collectively referred to as farmer awareness (Napier, 1991; Napier et al., 1991). Other factors are related to the context into which the innovation is introduced. These include, but are not limited to: farm size, land tenure, human capital, labor availability, credit availability, farmer attitudes toward risk and uncertainty, and market availability (Feder et al., 1985; Mercer, 1992; Raintree, 1983; Repollo and Castillo, 1989).

Organizing the Knowledge Base

One of the ways to organize knowledge regarding adoption of tree-based technologies, is to model how these factors influence farmer decision making. One of the simplest models is the decision tree. Decision trees are based on the psychological theory of elimination by aspects first discussed by Tversky (1972) and elaborated on by Gladwin (1983; 1989) and Gladwin and Murtaugh (1980). This theory states that a complex decision can be divided into discrete aspects each containing alternatives. Complex decisions are made on the basis of progressive elimination of alternatives in the decision. The decision criteria can be either orderings of alternatives on some aspect, dimension, or factor of the alternatives or they can be constraints that must be satisfied (Gladwin, 1983).

In Gladwin's examples (Gladwin, 1983; 1989), decision-tree models are constructed through directed interviews with farmers regarding the factors affecting their decision-making process. The information from these interviews is then synthesized by the researcher into a general model. In contrast, this analysis uses general information taken from the knowledge base regarding the adoption of tree-based innovations as the basis for model development (Robotham, 1996).

The major advantages of decision-tree models are simplicity and transparency. The reported accuracy of the site-specific models developed by Gladwin (1983; 1989) is also very high (80 to 95%). In addition, a decision-tree model can be easily transferred into a rule-based expert system. Since the primary goal of this analysis was the development of an expert system based on a simple and easily understandable model, a decision-tree format was adopted.

Accessing and Implementing the Knowledge Base

Even if a knowledge base is organized around an appropriate model, this does not guarantee that potential users of the knowledge will be able to access and implement the knowledge. With the increased availability of personal computers and advances in computer technology, computer programs that allow users to access a knowledge base in a convenient and understandable way have been developed. Expert systems have been used to capture and disseminate knowledge for a wide variety of situations related to agriculture and agricultural development (Plant and Stone, 1991) including the biophysical and environmental aspects of tree-based innovations (Warkentin et al., 1990; R. S. Yost, personal communication).

Table 14.1 Categories of Questions Used in the Model

Question category	No. of questions
Perception of the problem	4
Beliefs about tree-based technologies	4
Desires for the future	3
Market accessibility	3
Constraints	
Land	5
Knowledge	2
Labor	2
Materials	6
Other	2
Perceived future benefits	2
Total	33

Agroforestry Decision Tree

The decision-tree model developed for this analysis uses two major assumptions about farmer behavior: (1) farmers adopt a risk averse or "safety first" strategy, that is, a farmer must be reasonably assured of meeting subsistence production needs before taking the risk to incorporate trees into their system; and (2) farmers have the following possible courses of action: (a) plant traditional annual crops without changes in the existing system, (b) plant a fruit or nut tree orchard on all or part of their land, or (c) use an agroforestry technology based on contour planting of nitrogen fixing tree (NFT) hedgerows on their annual crop land.

The preconditions, opportunities, and constraints identified from the literature as well as the assumptions above were then used to formulate a set of 33 questions divided into 6 general categories (Table 14.1). These questions were combined to form the decision-tree that served as the basis for AGFADOPT, the expert system.

Mapping the Model into Agfadopt

Mapping the rules from the decision-tree model into the expert system shell appeared to be reasonably straightforward. Rule-based expert systems, such as AGFADOPT contain several components: choices, the potential system outcomes; rules, the if-then-else statements representing the knowledge; and qualifiers, the individual statements that are combined to make rules. Rules can be further divided into three types (Clancey, 1983): strategy rules, used to represent the plan or order of questioning; structural rules, used to reference a specific set of rules for a given conclusion; and support rules, that add conditions to the main conclusions. The THEN or ELSE conditions from one rule can serve as the IF conditions for subsequent rules. Through this process, rules are linked in a logical sequence that simulates reasoning.

The questions in the decision-tree model could readily be translated into qualifiers in the expert system. For example, question 1 in the decision tree, "Do farmers generally perceive that soil erosion is a problem?" became qualifier 1 in the expert system, "Farmers generally perceive soil erosion to be <a problem, not a problem>." These qualifiers were then combined to form a sequence of rules that reproduce the logic of the decision tree (Figure 14.1).

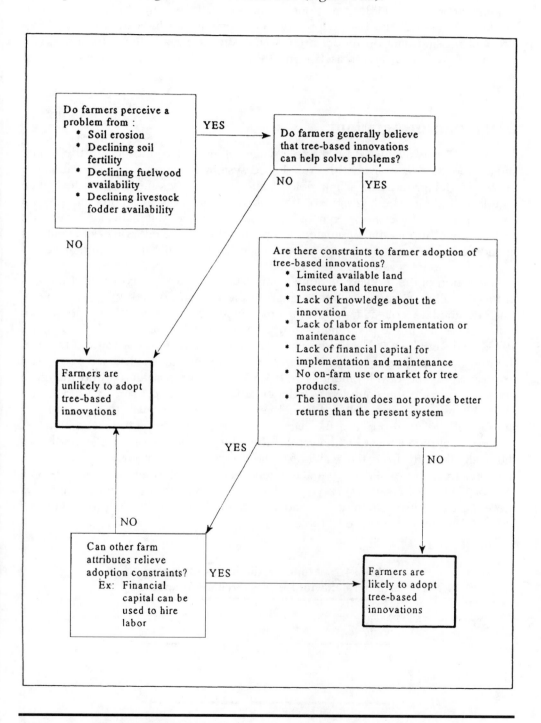

Figure 14.1 AGFADOPT decision logic.

Given the complexity of the decision tree, rule development and organization proved difficult. Although the majority of the rules used in the system were structural rules, the complexity of the decision tree necessitated several levels of strategy rules. In addition, a number of support rules were added to the system to identify missing data and to provide the user with an indication of the likely effect of missing data on model predictions. After these additions, the prototype version of AGFADOPT increased from 33 model questions to 87 qualifiers used in 72 different rules.

Evaluation of Agfadopt

AGFADOPT was evaluated on the following criteria: (1) verification of the model syntax and confirmation that the rule-based system is free of error; (2) validation of the model predictions using field data; and (3) assessment of performance criteria including speed, reasonableness of the interaction, and completeness.

Verification of the model syntax was done by using systematically developed sets of model inputs that simulated a large number of potential model situations including numerous unusual situations (such as large amounts of missing data) where the model was most likely to fail.

Validation of the model was done using data from a 1993 survey of 161 farmers in the Cordillera Central mountain range in the Dominican Republic. The survey was designed to collect baseline information on farming practices and farmer attitudes for use in the planning of future upland development efforts (Robotham, 1993; Witter et al., 1996). The survey contains questions that correspond to all of the questions used in the expert system. Although the survey data were not used extensively in the development of AGFADOPT, it was considered and therefore does not provide the completely independent data set necessary to validate the model. However, it is usable for model verification — to answer the question: "Is the model able to reproduce the data that were used in model development?"

Results of the verification analysis for the adoption of tree orchard technologies differ significantly from the results for the adoption of contour hedgerows. For tree orchards, the expert system accurately reproduces farmer decisions for 83% (134 of 161) of the farmers surveyed (Table 14.2). However, for contour hedgerows the expert system accurately reproduces farmer decisions for only 68% (110 of 161) of the farmers surveyed (Table 14.3).

Table 14.2 Model Prediction vs. Observed Farmer Behavior for Tree Orchard Adoption

Model prediction	Observed behavior		
	Yes	No	Total
Yes	83	15	98
No	12	51	63
Total	95	66	161

Table 14.3 Model Prediction vs. Observed Farmer Behavior for Contour Hedgerow Adoption

Model prediction	Observed behavior		
	Yes	No	Total
Yes	29	29	58
No	22	81	103
Total	51	110	161

The user interface and the overall utility of the model were tested through a survey that was included in copies of the AGFADOPT prototype that were distributed to colleagues in the U.S. and in other countries. These colleagues were asked to evaluate both system accuracy and performance. The initial responses regarding the AGFADOPT user interface have been positive. However, they represent only limited feedback and do not provide sufficient information for definitive conclusions.

Limitations of the Expert System

The accuracy of AGFADOPT for tree orchard technologies (83%) is similar to the accuracy (80 to 95%) presented for other decision-tree models (Gladwin, 1989). However, the accuracy of AGFADOPT for contour hedgerows (68%) is much lower. There are a number of potential reasons for the discrepancy between the results for trees, which are similar to other models, and for contour hedgerows.

First of all, the model is based on community-level generalizations from the literature while the data used in the model verification is from individual farmers. Scale may be an important issue in this type of analysis and may have different effects depending on the technology being considered. Since many of the benefits accrued from tree orchards result from the sale of products off-farm, tree orchard adoption may be more strongly affected by general, community-wide factors. In contrast, since the benefits from contour hedgerows are mainly confined to the farm itself (e.g., green manure, animal fodder, erosion control), site- and farmer-specific attributes may play a larger role in farmer decision making.

Second, the specific situation in the survey area where verification data was collected brought out other potential weaknesses of the expert system. One of the inaccuracies evident in Table 14.3 relates to a limitation of the model. The model does not incorporate the potential use of other contour barrier technologies such as grasses or dead plant materials. Of the 29 farmers who did not adopt contour hedgerows but were predicted to do so by the model, 15 used an alternative barrier technology.

The model also does not differentiate between farmers who use the practices on all of their land and those who use the practices on only a small area. Both circumstances are classified as adoption. By equating partial and complete adoption of technologies, this model may be masking important variations in farmer behavior that could have significant implications for land management.

Evaluation of the Approach

One goal of this analysis was to evaluate the utility and feasibility of using an expert system to organize and access a knowledge base of information related to farmer adoption of tree-based technologies. The development of AGFADOPT illustrates some difficulties but also some potential insights that come from incorporating even a relatively simple model into a deterministic, rule-based expert system. For example, comparison between system predictions and actual farmer behavior helped identify the importance of alternative soil erosion technologies in the survey area. This insight provides a potential area for further study.

Conclusions

This paper has presented AGFADOPT, a prototype expert system designed to model agroforestry adoption decision making. Verification of the model using data from a sample area in the Dominican Republic showed that the model replicated farmer decisions in 83% of cases for tree orchard adoption and 68% of cases for contour hedgerow adoption.

These results indicate that even a simple model based on general principles such as AGFADOPT may provide a useful way to represent knowledge regarding how a variety of factors influence farmer adoption of tree-based technologies. However, the system still requires significant additional evaluation, regarding both its applicability to other areas and its utility to potential users.

In addition to potentially helping technically trained planners and field personnel make decisions on specific projects or programs, evaluators of the prototype system have suggested that AGFADOPT has potential use as an interactive teaching tool. As such, it might provide a way to introduce students in technical fields to socioeconomic factors that affect adoption.

Expert systems such as AGFADOPT can also serve as the basis for much-needed research on those factors which affect agroforestry technology adoption under the wide variety of environmental, social, and political conditions where agroforestry is a potential land use. Comparisons between model predictions and field observations can help identify gaps in the knowledge base and facilitate hypothesis development. Additional research into areas where knowledge gaps exist can expand and improve the existing knowledge base. Effective use of this growing body of knowledge by planners, policy-makers and extension personnel is a prerequisite to the formulation and execution of effective upland development activities benefitting small farmers.

Acknowledgments

Collection of the survey data used in the model verification was supported by a Title 12 USAID Institutional Support Grant to Dr. Scott Witter, Michigan State University. The author's ongoing doctoral studies, including the development of AGFADOPT, are supported by a very generous grant from the East-West Center.

References

Allen, B.J. 1993. The problems of upland land management. Pages 225–241 in H. Brookfield and Y. Byron, Eds. *Southeast Asia's Environmental Future: The Search for Sustainability*. United Nations University Press, Tokyo.

Blair, G. and R. Lefoy. 1991. *Technologies for Sustainable Agriculture on Marginal Uplands in Southeast Asia*. Proceedings of a seminar held at Ternate, Cavite, Philippines, 10–14 December, 1990. ACIAR Proceedings No. 33. ACIAR, Canberra, Australia.

Clancey, W.J. 1983. The epistemology of a rule-based expert system — a framework for explanation. *Artif. Intelligence*. 20:215–251.

Dove, M.R. 1992. Foresters' beliefs about farmers: a priority for social science research in social forestry. *Agrofor. Syst.* 17:13–41.

Feder, G., R.E. Just and D. Zilberman. 1985. Adoption of agricultural innovations in developing countries: a survey. *Econ. Dev. Cult. Change*. 33(2):255–298.

Fujisaka, S. 1994. Learning from six reasons why farmers do not adopt innovations intended to improve sustainability in upland agriculture. *Agric. Syst.* 46:409–425.

Garrity, D.P. 1993. Sustainable land-use systems for sloping uplands of Southeast Asia. Pages 41–66 in *Technologies for Sustainable Agriculture in the Tropics*. ASA Special Publication 56. American Society of Agronomy, Madison, Wisconsin.

Gladwin, C.H. 1983. Contributions of decision-tree methodology to a farming systems program. *Hum. Organ.* 42(2):146–157.

Gladwin, C.H. 1989. *Ethnographic Decision Tree Modeling*. Sage, Beverly Hills, California.

Gladwin, C.H. and M. Murtaugh. 1980. The attentive-preattentive distinction in agricultural decision making. Pages 115–134 in P.F. Barlett, Ed. *Agricultural Decision Making: Anthropological Contributions to Rural Development*. Academic Press, New York.

Juo, A.S.R. 1989. New farming systems development for the wetter tropics. *Exp. Agric.* 25:145–163.

MacDicken, K.G. and N T. Vergara. 1990. Introduction to agroforestry. Pages 1–30 in *Agroforestry: Classification and Management*. John Wiley & Sons, New York.

Mercer, D.E. 1992. The economics of agroforestry. Pages 111–143 in W.R. Burch, Jr. and J.K. Parker, Eds. *Social Science Applications in Asian Agroforestry*. Winrock International, Petit Jean Mountain, Arkansas.

Napier, T.L. 1991. Factors affecting acceptance and continued use of soil conservation practices in developing societies: a diffusion perspective. *Agric. Ecosyst. Environ.* 36:127–140.

Napier, T.L., A.S. Napier, and M.A. Tucker. 1991. The social, economic, and institutional factors affection adoption of soil conservation practices: the Asian experience. *Soil Tillage Res.* 20:365–382.

Plant, R.E. and N.D. Stone. 1991 *Knowledge-Based Systems in Agriculture*. McGraw-Hill, New York.

Raintree, J.B. 1983. Strategies for enhancing the adoptability of agroforestry innovations. *Agrofor. Syst.* 1:173–187.

Repollo, Jr., A.Q. and E.R. Castillo. 1989. Agroforestry technology in hilly land households: factors influencing its adoption. Pages 117–132 in N.T. Vergara and R.A. Fernandez, Eds. *Social Forestry in Asia*. SEARCA, College, Laguna, Philippines.

Robotham, M.P. 1993. The Relationship between Plan Sierra Outreach Activities and the Adoption and Continued Use of Soil and Water Conservation Technologies by Upland Farmers. M.S. thesis, Michigan State University, East Lansing.

Robotham, M.P. 1996. Modelling socioeconomic influences on agroforestry adoption using a rule-based decision support system. In P.S. Teng, Ed. *Proceedings of the 2nd Symposium on System Approaches for Agricultural Development*. Kluwer Academic Publishers, AA Dordrecht, The Netherlands (in press).

Rogers, E. 1983. *Diffusion of Innovations*, 3rd ed. The Free Press, New York.
Tversky, A. 1972. Elimination by aspects: a theory of choice. *Psychol. Rev.* 79(4):281–299.
Warkentin, M.E., P K R. Nair, S R. Ruth, and K. Sprague. 1990. A knowledge-based expert system for planning and design of agroforestry systems. *Agrofor. Syst.* 11:71–83.
Watson, G.A. 1990. Tree crops and farming systems development in the humid tropics. *Exp. Agric.* 26:143–159.
Witter S.G., M.P. Robotham, and D.A. Carrasco. 1996. Plan Sierra, outreach programming, and sustainable adoption of soil and water conservation technologies by upland farmers. *J. Soil Water Conservat.* (in press).

Chapter 15

Developing an Integrated Nutrient Management Decision Aid

RUSSELL YOST AND JOHN Z.C. LI

Introduction

Nutrient management presents a complex challenge to modern agriculture and natural resource management. The complexity results both from complexity of processes determining nutrient availability and the competing goals. Management alternatives are needed that balance the competing goals of high crop productivity, environmental health, and economic viability. Knowledge-based computerized decision-aids such as ADSS (Acidity Decision Support System, Tropsoils Project, 1990), PDSS (Phosphorus Decision Support System, Yost et al., 1992), and NDSS (Nitrogen Decision Support System, Reid, 1996, personal communication) have been designed to solve complex problems in soil acidity, phosphorus deficiency, and nitrogen deficiency while PROPA (Professor Papaya, Itoga et al., 1990) and FACS (Fertility Advice and Consulting System, Yost et al., 1995) have been designed to solve nutrient, disease, insect, and cultural problems in papaya production and to make integrated fertilizer recommendations for samples submitted to the Agricultural Diagnostic and Service Center of the University of Hawaii. (ADSC), respectively. While these decision aids generate solutions that solve primary problems, the solutions may be in conflict and the competing goals may not be evenly considered. The example below illustrates conflicting solutions to addition of nutrient nitrogen to a cropping system. Integrated Nutrient Management Decision Aids must help plan the management of such conflicting solutions and competing objectives.

An Example

Applying the nutrient nitrogen (N) can correct soil N deficiency, increase crop yield, increase input cost, and increase likelihood of groundwater contamination. With regard to biophysical interactions such applications, depending on the chemical form, can increase, decrease, or not affect soil pH, each of which can strongly affect the availability of other nutrients. Similarly, liming to neutralize soil acidity can correct soil acidity, affect overall nutrient retention, and usually increase the amounts of nutrient calcium (Ca) and magnesium (Mg), possibly leading to an imbalance of Ca and Mg, again depending on the chemical form in which they are added. As suggested above, the biophysical complexity is augmented by the question of how to obtain the balance of amendments that provides the nutrients in just sufficient quantities to optimize crop productivity, environmental health, and ensures economic viability.

Clearly multiple expertise that deals with each of these specific requirements and adjusts for the multiple goals is needed to resolve conflicting solutions, optimize competing objectives, and find practical, workable solutions in an easy, timely fashion. Equally clear is the requirement that this expertise not be rigid and uncompromising in the face of direct conflicts.

Objectives of the Study

1. To identify the common features of individual nutrient problem-solving decision aids
2. To investigate system models for integrated problem solving, uncertainty handling, and conflict resolution
3. Outline a structure for an integrated nutrient management decision aid

Our survey on system models for integrated problem solving is by no means exhaustive. Uncertainty handling methods are discussed briefly.

Characteristics of Agricultural Knowledge and the Need for Integrated Knowledge Bases

We first describe some meta-knowledge (knowledge about the disciplinary knowledge) regarding agricultural systems that we have found helpful in framing our problem-solving effort. Agricultural systems are extraordinarily complex, appearing to bear similarity to the detail of a fractal process — the detail continues to increase as one looks closer such that there are infinite distances between two points a finite distance apart. So the complexity of biological systems seems to increase as one examines, separates, and analyzes the systems more closely. This observation illustrates why it may be futile to attempt to completely describe a biological system such as a soil, crop, and plant system that is the focus of the decision aids discussed here.

We view soil-crop-plant systems as including a large number of subsystems that contribute to the overall behavior, but any one of which can be the cause for specific failures that result in reduced systems properties: productivity, stability, resilience, and other properties suggested by Conway (1987). It is difficult to ensure specific levels of systems properties because they depend on nonlimiting conditions of subsystems. For example, ensuring crop productivity requires that nutrient status must be nonlimiting. Removing the constraints of nutrient deficiencies, however, does not ensure high productivity, rather productivity and the other systems properties depend on all the subsystems, such as water relations, disease, insect incidence, and cultural practices. Consequently, indicators and diagnostic techniques that accurately reflect nutrient status must be monitored by humans unless there are self-correcting nutrient supply mechanisms in place. Unfortunately, it is not possible to entirely predict system behavior because of the complexity. Consequently, one option is to monitor the system through systems of indicators and diagnostic measures so that remedial measures can take place before detrimental effects occur. Various aspects of nutrient, physiology, disease, and insect dynamics appear amenable to prediction and it is important to increase the breadth and depth of such predictability. However, at present, management cannot wholly depend on such predictability to ensure success in food production and security. For these reasons, diagnostic procedures are needed in order to improve and optimize food, fiber, feed, and forest production systems.

While diagnostic information is crucial to identify problems at the earliest opportunity, the prescription is also crucial because it determines the extent to which biophysical alternatives are identified and evaluated. The prescriptive step usually requires all available disciplinary information regarding processes that control and affect the deficient or excessive nutrient. This information is directly required to develop alternative ways to resolve the nutrient deficiency problem, be it by simple addition of the nutrient, the affecting of reactions controlling its availability, or identifying causes and remedies for the shortage or excess. This body of information is clearly disciplinary and requires the best available knowledge. Usually, however, decisions must be made quickly and have to be carried out with only the information that is accessible at the time of the decision. This clearly differentiates management decision making as opposed to research, or pursuit-of-knowledge decision making. The latter would wait until the new information becomes available whereas the former would clearly make the decision when circumstances required it to be made.

In addition to the meta-knowledge of how the domain knowledge and experience are used, certain characteristics of how the knowledge of each nutrient is shaped into a problem-solving framework also repeat for many of the nutrients.

A general pattern has appeared in our experience developing decision aids for the various nutrients — we often begin with questions about unmanageable factors and progress to manageable, and last to rapidly changing factors. Many of the unmanageable factors of the environment are often ascertained early in consultations by human experts. We found this to be a logical, efficient search sequence in problem solving. Often unmanageable factors such as soil elevation, slope, and physical characteristics set the bounds for subsequent possibilities of adjustment, change, or transformation to be exploited by management. In one of the systems that we have developed, we found it useful to identify the nonmanageable or difficultly manageable factors in considering landuse options (CSAS, 1992).

The sequence we found useful in soil-crop-economic systems progresses from soil characterization to crop requirements to cropping system management needed to minimize the mismatch between soil characteristics and crop requirements (Uehara, personal communication, 1982) and ultimately to the economic analysis of the management alternatives. We have considered three ways to link economic analyses of various systems. Initial systems such as ADSS considered biophysical factors then management that approximately optimized nutrient supply and then considered the economic benefit to that specific biophysical selection. In later systems such as PDSS, the management and economic factors are considered in the same form, leading to easier integration of biophysical optima and economic optima.

In a third type of system (Li and Yost, 1996), we seek to compare a reasonable set of management options that meet a user-selected set of goals. Intermediate results that are simulated are necessary to determine the probable consequences of alternative management choices.

We have found it useful to maintain a certain degree of disciplinary integrity in order to facilitate the maintenance and updating of knowledge bases of disciplinary information. Scientists that concentrate on developing disciplinary knowledge must be able to identify with the knowledge and take the initiative to indicate when updating of the knowledge base is needed. These people need specific rewards for their unique contribution to the maintenance and validity of the knowledge base.

As a final observation on the problem-solving sequence, we have found it useful to attempt to record and use information that sometimes is only qualitative at best. As further research develops, such qualitative information sometimes can be revised and updated to more quantitative relationships. One example from the development of the ADSS was an observation that aluminum toxicity was possibly reduced by the addition of organic materials in field research at Sitiung, Indonesia. We initially reported this in ADSS as only supplementary information. As more research results became available we were able to make more definite statements about the effect and ultimately quantitatively reduce calculated lime requirement by one tonne for every 10 tonnes of applied organic material. Additional information is needed — type of organic material varies the effect and in some regions the effect seems not to occur. The pattern is clear, however, and we can usefully record and transmit information even though it is not quantitative or absolutely certain.

Difficulties in Developing Decision Aids

It has long been recognized that knowledge acquisition — the gathering, structuring, documenting, and implementing in a computer-accessible format — is one of the more difficult aspects of decision-aid development. One difficulty in knowledge acquisition is that experts may have "compiled" their knowledge based on their experience and the usual range of variables in their geographic region of responsibility and no longer retain the fundamental stages or steps involved in the deduction or inference of the rationale (Etherington et al., 1989). This loss of the basic information can be unfortunate when a new set of circumstances present themselves and it is necessary to reevaluate the underpinnings of the logical procedures. Computer implementation makes it possible to store and replicate long complicated sequences of logic, experience, and rationale with a high degree of accuracy, thus retaining the detail that typically is lost when experts "compile" their knowledge into easily retrieved generalizations and rules-of-thumb. A few such examples suggest that computer-assisted memory can contribute to the quality of our information, particularly when it is time to reevaluate and reconsider old questions and knowledge in the light of new measurements or techniques.

Survey of System Models for Cooperative Problem Solving.

1. **Yellow page approach**, was originally proposed and designed for cooperative expert systems in distributed computing environments, in which each expert agent advertises its expertise on the global "yellow pages." To solve problems which require multidisciplinary knowledge, the user and other experts match their needs to the qualifications of the available experts on the pages and thus determine with whom to consult. Then a cooperation plan can be selected. The system structure is loosely coupled knowledge bases that operate autonomously, with initiative on the part of the user. By the term "loosely coupled," we also mean that these knowledge bases can be totally different from each other in their knowledge structures, inference mechanism, database organization, and input/output requirements. They do not necessarily share their data. Conflict resolution is the responsibility of the user.
2. **Blackboard approach**, consists of a "blackboard" where each expert agent contributes inferences based on initial input and posted inferences from other experts until the problem is solved or no further inferences are possible. This approach is useful for complex, ill-structured problems characterized by poorly defined goals and an absence of a predetermined decision path from the initial state to the goal. Conflicts can be resolved if they are identified by experts and satisfactory alternatives are posted on the blackboard. Structurally, all expert agents work on the same problem and respond opportunistically to changes on the blackboard, resulting in the incremental generation of partial solutions; no *a priori* order of experts. The blackboard serves as the common database, which is the only place that communication and interaction among the experts occurs. Obviously, these knowledge bases must have the same input/output requirements. Their inference engines, however, can be different from each other.
3. **General nurse approach**, by which the "general nurse" is described as the first agent to intercept a problem request and routes it to the experts most likely to solve the problem. In case multiple experts are needed, the "general nurse" identifies the primary problem and identifies the primary expert for the initial solutions. Then based on the primary experts's proposed solution to the primary problem, a reference link to the second relevant expert is made to investigate the implications of the proposed

solution on the secondary problem. Conflicts are resolved by the experts, not by the nurse. However, the nurse's role is important for effective problem solving. The nurse provides the user with guidance and interpretation of the problems. In implementation, the depth and breadth of this agent's knowledge base needs to be carefully determined so that it is broad enough to fit its role as a general nurse, but not so deep as to replace experts.

4   **Professor Papaya approach**, is a multidisciplinary knowledge-based computer program developed at the University of Hawaii at Manoa (Itoga et al., 1990). The program was developed from expertise from plant pathologists, soil fertilizer experts, entomologists, and general extension experts. To start the consultation, the user checks off observed symptoms on a general form with a fixed set of possible symptoms. The answers to the form are used to direct the problem request to the appropriate experts by means of a relevancy ranking. The most relevant expert starts a dialogue with the user. New information from the dialogue between the expert and the user updates the relevancy ranking and shifts the dialogue to the next most relevant expert once the current expert finishes the session. Finally, each expert gives his/her diagnoses and recommendations. This system primarily diagnoses problems and has no conflict resolution. Structurally, this system is an example of a closely-coupled DSS that uses a common inference engine and shares all data.

5   **Simple deterministic approach**, by which a fixed sequence of experts is followed, possibly determined by expert knowledge and experience. Conflict resolution is the responsibility of the user.

Our Experience

ADSS

We found it best to request information about unmanageable variables first in the system and manageable variables second (soil, crop, management, economics) in this sequence.

We had to assume that all other factors were nonlimiting in order to operationalize the system.

We tried to represent prediction uncertainty and found that some users ignored uncertainty estimates while others eagerly used the uncertainty estimates to quantify risk.

Propa

Interdisciplinary Relevancy factors proved to be a useful way to prioritize the various disciplinary knowledge domains. The form-based interface was easy to use and the graphic identification of insects illustrated the utility of photographs and drawings as part of the identification and characterization effort.

This system was, however, almost completely diagnostic. Very few recommendations or details of how to solve the problem were provided. In part, this was because a major portion of the system was devoted to disease and insect problems.

With disease knowledge, we learned that management solutions usually required developing resistant cultivars. Consequently, identification was quite important.

PDSS

We developed a strong diagnostic system with considerable emphasis on management remediation of P deficiencies, including economic evaluation and assessment. Forms were easy to use and facilitated the interaction between economic and biophysical management alternatives in developing the recommendations.

FACS

Multiple knowledge sources were integrated, however, there was very little integration of the consequences of one application on the other, with the exception of Ca and Mg. We did calculate the Mg contribution resulting from the pH adjustment by dolomite and the expected Ca increase resulting from both calcitic and dolomitic lime additions.

IITA-AFS

This decision aid was developed to contribute to the network (AFNETA, Agroforestry Network for Africa, supported by the International Institute for Tropical Agriculture [IITA]) by gathering collaborator data into one place that all could see, access, and share.

Distributed to assist formation and stimulate the network by providing a unique network product that was a synthesis of their knowledge.

N Manager for Windows

This system uses an integration of rule-based problem solving and simulation to determine strategies and management alternatives that together achieve user-specified goals.

Uncertainty Handling Methods

General options included the Bayesian cumulative probability, consideration of fuzzy logic in relation to the Bayesian approaches and, more recently, the Dempster-Shafer approach to evidence accumulation. The Bayesian evidence accumulation method, described by Pearl (1988), and implemented in PDSS, permitted the synthesis of disparate data and observations such as the combination of knowledge of location, with visual nutrient deficiency symptoms, values of soil nutrient analyses, and plant tissue analyses all into one index of availability. The Bayesian system, however, fails to distinguish between perfectly conflicting information (Prob. of P deficiency of 0.5) and no information at all — (Prob. of P deficiency of 0.5). The Dempster-Shafer method in such cases, however, clearly indicates that there is a lot of evidence and that it appears to be conflicting.

Conclusions

An ideal Integrated Nutrient Management Decision Aid must be able to (1) identify the possible conflicts of the problem which none of its components can solve by itself alone (need for cooperation), (2) use global knowledge in order to locate relevant experts and select the appropriate cooperation plan, and (3) modify the primary recommendation to enhance its positive effects and assure that it does not result in detrimental side effects. We have found that the blackboard approach was useful for diagnosis and information exchange, but does not provide a method of synthesizing observations into an overall conclusion.

A simple deterministic approach with user initiative has been successfully implemented in the FACS. However, conflict resolution is the responsibility of the users. A knowledge-base method of conflict resolution appears to be the most promising.

References

Conway, G.R. 1987. The properties of agroecosystems. *Agric. Syst.* 24:95–117.
CSAS. 1992. Crop Suitability Analysis System, Department of Agronomy and Soil Science, Department of Land Development, Deloitte and Touche Consultants, and the Hawaii State Department of Agriculture.
Dempster, A.P. 1967. Upper and lower probabilities induced by a multi-valued mapping. *Ann. Math. Stat.* 38(2):325–339.
Etherington, D.W., A. Borgida, R.J. Brachman, and H. Kautz. 1989. Vivid knowledge and tractable reasoning: preliminary report. Proceedings of the Eleventh International Joint Conference on Artificial Intelligence, August 20–25, 1989, Detroit, Michigan.
Gordon, J. and E.H. Shortliffe. 1985. A method for managing evidential reasoning in a hierarchical hypothesis space. *Artif. Intelligence*. 26:323–357.
Hayes-Roth, Barbara. 1985. A blackboard architecture for control. *Artif. Intelligence*. 26(1985):251–321.
Hayes-Roth, Barbara. 1983. The Blackboard Architecture: A General Framework for Problem-Solving? Tech. Rep. HPP-83-30. Knowledge Systems Laboratory, Computer Science Department, Stanford University.
Itoga, S.Y., C.L. Chia, R.S. Yost, and R.F.L. Mau. 1990. Propa: a papaya management expert system. *Knowledge-Based Syst.* 3:163–169.
Li, M. and R.S. Yost. 1996. N Manager for Windows: a two-way approach to the management of nutrient nitrogen. (In review).
Newell, A. 1969. Heuristic programming: ill-structured problems. Pages 360–414 in J. Aronofsky, Ed. *Progress in Operations Research*. John Wiley & Sons, New York.
Nii, H. Penny. 1986. Blackboard systems: the blackboard model of problem solving and the evolution of blackboard architectures. *AI Mag.* Summer 1986:38–53, August, 1986:82–106.
Pearl, Judea. 1988. *Probabalistic Reasoning in Intelligent Systems: Networks of Plausible Inference*. Morgan Kaufmann Publishers.
Shafer, G. 1976. *A Mathematical Theory of Evidence*. Princeton University Press, Princeton, New Jersey.
Shekhar, S. and C.V. Ramamoorthy 1989. Coop: a shell for cooperating expert systems. IEEE International Workshop on Tools for Artificial Intelligence. Architectures, Languages, and Algorithms. IEEE Computer Society Press, Los Alamitos, California. Pages 2–11.

Tropsoils Project. 1990. ADSS: Acidity Decision Support System. Tropsoils Project, University of Hawaii, Honolulu, Hawaii.

Yost, R.S., F.R. Cox, A.B. Onken, and S. Reid. 1992. The phosphorus decision support system. In *Proceedings Phosphorus Decision Support System*. Texas A&M University, College Station, Texas.

Yost, R.S., J.A. Silva, Y.N. Tamimi, N.V. Hue, and C.E. Evensen. 1995. Towards automating a nutrient management system. Paper presented at the Western Regional Nutrient Management Conference, March 9–10, 1995.

III

MODSS METHODOLOGIES, TOOLS, COMPONENTS, AND INTEGRATION

Chapter 16

Smart Pitchfork

PHILIP R. GOODRICH

A user-friendly expert system computer program SMART PITCHFORK has been developed to assist livestock producers utilize manure more efficiently for sustainable crop production, reduction of energy inputs, and pollution control. Animal manure nutrient information and nutrient application plans are determined from rules and calculations bypassing the normal time-consuming hand calculations which hinder the farmer or the custom applicator. SMART PITCHFORK allows the client to quickly see the options and understand how the advice was determined. It is also an educational tool which may be used by farmers, extension agents, farm managers, pollution control specialists, and agricultural consultants who deal with manure application to improve their skills in the analysis of manure application problems. The expert system provides information regarding handling animal manure including: what is the value of the manure, how much of each nutrient is available in the manure, how much nitrogen will be lost in the process of spreading and exposure to weather, how much credit can be claimed for previous manure applications, how much manure is in a storage structure, and how much manure should be applied to support the needs of a crop. Information for recommended nitrogen and phosphorus to produce goal levels of 12 crops is as specified by the Potash Institute of America. Created in using Level V™, the easy to understand program is available on DOS and Macintosh operating systems.

Introduction

Animal manure nutrient information and suggestions how to apply these nutrients are available in extension publications, but the calculations and decisions are complex relying on time-consuming sampling and testing of the nonuniform material collected and stored. Therefore, the calculations are usually not completed by the farmer or the custom applicator. Manure is often spread on the nearest field and placed so it looks good. Then no credit for the applied nutrients is taken and normal amounts of commercial fertilizer are applied to supply the crop planted in the field. Manure is often spread when the storage facility gets full and not on a planned schedule. Proximity to water is ignored and drainage systems sometimes are covered with manure.

Expert systems are knowledge-intensive computer programs. They use rules-of-thumb (or heuristics) to focus on the key aspects of particular problems and to manipulate symbolic descriptions in order to reason about the knowledge they are given. The best expert systems can solve difficult problems, within a very narrow domain, almost as well as human experts. This may be because knowledge systems (or symbolic systems) do not display biased judgments, nor do they jump to a conclusion and then seek to maintain those conclusions in face of disconfirming evidence. Or, in other words they are free from the behavioral change that may influence the judgment of an expert (Harmon and King, 1985).

Efficient and environmentally acceptable systems for application of animal manure can recycle large volumes of nutrients onto cropland. Animal manure is a large volume output of the animal industry and is usually associated with a land resource where the animal manure can be applied and the nutrients used by the crops. Approximately 123 million metric tons of waste on a dry basis are produced annually in the U.S., of which 57 million metric tons may be recovered economically (Van Dyne and Gilbertson, 1978). The managers of many animal production units are trained in the management and production of animals. They are less well trained in the crop production area and are not experts in the use of the byproducts of the animal production enterprise.

Why Use an Expert System in a Manure Application Program?

The major reason for developing and using an expert system is that it is the best tool to solve the problem.

1. It allows the manure applicator to deal effectively with complex, ill-defined, qualitative problems such as manure management systems.
2. It is an effective way to package technology and make it useful by "nontechnical" users.
3. It can result in better recommendations for solving the problem.
4. It can reduce the time spent by the manure applicator if the proper system is chosen for the needed situation.

Objective

The objective was to create an expert system for deciding where manure would be applied. The expert system would select an ideal system to transport and apply manure on specific fields in an efficient and environmentally sound manner. It would provide an educational tool which could be used by farmers, extension agents, farm managers, pollution control specialists, and agricultural consultants who deal with manure application to improve their skills in the analysis of manure application problems.

Constructing Smart Pitchfork

The expert system shell chosen to use in the development was Level V™. This expert system shell is rule-based and the knowledge base may be compiled to run on a variety of machines from small mainframe VAX's as well as the Macintosh and IBM/MS-DOS machines. Several expert systems have already been developed at the University of Minnesota in this shell and they may be chained together and run as a package (Goodrich et al., 1990). An earlier expert system that helped choose a field application system was constructed using Personal Consultant Plus™ (Goodrich and Kalkar, 1988).

The first step in building the expert system involved setting goals for the outcome of the system. What information regarding handling animal manure did the clients want to know?

- What is the value of the manure?
- How much of each nutrient is available in the manure?
- How much nitrogen will be lost in the process of spreading and exposure to weather?
- How much credit can be claimed for previous manure applications?
- How much manure is in a storage structure?
- How much manure should be applied to support the needs of a crop?

In reality, the applicator or farmer wants to know how much manure to apply to a field to provide sufficient nutrients to produce a yield goal. The other questions may be part of the process, but one single question gets to the center of the need for information.

Smart Pitchfork uses table information for recommended nitrogen and phosphorus to produce goal levels of 12 crops as specified by the Potash Institute of America (MWPS, 1985). The basic procedure for calculation of the residual nitrogen from previous applications is taken from the Midwest Plan Service Publication 18 (MWPS, 1985)

Operating the Expert System

New Information or Modifying an Old Data Set?

Initially the expert system (ES) asks (see Figure 16.1) if a new set of data or a data file is to be used. This means that previous data can be modified and all new data does not have to be input to run thus saving time and possibilities for error.

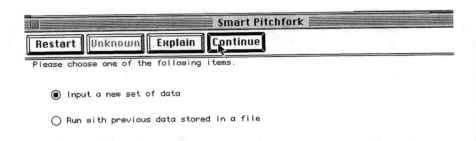

Figure 16.1 A screen with buttons and a choice. Keyboard or mouse input is allowed.

Type of Manure in the Storage?

Here either liquid or solid manure storage is selected. Calculations are quite different for the two types of manure. For liquid manure the user is asked if a chemical determination of the nutrient content is available. If available, the values for nitrogen, phosphorus, and potassium are entered as requested. If there has not been an analysis completed, a request for the species of animal is presented for the user to pick from the list. The ES then presents an estimate of the nutrients available in liquid manure (kilograms per 1,000 l). This is presented as total nitrogen, ammonium nitrogen, nitrate nitrogen, phosphate (P_2O_5), and potassium (K_2O). For solid manure the manure handling system asks a series of questions to determine how much bedding is in the manure.

Crop and Expected Yield

The user is asked for the crop to be grown and what the expected yield on that field will be. Crops are those that normally would be grown in Minnesota and Midwest conditions.

Method of Application

The planned method of application to the soil is requested next. Broadcast without incorporation, broadcast with incorporation, injection, and irrigation are the choices. Associated with each of these methods of application is a loss coefficient for nitrogen and phosphorus; irrigation producing large losses and injection producing minimal losses of nitrogen.

A summary screen informs the user of the loss coefficient and the percent losses to expect with the method of application. This is one of the important facets of expert systems, education of the user about the underlying assumptions of the solution. Previous year applications of manure are next determined. If the user affirms that manure has been applied in any of the three previous years, applied nutrients are requested for each year. Residual nutrients that are available will be included as credits.

How to Use Nutrients

The user now is asked how the nutrients are to be used. If the manure is applied to try and supply all the needed nitrogen, then phosphorus and potassium will normally be applied in excess. If phosphorus is chosen as the critical nutrient, there usually will not be enough nitrogen for the crop. This latter method is more efficient although nitrogen will have to be added later for most crops. Over-application of phosphorus is a problem with some soils especially if the excess application continues for many years.

The ES determines the amount of manure to apply per unit area. If additional nutrients are needed these are noted. Then the area available is requested. A summary table presents the amount that will be spread on the land. Suggestions on how to change to either make better use of the nutrients or to cover more land are provided. The user is able to save the results for further use and/or go back and change some of the currently used inputs. Thus several "what if" scenarios can be constructed and compared for the amount of manure available.

Determination of the Volume/Mass in Manure Storage

The ES can calculate the volume of manure in a variety of manure storages that are difficult for the average person to calculate. Cylindrical vertical storage, rectangular tanks, piles of trapezoid or pyramid shape, and horizontal cylindrical tanks are included. Dimensions for the chosen shape are requested and the volume presented. The area that can be fertilized by the volume contained in the storage facility is presented.

Explanation Feature

One of the important features of an ES is the ability to answer a users' question immediately. The "explain" function in Level V™ is accessed by a visual button displayed at the top of the screen. If there is an explanation of the question presented to the user, the button will be black; if it is dimmed, no explanation is available for this question. Triggering the button presents the explanation to the user. This is usually text that provides an expanded description or there may be a picture that explains more clearly what is requested, an indication of what terms are applied to an object that may be new to the user. Each explanation is customized by the creator to assist the user in understanding the knowledge included in the system and gaining increased capability to perform the task being undertaken. Conventional programs lack this immediate satisfaction or response to the user. The nonsequential form of ES allow the user to jump out to view messages and return. Figure 16.2 is a screen presented when an explain button is activated by the user.

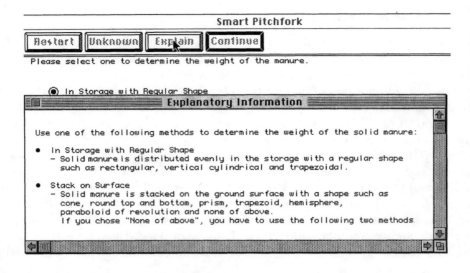

Figure 16.2 An "explain" screen that provides information to the user about the choices presented.

Use of Smart Pitchfork

Minnesota animal producers who also produce crops would be the main users of the ES to apply animal manure in an efficient manner. The population of Minnesota as a whole will benefit from the reduced pollution potential and improved environment. Crop farmers who wish to purchase excess manure from livestock farmers will benefit from an accurate measure of the nutrient composition and judge the value of the manure. Custom applicators could benefit the ES since manure may become a commodity that may be bought and applied.

Benefit to the Environment

The population of Minnesota as a whole will benefit from the reduced pollution potential and improved environment. Wise use of nutrients that also have pollution potential will protect the lakes and streams of Minnesota. The subsurface water supplies will be protected from nutrient and toxic substances when appropriate rates of fertilizers are matched to the soil conditions and the crop uptake of nutrients. Hazards will be considered each time the expert programs are used and the complex interactions of the environment accurately involved in the decisions made about application of animal manures to the soil. The ES for animal waste management will extend the experience of a specialist in animal waste management to help producers make excellent decisions. Thorough, correct, efficient and environmentally sound solutions will benefit both the producer and the environment.

References

Goodrich, P.R. and S.N. Kalkar. 1988. The manure application adviser. In Proceedings of the Second International Conference on Computers in Agricultural Extension Programs. Orlando, FL. February 10–11, 1988. Florida Extension Service. University of Florida, Gainesville. 7 pages. Paper No. 2,306 of the miscellaneous journal series of the Minnesota Agricultural Experiment Station.

Goodrich, P.R., B.J. Conlin, G.R. Steuernagle, and J.K. Reneau. 1990. The DAIRY MANAGER. Pages 542–546 in Proceedings of the 3rd. International Conference on Computers in Agricultural Extension Programs. Orlando, FL. February, 1990. Florida Extension Service, University of Florida, Gainesville.

Harmon, P. and D. King. 1985. Expert Systems, Artificial Intelligence in Business. John Wiley & Sons, New York.

Hayes-Roth, F., D.A. Waterman, and D.B. Lenat. 1983. Building Expert Systems. Addison-Wesley. Reading, Massachusetts. 444 pages.

MWPS (Midwest Plan Service). 1985. Livestock Waste Facilities Handbook. 2nd ed. Midwest Plan Service, Inc. Iowa State University, Ames.

Van Dyne, D.L. and C.B. Gilbertson. 1978. Estimating U.S. Livestock and Poultry Manure and Nutrient Production. Publication ESCS-12. U.S. Department of Agriculture Economics, Statistics, and Cooperatives Service. 145 pages.

Chapter 17

Using Desired Future Conditions to Integrate Multiple Resource Prescriptions: The Northeast Decision Model

MARK J. TWERY, SUSAN L. STOUT, AND DAVID L. LOFTIS

Abstract

Synthesis of scientific knowledge into forms useful to forest managers is essential. Integration of information needed by managers is complex. Tools that make integration feasible include decision support systems (DSS). The Northeast Decision Model (NED) is such a system and uses an original prescription design system to incorporate management goals for multiple objectives, analyze current forest conditions, recommend management alternatives, and predict future conditions under different alternatives. NED is designed to include a long-term, landscape-level view of the forest as an interconnected ecosystem that is too complex to understand at every level but which still must be managed. NED recommends potential treatments for all resources affected and provides options from which a manager may choose. The technique involves defining a management area of interest, defining goals for the area, identifying conditions necessary to meet each goal, and identifying conditions that can be met in conjunction with others, from most restrictive to least restrictive. The process begins with the selection of management objectives, or goals, for any or all of five resources: visual quality, wildlife, water, wood production, and general ecological objectives. These goals are defined for a management unit at a scale from one to many stands, generally within the range of 5 to 5,000 ha. Committees of experts in each of the specific resources have defined the conditions necessary to meet the specified goals, and have determined common variables to allow consistent evaluation of the conditions across goals. This integrated evaluation is a key element to the process of determining acceptable prescriptions and evaluating whether different alternative actions across the entire area will meet the desired conditions.

Introduction

Forests in the northeastern U.S. serve many purposes for many people including wood production, outdoor recreation, and wildlife habitat. There is great pressure to improve the management of forests, in part because of the history of highgrading and abuse of forest lands, and in part because of increased demands for a variety of benefits from forests. One of the biggest challenges is to manage a single piece of land for multiple benefits. The challenge is made even greater by the difficulty of assigning economic value to some of those benefits, such as scenic beauty and wildlife viewing. The pressure comes from many quarters, including corporate boardrooms and the general public. Most land managers know there are ways to improve their stewardship, and they seek help. Values other than timber also are gaining in importance, including the preservation of biological diversity and watershed protection. Active management of forest land is now practiced to increase all of these benefits. Production of wood fiber is still an important result of management, and the one that most often provides revenue to a landowner, but for many forest owners it is becoming less important than other resources. Landowners who do hold forests primarily for income from timber production often attempt to integrate other resources into their management as well.

Managing forests is difficult and complex. Managing forests for multiple values such as visual quality, wildlife habitat, wood products, and watershed protection while maintaining the sustainability of the full system in an integrated manner is even more difficult and complex (Marquis, 1992; Marquis and Stout, 1992). Much knowledge exists regarding management for various resources, but it is usually scattered in the literature, focused on a single resource (e.g., Davis and Clark, 1989; Manuel et al., 1993; McKinney et al., 1993; Loh et al., 1991), and not necessarily applicable to a manager's specific site. To address the need for solving complex forest management problems, the Northeastern Forest Experiment Station is developing a DSS called the Northeast Decision Model (NED) (Marquis, 1991; Twery, 1994; Kollasch and Twery, 1995). NED supports decision making by managing information about objectives, current forest conditions, silvicultural prescriptions to achieve the management objectives, and predictions of future conditions. A new conceptual design for making silvicultural decisions is the focus of this paper.

The NED is intended for application in the Northeastern U.S. and, with anticipated modifications, in other forests in the South and Midwest. Vastly different conditions exist within a particular forest type in different areas of a region, but these can be accounted for with site-specific information provided by the user, and by refining the expert prescriptions.

The intended clientele for NED includes all who are interested in management of forest land, principally those responsible for individual management decisions on specific units of land. To make full use of NED, however, will require some familiarity with both forest resource management and computer systems. Extensive explanation and help facilities will be included, but users will still need to be familiar with forest management terms and resource analyses. Most consulting foresters, industry woodland managers, or public natural resource managers should be comfortable using the system.

The NED Concept

The NED vision is that a variety of resource values can be achieved best by first determining the priorities of all management objectives, then resolving trade-offs among them, and finally selecting activities that are compatible with all goals and likely to produce the desired conditions. The intent is to produce a software system that cooperates with the user in the decision-making process without controlling that process.

The most original and complex part of NED is the prescription process, which uses expert systems technology to arrive at a prescription for management of the forest through silviculture. That is, it focuses on the implementation of actions on the ground and presumes that goals have been resolved previously. The process begins with selection of management objectives, or goals, from menus for any or all of five resources: visual quality, wildlife, water, wood production, and general ecological objectives. These goals are defined for a management unit at a scale from one to many stands, generally within the range of 5 to 5,000 ha. Committees of experts in each of the specific resources have identified a range of feasible goals and defined what future conditions are required to meet the specific goals. The desired future conditions (DFC) from all the selected goals are then grouped into compatible sets that can be implemented together. Each set is integrated into a consistent description of stands. These integrated descriptions are used to select a silvicultural system that will sustain the desired conditions over time. The integrated descriptions of target conditions are also used as the basis for an allocation step that identifies which stands are best suited to realize each combination of conditions. The prescription, in the form of a treatment recommendation, can be generated for each stand once the integrated description and the silvicultural system are available. A final integrated evaluation is a key element to the process of determining acceptable prescriptions and evaluating whether different alternative actions across the entire area will meet the desired conditions.

NED's Prescription Process

NED attempts to produce sustainable silvicultural prescriptions. "Sustainable" implies that the overall set of future conditions produced remains essentially stable over time, and this implies that within each attribute set, there will be a balanced age distribution within the stands assigned to that attribute set. Balance will be achieved across stands under even-aged silvicultural systems and within stands

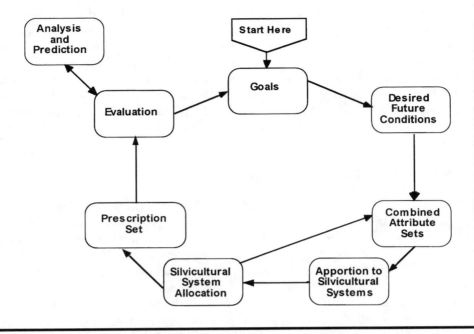

Figure 17.1 A schematic diagram of the NED prescription development process.

under an uneven-aged system. On small management units, this will have implications for the choice of the reentry period. A management unit with fewer stands will have a longer reentry period under even-aged systems.

Goal Selection

The steps of the prescription cycle are Goal Selection, DFC Identification, Attribute Set Integration, Allocation, Prescription, and Evaluation (Figure 17.1). Although direct entry of coded, multiple-resource goals will be possible for sophisticated users, NED will also include a Goals Interview for novice users to enter general goals and to translate them into a specific and compatible goal set. Examples of goals that may be selected from NED goal menus include enhancing fisher habitat, or creating a screening effect.

Desired Future Conditions

The DFC produced by the goal selection process describe the instantaneous conditions that must exist on the management unit to accomplish the desired goals. For the process to be sustainable, all these conditions must exist on the management unit at all times. All DFC are expressed in terms of measurable or derivable variables. This allows a test of whether the conditions exist at time of a field inventory or after a simulated prescription is implemented. Committees of experts in each of the disciplines addressed by NED have produced sets of conditions correlated to each of the goals. For example, the NED Wildlife committee has suggested that mature conifer stands and an abundance of dead and down woody material are elements of desired future conditions for fisher habitat; the Visual Quality committee has identified that 1,000 stems/acre more than 5 ft tall and less than the 3-in. diameter class creates a visual screen.

Attribute Set Integration

This is the most difficult step in the process. In operation it is actually an iterative step that includes cycles of identification of forest type and potential silvicultural systems. Whereas each DFC can be described with some precision, initially there are many differences and many unstated qualifications to each attribute set. The goal of the integration process is to produce mutually exclusive composite sets of compatible conditions such that all of the conditions in each set can be implemented on the same set of stands and be maintained over time by the use of the chosen silvicultural system. These sets, called attribute sets, reduce the number of condition sets which must be managed for to a minimum. The attribute set becomes the unit by which desired conditions are allocated to stands for realization.

The process first identifies which DFC may be unique and completely incompatible with others. For example, open water within a management area may be required for a wildlife habitat goal; however, growing timber or maintaining hiking trails cannot be accomplished in the same space as the open water. The second substep is to identify those conditions that are compatible with all remaining conditions. The best examples here are also wildlife habitat requirements, such as maintaining cavity trees or dead and down material. The next substep identifies size class and forest type combinations that need to be maintained. The final step adds special conditions that are relevant only to particular areas within a management unit, such as cold-water fish habitat or big-tree-appearance goals.

Allocation

The attribute set is the unit by which allocation of desired conditions to stands occurs. Allocation has two components: spatial and temporal. Spatial allocation identifies stands that will realize an attribute set at some period over time. Temporal allocation determines which stands within that group will realize which age classes at different times. If the integration process produces only a single attribute set, then spatial allocation is trivial because all stands will be assigned to that attribute set. However, temporal allocation is still necessary to determine whether and when silvicultural activities are required for each stand. If multiple attribute sets are needed, then stands must be assigned to each attribute set based on their compatibility with maintenance of the desired conditions. Note that stand development may move conditions within one stand from meeting one attribute set to another over time. Note also that current conditions within the management unit and the surrounding area may have a great influence on allocation.

Prescription Set Development

Prescription sets within NED are intended to address one planning cycle, typically a 5- to 10-year period. Within that period, the NED prescription process attempts to identify those DFC that are already met by current conditions, those that exist but must be actively maintained, and those that need active intervention to achieve. Because much time is often required to change conditions in a forest (such as changes from seedlings to mature trees, or creation of large den trees), prescriptions will often achieve only partial success in moving current conditions toward desired conditions in any given cycle. For example, if an entire management area is forested with 80-year-old stands, the creation of balanced age classes necessary for sustainable conditions can only be partially addressed in the first 10 years of planning. Thus, some of the key elements in developing a prescription set are recognition of adjacency constraints, limits of operational capability, and prioritization of stands for treatments.

Evaluation

The evaluation process within NED attempts to allow a user to evaluate alternatives based on current knowledge, without needing to wait decades to make the best choice. Evaluations are accomplished through use of simulations of forest growth and analyses of the management unit with respect to stated goals before and after proposed treatments. Current forest growth simulators included in NED are SILVAH, FIBER, NE-TWIGS, and OAKSIM. Analyses of the various resources addressed by NED produce evaluations of whether the conditions necessary to meet specified goals are met, nearly met, potentially reachable, or unattainable.

Conclusions

Integration of technology transfer tools with other systems that forest managers use (corporate databases and geographic information systems, for example) is crucial to their acceptance. Identification of further research needs is another important component of the development of a good, integrated, technology transfer tool. Through use, NED will clarify which parts of our knowledge base are weakest. The integrating links that identify compatibilities and conflicts among various parts of the forest ecosystem are clearly some of the more fruitful areas for additional knowledge development.

When recommendations are made by NED, the user will have the opportunity to override those suggestions or to customize the results. The system will expect that the user has some knowledge of silvicultural practice and, therefore, implementation of its prescriptions by persons without a resource management background may not be fully satisfactory. To make the system useful to more people, help facilities will include discussions for people with different levels of expertise from novice to expert, and a glossary of all technical terms will be accessible.

The prescription process developed for NED will provide a systematic way to integrate evaluation criteria for multiple resources within a managed forest. This integration should enable treatments to be identified for individual stands based on the conditions in the surrounding area, such that goals identified for the larger context can be met.

Because forest management is a long-term proposition and the future is uncertain and our knowledge is incomplete, we must monitor actions we take, determine whether they are having the desired effect, and adapt future actions to reflect our improved knowledge. Future versions of decision support software, such as NED, must be adapted as knowledge of the system increases and as better tools are developed they should be incorporated.

Acknowledgments

We gratefully acknowledge the considerable contribution of David Marquis, whose vision and foresight began the project and defined its scope. We are also indebted to Pete Kollasch, Deborah Bennett, Mike Rauscher, Linda Thomasma, Scott Thomasma, Laura Alban, and many others for their ideas and hard work on the Northeast Decision Model.

References

Davis, J.R. and J.L. Clark. 1989. A selective bibliography of expert systems in natural resource management. *AI Appl.* 3(3):1–18.

Kollasch, R.P. and M.J. Twery. 1995. Object-oriented system design for natural resource decision support: the Northeast Decision Model. *AI Appl.* 9(3):73–84.

Loh, D.K., M.D. Connor, and P. Janiga. 1991. Jack pine budworm decision support system: a prototype. *AI Appl.* 5(4):29–45.

Manuel, T.M., K.L. Belli, J.D. Hodges, and R.L. Johnson. 1993. A decision-making model to manage or regenerate southern bottomland hardwood stands. *South. J. Appl. For.* 17(2):75–79.

Marquis, D.A. 1991. The Northeast Decision Model: a multi-resource silvicultural decision model for forests of the northeastern United States. Pages 368–365 in Proceedings of the 1991 Symposium on Systems Analysis in Forest Resources. M.A. Buford, compiler. 1991 March 3–6; Charleston, South Carolina. General Technical Report SE-74. U.S. Department of Agriculture, Forest Service, Southeastern Forest Experiment Station, Asheville, North Carolina.

Marquis, D.A. 1992. Accommodating biological and social issues in hardwood management: the Northeast region. Pages 111–125 in Proceedings, 20th Annual Hardwood Symposium. 1992 June 1–3; Cashiers, North Carolina.: Hardwood Research Council, Memphis.

Marquis, D.A. and S.L. Stout. 1992. Multi-resource silvicultural decision model for forests of the northeastern United States. Pages 54–61 in Getting to the Future Through Silviculture — Workshop Proceedings. D. Murphy, compiler. 1991 May 6–9. Cedar City, Utah. General Technical Report INT-291. U.S. Department of Agriculture, Forest Service, Intermountain Forest and Range Experiment Station, Ogden, Utah.

McKinney, D.C., D.R. Maidment, and M. Tanriverdi. 1993. Expert geographic information system for Texas water planning. *J. Water Resour. Plan. Manage.* 119(2):170–183.

Twery, Mark J. 1994. Meeting tomorrow's challenges in silvicultural prescriptions: the northeast decision model. In Proceedings, 1993 SAF National Convention; 1993 November 7–10. Indianapolis, Indiana. Society of American Foresters, Bethesda, Maryland.

Chapter 18

A Multiattribute Tool for Decision Support: Ranking a Finite Number of Alternatives

D.S. YAKOWITZ

Abstract

Many decision-making problems can be expressed as the problem of deciding among a finite number of alternative management systems based on a finite number of criteria or attributes (impacts on the environment) of the alternatives. In this paper, a method is described for ranking the alternatives in order of preference without specifying weights on each criteria. The range from best to worst possible scores, given a priority order (ordinal) of the criteria, are computed by solving two simple linear programs and can be displayed by simple bar graphs. Strong or absolute dominance of one alternative over another can often be discerned immediately. The method can be easily incorporated into a decision support system (DSS) for a wide variety of applications including environmental decision making or as an enhancement to existing systems and methods. Adapting the method to include the consideration of a hierarchy of the criteria and the possibility of different states of nature with some probability are also developed. Use of the model within a sampling scheme to predicted unsampled subdivisions is also suggested. An example application and suggested display of the information are provided.

Introduction

Many decision-making circumstances involve the need to evaluate a finite number of possible choices (alternatives/candidates) based on a finite number of attributes (criteria). For example, a farmer is faced with choosing among several farm management practices which involve different tillage methods or chemical applications. The farmer is concerned with the economic impact of such a decision as well as those factors that affect the quality of both surface and groundwater in the vicinity of the farm. The questions are: (1) How should the feasible practices be evaluated? and (2) Once the practices are evaluated based on the criteria, how does one decide whether one management system is preferable over another if no system dominates all others with respect to all criteria?

Much of the multiple objective (MO) literature considers issues such as the one described above. Often the methods proposed seek to discern the decision maker's preference for a particular alternative in a highly interactive and subjective manner. In many cases, what is desired is a decision that is independent of the personal preferences of a single decision maker (DM). In this paper, a conceptually simple method is presented that requires a minimal amount of input from the DM once the criterion evaluations are commonly scaled. The method can be a valuable tool in the context of a DSS even if other MO methods are also used.

The criteria that the DM considers when evaluating the alternatives are generally noncommensurable, often expressed in quantities of differing units. Typically, the outcome of the evaluation process is expressed in terms of a payoff table or matrix in which an entry, v_{ij}, is the result of evaluating alternative j with respect to criterion i ($i = 1,2,...,m$ and $j = 1,2,...,n$). Based on these outcomes, a method is needed which objectively combines the criteria in order to recommend an alternative.

The most common evaluation scheme is to assign a numerical value (often unitless) for each v_{ij}, and a unitless positive weight, w_i, to each criterion, and then use the weighted average as the total utility, or worth, of alternative j. The alternatives can then be ranked based on the total utility. With this scheme, however, the DM is immediately faced with two tasks: (1) obtaining the values, v_{ij}; and (2) deciding on the weights, w_i.

The first task, the evaluation of each alternative for a single criterion, is often handled by direct examination of the impact of each alternative on the environment, or if this is not possible, by a best estimate of such impact or by using a simulation model to obtain it. The predicted outcomes, whether qualitative or quantitative can then be converted to unitless numbers, say from 0 to 1, by judgment, functional transformation, or other methods based on each single criterion. The lack of an objective, systematic method for performing the second task (multiobjective), i.e., determining the weights, is the prime motivation for this contribution. Numerous techniques for accessing weights have been proposed in the literature. Most solicit weights directly or indirectly (by a series of questions) from the DM. The resultant ranking of the alternatives can be very sensitive to the actual weights used. Specifying weights implies the ordering of the criteria based on importance. A method that does not depend on any particular set of weights was introduced by Yakowitz et al. (1993a) and presented in the context of a DSS by Yakowitz et al. (1993b). This method is proposed as a potentially useful component of DSS for addressing many problems that fall into the category of multiple attribute decision making. Rather than a single vector of weights, this method considers all weight vectors that are consistent with an importance order (ordinal ranking) of the criteria determined by the problem under consideration or imposed by the DM. The DM participates primarily in that part of the decision-making process for which he or she is best suited: the evaluations of v_{ij} based on a single criteria and determining the importance order of the criteria.

The purpose of this paper is to show how the proposed method, which eliminates much of the subjectivity and vagueness typically associated with multiattribute methods, can be incorporated into a DSS that would benefit from this analysis. The method is applicable to many problems including the one described above and others such as site selection, project selection, facility layout, medical diagnosis, and identification. Several adaptations of the method also are given to further illustrate its usefulness.

Method Outline

We begin by assuming that the DM has decided on the set of alternatives and the attributes to evaluate each alternative over. The ranking procedure is as follows.

- **Step 1**: the DM evaluates each alternative (e.g., by direct observation or simulation model) and obtains (assigns) the values v_{ij} for each alternative j and each criteria i.
- **Step 2**: the values from Step 1 are converted to a common unitless scale (0 to 1, for example) if needed.
- **Step 3**: an ordinal ranking (importance order) of the decision criteria/attributes is determined by the DM or by some default method.
- **Step 4**: an aggregation rule, which uses information from two simple linear programs, is used to produce a single total composite score for each alternative under consideration.
- **Step 5**: the alternatives are ranked based on the Total Composite Scores.

There are many methods/rules/procedures for obtaining and scaling the "raw" data. One scaling rule for Step 2 is the use of a set of generic scoring functions proposed in Wymore (1988) and adapted for evaluating hazardous waste containment caps and farm management systems in Lane et al. (1991) and Yakowitz et al. (1992), respectively. These functions are easy to use and are adaptable to many problem situations.

Transforming the range of each decision criterion to a common range of 0 to 1 is often referred to as scoring and accounts for the nonlinearity of the DM's values. For our purposes here we assume that the data have already been so converted (a process that should be a part of the DSS).

The aggregation rule involves computing best and worst composite scores for each alternative from the solutions of two simple linear programs. The final ranking is determined by a combination of these two composite scores.

The Aggregation Tool

The steps in the aggregation rule which lead to a ranking of the alternatives requires only a payoff (or score) matrix and an ordinal importance order of the decision criteria/attributes as input. For ease of exposition, we assume that the values in the score matrix are between 0 and 1 and that 1 is the preferred score. The aggregation rule consists of three parts: (a) determining an importance order (ordinal ranking) of the criteria, (b) determining the best and worst composite scores, and (c) aggregating the best and worst scores.

Part (a) can be determined by a number of ways usually dictated by the decision-making circumstances. For example, an ordering of the criteria/attributes can be given as the result of a group meeting. If other multiple attribute methods are used, such as the Analytic Hierarchy Process (Saaty, 1980) for example, then the weights, so determined, imply an ordinal ranking or importance order and this ordering (not the actual weights) can be used.

Once the importance order of the criteria is determined, best and worst total scores can be found according to Yakowitz et al. (1992, 1993a) without requiring the DM to set specific weights for each of the criteria (see also Salo and Hämäläinen, 1992).

We assume that if $i < j$ then criterion i has a higher priority than criterion j (i.e., criterion 1 has higher priority than criterion 2 and so forth). The priority order suggests that we should require that the weights, $w_i, i = 1, m$, have the following relation: $w_1 \geq w_2 \geq \ldots \geq w_m$. Therefore, given the priority order and the values in the scoring matrix or table and assuming an additive value function, the best (worst) composite score that alternative j can achieve is determined by solving the following linear programs (LPs).

Best (Worst) Composite Score:

$$\max(\min) \sum_{i=1}^{m} w_i v_{ij}$$

$$s.t. \sum_{i=1}^{m} w_i = 1$$

$$w_1 \geq w_2 \geq \ldots \geq w_m \geq 0.$$

The best composite score is found by maximizing the objective function while the worst composite score is found by minimizing the objective function. The first constraint is a normalizing constraint. The second fixes the importance order and restricts the weights to be positive. The above linear programs are solvable in closed form according to Yakowitz et al. (1993) and therefore, an LP solver is not required in a DSS for the above simple formulation.

For each alternative, the solutions to these two linear programs determine the maximum and minimum total scores possible for any combination of actual weights that any one could choose that are consistent with the given importance order of the criteria/attributes. Having these two objective values available immediately alerts the DM to the sensitivity of each alternative to the weights. These values can be displayed graphically (illustrated later in this paper in the context of an example) in the form of side by side bar graphs for each alternative with the best score at the top of each bar and the worst score at the bottom.

Ranking the Alternatives

The remaining task is to decide, based on the best and worst scores, how to rank the alternatives. An alternative that scores high with respect to both best and worst scores indicates that under the given importance order, that alternative is insensitive to any weights consistent with the importance order and scores highly. Additionally, if the worst composite score of one alternative is greater than the best composite score of another alternative, then that alternative strongly dominates (Yakowitz et al., 1993) or absolutely dominates (Salo and Hämäläinen, 1992) the other alternative.

An obvious rule for ranking the alternatives is one which simply uses the average of the best and worst scores for each alternative. Some justification for using the average is discussed in Yakowitz et al. (1993). The alternatives can then be ranked in decreasing order of this average score (highest average score ranks first, etc.).

An Illustrative Example

In this section we illustrate how the method described above could be used in aiding decision making concerned with environmental issues.

The DM, in this example, is concerned with the evaluation of different farming practices on crop land. Several other applications of the method with respect to selecting farm management systems can be found in Yakowitz et al. (1992, 1993b, 1993c). The DM is concerned with criteria that have an impact on surface and ground water quality. These include the amount of soil eroding from the field and the amount of pesticides and nutrients leaching below the root zone or running off the field. The criteria values are obtained from data or predicted values from a simulation model.

Example: Evaluating Alternative Farm Management Options

The conversion of a field in western Iowa to a conservation tillage practice and corn/soybean rotation are considered. The considered management system alternatives are:

- Alt. #1 deep disk tillage, continuous corn (DD,CC)
- Alt. #2 deep disk tillage, corn/soybean (DD,CB)
- Alt. #3 conservation tillage, corn/soybean (CT,CB)

Each of these alternatives were computer simulated and the results yielded the matrix of scores shown in Table 18.1. The criteria are listed in order of importance with nitrogen (N) losses in percolation topping the list, followed by N surface losses, sediment losses, net farm income, surfaces losses of four different pesticides, phosphorus (P) surface losses, and percolation and/or surface losses from two other pesticides. It is important to notice that Alt. #1, the current management system, was used as a baseline and scores of 0.5 were assigned for each criterion by design.

Table 18.1 Criteria and Scores for Three Farm Management Systems of Example 1

Criterion	Alt #1 (DD,CC)	Alt. #2 (DD, CB)	Alt. #3 (CT, CB)
N (p)	0.5	0.74	0.66
N (s)	0.5	0.51	0.98
Sediment	0.5	0.34	0.97
Net income	0.5	0.67	0.82
Alachlor (s)	0.5	0.00	0.22
Terbufos (s)	0.5	1.00	1.00
Atrazine (s)	0.5	0.92	0.81
Metolachlor (s)	0.5	0.94	0.79
P (s)	0.5	0.38	0.93
2,4-D (p)	0.5	0.50	0.50
Paraquat (s)	0.5	0.50	0.17
2,4-D (s)	0.5	0.50	0.50

(s) = sufrace losses, (p) = losses in percolation.

Composite scores are determined for each alternative by solving the best and worst LPs; the results are indicated in Figure 18.1. Notice that Alt. # 3 (CT,CB) strongly dominates the baseline alternative (DD,CC) indicated by the broken line in the figure. The results for Alternative #2 indicate a higher degree sensitivity to weights consistent with the priority order than Alt. #3. Ranking is as follows: Alt. # 3, Alt. #2, Alt. #1.

Using these criteria and ordering, conversion of the field to a corn/soybean rotation and a conservation tillage practice can be recommended. A DSS developed by the Agricultural Research Service and described by Yakowitz et al. (1992, 1993b) can be used to quickly evaluate the above and any additional alternatives or criteria. Other priority orders of the criteria can also be considered. The results of several decision making, or "what if," scenarios can be analyzed and the results displayed quickly and easily within a DSS contributing to an informed decision by the DM.

Decision Making Under Multiple States of Nature

The method described and illustrated above assumes that one can evaluate the payoff or score matrix with certainty (i.e., only a single state of nature). Multiple states of nature may need to be considered in order to make the best informed and minimum risk decision. For example, the DM in the above example may want to consider the alternatives under several different weather scenarios such as drought or flood conditions. In this case, a slight modification to the method above is suggested.

Assume first that there exists a finite and small number of possible scenarios, denoted by S_K, $k = 1,...K$, under which each alternative is to be evaluated and is assumed to occur with some known probability, p_K. The steps of the method above may be performed for all scenarios. Then, if we let T_{Kj} denote the total

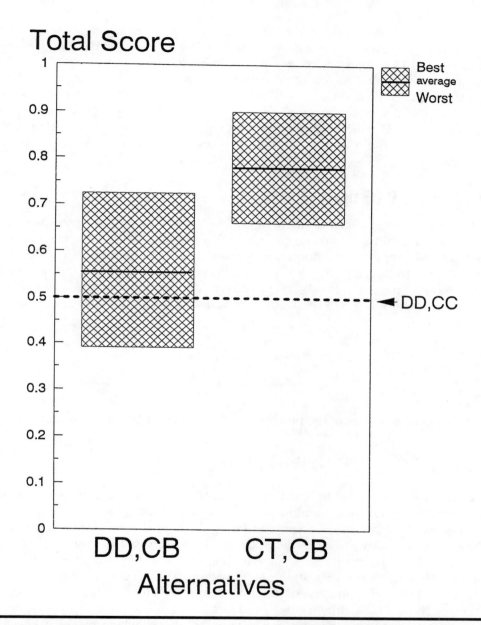

Figure 18.1 Graph of best and worst composite scores for the three farm management systems of the selected example.

score (average of best and worst composite scores) of alternative j under scenario k, alternative j can be ranked based on the following Total Value (TV) index:

$$TV_j = \sum_{k=1}^{K} p_k T_{kj}.$$

The ranking could then be obtained by ordering in decreasing order of **TV**. This ranking favors that alternative that has the highest probable total score with respect to the given importance order of the criteria. If one is not able to specify the probabilities explicitly, but if the scenarios can be ordered in descending order of likelihood of occurrence, then one can solve the best and worst Total Value by first determining T for each scenario and then solving two additional optimization programs obtained by substituting T for v and p (now assumed unknown) for w in the linear programs given earlier. The solution of these new linear programs will yield the best and worst probable total scores. The ranking could then be based on the average of these two resultant scores.

A Hierarchy of the Criteria

Additional constraints can easily be added to the LPs to account for a hierarchy of the criteria and still provide the range from best to worst composite scores. For example, suppose criterion i is composed of t subcriteria that are ordered in importance. Let $v_{i,k,j}$ and $w_{i,k}$, $k = 1,...,t$ indicate the values(scores) for alternative j and subweights (unspecified), respectively, associated with subcriteria k of criteria i. Then, the following two constraints are added to best/worst LPs to account for this hierarchy:

$$w_i = w_{i,1} + w_{i,2} + \ldots + w_{i,t}$$

$$w_{i,1} \geq w_{i,2} \geq \ldots \geq w_{i,t} \geq 0$$

The objective functions of the best/worst LPs for alternative j are then replaced by:

$$\max(\min) w_{i,1} v_{i,1j} + w_{i,2} v_{i,2j} + \ldots + w_{i,t} v_{i,tj}.$$

Again, there is no need to specify weights or subweights. Linear modifications due to other hierarchical considerations are easily made.

The example presented above can be modified to account for a hierarchy of the criteria as follows: Major criteria could be (1) subsurface losses, (2) surface losses, and (3) net income. The subcriteria from Table 18.1 would each fall into one of the above categories and ordered within each category by importance. In this case, since only one criteria fits into Category 3, four additional constraints (two for each of the first two categories) would be added to the best/worst linear programs. A solution procedure for solving the above, which makes it possible to easily examine alternative priority ordering in any level of the hierarchy, without the need of linear programs is under development.

Obtaining Indicator Values for Large Heterogeneous Sites

The MO method presented above can be used in the process of obtaining estimates of environmental indicators from large heterogeneous land areas. In this case it is assumed that a complete survey of the large land areas involved to obtain measurements of each indicator is prohibitive due to cost or time. Estimates of indicators can be obtained by combining a sampling strategy to obtain indicator values for randomly selected subdivisions of the land areas and the MO model presented above to predict the indicator values of unsampled subdivisions.

Simple sampling procedures on subdivisions of the area of concern are possible (Cochran, 1977), but unless the sample size is large, precision is sacrificed. To apply the methodology on a regional scale consisting of heterogeneous land areas, traditional estimates of the indicators from a small sample size (sample mean × number of subdivisions) can be improved if information on the location of the sampled subdivisions with respect to the unsampled subdivisions is incorporated into the calculations. Subdivisions not directly sampled can be evaluated based on any existing information (i.e., satellite images, etc.) as well as information provided by geostatistical analysis of, or other attributes of the nearest sampled segments. Subsequent sampling in the region can be used to update the estimates based on the best currently available information.

For example, estimates of indicators from unsampled subdivisions can be obtained as follows. Indicator values from sampled subdivisions that are closer in proximity to the unsampled subdivision of concern are allowed to have a greater influence on the indicator estimates for that subdivision than those samples that are far away. This is accomplished by treating the sampled subdivision indicator values as the criteria in the model. That is $v_{i,j}$ is interpreted as the value of indicator j obtained for sampled subdivision i. The sampled subdivisions can be ordered in increasing order of distance (say Euclidean) from the unsampled subdivision under consideration. (Note: other information such as similarity of topography might suggest a different ordering of the sampled subdivisions.) Once ordered, the two linear programs are then solved to obtain a range, from highest to lowest (in place of best to worst), for the indicator of concern. To illustrate the method, suppose a region has been divided into the subdivisions suggested in Figure 18.2 and a random sampling strategy is applied. Values of the indicator of concern are obtained by collecting data from the subdivisions indicated by cross-hatches. An estimate of the value of the indicator for subdivision U is desired. In this case, the sampled subdivisions are numbered so that a lower number indicates that the center of that subdivision is closer to the center of subdivision U than higher numbers. The range of values for the indicator for subdivision U is obtained by solving the two linear programs using this ordering of the sampled subdivisions. The average of highest and lowest value can be used as the predicted value for the indicator of concern for subdivision U.

As subsequent sampling and resampling over time are performed, the indicators can be updated. Additionally, the model is readily refineable, so that as additional information based on site specific and probability-based sampling data is made available, the evaluation process can be fine-tuned.

Conclusions

The conceptually simple method suggested here can be easily programmed into a generic DSS and can be a valuable tool in itself or can be incorporated with

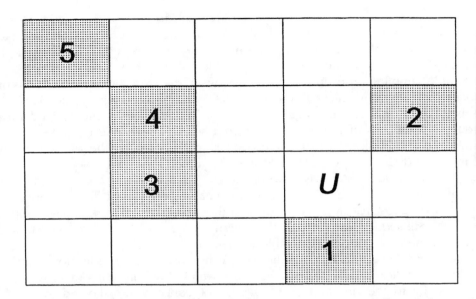

Figure 18.2 Region indicating the sampled subdivisions (dotted) and the unsampled subdivision U. The sampled subdivisions are ordered in increasing distance from U.

other MO methods or existing DSS applications. The method has particular appeal for environmental decision making since determining specific weights (which may satisfy only a few stakeholders) is eliminated. The methodology is easily adapted to decision-making problems under multiple states of nature or to include a hierarchy of the decision criteria. Additionally, other problems such as that illustrated above for obtaining the value of an indicator for unsampled sites, indicate that the model can be adapted and applied to many decision-making problems.

References

Cochran, W.G. 1977. *Sampling Techniques.* John Wiley & Sons, New York.

Lane, L.J., J.C. Ascough, and T.E. Hakonson. 1991. Multiobjective Decision Theory — Decision Support Systems with Embedded Simulation Models, Pages 170–176 in Proceedings of ASCE Irrigation and Drainage Conference, Honolulu, Hawaii.

Saaty, T.L. 1980. *The Analytic Hierarchy Process.* McGraw Hill, New York.

Salo, A.A. and R.P. Hämäläinen. 1992. Preference assessment by imprecise ratio statements. *Oper. Res.* 40(6):1053–1061.

Wymore, A.W. 1988. Structuring system design decisions, Pages 704–709 in C. Weimin, Ed. *Proceedings of International Conference on Systems Science and Engineering.* International Academic Publishers.

Yakowitz, D.S., L.J. Lane, J.J. Stone, P. Heilman, R.K. Reddy, and B. Imam. 1992. Evaluating Land Management Effects on Water Quality Using Multi-Objective Analysis Within a Decision Support System, American Water Resources Association 1st International Conference on Ground Water Ecology, April, Tampa, Fl. pp. 365–373.

Yakowitz, D.S., L.J. Lane, and F. Szidarovszky. 1993a. Multi-attribute decision making: dominance with respect to an importance order of the attributes. *Appl. Math. Computat.* 54:167–181.

Yakowitz, D.S., J.J. Stone, L.J. Lane, P. Heilman, J. Masterson, J. Abolt, and B. Imam. 1993b. A decision support system for evaluating the effects of alternative farm management practices on water quality and economics. *Water Sci. Technol.* 28(3–5):47–54.

Yakowitz, D.S., J. Stone, B. Imam, P. Heilman, L. Kramer, and J. Hatfield. 1993c. Evaluating Land Management Practices with a Decision Support System: An Application to the MSEA Site Near Treynor, Iowa, Soil and Water Conservation Society Conference on Agricultural Research to Protect Water Quality, Minneapolis, MN. Feb. pp. 404–406.

Chapter 19

Effects of Optional Averaging Schemes on the Ranking of Alternatives by the Multiple Objective Component of a U.S. Department of Agriculture Decision Support System

BISHER IMAM, DIANA S. YAKOWITZ, AND LEONARD J. LANE

Abstract

The effects of annual and long-term variations of climatological processes on the decision recommendations of a U.S. Department of Agriculture-Agricultural Research Service decision support system (DSS) and the effect that the point at which aggregation of information occurs in the system subprocesses are investigated. The objective of the system is to rank farm management alternatives in preference order consistent with their environmental and economical impacts. The decision-making process is divided into three subprocesses that include: (1) conversion of the simulation data values for each decision criteria into the unitless [0,1] domain through scoring, (2) computing best and worst composite scores, and (3) ranking the alternative management systems. In this study, four schemes were tested by changing the point at which data is aggregated using a stochastic ensemble of model input. In all cases, the results were insensitive to the point at which the aggregation was performed and thus are supportive of the currently implemented choice to aggregate the simulation output prior to the first decision model subprocess.

Introduction

Practical utilization of state-of-the-art water quality simulation models in identifying and aiding in the solution of nonpoint source pollution problems caused by agriculture was the motivation behind the development of a Water Quality Decision Support System (WQDSS), also known as MODeST (Multiple Objective Decision Support Tool) (Yakowitz et al., 1992; 1993c). The system, developed by the Agricultural Research Service (ARS) a branch of the U.S. Department of Agriculture, demonstrates the concept of coupling simulation models with multiobjective decision methods. The WQDSS integrates a field scale nonpoint source pollution and crop growth simulation models with a novel multiple criteria decision method in order to evaluate farm management systems and aid in the decision-making process. The WQDSS ranks a finite number of farm management systems in order of preference taking into account the effects of crop management, tillage operations, and nutrient and pesticide applications on water quality and economic attributes. These include sediment yield from the field, nutrient and pesticide loading, and net farm return. The current version of the WQDSS utilizes average annual values of the simulation data or historical record for each decision criteria to rank management alternatives. In this paper, we investigate the effects of stochastic decision criteria on the ranking of alternative systems using the WQDSS. The aim of the study is to evaluate the merit of using average annual values and suggest approaches that take into account the uncertainty in natural processes (e.g., precipitation, solar radiation, and temperature) in the decision-making process.

Description of the USDA-ARS WQDSS

The following description of the WQDSS components is rather brief and readers are referred to Yakowitz et al. (1992 and 1993b,c) for details regarding the general structure of the system, and to Yakowitz et al. (1993a) for details regarding the theoretical background of the decision component. Other references are noted where appropriate.

The Simulation Component

The main part of the simulation component of the WQDSS is the Groundwater Loading Effects From Agricultural Management Systems (GLEAMS) (Leonard et al., 1987). GLEAMS simulates daily values of runoff, sediment and water movement in the root zone, and the pesticide distribution in each of these processes. This model was linked to the nutrient submodel from the Chemical, Runoff and Erosion from Agricultural Management Systems (CREAMS) model (Knisel, 1980) and the crop growth component from the Erosion Impact and Productivity Calculator (EPIC) model (Williams et al., 1989). The modified GLEAMS accepts a daily precipitation input file and hydrology, sediment, pesticide, nutrient and crop growth parameter files. These parameter files must be designed to reflect management practices by using the user friendly input file builder in the system. The model's output consists of an annual summary file and a statistical summary file that includes minimum, maximum, and average annual values of several simulated processes.

The Decision Component

Multiple criteria decision making associated with water quality problems involves evaluating a set of management alternatives with respect to a multitude of attributes that describe the natural system's responses. Generally, these attributes must be transformed from their original units into a common unit or unitless range such that an abstract total measure of performance of each alternative can be quantified. Following the suggestion of Lane et al. (1991), a set of 12 scoring functions (value functions) suggested by Wymore (1988) are used in the system to scale the decision criteria from their original units into unitless values in [0,1]. For each criterion or attribute, a function type is chosen and constructed using information from the decision maker or the default settings based on the input data.

Although the WQDSS decision component uses an additive value function (see Goichochea et al., 1982, for example) to aggregate the scores of the decision criteria, alternatives are not ranked based on a single set of weights. In fact, if partial information regarding the importance of each decision criterion is available, alternative ranking is attained by utilizing two simple yet powerful linear programs in order to obtain the best and worst possible composite scores considering all feasible sets of weights (Yakowitz et al., 1993a). Suppose that there are n alternatives which must be ranked in order of preference with respect to a vector of m decision criteria. If qualitative partial information regarding the relative order of each criterion is available, a criterion matrix,

$$\mathbf{X} = \left[x_{ij}\right]_{\substack{i=1,2,\ldots m \\ j=1,2,\ldots n}}$$

may be defined by arranging the criteria in order importance. Hence, $x_{i,j}$ is the value of the *i'th* most important decision criterion with respect to the *j'th* alternative. Similarly, a scoring matrix,

$$\mathbf{V}(\mathbf{X}, \theta) = \left[v_{ij}\right]_{\substack{i=1,2,\ldots m \\ j=1,2,\ldots n}}$$

may be defined as a function of $\mathbf{X}$ as well as of the score function parameter vector θ. The best and worst composite scores of an alternative consistent with the importance order are found according to Yakowitz et al. (1993a) by solving the following linear programs for the weights $w_{i,i\ =\ 1,2,\ldots,m}$.

$$BSC_j(WCS_j) = \max(\min) \sum_{i=1}^{m} w_i v_{ij} \qquad (1)$$

$$st \quad \sum_{i=1}^{m} w_i = 1$$

$$w_1 \geq w_2 \geq \ldots \geq w_m \geq 0,$$

where in Equation 1: BCS_{jj} = best composite score of the *j'th* alternative, WCS_{jj} = worst composite score of the *j'th* alternative, and w = weight vector.

The first constraint of the above program normalizes the weights. The second constraint forces the solution to be consistent with the imposed importance order. Yakowitz et al. (1993a) showed that the extreme points of this program are defined by the following partial sums, s_{kj}, $k = 1,\ldots,m$:

$$S_{kj} = \frac{1}{k} \sum_{i=1}^{k} v_i \qquad (2)$$

The analytic solution of Equation 1 is then given by:

$$\mathbf{BCS}[\mathbf{V}(\mathbf{X},\theta)] = [BCS_j] = \max_{k=1,2,\ldots m}[s_{kj}]_{j=1,2,\ldots,n}, \qquad (3\text{-a})$$

$$\mathbf{WCS}[\mathbf{X},\theta] = [WCS_j] = \min_{k,1,2,\ldots m}[s_{kj}] j = 1,2,\ldots,n \qquad (3\text{-b})$$

Once the best and worst possible composite scores of each alternative are identified, management alternatives can be ranked based on the average value of their best and worst composite scores (Yakowitz et al., 1993a). The average composite score (**ABW**) is,

$$\mathbf{ABW}[\mathbf{V}(\mathbf{X},\theta)] = [ABW_j] = \left[\frac{BCS_j + WCS_j}{2}\right]_{j=1,2,\ldots,n} \qquad (4)$$

The solution of the multiple criteria decision-making problem is a vector of integers $\mathbf{R}$, whose *p'th* element r_p represents the index of the alternative holding the *p'th* rank. Or,

$$\mathbf{R} = \eta[\mathbf{ABW}(\mathbf{V}(\mathbf{X},\theta))] = [r_p]_{\substack{r \in \{1,2,\ldots n\} \\ p=1,2,\ldots,n}} \qquad (5)$$

The function η is a descending ordering function that may be defined as follows:

$$\eta = r_p \in \{1,2,\ldots,n\}; \; \forall \; p = \{1,2,\ldots n-1\} \; ABW_{r_p} \geq ABW_{r_{p+}}. \qquad (6)$$

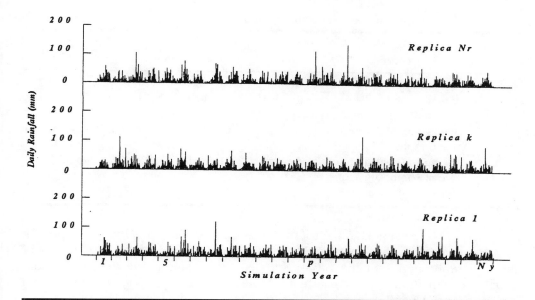

Figure 19.1 Example of a stochastic model input ensemble: daily precipitation. Generated using CLIGEN (Nicks and Lane, 1989) for Oakland, Iowa.

Since the method described above does not require specifying a weight vector, much of the subjectivity associated with multiple criteria decision making is eliminated.

Stochastic Decision Criteria Values

We now consider random decision criteria for two purposes. One purpose is to account for the influence of the stochastic nature of climate processes on the ranking of farm management alternatives. The second is to assess the effect on the decision results of the points at which each of the three decision making subprocesses: (1) conversion from quantitative to qualitative values, (2) aggregation, and (3) ranking, is performed. In this study, conversion is defined as the application of a set of scoring functions to convert a decision criteria matrix X into a scoring matrix $V(X,\theta)$. Aggregation is the process of calculating the best and worst composite scores of a set of alternatives using Equations 2 through 4. Finally, ranking is defined as the ordering of a set of alternatives in preference based on their average best and worst composite scores utilizing Equations 5 and 6.

Consider a population of N_r samples of random sequences of model input (rainfall, temperature, radiation) such that each sequence of the ensemble consists of N_y annual observations of the simulation model input (a sample can be seen in Figure 19.1).

Introducing each input sequence to the simulation model yields a sequence of model output. Hence, the annual decision criteria matrix (model output) arranged in order of importance is written as

$$\mathbf{X} = \left[X_{kp}\right]_{\substack{k=1,2,\ldots,N_y \\ p=1,2,\ldots,N_r}} = \left[x_{ijkp}\right]_{\substack{i=1,2,\ldots m \\ j=1,2,\ldots n \\ k=1,2,N_r \\ p=1,2,\ldots N_y}} , \quad (7)$$

Table 19.1 Summary of the Four Ranking Schemes

Scheme	Description	Conversion	Aggregation	Ranking	Averaging prior to
a	Full ensemble	Annual	Annual	Annual	—
b	Average annual decision criteria for each replica	Replica	Replica	Replica	Conversion
c	Average annual scores of each replica	Annual	Replica	Replica	Aggregation
d	Average annual best and worst composite scores for each replica	Annual	Annual	Replica	Ranking

where $[x_{ijkp}]$ is the simulation output value associated with the $i'th$ most important decision criterion with respect to the $j'th$ alternative occurring in the $p'th$ year of the $k'th$ replication sequence. To avoid lengthy notations we use the middle equality of Equation 7 to indicate the $p'th$ year of the $k'th$ replica of the random decision criteria matrix. Similarly a scoring matrix is written as

$$\mathbf{v}(\mathbf{X}, \theta) = \left[\mathbf{v}(\mathbf{X}_{kp}, \theta)\right] = \left[v_{ijkp}\right]_{\substack{i=1,2,\ldots m \\ j=1,2,\ldots n \\ k=1,2,N_r \\ p=1,2,\ldots N_y}} \quad (8)$$

The ranking vector obtained each year is then,

$$\mathbf{R}_{kp} = \eta\left[\mathbf{ABW}\left(\mathbf{v}(\mathbf{X}_{kp}, \theta)\right)\right] \quad (9)$$

To study the effects of random decision criteria on the WQDSS recommendation, we calculate the frequency of occurrence of each alternative at a given rank obtained by four different schemes. Table 19.1 summarizes the four ranking schemes based on the point of application of each decision subprocess (conversion, aggregation, and ranking) and also indicates explicitly when averaging of the data takes place.

In scheme (a), all replicas are considered as a single sample of $N_y \times N_r$ annual observations (simulations) of the random decision criteria. The alternatives are ranked annually and the frequency matrix **FR** is given by

$$\mathbf{FR} = \left[\mathbf{Fr}_{t,j}\right]_{\substack{t=1,2,\ldots,n \\ j=1,2,\ldots n}} = \left[\frac{N[r_t = j]}{NT}\right]_{\substack{t=1,2,\ldots,n \\ j=1,2,\ldots n}} \quad (10)$$

where N is the number of times at which alternative t occupies the rank j and NT is the total number of ranking vectors in the sample (in this case $NT = N_y \times N_r$). Clearly, in the above scheme, conversion, aggregation, and ranking are all implemented annually.

In the second approach, b), each replica is a realization of the random sequence. Thus, there will be N_r samples whose expected value (sample mean) of the decision criteria matrix is:

$$E[\mathbf{X}]_k = \frac{\sum_{p=1}^{N_y} \mathbf{X}_{kp}}{N_y}, k = 1, 2, \ldots N_r. \quad (11)$$

In this case, the scoring matrix and the average composite score vectors are functions of the sample mean of the decision criteria matrix for each replica of the ensemble.

$$\mathbf{v}(E[\mathbf{X}]_k, \theta) = [\mathbf{v}_{ij}]_k, k = 1, 2, \ldots N_r \quad (12)$$

$$\mathbf{ABW}_k = \mathbf{ABW}(\mathbf{v}(E[\mathbf{X}]_k, \theta)), k = 1, 2, \geq N_r. \quad (13)$$

Similarly, the ranking vector becomes a function of the scores of the replica's average annual criteria matrix.

$$\mathbf{R}_k = \eta\left[\mathbf{ABW}(\mathbf{v}(E[\mathbf{X}]_k, \theta))\right], k = 1, 2, \ldots N_r. \quad (14)$$

Applying Equation 14 to all replicas within the ensemble yields a sample of N_r ranking vectors. The ranking frequency matrix **FR** is given by Equation 10 with $NT = N_r$ (the number of replicas in the ensemble). Unlike the first scheme, this scheme averages the decision criteria matrix with respect to each replica. Hence, conversion, aggregation, and ranking are all performed on the replica level. In approach (c), the decision criteria are scored annually to produce a scoring matrix for each year p within a replica k. The average annual scoring matrix for each replica is then:

$$E[\mathbf{v}(\mathbf{X}_{kp}, \theta)]_k = \frac{\sum_{p=1}^{N_y} \mathbf{v}(\mathbf{X}_{kp}, \theta)}{N_y}, k = 1, 2, \ldots N_r. \quad (15)$$

The average composite score for each replica is determined by solving (3-a and b) using the replica's average annual scoring matrix. The alternatives are ranked according to:

$$\mathbf{R}_k = \eta\left[\mathbf{ABW}\left(E[\mathbf{v}(\mathbf{X}_{kp}, \theta)]_k\right)\right], k = 1, 2, \ldots N_r. \quad (16)$$

Again, a sample of N_r ranking vectors result by applying Equations 15 and 16 to each ensemble replica. Hence, ranking frequency can be obtained using Equation 10 with $NT = N_r$. This scheme differs from the two previous schemes in that conversion is implemented annually, while aggregation and ranking are both performed on the replica scale. The interface between the two temporal levels is provided by the sample expectation operation described in Equation 15.

In the last approach, (d), the decision criteria matrices are scored and the best and the worst composite score vectors are determined annually and the replica sample mean calculated by:

$$E\left[\mathbf{BCS}(\mathbf{v}(\mathbf{X}_{kp}, \theta))\right]_k = \frac{\sum_{p=1}^{N_y} \mathbf{BCS}(\mathbf{v}(\mathbf{X}_{kp}, \theta))}{N_y}, k = 1, 2, \ldots N_r . \quad (17\text{-}a)$$

$$E\left[\mathbf{WCS}(\mathbf{v}(\mathbf{X}_{kp}, \theta))\right]_k = \frac{\sum_{p=1}^{N_y} \mathbf{WCS}(\mathbf{v}(\mathbf{X}_{kp}, \theta))}{N_y}, k = 1, 2, \ldots N_r . \quad (17\text{-}b)$$

The alternatives are then ranked based on the average best/worst composite score:

$$E\left[\mathbf{ABW}(\mathbf{v}(\mathbf{X}_{kp}, \theta))\right]_k = \frac{E\left[\mathbf{BCS}(\mathbf{v}(\mathbf{X}_{kp}, \theta))\right]_k + E\left[\mathbf{WCS}(\mathbf{v}(\mathbf{X}_{kp}, \theta))\right]_k}{2}, \quad (18)$$

$$k = 1, 2, \ldots N_r$$

Thus, the ranking vector is,

$$\mathbf{R}_k = \eta\left[E\left[\mathbf{ABW}(\mathbf{v}(\mathbf{X}_{kp}, \theta))\right]_k\right], k = 1, 2, \ldots N_r \quad (19)$$

Applying Equation 19 to each replica within the ensemble yields a sample of N_r ranking vectors. The ranking frequency is then determined using Equation 10 with NT = N_r. From Equations 18 and 19, it is clear that in this case, conversion and aggregation are both performed on the annual scale, while ranking is performed on the replica scale. Similar to the latter two schemes, this transition between the two temporal scales is provided by the averaging operation described in Equation 17.

Case Study: Treynor, Iowa

Background

The above discussed schemes were applied to a 32-ha watershed monitored by the USDA-ARS Deep Loess Research Station near Treynor, Iowa. Historical records of 24 years including rainfall, runoff, percolation, sediment yield, nutrient and pesticide applications and crop yield are available. The current tillage practice on the watershed is deep disking with continuous corn [DD_CC]. To illustrate the above approach, a set of four alternative management systems are proposed for the field. These systems are (1) deep disking and corn — soybean rotation [DD_CB], (2) chisel plow and corn — soybean rotation [CP_CB], (3) ridge till and corn — soybean rotation [RT_CB], and (4) no till and corn — soybean rotation [NT_CB]. All five practices (including the conventional) must be evaluated with respect to a vector of sixteen decision criteria and two different importance orders. The first importance order considers five nutrient loading decision criteria as the most important. The first seven criteria in order are (1) nitrogen concentration in runoff, (2) nitrogen concentration in sediment, (3) nitrogen concentration in percolation, (4) phosphorus concentration in runoff, (5) phosphorous concentration in sediment, (6) sediment yield, and (7) net farm return. Among eight different pesticides applied in one or more of the five management practices, only four were predicted by the simulation model to show traces significant enough to be considered in the decision-making process. These pesticides were Alachlor,* Atrazine,* Bromoxynil,* and 2,4-D.* The importance order of the decision criteria associated with these pesticides is (8) Alachlor in runoff, (9) Alachlor in sediment, (10) Atrazine in runoff, (11) Atrazine in sediment, (12) Bromoxynil in runoff, (13) Bromoxynil in sediment, (14) 2,4-D in runoff, (15) Atrazine in percolation, and (16) 2,4-D in percolation. The second importance order considered net farm return to be the most important decision criterion and the order of all remaining criteria were shifted accordingly. The two importance orders are summarized in Table 19.2.

To parameterize the sixteen scoring functions historical climatological record was used in the simulation of the conventional management system. The average annual value of each decision criterion was used to determine the baseline parameters for each corresponding scoring function. Farm returns were calculated using a simplified cost benefit equation in which benefits were assumed to result from the sale of crop at the average prices estimated for the period between 1988 and 1990 (USDA, Agricultural Statistics, 1991). Average values of production cost were estimated using the Cost and Return Estimator (CARE) model (Midwest Agricultural Research Associates, 1988). Input to the model consisted of farm management operations including tillage, nutrient and pesticide applications, and labor costs (Heilman et al., 1993). The CARE model input was obtained from the Iowa Soil Conservation Service. Table 19.3 lists the sale prices and the costs associated with a typical crop year given the tillage practice as estimated by CARE.

Historical Weather Record

The above management alternatives were evaluated using the available historical record prior to performing the stochastic experiment. To do so, the 24-year historical climatological record was used in the simulation of the four alternative and the conventional management systems. The five resulting alternative output sequences were considered as a single replica and the four ranking schemes

---

* The USDA neither guarantees nor warrants the standard of the products mentioned above, and the use of the names by the USDA implies no approval of the product to the exclusion of others that may also be suitable.

Table 19.2 Two Importance Orders of the Decision Criteria Considered in the Experiment

	Importance order	
	#1	#2
1	N in runoff	Net returns
2	N in sediment	N in runoff
3	N in percolation	N in sediment
4	P in runoff	N in percolation
5	P in sediment	P in runoff
6	Sediment yield	P in sediment
7	Net returns	Sediment yield
8	Lasso in runoff	Lasso in runoff
9	Lasso in sediment	Lasso in sediment
10	Aatrex in runoff	Aatrex in runoff
11	Aatrex in sediment	Aatrex in sediment
12	Buctril in runoff	Buctril in runoff
13	Buctril in sediment	Buctril in sediment
14	Dacamine in runoff	Dacamine in runoff
15	Aatrex in percolation	Aatrex in percolation
16	Dacamine in percolation	Dacamine in percolation

Table 19.3 Estimated Production Costs and Sales Prices for Treynor, Iowa

Crop	Sales price $/mg	Production costs ($/HA)			
		Deep disk	Chisel plow	Ridge till	No till
Corn	90.75	180.00	178.00	164.00	146.00
Soybean	228.43	375.00	350.00	289.00	281.00

described in Table 19.1 were used to rank the alternatives in order of preference. Since only one replica is available (Nr = 1, Ny = 24), only scheme (a) yielded frequency data given in Figure 19.2. For this case schemes (b), (c), and (d) yielded a single observation of the ranking vector for each scheme and each importance order results are listed in Table 19.4. Notice that in Table 19.4, the conventional practice [DD_CC] has an average best and worst composite score (ABW) of 0.5 for both importance orders using scheme (b). This occurs since the average annual value of each decision criterion is used to define the baseline parameter for each corresponding scoring function. On the other hand, when ranking schemes (c) and (d) are used the conventional practice's composite score varied from 0.5.

When using ranking scheme (a), it is reasonable to recommend as best that alternative that has the maximum frequency of ranking first. From Figure 19.2, alternative [RT_CB] is recommended when nutrient related decision criteria are most important (importance order I) while the alternative [NT_CB] is recommended when the economical criterion supersedes the environmental criteria (importance order II). However, from Table 19.4 we see that the numerical values of the

Effects of Optional Averaging Schemes 227

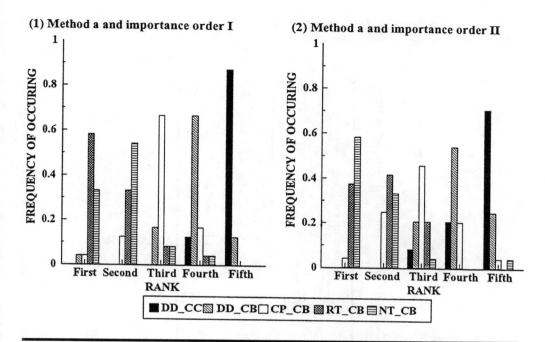

Figure 19.2 Ranking frequency for five management alternatives simulated for 25 years of historical climatologic record. Treynor, Iowa. N_y = 24 years, N_r = 1 replica.

average best/worst composite scores of the first two alternatives [RT_CB] and [NT_CB] were quite close for both importance orders when schemes (b), (c), and (d) were applied indicating the competitive nature of these two management systems.

Stochastically Generated Weather

As mentioned, GLEAMS accepts daily rainfall data, monthly minimum and maximum temperature, and monthly mean radiation. In this part of the experiment we use the weather generator CLIGEN (Nicks and Lane, 1989) to generate 125-year replications of data for the station nearest to Treynor (Oakland, Iowa). Input sequences (125-year) were deemed more desirable than a single 2,500-year sequence for the following reasons. First, independent sequences are needed to understand the behavior of each management alternative under varying condition. Second, the modified GLEAMS simulation model is limited to a 40-year simulation. For each sequence the following steps were applied. First, a randomly generated seed was introduced to CLIGEN with the necessary information. Second, generated daily rainfall data was then read and written to a GLEAMS precipitation file. Temperature and radiation data were averaged for each month during every year and then used with the GLEAMS hydrology parameter files corresponding to the five management systems. Third, the simulation was performed for each alternative system using the stochastic precipitation and hydrology files along with the erosion, nutrient, pesticide, and crop growth files. The annual crop yield was used to calculate the annual farm net return based on the information provided

Table 19.4 Alternative Ranking Based on Three Averaging Schemes of Decision Criteria Simulated using 24 years of Historical Record in Treynor, Iowa

Scheme	Importance order I						Importance order II					
	b)		c)		d)		b)		c)		d)	
#	(1)	(2)	(3)	(4)	(5)	(6)	(7)	(8)	(9)	(10)	(11)	(12)
Rank	Alt	ABW	Alt	ABW	Alt	ABW	Alt	ABW	Alt	ABW	Alt	ABW
1	RT_CB	0.826	RT_CB	0.798	RT_CB	0.785	NT_CB	0.789	NT_CB	0.768	NT_CB	0.742
2	CP_CB	0.817	NT_CB	0.793	NT_CB	0.782	RT_CBB	0.732	RT_CB	0.732	RT_CB	0.718
3	NT_CB	0.716	CP_CB	0.712	CP_CB	0.700	CP_CB	0.687	CP_CB	0.689	CP_CB	0.663
4	DD_CB	0.621	DD_CB	0.653	DD_CB	0.644	DD_CB	0.519	DD_CB	0.615	DD_CB	0.591
5	DD_CC	0.500	DD_CC	0.542	DD_CC	0.527	DD_CC	0.500	DD_CC	0.546	DD_CC	0.538

Ny = 24 year, Nr = 1 replica.

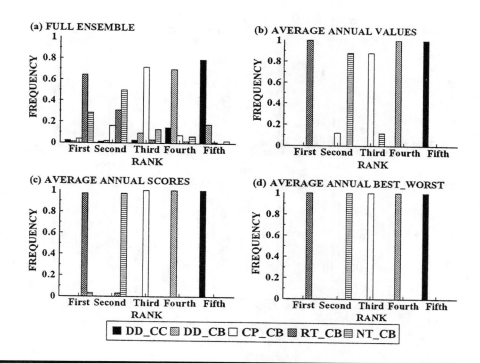

Figure 19.3 Frequency of alternative ranking for four different ranking schemes when nutrient impact on water quality is more important than farm return.

in Table 19.3. Once all simulation runs were completed, the ranking experiments described by Equations 9 thorough 19 were performed.

Equation 10 was used to determine the frequency of occurrence of an alternative at a given rank while Equations 11, 15, and 18 were used to estimate the mean values for the last three experiments with N_y = 25 years per replica and N_r = 100 replicas in the ensemble.

Figure 19.3 illustrates the results of the frequency analysis when the first importance order is imposed. As indicated, alternative [RT_CB] dominates the first position throughout the experiment. However, there were instances in which other alternatives were ranked first, especially when the whole ensemble was treated as a single sample (Figure 19.3-a). The maximum frequency of occurrence with respect to this case would yield the following order: [RT_CB], [NT_CB], [CP_CB], [DD_CB], and finally [DD_CC]. The remaining schemes are consistent with this order. The ranking indicated by Figure 19.3-d provides the most evident ordering since only a single alternative occupies a given rank.

Similar behavior was also observed with respect to the second imposed importance order (Figure 19.4). (Recall that, in this case, net returns has the highest priority.). All schemes produced the following ranking: [NT_CB], [RT_CB], [CP_CB], [DD_CB], and [DD_CC]. However, there were instances in which alternative [RT_CB] ranked first for schemes (a) (b) and (c) indicating the competitive nature of the first two alternatives. Again, scheme (d) provided the most decisive ranking.

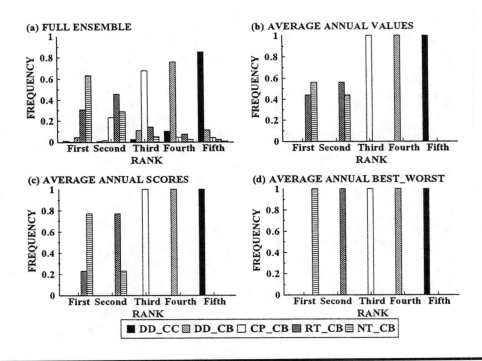

Figure 19.4 Frequency of alternative ranking for four different ranking schemes when farm return is considered the most important decision criterion.

Summary and Conclusion

Four different schemes for ranking a discrete set of management systems were used in conjunction with the USDA-ARS WQDSS. The four schemes differ among each others with respect to the temporal level at which the quantitative decision criteria matrix is converted into a qualitative scoring matrix and aggregated using the best and worst possible composite scores for each alternative. Two temporal levels were considered, the annual level and the replica level.

The four schemes were tested for two different importance orders reflecting two possible resource management strategies to be used to evaluate five different farm management systems. Then, 100 sequences of stochastically generated climatological data were used to provide 100 replicas of simulation output for the same five management systems. In this case, the final ranking vector corresponding to each ranking scheme was determined by observing which alternative had the maximum frequency of occurrence at a given rank.

With respect to the first importance order, all schemes produced the same ranking of the alternatives for both the historical record and the stochastically generated record. Similarly, the ranking produced by applying the above four schemes to the second importance order produced identical results for both the historical and the stochastic experiments. The consistency of the schemes suggests that the decision model is insensitive to the point at which the expectation (averaging) operation is performed.

Similarities between the results of each scheme when applied to the historical record and when applied to the stochastic record indicate that the random generator, CLIGEN, successfully reproduced the general behavior of the climatological record for the area under study. Furthermore, similarities between the ranking vectors obtained from all ranking schemes with respect to each importance order suggests that the currently used approach in the WQDSS may be adequate (ranking is currently based on the average annual values of the decision criteria, the least computationally cumbersome approach since averaging is done prior to conversion). Nevertheless, the information generated by each of the four schemes can be of value to a decision maker. For example, scheme (a) when used for a reasonably long historical record, allows the decision maker to evaluate the impact of the annual variation of the climatological processes on the decision-making process. These variations may include extreme events for which management alternatives respond differently. Schemes (b) and (c), on the other hand, exert a smoothing effect on the average best/worst composite scores of competing alternatives, hence, providing a more discerning evaluation of competitive alternatives than scheme (a). Finally, scheme (d) appears to produce the most conclusive ranking.

The above discussion exemplifies the benefits of considering more than one ranking strategy when the decision-making problem includes closely competing alternatives. Decision makers can gain different perspectives by using all of the above schemes.

If the historical record is reasonably long, as in this example, applying the above four schemes to the historical record alone may exhibit sufficient information to assess the effects of the annual variation in the climatological processes on the decision recommendations, hence avoiding the burden of a stochastic experiment.

References

Goicoechea, A., D. Hansen, and L. Duckstein. 1982. Multiobjective Decision Analysis with Engineering and Business Applications. John Wiley & Sons, New York.
Heilman, P., D.S. Yakowitz, J.J. Stone, L.A. Kramer, L.J. Lane, and B. Imam. 1993. An Exploration of the Economics of Farm Management Alternatives to Improve Water Quality, Application of Advance Information Technologies: Effective Management of Natural Resources. Proc. IETS/ASAE Spokane, Washington. June 18–19. pp. 193–205.
Knisel, W.G., Ed. 1980. A Field-Scale Model for Chemicals, Runoff, and Erosion from Agricultural Management Systems, Conserv. Res. Report No. 26. U.S Department of Agriculture, Science and Education Administration.
Lane, L.J., J.C. Ascough, and T.E. Hakonson. 1991. Multiobjective Decision Theory Decision Support Systems with Embedded Simulation Models, Irrigation and Drainage Pro. IR Div/ASCE conference in Honolulu, Hawaii. July 22–26. pp. 445–451.
Leonard, R.A., W.G. Knisel, and D.A. Still. 1987. GLEAMS: Ground water loading effects of agricultural management systems, Trans. ASAE. 30(5):1403–1418
Midwest Agricultural Research Associates. 1988. User Manual — Cost and Return Estimator. USDA-ARS Soil Conservation Service. Contract no. 54-6526-7-268.
Nicks, A.D. and L.J. Lane. 1989. Weather generator. Pages 2.1–2.19 in L.J. Lane and M.A. Nearing, Eds. USDA-Water Erosion Prediction Project: Hillslope Profile Model Documentation. BSERL Report No. 2. USDA-ARS National Soil Erosion Research Lab., W. Lafayette, Indiana.

Williams, I.R., C.A. Jones, J.R. Kiniry, and D.A. Spanel. 1989. The EPIC crop growth model. Trans. ASAE. 32(2):497–511.

Wymore, A.W. 1988. Structuring system design decisions. Pages 704–709 in C. Weimen, Ed. Proceedings of the International Conference on Systems Science and Engineering, Beijing, China July 25–28. International Academic Publishers,

Yakowitz, D.S., L.J. Lane, J.J. Stone, P. Heilman, R.K. Reddy, and B. Imam. 1992. Evaluation of Land Management Effects on Water Quality Using Multiobjective Analysis within a Decision Support System, Proceedings of the AWRA First International Conference on Ground Water Ecology. Miami, Florida. April. pp. 265–374.

Yakowitz, D.S, L.J. Lane, and F. Szidarovszky. 1993a. Multi-attribute decision making: dominance with respect to an importance order of the attributes. *Appl. Math. Computat.* 54:167–181.

Yakowitz, D.S, J.J. Stone, B. Imam, P. Heilman, L.A. Kramer, and J.L. Hatfield. 1993b. Evaluating Land Management Practices with a Decision Support System: an Application to the MSEA site near Treynor, Iowa. Agricultural Research to Protect Water Quality, Minneapolis, Minnesota. February. pp. 404–406.

Yakowitz, D.S, J.J. Stone, L.J. Lane, P. Heilman, J. Masterson, J. Abolt, and B. Imam. 1993c. A decision support system for evaluating the effects of alternative farm management systems on water quality and economics. Water Sci. Technol. 28(3–5):47–54.

Chapter 20

Integrating Agricultural Expert Systems with Databases and Multimedia

AHMED RAFEA

Abstract

This paper presents the needs for integrating agricultural expert systems with databases and multimedia, and gives examples of typical expert systems already developed and deployed that include a database and multimedia components. The needs for integration with database are the storage of large number of static data that should be entered when consulting an expert system, and the unavailability of some of these data to users. The main problem identified as a result of this integration is the maintenance of both the knowledge base and the database assuming that the expert system tools support calling a database retrieval program. The needs for integration with multimedia are enhancing the symptoms acquisition, disorder verification, and explaining agriculture operation. The main problem identified to accomplish this integration is proper identification of the media to be integrated assuming that large storage area and input/output devices are available.

Introduction

This paper uses an expert system of crop management as an example of agricultural expert systems. Crop management can be defined as the set of agricultural operations done to produce a crop. Advice about these operations are given by specialists in soil and water management, plant pathology, entomology, production, breeding, and horticulture. Analysis of these operations has revealed that an expert system for crop management is a family of expert systems that works together to generate a schedule for agricultural operations. It is a complex scheduling problem. In order to break down this complex problem into a less complex problem, the functionality of these operations was analyzed and consequently classified into these categories: irrigation, fertilization, plant caring, and disorder remediation. The analysis also revealed that knowledge behind advising about these operations is also dependent on a certain crop though the same categories stand for any crop. As these expert systems use some common data and knowledge about the domain, a common knowledge base was built to be shared by all these expert systems. Database and multimedia are identified as essential components for the success of those expert systems.

The paper consists of three sections in addition to the introduction and conclusion. The first section reviews expert systems developed in Agriculture. The second and third sections discuss the needs for integrating databases and multimedia with expert systems.

Expert Systems in Agriculture

Knowledge-based expert system technology has been applied to a variety of agricultural problems. EPINFORM (Caristi et al., 1987) is an expert system developed with the INFER inference engine and the EXPLAIN program for predicting the effects of stripe rust and *Septoria nodorum* blotch on wheat yields relatively early in the growing season. PMDS (Rauscher and Hacker, 1989) is an expert system to facilitate insect pest population control. An expert system for wheat using Generic Task methodology has also been developed in collaboration between the Central Laboratory for Agricultural Expert Systems, Egypt, and Michigan State University (Kamel et al., 1994).

COMAX (Lemon, 1986) the Crop Management Expert, developed by the U.S. Department of Agriculture (Mississippi State University and Clemson University), is an expert system which can predict crop growth and yield in response to external weather variables, soil physical parameters, soil fertility, and pest damage. Another system has been developed for cotton (Plant et al., 1988). Palmer (1986) and other researchers at Purdue University have developed PLANTER, which suggest whether and when to plant or replant corn or soybeans, and the GRAIN MARKETING ADVISOR, which utilizes information and knowledge about prices, drying and storage facilities, and transportation to make marketing decisions. A set of expert systems for cucumber production management have been developed (Rafea et al., 1991; Rafea et al., 1992; El-Dessouki et al., 1993).

The United Nations University (UNU) Agroforestry Expert System (AES) was designed to support land-use officials, research scientists, farmers, and other individuals interested in maximizing benefits gained from applying agroforestry management techniques in developing countries (Warkentin et al. 1990). It provides recommendations for tree species and interspace parameters for tree-crop symbiotic plantings using alley cropping. DBL-CROP (Halterman et al., 1988) was developed for the double crop winter wheat and soybean domain. It provides recommendations for fertilization, herbicide selection, variety selection, seeding rate, and residue management.

CHAMBER (Jones and Haldeman, 1986) was designed for the diagnosis of failures in controlled environment plant research chambers. NERISK (Rauscher and Hacker, 1989) assists users in assessing the impact of pesticides on beneficial arthropod predators and parasitoids in agricultural systems.

Expert systems have also been applied to the problems of diagnosing soybean diseases (Michalski et al., 1983), pest and drought control for apple orchard management (Buchanan, 1986), and pine seedling management (Rauscher and Hacker, 1989). Expert system technology can also be applied to tactical management of agricultural systems and for hybrid corn drying control (Peart et al., 1986).

Integration with Database

This section presents the need for integrating expert systems with databases, and gives an example of this integration using an irrigation expert system.

Need for Databases

Database technology is one of the fields that extensively makes use of computers. It is the heart of any information system to be developed within any organization. In agriculture, information systems have been used, as in any other discipline, in management, research, finance, and other areas. At some sites where expert systems may be needed, information system may already be there. Therefore, it may be necessary to integrate the expert system with the working environment, especially if the data needed by the expert system are part of an existing database. If there is no previous information system within any location, we found that expert systems need databases to store static data of a specific plantation in order to be used by the inference engine of the expert system, otherwise the system has to ask the user each time they run the system to enter these data. For sure, this is inconvenient to the user who is, in our system, the extension worker giving advice to several growers who may have more than one plantation. Although one may claim that this problem can be solved by simply keeping these data on a file and retrieve them when needed, this claim is untrue because managing the set of files for all plantations, memorizing the name of each plantation file, maintaining the integrity of the knowledge base with the database, and other reasons require the usage of a knowledge base and database management system.

Database Integration Example

We consider irrigation expert system within a larger framework which is the crop management. Three expert systems were developed for irrigation of orange, cucumber, and lime. The method used for the first two systems (orange and cucumber) was based on model approach implemented using expert systems techniques. The method used for the third system (lime) was based completely on heuristic approach. In this section, the expert system which was based on model approach will be described as it is the approach which needs more data than the heuristic approach.

Table 20.1 Example of an Irrigation Schedule

Month	Quantity (m³/feddan/irr)	Frequency
January	0	0
February	360	1
March	290	1
April	280	2
May	310	2
June	350	2

"Irrigation system" can be defined in terms of inputs and outputs. The inputs are the properties of soil, water, and climate of a certain plantation in addition to other factors related to the plantation such as the irrigation method, number of trees, drainage status, and others. The outputs are essentially the water quantity and the application frequency. The inputs can be classified into two broad categories: static data and dynamic data. The static data include data which do not change in short period while the dynamic data include the changeable data in the field. The static data consist of farm data, soil data, data of water used for irrigation, and climate data. Farm data include the area, number of trees, distance between trees and between rows, used irrigation system, source of irrigation water, and drainage system. The soil data include soil texture, field capacity, and permanent wilting point. The water data include electrical conductivity of irrigation water. The climate data include monthly average of daily temperature, daily relative humidity, daily duration of maximum sunshine hours, actual sunshine hours, and extra terrestrial radiation. On the other hand, dynamic data include status of weeds in the farm.

The irrigation schedule is produced using an irrigation model (Hargreaves, 1983). This model is based on calculating the potential evapotranspiration (Etp) and then the model considers the soil and calculates soil moisture deficit for each crop, taking into account the individual crop requirement. A typical case of using this model to the citrus expert system (Salah et al., 1993) is as follows:

Site:	*Toukh*	*Grower:*	*Khaled*
Plantation date:	01/01/1978	Area:	20 feddan
Distance between rows:	5 m	Distance between trees:	5 m
Irrigation system:	Flooding	Water source:	River

It should be noticed that these data are not enough to compute the irrigation schedule as the data for soil, water, and climate are not known by the grower. The system, in this case, assumes default values stored in the system database to respond to this consultation. The generated schedule for the first 6 months of the year is shown in Table 20.1.

The integration of this expert system with the database has necessitated the development of a knowledge and data management system in order to maintain both of them. For example, if we have an attribute like "texture" in an object "soil," where this attribute gets its value from the database when consultation begins, deleting this attribute from the database should be prohibited in order to keep the integrity of the system.

Integration with Multimedia

This section presents the need for integrating expert systems with multimedia and gives an example of this integration using a disorder remediation expert system.

Need for Multimedia

The need for integrating expert systems with multimedia will be done through discussing where each multimedia type could be used to enhance the utilization and performance of the expert system. Providing explanations during consultation and/or after reaching a conclusion can also be enhanced using all types of multimedia.

Images

It was found that describing symptoms in words is very difficult and sometimes is very confusing. Therefore, images are identified to be used for two main purposes: describing a disorder symptom and confirming the diagnosis of the cause of a certain disorder. Detailed images for all symptoms, and unique images that confirm the occurrence of disorders at different stages, should be collected.

Although images are very useful in acquiring the user inputs, the uncertainty problem is still there. Therefore, giving the user the option to select an image with a degree of certainty should be provided. Providing more than one picture for the same symptom can reduce the user uncertainty, but this will lead to exerting more efforts in collecting and classifying the images.

Video

As already explained, the output of an expert system for crop production management is a set of agricultural operations. Describing how to perform an agricultural operation in words is very hard and one can never guarantee that the user can understand what has been written. Displaying a video for a professional doing the recommended operation would be very informative.

Sound

Sound is essential because sometimes it is not easy to write terminologies used by growers in daily life. In addition, combining the video with sound is also recommended to comment on how the operation is done.

Multimedia Integration Example

The main function of the disorder remediation expert system is to generate a prescription to remediate a certain disorder or a set of disorders. If the user suspects the cause of disorder(s), they can provide the system with their suspicion, and the system confirms or rejects this suspicion. If the user has no suspicion, they can provide the system with the symptoms of the disorders, and the system identifies the cause(s) of the disorder(s). Figure 20.1 depicts a typical output screen

Treatment Operation
Detailed Operation Information

Date :   28/6/94
Disorder :   white_fly
Material Name :   actellic 50%
Mode of entry :   contact
Quantity :   300   ml/100 L
Method :   foliar application
Tool :   sprayer
Application   avoid high temperature
 Time   during spraying
Advice :   install nets before transplanting
                    spray only when number of
                    insects reaches 2-3 per leaf,
                    make deep harvest before
                    spraying

Figure 20.1   Typical output screen of the disorder remediation expert system.

of disorder remediation expert system for cucumber production under a plastic tunnel (El-Dessouki et al.,1993).

So far, we have only integrated images with the expert system. The integration process has passed into several steps: the identification of images to be included, collection and scanning of images, and modifying the knowledge base to integrate the images.

Images Identification

The identification of images was done by studying the relation between the knowledge representation and its presentation. For example, the value of the attribute "leaf spot color" of the object "leaf spot" has a set of images for different colors. For each color, there may be more than one image pending on other attributes of the spot such as its shape, its position, etc. Another example is the images of a diagnosed disorder which may differ according to the severity of this disorder. Therefore, thorough examination of typical observation has been conducted to identify proper images.

Collection and Scanning of Images

Four sources are recognized to get the identified images: the slides used by domain experts in their presentation, the extension documents, books, and pictures taken from the field.

A combined slide and flat-bed color scanner with a resolution up to 1200 dpi was used for scanning pictures and slides. However, scanning with this high resolution needs a lot of disk storage. We have found, practically, 300 dpi is sufficient to produce a good image. A typical size of the images used in the system ranges from 60 to 369 kB. This difference is due to the size of the image to be displayed. In order to solve the problem of the disk storage space, we decided to distribute the expert systems with images on CD-ROMs in the future.

Knowledge Base Modification

The observation class was modified to include links to images, and additional rules were added in order to enable the image display method to select the appropriate image out of a set of related images and then present it to the user.

For example, when the system is to ask about the leaf spot color, and the user wants to retrieve the image related to a white spot, there should be a method to select the appropriate image among the set of images linked to this attribute value. If the shape of the spot is irregular, and we have four images of white spot, then the system should select the irregular white spot image.

Conclusion

This paper has revealed that integration of expert system with databases is needed to store static data of a certain plantation, and to have default values of different locations where the system is to be installed. A more general solution of accessing default values is to integrate the expert system with a geographic information system (GIS). The main technical problem that can be raised due to integration with databases is the maintenance of both the knowledge base and the database assuming that the expert system tools supports calling a database retrieval program. The unavailability of a such retrieval program is a major problem which should be taken care of from the very beginning of an expert system project. The maintenance problem could be solved, either manually in case that the developer uses A ready-made package, by taking care of the database, when making any modification to the knowledge base, or by building a knowledge base and a database management system in case that the developer uses a tool built in house.

The needs for integration with multimedia are the enhancement of the symptoms acquisition, disorder verification, and the explanation of agriculture operations. The main problem identified to accomplish this integration is the proper identification of the images, video tapes to be integrated, and the knowledge modification to link the different attributes, and values with proper media, assuming that large storage area, and input/output devices are available.

References

Buchanan, G.G. 1986. Expert systems: working systems and the research literature. *Expert Syst.* 3:32–51.

Caristi, J., A.L. Scharen, E.L. Sharp, and D.C. Sands. 1987. Development and preliminary testing of EPINFORM, an expert system for predicting wheat disease epidemics. *Plant Dis.* December:1147–1150.

El-Dessouki, A., S. Edrees, and S. El-Azhari. 1993. CUPTEX: an integrated expert system for crop management. In Proceedings of the Second Expert Systems and Development Workshop ESADW-93. Ministry of Agriculture, Cairo, Egypt.

Halterman, S.T., J.R. Barrett, and M.L. Swearingin. 1988. Double cropping expert system. *Trans. ASAE.* January-February:234–239.

Hargreaves, G.H. 1983. *Practical Agroclimate Information System.* Westview Press, Colorado.

Jones, P. and J. Haldeman. 1986. Management of a crop research facility with a microcomputer-based expert system. *Trans. ASAE.* January-February:235–242.

Kamel, A., K. Schreder, J. Sticklen, A. Rafea, A. Salah, U. Schulthess, R. Ward, and J. Ritchie. 1994. An integrated wheat crop management system based on generic task knowledge based systems and CERES numerical simulation. *AI Appl.* 8(3).

Lemon, H. 1986. Comax: an expert system for cotton crop management. *Science.* 233:29–33.

Michalski, R.S., J.H. Davis, V.S. Bisht, and J.B. Sinclair. 1983. A computer-based advisory system for diagnosing soybean diseases in Illinois. *Plant Dis.* 67:459–463.

Palmer, R.G. 1986. How expert systems can improve crop production. *Agric. Eng.* September/October:28–29.

Peart, R.M., F.S. Zazueta, P. Jones, J.W. Jones, and J.W. Mishoe. 1986. Expert systems take on three tough agricultural tasks. *Agric. Eng.* May/June:8–10.

Plant, R., L. Zelinski, P. Goodell, T. Kerbey, L. Wilson, and F. Zalom. 1988. CALEX: an integrated expert decision support system for farm management. Pages 196–201 in F. Zazueta and A. Bottcher, Eds. Proceedings of the International Conference on Computers in Extension. Volume 2. Gainesville, Florida.

Rafea, A., M. Warkentin, and S. Ruth. 1991. An expert system for cucumber production in plastic tunnel. In Proceedings of the World Congress on Expert Systems. Orlando, Florida.

Rafea, A., A. El-Dessoki, S. Nada, and M. Youssef. 1992. An expert system for cucumber production management under plastic tunnel. In Proceedings of the First International Conference on Expert Systems and Development (ICESD-92). Ministry of Agriculture, Cairo, Egypt.

Rauscher, H.M. and R. Hacker. 1989. Overview of artificial intelligence applications in natural resource management. *J. Knowl. Eng.* 2(3):30–42.

Salah, A., H. Hassan, K. Tawfik, M. Mahmoud, and I. Ibrahim. 1993. CITEX: an expert system for citrus crop management. Second National Expert Systems and Development Workshop (ESADW93). Ministry of Agriculture, Cairo, Egypt.

Warkentin, M., P.K. Nair, S. Ruth, and K. Sprague. 1990. A knowledge based system for planning and design of agroforestry systems. *Agrofor. Syst.* 11:71–83

Chapter 21

WATERSHEDSS©: A Decision Support System for Nonpoint Source Pollution Control in Predominantly Agricultural Watersheds*

D.L. OSMOND, D.E. LINE, J.A. GALE, R.W. GANNON, C.B. KNOTT, K.A. PHILLIPS, M.H. TURNER, S.W. COFFEY, J. SPOONER, P.D. ROBILLARD, M.A. FOSTER, AND D.W. LEHNING

In order to adequately control nonpoint source (NPS) pollution, it is necessary to accurately determine the water quality problem. Once the water quality problem has been defined and the pollutant types and sources identified, a system of NPS controls can then be selected and installed within the watershed. Selection and positioning of the most appropriate best management practices (BMPs) and landscape features are important in order to minimize or mitigate the identified pollutant(s). WATERSHEDSS (*WATER*, *So*il, and *H*ydro-*E*nvironmental *D*ecision *S*upport *S*ystem) has been designed to aid managers of predominantly agricultural watersheds in determining their water quality problems and selecting appropriate land treatment practices. WATERSHEDSS is mounted on the World Wide Web

---

* Developed under the U.S. Environmental Protection Agency Grant, Cooperative Agreement #CR822270, Understanding the Role of Agricultural Landscape Feature Function and Position in Achieving Environmental Endpoints.

(Web) at http://h2osparc.wq.ncsu.edu. The two primary objectives of the decision support system (DSS) are (1) to transfer water quality and land treatment information to watershed managers in order to assist them in making appropriate land management/land treatment decisions and (2) to assess and evaluate NPS pollution at the watershed scale, based on user-supplied information.

Introduction

Most of the contamination of surface waters in the U.S. is due to NPS of pollution. The U.S. Environmental Protection Agency (USEPA) estimates that approximately 60% of the total NPS pollution load on assessed surface waters is due to agricultural runoff (USEPA, 1990). The primary agricultural pollutants are sediment, nutrients, pathogens, and pesticides (USGAO, 1990).

When protecting or restoring a surface water resource, the area of concern is defined by the watershed that drains into the receiving surface water. The water quality within the watershed is affected by climate, topography, hydrology, soils, agricultural practices, and other human activities. Thus, to protect a water resource or remediate a water quality problem, the appropriate scale of management is the whole watershed. In an attempt to better understand agricultural NPS pollution control at the watershed scale, the U.S. Congress authorized the 15-year, voluntary, $64 million experimental Rural Clean Water Program (RCWP) (Federal Register, 1980).

The RCWP was unique in that federal and state agency collaboration was mandated for this program. The program was administered by the U.S. Department of Agriculture (USDA) with concurrence by the USEPA. Participating federal agencies included the USDA Consolidated Farm Services Agency (formerly the Agricultural Stabilization and Conservation Service), the USDA Extension Service, the USDA Natural Resource Conservation Service (formerly the Soil Conservation Service), the USDA Economic Research Service, the USDA Agricultural Research Service, and the U.S. Geological Survey.

Throughout the U.S., 21 watershed-scale projects were selected to participate in the RCWP (Gale et al., 1993). These projects represented a range of agroecological conditions and water quality problems. The projects ran for 10 to 15 years (1980 to 1995) and included a multiyear, preimplementation planning and water quality monitoring phase, followed by BMP implementation with continued water quality monitoring. Up to 75% federal cost-share was provided to farmers for recommended land treatment structures and management practices, with a maximum rate of $50,000 per participant.

In the final RCWP evaluation, investigators discovered common lessons learned about managing agricultural watersheds, despite the diversity of RCWP projects (Gale et al., 1993; Coffey et al., 1992). Those watershed projects managed successfully followed similar technical and organizational processes. The technical processes, which will be referred to throughout the paper as watershed assessment and evaluation, included identification of (1) the water resource of concern, (2) the water quality impairment or threat of impairment, (3) the pollutant and the pollutant source, (4) the critical area within the watershed, and (5) the appropriate systems of BMPs.

One way to transfer the knowledge gained from the RCWP and other NPS pollution control projects is to incorporate it into a DSS. In order to aid watershed managers, the North Carolina State University (NCSU) Water Quality Group has developed the prototype DSS (WATERSHEDSS) to transfer lessons learned from the RCWP experience and other agricultural watershed projects to watershed managers, land treatment specialists, and others working in the field of water quality and agricultural watershed management.

WATERSHEDSS serves varied technical audiences by (1) educating interested individuals about water quality and land treatment information pertinent to watershed management and (2) providing decision support technology on problem definition, pollutant transport modeling, and land treatment planning.

WATERSHEDSS is an expert system-like interface that links predetermined linear paths to a solution end-point; an extensive education component accessible throughout the interface by hypertext links; forms linked to databases; and a link into a modeling environment. The structure of the DSS allows users to access narrative descriptions, water quality criteria and standards, and models. WATERSHEDSS is mounted on the Web and can be accessed at http://h20sparc.wq.ncsu.edu or http://www.bae.ncsu.edu/bae/programs/extension/wqg.

The two primary objectives of WATERSHEDSS, which are complementary and simultaneous because of the computing environment, are (1) to transfer water quality and land treatment information to watershed managers in order to assist them in making appropriate land management/land treatment decisions and (2) to assess and evaluate NPS pollution in a watershed based on user-supplied information and decisions.

The Structure of Watershedss

WATERSHEDSS is comprised of the following six components (Figure 21.1):

1. A hypertext expert systems-like user interface that serves as the watershed assessment and evaluation framework for the DSS
2. An education component
3. An annotated bibliography of NPS literature
4. An agricultural BMP database
5. A modeling tool
6. A pollutant budget spreadsheet

The hypertext expert systems-like user interface, the education component, the annotated NPS bibliography, and the modeling tool can be accessed from the home page of WATERSHEDSS. The education component, the modeling tool, the annotated bibliography can also be accessed from a list of options included on almost every page of WATERSHEDSS. The agricultural BMP database and the pollutant budget spreadsheet can only be accessed by traversing a series of links within the DSS.

Component 1 — The Expert System-Like User Interface

The interface serves as the framework of the DSS, is authored in the HTML (HyperText Markup Language), and is browsable on the Web. This interface serves

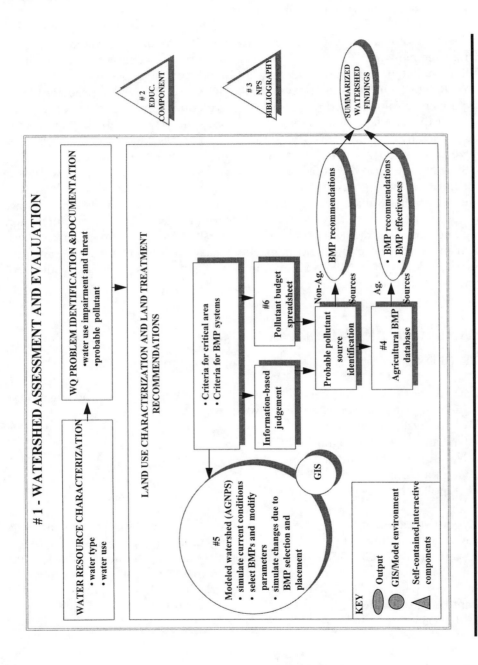

Figure 21.1 Diagram of WATERSHEDSS.

as the watershed assessment and evaluation portion of the DSS. The interface guides the user through a series of questions that are based on the watershed assessment and evaluation process developed from findings of the RCWP. Response choice and the resulting linear path links take the user through the watershed assessment and evaluation process in a step-wise fashion. This process starts with the water resource identification and finishes with the BMP selection. Three of the other five components of the DSS (agricultural BMP database, WATERSHEDSS modeling tool, and pollutant budget spreadsheet) are also part of the watershed assessment and evaluation. The remaining two components, education and annotated NPS bibliography (although not part of the watershed assessment and evaluation), are accessed throughout the user interface by hypertext links and can be utilized by the user to help make decisions.

The purpose of the watershed assessment and evaluation process within the DSS is to guide users in water quality problem identification and eventually in selection, and in some cases positioning, of appropriate BMPs in order to minimize or mitigate pollutant delivery to the water resource. Watersheds can be assessed and evaluated at different levels of detail within the DSS dependent upon the water quality problem, the amount of data available, and the user's ability to utilize the modeling tool.

However, all users are asked the same series of generalized questions that reflect the steps necessary for successful watershed management. Questions address water resource characterization, problem documentation, and land use characterization. Based on their answers to the questions, users are able to obtain a summarized set of water quality and watershed findings that include: water resource type and use, water quality impairment, probable pollutants, the sources of the pollutant and recommended BMPs. If the pollutant is derived primarily from agricultural sources, then the recommended BMPs are accessed from the agricultural BMP database. All other recommended BMPs are based on the best-available information about each particular pollutant and source.

While the users ultimately provide their own answers to all of the questions in Component 1, the value of this watershed assessment and evaluation tool lies in the problem-solving process, which has evolved from analysis of the most extensive watershed experience to date in the U.S., the RCWP. This step-wise guidance will prompt users, when necessary, to seek additional information, either from the education and annotated NPS bibliography components of WATERSHEDSS, or from water quality sampling data and/or land use information for their own watersheds.

Component 2 — The Education Component

This component includes information on 24 broad water quality categories, ranging from water use characterization to fish habitat assessment. There are also five land treatment information sections available in the education component: ten fact sheets on NPS control topics, NPS pollution control programs, descriptions of BMPs that control different NPS pollution problems, extensive information on wetlands (both natural and constructed), and two case studies designed to lead the user through specific examples of successful agricultural NPS pollution control projects managed on a watershed scale. Concepts discussed within the context of watershed project case studies include water quality problem identification, establishment of objectives and goals, critical area determination and selection, and water quality monitoring and evaluation.

Component 3 — A Link from the User Interface of WATERSHEDSS to the NCSU Water Quality Group's Annotated Bibliography of NPS Literature

This link can be used to assess more than 6,000 annotated articles on NPS water pollution and its control. This part of the DSS allows easy access to, and search capabilities of, this literature resource.

Component 4 — An Agricultural BMP Database

The database was started by the NCSU Water Quality Group and further developed at The Pennsylvania State (PSU) (Foster et al., 1995). The database consists of most USDA Natural Resource Conservation Service (NRCS)-approved BMPs for controlling sediment, phosphorus, nitrogen, fecal coliform, and oxygen demanding compounds.

Information in the BMP database is accessed from a form that lists pollutant sources for particular pollutants. The output from the database search is a list of recommended BMPs for a pollutant from a specific source. Efficiency ranges for the BMPs in the database, which remediate a particular pollutant, are included for some, but not all, of the BMPs. These ranges are not absolute and BMP efficiency varies with location. References from which the database has been developed and NRCS BMP codes and descriptions are available. The database has been further expanded by the NCSU Water Quality Group to include natural and constructed wetlands and riparian areas as land treatment resources that can mitigate agricultural NPS pollution.

Component 5 — A Modeling Tool

This component consists of a GIS linked to a spatially distributed pollutant runoff model — Agricultural NonPoint Source, AGNPS (Young et al., 1987), and is available to for viewing an example digitized watershed located in Gaston County, North Carolina. This example enables users to view different best management practice decisions.

The public domain GIS software, Geographical Resource Analysis Support System — GRASS (USACERL, 1993), was selected for use in WATERSHEDSS. The GRASS/AGNPS link was developed by Srinivasan and Engel (1994), but has been modified to include point source, pesticide, and channel information. The new WATERSHEDSS modeling tool is available for users to FTP to their sites. Maps generated by the modeling tool will allow users to make a detailed comparison of the effects on water quality of various land uses, landscape feature placement, and land management scenarios in their watershed. The WATERSHEDSS modeling tool is available throughout the DSS.

Component 6 — A Pollutant Budget Spreadsheet

The spreadsheet is available to aid in pollutant source determination if the pollutant is either nitrogen or phosphorus. The spreadsheet, only accessible from the Agricultural Pollutant Source page, is a simple form onto which the user enters land use data, expressed as number of acres. The pollutant budget can only be accessed after traversing the watershed assessment and evaluation links in the DSS.

An Example of the Use of Watershedss

From the home page of WATERSHEDSS, a user clicks on WATERSHEDSS to start the assessment and evaluation process. At the next page, the user selects water use under the appropriate water resource type. In this example, the user selects *River/Stream* as the water resource and *Aquatic Life Support* as the impaired or threatened water use.

At this juncture, the user will be asked to either further identify the water use or select the pollutant of concern. For this example, the user must identify the water use by choosing between *Salmon, Trout,* or *Bass*. Selection of *Trout* moves the user to a question about the water quality problem (*ammonia, dissolved oxygen, metals, organics, pesticides, sediment, temperature,* or *turbidity*). Depending on which water quality problem the user selects, he or she will be led to one of several paths that almost always ends at the *Pollutant Source* page. For this example, if the user selects turbidity as the problem, caused by algal respiration that in turn is due to high total phosphorus concentrations (>0.05 mg/l), then the user will end up on the *Phosphorus: Pollutant Sources* page.

The *Pollutant Source* page is designed to assist the user in pollutant source identification, and to illustrate fundamental concepts in watershed management and in the identification of potentially appropriate BMPs. From the *Pollutant Source* page users can access information on important watershed management concepts: *identification of pollutant sources, identification of critical areas,* and *systems of BMPs*.

Best management practices will be recommended based on the pollutant source selected by the user. If the user is unsure of the pollutant source, and if the pollutant is phosphorus or nitrogen, the user will be directed to utilize the pollutant budget spreadsheet option. In this example, the user would enter the acreage of the different types of land uses into the pollutant budget spreadsheet: 1,000 acres corn, 1,000 acres soybeans, 1,000 acres pasture, 100 acres residential, and 50 acres business. The results, under percentage of phosphorus load, indicate that the land use that has the potential to contribute the largest amount of phosphorus to the water resource is from the cropland. Using the results from the pollutant budget spreadsheet, the user would return to the *Phosphorus Agricultural Pollutant Source* page and click on *cropland, dry* as the pollutant source.

In the example, all of the pollutant listed from the *Phosphorus: Pollutant Source* page, except agricultural pollutant sources, are HTML links to descriptions of the recommended BMPs for that particular pollutant source. For example, if *Boats/Marinas* is chosen as the pollutant source, the user makes that selection and the recommended BMPs that control phosphorus (*boat operation; liquid waste and fuel disposal; sewage disposal, and; solid waste generation and disposal*) are listed and described.

The agricultural pollutant sources are coded as a form that is linked to the PSU agricultural BMP database. After the user selects a pollutant source (e.g., *cropland, dry*) and clicks on *Submit* button, recommended BMPs are listed. The information included as part of the recommended BMPs are BMP efficiency data for some of the practices, descriptions on each recommended BMP, NRCS BMP code numbers, the references from which the information was collected, and a link to the NRCS *National Handbook of Conservation Practices* web site.

Results of the watershed assessment and evaluation are presented in a watershed summary table that precedes the recommended BMPs. If the user has digitized information and chooses to use the WATERSHEDSS modeling tool, then the GIS-generated maps will suffice as the watershed summary. The watershed summary table, used in conjunction with the recommended BMPs and the information about critical areas and systems of best management practices, will allow users to make decisions about BMP selection and placement.

Conclusion

WATERSHEDSS is a prototype watershed management DSS that enables users to assess and evaluate water quality impairment or threats of impairment. The DSS is also designed to provide both water quality and land treatment information to users. WATERSHEDSS accomplishes the assessment and evaluation and information transfer through its six interacting components: (1) a hypertext expert systems-like user interface that aids in watershed assessment and serves as an evaluation framework for the DSS, (2) an education component, (3) annotated bibliography of NPS literature, (4) an agricultural BMP database, (5) a linked GIS/water quality model, and (6) a pollutant budget spreadsheet. The intended audience for the DSS includes watershed managers and land treatment personnel who need both information and assessment tools to assist them in the decision making necessary to effectively protect water quality.

REFERENCES

Coffey, S.W., J. Spooner, D.E. Line, J.A. Gale, J.A. Arnold, D.L. Osmond, and F.J. Humenik. 1992. Elements of a Model Program for Nonpoint Source Pollution Control, In 10 Years of Controlling Agricultural Nonpoint Source Pollution: The RCWP Experience, The National Rural Clean Water Program Symposium, U.S. Environmental Protection Agency, Washington, D.C. EPA/625/R-92/006.

Federal Register. 1980. Rural Clean Water Program (RCWP): Proposed rule, Fed. Reg. 40:14006.

Foster, M.A., P.D. Robillard, D.W. Lehning, and R. Zhao. 1996. STEWARD, a knowledge based system for selection, assessment, and design of water quality control practices in agricultural watersheds, Water Resour. Bull. in review.

Gale, J.A., D.E. Line, D.L. Osmond, S.W. Coffey, J. Spooner, J.A. Arnold, T.J. Hoban, and R.C. Wimberly. 1993. Evaluation of the Experimental Rural Clean Water Program, National Water Quality Evaluation Project, NCSU Water Quality Group, Biological and Agricultural Engineering Department, North Carolina State University, Raleigh. EPA-841-R-93-005.

Line, D.L., S.W. Coffey, and D.L. Osmond. 1996. WATERSHEDSS GRASS-AGNPS Modeling Tool. Trans. ASAE. in review.

Srinivasan, R. and B.A. Engel. 1994. A spatial decision support system for assessing agricultural nonpoint source pollution. Water Resour. Bull. 30(3):441-462.

USACERL. 1993. GRASS 4.1 User's Reference Manual. U.S. Army Corps of Engineers, Construction Engineering Research Laboratories, Champaign, Illinois.

USEPA. 1990. National Water Quality Inventory: 1988 Report to Congress. U.S. Environmental Protection Agency, Washington, D.C. EPA 440-4-90-003.

USGAO. 1990. Water Pollution: Greater EPA Leadership Needed to Reduce Nonpoint Source Pollution. U.S. General Accounting Office, Gaithersburg, Maryland. GAO RCED-91-10.

Young, R.A., C.A. Onstad, D.D. Bosch, and W.P. Anderson. 1987. AGNPS, Agricultural Non-Point-Source Pollution Model. A Watershed Analysis Tool. USDA, Conservation Research Report.

Chapter 22

A Knowledge-Based Reasoning Toolkit for Forest Resource Management

STEPHEN B. WILLIAMS AND DAVID R. HOLTFRERICH

Abstract

Forest resource managers on two U.S. Forest Service (USFS) Ranger Districts are using a decision support system (DSS) known as INFORMS-R8 to support common district planning activities. One component of this system, the knowledge base component, is being used to address a broad range of issues, ranging from selection of public firewood sites to assessment of wildlife habitat. Until recently, the expert-defined rulebases that are the basis of this knowledge base component were converted into C Language Integration Production System (CLIPS) code by highly skilled programmers. The CLIPS-encoded rulebases were then integrated into INFORMS-R8 for use. In order to enable distribution of this technology across many Ranger Districts, a knowledge-based reasoning toolkit, or "rulebase toolkit," is being built to promote self-sufficiency by Ranger District staff in building and maintaining CLIPS-encoded rulebases. The history of rulebase use on this project and preliminary feedback from users of the initial beta version of the toolkit highlights the value of rulebase technology in natural resource management and the need for this knowledge base reasoning toolkit.

Introduction

In the U.S., public scrutiny of public land management decisions has intensified in recent years. Recent public debate concerning the protection of the endangered northern spotted owl and protection of old growth forests in the Northwestern U.S. is a clear example of this scrutiny. As the land manager of 191 million acres of some of the nations most scenic and ecologically valuable lands, the USFS is in the middle of many of these public land management controversies. Beyond these newsworthy land management controversies, USFS resource managers deal with hundreds of lesser issues (Perisho et al., 1995) annually, which all have potential for controversy at either the local, regional, or national level. More than ever, the USFS is expected to manage National Forest lands in a consistent, scientifically reasonable, and ecologically sustainable manner.

To foster consistent, scientifically based decision making that utilizes the wealth of forestry knowledge available, the USFS has directly developed or sponsored development of dozens of computer software decision-support tools in the last decade (Schuster et al., 1993). Many of these tools are oriented toward simulation modeling, and many are used to address specific issues (e.g., wildlife, fisheries, or timber growth). One tool, INFORMS-R8 (Integrated Forest Resource Managment System — Region 8), is somewhat unique in that it is a DSS that offers access to multiple models and a knowledge base component to support a great variety of resource planning tasks (Loh and Saarenmaa, 1992; Holtfrerich et al., 1992) (see Figure 22.1). This integrated software package was developed by Systems Technology Applications in Renewable Resources Laboratory (STARR Lab) at Texas A&M University in cooperation with Forest Health Protection in the Southern Region (Region 8) and Forest Health Protection's Methods Application Group of the USFS (Oliveria and Swain, 1991; Williams, 1992).

A central component of INFORMS-R8, and the component which lends itself particularly well toward addressing a variety of emerging and fast changing resource management issues in a consistent manner, is the knowledge base component. This component offers resource managers a mechanism to store forestry knowledge in the form of rulebases. Resource managers can then analyze forest land data using these rulebases. Outputs generated by these natural resource-related rulebases help define practical land management activities that will promote sustainable forest ecosystems.

As resource managers from two Ranger Districts in the Southern Region of the USFS have embraced and subsequently expanded the use of the knowledge base component of INFORMS-R8, Forest Health Protection staff in the Southern Region (i.e., the project sponsors) and STARR Lab recognized the need to facilitate more rapid and efficient development of individual rulebases. A new cooperative effort was established in 1994 between STARR Lab and the USFS to build a knowledge base reasoning toolkit, or "rulebase toolkit." This toolkit application will eliminate the need for the highly skilled application programmers now required to transfer expert-defined rules into logic that can be processed by the knowledge base engine used within INFORMS-R8. The Rulebase Toolkit will empower resource managers to more easily capture natural resource knowledge into rulebases to support planning tasks and then utilize those rulebases via INFORMS-R8. The following sections describe both the needs driving the development of the Rulebase Toolkit and the toolkit's functionality.

Defining a project area through the "clipping" function.

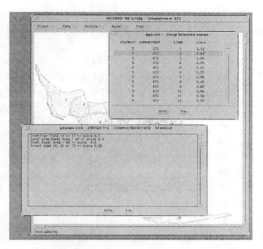

Analyzing conditions with rulebases.

Building a project alternative.

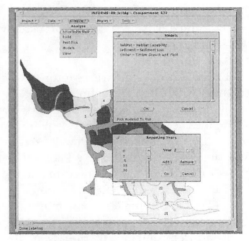

Analyzing long-term management effects via models.

Figure 22.1   Sample screens demonstrating major functions of INFORMS-R8.

Project Background

The planning tools of INFORMS-R8 were first available to resource managers in 1990, when the software was piloted on the Neches Ranger District of the Davy Crockett National Forest in east Texas (Williams, 1992). The original focus of the knowledge base component of INFORMS-R8 was on defining expert-based rules, commonly called rulebases, to rate the practicality of applying various silvicultural treatments (i.e., vegetation management activities) to forest stands in the District. USFS subject "experts" skilled in planning, silviculture, wildlife, soil science, and general forestry met over the course of three 1- to 2-day meetings in early 1990 to determine what rulebases would be built and to define the actual rules for each rulebase. Twelve rulebase subjects representing commonly utilized silvicultural treatments of the time were selected for testing rulebase technology:

- Prescribed Burning Suitability
- Overstory Removal Suitability
- Commercial Thinning Suitability
- Precommercial Thinning Suitability
- Release Cutting Suitability
- Regeneration Harvest Suitability
- Clearcut Harvest Suitability (regeneration harvesting method)
- Seedtree Harvest Suitability (regeneration harvesting method)
- Shelterwood Harvest Suitability (regeneration harvesting method)
- Uneven-Aged Management Suitability (regeneration harvesting method)
- Group Selection Harvest Suitability (uneven-aged stand management method)
- Single Tree Selection Harvest Suitability (uneven-aged stand management method)

Expert knowledge was captured to determine the need for and applicability of these silvicultural treatments as they related to forest stand conditions. The resulting set of rulebases reflected input from USFS planning standards and guidelines, court orders, a Vegetation Management Environmental Impact Statement, Best Management Practices for the state of Texas, the Red Cockaded Woodpecker Comprehensive Plan, other USFS manuals and handbooks, and many years of professional resource management experience.

Subject rulebases such as the commercial thinning rulebase are built around the EMYCIN formula (Buchanan and Duda, 1983) and the concepts of facts, rules, and numerical ratings (see Figure 22.2). A rulebase is composed of one or more 'facts' — determinants of the issue. For the commercial thinning rulebase, five facts were identified as being important determinants as to whether or not commercial thinning would be an applicable silvicultural treatment for a particular forest stand.

One of the five facts is the hazard rating value for southern pine beetle (SPB) damage. One or more 'rules' are defined for each fact. These rules are used to assign a numerical rating (i.e., quantitative value) based on the rule logic. For the SPB hazard rating fact, one rule assigns a numerical rating of +0.75 if the value of the SPB hazard rating is "high;" another rule assigns a numerical rating of +0.25 if this rating is "medium;" and so on for a "low" rating. These numerical ratings largely represent qualitative statements. The following scale shows the framework used to convert qualitative statements into numerical ratings:

Qualitative scale	Numerical conversion
Absolutely favorable	+1.00
Strongly favorable	+0.75
Moderately favorable	+0.50

Fact 1: Basal area
    Basal area exceeds the thinning guide value in the Land Management Plan?
        Rule 1: Yes = +.75
        Rule 2: No = 0

Fact 2: Sawtimber operability
    Total removal > 1000 board feet?
        Rule 1: Yes = 0
        Rule 2: No = -.75

Fact 3: Poletimber operability
    Operability > 2.5 cubic units?
        Rule 1: Yes = 0
        Rule 2: No = -.75

Fact 4: Crown closure
    Crowns generally touching or interwoven?
        Rule 1: Yes = 0
        Rule 2: No = -.5

Fact 5: Southern pine beetle rating
        Rule 1: High = +.75
        Rule 2: Medium = +.25
        Rule 3: Low = 0

Figure 22.2 Commercial thinning rulebase facts, rules, and numerical ratings.

Slightly favorable	+0.25
Inconclusive	0
Slightly unfavorable	−0.25
Moderately unfavorable	−0.50
Strongly unfavorable	−0.75
Absolutely unfavorable	−1.00

A rule score (i.e., numerical rating) for a given fact that approaches +1.0 suggests that commercial thinning is an increasingly favorable silvicultural treatment for a given forest stand. Scores approaching −1.0 suggest that commercial thinning is an increasingly unfavorable treatment option, and scores near zero have little or no effect on the favorablility of the treatment. The numerical ratings assigned by the appropriate rule for each fact are combined across all relevant facts of the rulebase using the EMYCIN formula to derive an overall score (see Figure 22.3). For commercial thinning, the five fact-related rule scores are combined to derive the overall score. (Note that each fact score, as well as the overall score, always falls somewhere between −1.0 and +1.0.)

Using the notes from the three "expert" sessions, knowledge engineers at Texas A&M University developed the knowledge bases using a modified version of the National Aeronautics and Space Administration's (NASA) CLIPS expert system building tool. CLIPS was modified by Texas A&M to allow storage and retrieval of data from the Oracle relational database management system (RDBMS). Oracle is the RDBMS used by the USFS agency-wide and is used to store USFS resource data.

EMYCIN Formula

A rulebase is composed of various FACTS. Each FACT may have one or more RULES. Each RULE represents a specific condition, and each condition is assigned a numeric score. The following is the EMYCIN algorithm used to derive a Numeric Rating (NR) for the rulebase.

if $0 \leq A \leq 1$ and $0 \leq B \leq 1$ *then* $NR = A + (B \cdot (1 - A))$
if $-1 \leq A < 0$ and $-1 \leq B < 0$ *then* $NR = A + (B \cdot (1 + A))$
if $(A \cdot B) < 0$ *then* $NR = (A + B) / (1 - \text{lesser of the absolute values for A and B})$

where:

"A" is the initial starting value (0) or the value from the previous processing step, and
"B" is the value from the rule being currently considered.

Example: Commercial Thinning

Rules fired	Values processed
Fact 1: Basal area exceeds guide value = +.75	$0 + (+.75 \cdot (1 - 0)) = +.75$
Fact 2: Saw timber removal > 1000 bf = 0	$+.75 + (0 \cdot (1 - .75)) = +.75$
Fact 3: Pole timber > 2.5 cubic units = 0	$+.75 + (0 \cdot (1 - .75)) = +.75$
Fact 4: Crown closure complete = 0	$+.75 + (0 \cdot (1 - .75)) = +.75$
Fact 5: Southern pine beetle rating—medium = +.25	$+.75 + (+.25 \cdot (1 - .75)) = +.8125$

Overall rulebase score (Numerical Rating) = +.8125

Figure 22.3 Cumulative score derived through the EMYCIN formula.

Texas A&M staff completed development of 12 CLIPS encoded rulebases over the course of several weeks. Texas A&M staff also integrated these finished products within INFORMS-R8 so that each rulebase could then be utilized within the INFORMS-R8 framework to support forest planning activities (see the upper right-hand screen in Figure 22.1). Final adjustment of these rulebases occurred over several months, as USFS staff utilized the rulebases in planning. These adjustments reflected either minor problems found during use or new knowledge to address unusual site conditions. Initially, any adjustments to the rulebase program logic were accomplished by Texas A&M staff. Over time, simple rule score changes were handled by USFS resource specialists, but changes to rulebase logic and development of new rulebases still required the technical expertise of computer programming specialists.

The 12 silvicultural rulebases developed by Texas A&M staff improved the Neches Ranger District's ability to perform consistent, thorough, and scientifically based project planning. Previous planning procedures were based on a resource specialist visiting the project site to collect data, and subsequently "walking" the site to assess site conditions and formulate options on how to treat forest stands within the project area. Upon returning to the office, the resource specialist would document current conditions and list treatment alternatives while considering various documents (such as the Forest Plan), their own professional knowledge, and, if necessary, advice from other available specialists.

The quality of this analysis depended to a large extent on the experience of the individual, the availability and skill of other specialists, and the individual's ability to account for guidelines and procedures scattered throughout many different documents. The rulebases improve this process, as all relevant knowledge concerning application of these silvicultural treatments is embedded in the rulebases. Anyone who runs the rulebases via INFORMS-R8 will get the same results. In addition, for any project or forest stand that is evaluated, the same analysis criteria are used every time by the rulebases.

Over 260 domain-specific rules are embedded in these 12 rulebases. About half of the data needed to drive these rulebases is located in the Region-supported forest stand inventory database. The other needed data describing stand conditions is located in a separate Oracle table. This other data represents information that had been used in decision making but was not previously kept in electronic form. For these initial rulebases, all needed data represented attribute data for forest stands. None of the data required spatially oriented information, such as distance to a road, etc.

The knowledge base component of INFORMS-R8 has been used successfully for several years now on the Neches Ranger District. Sound forestry knowledge tailored to the local conditions of this east Texas site have yielded a set of rulebases that foster consistent and reasonable management decisions on District lands. However, the difficulty in applying this technology on all Forest Service lands is in the number of knowledge bases needed.

In a more recent implementation of INFORMS-R8, on the Jessieville Ranger District of the Ouachita National Forest in Arkansas, an entirely new set of silvicultural rulebases were defined. In addition, the technology was further extended to a variety of other nonsilvicultural analysis tasks, such as wildlife habitat improvement opportunities and location of historic resource sites (Williams et al., 1995). This District's use of rulebases required enhancements to the CLIPS component of INFORMS-R8 in order accommodate the use of spatially oriented rules. A list of the current rulebases in use on the Jessieville Ranger District is enlightening:

Nonspatially oriented rulebases	Description
Early seral habitat	Rates forest stands for providing early seral habitat
Mature growth	Rates forest stands for providing mature habitat conditions
Prescribed burning	Rates forest stands for appropriateness of prescribed burning
Cerulean Warbler	Rates forest stands as to quality of foraging and nesting habitat
Even aged management	Rates forest stands for appropriateness of performing seedtree or shelterwood regeneration harvesting
Overstory mast development	Rates forest stands for appropriateness of overstory mast development as a vegetation treatment
Pine bluestem	Rates forest stands for suitability for conversion to a pine/bluestem ecosystem
Commercial thinning	Rates forest stands for appropriateness of commercial thinning

Nonspatially oriented rulebases	Description (Continued)
Timber stand improvement	Rates forest stands for need of release, weeding, or precommercial thinning
Single tree selection	Rates forest stands for appropriateness of performing a single-tree selection harvest cut (uneven-aged stand management)
Group selection	Rates forest stands for appropriateness of performing a group selection harvest cut (uneven-aged stand management)
Mast production potential	Rates forest stands as to potential for hard mast production
Spatial rulebases	Description
Erosion potential	Determines the overall potential for soils to erode
Suitability for firewood	Determines if areas are suitable for public firewood production
Historic site survey intensity	Through favorability rating, determines recommended survey intensities to locate historic heritage resources
Prehistoric site survey intensity	Same as above but for prehistoric resources
Potential road damage	Determines potential of road segments to damage the ecosystem, includes road density
Trail/road vista	Rates road or trail segments for suitability for vistas
Management indicator species	Rates areas as to suitability for providing scarlet tanager forage areas

Note that the Jessieville Ranger District (RD) has developed two types of rulebases: spatial and nonspatial. Spatial rulebases are composed of facts, wherein at least one of the facts represents a spatial attribute; an example would be the distance to a perennial stream fact. The value of this fact is derived through a spatial operation spawned by the executed rulebase, with the results fed back to the rulebase in order to attach a numerical rating (see Figure 22.4). Nonspatial rules are like those used on the Neches RD, wherein fact values are all attributes of a specific spatial feature such as a forest stand polygon.

The Jessieville RD staff have expressed a potential need for additional rulebases as resource issues emerge and evolve. In addition, both the Jessieville and Neches RD staffs have already been faced with adjusting existing rulebases as legal and agency guidelines change, and as scientific knowledge is refined. Considering that there are hundreds of USFS Ranger District offices nationwide managing a diversity of landscapes and grappling with hundreds of issues, the methods used to date to define, build, and maintain knowledge bases would be extremely costly if implemented across the Forest Service

Requirements for A Rulebase Toolkit

Development of a Rulebase Toolkit has been scoped to address many of the bottlenecks that could restrict the applicability of rulebase technology within the

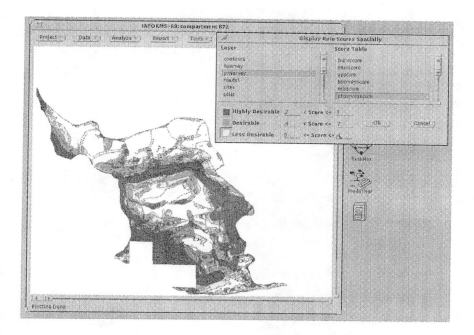

Figure 22.4 INFORMS-R8 screen displaying spatial rulebase results.

USFS: specifically, that could restrict the full utilization of the knowledge base component of INFORMS-R8. The overall objective is to create a user-friendly application through which field-level forest resource managers can construct and implement natural resource-related, CLIPS-encoded rulebases. The requirements specifications driving development of the toolkit are based on lessons learned over 5 years on INFORMS-R8 user sites, supplemented by several meetings in which ideas were captured from involved USFS staff as well as other USFS research scientists. Some of the major requirements for the toolkit are:

- Usuable by staff with no knowledge of CLIPS.
- CLIPS-encoded rulebases automatically generated.
- Capability to automatically add completed CLIPS-encoded rulebases to INFORMS-R8 for immediate use.
- When defining facts and rules, directly utilizes USFS resource data elements stored in ORACLE (the official corporate USFS Relational Database Management System).
- When defining facts and rules, users can select data elements from multiple ORACLE tables.
- When defining facts and rules, users can utilize spatially derived data.
- When defining facts and rules, users can utilize the results of other rulebases.
- Integrity of rulebases must be protected through security mechanisms.
- Must track rulebase history (author, cited references, change dates, etc.).
- Must include mechanisms to perform rulebase cross-checking or "truthing" (needed to protect credibility and validity of rulebases).

- Need mechanism to help organize or catagorize rulebases (i.e., recreation, wildlife, etc.).
- Mechanism to update and delete existing rulebases.
- Rulebase testing tools.
- Help facilities and supporting documentation.
- Operable using USFS corporate hardware and software.

Toolkit Design

In light of the preceding requirements, there are three technical areas of concern in designing the Rulebase Toolkit:

1. The type of CLIPS rules that must be built
2. The metadata tables (Holtfrerich et al., 1993) needed to support the building of these rules
3. The user interface and interaction needed to obtain information for building rules

Each of these is discussed in detail, below.

Rule Types

INFORMS-R8 users have built rules within rulebases requiring distinct operations in terms of database manipulations, spatial operations, and overall CLIPS coding. These rules, each taking into account the three processes, were categorized by the developers as: entity rules, query rules, and spatial rules.

"*Entity rules*" are rules wherein the condition of a fact (a data value) is contained in a table linked directly to a table representing a spatial feature. For example, a rule that translates an SPB hazard rating of "high" (an attribute value) to the numerical rating of +0.75 is an entity rule when that attribute value ("high") is stored in a table linked directly to the spatial table representing the forest stand polygon. Many nonspatial rules defined in the USFS Southern Region are entity rules as stand inventory data (often used in rulebases) are stored in a fairly simple set of tables linked directly to the stand polygons.

"*Query rules*" are those rules wherein the condition of the fact is available only by searching through multiple linked tables. In this case, the fact condition or attribute for a spatial feature is not stored in a table linked directly to the spatial feature. This situation is common in the USFS, especially in Regions where highly normalized ORACLE data structures are being designed to store natural resource data. Handling query rules within the toolkit involves two steps: query and score. The *query* step builds the CLIPS rules needed to query the database through the linking of key columns to obtain the needed data field and data value; the *score* step builds a rule just like the entity rule to assign a score based on these data attributes.

"*Spatial rules*" allow for the manipulation of spatial information in order to retrieve a fact condition. The spatial rule construction function has not been implemented yet. Examples of spatial rules include derivation of the distance of

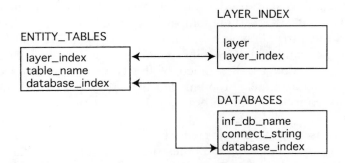

Figure 22.5 Metadata tables to support entity rule construction.

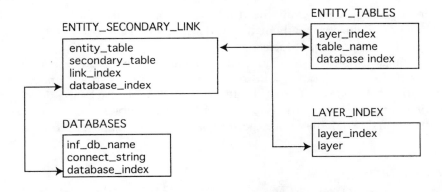

Figure 22.6 Metadata tables to support query rule construction.

a stand from the nearest stream, the condition of neighboring stands, and the aspect of a stand. Currently, spatial rulebases spawn ARC/INFO AMLs (Arc Macro Language queries) when executed, which derive these spatial attributes (see FUTURE DEVELOPMENTS section for further detail).

Metadata and Other Data Tables

Metadata tables store all information related to available data and their location(s), rule templates, and computed fact information. Figures 22.5 and 22.6 illustrate how data are located through metadata tables. Figure 22.5 illustrates the relationship between metadata tables that support entity rule construction. These three tables provide the system with information about available spatial layers (through LAYER_INDEX), the table names for layer attribute data (through ENTITY_TABLES), and how the system can access the attribute tables (through DATABASES).

Figure 22.6 illustrates the relationship of metadata tables needed to support query rules, in which tables containing attribute data are not directly linked to spatial tables. In this figure, a secondary "lookup" table (ENTITY_SECONDARY_LINK) is used to guide the system to the specific table wherein the attribute data is located.

Two metadata tables are used to describe available rule templates, TEMPLATE and TEMPLATE_VARIABLES. These templates are used to store predetermined rule structures and the variable syntax within these structures. Basically, these tables store much of the standard CLIPS code required beyond the specific issue-related rules. This "extra" code handles much of the control logic needed to form a completely executable CLIPS program. Much of the code represented in these templates is added to the front end of the CLIPS program, followed by the user-defined rules.

One table is used to store facts created by mathematical computations on one or more database columns (e.g., attributes such as stand age or stand basal area). For example, in the Southern Region, the stand inventory database contains only the "birthdate" of the stand — such as the year 1915; to derive "Forest Stand Age," this year must be converted to current age by subtracting the birthdate from the current year (e.g., 1995 – 1915 = 80). Thus, the condition of this fact can not be determined by simply retrieving the database attribute, but must be calculated by manipulating the attribute mathematically. The COMPUTED_FACTS table supports definition of such computed facts.

The data tables shown in Figure 22.7 contain rule and rulebase information that support easy development and transformation of facts, rules, and numerical ratings into CLIPS code. Note that the RULE_TRACK table tracks the development and maintenance history of the rulebases. All of these tables may evolve as more of the functional requirements of the toolkit are put in place in the next year.

User Interface

The targeted users of this toolkit system are forest resource managers and staff. As their computer expertise may be minimal, the system should be simple to use. Therefore, a point-and-click interface was developed using the X-Window System and XView. All rule information, as created interactively by the user, will be displayed in English syntax to foster ease of interpretation. Although the process of implementing a rulebase from initial inception to fully tested finished product is a complicated process requiring a healthy learning curve, the toolkit will dramatically shorten the amount of time needed to develop a rulebase, since coding in the CLIPS language is eliminated. Users of the INFORMS-R8 knowledge bases can thus concentrate on other less technical aspects of the process.

Toolkit Role in Rulebase Development

Developing a rulebase using the toolkit generally follows the methods used for developing typical expert systems. First, a knowledge engineering session is used to elicit knowledge from experts in the field of study. This knowledge is then transferred into facts and rules. The rules are then coded into the rulebase engine's language — in this case, CLIPS. The Rulebase Toolkit can aid in all three steps; however, its obvious role is to facilitate the coding of CLIPS-encoded rulebases.

A Knowledge-Based Reasoning Toolkit for Forest Resource Management

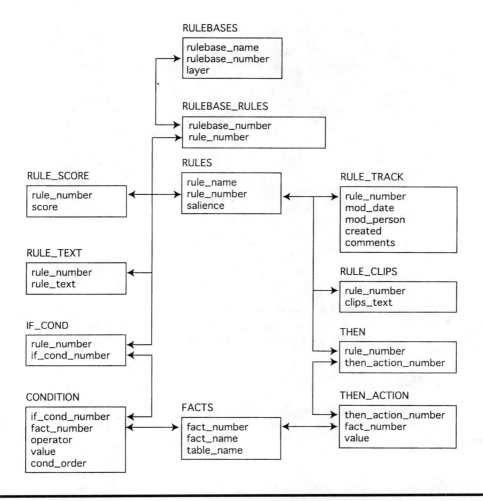

Figure 22.7 Data tables to support overall construction of rules and rulebases.

Eliciting Knowledge

Eliciting knowledge from experts is a key concern in developing rulebases, and select USFS employees have been trained in the methods for eliciting such knowledge. These individuals (i.e., knowledge engineers) will organize meetings with resource specialists in the subject areas of interest. In these meetings, the process for building, testing, and using rulebases is discussed. Once all involved parties have an understanding of this process, knowledge extraction begins. The knowledge engineers will get everyone to discuss their solution methods (e.g., how do you decide what silvicultural treatments are needed) and how they attack the targeted problem. Then a consensus is built. Once this consensus is reached, rules are constructed.

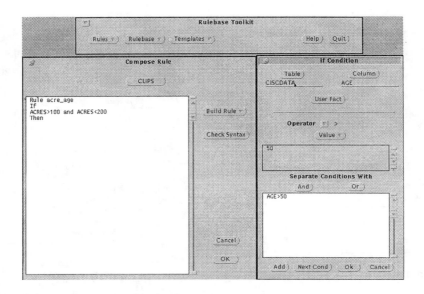

Figure 22.8 Rulebase toolkit. First steps in building rules: clicking on the "Rules" button and then the "Build Rule" button yields the above windows.

Constructing Rules

After a consensus solution method is developed, the knowledge engineer will build if-then rules. These rules will then be agreed upon by the group that developed them. The next step is to code the rulebase using the toolkit.

Encoding Rulebases

Once the rules are developed, the knowledge engineer can start to build the rulebase using the toolkit. There are two main processes involved in building a rulebase:

1. Building the individual rules
2. Composing the rulebase from the rules and templates so that a complete functional CLIPS rulebase is generated

Figure 22.8 shows the rule composition function screen used to build individual rules for a rulebase. The left window contains the English text of the rule being built although the user can toggle a button to see the CLIPS code associated with this text. The right window shows the if-condition building window. Building a condition consists of selecting a fact (i.e., table and table column from a database or a user-defined fact) and some logical condition for this fact (e.g., ACRES > 100 and ACRES < 200). Multiple facts can be used in each condition. Consecutive conditions are connected with an implicit "and." Once the if-condition is built, the user clicks on "OK," which inserts the conditional information into the appropriate data tables.

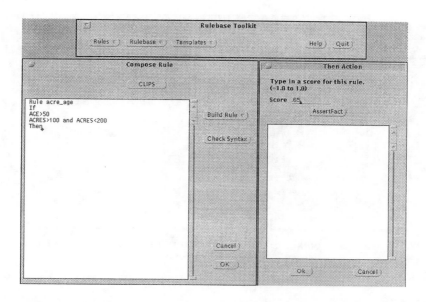

Figure 22.9 Rulebase toolkit. To complete a rule, the "Then" action must be defined.

After clicking the "OK" button above, the user must complete the "then" part of the rule (i.e., what to do if the condition is true). This is facilitated by the screen shown in Figure 22.9, which automatically appears after clicking "OK" on the previous screen. In this step, the user enters the rule score, set by consensus during the knowledge engineering sessions. Thus, if the rule is true, then this score will be asserted to the fact base. Beyond defining a score as part of the then-condition, the user also has the opportunity to define an additional fact and fact value. The result of this definition is to assert a fact tuple (i.e., a row in the fact base that includes a fact name and fact value) if the rule is true. These rule generated facts are sometimes used to set fact values that are used by other rules. Clicking "OK" after defining the then-condition inserts this information into the appropriate data tables.

At this time, the user can check the syntax of the rule by clicking on the "check syntax" button. The toolkit uses the CLIPS engine to verify correct syntax and reports problems to the user. If the rule is syntactically incorrect, the user must edit the rule to correct it. Once the rule is syntactically correct, clicking the "OK" button inserts the rule text into the appropriate data table. The user will then be prompted for rule tracking information.

Users can also copy and edit rules using the toolkit. Copying and then editing an existing rule saves time when making new rules that differ only slightly from an existing rule. The new rule may judge the same facts for different values and/or return a different score than the existing rule. For forest resource management rules, this is a common situation. Rules dealing with stand basal area, stand age, and dominant tree species are common in many current silvicultual treatment rulebases. The toolkit rule editing feature is depicted in Figure 22.10. The user can change values and operators during an editing session through the "Edit Condition" window. In the Figure 22.10 example, the user has clicked on the "Value" button, selected a value type (in this case, "Integer") from a pop-up menu

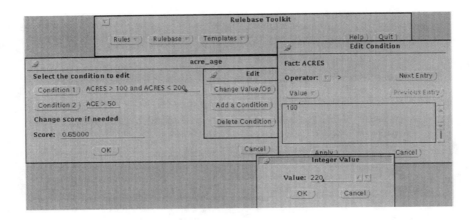

Figure 22.10 Rulebase toolkit. Existing or copied rules can be easily edited using the above windows.

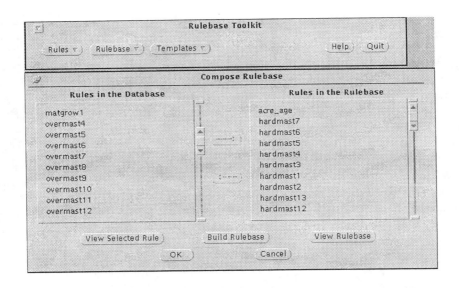

Figure 22.11 Rulebase toolkit. To compose a rulebase, users select the needed rulebase rules from a scrollable list on the left side of the above window.

(not shown), then adjusted the value for ACRES from 100 to 220. The user can also insert or delete rule conditions at this point.

The rulebase composition screen is shown in Figure 22.11. A rulebase is basically a collection of rules. The toolkit creates a CLIPS encoded rulebase by combining the CLIPS code for a set of chosen rules with the CLIPS code from a template of a typical rulebase. The template contains CLIPS code related to score-combining functions as well as CLIPS code related to rules that retrieve data from a database and rules that send results to a database. During rulebase composition, the system automatically checks that each rule is able to be fired: that is, that all facts in the conditional part of the rule have previously been asserted somewhere in the action ("then") part of another rule within a rulebase. Finally, the user is able to view individual rules and to view the entire newly constructed rulebase.

A final module within the toolkit allows completed rulebases to be installed automatically within INFORMS-R8. As a result, these rulebases will appear as options in the appropriate INFORMS-R8 menus listing available rulebases, and can be selected and executed for immediate use. (This assumes that INFORMS-R8 has been configured to recognize the same spatial and nonspatial resource databases that were used by the toolkit to build the rulebase, as will usually be the case with current and potential USFS users.)

Conclusions

Rulebase technology is providing a valuable analysis tool in analyzing complex forest resource management issues. In early testing of this technology in a precursor to the INFORMS-R8 application, some USFS staff were skeptical of its utility. In the early 1990s, a limited but useful set of silvicultural treatment rulebases were embedded within INFORMS-R8 and used operationally to support day-to-day silvicultural prescription planning. Today, a variety of specialists at the Ranger District level of the USFS have embraced the knowledge base component of INFORMS-R8 and have applied this technology well beyond the original silvicultural rulebases. Current users are applying this technology in creative ways, and are anxious to extend its use to other Ranger Districts across the USFS. Users have found that the use of rulebases:

- Simplifies analysis by organizing complex decision logic
- Provides easy access to expert knowledge when needed
- Promotes consistent decision making across staff, time, and space
- Facilitates improved documentation of decision making
- Facilitates quick analysis on large project areas
- Provides easy access to and consideration of important data
- Ensures important factors are consistently used in the decision-making process
- Provides a training vehicle for district trainees
- Provides a vehicle for capturing knowledge from subject experts
- Highlights research needs by defining knowledge gaps

However, to extend this technology operationally to even a small set of the hundreds of Ranger District offices requires a tool that will promote self-sufficiency in building and maintaining CLIPS-encoded rulebases. The Rulebase Toolkit is intended to do just that.

Early feedback from users of the beta version of the Rulebase Toolkit has validated the need for this application as originally envisioned by the project sponsors and developer. Several of the rulebases listed earlier in this paper were, in fact, developed using the Rulebase Toolkit. The toolkit, by being blended in with and utilizing existing USFS computer technology — workstations, ORACLE, Arcinfo, existing resource databases, and INFORMS-R8 — has facilitated easy construction of rulebases by USFS resource specialists. These users can construct the rules directly using the resource data they are familiar with.

Beyond simply facilitating the creation of CLIPS-encoded rulebases, the Rulebase Toolkit has had some unexpected benefits. The empowerment of USFS staff to create their own rulebases has spawned exceptional creativity in applying the technology to a wide variety of forest resource management issues. With this tool on-site, with their intimate knowledge of the resources they manage, and after generating some initial rulebases, these users have begun to realize the potential of this technology far beyond what CLIPS programmers could relate to and beyond what USFS staff, previously dependent upon those programmers, thought possible. In addition, with the capability to generate executable rulebases on-site, USFS staff have been able to quickly encapsulate decision logic and then, clarify problems or conflicts among a cross section of resource specialists early in the planning process. These quickly generated rulebases serve as a catalyst in communication.

In the next 2 years, the development team at Texas A&M and the project sponsors from the Forest Health Staff in the Southern Region and the Methods Application Group of Forest Health Protection at the national level hope to bring this project to fruition with the addition of the spatial rule composition function, and the other functions discussed. Other USFS staffs and nonUSFS entities who have seen the utility of this implementation of rulebase technology have expressed a strong interest in utilizing this technology.

References

Buchanan, B.G. and R.O. Duda. 1983. Principles of rule-based expert systems. Pages 164–215 in M.C. Yovits, Ed. *Advanced in Computers*. Volume 22. Academic Press, New York.

ESRI. 1995. ARC News. Volume 17. Environmental Systems Research Institute, Inc., Redlands, California.

Holtfrerich, D.R., Y.K. Choo, B. Stiles, and D.K. Loh. 1992. Incorporating Software Integration into Project Planning. ASPRS/ASCM/RT 92 Technical Papers, Volume 5. American Society for Photogrammetry and Remote Sensing and American Congress on Surveying and Mapping. Bethesda, Maryland. Pages 464–470.

Holtfrerich, D.R., Y.K. Choo, and S.B. Williams. 1993. Integration of a commercial GIS in INFORMS-TX. In Proceedings: GIS 93, Vancouver, British Columbia, February 15–18, 1993.

Loh, D.K. and Hannu Saarenmaa. 1992. The Design of Integrated Resource Management Systems. Texas A&M University, Department of Rangeland Ecology and Management, College Station, Texas, and the Finnish Forest Research Institute, Department of Mathematics, Silva, Finland.

Oliveria, F. and K. Swain. 1991. Progress Report for Implementation of INFORMS on the National Forests in Texas-Technology Development Project R8-90-13. Internal Report, Forest Pest Management Washington Office. October 1991. 6 pages.

Perisho, R.J., F.L. Oliveria, and D.K. Loh. 1995. INFORMS-R8 — a tool for ecosystems analysis. In 1995 ESRI User Conference Proceedings. May 22–26, Palm Springs, California. Environmental Systems Research Institute, Inc., Redlands, California.

Schuster, E.G., L.A. Leefers, and J.E. Thompson. 1993. A Guide to Computer-Based Analytical Tools for Implementing National Forest Plans. Gen. Tech. Rep. INT-296. Ogden, Utah. U.S. Department of Agriculture, Forest Service, Intermountain Research Station. 269 pages.

Williams, S.B. 1992. INFORMS-TX Overview. In Proceedings: Spatial Analysis and Forest Pest Management. Mountain Lakes, Virginia. April 27–30, 1992.

Williams, S.B., D.J. Roschke, and D.R. Holtfrerich. 1995. Designing configurable decision-support software: lessons learned. *AI Appl*. 9(3): in press.

Chapter 23

Subjective Evaluation of Decision Support Systems Using Multiattribute Decision Making (MADM)

JAMES C. ASCOUGH, II, LOIS ANN DEER-ASCOUGH, MARVIN J. SHAFFER, AND JON D. HANSON

Abstract

Structured evaluation of decision support system (DSS) development stages may be found in development plans, however, formal and systematic evaluations of "real world" DSS are rarely performed throughout the creation process. As a result, we relinquish the chance to gain valuable information about what potential customers think about the system, how well the code is written, the extent to which the system actually supports decision making, etc. Andriole (1982) estimated that reliable evaluations were conducted on about only 10% of government-sponsored DSS development projects. Multiattribute decision making (MADM) is presented herein as a mechanism to account for different attributes or system effectiveness measures (SEMs) considered important when evaluating a DSS and as a means for combining the SEMs into a comprehensive measure of value, that is, a single overall assessment of system utility. MADM is an appropriate tool for evaluating DSS in that it is capable of distinguishing between satisfactory and unsatisfactory systems, or components of a system. A case study is presented showing how MADM was used to evaluate two potential Graphical User Interfaces (GUIs) for the USDA-ARS Great Plains Framework for Agricultural Resource Management (GPFARM) DSS.

Introduction

This paper is divided into two major parts. The first part presents a generic framework for considering evaluation issues and a multipurpose approach for selecting categories of evaluation techniques. The approach is based upon that suggested by Adelman (1992) and includes technical, experimental, and subjective evaluation categories. The second part of the paper suggests a MADM evaluation hierarchy for DSS as: (1) a formal structure for conceptualizing SEMs or attributes, (2) a technique for decomposing the global SEM (i.e., the overall utility of the DSS) into upper and lower (bottom) level component SEMs, and (3) a means for combining the SEMs into a comprehensive measure of value, that is, a single overall assessment of system utility. Development of a multiattributed hierarchy is a subjective evaluation process that should account for all the different SEMs considered important when evaluating the DSS. This type of hierarchy is particularly useful for: (1) gauging the development progress of a single DSS, (2) performing an evaluation on some component of a DSS (e.g., the GUI), and (3) evaluating and comparing two or more DSS that are similar in nature (e.g., DSS for agricultural watershed management). Two well-known techniques are discussed for identifying evaluation attributes and structuring them into the MADM hierarchy: the descending (hierarchical) approach (Keeney and Raiffa, 1976) and the ascending (attribute listing) approach (Kelly, 1979). A case study is given detailing the use of MADM in evaluating two potential GUI's for the USDA-ARS GPFARM DSS.

Why Evaluate Decision Support Systems?

DSS can be defined as interactive computer programs that may utilize numerical methods, such as water quality or groundwater models; analytical methods, such as decision analysis tools or optimization algorithms; and spatial analysis methods, such as geographic information systems, for formulating a framework to help decision makers rank management alternatives, analyze their impacts, and interpret and select appropriate options for implementation. Morton (1980) states, "The term 'decision support systems' (DSS) refers to the use of computer-based systems, often interactive, to support humans as they make certain types of partially structured decisions."

It is easy to study these definitions and see why successful DSS are so difficult to build. The development of software that: (1) typically structures problems differently than people do and (2) may contain numerical, analytical, and spatial analysis methods, is not a trivial task. Unfortunately, the "state of the art" has not yet matched the "state of the expectation" (Andriole, 1989). Although there are occasional successes, many, if not most, DSS that are developed are never used. Federal agencies have spent large sums of money on DSS technology with minimal benefits to operational personnel. The USDA-ARS is currently allocating considerable human and financial resources towards the development of DSS for ranking agricultural best management practices on farm, ranches, and watersheds. Private industry has spent large sums of money developing DSS with minimal impact towards enhancing the decision-making processes of targeted users. Why does this keep occurring? One explanation may be that formal evaluation procedures probably were not used in the development cycles, thus, there was no way of knowing if user requirements were really being met. Another problem is that DSS development, and indeed most software development, is currently technology driven instead of requirements driven. Fortunately, there has been considerable progress in development of methods for increasing the probability that the system design process will be requirements driven, and that, in turn, the system will be be used. Evaluation is an important link in the application of a requirements driven development cycle, for it provides the information that keeps the development process on track (Adelman, 1992).

Structured evaluation is typically absent from "real world" DSS development cycles. As a result, we exclude an important step of the development process that will have a significant impact on whether or not the DSS is to be accepted and used. Moreover, it is important we acquire this information early in the development cycle when changes are less expensive to make than at the final, more costly stages of development. Andriole (1982) estimated that reliable evaluations were conducted on about only 10% of government-sponsored DSS development projects. Unfortunately, the last 13 years have not brought much improvement. There are a variety of reasons for this, as Andriole (1989) states, "First, there are always those who are inherently distrustful of structured analyses of any kind, particularly when they challenge conventional wisdom. Many decision support systems have not been evaluated because their designers and users felt that the systems were working fine and that everyone liked them. Other systems have gone unevaluated because the projects simply ran out of money." Finally, as Riedel and Pitz (1986) point out, "All of this suggests that utilization of evaluation results should be explicitly addressed in the evaluation design. By utilization of evaluation results we mean that the evaluation should have some impact on future development or deployment of the DSS." Simply put, this statement means DSS should be evaluated to help ensure that they will be used, and most importantly, that they will be useful.

Who Should Evaluate What, and How?

Evaluations are typically required by both the DSS development and sponsoring groups. The development group includes users, customers, designers, and programmers. The sponsoring group may or may not be users. It is difficult, yet crucial, to get both sponsors and users to participate in the development process. Evaluators need to be aware of this problem and take whatever steps necessary to solve it in order to increase the probability of successful system development and implementation. Although users and sponsors have specific criteria for the DSS under development, each is primarily concerned with overall system performance based on meeting their requirements. In other words, the users and sponsors are interested in knowing: "Is the system working well and is it meeting our requirements?" Answering this question is deceptively difficult because developers, users, and sponsors have different criteria and ideas of what "their" system should contain to be successful. They may have different assumptions of what constitutes success on one or more of the criteria; they may disagree on definitions for the criteria themselves (Adelman, 1992). The answer to the above questions must be found since they provide critical information regarding the direction that development will proceed. By integrating technical, experimental, and subjective evaluation methods, the development team can navigate the development process on a course that will improve the probability that the DSS will be effectively used.

Specific evaluation or measurement approaches depend on selected criterion. Evaluators must know how to implement different evaluation methods in order to provide the required information on different criteria. It is important to realize that identifying evaluation criteria is both a subjective and objective process. All users have different criterion important to them, and many are uncertain of exactly what their decision requirements are. As a result, the identification of evaluation criteria is often neglected early in the DSS development process. When this happens, the development team leaves itself without measure(s) for defining a successful system. Criterion for evaluation may include many factors, such as user satisfaction, the quality of the decisions made with the system, the completeness of its databases, the logical soundness and appropriateness of its numerical, analytical, or spatial methods, and so forth. User satisfaction alone is not a sufficient criterion for evaluation, however, user satisfaction is certainly a critical requirement for use of the system. A system that is technologically superior is worth little or nothing if the users are dissatisfied with it. Similarly, user satisfaction is a necessary requirement for assessing the adequacy of an evaluation. System evaluations must provide information that will help the development team determine if the users of the system will be satisified with it (Adelman, 1992). A fundamental requirement is to adopt a flexible approach that is based on the evaluation purpose and situation.

Adopting a Multipurpose Evaluation Format

Evaluation is an iterative process that goes hand in hand with the prototyping approach commonly used in DSS development. The evaluator needs to be able to use different methods to answer different questions at various stages of the development process. This requires a multipurpose evaluation format. The format as described by Adelman (1992) has three categories: technical, experimental, and subjective. The technical evaluation category focuses on evaluating the DSS from both internal (algorithmic) and external (I/O) perspectives. For example, someone considering the technical evaluation of a DSS might focus on testing the validity of the numerical models it might contain. That is, do the models accurately represent the processes they are modeling? Other concerns lie with conventional test and evaluation issues. For instance, can the system be effectively and efficiently integrated with the software and hardware components of the organization? Was it developed with the design and coding standards of the organization in mind? Can system data requirements be met?

The experimental evaluation category focuses on obtaining objective measures of system performance. The goal is to ascertain whether users make significantly better or faster decisions or use significantly more information working with rather than without the system, and to identify mechanisms for improving performance if the system fails to measure up based on evaluation criteria (Adelman, 1992). If possible, experts should also participate in the experimental evaluation in order to systematically assess whether system performance is a function of user type.

The subjective evaluation category focuses on evaluating the DSS from the perspective of potential users. The goal is to determine whether the users generally like the DSS. Most of the user feedback will concern improving the system interface, both in terms of its ease of use and compatibility with the background, training, and needs of the user. In addition, subjective evaluation can be accomplished through the use of MADM by identifying SEMs that will provide the information required by the development team to determine system utility. The SEMs may be either subjective or objective attributes and should be correlated with the overall utility of the DSS.

Adelman (1992) compares the multipurpose approach to the traditional concepts of verification and validation. Technical evaluation methods, which are directed toward looking inside the "black box," represent verification methods. Experimental evaluation methods represent validation methods; they are directed towards determining whether the system actually improves decision-making performance. Subjective evaluation methods have a role in both verification and validation, although they are more effective for validation. Frequently, there is a lack of agreement concerning the definitions for verification and validation and one must be cautious when using these terms when referring to DSS.

MADM as a Subjective Evaluation Tool

MADM evaluates utility or value functions intended to accurately express a decision maker's outcome preferences in terms of multiple attributes. MADM decomposes the complex problem of assessing a multiattribute utility function into one of assessing a series of unidimensional utility functions. When applied to subjective evaluation, the DSS is decomposed into evaluation attributes (or SEMs) against which it can be measured. Before continuing, a few basic concepts of MADM are defined.

Basic MADM Concepts

Attributes, Alternatives, and Value Functions

An attribute is a characteristic of the alternatives that the users consider important. SEM is used in this paper as a synonym for attribute. Let x_{ij} be the value of attribute j for alternative i and X_i be the vector of values of attributes j = 1,..., n for i. Value scaling is the creation of unidimensional attribute value or utility functions $v_j(x_{ij})$ that convert an attribute into a measure of worth. Attribute value functions include just the user's evaluations of different levels of the criterion; utility functions, in addition, capture user attitudes toward risk. Combination or reintegration rules combine several single attribute value functions $v_j(x_{ij})$ into an overall index of worth or utility $V(X_i)$. An example is the simple additive weighting (SAW) value function, in which the overall worth is the weighted sum of scaled attributes

$$V(X_i) = V_i = \sum_{j=1}^{n} w_j v_j(x_{ij}) \qquad i = 1,...,m \qquad (1)$$

where w_j are weights used to combine the attributes and m is the number of alternatives.

Attribute Generation

The attributes need to be defined well enough so that one can obtain measures (either subjective or objective) of how well the DSS performs on each of them. The definition of the attributes should be clear and easily measurable, and the attributes should contribute to the overall worth given to the system by key users or decision makers. One way of ensuring this is to derive the attributes hierarchically from the overall goal. Few techniques for generating evaluation attributes and structuring them into a hierarchy tree have been proposed by decision scientists. Two techniques are typically applied by decision analysts using MADM: (1) the descending (hierarchical) approach (Keeney and Raiffa, 1976) and (2) the ascending (attribute listing) approach (Kelly, 1979). In the descending approach, the upper level attributes of each branch of the tree are listed first; then each branch, in turn, is subdivided into its component attributes. The process continues until the lowest level attributes in the tree are identified. The ascending approach proceeds by first creating a list of all lower level attributes in the tree without any concern for the hierarchical nature of the problem.

Attribute Weighting

Weights should be elicited from the users and represent subjective judgments that can be used to measure the system's value to the users and the sponsors. That is, the role of weight serves to express the importance of each attribute relative to the others. There are many different techniques for obtaining relative importance weights across attributes. Good reviews can be found in Hobbs (1980) and Hwang and Yoon (1981). Common techniques include weighting by rank (e.g., rank reciprocal weighting and rank sum weighting) and ratio weighting. The inherent difficulty in getting a consensus on the value of the weights and the impact of the weights on the overall index of worth or utility are weaknesses of the MADM method.

Normalization of Attributes

Attribute values are frequently normalized to eliminate computational problems caused by incommensurable measurement units in a decision matrix. This is not always necessary but may be required for many MADM methods. Normalization aims at obtaining comparable scales, which allow interattribute as well as intraattribute comparisons. Consequently, normalized ratings have dimensionless units and, the larger the rating becomes, the more preference it has (Yoon and Hwang, 1995). Linear normalization is a simple and common normalization procedure that divides the ratings of a certain attribute by its maximum value. The normalized value of x_{ij} is given as

$$r_{ij} = \frac{x_{ij}}{x_j^*} \qquad i = 1,\ldots,m; \ j = 1,\ldots,n \qquad (2)$$

where r_{ij} is the normalized value and x_j^* is the maximum value of the jth attribute.

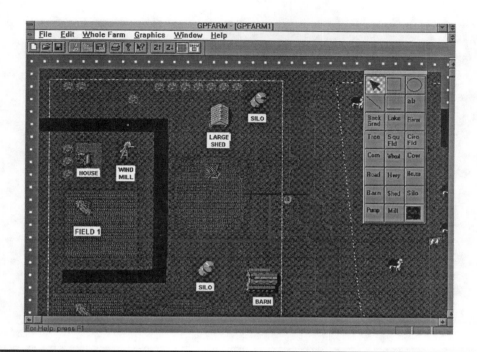

Figure 23.1 GPFARM symbolic graphical user interface (GUI).

A MADM Case Study

Two potential GUIs for the GPFARM DSS (Ascough et al., 1995) were evaluated using MADM. The evaluation was performed both as an objective test of MADM to distinguish between the preferred GUI and to assess the applicability of MADM for future whole system evaluation. The first interface was called the symbolic GUI and is shown in Figure 23.1. The second interface was called the schematic GUI and is shown is Figure 23.2. Evaluation was based on judgments concerning the input screens, output screens, help screens, miscellaneous characteristics, and operational concerns associated with the two GUIs.

Evaluation Methodology

The evaluation team consisted of customer focus group members for the GPFARM DSS project; including farmers and ranchers, extension personnel, and agricultural consultants. The initial step in evaluating the two GUIs was the development of a hierarchical attribute tree by the evaluation team that reflected interface characteristics that they considered important for the GPFARM GUI. The hierarchical attribute tree is shown in Figure 23.3. The next step was to generate the weights for the attributes. The pairwise ranking approach of Morris (1964) was used which requires the decision maker(s) to make pairwise preference judgments between attributes. In this case, the pairwise judgments reflected an open consensus between the evaluation team members as to the preference relationships between

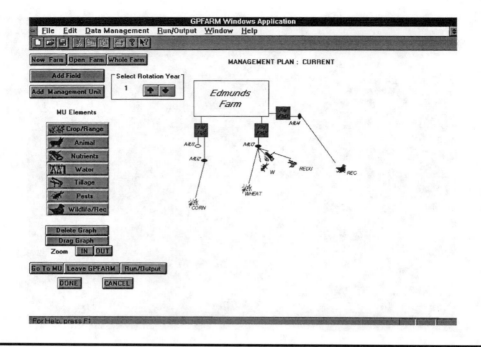

Figure 23.2 GPFARM schematic graphical user interface (GUI).

the attributes. Once a complete ranking order was generated for each level of the attribute tree, quantitative values for the weights were arbitrarily agreed upon by the evaluators. The weighting data for the attribute tree hierarchy are shown in Figure 23.4.

A questionnaire containing Likert-type qualitative judgment response questions for each of the attributes was given to the evaluation team members. The Likert scale ranged from 1 (a strongly negative response such as dislike or disagree) to 10 (a strongly positive response such as like or agree). The evaluation team responses (scores) for each attribute were then averaged for the GUIs. The mean response score evaluation data are presented in Table 23.1.

A comparable rating was obtained through the use of linear normalization. The partial normalized decision matrix is given as

	X_{111}	X_{112}	X_{113}	$X_{114}, \ldots,$	X_{42}	X_{43}
Symbolic	1.00	1.00	1.00	0.95	1.00	1.00
Schematic	0.84	0.96	0.97	1.00	0.88	0.84

where the decision matrix values are the normalized r_{ij}s.

Evaluation Results

The overall utility or value of the symbolic GUI was then calculated using a simple additive weighting (SAW) value function as

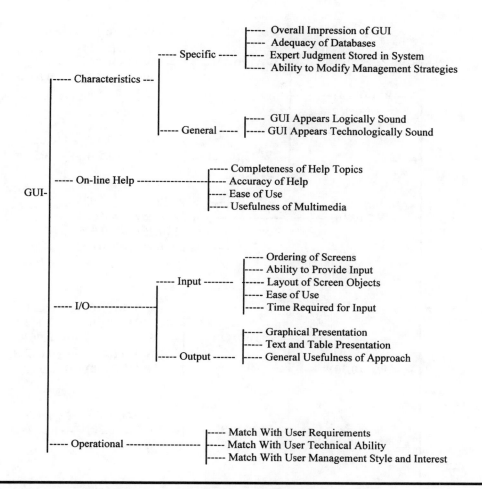

Figure 23.3 GUI evaluation hierarchical attribute tree.

$$V(Symbolic) = \sum_{j=1}^{21} w_j r_{ij} \qquad (3)$$

$$= 0.03(1.0) + 0.03(1.0) + 0.02(1.0) + 0.02(0.95)$$

$$+ \ldots 0.12(1.00) + 0.08(1.00) = 0.703$$

The overall utility or value of the schematic GUI was calculated using Equation 3 as

V(Schematic) = 0.03(0.84) + 0.03(0.96) + 0.02(0.97) + 0.02(1.0)
+ ..0.12(0.88) + 0.08(0.84) = 0.647

Based on the calculated overall values, the evaluation team preferred the symbolic GUI to the schematic GUI.

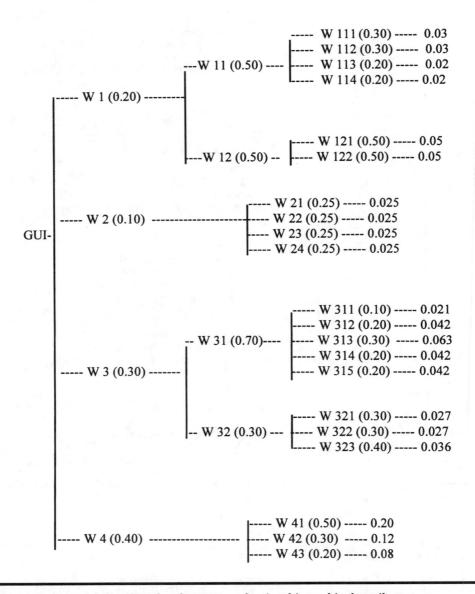

Figure 23.4 Weighting data for the GUI evaluation hierarchical attribute tree.

Summary and Conclusions

This paper discusses three different types of evaluation techniques: subjective, technical, and experimental. The initial resolution to build a DSS is essentially a hypothesis that the system will improve the decision-making performance of an organization. If evaluators can help members of the sponsoring team assess the adequacy of this hypothesis at the beginning of the development cycle, then they can possibly reduce costs and better define the requirements of the DSS. Once the development process is under way, the job of the evaluator is to systematically collect, filter, and aggregate data about the system in order to test the hypothesis (Adelman, 1992). The application of formal evaluation methods helps members

Table 23.1 GUI Evaluation Attribute Judgment Data

Attribute or SEM	Weight	Symbolic GUI score	Schematic GUI score
1. Characteristics			
1.1 Specific			
1.1.1 Overall impression of GUI	0.03	8.2	6.9
1.1.2 Adequacy of databases	0.03	5.7	5.5
1.1.3 Expert judgment stored in system	0.02	6.2	6.0
1.1.4 Ability to modify management strategies	0.02	7.3	7.7
1.2 General			
1.2.1 GUI appears logically sound	0.05	7.1	7.0
1.2.2 GUI appears technically sound	0.05	6.0	6.6
2. On-line help			
2.1 Completeness of help topics	0.025	4.8	5.7
2.2 Accuracy of help	0.025	7.8	8.0
2.3 Ease of use	0.025	8.0	7.4
2.4 Usefulness of multimedia	0.025	3.2	4.1
3. I/O			
3.1 Input			
3.1.1 Ordering of screens	0.021	7.9	6.6
3.1.2 Ability to provide input	0.042	5.3	5.9
3.1.3 Layout of screen objects	0.063	8.3	6.8
3.1.4 Ease of use	0.042	7.7	7.6
3.1.5 Time required for input[a]	0.042	83	67
3.2 Output			
3.2.1 Graphical presentation	0.027	5.6	5.2
3.2.2 Text and table presentation	0.027	6.1	6.4
3.2.3 General usefulness of approach	0.036	7.1	6.7
4. Operational			
4.1 Match with user requirements	0.20	6.4	6.2
4.2 Match with user technical ability	0.12	6.8	6.0
4.3 Match with user management style and interest	0.08	8.3	7.0

[a] Time required for input is a quantitative value, all other values are qualitative determined from a Likert scale (1–10) questionnaire.

of the sponsoring and development teams monitor the perceived utility of the system during its development and take corrective action if needed to increase the probability of use.

A case study was presented showing how MADM was successfully used to evaluate symbolic and schematic GUI development approaches for the USDA-ARS GPFARM DSS. The evaluation team generated a hierarchical tree of attributes deemed important in evaluating the GUIs, determined the attribute weights, and judged the attributes using a ten point Likert scale questionnaire. A simple additive weighting (SAW) value function was then used to calculate the overall utility or value of each GUI. It was concluded that the evaluation team preferred the symbolic GUI over the schematic GUI on the basis of the higher overall value calculated for symbolic GUI.

References

Ascough, II, J.C., J.D. Hanson, M.J. Shaffer, G.W. Buchleiter, P.N.S. Bartling, B.C. Vandenberg, D.A. Edmunds, L.J. Wiles, G.S. McMaster, and L.R. Ahuja. 1995. The GPFARM decision support system for whole farm/ranch management. Proceedings of the Workshop on Computer Applications in Water Management, Great Plains Agricultural Council, May 23–25, 1995. Fort Colllins, Colorado.

Adelman, L. 1992. *Evaluating Decision Support Systems.* John Wiley & Sons, New York.

Andriole, S.J. 1989. *Handbook for the Design, Development, Evaluation, and Application of Interactive Decision Support Systems.* Petrocelli, Princeton, New Jersey.

Andriole, S.J. 1982. The design of micro-computer based personal decision aiding systems. *IEEE Trans. Syst. Man Cybern.* SMC-12:463–469.

Hobbs, B.F. 1980. A comparison of weighting methods in power plant siting. *Decision Sci.* 11:725–737.

Hwang, C.L. and K.P. Yoon. 1981. *Multiple Attribute Decision Making: Methods and Applications.* Springer Verlag, Berlin/Heidelberg/New York.

Keeney, R.L. and H. Raiffa. 1976. *Decisions with Multiple Objectives.* John Wiley & Sons, New York.

Kelly, C.W. 1979. Program Completion Report: Advanced Decision Technology Program (1972–1979). Tech. Report TR 79-3-3. Decisions and Designs, Inc. McLean, Virginia.

Morris, W.T. 1964. *The Analysis of Management Decisions.* Irwin Press, Homewood, Illinois.

Morton, M.S. Scott. 1980. Book review of "Decision Support Systems: Current Practice and Continuing Challenges" by Stephen L. Alter. *Sloan Manage. Rev.* 21:77.

Riedel, S.L. and G.F. Pitz. 1986. Utilization-oriented evaluation of decision support systems. *IEEE Trans. Syst. Man Cybern.* SMC-16:980–996.

Yoon, K.P. and C.L. Hwang. 1995. *Multiple Attribute Decision Making.* Sage Publications, Inc. Thousand Oaks, California.

Chapter 24

CERES Models in Multiple Objective Decision-Making Process

G.J. Kovács, J.T. Ritchie, T. Németh

The goal of this paper is to show a special application of CERES models in the preparation of a multiple objective decision-making tool in Hungary. The CERES-CEREAL generic model was used. Despite of this multipurpose capacity CERES is still a robust model. The applied **objectives** were to maximize production or profit and minimize nitrate leaching. One step of doing that was to simulate production and leaching in an existing long-term experiment. The nitrate nitrogen accumulation was analyzed in the soil profile in the 20th year of the field experiment. After having proven the right estimation on the long run, simulation experiments were conducted to gain estimates of leaching for each year. Then hypotheses were generated about the possible factors of the differences between nitrate leaching during the rotation of maize and wheat crops. This is a preparatory part of the long-term multicriteria optimization process. Advantages of a simulation technique is shown when expensive field experiments and analyses can be reduced greatly in number.

Introduction

The process of decision making consists of four major points: (1) information on the present status of the system in question, (2) knowledge about the cause and effect relationships within the system, (3) ability to create alternatives, and (4) setting priorities to choose the most appropriate solution of a problem. Each one of these phases may be helped by packages of information. To aid the point 1 a database should be produced, managed, and maintained. Point 2 can be summarized best in the format of a computer (simulation) model. Point 3 can be aided by programs (visualization, optimization, etc.) to manage the model outputs. For point 4 limit values or other conditions can be set in the package to help prioritize among alternatives.

The heart of a decision support syster (DSS) is the model (point 2) that describes the relationships between the elements of the system. The reliability of the DSS first depends on the reliability of the model. In this paper the focus is on the applicability of the CERES models in multiple objective DSS. For this purpose a long-term experimental database has been developed. The CERES models were modified to do continuous runs and to simulate maize and wheat in the same run of crop rotation.

The goal of this paper is to show a special application of CERES models in the preparation of a multiple objective decision-making tool in Hungary. The applied objectives were to maximize production or profit and minimize nitrate leaching. One step of doing that was to simulate production and leaching in an existing long-term experiment. The nitrate nitrogen accumulation was analyzed in the soil profile in the 20th year of the field experiment. After having proven the right estimation on the long run, simulation experiments were conducted to gain estimates of leaching for each year.

CERES Models

CERES models were designed to describe the system of the crop and its environment (Jones and Kiniry, 1986). One of the emphases of the model developers was to harmonize the scales of the parts of the model. This is why it can be used equally well to estimate nitrate leaching and the grain yield and also transpiration or nitrogen uptake. Despite this multipurpose capacity, CERES is still a robust model and concentrates only on the most important functional relationships between the main components of the ecosystem to model the production and pollution processes on the field. Some earlier works help to trace the development of the subject (Ritchie, 1989; Kovács et al., 1989, 1994, 1995; Pethö et al., 1994).

The version of CERES used here corresponds to the CERES-Maize and -Wheat included in DSSAT v.2.1.

Field Experiment

A conventional field experiment was used to evaluate and adopt the CERES models for the Hungarian conditions. A 20-year period (1969 to 1988) has been used from a long-term fertilizer experiment at Nagyhörcsök, Hungary. The soil is a loamy calcaric chernozem with a level of groundwater of 13 to 15 m from the surface. It is important to note that the deepest accumulation of nitrate that exceeded the "natural" nitrate concentration was at 530 cm from the surface. (Nitrate concentration under the long-term control plots were considered being "natural" concentration.) In the experiment nitrogen, phosphorus and potassium fertilizers were applied in different rates. For our purposes only nitrogen treatments were selected: 0, 50, 100, 150, 200, and 250 kg/ha/year. The experimental design was split-split-plot. Nitrate accumulation was analyzed by taking soil cores by hand augers until the depth of measurable nitrate enrichment. According to earlier works it was about 5 m soil depth (Németh at al., 1988, 1989).

Simulation Experiment

To create a DSS for nitrogen fertilization requires us to learn more about the long-term conditions of nitrate leaching. Field measurements of this kind are very difficult, expensive, and time consuming. Models have been developed to simulate nitrogen behavior in the soil and plant (Addiscott and Whitmore, 1987; Frissel and van Veen, 1981; Godwin and Jones, 1991; Fehér at al., 1991). This special application of CERES was made using the long-term experiment, as it is described above, to answer this problem. After the model was adapted to the local conditions and genotypes, yearly estimates of nitrate leaching were simulated, then analyzed. The study showed large differences in leaching between years. It gives a possibility to analyze the relationships between leaching and its possible factors included in the simulation. The following step is an optimization of the treatments to maximize production and minimize contamination (Alocilja and Ritchie, 1993)

Discussion

Observed experimental data were compared with simulated ones for the yearly averages of yields (Figure 24.1), summed nitrogen balance for the 20 years (Figure 24.2), total leached nitrate nitrogen in the rooted zone and under it at the end of 20 years (Figure 24.3). CERES simulations gave good estimations for these components of the agricultural ecosystem. It was found that the amount of N-accumulated under the root zone in case of overfertilization was about the same as the nitrogen balance of that treatment (in 250 kg/ha/year N treatment: 1200 to 1300 kg/ha) both in the experiment and the simulation.

The estimation satisfied the statistical requirement (Figure 24.4) that the values of leaching measured in the deep soil profiles (under 170 cm depth) of the fertilizer treatments were within the 0.95 confidence intervals of the means of estimations ran for the 20 years. The estimated yearly nitrate nitrogen leached under 170 cm of soil depth is shown on Figure 24.5. The differences of N-leaching are great among years (0 to 240 kg/ha/year). In order to use the results for decision making it is necessary to identify the factors of the yearly leaching.

The first factor could be the yearly precipitation (Figure 24.6). Comparing Figures 24.5 and 24.6 shows a very light relationship only. Yearly drainwater (Figure 24.7) shows a stronger connection to leaching, but still there are contradictions (i.e., 1970 and 1986). Taking into consideration the yearly simple balance of nitrogen (given by fertilizer and taken off by plant harvest, Figure 24.8) helps partly and also confuses the picture in other years.

Figure 24.9 shows the summed balance for any given year (nitrogen added as fertilizer and taken off at plant harvest and leached by penetrating water). This shows the massive accumulation of nitrate after 1981 and especially from 1985 to 1986. Refering to Figure 24.7, it is obvious that the simulated summed N-balance and drainwater together explain nitrogen leaching (Figure 24.5).

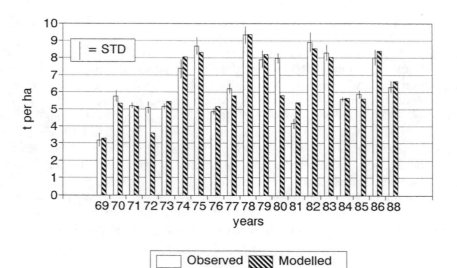

Figure 24.1 Observed and modelled yields.

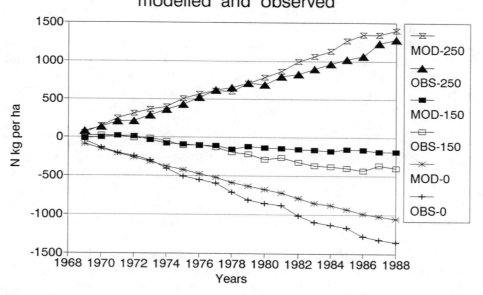

Figure 24.2 N-balance of 3 nitrogen levels.

CERES Models in Multiple Objective Decision-Making Process   285

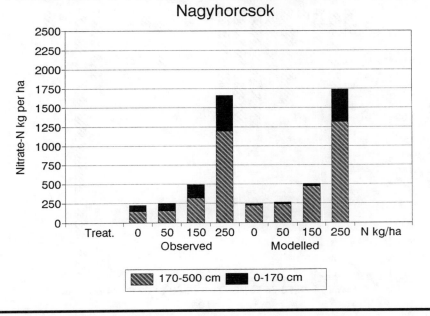

Figure 24.3   Nitrate in 2 depths after 20 years.

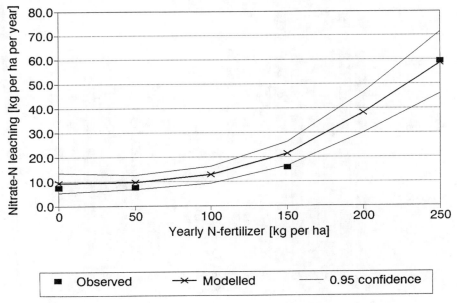

Figure 24.4   Nitrate leaching.

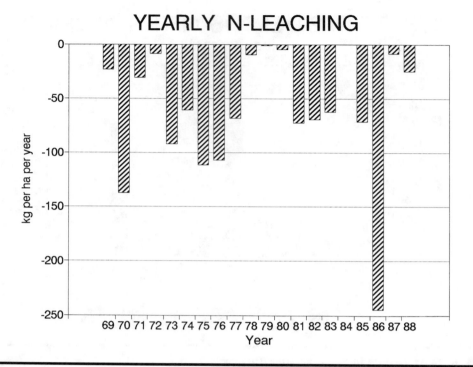

Figure 24.5 Yearly N-leaching.

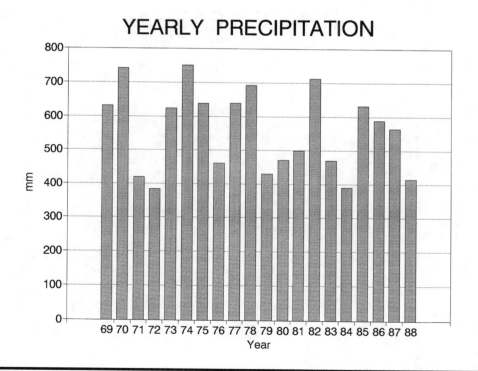

Figure 24.6 Yearly precipitation.

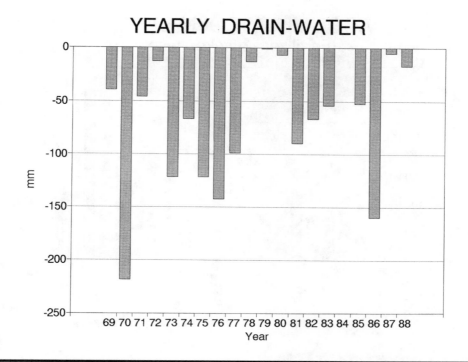

Figure 24.7 Yearly drain-water.

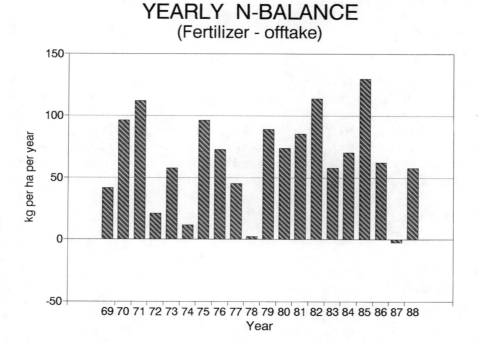

Figure 24.8 Yearly N-balance.

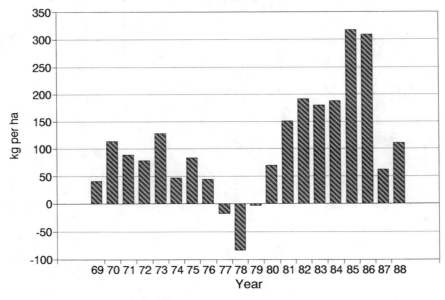

Figure 24.9 Accumulated N-balance.

The decision support tool can be put together using locally parametrized CERES models and based on the relationships demonstrated above.

Summary

CERES-Maize and -Wheat models are applicable to simulate yields, nitrogen transformation, and water and nitrogen transport processes running continuously through decades of years giving good estimates. Using long-term experiments to parametrize the models locally, reliable tools can be developed for decision makers for environmental as well as economical studies. The number of expensive field experiments and analyses can be reduced significantly by simulation experiments.

References

Addiscott, T.M. and A.P. Whitmore. 1987. Computer simulation of changes in soil mineral nitrogen and crop nitrogen during autumn, winter, and spring. *J. Soil Sci.* 34:709–721.

Alocilja, E.C. and J.T. Ritchie. 1993. Multicriteria optimization for a sustainable agriculture. Pages 381–396 in F. Penning de Vries, P. Teng, and K. Metselaar, Eds. *Systems Approaches for Agricultural Development*. Kluwer Academic, Dordrecht, The Netherlands.

Fehér, J., M.Th. van Genuchten, G. Kienitz, T. Németh, Gy. Biczók, and G. Kovács. 1991. DISNIT2, a root zone water and nitrogen management model. Pages 197–205 in G. Kienitz et al., Eds. *Hydrological Interactions Between Atmosphere, Soil and Vegetation*, Proc. of the Vienna Symp., August 12–25 1991. IAHS Publication No. 204.

Frissel, M.J. and J.A. van Veen, Eds. 1981. *Simulation of Nitrogen Behaviour of Soil-Plant Systems.* PUDOC, Wageningen, The Netherlands.

Godwin, D.C. and C.A. Jones. 1991. Nitrogen dynamics in soil-plant systems. Pages 287–321 in J. Hanks and J.T. Ritchie, Eds. *Modeling Plant and Soil Systems. Agronomy,* No. 31. ASA-CSSA-SSSA, Madison, Wisconsin.

Jones, C.A. and J.R. Kiniry, Eds. 1986. *CERES-Maize. A Simulation Model of Maize Growth and Development.* Texas A&M University Press, College Station.

Kovács, G.J., J.T. Ritchie, A. Werner, G. Máthé-Gáspár, and P. Máthé. 1989. Modeling the Leaf Area Development of Different Maize Genotypes. IBSNAT Symposium 1989: Proceedings of 81st Annual Meeting of the American Society of Agronomy, Las Vegas, Nevada, October 1989. Part II.

Kovács, G.J. and J.T. Ritchie. 1994. Using Simulation Models to Estimate Nitrate Leaching and Crop Production at the Farm Level in Hungary. Invited paper at ASA-CSSA-SSSA Annual Meeting. Seattle, Washington, November 13–18. *Agronomy Abstract.* Page 21.

Kovács, G.J., J.T. Ritchie, and J. Nagy. 1995. Optimization of Agricultural Technology Applying CERES Models. MALAMA 'AINA 1995. International Conference on MODSS for agriculture and environment, in print: *Multiple Objective Decision Making for Land, Water, and Environmental Management.*

Németh, T., G.J. Kovács, and I. Kádár. 1988. NO_3^-, SO_4^- and "water soluble salts" accumulation in soil profile of a long term fertilization experiment. *Agrokém. Talajtan.* 36–37:109–126.

Németh T., G.J. Kovács, J. Fehér, and Z. Simonffy. 1989. Nitrate leaching studies in lysimeters with sandy soils. *Agrokém. Talajtan.* 38(2):239–242.

Petö, K., L. Huzsvai, and G.J. Kovács. 1994. A method to test the effect of fertilization using CERES MAIZE 2.1 model in maize monoculture. *Növénytermelés.* 43(6):521–531.

Ritchie, J.T. 1989. Crop models and decision making: scenario for the future. In A. Weiss, Ed. *Climate and Agriculture Systems Approaches to Decision Making.* Proc. Am. Meteor. Soc. 5–7 March 1989, Charleston, South Carolina.

Chapter 25

Advances in Ration Formulation for Beef Cattle Through Multiple Objective Decision Support Systems

STEVEN S. VICKNER AND DANA L. HOAG

Introduction

Systems analysis has been widely and successfully applied to many problems in agricultural and resource economics. Agricultural economists have used mathematical programming techniques, such as linear programming (LP), in the formulation of least cost rations. Perhaps the earliest documented application of LP to the formulation of least cost livestock rations is by Waugh (1951). Waugh's dairy cattle model minimized total feed cost subject to maintaining various minimum animal nutritional requirements. While LP is a powerful analytical approach to solving this complex problem, agricultural economists have long recognized that the traditional ration formulations may yield unacceptable results. Examples of unacceptable rations include nutritionally unbalanced rations and rations that are too unique to be practical. To counter this, more realism is introduced through a series of lower and upper bound constraints on feeds or vitamin and mineral levels. Infeasible solutions, however, may be encountered using this approach.

More recently, agricultural economists have employed multiobjective techniques such as goal programming (GP) to provide the flexibility necessary to overcome the difficulties encountered with the rigid LP formulations. Rehman and Romero (1984) used variants of GP and multiple objective programming in the formulation of dairy cattle rations. By converting rigid nutritional constraints into "softer" goals they were able to circumvent the frequently encountered problem of infeasibility. Also, they utilized penalty functions to avoid violating important nutritional goals of energy and crude protein. Rehman and Romero (1987) used GP with penalty functions to develop feed rations for beef cattle. They incorporated additional nutritional goals and added a cost goal. Lara (1993) used multiple objective fractional programming in the formulation of dairy cow diets. The fractional programming approach enabled Lara to include the goal of maximizing the percent of on-farm surplus feeds in the ration. Lara and Romero (1994) used the STEM method to soften rigid nutritional constraints in dairy cattle rations.

Many analysts in the fields of systems engineering and operations research have begun to leverage the power of the decision support system (DSS) concept. A DSS is an integrated computing framework that combines a database, model base, and user interface to facilitate the development and evaluation of alternative courses of action. Often, an optimization model is one component of the model base in the DSS. Crabtree (1982) adopted this methodology by building an interactive computer system for the development of beef and dairy cattle least cost rations. At the heart of the model base was a standard Waugh-like LP model. Ranchers in need of feeding assistance would call the central agricultural extension office that maintained and used the system. Extension economists would then assemble the necessary inputs, such as animal type, weight, rate of gain and available feeds, run the model and, with delay, report the alternative rations to the rancher.

This paper combines and extends the aforementioned research to develop a GP-based, PC-based DSS for developing and evaluating alternative beef cattle ration formulations. Within the model base of the DSS is a GP model similar to Rehman and Romero's, less the penalty functions. A new dimension added to the GP model is time. The linear GP model optimizes over four sequential feeding periods to ensure smooth transitions from one ration to the next. Like the Crabtree model, this model is not a stand alone, rather it resides in a DSS. However, the DSS can be run directly by agricultural economists, animal scientists, veterinarians, feedlot operators, feed store operators and independent farmers and ranchers in a user-friendly, real-time PC environment. The DSS has an extensive database of feeds and feed nutritional profiles. Finally, the DSS includes a user interface to collect model inputs and to deliver output reports.

The DSS Prototype

Model Base

At the core of the DSS, in the model base, is a linear GP model. The GP model was built using standard GP methodology (Goicoechea et al., 1982). The model determines a mix of feeds per head of cattle per day that minimizes the weighted sum of deviations from the goals in the GP model. Stated compactly the GP model is:

$$\text{minimize:} \quad \sum_{i=1}^{5}\sum_{t=1}^{4} P_{it}\left(w_{it}^{+}d_{it}^{+} + w_{it}^{-}d_{it}^{-}\right) + \sum_{t=1}^{4} P_{t}\left(w_{t}^{+}d_{t}^{+} + w_{t}^{-}d_{t}^{-}\right) \qquad (1)$$

subject to:

$$\left(100/G_{it}\right)\left(\sum_{j=1}^{10} a_{ij}x_{jt}\right) - d_{it}^+ + d_{it}^- = 100 \qquad \forall i,t \qquad (2)$$

$$\left(100/G_{t}\right)\left(\sum_{j=1}^{10} c_{j}x_{jt}\right) - d_{t}^+ + d_{t}^- = 100 \qquad \forall t \qquad (3)$$

$$(1-\alpha)x_{jt} - x_{j(t+1)} \leq 0 \qquad \forall j,t \qquad (4)$$

$$(1+\beta)x_{jt} - x_{j(t+1)} \geq 0 \qquad \forall j,t \qquad (5)$$

$$x_{jt} \geq 0; d_{it}^+ \geq 0; d_{it}^- \geq 0; d_{t}^+ \geq 0; d_{t}^- \geq 0 \qquad \forall i,j,t \qquad (6)$$

Equation 1, the objective function, represents the weighted sum of deviations from the goals. The first terms in the objective function (e.g., the double summation) represent weighted deviations from the nutritional goals in each time period. The parameter P_{it} is the weight on the ith nutritional goal in the tth time period. The variable $d_{it}^+ (d_{it}^-)$ is the positive (negative) deviational variable in the ith nutritional goal in the tth time period. The variable $d_{it}^+ (d_{it}^-)$ measures the percent deviation above (below) the ith nutritional goal in the tth time period. The parameter $w_{it}^+ (w_{it}^-)$ is the weight of $d_{it}^+ (d_{it}^-)$ in the ith nutritional goal in the tth time period.

The second terms in the objective function (e.g., the single summation) represent weighted deviations from the cost goal. The parameter P_t is the weight on the cost goal in the tth time period. The variable $d_t^+ (d_t^-)$ is the positive (negative) deviational variable in the cost goal in the tth time period. The variable $d_t^+ (d_t^-)$ measures the percent deviation above (below) the cost goal in the tth time period. The parameter $w_t^+ (w_t^-)$ is the weight of $d_t^+ (d_t^-)$ in the cost goal in the tth time period.

The four time periods corresponding to the subscript t are very flexible. They can be set to equal duration, such as 30 days each, or each can be set at a different duration. This allows the rancher to customize an entire feeding schedule according to an animal's physiology.

Equation 2 represents the i = 5 nutritional goals. The five nutritional goals are net energy for maintenance (NEm), net energy for gain (NEg), crude protein (CP), calcium (Ca) and phosphorous (P). Each nutritional goal is also represented in each time period. Since there are 4 time periods in the planning horizon, there are 20 nutritional goals in total. Equation 3 represents the cost goals. The cost goal is represented in each time period, resulting in 4 cost goals in total. Combined, there are 24 positive deviational variables and 24 negative deviational variables. The parameter a_{ij} represents the units of the ith nutrient in the jth feed. The decision variable x_{jt} represents the dry kilograms of the jth feed used in the tth time period in the ration. Since the model accommodates 10 feeds and 4 time periods, there are 40 decision variables. The constant G_{it} is the target level for the ith nutritional goal in the tth time period. The constant G_t is the target level for the cost goal in the tth time period. As each goal is measured in incommensurate units, it was necessary to convert them into percentages by dividing both sides of each goal equation by the target level and then multiplying both sides by 100. A 1 unit deviation in a goal represents a 1% deviation in that goal. The deviational variables are exploited in the user-requested output reports as feedback mechanisms to indicate which goals are being under-, exactly-, and over-achieved.

Equations 4 and 5 represent the [(2)(j)(t-1)] or 60-ration composition continuity constraints for the decision variables. These constraints ensure a smooth transition from period to period for individual feeds in the ration. As the beef cattle mature over the feeding horizon, their nutritional requirements also change. Myopically applying static ration models ignores ration composition continuity. Animal productivity and health are threatened when diets radically change as is often the case when successively applying static models through time. The continuity constraints prevent this "bang-bang" phenomenon from occurring. The parameters α and β, exogenously specified by the user, belong to the interval [0,1]. The smaller they are, the tighter are the bands of tolerance. For example, if $\alpha = \beta = 0.2$, then the amount of the jth feed used in next period's ration must lie between 80 and 120% of the jth feed used in this period's ration. These parameters can be varied in a sensitivity analysis as the DSS user deems appropriate. These constraints are similar to the harvest flow constraints used in LP models by forest planners (Dykstra, 1984). Other nutritional constraints and stomach capacity constraints were omitted from the model due to the model size limitations of the spreadsheet software used in the development of the DSS prototype. Finally, Equation 6 represents the usual nonnegativity constraints.

Goal programming minimizes the sum of weighted deviations from the targets of respective goals. The target levels, the weights of each goal and the weights of the deviational variables simultaneously influence the direction of the solution. These parameters are supplied by the DSS user. One of the key aspects to using GP successfully is knowing *a priori* what the target levels should be. Animal scientists have developed regression formulas defining what the minimum daily nutritional requirements are for a given starting weight and desired rate of live-weight gain for nine different beef cattle types (Nutrient Requirements of Beef Cattle, 1976). Deficiencies in the nutritional goals can cause expensive animal ailments, weight loss, and even death. Depending on the sophistication of the feeder, other minerals and vitamin levels may be targeted in the rations, but NEm, NEg, CP, Ca, and P are the most commonly cited needs for animal health and productivity. Of these, the most important nutritional requirements for beef cattle are NEm, NEg, and CP (Perry, 1984). This information can be used to develop weighting strategies for the goals.

The regressions for each body type are the second component of the model base. They are used to determine the targets as shown in the following example for NEg. The NEg target, GS_{it}, for a medium-frame steer calf in the first feeding period is given by:

$$NEg = 0.0557 W^{0.75} LWG^{1.097}$$

NEg is the minimum daily megacalories (Mcal) of net energy for gain, W is the animal's weight in kilograms and LWG is the desired daily live-weight gain in kilograms. LWG is a chosen management decision pursued by the farmer or rancher. A 200-kg medium-frame steer calf with a desired LWG of 0.8 kg would require a minimum of 2.32 Mcal/day. Since the GP model is a dynamic four-period model, a separate NEg target level must be calculated for the next three planning periods. If there were 30 days in each period, the steer would grow at a rate of 24 kg per period, assuming that the LWG was desired to stay fixed at 0.8 kg/day. The successive NEg targets given by the nonlinear function would be 2.52, 2.73, and 2.92. Updating the nutritional targets each period based on the weight of the animal reduces the error that would be imposed in a single period model. The DSS automates these calculations for the user and accesses the regression equation coefficients using spreadsheet lookup functions.

In Waugh's least cost ration, cost is the objective function to be minimized. Cost is incurred when purchasing the various feeds that compose a ration. In the GP model cost is a goal. The target level for this goal can be set in two ways. The most direct approach is to set the target using historical cost accounting data from the user or from enterprise budgets. The second approach involves finding a solution to Waugh's least cost ration and using it as the target in the GP model. Both of these methods, however, ignore the opportunity cost of the value of foregone production when a limiting nutritional goal is under-achieved. This issue can be addressed through the weights in the objective function as discussed next.

There is little *a priori* information for determining the relative weights for each goal. Most producers, perhaps, would weight cost more heavily than the various nutritional goals. However, forcing the GP to exactly satisfy a relatively low cost target through the use of relatively heavy weights may result in a severe violation of a nutritional target. This violation may result in lost production and, hence, lost revenue. Thus, the DSS imposes equal weights on the goals initially and utilizes sensitivity analysis to evaluate alternative weighting schemes. Weights on the positive and negative deviational variables may be easier to set. For example, it is natural to assume the weight on the negative deviational variable in the cost goal is zero. In other words, there is no penalty for being below the cost target. Similarly, the weights on the positive deviational variables in certain nutritional goals may be low relative to the weights on the negative deviational variables in those goals.

Database

The livestock feeding application lends itself well to the DSS methodology. Animal scientists, nutritionists and veterinarians have studied livestock nutrition for decades and have compiled a wealth of knowledge on the subject. A common reference used for feeding beef cattle is the National Academy of Sciences' Nutrient Requirements for Beef Cattle publication (1976). The publication contains the aforementioned regression formulas used in the calculation of minimum daily nutrient requirements for different classes of beef cattle. The second major component of the publication is the table of feeds. There are over 250 feeds for beef cattle recorded in the feed table. Each feed has a corresponding value for the energy, protein, mineral, and vitamin parameters. These represent the a_{ij} technological coefficients in the GP model. The DSS leverages this database using spreadsheet lookup functions to enter and update the a_{ij} technological coefficients in the GP model. This accelerates the process of developing and evaluating alternative ration formulations as a DSS user can choose various feeds while the GP model is updated instantly.

These a_{ij} values are only averages and are influenced by many factors. Growing conditions, such as soil type, climate and slope, as well as management actions, such as irrigation and chemical applications, will influence the values of the technological coefficients. Some ranchers test each feed prior to feeding to get a more accurate reading of its nutritional contents. Moreover, the c_j cost coefficients change as commodity market conditions change. For these reasons the database has an option to override the hard-coded values for nutrients in feeds and the prices of feeds.

User Interface

The user interface was the first component developed in the DSS. This was necessary to identify all of the input and output requirements of the decision maker and the decision to be made. Beef cattle nutritionists, animal scientists, extension agents, and agricultural economists were consulted prior to building the DSS to determine these requirements. The relevant user-supplied inputs are animal type, starting animal live-weight, desired daily rate of live-weight gain, number of days in each feeding period, cost targets, ten relevant feeds, the weights for the goals, and the parameters for the ration composition continuity constraints. All of this information is readily available and easily input for most DSS users.

There are several user requested outputs. The rancher receives a report indicating, on both a dry and as-fed basis, the volume of each feed to use in each time period, as well as a table summarizing the costs of the ration. Information in the goal analysis section is also captured in a table indicating the percent deviation above and below the nutritional and cost goals for each time period. The deviational variables measure this and, thus, provide feedback on the expected nutritional health of the cattle as well as the cost of the operation. If the deviations appear unreasonable or sensitivity analysis is desired, the rancher can adjust the weights of the respective goals.

Case Study

The best approach to describing the benefits of this DSS is an actual demonstration of its potential. The DSS was run for a rancher with 20 large-frame bull calves. The calves start at a live-weight of 400 kg at the beginning of a 120-day feeding schedule. During each of the four 30-day periods in the schedule, the desired LWG is 1.4 kg/day. The finished weight of the bull calves is expected to be 568 kg. The rancher has also selected 10 of the most accessible feeds from the 25 feeds available in the DSS's database. These feeds include fresh alfalfa, alfalfa hay, alfalfa silage, barley grain, corn grain, corn silage, corn stover, sorghum grain, soybean meal, and urea. The initial values for the ration composition continuity constraints are $\alpha = \beta = 0.2$. The weights on the goals, the P_{it}, are set equal initially to show what would happen if there were no priorities among the goals. The weights on the deviational variables, however, reflect that it is not preferred to under-achieve the nutritional goals of NEm, NEg, and CP or over-achieve the cost goals. Finally, the targets for the cost goals are set to $0.38/head/day based on historical cost accounting data.

Table 25.1 Optimal Baseline Ration

Feed[a]	Period 1	Period 2	Period 3	Period 4
Corn silage	0.53	0.63	0.76	0.91
Corn stover	6.10	6.03	5.95	5.87
Sorghum grain	2.10	2.21	2.30	2.39
Urea	0.13	0.14	0.15	0.15
Total	8.87	9.01	9.16	9.32

[a] All quantities of feed are measured in dry kilograms.

Table 25.1 summarizes the optimal baseline ration. Although the ration is composed mainly of corn stover and sorghum grain, it also includes small amounts of corn silage and urea. The amount of corn stover steadily declined throughout each successive feeding period while the amounts of the other feeds steadily increased. The total dry kilograms of feed in each ration in the four periods are, respectively, 8.87, 9.01, 9.16, and 9.32. A goal analysis indicates that the NEm goal is always over-achieved and the CP, Ca, and P goals are always exactly-achieved. The NEg goal is over-achieved in the first three periods and under-achieved in the fourth period. This fourth period deficiency may be an area of concern for animal health and productivity, especially in the critical final 30-day finishing period. Finally, the cost goal is under-achieved in the first two periods, exactly-achieved in the third period, and over-achieved in the final finishing period. This cost overrun is viewed as potential financial control problem and may be investigated using sensitivity analysis.

Summary

A prototype DSS was built in a spreadsheet to determine the optimal mix of feed per head per day for various cattle body types. The DSS relies on a linear GP model that builds on earlier efforts by other authors. The weighted sum of deviations from five nutritional goals and a cost goal are minimized over four feeding periods that compose the holding period of the animal. An extensive database holds technical information needed by the model. A user interface elicits subjective information about user goals and their weights, as well as information about animal types and feeds to consider in the optimization. The model also allows a user to specify continuity requirements for feeding patterns from one period to the next.

The GP solution is a more general solution than Waugh's least cost ration solution. It includes, as a special case, a rigidly defined LP solution. The GP solution, however, increases tractability by allowing various constraints to be violated. A case study showed how the program could be used. The GP model enhances the chance that a rancher will find a ration that meets his needs. The DSS in a spreadsheet increases the access to this sophisticated modeling technique by making it relatively user-friendly. An extension application or private market version could improve the DSS by avoiding the limitations of the spreadsheet software.

References

Crabtree, J.R. 1982. Interactive formulation system for cattle diets. *Agric. Syst.* 8:291–308.

Dykstra, D.P. 1984. *Mathematical Programming for Natural Resource Management.* McGraw-Hill, New York. 318 pages.

Goicoechea, A., D.R. Hansen, and L. Duckstein. 1982. *Multiobjective Decision Analysis with Engineering and Business Applications.* John Wiley & Sons, New York. 519 pages.

Lara, P. 1993. Multiple objective fractional programming and livestock ration formulation: a case study for dairy cow diets in Spain. *Agric. Syst.* 41:321–334.

Lara, P. and C. Romero. 1994. Relaxation of nutrient requirements on livestock rations through interactive multigoal programming. *Agric. Syst.* 45:443–453.

Nutrient Requirements of Beef Cattle. 1976. 6th rev. ed., National Academy of Sciences, Washington, D.C. 90 pages.

Perry, T.W. 1984. *Animal Life-Cycle Feeding and Nutrition.* Academic Press, San Diego. 319 pages.

Rehman, T. and C. Romero. 1984. Multiple-criteria decision-making techniques and their role in livestock ration formulation. *Agric. Syst.* 15:23–49.

Rehman, T. and C. Romero. 1987. Goal programming with penalty functions and livestock ration formulation. *Agric. Syst.* 23:117–132.

Waugh, F.V. 1951. The minimum cost dairy feed. *J. Farm Econ.* 33:299–310.

Chapter 26

Conflict Resolution Among Heterogeneous Reusable Agents in the PARI DSS

MURRAY J. BENTHAM AND JIM E. GREER

Introduction

Farmers and ranchers in the Prairie Region of Western Canada are currently facing many challenges of which soil degradation is one of the key issues. Conservation production systems must be implemented to ensure the long-term sustainability of the soil resource and in turn the sustainability of agriculture. Farm managers need sound advice and accurate information on all aspects of their farm operations in order to make appropriate decisions that will conserve the soil, land, and water resources while also maintaining the economic viability of their operations. In order to deal with the many complex issues and numerous alternatives available, they need access to various knowledge sources to support their production decisions.

A vast amount of technical information, both published and unpublished, resides in various locations, but may not be readily available nor utilized at present. In addition, many "experts" in various disciplines, at research and educational institutions and at extension agencies, have considerable knowledge and experience that is not being fully utilized. Innovative multiple objective support systems are required to make information and knowledge available to farm managers to support their production decisions.

Agriculture and Agri-Food Canada, Research Branch, under the Parkland Agriculture Research Initiative (PARI), is developing a conservation production decision support system (DSS), comprising multiple agents to facilitate multiobjective decision making focusing on encouraging adoption of sound economically viable soil conservation practices. The primary client of the PARI DSS will be farm managers interested in adopting conservation farming practices, although other clients may include extension specialists, research scientists, and private industry.

The PARI DSS is being collaboratively developed with producers, governments, universities, and private industry. A 4-level framework, 0 to 3, enables flexible natural growth while providing a "single window" to decision support. Level-0 acts as a clearing house, providing services and administrative support to PARI DSS clients. Level-1 consists of basic decision support based on databases and management information systems. Level-2 provides advanced decision support consisting of standalone analyzers. Level-3, which provides the most sophisticated support through the integration of multiple heterogeneous reusable expert systems utilizing a blackboard architecture, is the focus of this paper.

Background

Present day problem-solving systems are large and complex reflecting rapid and constant changes to objectives in application domains. Often these systems require the expertise of multiple expert agents cooperating together to produce comprehensive solutions to complicated problems. These agents might be developed from scratch for each application system, a costly and timely process, or more recently, application systems are being built with heterogeneous and reusable agents. The context of our application necessitates a further challenge to utilize existing "off-the-shelf" heterogeneous reusable agents, not only to minimize costs and development time, but to create an open system. *Open systems* receive new knowledge and information from outside themselves at any time, and hence, their operation is always subject to unanticipated outcomes (Hewitt, 1991).

An *agent* is defined as a logically independent computational process representing a specific area of expertise, such as an expert system, developed to deliver a specific knowledge set (Bond and Gasser, 1988). Information represented in a modular and logically independent method is potentially *reusable* (Neches et al., 1991). A reusable agent system is computationally analogous to a pool of human specialists. Hence, a *multiagent system* is composed of agents dynamically selected from an existing library and integrated with no, or minimal, customized implementation (Lander, 1994).

Heterogeneity among agents can be classified as logical and implementational. Logical, or knowledge (as they are sometimes called), heterogeneous agents are characterized by differences in declarative knowledge, solution evaluation criteria, goals, capabilities, and priorities (Lander, 1994). Implementational heterogeneous agents can be characterized by differences in architectures, algorithms, languages, inference engines, or hardware requirements. These differences may also be classified as representation heterogeneity (Lander, 1994).

Our goal of an open system not only strives to use existing "off-the-shelf" agents, but seeks to provide collaborators the freedom to develop computer-based agents for their own specific use. Furthermore, integrating heterogeneous reusable agents provides a natural and flexible development methodology whereby new agents can be added and existing agents deleted in response to system requirements. Other advantages to using "off-the-shelf" heterogeneous reusable agents which fulfill our goals include (Lander, 1994):

- The uniformity of expertise within each agent makes them easier to design, develop, and maintain than single large systems.
- Consistency of knowledge can be maintained within local boundaries without necessitating agreement across boundaries.
- Software modifications can be focused at the problem or enhancement within the agent, without requiring changes to be disseminated throughout the entire system.

In the paradigm of cooperating expert systems, heterogeneous reusable agents must be integrated to provide effective technical interaction. Within these systems, the state of problem solving must be communicated and agents' actions must be coordinated to arrive at mutually acceptable solutions. However, in these cooperative environments, conflicts must be resolved as a result of incomplete or inconsistent knowledge and/or incorrect assumptions, different problem-solving techniques, and different solution evaluation criteria (Lander and Lesser, 1994). Anticipating and removing potential conflicts through software engineering at agent-development time is not possible since it is not known what knowledge will be contained in an open system. Software reuse in any form is hindered by the absence of technical tools and techniques to support information sharing (Neches et al., 1991). However, collectively joining diverse knowledge from the viewpoints of cooperating experts provides an extremely important source of balance and robustness in many real-world situations. This is characterized by carefully selected human project teams and work groups. Teams and groups can solve problems which are normally beyond the comprehensiveness of individual experts, and in doing so, provide potentially creative and innovative solutions resulting from a rich and varied body of knowledge (Lander and Lesser, 1989a).

In this paper, we discuss distributed artificial intelligence, focusing on multi-agent systems in terms of the formulation, description, and allocation of problems, and the interaction, communication, coordination, and conflict resolution among agents as it relates to the application context of the agricultural DSS called the PARI DSS.

Distributed Artificial Intelligence

Historically, researchers divided Distributed Artificial Intelligence (DAI) into two main areas (Bond and Gasser, 1988). *Distributed Problem Solving* (DPS) undertakes the work of problem solving by dividing the tasks among a number of cooperating modules that share knowledge about the problem and the developing solution (Smith and Davis, 1981). Alternatively, research in *Multiagent Systems* concentrates on coordinating the knowledge, goals, skills, and plans of autonomous intelligent agents to facilitate problem solving through their joint intelligent behavior.

Multiagent Systems

The agents in a multiagent system might work together toward a single global goal or toward separate individual goals that cooperatively interact to meet a common goal. In the PARI DSS, the agents are heterogeneous, and in most cases "off-the-shelf" (i.e., reusable) with separate individual goals, which are integrated through a cooperative environment to provide comprehensive decision support. As in a DPS system, agents in a multiagent system need to share knowledge about problems and solutions but must also reason about the coordinating processes used among the agents in the cooperative environment. Coordination in a multiagent system, particularly an open system (Hewitt, 1991), can be difficult, as situations can arise where there is no possibility for global control, globally consistent knowledge, globally shared goals, or global success criteria, or sometimes, not even a global representation of a system (Hewitt, 1991). Many of the coordination difficulties in multiagent systems can be attributed to several main problems which are identified in the following subsections.

Formulation, Description, and Allocation of Problems

In the same way that work is divided among human experts, there must be a division of labor and organization in order to distribute tasks among agents (Bond and Gasser, 1988). In order to accomplish this distribution, tasks must be formulated and described in such a way as to facilitate their distribution (Bond and Gasser, 1988). Complicated tasks requiring more resources or knowledge than is held by a single agent must be decomposed in order that they can be accomplished. Tasks must be allocated or assigned to particular agents that are capable of performing them. All these requirements are very interdependent.

Interaction, Communication, and Coordination Among Agents

In order for several intelligent agents to combine their expertise and efforts, interaction and communication are necessary underlying concepts of multiagent systems. Interaction in a multiagent system means a type of collective action in which agents take action or make decisions based on influence from the presence or knowledge of other agents (Bond and Gasser, 1988). Unlike agents' perceptions, beliefs, and goals, which may or may not be distributed, interaction is inherently distributed and is necessary for the successful coordination of agents' actions.

Conflict Resolution Among Agents

Intelligent problem-solving agents must externalize parts of their world in order to reason about them (Bond and Gasser, 1988). This externalization is subject to problems of abstraction and incompleteness due to the problem of attempting to fully represent an object or process. Therefore, intelligent agents need to contend with differences and uncertainty between their externalized representations and the affairs actually represented. When multiple agents are integrated they must externalize one another into representations which must then be aligned to coordinate with each other. Alignment does not mean that agents share representations but rather that the representations must allow them to act so as to fulfill their individual goals. Developing common representations, such as standards for communication protocols or information exchange (Durfee, 1987), appear to be workable, but it depends very much on the agents' compatibility to interpret these protocols (Winograd and Flores, 1986). Furthermore, agents may have knowledge bases which contain differing beliefs, or they might have the same beliefs but with different confidence levels. Differences like these lead to conflicts among the agents.

The Paradigm of Cooperating Experts

When human experts cooperate on a single problem, they contribute their multiple diverse viewpoints in the problem-solving process. Merging diverse knowledge is common in real-world situations and brings with it a source of robustness and balance (Lander and Lesser, 1989a) which is required to develop comprehensive credible solutions. A plant breeder and an agrologist work together to develop crops with high quality and good yields. A chemist and a machinery engineer will cooperate to produce effective and environmentally safe herbicides. In general, the greater the complexity of problem solving, the more cooperating experts are required. Cooperation and diversity can be advantageous, providing increased problem-solving abilities beyond the individual expert and promoting creativity and innovation (Lander and Lesser, 1989b). However, conflicts result from cooperation and these must be resolved among the experts through information exchange. Conflict resolution among heterogeneous reusable agents in the PARI DSS poses some special problems which are addressed in the following section.

Cooperating Expert Systems

The PARI DSS is being developed collaboratively with cooperation of specialists from federal and provincial governments, universities and private industry. All of the expert systems within the PARI DSS either exist "on-the-shelf" as reusable agents or are being developed as reusable agents. A specialized multiagent system which shares expertise has several advantages over a large monolithic system developed to solve the same problem(s) (Lander and Lesser, 1989b).

The multiagent system approach provides the PARI DSS with a natural, flexible development framework in which to integrate heterogeneous reusable agents. This approach provides the capability to implement agents independently, with minimal time and resource requirements, add agents to the PARI DSS as they become available, and to delete or substitute agents in response to changes in specifications and technology. Agents developed for multiple uses are generally more reliable and their development cost can be amortized over multiple uses (Lander and Lesser, 1994). In addition, a modular expert approach provides the ease and simplicity associated with developing, debugging, testing, and maintaining small systems.

The PARI DSS is to utilize reusable expert systems, and with minimal modification to them, integrate them into a cooperating multiagent system. There are two main approaches to developing cooperative expert system environments. A popular approach is to establish a global structure containing information specifically about each agent and corresponding conflict strategies (Klein and Lu, 1990; Lander and Lesser, 1989b; Werkman et al., 1990). A contrasting approach is to develop each agent with its own conflict knowledge, separate from its domain level knowledge. This conflict management knowledge is not accessible or known to other agents (Polat et al., 1993). There are tradeoffs between these two approaches. With the global structure approach, information must be specifically defined about each agent and corresponding conflict strategies. However, with the contrasting approach, each agent must be developed specifically with its own set of conflict resolution schemes and translation capabilities, in order for the system to interact. Dependencies exist in both approaches. One approach has dependence on the global conflict structure and the other approach has dependence on the agent itself.

The approach, in which each agent has a built-in set of conflict resolution schemes and translation capabilities, defies the term *reusable agent,* and hence, is not applicable to the PARI DSS. The global conflict structure approach holds more promise for conflict management within the PARI DSS. Although each agent in a cooperating expert system is independent and a fully functional knowledge-base system, Lander and Lesser (1989a), do not believe that "off-the-shelf" agents will work effectively, even with minor modifications. Some of their major reasons include (Lander and Lesser, 1989a):

- A shared communication language does not exist within a reusable agent.
- There is no mechanism to handle inconsistency and unique beliefs among reusable agents.
- Inadequate internal knowledge representations do not permit sufficient goal and history information (i.e., cases) to be retained for cooperative solution revision.
- There is no mechanism for sharing timely information during problem solving.
- There is no provision for resolving conflicts.

Section 4 addresses these concerns and describes the application of conflict management techniques in the PARI DSS in an attempt to accommodate the use of "off-the-shelf" reusable agents.

Application of Conflict Management Techniques in the PARI DSS

Within the PARI DSS, it is necessary to integrate several conflict management techniques to use the "off-the-shelf" reusable intelligent agents in this cooperative problem-solving system. This is not an easy task. As stated by Lander and Lesser (1994) "There is a high degree of complexity inherent in building heterogeneous agents that can understand each other well enough to positively affect mutual work." Figure 26.1 illustrates the general system architecture for level-3 of the PARI DSS. A Global Control Expert (GCE) is being developed to interact with a Global Knowledge Base (GKB), which will be implemented using a blackboard-based problem-solving approach. The blackboard provides the communication medium since a common communication language among reusable agents is not possible. Meta-knowledge about the reusable agents, accessible by the GCE, will be utilized to handle inconsistency and unique beliefs among the agents. The GKB will contain the meta-knowledge about each reusable agent including

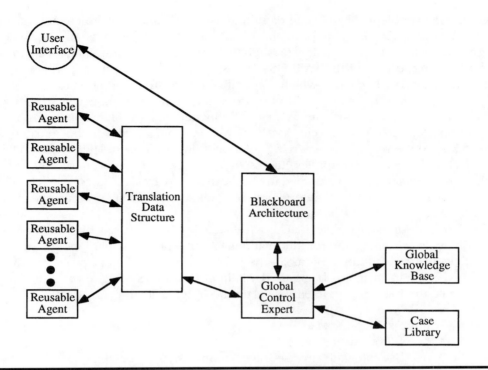

Figure 26.1 PARI DSS level-3 system architecture.

declarative knowledge, assumptions, beliefs, problem-solving techniques, solution evaluation criteria, minimum and maximum constraints, if applicable, conflict threshold limits, and other relevant meta-knowledge. Reusable agents post their information to the blackboard via a translation data structure which temporarily holds the data while the GCE, utilizing agent meta-knowledge, determines the required subset of data required to be placed on the blackboard for each agent and the problem-solving process, and then distributes this data to the blackboard. Case-based reasoning, utilizing a case history database, may be incorporated to solve future related problems. Compromise negotiation and weighted constraints are evoked, utilizing the GKB, when conflict thresholds are reached. In some cases, an agent may need to relax some requirement on the solution in order to resolve conflict. Flexibility values are used which indicate the degree of relaxation permitted. The following discusses the combination and extension of these approaches to conflict management, in order that a successful methodology can be developed to use "off-the-shelf" reusable intelligent agents in the PARI cooperative problem-solving paradigm.

The Blackboard Model Approach to Problem Solving

The blackboard model is a relatively complex problem-solving model providing the conceptual organization of information and knowledge, and prescribing the dynamic control and use of it pertaining to the problem-solving behavior within the overall system (Engelmore and Morgan, 1988; Jagannathan et al., 1989). The blackboard data structure, in its most general form, consists of a global data area containing problem-solving state data. Agents incrementally make changes to the blackboard based on their problem-solving ability at a given time. The agents communicate and interact solely through the blackboard. A controlling mechanism, the GCE in the case of the PARI DSS, is responsible for determining which agent to activate, when, and using which part of the blackboard. In addition to organizing information and knowledge, blackboard systems have a particular reasoning behavior associated with them as a result of incremental problem solving. At each solution step or control cycle, any type of reasoning step can be initiated in response to the part of the emerging solution being addressed. Hence, the control mechanism selects and applies agents dynamically and opportunistically rather than by fixed and preprogrammed methods. In some cases, the control mechanism might activate an *a priori* determined set of knowledgeable sources, as in the PARI DSS.

The development of the blackboard model was motivated by the need for flexibility in reasoning and for information sharing. The blackboard model is useful in a broad range of applications and is able to facilitate different problem-solving methods. In particular, it is being used within the PARI DSS because of its capability to handle the following problem characteristics (Jagannathan et al., 1989):

- A huge solution space
- Various input data
- A need to integrate diverse information
- A need for cooperation and coordination among heterogeneous reusable agents in forming credible solutions or recommendations
- The requirement for an evolutionary solution

Recognition-Oriented Blackboard System

Within the problem-solving paradigm of the blackboard model, there are two methods for determining the next state in the solution process: search and recognition. At any solution step or state, the search method generates and evaluates the possible next states, whereas the recognition method simply knows what the next state should be through a pattern matching process that involves scanning the knowledge base for knowledge that can be applied to a state (Jagannathan et al., 1989). In the case of the PARI DSS, the recognition-oriented approach is taken (i.e., the knowledge base is the GKB and the knowledge that can be applied to a state is either activation rules or conditional rules). Naturally, there may be several pieces of knowledge (i.e., rules) applicable for a given state and a decision is necessary to determine what knowledge (i.e., rules) to apply. This decision is also knowledge-based, or in the case of the PARI DSS, rule-based. In a recognition-oriented blackboard system, this decision results in triggering either an activation rule or a conditional rule. The triggering of an activation rule will fire and control a set of agents, based on the predetermined rules for a given situation, resulting in a change to the solution space on the blackboard. The triggering of a conditional rule will fire a predetermined rule set for a given situation resulting in a contribution to the solution space on the blackboard. In the PARI DSS, the activation rule sets, located in the GKB, trigger and control the reusable agents in their search for a solution. The conditional rule sets contribute to the solution by bridging gaps and solving "grey" areas between the agents as they are being fired.

The Global Control Expert

The GCE is intended to provide the global control necessary for interaction, communication, and coordination among agents. In addition, the GCE has access to meta-knowledge about the reusable agents, which is contained within the GKB, and used to handle inconsistency and unique beliefs among the agents. In this way, the GCE should be able to reason about the coordinating processes used among the agents in the cooperative environment. The GCE, by utilizing the GKB, allocates tasks to the agents by matching the task requirements with the agent's knowledge and resources. The GCE triggers agents into action based on influence from the agents' contributions, the knowledge of other agents, and the problem-solving process. Since there is no common communication language shared among the agents, the GCE monitors the agents' processing of tasks, utilizing the GKB, providing the necessary interaction and communication among the agents via the blackboard and the translation data structure. Reasoning on the part of the GCE is aided by meta-knowledge about agents, and possibly case-based histories, enabling the GCE to partially predict future actions of an agent.

As the agents interact through the GCE, they post contributions to the shared portion of the blackboard. The GCE detects conflicts when agents post contributions judged to be in disagreement with each other based on the constraints and goals of the original task. In resolving conflict among heterogeneous reusable agents, the GCE needs sufficient artificial intelligence and sufficient access to required knowledge, so as to have a global view of the problem-solving activity with the power to make global decisions that are coherent and in the best interest of the agents involved. In so doing, it may be necessary to settle for only partial goal satisfaction. On the other hand, possibly goals can be relaxed, so that all goals can be fulfilled. In the extreme case of conflict resolution, perhaps some goals need to be fully relinquished in order to find a solution. The GCE must be able to make decisions about what goals to emphasize, and localize revisions to these goals as much as possible.

Meta-Knowledge

As the PARI DSS evolves, with the addition of new agents and deletion of agents no longer required, the meta-knowledge within the GKB needs to be updated. The objective of the GCE, when using meta-knowledge, is not to locate a specific solution, but to focus the GCE in its search for a solution (as in Lander and Lesser, 1994). In order for the GCE to guide the processing towards an acceptable result, the GCE must be able to determine the agents' requirements for solutions. This is accomplished by utilizing the agents' meta-knowledge. Agent meta-knowledge will consist of knowledge including:

- What are the agent's goals?
- What implicit heuristics are employed?
- What results is the agent capable of providing?

- What are the agent's input parameters?
- What are the agent's conflict thresholds?
- What are the agent's weighted constraints?
- How are constraints imposed?

One of the key entities of the meta-knowledge will be constraints. The GCE must be able to assimilate constraining information of each agent in order to guide the problem-solving activities. Constraints are discussed in more detail in the next section.

Compromise, Negotiation, and Relaxation

In heterogeneous reusable agent problem solving, each agent can determine the status of its local solution, but there is no effective method for an agent to determine the status of the global solution, as no agent has knowledge of the constraints of other agents (Lander and Lesser, 1992). Although researchers (Sycara et al., 1991) have addressed this problem to some extent in their literature on constraint-directed negotiation, the investigations they conducted on this subject in a multiagent environment necessitated that the agents share a global consistent problem-solving methodology and agent architecture. These agents are heterogeneous, based on their resource requirements, but homogeneous in their implementation, and implicitly use their homogeneity for control.

Although conflict resolution can be attempted using any of several methods of negotiation, compromise negotiation holds the most promise for dealing with negotiation among heterogeneous reusable agents in the PARI DSS. Compromise negotiation uses a methodology where values are iteratively revised by sliding them along some dimension until a mutually agreeable position is found (Lander and Lesser, 1989b). An example of compromise negotiation is demonstrated when a customer is purchasing a car. The car buyer and salesman iteratively propose prices (i.e., slide a value along a monetary scale) until the two proposals converge. For compromise negotiation to be effective, a small number of dimensions should be involved, a method should exist for determining whether the proposed values are moving toward each other, and the values should be within an acceptable range (Lander and Lesser, 1989a). The GCE will utilize weighted constraints from the agent meta-knowledge in the GKB and hence, will know the agents' "bottom line" constraints, leaving the main task of identifying and compromising the salient constraints. By establishing a set of conflict attributes as part of the meta-knowledge for each agent, and defining threshold values for these attributes, the GCE can reason as to when to try compromise negotiation.

An alternate approach to conflict resolution among heterogeneous reusable agents in the PARI DSS is relaxation. Relaxation methods are invoked when the problem-solving activity is believed to be over constrained (Lander and Lesser, 1992) resulting in no or little progress in solution building. Relaxation is invoked when one or more agents relax some requirement on a goal or solution, thereby enabling problem solving to advance. Sometimes through relaxation, a case stored in the case history database which previously failed, may become immediately acceptable. Within each agent's meta-knowledge, weighted constraints are assigned to each information type or variable, which define the degree to which that agent is willing to relax that information (Lander and Lesser, 1992). When managing conflict, the GCE will evaluate the weighted constraints of the agents and the constraints being imposed by the solution analysis and thereby determine if relaxation is feasible.

A PARI DSS Scenario

Through the use of a third-party agent, a shared global blackboard, agent meta-knowledge, and several conflict resolution techniques, we are attempting to utilize "off-the-shelf" reusable agents. Integrating heterogeneous reusable agents in the collaboratively developed PARI DSS provides a natural, flexible development methodology, providing the capability to add agents as they become available and to delete or substitute agents in response to changes in specifications and technology. Agents developed for multiple uses are generally more reliable and their development cost can be amortized over multiple uses (Lander and Lesser, 1994). Conflict resolution plays a major role in the success of integrated reusable agents and forms the basis for communications and coordination protocols. The following scenario will help to illustrate the role of conflict resolution among heterogeneous reusable agents in the PARI DSS. The following examples of PARI DSS reusable agents will be used: (1) CROPMAN, a crop management expert system (ES), which focuses on crop production including crop selection and rotation, herbicide requirements, fertility, and potential plant disease (de Gooijer, 1995); (2) a Stepwise Technology Adoption Risk Reduction Tool ES called STARRT, which evaluates the economic impact of different cropping decisions (Alexiev et al., 1995); and (3) a planting and residue ES called PARMS, which determines the level of success of planting a crop given the planting equipment, current production practices, and conservation goals (Pavlik et al., 1995).

Even without considering other reusable agents in the PARI DSS, such as weed and insect agents, the complexity of integration and conflict resolution can be illustrated. Suppose that a producer, during pregrowing season planning, requests the PARI DSS to advise on the crop(s) to grow with the best economic net return. The producer's field-specific database is queried and required data is placed on the global blackboard. The GCE triggers CROPMAN which determines crop candidates that can be grown based on crop rotations, fertility, soil moisture, potential plant disease, herbicide residue, and agroclimatic zone. These results are posted to the blackboard via the translation data structure, and the GCE triggers STARRT which uses the information on the blackboard and determines the best crop(s) to grow on each field based on present market outlooks and input costs, and posts its results to the blackboard. The GCE invokes PARMS which uses the information from the blackboard, machinery inventory, and current levels of crop residue to determine that the crop(s) recommended cannot be seeded without equipment modifications due to the high level of crop residue present. PARMS posts the applicable information to the blackboard including the costs for equipment modification. The GCE recognizes a conflict and accesses the case history database to realize that this decision process and results have not occurred previously and updates the case-based history database. The GCE, with the aid of the GKB, determines that STARRT needs to reevaluate the net return given the cost of equipment modifications. If the net return indicates the recommended crops are not cost effective, then CROPMAN is triggered again with imposed crop constraints. The process continues until a viable net return results with no conflicts.

Conclusions

Lander (1994) states; "An agent can be either an existing piece of software, modified to work within an agent set, or it can be a piece of software specifically designed as a reusable agent to work within an agent set. The first case is more problematic." Neches et al. (1991) indicate that sharing and reusing knowledge sources in any form is difficult, but feasible, if knowledge sharing technology and infrastructure tools can be created with which to facilitate knowledge-based system development and operation. The PARI DSS is challenged with the integration and utilization of "off-the-shelf" reusable intelligent agents in this cooperative problem-solving system.

We believe that the PARI DSS has an advantage when utilizing "off-the-shelf" reusable agents in that the recognition-oriented blackboard system method being employed will simplify the problem-solving paradigm over the search method. The GCE, currently under development, plays a major role in the PARI DSS. As the control mechanism, the GCE is comprised of problem-solving heuristics as opposed to application heuristics. The GCE's responsibilities include monitoring the status of the blackboard to determine if the current solution state necessitates the triggering of an activation rule, or a conditional rule, and then controlling these predetermined rule sets, including the predetermined set of agents, if an activation rule is fired. The GCE, as the knowledge source control executive, must be able to assimilate and apply information from agents in order to guide the problem-solving activities. Therefore, agent information must be consistent with the GCE's local knowledge and thereby, the GCE relies heavily on agent meta-knowledge. The GCE's most difficult task will be guiding the problem-solving activities of "off-the-shelf" reusable agents in a cooperative environment to provide comprehensive decision support. We have discussed the role of conflict resolution in the integration of heterogeneous reusable agents and how it forms the basis for communication and coordination protocols.

Although there are several conflict resolution approaches and techniques that exist and are currently being researched, we believe that our investigation and consideration of the available conflict management methodologies has resulted in the development of a design which holds the most promise for integrating heterogeneous reusable agents. In addition to our current complement of conflict resolution methodologies and techniques, we believe the incorporation of case-based reasoning would assist in the solution of future related problems. The GCE, GKB, and blackboard architecture are currently being developed. Our plan is to utilize case-based reasoning in the future depending on the result of our prototype development. The PARI DSS is being built as an open system, and hence, there will be ongoing development.

Acknowledgments

We would like to acknowledge the financial support of the Agriculture and Agri-Food Canada, Research Branch, Parkland Agriculture Research Initiative (PARI).

References

Alexiev, V., K. Duhaime, and P. Dzikowski. 1995. *STARRT — Stepwise Technology Adoption Risk Reduction Tool*. (v1.3) [Computer software], Alberta Agriculture, Edmonton, Alberta.

Bond, A.H. and L. Gasser, Eds. 1988. *Readings in Distributed Artificial Intelligence,* Morgan Kaufmann Publishers, San Mateo, California.

de Gooijer, H.C. 1995. *CROPMAN — Crop Management*. (v2.0) [Computer software], 20-20 Agricultural Services, Kelliher, Saskatchewan.

Durfee, E.H. 1987. A Unified Approach to Dynamic Coordination: Planning Actions and Interactions in a Distributed Problem Solving Network. Doctoral dissertation. University of Massachusetts, Amherst.

Engelmore, R. and T. Morgan, Eds. 1988. *Blackboard Systems*. Addison-Wesley, New York.

Hewitt, C. 1991. "Open information systems semantics for distributed artificial intelligence." *Artif. Intell.* 47(1):79–106.

Jagannathan, V., R. Dodhiawala, and L.S. Baum, Eds. 1989. *Blackboard Architectures and Applications*. Academic Press, San Diego, California.

Klein, M. and S.C.-Y. Lu. 1990. Conflict resolution in cooperative design. *Int. J. AI Eng.* 4(4):168–180.

Lander, S.E. 1994. Distributed Search and Conflict Management Among Reusable Heterogeneous Agents. Doctoral dissertation. University of Massachusetts, Amherst.

Lander, S.E. and V.R. Lesser. 1989a. A framework for cooperative problem-solving among knowledge-based systems. Pages 1–14 in *Proceedings of the 1989 MIT-JSME Workshop on Cooperative Product Development*.

Lander, S.E. and V.R. Lesser. 1989b. A framework for the integration of cooperative knowledge-based systems. Pages 472–477 in *Proceedings of the 1989 IEEE International Symposium on Intelligent Control*. IEEE Computer Society Press, Washington, D.C.

Lander, S.E. and V.R. Lesser. 1992. Customizing distributed search among agents with heterogeneous knowledge. Pages 335–344 in *Proceedings of the 1992 First International Conference on Information and Knowledge Management*.

Lander, S.E. and V.R. Lesser. 1994. Sharing Meta-Information to Guide Cooperative Search among Heterogeneous Reusable Agents." Tech. Report CMPSCI 94-48. Department of Computer Science, University of Massachusetts, Amherst.

Neches, R. et al. 1991. Enabling technology for knowledge sharing. *AI Mag.* 12(3):36–56.

Pavlik, C., E. Smith, and C.W. Lindwall. 1995. *PARMS — Planter and Residue Management System*. (v2.1) [Computer software], Agriculture and Agri-Food Canada, Lethbridge, Alberta.

Polat, F., S. Shekhar, and H.A. Guvenir. 1993. Distributed conflict resolution among cooperating expert systems. *Expert Syst.* 10(4):227–236.

Smith, R.G. and R. Davis. 1981. Frameworks for cooperation in distributed problem solving. *IEEE Trans. Syst. Man Cybern.* 11(1):61–70.

Sycara, K.P. et al. 1991. Distributed constrained heuristic search. *IEEE Trans. Syst. Man Cybern.* 21(6):1446–1461.

Werkman, K.J. et al. 1990. Design and Fabrication Problem Solving Through Cooperative Agents. Report No. 90-05. ATLSS Engineering Research Centre, Lehigh University, Bethlehem, Pennsylvania.

Winograd, T. and F. Flores. 1986. *Understanding Computers and Cognition*. Alblex, Norwood, New Jersey.

Chapter 27

Integrated Decision Making for Sustainability: A Fuzzy MADM Model for Agriculture

ELIZABETH G. DUNN, JAMES M. KELLER, AND LEONIE A. MARKS*

Abstract

Fuzzy logic is an approach that is particularly useful for formulating and solving multiple attribute decision making (MADM) problems related to agricultural sustainability. The specific decision of interest is the selection among alternative farming systems in order to achieve the highest degree of sustainability. The fuzzy logic approach offers several advantages including the ability to (1) integrate mixed qualitative and quantitative data, (2) utilize data expressed in noncommensurate units without the need for rescaling, (3) include information that is relevant to the decision but may be vague or imprecise, (4) avoid inappropriately high levels of discrimination between alternatives that differ only slightly in their attributes, (5) model the interrelationships between the different sustainability criteria, and (6) provide a full ranking of the alternatives. To illustrate the application of fuzzy logic in MADM, a prototype model for selecting between two farming systems in northern Missouri is described and a decision support system (DSS) based on fuzzy logic is proposed.

---

* The authors wish to acknowledge the helpful suggestions of John E. Ikerd and Paul D. Gader, the programming assistance of Larry D. Godsey, and the information on forage-livestock systems provided by Kevin C. Moore.

Introduction

As the sustainability of agriculture becomes a topic of increasing interest, there is a need for DSS which can be used by farmers and public decision makers. Because of the multidimensional nature of agricultural sustainability, MADM provides an appropriate model for the development of DSS. However, the nature of the decision criteria and issues associated with the measurement of these criteria create several challenges in the development of DSS. In this paper it is argued that a fuzzy logic approach to MADM can overcome several key obstacles inherent in modeling decisions related to agricultural sustainability.

This paper is divided into four sections. The first section briefly describes the decision-making context. The second section is concerned with the application of MADM to the problem of sustainability. Several key terms used in MADM are defined. An example is provided to illustrate the use of MADM in modeling the selection of a sustainable farming system. At the end of the section, key issues in modeling agricultural systems with MADM are delineated. Section three provides a general introduction to fuzzy (multivalent) logic. The use of fuzzy logic to alleviate some of the problems in decisions related to sustainability is also discussed. Finally, section four describes a prototype model for applying fuzzy logic to the problem of agricultural sustainability. The paper concludes with a discussion of several important modeling issues that need to be addressed in order to develop a multiple attribute DSS based on fuzzy logic.

The Decision-Making Context

This paper is concerned with the decision of selecting among alternative farming systems in order to choose the system that has the highest degree of sustainability. A farming system typically consists of several smaller component systems, such as different types of crops, livestock, woods, and other subsystems and enterprises. The components of the farming system are arranged spatially and temporally, and they are interrelated to greater or lesser degrees. Within this context, the decision maker is a farmer operator or farm household.

Agricultural Sustainability

The concept of agricultural sustainability is derived from the more general concept of sustainable development. The idea of sustainable development first surfaced in the early 1980s and gained prominence with the publication of the Brundtland report, which defined sustainable development as "...development that meets the needs of the present without compromising the ability of future generations to meet their own needs" (WCED, 1987). There are a number of context-specific definitions of sustainability, but all existing definitions tend to contain at least some elements that relate to the three dimensions of sustainability: economic, environmental, and social.

The sustainability of agricultural systems can also be characterized as having these three dimensions. Agricultural production processes involve the manipulation of the natural environment to create products that are useful to people. The productivity of farming systems depends on the quality and quantity of inputs, such as water and soils. Agricultural productivity, in turn, is a major determinant of the profitability of farming. Likewise, farming is an activity which is anchored in the rural landscape and is a major component in the vitality of rural communities.

In order to ensure the sustainability of agricultural systems, decision makers need to incorporate information from all three dimensions of sustainability into their decisions. MADM approaches can form a useful basis for supporting the complex decision environment in agriculture. The following section considers MADM and places it within the context of agricultural sustainability through the introduction of an example. The same example is used to help illustrate some of the key issues that arise using MADM to model agricultural systems.

MADM and Agricultural Sustainability

The MADM approach is a useful way to model the problem of agricultural sustainability for several reasons. First, as already mentioned, sustainability is a multidimensional concept, encompassing numerous decision criteria that relate to economic, environmental, and social goals. Second, the goals of sustainability often conflict in the short run. MADM approaches are specifically designed to handle multiple, conflicting goals. Finally, it is likely to be difficult to draw conclusions about the overall sustainability of a farming system from an informal comparison among alternative farming systems. Farming systems that maximize the performance on one decision criterion may have mediocre or poor performance on another criterion.

MADM

MADM is the process of selecting a preferred alternative from a finite, discrete set of alternatives, where each alternative is described in terms of multiple attributes. The alternatives are the choices that the decision maker is considering. For the problem discussed in this paper, the alternatives are the different farming systems that the decision maker is considering for implementation.

The attributes in the MADM problem are the variables that the decision maker uses as the basis for evaluating alternatives. In other words, the attributes represent the decision criteria of the decision maker. The measurements (data) for the MADM problem are the specific realizations of the attributes for each alternative.

These data can be represented in an nxm decision matrix whose elements consist of the data for the problem. There are n rows in the decision matrix, one for each alternative A_i (I = 1,...,n), and m columns, one for each attribute X_j (j = 1,...,m). Each element of the matrix, x_{ij}, represents the measurement on the ith alternative with respect to the jth attribute. The decision matrix for the problem discussed in this paper is illustrated in a specific example below.

An Agricultural Example

A typical farming system in the north central region of Missouri is used to generate decision alternatives. The farmer operator works full time and cultivates 640 acres

	Profit	Cash-Flow	ROA	Erosion	N-Loss	Landscape	Safety	Esteem	Lifestyle
FS A_1	40	med	8	6.6	40	high	low-med	med-high	high
FS A_2	45	med-high	9	5.05	35	high-vhigh	med	high	high

Figure 27.1 Decision matrix for agricultural example MADM problem (FS = farming system; med = medium; vhigh = very high).

of wheat-soybean-corn in rotation. In addition, the farm has a cow-calf operation with 30 head (29 cows, 1 bull). The overall size of the farm is 800 acres, of which 360 are owned and 440 are rented. All of the rented land is used for the wheat-soybean-corn crop rotation.

The farmer is considering whether to continue with the current system of continuous grazing or to switch to a system of rotational grazing. The distinguishing feature of rotational grazing is that grazing access is limited to only certain parts of the pasture at certain times. The two alternatives are represented by A_1 and A_2, where A_1 denotes the farming system that includes continuous grazing and A_2 denotes the farming system with rotational grazing. In order to select between the two farming systems on the basis of sustainability, appropriate decision criteria need to be identified.

Under the economic dimension of sustainability, three subgoals are included in the model as follows: (1) return to unpaid labor, management, and equity (PROFIT); (2) cash-flow feasibility (CASH-FLOW); and (3) return on assets (ROA). Under the environmental dimension of sustainability the following subgoals are included in the model: (1) soil erosion (EROSION), (2) nitrogen losses (N-LOSS), and (3) landscape attractiveness (LANDSCAPE). The decision criteria PROFIT, ROA, EROSION, and N-LOSS are quantitative variables, while CASH-FLOW and LANDSCAPE are qualitative variables. Under the social dimension of sustainability, the three subgoals considered are (1) occupational safety (SAFETY), (2) self-esteem of the farm operator (ESTEEM), and (3) quality of lifestyle (LIFESTYLE). All three social subgoals are measured qualitatively. The data for this example were drawn from expert opinion. The problem is summarized in the decision matrix shown in Figure 27.1.

Key Issues in MADM Models

There are several issues that generally arise when using MADM; some of these issues are particularly important when MADM methods are used to model agricultural sustainability. Six of these issues are discussed below. The purpose of this discussion is to lay the foundation for understanding the potential contribution of fuzzy logic to MADM.

Mixed Data

The decision variables that underlie the three dimensions of sustainability are measured using a variety of scales, both qualitative and quantitative. When a decision matrix contains a mixture of qualitative and quantitative criteria, there are usually two choices available to the decision maker: (1) use a MADM method which can handle mixed data (Hwang and Yoon, 1981) or (2) transform all of the existing data to some common numerical scale.

Conversion of the attributes to a numerical scale usually involves two steps. The first step is to convert the qualitative data to some bipolar interval scale. Thus, a variable such as landscape attractiveness can be converted to a corresponding numerical scale. Once the qualitative variables have been converted to a quantitative scale, the next step is to convert the data measured on different quantitative scales to a commensurate scale.

Noncommensurate Data

Some type of normalization procedure is usually used to transform quantitative data that have been measured on different interval and ratio scales. Once the data have been rescaled, or normalized, the resulting measurements can then be used to rank the alternatives. The three most commonly used normalization methods are vector normalization, linear scale transformation, and range transformation (Howard, 1991; Hwang and Yoon, 1981). The choice of normalization procedure can considerably affect the ranking of alternatives (Howard, 1991).

Vague and Imprecise Decision Criteria

Decision variables that are vague, imprecise, or ambiguous may also be included in the decision matrix. This presents a problem because such variables are not easily modeled with conventional MADM techniques. Variables that are vague are sometimes referred to as intrinsically fuzzy. They are nondichotomous in nature and cannot be delimited by sharp boundaries. The measurement of such variables can be problematic as they do not relate to any known quantitative variables. Imprecise variables are usually composite variables, which could, in principle, be broken down into a large number of more precise variables. Ambiguous variables are quantitative in nature but there are various possible interpretations of the resulting measurement. One of the most common problems in measuring sustainability is when the interpretation of a measurement is a matter of disagreement.

Inappropriate Discrimination between Attribute Values

One problem which presents itself in MADM is the degree to which the distance between intraattribute and interattribute values is accurately reflected in the final ranking of alternatives. The severity of this problem often varies with the type of method chosen to evaluate the alternatives. This problem can be considered a considerable drawback when two or more of the attribute values are close together.

Modeling Interrelationships between Sustainability Criteria

An important distinction between methods for solving MADM problems is whether compensation, or tradeoffs, are allowed among criteria. A choice strategy is compensatory if trade-offs between attribute values are permitted, otherwise it is noncompensatory. In the case of evaluating alternative farming systems on the basis of sustainability, some tradeoff among criteria is desirable.

Ranking of Alternatives

MADM methods vary with respect to how much information they provide in the final ranking of alternatives. Some methods provide a full ranking of alternatives, which is desirable in the case of the evaluation of alternative farming systems.

An Introduction to Fuzzy Logic

Fuzzy logic refers to an inference system that is based on the concept of multivalence. In contrast to bivalent logic, in which a statement must be either true or false, multivalent logic allows for a statement to have some degree of truth. Unlike probability theory, which plays an important role in modeling the uncertainty associated with future or unknown events, fuzzy logic can deal with ambiguities that exist in known or realized events.

Fuzzy Sets

Fuzzy sets are the basic building blocks in fuzzy logic. A fuzzy set, A, can be described as a set of ordered pairs (Zimmerman, 1987):

$$A = \{(x, \mu_A(x)) | x \in X; 0 \leq \mu_A(x) \leq 1\}$$

where x is an object in the collection of objects, **X**, and $\mu_A(x)$ indicates the degree of membership of x in A. The mapping μ_A is called the membership function for the fuzzy set A. For every object in X, the membership function for A returns a value from zero to one. Membership values are independent in the sense that they do not have to sum to one. Fuzzy set theory is based on the idea that an object can have grades or degrees of partial membership in the fuzzy set. While classical (Boolean) sets allow for the statement that "x is an element of A" to be either (totally) true or (totally) false, fuzzy sets allow for the statement to be partially true and partially false (at the same time). The higher the degree of membership of x in A, the more closely x corresponds to the idea expressed by the fuzzy set A.

Linguistic Variables

A linguistic variable is a variable with a range which consists of linguistic terms. Linguistic variables can be completely described by the triple (X,T,b), where X is

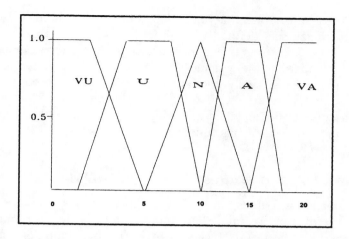

Figure 27.2 Landscape attractiveness as a linguistic variable.

the name of the variable, T is the term set for X, and b is the base variable for X. Linguistic variables can be thought of as being constructed of fuzzy sets since each of the terms in the term set (T) is a fuzzy set. Figure 27.2 provides an example of a linguistic variable to illustrate the concept.

The name (X) of the linguistic variable illustrated in Figure 27.2 is "landscape attractiveness." The term set (T) for this variable consists of five terms, which are (1) VU: very unattractive, (2) U: unattractive, (3) N: neutral, (4) A: attractive, and (5) VA: very attractive. The base variable (b) is an interval scale from 0 to 20. Note that each of the terms in the term set is a fuzzy set. That is, each of the five terms is described by a membership function. For example, the membership function for "neutral" has the shape of a triangle with a base that is 10 units wide and centered on 10. Consider a landscape that receives a measurement of 11 on the base variable. Its degree of membership in the fuzzy sets VU, U, and VA would be zero, while it would have a degree of membership of 0.5 in A and 0.75 in N.

Fuzzy Logic

Fuzzy logic is an inference mechanism built on fuzzy set theory. In a fuzzy logic rule-based system, knowledge is represented by IF-THEN rules. The symbols in the rules are values of linguistic variables modeled by fuzzy sets.

An example of two rules relating soil erosion (EROSION) and landscape attractiveness (LANDSCAPE) to environmental sustainability are as follows:

IF (EROSION is A1) and (LANDSCAPE is B1)

THEN environmental sustainability is C1

IF (EROSION is A2) and (LANDSCAPE is B2)

THEN environmental sustainability is C2

In these rules, A1 and A2 are fuzzy sets defined over the range of "soil erosion," B1 and B2 are fuzzy sets defined over the range of "landscape attractiveness," and C1 and C2 are fuzzy sets over a normalized scale for environmental sustainability. In the standard implementation of fuzzy logic, the minimum degree to which the input values satisfy the antecedent clause is used to produce a modification of the consequent clause fuzzy set (by truncation) for each rule. The modified consequents for all rules are added, and the actual numeric value of the output is the centroid of this function.

Use of Fuzzy Logic to Address Issues in MADM

Fuzzy sets, linguistic variables, and inference mechanisms based on fuzzy logic can provide useful tools for modeling and solving problems formulated in the MADM framework (Chen and Hwang, 1992). The opportunities for combining fuzzy set theory and MADM in order to model the specific problem of sustainability appear very promising, although few applications have been made to date. The few notable exceptions include Munda (1993), Munda et al. (1994), Molder et al. (1991), and Kiker (1994). Fuzzy techniques can help to address the six issues raised in the previous section.

First, linguistic variables can be used to incorporate vague, imprecise, and ambiguous variables into decisions regarding sustainability. Second, linguistic variables can be used to model mixed data in the decision matrix. The qualitative data can be manipulated with fuzzy operators defined on nominal or ordinal scales, or it can be converted to an interval scale through the assignment of an artificial base variable. Third, fuzzy logic allows the model to be solved without normalization of the quantitative data. The fuzzy sets associated with the terms of linguistic variables are already commensurate in the sense that as they all take on values from zero to one, reflecting degrees of membership.

The problem of inappropriately large discrimination in the ranking of alternatives can be avoided through the use of fuzzy techniques. Fuzzy sets allow for smooth, or continuous, changes in the degree of membership. This prevents the common problem of having to define a discrete threshold measurement for a variable. Finally, the application of fuzzy logic to MADM problems permits a full ranking of the alternatives, which is helpful in decision support. In the following section, a prototype MADM model of agricultural sustainability is discussed. The paper concludes with a discussion of model validation and issues involved in creating an effective DSS.

A Fuzzy MADM Model of Agricultural Sustainability

In order to experiment with applying fuzzy logic to a MADM formulation of the problem of agricultural sustainability, a prototype model was constructed. The prototype model was constructed using linguistic variables for each of the nine subgoals and the three major goals. A hierarchical fuzzy rule base was used as the inference engine in the model. For simplicity, the attributes were weighted equally in the rule base.

Table 27.1 Agricultural Example Results, both Levels of Fuzzy MADM

	Economic dimension	Environmental dimension	Social dimension	Overall sustainability
A_1	0.562	0.672	0.618	0.591
A_2	0.564	0.803	0.907	0.765

The fuzzy MADM model was solved using commercially available software (CubiCalc by HyperLogic). The results for both stages of the problem are provided in Table 27.1. The measures of sustainability that result from the application of the model provide a ranking that indicates that the second alternative is more sustainable than the first. This result is consistent with *a priori* expectations, since alternative 2 was at least as good as or better than alternative 1 on each of the decision criteria.

Conclusions

We have shown how fuzzy logic can improve the modeling of agricultural sustainability under a MADM framework. There are several remaining issues that must be addressed in order to refine the existing prototype model.

First, care should be exercised in the selection of decision criteria so that the criteria provide full coverage of the concept of sustainability while adequately reflecting the concerns of actual decision makers. A participatory research process would help to ensure that the model includes criteria that are relevant to actual decision makers.

Second, additional research is needed on the proper construction of the membership functions. Applications outside of agriculture indicate that model results are not overly sensitive to the shape of the membership functions. However, the placement of these functions on the base variable scale is an implicit form of weighting.

Third, experimentation is needed with alternative aggregation operators in the rules. Likewise, the implication of using a hierarchical structure in the fuzzy rule base needs to be explored. In addition, the performance of fuzzy MADM approaches should be evaluated relative to existing, nonfuzzy MADM approaches.

Finally, there are several stages in the modeling process at which there is either an implicit or explicit weighting of the decision criteria. Weighting is extremely important in influencing the outcome of MADM models. The weighting of criteria should clearly reflect the preference structure of the decision maker relative to the goal of sustainability. This would help to ensure the relevance of the DSS and its results.

References

Chen, S.J. and Ching-Lai Hwang. 1992. *Fuzzy Multiple Attribute Decision Making*. Springer-Verlag, New York.

Dunn, E.G., J.M. Keller, L.A. Marks, J.E. Ikerd, P.D. Gader, and L.D. Godsey. 1995. Extending the application of fuzzy sets to the problem of agricultural sustainability. Pages 497–500 in *Proceedings of ISUMA-NAFIPS '95*. IEEE Computer Society Press, Los Alamitos.

Howard, A.F. 1991. A critical look at multiple criteria decision making techniques with reference to forestry applications. *Can. J. For. Res.* 21(11):1649–1659.

Hwang, Ching-Lai and K. Paul Yoon. 1981. *Multiple Attribute Decision Making Methods and Applications: A State-of-the-Art Survey*. Lecture Notes in Economics and Mathematical Systems Series 186. Springer-Verlag, New York.

Kiker, Clyde F. 1994. *Adaptive Policy in Ecological Certification Processes: A Multi-Valued Logic Approach*. SP 94-17. Food and Resource Economics Department, University of Florida, Gainesville.

Molder, Pauline J., Patti D. Negrave, and Richard A. Schoney. 1991. Descriptive analysis of Saskatchewan organic producers. *Can. J. Agric. Econ.* 39(4):891–899.

Munda, G. 1993. Fuzzy Information in Multicriteria Environmental Evaluation Models. Ph.D. dissertation. Free University of Amsterdam, The Netherlands.

Munda, G., P. Nijkamp, and P. Rietveld. 1994. Qualitative multicriteria evaluation for environmental management. *Ecol. Econ.* 10:97–112.

WCED (World Commission on Environment and Development). 1987. *Our Common Future*. Oxford University Press, Oxford.

Zimmermann, Hans J. 1987. *Fuzzy Sets, Decision Making, and Expert Systems*. Kluwer Academic Publishers, Boston.

Chapter 28

Elements of a Decision Support System: Information, Model, and User Management

D.E. MOON, S.C. JECK, AND C.J. SELBY

Introduction

Land resource issues tend to be large and complex and involve multiple disciplines with conflicting goals. Plans and decisions are made frequently in an atmosphere of urgency and conflicting values compounded by inadequate access to either the necessary data and information tools or a well-defined decision process. This is because land resource issues, especially those arising unexpectedly, rarely provide the luxury of a long lead time in which to compile and process the necessary data or to develop the necessary decision support tools and criteria.

Frequently, limited resources are used to develop decision support systems (DSS) which address only specific subsets of larger problems and few resources are available to support data sharing, multipurpose model development, or integration into the larger decision process. This approach, while expedient for the problem subset, is still expensive and leads to inefficiencies in the overall decision process through overlap in data acquisition, processing, and management, overlap in model development and testing, and unnecessary difficulty in integrating the subset solutions into the overall decision process. It also sacrifices the long-term utility and adaptability of the tools used to solve the subset problems. Compounding the problem of inefficiency are rapidly changing technologic, economic, social, and political conditions, an increasing need to integrate new parameters and goals into the decision process, and decreasing time frames which may quickly make the application obsolete.

One approach to meeting this challenge is to design DSS which can be adapted, modified, or expanded efficiently to meet changing needs, which can integrate other subset tools or be easily integrated into other DSS, and which can access and share existing data to save redundant development and data costs.

Fortunately, land resource problems can be represented by a limited number of generic problem types. These can be used to develop generic solutions which provide an efficient, flexible, and highly adaptable land analysis and DSS shell. A generic decision support shell, which supports most or all aspects of decision support, will enable more rapid application development and solution delivery than "one of a kind" approaches and simplify the integration of multiple decision support tools.

The following discussion presents the elements of a DSS required to meet rapidly changing demands. The discussion also makes reference to the LANDS (Land Analysis and Decision Support) system (Moon and Jeck, 1994a; Moon and Jeck, 1994b). LANDS manages and integrates the data and models required to meet a broad spectrum of land management and planning needs. LANDS has been used to evaluate single objective problems ranging from large- and small-scale allocation of production systems to land, through manure management regimes to meet environmental constraints for nitrate contamination of groundwater, to the evaluation of effectiveness and feasibility of conservation tillage practices. We make reference to the LANDS system to demonstrate that what we discuss is feasible. We emphasize, therefore, not the specific tools and applications incorporated into the shell for any specific application, but rather the integration of diverse data, problems, models, and user requirements into a single system.

The Nature of the Problem

A land use decision problem will exhibit one or all of the following characteristics.

1. **Diverse data** — these data will range from biophysical (soil, terrain, climate, ecology, biology), through land use (current and potential), economics (production and sales), and policy (zoning, environmental and health regulations) to social preferences and goals.
2. **Variable levels of data detail, analytical detail, and data volumes** — the size of the decision problem, the number of options, and the level of analytical detail required will dictate the size of the database. For example, production schedules could be defined as annual summaries or as daily activities with corresponding impacts on the size of the database required.
3. **Model diversity** — the potential models required to address a complex decision may encompass a wide range of type and sophistication. The DSS may use expert systems (simple rule-based algorithms or more sophisticated fuzzy logic and uncertainty-based algorithms), empirical models (more or less statistically based), or sophisticated deterministic or process-based models. This progression generally corresponds to an increase in the data required to support the models. The models may be used independently or as linked components of the decision process.

4. **Changing needs** — conditions, knowledge, and attitudes change and new technology will create new options. These require the DSS to be adaptable. It will have to cope with new attributes, new things about which we must store data, and new models. One danger of computer-based DSS is that we will use them past their applicable life because of the costs and effort required to adapt them to new conditions.

Challenges

Information Management

Information management encompasses several issues. These issues range from data modeling through data entry, editing, and retrieval, to data integrity and concurrency. Changing the data model of an existing database is demanding and expensive. A generic and adaptable system will meet new conditions and demands without modification. The same is true of the user interface. It should support changing demands without requiring modification. The information management system must support the following tasks.

1. **Manage decision support data** — the DSS will support data entry, editing, maintenance, and retrieval. The underlying data model clearly defines the following:
   a. Data entities (things about which we store data)
   b. The attributes used to describe them
   c. The domains of these attributes (valid values, formats, etc.)
   d. Relationships between entities
   e. Attributes of these relationships
2. **Data integrity** — the DSS must ensure data integrity by enforcing both relational integrity and business rules. For example, a soil must have a minimum set of attributes, determined using specific methods before the system will accept the data. The user interface or database triggers perform this task.
3. **Unanticipated types of data** — decision support needs will change. Existing models will be refined, new models will be developed, and new technologies will be available. These will require the creation of new entities, relations, and attributes. Without a robust data model and user interface, these changes will require expensive and time-consuming changes to the underlying data model and user interface.
4. **Large data volumes** — the various interpretive needs and models used to support decisions require large volumes of data. Different interpretive models will require different levels of data generalization. The system should store the greatest level of information detail available. If necessary, applications or models should use generalizations of stored data rather than forcing generalization before data storage.
5. **Data redundancy** — information should be stored in one and only one place. Duplication of data wastes storage space and imposes an overhead to provide manual or programmatic concurrency updates. It may also lead to inconsistent forms of the same data. However, when performance is important storing data derived from existing data may be appropriate.

Security and Distribution

The large number of individuals and organizations requiring input and/or access to the database imposes serious demands on security and distribution protocols. The distribution of data files to individual users imposes the challenge of ensuring data concurrency while using distributed databases imposes the challenge of ensuring data security.

1. **Data access and protection** — multiple users and agencies will require access to the database. The database should therefore provide different levels of access to different sets of data. For example, the agency or person responsible for the data may be the only one with write access to its particular data. Similarly some data may be confidential and should have only programmatic access to ensure confidentiality. Roles and priviledge assignments can meet this requirement.
2. **Data sharing** — many applications use the same core data. These data are shared among all applications to be sure that model results are comparable. In addition the output from one model may provide input to another model. It may in turn provide input to yet another model. The management and storage of intermediate results and the sharing of data between models are roles of the information manager.

Land use planning decisions require data from a number of agencies. Each agency maintaining a set of duplicate data is inefficient, especially if it is logical for one agency to be ultimately responsible. These data should be shared via distributed databases rather than by replicating many copies of the data across many systems. Replication breeds lack of concurrency and increased maintenance costs.

Structuring the Process

The decision process consists of a series of steps some of which will be dependent on the results or preparations of a previous step and will in turn have other tasks dependent on their successful completion. Modification of input or conditions at precursor steps may invalidate all subsequent analysis. The system must guide the decision maker through the appropriate procedures. It must ensure that task dependencies are enforced and it must recognize and enforce cascading changes when intermediate steps are modified.

1. **Guiding the user and enforcing the structure** — the system provides a logical ordering of tasks to be done. Menus and forms present the tasks in the required order. The system also enforces sequential entry using triggers that test for precursor steps before allowing the user to proceed. It also needs to check for cascading effects should any intermediate step be modified after initial processing. The system therefore requires a mechanism for defining, storing, and enforcing task dependencies and dependency status.
2. **Ensuring minimum requirements** — many applications require minimum (mandatory) data but may use optional data. The addition of optional

data may increase the reliability or the precision of the prediction. The DSS should accommodate two approaches to handling missing data. The first is to use a more general model (usually empirical or stochastic). The second is to use default values (either generated from secondary models, chosen by the user as representative, or chosen by the application developer as a good compromise). The impact of model generated input data, using defaults, or using less accurate empirical models should be clearly identified and understood.

Model Management

The array of problems and issues associated with land use decisions requires an array of predictive models. The array of models required imposes an array of data inputs, run parameters, and output management needs. In addition, models are interdependent. Output from one model provides input to another that in turn provides input to a successor model. Because the DSS presents these models with a common user interface, the user need learn only one interface. The system manages the interaction of models used in the decision process.

The DSS should adapt itself to the multiplicity of available decision support models and those to be developed. It is not the responsibility of the model to adapt itself to the DSS. Adaptation to external models requires an input/output facility to feed external models their data in the required format. Adaptation also requires an import facility to convert results to the appropriate format for storage, reporting, or subsequent analysis.

User Management

A DSS should support at least five types of user. For a particular DSS, a single person may perform all five functions. However, as the complexity of the application increases this becomes less likely. The users identified below will have different levels of computer, systems, and domain expertise. This requires that the DSS exhibit different characteristics for the different users. The following list of users represents an increasing need to protect the system from inadvertent damage. We should also remember that, as a DSS become more user-friendly and easier to execute, we have an increasing obligation to protect the user from the system.

1. **Application developers** — the DSS should support one or both of the following functions. It should free the developer from having to conform to the DSS data model or user interface by being able to generate whatever data format the developer has chosen to use. It should also provide a development environment. If developers choose to use the DSS development tools, they will be free from the system tasks of managing data and building user interfaces.
2. **Domain specialists** — domain or discipline specialists will be responsible for applying the models and validating data and results. Once applications are in place domain specialists will use them to construct scenarios reflecting

the various options available to the decision maker. The DSS must make the necessary data available to both the domain specialist and the applications. It should also manage the linkage and execution of multiple models necessary to construct the scenarios. Finally, it should manage the results for display or subsequent analysis. The domain specialists should not have to understand the underlying data model or application code.

3. **Evaluators** — scenarios constructed by the domain specialist must be evaluated in context. Domain specialists may not be able to evaluate options that are outside their discipline or domain and may not be able to place the scenarios in full context. This falls to the evaluator or evaluation team. An evaluation team may well be comprised of domain specialists. The evaluation may take the form of some single or multiple objective optimization or conflict resolution processes. The DSS should support "what if" scenarios necessary to identify the tradeoffs associated with the various scenarios and/or multiple objectives related to the decision problem.

4. **Decision makers** — ultimately, the appropriate decision authority will make the decision. The results of the analysis must be clear and understandable by the decision makers. The decision makers will be responsible for the decision and should have confidence in the information on which they based the decision. The DSS should distill the necessary information for the decision maker but should also do at least two things. It should provide formal documentation of all data, algorithms, and assumptions used in the distillation process and it should provide some measure of confidence for the distilled information. It will fall to the decision maker to decide but it falls to the DSS (including the people) to identify clearly the risks and uncertainties associated with the possible decisions.

5. **Monitors** — once taken, implementation of the decision will require monitoring. The DSS should support analysis of the impact of variances from the proposed implementation plan and allow feedback for further analysis or modification.

A Layered Technologies Approach

Building a DSS to meet these challenges is a nontrivial exercise. Meeting each challenge is a major task itself. Fortunately we can take advantage of existing technologies to ease the task. We can layer existing technology on top of the data used to support decisions (Figure 28.1). Each layer performs a given set of operations against the data. Successive layers provide more sophisticated analysis and provide access by less skilled users to the underlying technologies. Each successive layer is also dependent on the underlying technologies for access to the data. The goal is for the user to concentrate on applying their domain expertise not on learning the technology. This approach has both advantages and disadvantages. The principal advantage is that commercial vendors of enabling technologies such as GIS (geographic information system) and RDBMS (relational database management system) produce more powerful and flexible tools than we are likely to build ourselves. The principal disadvantage is that we become dependent on the underlying technologies over which we have little control.

Elements of a Decision Support System: Information, Model, and User Management

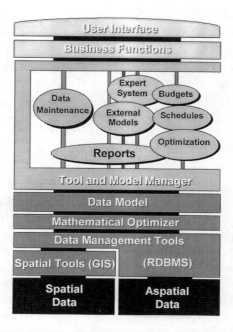

Figure 28.1 Layered technologies (adapted from Hunt and Jones, 1995).

Database

The database is the foundation of the entire structure. Without a clear understanding of the nature and limitations of the underlying data, there can be no basis for assigning confidence to the results of the DSS. Most applications will work with both spatial and aspatial data.

1. **Spatial data** — spatial data represent points, lines, areas, or networks. Most spatial data are presented in map form. At a minimum, spatial elements will have attributes of location. They may also have attributes of length, area, topology, connectivity, contiguity, and proximity. Biophysical conditions (soil, terrain, climate, vegetation types), land use (present and potential), land zoning (parks, reserves), and farm or industry locations are commonly represented as spatial entities.
2. **Aspatial data** — aspatial data encompasses the nonspatial attributes (timber volume, crop type, crop needs, production levels, zoning regulations, etc.) of spatial entities.

Spatial Tools (GIS)

The GIS occupies the first layer of enabling technologies. It captures, manages, analyzes, and displays the spatial nature of project data. The GIS provides several very useful functions. For example, it can provide integration through spatial overlays of things like climate, soil, and terrain to define areas of homogeneous biophysical conditions for crop growth. Evaluating spatial entities in context with other spatial entities is often necessary. The GIS can identify contiguous areas, minimum contiguous areas, boundary conditions, proximity, buffer zones, and linear connectivity and flow direction. Many well-tested and supported commercial GIS are on the market. Most will have an interface to an RDBMS as described below. The nature of the linkage between the GIS and RDBMS will strongly influence the flexibility and ease of maintenance of spatial entities (Moon et al., 1993).

Data Management Tools (RDBMS)

The RDBMS occupies the first and second levels of enabling technology. It can manage and analyze attributes of either spatial or aspatial entities and in many cases encompasses the GIS. The RDBMS frees the developer and/or user from file management and complex input/output programming. If available in the RDBMS, structured query language (SQL) facilitates data retrieval and analysis from multiple related tables. SQL uses a syntactically simple language that frees the user from specifying the retrieval procedures. While powerful and easily learned, RDBMS generally lack the speed required for *ad hoc* retrieval and display of spatial features. Therefore, spatial attributes are generally managed using proprietary file structures in the GIS. Recent developments are however closing this performance gap. In addition, procedural languages are more efficient for some applications. The RDBMS should, therefore, have either a procedural option or a precompiler for procedural languages like "C," "C++," "FORTRAN", etc. Except for spatial analysis and display, it is more efficient and easy to manage or execute most analytical functions within the RDBMS rather than the GIS.

Mathematical Optimizer

A useful, if not essential, enabling technology is a mathematical optimizer. It should support linear and/or nonlinear, integer and/or noninteger optimization. If the system generates MPS code, it can use an array of commercial or public domain optimizers.

Data Model

Nothing will influence the efficiency and flexibility of the DSS more than the data model. If the data model cannot cope with new conditions or data, the user must ultimately resort to expensive and time-consuming adaptations to the existing data model and user interface. Entity-relation modeling, combined with data normalization and abstraction, offers a powerful and flexible data modeling paradigm (Jeck et al., 1995, and Moon et al., 1993).

All information related to the decision problem is defined in terms of **entities** or subjects (things relevant to the problem), **relations** (relationships between entities), and **attributes** (properties of the entities or relations relevant to the problem). For example, crops are a type of entity and have biophysical requirements as attributes. Soils, climates, and terrain have biophysical attributes. The suitability of a crop on a piece of land is an attribute of the relationship between crop, soil, terrain, and climate. There are different kinds of relationships. These may be associative (simple one to one associations), hierarchical (parent-child-grandchild relationships), or definitional. Hierarchical relationships may be strict (an entity may belong to one and only one higher group) or general (an entity may belong to more than one higher group). Hierarchical relationships can also support "fuzzy" memberships by using probability of membership as an attribute of the relation. Definitional relationships group different types of entities (e.g., land use, soil, terrain, and climate combine to define a land management unit).

As an aside, the evolution of object-oriented data models shows real promise for application modeling. Unfortunately, object-oriented data modeling has neither developed a set of consistent standards that will allow true portability of objects nor provided an effective replacement for relational problems. As well, commercially supported object-oriented database systems that offer the range of information management functions required to meet large-scale DSS are not available. Despite this, some combination of object-oriented systems and RDBMS will become dominant. In anticipation of this evolution, the data model should incorporate many characteristics of object-oriented data models such as entity hierarchies, inheritance, and associated processes that should make the transition to a hybrid object/relational database management system relatively painless.

Tool and Model Manager

When combined with the data model, the enabling technologies described present a powerful and flexible development environment. However, the application of these enabling technologies to decision support still requires significant programming expertise. Making this powerful capability available to users who lack programming and systems skills requires a tool or model manager to make these tools or models available to them. The ability of this manager to incorporate internal and external models will greatly increase the flexibility and power of any DSS.

Data Maintenance (Entities, Attributes, Integrity)

Key to any decision support application is the management of data. The data maintenance tools perform four key functions.

1. The definition entities (the things or types of things of interest). These may include such things as soils, map delineations, crops, wells, parking lots, models, people, agencies, etc.
2. The definition of relationships (how different things are related). These may include such things as specific crops on specific soils, members of a taxonomic class, soil series found in a map delineation, etc.
3. The definition of attributes (properties) which describe both the entities and the relationships between entities.
4. The definition of domains and integrity constraints that the system will enforce and make available to the user as on-line help.

Expert System

Expert system shells provide a means of encoding expert knowledge and making it available to nonexperts. The expert system has the added advantage of formalizing the knowledge base and making it available to scrutiny, testing, and modification. The system stores and maintains the knowledge base and rule sets within the database and exports them to the expert system shell for inferencing. The system then imports the results back to the database for subsequent reporting or processing.

Budgets, Schedules, and Optimization

- **Budgets** — the budgeting tool constructs detailed or summary budgets and cash flows. It can report the timing of costs and returns for individual or grouped operations, resources, or commodities involved in a land use option. It uses the resource quantities attached to each operation in the schedule (see below) multiplied by the prices attached to each resource or product to generate a budget on a per unit area basis for each feasible option. After choosing a land use plan, the system uses individual budgets to report an enterprise wide budget broken down by each land use option.
- **Schedules** — the scheduling tool applies the project management concepts of operations (tasks), resource assignments to operations, and task dependencies. A production schedule should be as general or as detailed as desired. Besides the more traditional project management concepts, the LANDS system supports specification of phenologically driven plus resource and time driven operations. It also allows specification of the biophysical conditions necessary to complete the operation (for example, minimum water stress to support sugar set in sugar cane or soil and water conditions necessary to support trafficability).
- **Optimization** — the land allocation model considers all feasible options and start times. It recognizes seasonal availability of resources and the availability and cost of borrowed capital. It optimally allocates constrained resources (seasonal if appropriate) to land use options while respecting all other user specified constraints such as monthly cash flow, soil loss by management area or soil type, nitrate leaching, production goals, or minimum or maximum areas allocated to a specific land use to maximize net income. If appropriate, the system can generate a detailed project schedule. The schedule can include the start dates and durations of all tasks, the timing, quantity, and costs of all resources, and the timing and return of yields.

External Models

There may be external models which address aspects of the decision issue at hand. If suited to the problem and integrated into the DSS and if the necessary data are available these models can save time and money. The system therefore facilitates the integration of external models.

Business Functions

Business functions are the actual uses to which the DSS tools or models are applied. The determination and allocation of animal production quotas and resultant animal waste consistent with groundwater quality may be a business function. While performance of this function may use most or all of the enabling technologies, models, and tools, the person performing the function should only need to concern themselves with supplying the necessary data and parameters to perform the allocation, not with linking and integrating the various models and tools to produce the result. The business functions define and implement the linkages and invoke the necessary models or tools.

User Interface/Task Manager

The user interface provides access by system specialists, data specialists, and domain specialists (soil scientists, crop scientists, economists, policy analysts, etc.) to the underlying tools, technologies, and data. It also provides access by the end user or decision maker to the business functions prepared by the system and domain specialists. For example, the allocation of land use options to specific land area under political, social, and resource constraints requires the following tasks. The expert system supports specification of biophysically capable options (including needed land improvements) but suitability and feasibility require more detailed specification of resource use, production, and budget schedules. Budgeting and scheduling tools support these phases. The land allocation model supports the definition of economic, ecological, social, or personal constraints on the solution, the optimizer solves the problem, and the reporting tool reports the results. The user interface and task manager ensure and facilitate the performance of the necessary tasks in the appropriate order and ensure that subsequent changes to precursor tasks will be propogated through the rest of the decision process.

Ideally, the user interface and task manager should free the user from the need to understand the complexities of operating systems, data models, input/output definitions, and model parameterization. In essence, the user interface provides for the care and feeding of the database, the models or procedures that call the data, and the business functions used to support the "what if" analysis necessary for decision making.

Conclusions

To be an effective aid to decision support a DSS must have the following characteristics.

1. It must offer an evaluation interface that is accessible to the decision maker when needed and that does not require a team of specialists to operate. The DSS will almost certainly require a team of specialists to set up but should not require the team to be present for the evaluation of alternative options.
2. It must be quickly adaptable to new problems. This requires a tool kit allowing the DSS domain specialists to construct and place the options to

be evaluated into an evaluation framework that is accessible to the decision maker.
3. It must operate against clean and complete data. This requires the ability to integrate and reconcile disparate data and data sources.
4. It must use validated models, applied at the appropriate scale, using adequate data, and interpreted with due caution.

With two notable exceptions, the LANDS system shell provides the framework and tools needed to build an effective DSS. The exceptions are the user interface and the model integrator. The LANDS user interface is not intuitive enough or easily enough accessible to support nonspecialist decision makers. The advent of powerful integrated graphical user interface tools will simplify meeting the first criteria. The current model integrator requires programmer intervention to integrate new models. A model integrator which will allow model integration without programmer intervention is in the design phase.

References

Hunt, H. and R.K. Jones. 1995. Building a forest management decision support system product: challenges and lessons learned. Pages 92–102 in M.J. Power, M. Strome, and T.C. Daniel, Eds. *Proceedings: Decision Support 2001*. Publ. American Society of Photogrammetry and Remote Sensing.

Jeck, S.C., D.E. Moon, C.J. Selby, and M.-C. Fortin. 1994. A Universal Data Model for Biophysical and Land Evaluation Data Management. Paper presented to American Society of Agronomy, Crop Science Society of America, Soil Science Society of America Annual Meetings, Seattle, Washington. 15 pages. (Available from the authors).

Moon, D.E. and S.C. Jeck. 1994a. LANDS, A Land Analysis and Decision Support System: Concepts and Procedures. Centre for Land and Biological Resources Research, Agriculture and Agri-Food Canada, Vancouver, British Columbia. 149 pages.

Moon, D.E. and S.C. Jeck. 1994b. LANDS, A Land Analysis and Decision Support System: Operations Manual. Centre for Land and Biological Resources Research, Agriculture and Agri-Food Canada, Vancouver, British Columbia. 352 pages.

Moon, D.E., S.C. Jeck, and C.J. Selby. 1990. The LANDS system, a land analysis and decision support system. Pages 361–369 in *Proceedings of GIS90. Making It Work*. Reid Collins, Van. and Forestry Canada, Vancouver, British Columbia.

Moon, D.E., C.J. Selby, S.C. Jeck, and J. Silver. 1993. Extending the power of the GIS using relational data models. Pages 1031–1038 in *GIS-93: Eyes on the Future, 7th Annual Symposium on Geographic Information Systems, Feb 15–18, 1993*. Forestry Canada, BCMOF, BCMOELP, Vancouver, British Columbia.

Moon, D.E., S.M. Ulansky, S.C. Jeck, and B. Stennes. 1994. Abbotsford Aquifer Agriculture Information System Pilot Study. A Report Prepared under Contract to the B.C. Ministry of Agriculture, Fisheries and Food as part of the Canada-British Columbia Soil Conservation Program. Victoria, British Columbia.

IV APPLICABILITY TO ECONOMIC, SOCIAL, POLICY, RISK, AND SUSTAINABILITY ISSUES

Chapter 29

Adoption of Environmental Protection Practices in the Scioto River Watershed: Implications for MODSS*

TED L. NAPIER AND SILVANA M. CAMBONI

Data were collected from 1,305 land owner-operators in the Scioto River watershed in central and south-central Ohio in 1991. Comparable data were collected from 245 land owner-operators in the Darby Creek watershed located in central Ohio in 1994. Findings revealed that socioeconomic factors, farm structure variables, personal characteristics of the primary farm operator, and institutional factors were useful for explaining variance in conservation attitudes but not conservation behaviors. Study findings suggest that conservation programs designed to stimulate awareness of environmental problems and to develop favorable attitudes toward environmental issues will have little impact on land owner conservation behaviors. Study findings suggest that land owner-operators respond to economic incentives even when they do not value the type of conservation system being advanced. It is argued that voluntary conservation programs can be effectively implemented, if adequate economic resources are provided as incentives to cooperating land owners. Alternative intervention approaches will have to be explored, if economic resources are not available to use as incentives.

---

* Financial support for salaries and data collection were provided by the Ohio Agricultural Research and Development Center of the Ohio State University, and the Management Systems Evaluation Area project funded by several agencies of the U.S. Department of Agriculture and the U.S. Environmental Protection Agency. The positions advanced in the paper are solely those of the authors and do not represent official positions of the funding agencies.

Introduction

Lack of adoption and/or rejection of environmental protection practices at the farm level constitutes a significant socioenvironmental issue in the U.S. because production agriculture has been shown to be the primary source of nonpoint pollution (Clark et al., 1985; Halcrow et al., 1982; Lovejoy and Napier, 1986; Napier et al., 1994; Page, 1987; Swanson and Clearfield, 1994). Use of farming systems that exacerbate soil loss from agricultural land will produce on-site damages such as loss of future agricultural productivity of land resources, loss of future resale value of farm land, reduced aesthetic value of land resources, and loss of agricultural production inputs. Environmentally unfriendly farming systems also contribute to off-site damages such as loss of wildlife habitat, disruption of transportation systems (highway, waterway, and rail) via sedimentation, increased costs of making water potable, contamination of surface and subsurface water supplies, and loss of recreational use of eroded land and polluted water resources (Clark et al., 1985; Halcrow et al., 1982; Lovejoy and Napier, 1986; Napier, 1990; Swanson and Clearfield, 1994).

Socioenvironmental costs of using inappropriate farming systems have been documented in many geographic regions of the U.S., and numerous conservation programs have been implemented to encourage adoption of conservation practices. While conservation initiatives have significantly reduced soil loss in the U.S. during the past 60 years, many land owner-operators continue to employ agricultural production systems that contribute to environmental degradation. Land owner-operators often continue to use farming systems that they know are contributing to on-site and off-site damages.

Many attempts have been made to understand why land owner-operators use land-degrading practices and a multitude of social science theories have been formulated to explain conservation behaviors of U.S. farmers. While some of the theoretical perspectives have been shown to explain considerable variance in conservation attitudes and perceptions, explanation of actual use of conservation behaviors has proved to be problematic. The purpose of this paper is to report the findings of two studies conducted in Ohio that were designed to identify covariates of adoption of conservation practices at the farm level. Study findings revealed that a number of factors were significantly correlated with attitudes and conservation behaviors. Statistical models demonstrated that it was possible to predict attitudes relatively well, however, the amount of explained variance for actual adoption behaviors was very low. Implications for future intervention programs and multiple decision making within the study regions are discussed.

Study Region

Findings reported in this paper are synthesized from statistical modeling of data collected from land owner-operators within the Scioto River watershed. Data were collected at two time periods. The first data set was collected in 1991 via personal interviews with 1,305 farm owner-operators within the Scioto River watershed, who were actively engaged in production agriculture. Data were collected from 245 land owner-operators in the Darby Creek watershed who were personally interviewed during the winter and spring of 1994. Since the Darby Creek hydrologic unit is a tributary of the Scioto River watershed, farmers had been interviewed within the watershed in 1991 as part of the Scioto River study. Both cross-sectional and time series findings are reported for the Darby Creek study group.

1991 Study Region

Data were collected in 1991 from a stratified random sample of land operators within the **Scioto River** watershed. The watershed encompasses all or portions of 16 counties in central and south-central Ohio. The topography varies from flat farm land in the north to rolling hills in the south. Soils are deep, fertile, and generally well-drained.

Land operators within the Scioto River watershed employ technology-intensive and chemical-intensive production systems that can contribute to environmental degradation if conservation production systems are not employed. While groundwater pollution from agricultural sources is not a significant issue within the study area (Baker, 1990; Baker et al., 1994), surface water is subject to pollution via farm chemicals and displaced top soil (Napier and Brown, 1993). Occasionally, nitrate levels in the Scioto River watershed are elevated to the point that health alerts are issued and regional residents are cautioned to take preventive action.

Corn and soybeans constitute the primary farm products produced within the study region. Approximately 60% of all farm income in the Scioto River watershed is derived from these two crops. Wheat contributes about 10% of total farm income and livestock about 25%. The primary crop rotation system is corn and soybeans on alternate years. When there is variance from this system, farmers include wheat in the rotation (Napier and Brown, 1993; Napier and Sommers, 1996).

1994 Study Region

Data were collected in 1994 from a stratified random sample of land owner-operators within the **Darby Creek** watershed which is composed of the Big Darby Creek and the Little Darby Creek. The study region is located in central Ohio and is in the northwest portion of the Scioto River watershed. The hydrologic unit occupies portions of six counties.

The watershed is located adjacent to Columbus and has been subject to extensive land use change during the past 3 decades as a result of metropolitan expansion. Suburbanization is beginning to change the human ecology of the Darby Creek watershed (Napier et al.,1995), even though agriculture remains the primary economic activity in most areas of the watershed.

The topography within the Darby Creek hydrologic unit is flat to gently rolling. Soils are deep, fertile, and generally well-drained.

Revenue from the sale of corn, soybeans, and wheat constitutes about 60% of all farm income within the study area. Sale of animal products contributes another 15% to total farm income (Napier and Camboni, 1994).

Production agriculture within the Darby Creek watershed is primarily large-scale and technology-intensive. Mennonite-Amish farmers compose a significant minority of land owner-operators within the Darby Creek hydrologic unit and their farm production systems vary considerably from their nonMennonite neighbors. Mennonite farmers, especially Amish farmers, employ more diversified farm production systems and use less technology-intensive and less chemical-intensive production systems (Napier and Sommers, 1996; Sommers and Napier, 1993).

While production agriculture remains the primary mode of economic activity within the watershed, the area is used extensively for recreational purposes by central Ohio sportspersons. Concern is frequently expressed by central Ohio residents about the environmental integrity of the watershed.

Extensive public and private resources have been focused on the Darby Creek watershed to prevent environmental deterioration of water quality and to protect existing wildlife habitat. A number of public and private conservation groups have implemented soil and water conservation programs within the study area during the past 5 years (Napier, 1993; Napier, 1995). Due to the concentration of economic and human resources within the watershed, the study region provides a natural laboratory for assessing the role of conservation programs as a means of encouraging adoption of soil and water conservation behaviors.

Descriptive Findings

Descriptive findings derived from data collected among farmers in the **Scioto River** watershed demonstrated that respondents perceived fertilizers and pesticides in groundwater as posing little threat to human health (Napier, 1993). Approximately 47% of respondents indicated that fertilizers in groundwater posed little or no threat to human health, and 37% reported that pesticides in groundwater posed little or no threat to human health (Napier and Brown, 1993).

Respondents believed that farmers should be punished for using excessive farm chemicals and that land owners should be forced to periodically test groundwater for chemical pollution (Napier and Camboni, 1993). Study participants did not believe that farmers should be forced to reduce chemical application rates to protect groundwater resources (Napier, 1993; Napier and Brown, 1993).

Most respondents perceived that they had relatively little knowledge of groundwater pollution within their county of residence even though a majority believed that groundwater pollution was an important environmental issue. Only 27.6% of the respondents believed that they had considerable or complete knowledge about groundwater pollution in county of residence (Napier, 1993; Napier and Brown, 1993).

Respondents indicated that farm magazines were the most frequently used source of information about groundwater pollution. Approximately 80% of the respondents indicated that they had used farm magazines for information about groundwater pollution. Less than half of the respondents reported using other sources of information such as government conservation agencies or the extension service. Neighbors, television, radio, crop consultants, and agricultural business representatives were used infrequently by respondents as sources of information about groundwater pollution. Only 4.5% of the respondents indicated that they had never received information about groundwater pollution (Napier and Brown, 1993).

Respondents were requested to provide the number of pounds of nitrogen, phosphate, and potassium usually applied to corn, soybeans, and wheat on a per acre basis. Study findings revealed that land operators within the Scioto River watershed were applying fertilizers in an environmentally responsible manner (Napier and Sommers, 1994). While some farmers were over-applying chemicals, a large majority of farmers were applying nutrients consistent with industry recommendations.

Land operators were asked to evaluate ten criteria that could be used when making adoption decisions about farm technologies and farming systems. The possible responses ranged from **not important** to **very important**. Study findings revealed that all factors examined were shown to be important or very important to respondents (Lang, 1993). The most important considerations in the decision-making process for respondents were *profitability* and *saving on input costs* followed closely by *maintenance costs*, *potential loss of production*, and *size of investment to adopt*. Factors such as access to information, awareness of need for innovation, saving in time, and ease of use were shown to be less important, even though these factors were also perceived to be important. The least important factor was disruption of the existing farming system to adopt an innovation. However, over 60% of the respondents considered disruption of farming system to be important or very important in the decision-making process.

The perceived impact of adoption of soil and water conservation practices on input costs was assessed. Study findings revealed that about 47% of the respondents believed production costs would not change or change very little, if land owner-operators adopted groundwater protection practices (Napier, 1993; Napier and Brown, 1993).

Frequency of use of ten agricultural practices was examined in the study. Practices representing both conservation and potentially degrading approaches to farming were assessed. Approximately 30.4% of the respondents reported they frequently or always used conservation tillage with one-third ground cover at planting time. Another 25.1% indicated they frequently or always used no-till. Unfortunately, the environmental benefits of these minimum tillage systems could have been negated by adoption of environmentally unfriendly practices. For example, about 53% of the respondents indicated they used fall tillage and 43.3% applied manure during winter frequently or always. About 17% indicated that they frequently or always applied fertilizer in the fall (Napier, 1993; Napier and Camboni, 1993).

Data collected in 1994 from land owner-operators in the **Darby Creek** hydrologic unit revealed that many of the descriptive findings observed in the Scioto River watershed were replicated. Darby Creek watershed farmers also reported use of farming systems that simultaneously contributed to protection of the environment and potential degradation. While 56.7% of the respondents indicated they used no-till, 62.9% used banded fertilizer application, and 45.7% reported they used conservation tillage every other year or every year, many farmers also used environmentally unfriendly practices. For example, approximately 43% of the respondents reported using fall tillage, 23.7% used deep plowing (moldboard plowing), 38.4% reported fall application of fertilizer every year or every other year (Camboni and Napier, 1994; Napier, 1995).

Like their Scioto River counterparts, Darby Creek respondents did not perceive groundwater contaminated by farm chemicals as posing an environmental problem and that human health was not threatened by agricultural pollution. Land owner-operators believed that farmers should be punished for over-applying farm chemicals and that land owners should be required to periodically test for groundwater pollution.

Respondents did not believe that farmers should be required to reduce fertilizer and pesticide application rates to protect the environment. A large minority (41.2%) of respondents believed that agriculturalists could reduce chemical application rates without reducing productivity (Camboni and Napier, 1994; Napier, 1995; Napier and Camboni, 1994; Napier et al., 1995).

When there were differences between the descriptive findings for the Darby Creek and Scioto River studies, however, the differences were relatively slight. For example, respondents in the Darby Creek watershed believed that adoption of soil and water conservation practices would slightly increase production costs and result in a slight decline in output (Camboni and Napier, 1994; Napier, 1995; Napier and Camboni, 1994). As noted earlier, Scioto River respondents believed that production costs would not change significantly.

Several variables were included in the Darby Creek study that were not examined in the Scioto River project. Some of the findings from those data are presented as follows.

Level of perceived risk associated with use of farm chemicals was assessed in the Darby Creek study. Perceived level of risk associated with such factors as water quality, food safety, food quality, health of applicator, animal health, wildlife, beneficial plants, beneficial insects, and human health was evaluated. Possible responses ranged from *no risk* to *serious risk*.

Descriptive findings revealed that respondents perceived little or no threat to any of the things assessed (Napier, 1995; Tucker and Napier, 1995). Health of applicator was perceived to be subject to the highest level of risk; however, the level of perceived risk was relatively low. Adverse impact on food quality was perceived to have the lowest level of risk (Camboni and Napier, 1994; Napier, 1995; Napier and Camboni, 1994; Tucker and Napier, 1995).

Perceptions of wetlands were examined using a seven-item semantic differential scale. Descriptive findings revealed that respondents were neutral, to slightly negative, toward wetlands. Wetlands were perceived to be unprofitable, slightly ugly, and slightly wasteful. Wetlands were also perceived to be slightly good, and slightly helpful (Napier, 1995; Napier et al., 1995).

Perceptions about the Darby Creek watershed as a resource to be protected and valued were examined using a seven-item Likert-type attitude scale (Edwards, 1957; Nunnally, 1978). Descriptive findings revealed that study respondents perceived the watershed positively and placed considerable value on the Big Darby and Little Darby creeks. Land owner-operators, however, did not believe that farmers should assume the costs and/or responsibility for protecting environmental quality within the watershed. Descriptive findings strongly suggest that resident land owners would not be supportive of environmental action if they had to internalize costs of protecting the resource.

Attitudes toward environmental action were examined by asking respondents to rate the level of emphasis placed on protecting environmental quality within the Darby Creek watershed. About 82% of the study participants reported that too much emphasis had been placed on environmental action in the recent past (Napier et al., 1995).

Data for Mennonite farmers and nonMennonite farmers were compared within the Darby Creek watershed in 1991 and again in 1994 (Napier and Sommers, 1996; Sommers and Napier, 1993). Descriptive findings revealed that Mennonite farmers differed from nonMennonite farmers in several ways. Mennonite farmers were more concerned about groundwater pollution. This finding was explained in the context of Amish farmers not using modern technologies to protect family from groundwater contamination by agricultural chemicals. Mennonite farmers used fewer agricultural chemical inputs and fewer technology-intensive production systems than their nonMennonite counterparts. Unfortunately, it was also observed that Mennonite farmers used *fewer* environmentally friendly tillage systems than nonMennonite farmers. Mennonite farmers used conservation agencies much less frequently than nonMennonite farmers; however, Mennonite farmers indicated a willingness to participate in education-information programs focused on soil and water conservation. Mennonite and nonMennonite farmers used profitability and efficiency criteria when making adoption decisions about technologies and techniques at the farm level.

Multivariate Findings

Multiple regression analysis, multiple discriminant analysis, and analysis of variance were used to examine the relative predictive abilities of a number of social, economic, and institutional factors when they were considered simultaneously. These findings are summarized below.

Attitudes toward groundwater pollution were examined using multiple regression modeling. Findings revealed that individuals who perceived contamination of groundwater from agricultural sources as constituting a threat to the health of family members also believed that pollution of groundwater was a significant issue and were more willing to force land owner-operators to adopt groundwater protection practices. Farmers who were more knowledgable of the extent of groundwater pollution in county of residence were less concerned about groundwater pollution and were less willing to force land operators to adopt groundwater protection practices. Respondents were less concerned about groundwater pollution from agricultural sources and less willing to force farmers to adopt groundwater protection practices, if they were grain farmers with high debt loads who believed that adoption of conservation practices would result in higher production costs.

Several other farm characteristics were shown to be significantly correlated with attitudes toward groundwater pollution, however, they explained very little additional variance. A total of 33.8% of the variance in attitudes toward groundwater pollution was explained by the statistical model, which is considered to be quite good using cross-sectional data (Napier, 1994; Napier and Brown, 1993).

Frequency of use of tillage practices was assessed using multiple regression analysis and the findings revealed that none of the predictive variables included in the models were useful in explaining a large percentage of the variance in the tillage practices examined (Camboni and Napier, 1994; Napier and Camboni, 1993). The findings for the Scioto River watershed study group revealed that none of the regression models explained more than 16% of the variance in tillage practices assessed, except winter application of manure, which explained 29% of the variance. Most of the models explained between 2 and 10% of the variance in specific tillage practices (Napier and Camboni, 1993). Comparable modelling for the Darby Creek study group revealed similar results. Statistical models explained from 2% of the variance for ridge tillage to 19% for winter application of manure

(Camboni and Napier, 1994). None of the variables included in the statistical modelling consistently entered the models.

Nutrient application rates were examined in the context of pounds of nitrogen, phosphate, and potassium per acre in the Scioto River study, and as pounds of nutrient per bushel of output per acre for the Darby Creek watershed study group (Napier and Sommers, 1994; Napier and Camboni, 1994). Separate regression models were calculated for corn, soybeans, and wheat for both study groups. The regression findings for the two studies were almost identical. Farm structure factors, institutional variables, personal characteristics of primary farm operator, and a host of other variables were shown to be very poor predictors of the variance in nutrient application rates. None of the statistical models explained more than 10% of the variance in nutrient use for corn, soybeans, and wheat.

Comparison of Mennonite and nonMennonite farmers was accomplished by disaggregation of Mennonite farmers and nonMennonite farmers (at both time periods) into separate groups and comparing characteristics using discriminant analysis. Mennonite farmers in 1991 were shown to be more concerned about groundwater pollution than were nonMennonites; however, nonMennonites were less willing to participate in educational programs designed to reduce farm chemical use. Mennonite farmers were less well educated, participated less frequently in government farm programs, and worked off the farm less than non-Mennonite farmers (Sommers and Napier, 1993).

Data collected in 1994 revealed that Mennonite farmers were significantly different from nonMennonite farmers in terms of off-farm employment for primary farm operator and mate, and participation in government programs. Mennonite farmers used conservation tillage systems much less frequently than nonMennonite farmers (Napier and Sommers, 1996). The statistical models developed using discriminant analysis were shown to be relatively good using cross-sectional data.

Willingness to participate in wetlands trading system was assessed in the Darby Creek watershed by asking land operators if they would be willing to sell rights to build wetlands on their farms, if they received market price for the land. Land owners were informed they would not be permitted to modify the wetland once it was constructed but would retain all other rights to the wetland resource. Respondents were informed that any person or organization wishing to remove an existing wetland could contract with a land owner to build a comparable wetland at another site. Such an arrangement would meet the legislative intent of Section 404 of the Clear Water Act. Respondents were also informed that persons wishing to construct a wetland on their farms would pay for all construction costs. Land owner-operators who indicated they would participate in such a system, and those who were undecided, were compared with respondents who indicated they would not be willing to participate in such a trading system.

Discriminant analysis was used to compare the two study groups. Study findings revealed that land owners who valued Darby Creek as a resource more highly, valued wetlands more highly, operated larger farms, were less well educated, owned more wetlands, worked more days off the farm, were more willing to participate in the wetlands trading system. The discriminant model revealed that variables shown to be significant in differentiating the two groups of farmers were relatively good predictive factors (Napier et al., 1995).

Comparison of adoption behaviors over time was accomplished by disaggregation of Darby Creek respondents from the 1991 Scioto River sample, and comparing selected variables with comparable data collected in 1994. Since the Darby Creek watershed is a tributary of the Scioto River, farmers within the Darby Creek hydrologic unit had been interviewed in 1991.

Comparison of conservation behaviors over time demonstrated that change had occurred within the Darby Creek watershed from 1991 to 1994; however, some of the changes are not desirable from the perspective of environmental quality. Comparison of the distributions of tillage practices over time, demonstrated movement toward the ends of the scales measuring frequency of use of tillage practices. Such a finding indicates that fewer farmers were experimenting with alternative tillage systems in 1994 compared with 1991. This finding suggests that use of specific tillage practices in this region is becoming more firmly entrenched over time.

If the patterns discovered in 1994 continue over longer time frames, it is highly likely that it will be very difficult to motivate land owner-operators, who have adopted environmentally degrading practices, to change behaviors. This is especially true if voluntary approaches are used as they have been in the last 60 years (Napier and Johnson, 1996).

It was also observed that environmental attitudes had not significantly changed within the watershed, even though many conservation programs had been implemented within the study region between 1991 and 1994. Fertilizer application rates had not changed between the two study periods, although extensive resources had been expended to affect chemical application rates (Napier and Johnson, 1996).

Conclusions and Implications

The most important conclusion derived from study findings is that access to education-information, farm structure variables, economic factors, personal characteristics, and institutional factors are relatively good predictors of attitudes toward soil and water conservation; however, they are very poor predictors of actual conservation behaviors. Research findings also demonstrated that conservation behaviors are basically independent of soil and water conservation efforts within the study watersheds, especially intervention programs implemented by The Ohio State University Cooperative Extension Service within the Darby Creek watershed.

Findings also revealed that favorable conservation attitudes were not translated into adoption of soil and water conservation practices at the farm level. The implication of these findings is that soil and water conservation programs designed to create awareness of environmental issues and to develop positive attitudes toward the environment will probably have little impact on farm-level decisions concerning use of conservation production systems. While awareness of conservation problems and the creation of positive attitudes toward the environment, may enhance the probability that conservation practices will be adopted, they are only the initial step in a very complex decision-making process.

Research findings focused on the wetlands trading system strongly suggest that economic incentives can motivate land owner-operators to participate in soil and water conservation programs even when they have not internalized favorable attitudes toward the outcome of the conservation effort. A significant minority of land owners within the Darby Creek watershed indicated a willingness to participate in a wetlands trading system that would create wetlands on their property, when assured they would receive market price for land resources while retaining ownership of the resource. This was shown to be true even when land owners did not perceive wetlands positively. The implication of this finding is that land owners can be motivated to implement conservation programs that they do not value, if economic incentives are sufficiently large to compensate them adequately for participating.

Study findings revealed that land owner-operators had adopted a number of soil and water protection practices. Unfortunately, they also simultaneously used production practices that could contribute to environmental degradation. The implications of these findings is that mechanisms must be developed to ensure that all components of the farm production system are compatible and contribute to maintenance of environmental quality.

In sum, study findings strongly suggest that voluntary programs designed to motivate land owner-operators to adopt soil and water protection programs must rely heavily on economic incentives to be successful. Education-information-technical assistance can facilitate adoption but will be basically ineffective without economic incentives. If adequate economic resources are available for use as incentives to adopt, voluntary programs can be effective. If adequate economic resources are not available, other approaches will have to be considered such as regulatory approaches or private market systems (Napier, 1994; Napier et al., 1995). Continued reliance on the education-information-technical assistance approach will probably not be very effective and will remain costly to U.S. taxpayers.

References

Baker, David B. 1990. Groundwater quality assessment through cooperative private well testing: an Ohio example. *J. Soil Water Conserv.* 45(2):230–235.

Baker, David B., Laura K. Wallrabenstein, and Peter R. Richards. 1994. Well vulnerability and agricultural contamination: assessments from a voluntary well testing program. In Diana Weigman, Ed. *Proceeding of the Fourth National Conference on Pesticides: New Directions in Pesticide Research, Development and Policy.* Virginia Water Resources Center at the Virginia Polytechnic Institute and State University, Blacksburg, Virginia.

Camboni, Silvana M. and Ted L. Napier. 1994. Socioeconomic and Farm Structure Factors Affecting Frequency of Use of Tillage Systems. Invited plenary paper presented at the Agrarian Prospects III symposium (Prague, Czech Republic).

Clark, Edwin H. II, Jennifer A. Haverkamp, and William Chapman. 1985. *Eroding Soils: The Off-Farm Impacts.* Conservation Foundation, Washington, D.C.

Edwards, Allen. 1957. *Techniques of Attitude Scale Construction.* Appleton-Century-Crofts, New York.

Halcrow, Harold, E.O. Heady, and M.L. Cotner. 1982. *Soil Conservation Policies, Institutions, and Incentives.* Soil and Water Conservation Society Press, Ankeny, Iowa.

Lang, Keith S. 1993. Factors Influencing the Criteria Farm Operators Use in Making Adoption Decisions: A Diffusion-Farm Structure Perspective. Masters thesis. The Department of Agricultural Economics and Rural Sociology. The Ohio State University, Columbus.

Lovejoy, Stephen B. and Ted L. Napier, Eds. 1986. *Conserving Soil: Insights from Socioeconomic Research*. Soil and Water Conservation Society Press, Ankeny, Iowa.

Napier, Ted L. 1990. *Implementing the Conservation Title of the Food Security Act of 1985*. Soil and Water Conservation Society Press, Ankeny, Iowa.

Napier, Ted L. 1993. The socioeconomics of groundwater protection in the Scioto River Watershed of Ohio. Pages 242–243 in *Agricultural Research to Protect Water Quality*. Soil and Water Conservation Society Press, Ankeny, Iowa.

Napier, Ted L. 1994. Regulatory approaches for soil and water conservation. Pages 189–202 in L.E. Swanson and F.B. Clearfield, Eds. *Agricultural Policy and the Environment: Iron Fist or Open Hand*. Soil and Water Conservation Press, Ankeny, Iowa.

Napier, Ted L. 1995. Soil and water conservation in the Darby Creek Hydrologic Unit of Ohio. Pages 203–206 in *Clean Water-Clean Environment-21st Century*. American Society of Agricultural Engineers, St. Joseph, Michigan.

Napier, Ted L. and Deborah E. Brown. 1993. Factors affecting attitudes toward groundwater pollution among Ohio farmers. *J. Soil Water Conserv.* 48(5):432–439.

Napier, Ted L. and David G. Sommers. 1994. Correlates of plant nutrient use among Ohio farmers: implications for water quality initiatives. *J. Rural Stud.* 10(2):159–171.

Napier, Ted L. and David G. Sommers. 1995. Farm production systems of Mennonite and non-Mennonite land owner-operators in Ohio. *J. Soil Water Conserv.* 51(1):71–76.

Napier, Ted L. and Eric Johnson. 1996. Conservation Initiatives within the Darby Creek Hydrologic Unit of Ohio: An Assessment of Impacts. Presented at the 1996 Soil and Water Conservation Society Meeting (Keystone, Colorado).

Napier, Ted L., Sam E. McCarter, and Julia R. McCarter. 1995 Willingness of Ohio land owner-operators to participate in a wetlands trading system. *J. Soil Water Conserv.* 50(6):648–656.

Napier, Ted L. and Silvana M. Camboni. 1994. Correlates of Nutrient Application Rates: The Role of Socioeconomic Factors. Invited plenary paper presented at the Agrarian Prospects III symposium (Prague, Czech Republic).

Napier, Ted L. and Silvana M. Camboni. 1993. Use of conventional and conservation practices among farmers in the Scioto River Basin of Ohio. *J. Soil Water Conserv.* 48(3):231–237.

Napier, Ted L., Silvana M. Camboni, and Samir A. El-Swaify, Eds. 1994. *Adopting Conservation on the Farm: An International Perspective on the Socioeconomics of Soil and Water Conservation*. Soil and Water Conservation Society Press, Ankeny, Iowa.

Nunnally, J.C. 1978. *Psychometric Theory*. McGraw-Hill, New York.

Page, G. William, Ed. 1987. *Planning for Groundwater Protection*. Academic Press, San Diego, California.

Sommers, David G. and Ted L. Napier. 1993. Comparison of Amish and non-Amish farmers: a diffusion-farm structure perspective. *Rural Sociol.* 58(1):130–145.

Swanson Louis E. and Frank B. Clearfield. 1994. *Agricultural Policy and the Environment: Iron Fist or Open Hand*. Soil and Water Conservation Society Press, Ankeny, Iowa.

Tucker, Mark and Ted L. Napier. 1995. Perceptions of Risk Associated With Use of Farm Chemicals: Implications for Conservation Initiatives. Presented at the annual meeting of the Soil and Water Conservation Society (Des Moines, Iowa).

Chapter 30

Soilcrop — A Prototype Decision Support System for Soil Degradation-Crop Productivity Relationships

S. GAMEDA AND J. DUMANSKI

Abstract

Challenges related to analytical tools and methodologies are involved in addressing sustainable land management based on multiobjective criteria. Integration of a number of analytical tools is required, including the use of geographic information systems (GIS), expert systems, biophysical and economic models. Methodologies are also required for eliciting farmer knowledge, abstraction of data into indicators, and integration of databases from disparate domains. A modular approach suited for incorporating analytical tools and methodologies was used for developing a decision support system (DSS) for sustainable land management (SLM), based on the framework for evaluation of SLM. A method was developed to integrate farmer knowledge with scientific research into a knowledge base on soil degradation-crop productivity relationships. The knowledge base was used in a prototype soil conservation expert system which was developed as the basic component of the DSS. The expert system determines the type and level of farm-level soil degradation, associated impacts on crop yield, and the remedial management practices required. The expert system is currently being enhanced with links to geo-referenced data and a biophysical process model. DSS components which include socioeconomic criteria will be developed based on data and knowledge bases obtained from case studies conducted in a number of countries. The overall DSS will provide biophysical and socioeconomic criteria for determining the sustainability of cropping practices in different agroenvironments.

Introduction

Addressing sustainability issues based on multiobjective criteria poses a number of challenges. These can be categorized as: (1) lack of integrated analytical tools; (2) disparities in the temporal and spatial scales at which the different objectives function; (3) need for appropriate supporting land resource, land use, and economic performance databases; and (4) need for integration of qualitative knowledge and quantitative data on sustainability. A number of primarily biophysical and economic models are currently in use that provide powerful analytical tools for assessing aspects of sustainability. However, their applicability to the assessment of SLM is limited by their lack of integration of multiple objectives and their requirements for exhaustive data sets.

The challenge of scale from both spatial and temporal perspectives has also not been addressed adequately. While issues of sustainable agriculture are often addressed at a macro policy level, activities affecting sustainability are most often carried out at micro or farmer levels. Furthermore, the impacts of certain objectives are dynamic, while those of others occur over much longer time spans. These dichotomies require the structuring of problems of scale into hierarchies, with associated methods of spatial and temporal aggregation and analysis.

A further challenge is disparity in databases and the difficulties associated with integrating disparate databases. An example is the lack of linkages between biophysical and economic databases. Integrating such data requires the development of functions that can translate the impacts of land resource use into economic terms, or alternately translate economic activities into their impacts on biophysical resources. Moreover, data need to be abstracted into appropriate indicators for assessing the sustainability of land management practices. Methodologies are, therefore, required for compiling such indicators and structuring them into databases for use in DSS.

Farmer Knowledge and Indicators of SLM

The development of indicators for SLM is still in its early stages. Consequently, there are many areas where sustainability assessments need to be carried out with sparse and incomplete information, which often requires that researchers tap into the qualitative knowledge of farmers. The importance of qualitative farmer knowledge has always been recognized but rarely used for scientific research, because the methodologies to make use of it were not widely available.

This can be attributed to the difference between scientific data and qualitative knowledge. Data consist of facts or figures from which specific cause-effect relationships can be determined. Qualitative knowledge encompasses implicitly aggregated facts and information, incorporates uncertainty, and draws from experience, resulting in intuitive, general relationships or correlations between different factors of interest. For example, researchers can obtain data on actual topsoil depth and crop yield reductions to derive explicit relationships between them. Qualitative knowledge, however, would entail abstraction of topsoil depth into soil color (appearance of subsoil) and aggregation of crop yield reductions across seasons, soil types, and landscapes to provide heuristic relationships between soil color and crop response. The precision obtained is more coarse than can be derived from actual data, but this is counter-balanced by a greater ability to apply the knowledge base to a wider set of conditions.

The use of farmer knowledge in conjunction with scientific principles has led to breakthroughs in such areas as soil classification and land evaluation in developing countries (Pawluk et al., 1992). Similarly, the World Bank has developed guidelines that incorporate the use of farmer-provided production estimates as a source of valid data (Murphy et al., 1991). Such successes have led to greater interest in devising methodologies for incorporating farmer knowledge with scientific research, and using the combined knowledge base for developing DSS.

Methodology

The advent of expert systems, coupled with shells facilitating their development, has expedited the integration of qualitative knowledge with quantitative information obtained from scientific research. To date, however, such systems have tended to be specific to production concerns within a given environment, e.g., production of commodities such as cotton under high capital and material input systems in North American agriculture (McKinion et al., 1989).

Because SLM is more complex than simple production systems, a mix of approaches needs to be used to develop an appropriate DSS. These consist of agroenvironmental databases for supporting decisions and input to analyses, establishment of relationships between qualitative farmer knowledge and scientific data, and identification of biophysical and economic models to link various levels of interaction between land management practices, biophysical responses, and socioeconomic impacts.

Studies are being conducted at Agriculture and Agri-Food Canada (AAFC) to develop a DSS for SLM. A modular approach is being taken to individually develop each modular component of the overall DSS. Each component is designed to function separately, and the aggregate of the components will comprise the overall DSS. The objective of this study is to integrate the disparate tools and methodologies identified above into the DSS for SLM.

Framework for Evaluation of SLM

A series of steps has been taken to address the issues of SLM within a DSS that encompasses multiple objectives. An International Working Group headed by the International Board for Soil Research and Management (IBSRAM) and the Centre for Land and Biological Resources Research (CLBRR), in collaboration with FAO and others, has developed a scientifically based, international Framework for the Evaluation of Sustainable Land Management (FESLM) (Smyth and Dumanski, 1993). SLM is defined as a system that combines technologies, policies, and activities aimed at integrating socioeconomic principles with environmental concerns so as to simultaneously maintain or enhance production/services, reduce the level of production risk, protect the potential of natural resources, be economically viable, and be socially acceptable (Dumanski and Smyth, 1994). These objectives constitute the five basic pillars of SLM, respectively, **productivity**, **security**, **protection**, **viability,** and **acceptability**. The FESLM is a logical pathway analysis procedure to guide the evaluation of land use sustainability through a series of scientifically sound steps (Smyth and Dumanski, 1993). It is comprised of three main stages: (1) identification of the purpose of evaluation, specifically land use systems and management practices; (2) definition of the process of analysis, consisting of the evaluation factors, diagnostic criteria, indicators, and thresholds to be utilized; and (3) an assessment endpoint that identifies the sustainability status of the land use system under evaluation.

The FESLM is used as the basis for developing components of the SLM-DSS. The basic component is a prototype expert system for soil conservation. Further enhancements will be achieved through (1) development of information and knowledge bases by means of case studies, (2) structuring the information and knowledge bases into relationships between pertinent FESLM parameters, and (3) coding information and knowledge bases, plus associated relationships into a computerized DSS.

Prototype Expert System for Soil Conservation

The objective of the prototype expert system for soil conservation was to integrate farmer knowledge with scientific research for soil degradation-crop productivity relationships. The prototype expert system, SOILCROP, was developed on the premise that innovative conservation farmers are adept in early identification of soil degradation problems and implementing remedial measures. It is recognized that these innovative farmers have intuitively developed and made use of indicators of the status of soil quality on their land to assist them in their diagnosis of degradation problems and choice of management practices.

Establishing relationships between qualitative farmer knowledge derived from these innovative farmers and quantitative scientific data, involves significant methodological challenges. In the development of SOILCROP, the qualitative knowledge was dispersed among a number of expert farmers and there were gaps or conflicts in the knowledge base. This required extensive knowledge elicitation through interviews, resolution of conflicting information, and validation of the knowledge base with experience from local agronomists. Information for the knowledge base was compiled by administering a structured questionnaire to identified innovative conservation farmers. Information sought in the questionnaire consisted of (1) identification of indicators of the level of degradation, (2) expected yield reductions due to each level of degradation, and (3) the management practices implemented for soil amelioration at each level of degradation.

The indicators identified by innovative conservation farmers (Table 30.1) were not differentiated according to their relative diagnostic importance, although it was clear that the farmers had implicit criteria for selective use of the indicators. There was, therefore, a need to classify and rank indicators in order to signify their relative importance. A relative weight was associated with each degradation indicator using a two-stage weighting system. Initially, indicators were classified as either strategic, cumulative, or tentative. Strategic indicators are those from which definite identification of type and level of degradation can be deduced without question. Cumulative indicators identify degradation additively, i.e., several are necessary before a definitive statement can be made. Tentative indicators suggest soil degradation, but they need to be in combination with other indicators

Table 30.1 Indicators of Soil Degradation Severity Identified by Innovative Conservation Farmers

	Wind erosion	
Slight	Moderate	Severe
Lighter colored soil	Large, heavy dust clouds	Very stunted to no plant growth
Soil accumulation in ditches, other low areas, fence lines	Soil accumulation along fence lines and shelterbelts	Crusting on more than half of field
Slightly shorter crops	Mixing of top two soil layers	Very heavy dust clouds
Increased stoniness on knolls and hilltops	Soil crusting covering half the field	White-topped knolls visible from across fields
Cultivated soil appears coarse	Texture becomes more coarse	Increased stoniness over most of field
Crusting	Uneven plant growth	Heavy soil accumulation at field edges and fence lines
Some dust in the air	Increased stoniness over parts of field	Very small soil aggregates
	Hilltops light-brown color	Disappearance of fence lines
	Thinner, shorter plants	
	Decrease in soil aggregate size	

	Water erosion	
Slight	Moderate	Severe
Lighter colored knolls	Increased stoniness	No topsoil left
Thinner plant growth	Poor crop growth	Water runs visible from across fields
Soil crusting on knolls and deposition areas	Topsoil subsoil mixing	Fully formed gullies that cut down to subsoil
Appearance of cultivable rills and small gullies	Soil crusting	Increased number of rocks on ridges
Poor germination	Rills forming around field edges	Very uneven crop growth
Increased stoniness on hilltops	Appearance of larger rills and small gullies	Difficulty establishing forages and other crops
	Soil deposition at slope bottoms	

to identify degradation type and level. Indicators were subsequently ranked according to their relative diagnostic merit (Table 30.2). A portion of the expert system rule base was structured on the basis of these indicators and their associated certainty factors.

Scientific research has generated extensive data and information in the domain of erosion based degradation. However, conflicting evidence exists as to the relative effect of different factors or parameters on soil degradation. There is, also, a large dependence on locational influences. Furthermore, as the relationships developed tend to be empirical, it is difficult to quantify these effects parametrically. For example, relationships developed between the depth of topsoil eroded and crop yield losses vary depending on such factors as depth of topsoil layer, quality of organic matter in the topsoil, nature of the subsoil, and climatic conditions.

During expert system development, it was necessary to organize and synthesize the underlying symbolic relationships behind the complexity of such information. This entailed identifying the parameters to be considered, establishing criteria for categorizing identified parameters, indicating the assumptions made in developing

Table 30.2 **Examples of Degradation Indicators with Associated Class, Rank, and Certainty Factors**

Wind erosion indicators	Degradation level	Class Type	Class Score	Rank	Value
Crusting on more than half of field	sv	2	7	7	49
Very heavy dust clouds	sv	1	10	10	100
White-topped knolls visible from across fields	sv	1	10	10	100
Increased number of stones on knolls and hilltops	sl	1	10	10	100
Very small soil aggregates	sv	3	3	7	21
Large, heavy dust clouds	md	1	10	10	100
Mixing of top two soil layers	md	2	7	10	70
Difficulties incorporating chemicals	md	3	3	1	3
Hilltops light-brown color	md	2	7	4	28
Soil accumulation in ditches, other low areas, fence lines	sl	1	10	10	100
Cultivated soil appears coarse	sl	2	7	7	49
Some dust in the air	sl	1	10	10	100
Slightly shorter crops	sl	2	7	4	28
Very stunted to no plant growth	sv	2	7	10	70
Uneven plant growth	md	2,3	7,3	7	21–49

Water erosion indicators	Degradation level	Class Type	Class Score	Rank	Value
Gullies that cut down to subsoil	sv	1	10	10	100
Appearance of larger rills and small gullies	md	1	10	7	70
Appearance of cultivable rills	sl	1	10	7	70
Increased stoniness on parts of field	md	1	10	7	70
Soil deposition at slope bottoms	md	1	10	10	100
Increased soil hardness	md	2	7	4	28
Greater frequency of hardpan areas	sl	3	3	4	12
Decreased soil moisture	md	3	3	1	3
Increased flooding of sloughs	sl	2	7	4	28
Very uneven crop growth	sv	2	7	7	49
Buried seedlings in slope bottoms	sv	1	10	10	100

Strategic, cumulative, and tentative indicators were assigned scores of 10, 7, and 3, respectively, in accordance with their relative diagnostic merit.

Legend: Degradation level — sl = slight, md = moderate, sv = severe; Class — Type: 1 = strategic, 2 = cumulative, 3 = tentative; Score: 10, 7, 3; Rank — Relative ranking, 10 = highest, 1 = lowest; Value — Certainty level for use by expert system shell, Score x Rank = Value.

criteria, and stating the symbolic relationships as a set of rules for use in the expert system. Such a process revealed the underlying links between the qualitative knowledge base of innovative conservation farmers and the quantitative soil quality information from scientific research (Tables 30.3, 30.4). The principles used in developing symbolic relationships for use in SOILCROP were based on procedures originally described by Hackett (1988).

Table 30.3 Criteria for Categorizing Selected Parameters into Symbolic Classes for Use in SOILCROP

Table 3a

Examples of parameters used*				Rating developed for SOILCROP
Topsoil loss (%)	Organic C (%)	Normal yields**, Spring Wheat (kg/ha)	Yield loss (%)	
<25	<2	<1,500	<10	Low/slight
25–50	2–3	1,500–2,500	10–35	Moderate
>50	>3	>2,500	>35	High

Table 3b

Yield recovery (% of uneroded)	Rating developed for SOILCROP
<85	Highly insufficient
85–90	Moderately insufficient
90–95	Slightly insufficient
95–105	Restored
105–110	Slightly increased
110–115	Moderately increased
>115	Highly increased

Table 3c

Location	Erosion-yield relationships (y = yield; x = depth of topsoil removed)	Yield response to incremental topsoil loss (SOILCROP equivalent)
Taber	$y = 1,121 - 47.5x + 0.74x^2$	Low
Lethbridge	$y = 1,289 - 93.6x + 1.48x^2$	Low to moderate
Hill Spring	$y = 1,554 - 117.6x + 2.65x^2$	Moderate
Cooking Lake	$y = 2,578 - 164.1x + 2.73x^2$	Severe
Josephburg	$y = 2,665 - 45.9x - 2.05x^2$	Severe

* Each parameter is rated individually
** Example of assumptions used in developing criteria for normal spring wheat yields.

The average yield for the region is 2,000 kg/ha. Yields within 25% of this value are considered moderate; yields below or above this threshold are considered low or high, respectively. Similar procedures were used for ranking other parameters.

Based on information from soil erosion crop productivity research (Larney et al., 1992; 1994).

The symbolic relationships obtained from combined farmer knowledge and scientific research sources were used to develop the set of rules which drive and control the expert system. Some examples of the types of rules derived from the criteria and classifications described in Tables 30.3 and 30.4 are:

    If: soil zone is Brown Chernozemic
    Then: yield potential is low (Table 30.4)
    and   erosion impact on yields is low (Table 30.3c)
    and   response to amendments is low to moderate (Table 30.4)

Table 30.4 Examples of Developing Symbolic Classes used in SOILCROP in Relation to Different Agroenvironments (Location)

Location	Soil zone	Common topsoil depth (cm)	Biological characteristics					Erosion effects					Productivity response			
			Organic C			Yields			Erosion		Yield loss					
			(%)	ES** equiv.		(kg/ha)	ES equiv.		Depth (cm)	ES equiv.	(%)	ES equiv.	Amend't	% of uneroded	Yield recovery	SOILCROP equivalent
---	---	---	---	---	---	---	---	---	---	---	---	---	---	---	---	---
Taber	BC*	10–15	1.40	Low		1,146	Low		5	Mod	23	Mod	Fertilizer	105	Restored	
													Manure	119	Moderately increased	
									10	Sev	44	Sev	Fertilizer	80	Moderately insufficient	
													Manure	97	Restored	
Lethbridge	DBC	15	1.58	Low		1,205	Low		5	Mod	12	Low	Fertilizer	125	Highly increased	
													Manure	139	Highly increased	
									10	Sev	67	Sev	Fertilizer	84	Moderately insufficient	
													Manure	132	Highly increased	
Hill Spring	BIC(tn)	10–13	2.97	Mod		1,522	Mod		5	Mod	28	Mod	Fertilizer	144	Highly increased	
													Manure	202	Highly increased	
									10	Sev	59	Sev	Fertilizer	118	Moderately increased	
													Manure	197	Highly increased	
Cooking Lake	GL	15–18	3.47	High		2,585	High		5	Mod	32	Mod	NA			
									10	Sev	48					
Josephburg	BIC(tk)	30–33	4.00	High		2,653	High		5	Slight	11	Mod	NA			
									10–15	Mod	19–50	Mod-Sev				
									20	Sev	63	Sev				

* BC = Brown Chernozemic; DBC = Dark Brown Chernozemic; BIC(tn) = Black Chernozemic (thin); GL = Gray Luvisol; BIC(tk) = Black Chernozemic (thick).
** ES = expert system (SOILCROP).

If: soil zone is Black Chernozemic (thin)
Then: yield potential is moderate (Table 30.4)
and erosion impact on yields is moderate (Table 30.3c)
and response to amendments is high (Table 30.4)

If: soil zone is Brown Chernozemic, and erosion level is moderate
Then: fertilizer amendments to eroded areas will restore yields (Table 30.4)
and manure amendments to eroded areas will moderately increase yields (Table 30.4)

If: soil zone is Black Chernozemic (thin), and erosion level is moderate
Then: fertilizer amendments to eroded areas will highly increase yields (Table 30.4)
and manure amendments to eroded areas will highly increase yields (Table 30.4)

These examples, although based on limited research, illustrate the symbolic relationships that can be drawn and utilized in an expert system. Such relationships are easily changed as more reliable information becomes available.

SOILCROP is currently being enhanced with links to extensive land resource databases through a GIS. This will enable input of location specific physical attributes to better define the environmental envelope within which the land use systems operate. Another enhancement is to incorporate a prognostic component which draws on modelled output from process models such as EPIC to provide the long-term impact of chosen management practices on soil degradation and crop productivity. A schematic of the overall structure is given in Figure 30.1.

Application of SOILCROP Experience to FESLM-Based DSS

Information provided by innovative conservation farmers often contains implicit multiobjective criteria for sustainability, as for example, the level of degradation and associated yield loss that farmers are willing to internalize before implementing ameliorative practices, the number of years that drought can be tolerated before the systems fail, etc. However, explicit criteria are required for these principles to be applied outside the domains in which they were obtained. The FESLM-based DSS being developed is aimed at establishing explicit tradeoff criteria for a range of agricultural production systems in different agroenvironments. This requires that land resource databases be established and mobilized, expert farmer knowledge be obtained, and that biophysical and economic impact models be integrated within an interactive structure. The following discussion describes the process to develop the knowledge base and computer coding components of the DSS.

Indicator Database

The key to the success of a multiobjective DSS is a comprehensive, purpose-oriented database. Extensive soil inventories and considerable volumes of data on agricultural land resources are necessary. These data should be structured into natural biophysical units such as land resource areas (LRAs), whereby each area consists of landscapes with similar ecoclimatic characteristics, land potentials,

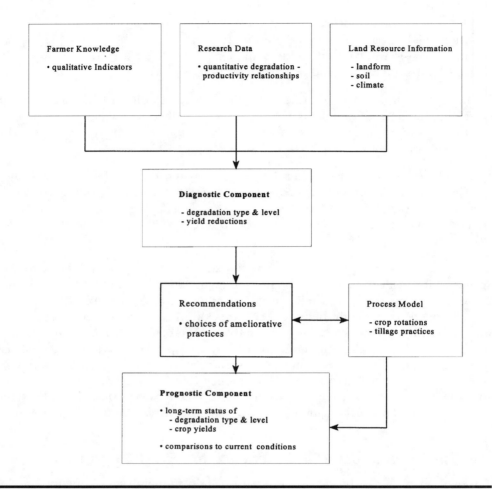

Figure 30.1 Schematic of SOILCROP expert system structure.

degradation risk, etc. This store of land-related information should be complemented by long-term research trials on the effects of soil degradation on crop production, and on the effects of various management practices combining crop rotation, tillage, etc. on sustainable agricultural production. Such databases have already been compiled at AAFC. These kinds and detail of data are not, however, available in many parts of the world. Moreover, even when available, they need to be aggregated into suitable indicators for purposes of assessing and monitoring the status of SLM objectives. In these situations, much can still be achieved through farm-oriented case studies where each farm is treated as the experimental unit.

Accordingly, an FESLM-based case study was conducted on a rain-fed agricultural production system in Alberta, Canada to determine farm level FESLM evaluation factors, diagnostic criteria, indicators, and thresholds (Gameda and Dumanski, 1995). The study provided information on the different types of data required to conduct an assessment of sustainability (Table 30.5). Based on this experience, more exhaustive case studies are being conducted on nine different farming systems in Saskatchewan, Canada. The information and knowledge from 31 participating farmers is being elicited using participatory rural appraisal principles

Table 30.5 Evaluation Factors, Diagnostic Criteria and Indicators for Evaluating the Sustainability of a Rain-Fed Crop Production System

Evaluation factors		Diagnostic criteria	Indicators
Physical	Moisture availability (low seasonal precipitation)	Soil moisture conservation	Percent land in fallow
	Growing season length	Soil cover maintenance for erosion control	Method of fallow
	Hail storm hazards		Depth of soil moisture at seeding
	Erosion hazards		Percent and trends of degradation
	Salinization hazards		Type of tillage practice
Biological	Soil fertility management	Maintenance of adequate nutrients	Changes in fertilizer and pesticide use
	Weed control in fallow		Length and diversity of rotations
			Crop yield trends
Economic	Operating costs	Production cost reductions	Trends in cost of production
	Revenues	Risk minimization	Gross margin
	Government subsidies		Number and types of government programs
	Management objectives (profit/utility maximization; risk reduction, etc.)		Farmer's management strategies
Social	Viability of rural communities	Proximity to available services	Distance to services
	Acceptability of farming practices	Off-farm impacts and perceptions	Social perceptions of conservation practices
			Reasons for participation in government programs

(Chambers, 1994). Concurrently, on-site biophysical and socioeconomic studies are being conducted to develop the databases from which strategic indicators will be developed for each of the SLM objectives. The information obtained is geo-referenced to the farm unit, and appropriate relationships between the data layers will be drawn for the analysis of each production system.

Additional case studies are being conducted in a number of other countries in order to compile databases of factors, indicators, and thresholds for FESLM assessments of a wide range of farming systems. Guidelines outlining the methodology and expected outcomes for the case studies have been drawn up with assistance of the Rockefeller Foundation (Dumanski et al., 1995) and are already implemented in several countries including Australia, Zimbabwe, Mexico, and the Phillipines. Further case studies will begin in Indonesia, Thailand, and Viet Nam.

The data and information obtained from the case studies identified above will be compiled into databases comprising region-specific evaluation criteria, indicators, and thresholds for local applicability. These would then serve as a basis for the development of a prototype FESLM-DSS (Figure 30.2) for rain-fed upland soils. Methods for compiling and structuring the database will be similar to those used in the development of SOILCROP. The DSS will be designed as a generic system applicable globally, but structured to draw on region-specific data and knowledge bases.

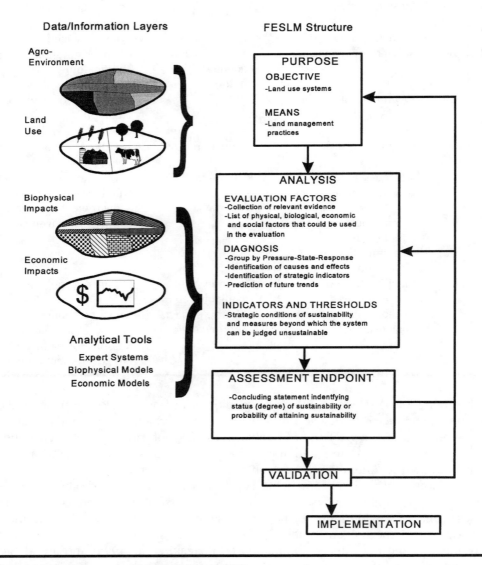

Figure 30.2   Schematic of FESLM-based DSS.

Conclusions

Several issues need to be taken into consideration for development of a DSS for SLM based on multiobjective criteria. These include the use of various analytical tools, e.g., GIS, expert systems, biophysical and economic models. In addition, a number of methodologies are required for drawing on hierarchy concepts to address spatial and temporal scale disparities, developing transfer functions for linking databases from disparate domains and, most importantly, combining qualitative farmer knowledge with scientific research.

The SOILCROP prototype expert system illustrates the feasibility of developing a DSS that integrates farmer knowledge and scientific research. Although SOIL-CROP is an operational tool for diagnosis and prognosis of soil conservation issues, it will become the foundation for the development of the DSS for SLM based on the FESLM. The system will mimic the management decisions and practices used by farmers, the basis of which are the necessary tradeoffs between maintenance of resource quality, the cost of remedial practices, and assessments of short-term economic returns and long-term investment.

Experience with the SOILCROP expert system indicates that a modular approach is necessary to develop the components of a FESLM-based DSS. Each component should be designed to address a specific objective, but also linked together so that in aggregate, the multiple objectives of the DSS are properly treated.

A reliable and scientifically credible knowledge base integrating farmer knowledge with scientific data is essential to the development of a FESLM-based DSS. It is important, however, that the knowledge base be validated by domain experts to ensure its credibility and acceptance.

Case studies provide an important method to build up the necessary knowledge base for decision support. Several of these are being conducted in different countries according to FESLM guidelines. Results from these case studies, featuring combined biophysical and socioeconomic components, will enable comparative assessments of economic, biophysical, and social sustainability based on the FESLM.

In today's world, the development of DSSs for SLM (or other applications) is not limited by computer technology; on the contrary the tools are readily available. The larger constraints are the lack of knowledge of the interactions and tradeoffs between the objectives of SLM, and an incomplete understanding of the concepts and dynamics of sustainability. Improvements in the knowledge base will come about through more active participation of the scientific community, field evaluations through case studies, and farm level experimentation.

References

Chambers, R. 1994. The origins and practice of participatory rural appraisal. *World Dev.* 22(7):953–969.

Dumanski, J., S. Nortcliff, H. Eswaran, R. Fuentes, and J.K. Syers. 1995. Guidelines for the Implementation of Case Studies in Sustainable Land Management. Unpublished Report. 13 pages.

Dumanski, J. and A.J. Smyth. 1994. The issues and challenges of sustainable land management. In R.C. Wood and J. Dumanski, Eds. Proceedings of the International Workshop on Sustainable Land Management for the 21st Century, Volume 2. Plenary Papers. Agricultural Institute of Canada, Ottawa.

Gameda, S. and J. Dumanski. 1995. Framework for evaluation of sustainable land management: a case study of two rain-fed cereal-livestock farming systems in the Black Chernozemic soil of southern Alberta, Canada. *Can. J. Soil Sci.* (in press).

Hackett, C. 1988. Matching Plants and Land: Development of a General Broadscale System from a Crop Project for Papua New Guinea. Natural Resources Series No. 11. Division of Water and Land Resources. CSIRO, Australia. 82 pages.

Larney, F.J. 1992. Productivity of artificially eroded soil: the Lethbridge "scalping" studies. Pages 51–56 in Soil Conservation: Many Ways to Make it Work. Proceedings of the Soil Conservation Workshop & Alberta Conservation Tillage Soc. 14th Annual Meeting, January 13–15, 1992, Edmonton, Alberta.

Larney, F.J., R.C. Izaurralde, H.H. Janzen, B.M. Olson, E.D. Solberg, C.W. Lindwall, and M. Nyborg. 1994. Soil erosion-crop productivity relationships for six Alberta soils. *J. Soil Water Conserv.* 50(1):87–91.

McKinion, J.M., D.N. Baker, F.D. Whisler, and J.R. Lambert. 1989. Application of GOSSYM/COMAX system to cotton crop management. *Agric. Syst.* 31:55–65.

Murphy, J., D.J. Casley, and J.J. Curry. 1991. Farmers' Estimations as a Source of Production Data: Methodological Guidelines for Cereals in Africa. World Bank Technical Paper No. 132. 71 pages.

Pawluk, R.R., J.A. Sandor, and J.A. Tabor. 1992. The role of indigenous soil knowledge in agricultural development. *J. Soil Water Conserv.* 47(4):298–302.

Smyth, A.J. and J. Dumanski. 1993. FESLM: An International Framework for Evaluating Sustainable Land Management. World Soil Resources Report 73. FAO, Rome. 74 pages.

Chapter 31

Evaluation of a Prototype Decision Support System for Rangelands in the Southwest United States

PAUL A. LAWRENCE, LEONARD J. LANE, JEFFRY J. STONE, AND DIANA S. YAKOWITZ

Abstract

The appearance of rangelands in parts of the southwestern U.S. is changing from grasslands with scattered shrubs to sparse grass populations under dense stands of mature woody shrubs, mostly of mesquite (*Prosopis velutina*). Decisions on whether to control the mesquite, coupled with a variety of grazing systems, make it appropriate to use decision support systems (DSS) as an advisory tool. This paper presents a preliminary evaluation of grazing and vegetation management systems for rangelands using a prototype multiobjective DSS (P-MODSS) developed by the USDA Agricultural Research Service at the Southwest Watershed Research Center in Tucson, Arizona. Sixteen years of measured and calculated data from four experimental watersheds were used to quantify physical resource decision variables of runoff depth, sediment yield, aboveground net primary production (ANPP), and peak rate of runoff. The results suggested that when it was desirable to maximize ANPP and minimize runoff and sediment yield, yearlong grazing with mesquite removed was preferred to the conventional system of yearlong grazing with mesquite retained and rotation grazing with and without mesquite. However, when both runoff and ANPP were maximized and sediment yield was the most important criteria, no alternative was better than the conventional system. The importance order of the decision variables and the scoring function shape are highlighted to demonstrate the utility of the P-MODSS to evaluate management systems on rangelands.

Introduction

Rangelands cover a large portion of the land surface of the world. Depending on the definition of rangelands, 25 to 50% of the earth's land is used as rangelands (van Gils, 1984). Characteristics that link rangeland ecosystems worldwide include a complex biology, large and heterogeneous management units, variable climate, and socioeconomic pressures for change or modifications (Stuth and Stafford Smith, 1993). These special features of rangeland management necessitate that a broad perspective be taken in the planning process and that information be well organized.

While the major commercial use of rangelands in the U.S. is livestock grazing to produce food and fiber, rangelands provide other less tangible values such as natural beauty, open spaces, wildlife habitat, and the ecological study of natural ecosystems (National Research Council, 1994). If rangelands are to be sustainable, methodologies are needed to facilitate informed decision making. Such methodologies should consider the whole system rather than concentrate on individual components. DSS provide a means of integrating databases, computer simulator models, economic analyses, and geographical information systems in a package that is practical and informative to the land use planner. In addition, the DSS needs to fulfil user requirements, be technically correct, consider diverse objectives, and present the user with information on feasible alternatives to the current grazing system.

This paper presents preliminary results of the application of the P-MODSS developed by the USDA Agricultural Research Service, Southwest Watershed Research Center, in Tucson, Arizona, to evaluate four current and alternative rangeland management practices in southwestern U.S. Specifically, the work demonstrates the effects of an importance order on the outcomes from the DSS for four decision criteria. The P-MODSS (Yakowitz et al., 1992a,b) was originally developed to assess the effects of alternative management practices on surface and groundwater quality. Use of the P-MODSS to evaluate farming practices in cropping lands (Yakowitz et al., 1992a; Yakowitz et al., 1993; Heilman, 1995) and the design of trench caps for shallow landfill waste (Lane et al., 1991; Paige et al., 1994) are well established, however it is only recently that the application to rangelands has been examined (Renard and Stone, 1993). Although the P-MODSS has the capacity to incorporate continuous simulation models, the analyses reported in this paper were restricted to using measured data from four instrumented watersheds on the Santa Rita Experimental Range to quantify the decision variables.

Background to the Problem

Increases in velvet mesquite (*Prosopis velutina* Woot.) have changed the general appearance of much of southern Arizona from grasslands with scattered young mesquite to sparse grass populations under dense stands of mature woody shrubs (Parker and Martin, 1952; Cable and Martin, 1973). When the mesquite is controlled, herbaceous cover and production have increased (Parker and Martin, 1952; Martin, 1963) while soil loss and runoff have declined (Renard et al., 1991; Martin and Morton, 1993). Methods to control the invasion of mesquite include mechanical or chemical means, burning, or a combination of each. However, costs to control the mesquite may not be economically justified from the net returns of the cattle alone. In addition, increased awareness of wider, socioeconomic aspects of desert grasslands and recognition of the continued degradation of natural resources have engendered an ecological perspective into vegetation management.

In addition, grazing management is linked to vegetation management. Options for grazing are continuous yearlong or a system of seasonal grazing that employs one or more types of grazing and nongrazing periods. The intended benefit of rotational grazing systems is improved range condition. Although rotational grazing has lead to increased grass production (Martin, 1973), these have not necessarily transferred to positive livestock responses (Driscoll, 1967) despite the increased input of management. While many studies have focused on the health of the plant and animal, there is a need to consider the effects of grazing from the perspective of soil, water, and air quality. To this end, some grazing practices may create an unsustained and unbalanced structure between the abiotic and biotic components of the ecological system.

Methods

Description of Watersheds and Data Collected

Eight experimental watersheds were established by the USDA Agricultural Research Service during 1975 on the Santa Rita Experimental Range, located 50 km south of Tucson, Arizona. The purpose of the watershed study was to examine the impact of grazing and vegetation manipulation methods in the semidesert (300 to 400 mm annual precipitation) regions of the southwestern U.S. Four of these watersheds were selected for this analysis. One pair of watersheds were grazed yearlong, while the other pair of watersheds were grazed using the Santa Rita rotation method (Martin, 1978). This method involved grazing once during March to October and once during November to February in a 3-year rotation, with 12-months rest between grazing periods. The dominant feature of the method is that rangeland to be grazed in the spring is rested during the preceding summer and winter and is an important component for proper grazing management in this environment (Arizona Interagency Range Committee, 1973). In 1974, two watersheds (WS2 and WS4) were treated with basal applications of diesel oil to control the invasion of mesquite and retreated when needed to keep the watersheds mesquite free. The other two (WS1 and WS3) remained unchanged. In 1994, the aerial coverage of mesquite and dense woody shrubs in WS1 and WS3 was 17 and 22%, respectively. Treatments and physical characteristics of the watersheds are given in Table 31.1.

Each watershed is equipped to measure precipitation (rate and depth), surface runoff (rate and depth), and sediment yield. On the occasions when there were insufficient sediment samples, sediment yield for the storm event was estimated

Table 31.1 Summary of Physical Characteristics of the Experimental Watersheds

WS	Area (ha)	Land Use Treatment		Soil type	Average (%)	Stream (m)
		Grazing	Vegetation			
1	1.63	Yearlong	Mesquite +	Sandy	3.43	329
2	1.76	Yearlong	Grass	Sandy	4.21	256
3	2.76	Rotation	Mesquite +	Fine sandy	3.01	298
4	1.97	Rotation	Grass	Fine sandy	4.01	306

using the measured depth and peak rate of runoff. The precipitation and runoff data are considered to be excellent (Renard et al., 1991). Periodic measurements were also made of channel cross-sections, grass density, and cover.

Summary of Hydrological Responses

Precipitation at the Santa Rita watershed study varies considerably from year to year, and from season to season. From 1976 to 1991, annual precipitation varied from 177 to 641 mm. Mean annual precipitation across the four watersheds was 373 mm, with a coefficient of variation of 29%. The bimodal distribution of monthly precipitation for WS1 is shown in Figure 31.1. Surface runoff is generated by short-duration, high-intensity summer storms. Between June and August, runoff producing storms represented 26% of the annual precipitation, but produced 66% of the total annual runoff. The temporal and spatial variability of storms in the Southwest has important implications for range management (Renard and Stone, 1993). For example, when spatial variability is high, summing storm events from a single raingauge to determine monthly or seasonal precipitation for water supply and forage management may unknowingly be misleading. In addition, partial wetting of the pasture will generate heterogeneous growth that may influence the degree of pasture utilization, and possibly species vigor and composition.

In this environment, essentially all the infiltrated moisture is partitioned as either bare soil evaporation or transpired by plants (Renard et al., 1993). Percolation below the root zone is infrequent and represents a negligible component in the average annual water balance.

Overview of the P-MODSS

The major components of the P-MODSS are (1) a modified version of the GLEAMS (Groundwater Loading and Evaluation of Agricultural Management Systems, Leonard et al., 1987) simulation model, (2) databases and default values for parameterizing the simulation model, (3) a decision model with embedded scoring functions and an algorithm for ranking alternatives, (4) a system driver, and (5) a user interface and report generator. Further details of the P-MODSS are given by Yakowitz et al. (1992a,b).

The decision model is based on multiobjective decision theory that combines the dimensionless scoring functions of Wymore (1988) with the decision tools presented in Yakowitz et al. (1992b;1993). The scoring functions convert predicted

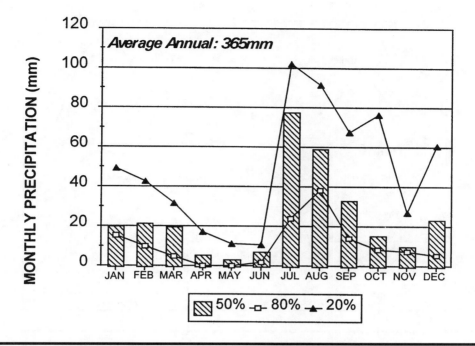

Figure 31.1 Distribution of monthly precipitation for 50, 80, and 20% probability of exceedence, watershed WS1 (1976 to 1991).

or observed data to a unitless scale of 0 to 1, where 0 is the worst and 1 is the best possible score. The scoring functions are a means of scaling each decision criteria by which the current and alternative management systems are evaluated. The four generic scoring functions are more is better, more is worse, a desirable range, and an undesirable range (Figure 31.2). Net returns and productivity are examples of decision criteria that would be associated with the more is better scoring function shape, while erosion would be associated with the more is worse score shape. The functions are constructed such that the average annual value of the management system conventionally used is assigned the score of 0.5 for all decision variables. The slope of the function at the score of 0.5 is determined by the annual minimum and annual maximum values of the decision criteria. The scores of the decision criteria for alternative management practices are computed in relation to the conventional system.

The importance order of the decision variables can be specified by the user or computed by the normalized slope of the function for the conventional management system. This latter method is the default importance order and assigns more value to those decision criteria for which small differences in the values of the alternative criteria make a large difference in the score. After the importance order has been determined, a best and worst composite score is computed by the method developed by Yakowitz et al. (1992b) using two linear programs for each alternative. This method is a feature of the P-MODSS which eliminates some of the subjectivity associated with assigning weights to the decision criteria. Finally, the alternatives are ranked in descending order according to the average of the best and worst composite scores. The preferred or "best" alternative is the one with the highest average score.

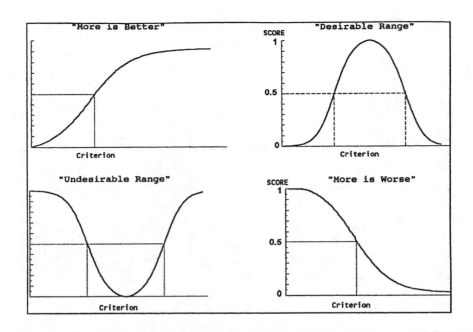

Figure 31.2 Generic scoring function types and shapes (based on Wymore, 1988).

Selection of Decision Variables

The National Research Council (1994) Committee on Rangeland Classification has developed a number of indicators to judge rangeland health. Among these, soil stability, soil erosion, and watershed function are considered to be important indicators of rangeland health. Using the 16 years of recordings from the Santa Rita watersheds, four measurements were selected as decision variables to evaluate the grazing and land management systems. These decision variables were annual runoff (mm), maximum annual peak rate of runoff (mm/hr), annual sediment yield (t/ha), and annual aboveground net primary production (g/m^2). Annual runoff, peak rate of runoff, and total annual sediment yield were calculated from measured data for the period 1976 to 1991 for each watershed and used to quantify the decision variables. ANPP is the total aboveground dry weight biomass produced per unit area in a growing season. Relationships between annual actual evapotranspiration and ANPP are given by Rosenzweig (1968), Webb et al. (1978), Lane and Stone (1983), and others. Lane and Stone (1983) showed that, in the absence of detailed information, annual actual evapotranspiration can account for 80% of the variation in ANPP. Using the discrete form of the water balance equation, annual actual evapotranspiration was derived as the difference between annual precipitation and annual runoff, assuming negligible percolation losses. The estimate of actual evapotranspiration for WS1 and WS3 was reduced by the coverage of mesquite and other woody shrubs (17 and 22%, respectively) to reflect the ANPP for grass forage. In this respect, ANPP was intended to represent a surrogate for productivity, although not a direct indicator of economic returns to the rancher.

Hence, the impact of each grazing and vegetation management system on factors that affect the volume and discharge of surface water, soil loss, and productivity were considered through the selection of the decision variables. It is emphasized that the analysis represented only a preliminary assessment of rangelands using measured data. As multiobjective databases are extended and more complex rangeland and ecosystem simulation models are considered, more decision criteria will need to be included in the evaluation. Further, the decision variables are not reflective of off-site effects nor do they address the objectives of broader issues of natural resource conservation.

Scoring Functions

The scoring function type more is worse was chosen for the decision variables of depth of runoff, peak rate of runoff, and sediment yield. In this sense, the preferred management system should minimize these physical resource decision variables. ANPP, as a surrogate for productivity, used the scoring function type "more is better."

Management Systems Evaluated

The four grazing and vegetation management systems evaluated were yearlong grazing (YL) and rotation grazing (ROT), each with (+m) and without (-m) mesquite vegetation.

Results and Discussion

The evaluation of rangeland management systems is discussed first using only the measured and estimated data from the experimental watersheds. This was done to demonstrate that it is possible to identify preferred management systems without the use of a computerized DSS. Following this initial analysis, an evaluation was performed by incorporating the information from the measured data into the P-MODSS. This evaluation takes advantage of two features of the P-MODSS, namely the importance order of the decision variables and the type of scoring function, to assess practical, multiobjective decision making scenarios that are normally not possible from a simpler analysis of measured data.

Evaluation without a Decision Model

Average annual values for the four decision variables (runoff depth, peak rate of runoff, sediment yield, and ANPP) for each grazing and vegetation treatments are given in Table 31.2. The results showed that the yearlong grazing treatments produced less annual runoff and sediment yield and lower peak discharge compared to the rotation grazing systems. This finding may be associated with the higher density of perennial grasses in the yearlong grazed watersheds. Martin and Morton (1993) reported that during 1985 and 1986, low summer rainfall, coupled

Table 31.2 Average Annual Values for Physical Resource Decision Variables. Coefficient of Variation (%) shown in brackets.

Decision variable	Grazing and vegetation practice			
	Yearlong + mesquite WS 1 Conventional	Yearlong + no mesquite WS 2 Alternative 1	Rotation + mesquite WS 3 Alternative 2	Rotation + no mesquite WS 4 Alternative 3
Annual runoff	18.7	13.2	40.8	37.3
Peak runoff rate	24.3	21.8	32.6	35.7
Sediment yield	1.4	0.8	7.3	7.1
ANPP	74.1	103.7	66.5	85.6

with March to October grazing in 1986, resulted in the grass density in the rotation grazing watersheds being less than half that measured in the yearlong grazing watersheds. Rainfall during 1985 was almost 20% below the annual mean. Martin (1973) observed that recovery of grass density after dry conditions was slow if March to October grazing was imposed during drought or in the summer following drought. The lower density and coverage of grass, in association with the soil texture, may be responsible for the greater runoff from the rotation grazed watersheds (WS3 and WS4) compared to the yearlong grazed watersheds (WS1 and WS2). To support these differences, an examination of the frequency of runoff events showed that the yearlong grazed watersheds averaged almost half the number of runoff events (11 per year) compared to the rotation grazed watersheds (19 events per year).

When the results in Table 31.2 were grouped according to the vegetation management, the mesquite-free watersheds (WS2 and WS4) produced less runoff and sediment yield and more ANPP than the mesquite-invaded watersheds (WS1 and WS3). The estimated ANPP was 35% greater on the mesquite-free watersheds than on the mesquite watersheds. The effect of vegetation manipulation on the mean annual peak runoff rate and the mean annual frequency of runoff events was indistinguishable.

This examination of the effect of grazing and vegetation treatments on the selected physical resources yielded several outcomes. First, it suggested that yearlong grazing was preferred to rotation grazing if the desired intention was to maximize ANPP and minimize runoff and sediment yield. Second, controlling mesquite satisfied both goals of natural resource conservation and production. Hence, the results indicated that the practice of controlling mesquite and yearlong grazing represented the preferred management system for southern Arizona rangelands. However, this outcome was somewhat limited in terms of considering other important factors associated with grazing and ranch management, such as economics, labor and management for stock handling, fencing, herd composition, and the biophysical limitations of the land. In addition, the outcome was independent of the impact on broader, multiobjective issues such as wildlife habitat, human perception of the use of rangelands, and the general concepts of integrated resource management. Finally, the outcome was based on the implicit assumption that the four resource decision variables were of equal importance.

Evaluation Using a Decision Model

The application of a DSS is designed to enhance and assist the land manager make decisions on effective land management. This is normally done by comparing the current conventional practice to a number of feasible alternative management practices. For this study, yearlong grazing with mesquite (YL+m) was selected as the conventional practice and compared to the other management systems on the basis of the four physical resource decision criteria. Three analyses were performed to demonstrate some of the advantages of using the P-MODSS to evaluate the sustainability of rangelands in the Southwest. These analyses were

1. An evaluation with equal importance attached to the four decision variables.
2. An evaluation using the default importance order attached to the decision variables.
3. An evaluation when the scoring function for depth of annual runoff is changed from more is worse to be more is better. This scenario represented the basis for earlier research into vegetation manipulation for water yield enhancement in Arizona (Hibbert, 1965; Ffolliott and Thorud, 1974).

Decision Variables with Equal Importance

The results of the analysis using the P-MODSS to evaluate the four management systems when equal importance was placed on the decision criteria are given in Figure 31.3. Clearly, the YL-m was the most preferred system. The two rotation systems (ROT+m and ROT-m) were less preferred to the two yearlong systems. This outcome is consistent with the earlier examination of the measured data.

Decision Variables with an Importance Order

The P-MODSS allows the user to select an importance order for the decision variables. This is a realistic feature designed to accommodate a particular preference associated with the decision criteria. For example, the user may wish to place higher importance on surface runoff than on the other decision variables. The default ordering of the criteria ranks highest that criterion which has the potential for the greatest change in score when a small change in the criteria near the conventional practice is observed (Yakowitz et al., 1992b). The score matrix for the decision variables with the default importance order is given in Table 31.3. Ranking the decision criteria by the normalized value of the slopes of the scoring functions resulted in a default importance order of (ranked from most to least importance): ANPP > sediment yield > runoff depth > peak rate of runoff.

Figure 31.4 shows the best (top line of bar) and worst (bottom line of bar) composite scores for the default importance order. When greatest importance was placed on ANPP, YL-m dominated the conventional (YL+m) and the alternative grazing and vegetation management systems. On the basis of the average composite score, YL+m was preferred to ROT+m and ROT-m. However, the length of the bars in Figure 31.4 suggested the outcome was highly sensitive to a particular weight vector consistent with the importance order. For some possible weighting schemes, ROT-m was equal to or better than YL+m, but for the majority of the weighting

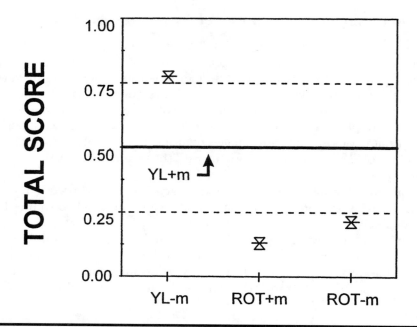

Figure 31.3 Composite scores for yearlong (YL) and rotation (ROT) grazing with (+m) and without (-m) mesquite vegetation with an equal importance order for the decision variables.

Table 31.3 Score Matrix for Decision Variables with Default Importance Order

	Grazing and Vegetation System			
	Yearlong + mesquite WS 1	Yearlong + no mesquite WS 2	Rotation + mesquite WS 3	Rotation + no mesquite WS 4
Decision variable	Conventional	Alternative 1	Alternative 2	Alternative 3
Annual runoff	0.5	0.787	0.004	0.011
Peak runoff rate	0.5	0.611	0.178	0.110
Sediment yield	0.5	0.524	0.291	0.299
ANPP	0.5	0.931	0.341	0.729

schemes, this alternative was less preferred to the conventional system of yearlong grazing with mesquite. The ROT+m alternative was the least preferred system.

Modifying the Scoring Functions

In the above analyses, the depth of runoff was associated with a scoring function type of "more is worse." However, in an environment where water is a limiting factor, a rancher may be interested in grazing systems that generate runoff for water harvesting projects and water supply for stock dams.

To evaluate this option, the scoring function type for runoff was changed to "more is better" and the P-MODSS used to reevaluate the management systems.

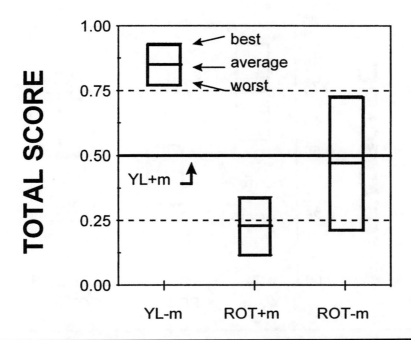

Figure 31.4 Composite scores for yearlong (YL) and rotation (ROT) grazing with (+m) and without (-m) mesquite with a decision variable importance order of ANPP > sediment > runoff > peak runoff rate.

This changed the score matrix for runoff for YL-m to 0.213, for ROT+m to 0.996, and for ROT-m to 0.990. All other values of the score matrix were unaltered from the values shown in Table 31.3. For the initial evaluation, the default importance order (ANPP > sediment yield > runoff > peak runoff rate) was used. The scoring functions for ANPP, sediment yield, and peak runoff were unchanged.

The results of best and worst scores following the adjustment to the runoff scoring function are shown in Figure 31.5. When the average composite score is used to rank the alternatives, YL-m was again found to be the preferred system. The ROT-m system was also preferred to the conventional system of YL+m. Changing the scoring function for runoff from "more is worse" (Figure 31.4) to "more is better" (Figure 31.5) increased the sensitivity of the outcome for YL-m but reduced the sensitivity of the outcome for ROT-m.

Next, an importance order was imposed so that runoff and sediment yield were of equal importance, but with a higher importance order than ANPP and peak runoff rate. With this imposed importance order, the rotation grazing systems (ROT+m and ROT-m) were marginally preferred to the yearlong grazing systems (Figure 31.6). No one alternative completely dominanted the other two. Given the importance order of the decision variables, the ROT-m was the most preferred system, and the YL-m the least preferred system.

As a final examination of grazing and vegetation systems, the importance order of the decision variables was adjusted to give sediment yield the greatest importance (i.e., sediment > runoff > ANPP > peak runoff rate). In Figure 31.7, the results showed that when sediment was the primary concern ahead of runoff, ANPP, and peak discharge, there was little to distinguish between the alternatives

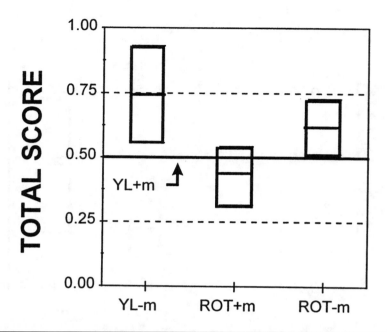

Figure 31.5 Composite scores for yearlong (YL) and rotation (ROT) grazing with (+m) and without (-m) mesquite with a decision variable importance order of ANPP > sediment > runoff > peak runoff rate. Scoring function for runoff is "more is better."

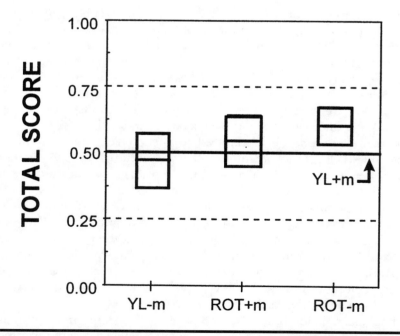

Figure 31.6 Composite scores for yearlong (YL) and rotation (ROT) grazing with (+m) and without (-m) mesquite with a decision variable importance order of runoff = sediment > ANPP > peak runoff rate. Scoring function for runoff is "more is better."

Evaluation of a Prototype Decision Support System for Rangelands

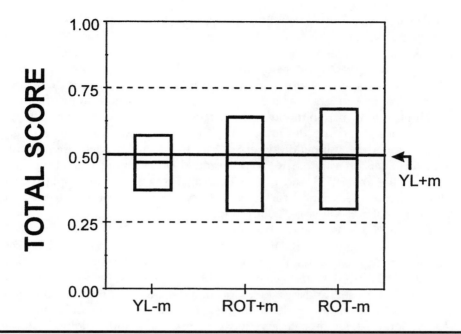

Figure 31.7 Composite scores for yearlong (YL) and rotation (ROT) grazing with (+m) and without (-m) mesquite with a decision variable importance order of sediment > runoff > ANPP > peak runoff rate. Scoring function for runoff is "more is better."

and the conventional management systems. The average score for all alternatives ranged from 0.47 to 0.49, and all were less than the 0.5 score associated with the conventional treatment. Figure 31.7 showed that the alternatives differed in the length of the bar, suggesting that the composite score for YL-m was the least sensitive to the weights consistent with the importance order.

Summary and Conclusions

A preliminary analysis was undertaken of the extensive rangeland areas represented by the grazing and vegetation management systems monitored on the Santa Rita Experimental Range in southwestern Arizona using a prototype DSS. Of the four rangelands management systems considered, the P-MODSS identified yearlong grazing with mesquite removed as a preferred system when it was desirable to maximize ANPP and minimize runoff depth, sediment yield, and peak discharge. This treatment produced the highest average composite score regardless of the importance order given to ANPP, sediment yield, and runoff. However, when the scoring function for runoff was changed from "more is worse" to "more is better" to reflect the limitations of water in a dry environment, and sediment yield and runoff were given equal importance, then the rotation grazing systems were preferred to the yearlong grazing system. Alternatively, when sediment yield was given the highest importance, all four systems considered in this evaluation were almost identical, but yearlong grazing with mesquite removed displayed the least sensitivity of the outcome to a given weighting vector. Based on the conditions and data set from the experimental watersheds, these evaluations suggested that yearlong grazing with no mesquite was a management system which reflected a stable balance of conservation and productivity objectives.

There are two important considerations for the use of a MODSS for evaluating rangelands. First, a quantitative framework is available to assess current and alternative practices of grazing and vegetation control. This framework incorporates technical information about the management system by quantifying the decision variables and the desirable utility of the decision variable through the selection of the scoring function type. Second, the user can adjust the importance order of the decision variable without the need of assigning weightings. This is a feature of the P-MODSS and one which encourages the full application of a MODSS to explore scenarios for natural resource conservation. Hence, the P-MODSS can make a valuable contribution towards the nation's assessment of rangeland health and ecological condition.

Future work using the P-MODSS on rangelands needs to address the limitations encountered with this preliminary work. First, the decision variables need to be quantified using a continuous, dynamic simulation model. This would allow long-term values to be used and consideration may be given to increase the number of decision variables in the evaluation (e.g., seasonal timing of soil water storage, alternative land uses, interactions between livestock and wildlife, vegetation composition, a more detailed economic analysis). Second, data are needed to expand the scope of decision variables to integrate off-site, sustainability, and socioeconomic aspects of rangelands.

Acknowledgment

The support of the Queensland Department of Natural Resources and the Rural Industries Research and Development Corporation, Australia, to undertake graduate studies at the University of Arizona, Tucson, is gratefully acknowledged by the first author (PAL).

References

Arizona Interagency Range Committee. 1973. Grazing Systems for Arizona Ranges. University of Arizona, Tucson. 36 pages.

Cable, D.R. and S.C. Martin. 1973. Invasion of semidesert grassland by velvet mesquite and associated vegetation changes. *Arizona Acad. Sci.* 8:127–134.

Driscoll, R.S. 1967. Managing Public Rangelands: Effective Livestock Grazing Practices and Systems for National Forests and National Grasslands. USDA AIB-315.

Ffolliott, P.F. and D.B. Thorud. 1974. Vegetation Management for Increased Water Yield in Arizona. Technical Bulletin 215. Agricultural Experiment Station, University of Arizona, Tucson. 38 pages.

Heilman, P. 1995. A Decision Support System for Selecting Economic Incentives to Control Nonpoint Source Pollution from Agricultural Lands. Ph.D. dissertation. University of Arizona, Tucson. 211 pages.

Hibbert, A.R. 1965. Forest treatment effects on water yield. Pages 527–543 in *International Symposium on Forest Hydrology. National Science Foundation Advances in Science Seminar Proceedings*. Pergamon Press, New York.

Lane, L.J. and J.J. Stone. 1983. Water balance calculations, water use efficiency and aboveground net production. Hydrology and Water Resources in Arizona and the Southwest. Office of Arid Land Studies, University of Arizona, Tucson. 13:27–34.

Lane, L.J., J. Ascough, and T.E. Hakonson. 1991. Multiobjective decision theory — decision support systems with embedded simulation models. Pages 445–451 in ASCE Irrigation and Drainage Proceedings, July, Honolulu, Hawaii.

Leonard, R.A., W.G. Knisel, and D.A. Still. 1987. GLEAMS: groundwater loading effects of agricultural management systems. *Trans. ASAE*. 30(5):1403–1418.

Martin, S.C. 1963. Grow more grass by controlling mesquite. College of Agriculture, University of Arizona. *Prog. Agric. Ariz.* 15:15–16.

Martin, S.C. 1973. Responses of semidesert grasses to seasonal rest. *J. Range Manage.* 26:165–170.

Martin, S.C. 1978. The Santa Rita grazing system. Pages 573–575 in D.N. Hyder, Ed. *Proceedings of the 1st International Rangeland Congress*. Society Range Management, Denver, Colorado.

Martin, S.C. and H.L. Morton. 1993. Mesquite control increases grass density and reduces soil loss in southern Arizona. *J. Range Manage.* 46:170–175.

National Research Council. 1994. *Rangeland Health — New Methods to Classify, Inventory and Monitor Rangelands*. National Academy Press, Washington, D.C. 180 pages.

Paige, G.B., T.E. Hakonson, D.S. Yakowitz, L.J. Lane, and J.J. Stone. 1994. A prototype decision support system for the evaluation of shallow land waste disposal trench cap designs. Pages 111–117 in Proceedings of ER'93, Meeting the Challenge, Environmental Remediation Conference, Augusta, Georgia.

Parker, K.W. and S.C. Martin. 1952. The Mesquite Problem on Southern Arizona Ranges. U.S. Department of Agriculture Circular 908.

Renard, K.G., F.A. Lopez, and J.R. Simanton. 1991. Brush control and sediment yield. Pages 12.38–12.45 in Proceedings of the 5th Federal Interagency Sedimentation Conference, Las Vegas, Nevada.

Renard, K.G. and J.J. Stone. 1993. Integrated watershed management. Pages 355–379 in Water Harvesting for Improved Agricultural Production. Proc. FAO Expert Consultation, Water Report 3.

Renard, G., L.J. Lane, J.R. Simanton, W.E. Emmerich, J.J. Stone, M.A. Weltz, D.C. Goodrich, and D.S. Yakowitz. 1993. Agricultural impacts in an arid environment: Walnut Gulch studies. *Hydrol. Sci. Technol.* 9(1–4):145–190.

Rosenzweig, N.L. 1968. Net primary productivity of terrestrial communities: prediction from climatological data. *Am. Nat.* 102:67–74.

Stuth, J.W. and M. Stafford Smith. 1993. Decision support for grazing lands: an overview. Pages 1–35 in J.W. Stuth and B.G. Lyons, Eds. *Decision Support Systems for the Management of Grazing Lands — Emerging Issues*. Man and the Biosphere Series, Volume 11. UNESCO (Paris) and the Parthenon Publishing Group (New York).

van Gils, H. 1984. Rangelands of the world: unifying vegetation features. In W. Siderius, Ed. Proceedings. Land Evaluation for Extensive Grazing (LEEG). International Institute for Land Reclamation and Improvement, The Netherlands. Pub. No. 36.

Webb, W., W. Szarek, R. Lavenroth, R. Kinerson, and M. Smith. 1978. Primary productivity and water-use in native forest, grassland and desert ecosystems. *Ecology*. 39(6):1239–1247.

Wymore, A.W. 1988. Structuring system design decisions. Pages 704–709 in C. Weimin, Ed. *Systems Science and Engineering*. Proceedings of the International Conference on Systems Science and Engineering (ICSSE '88), July 1988. International Academic Publishers.

Yakowitz, D.S., L.J. Lane, J.J. Stone, P. Heilman, and R. Reddy. 1992a. A decision support system for water quality modeling. Pages 188–193 in *ASCE Water Resources Planning and Management*. Proceedings of the Water Resources Sessions/Water Forum '92, Baltimore, Maryland.

Yakowitz, D.S., L.J. Lane, J.J. Stone, P. Heilman, R. Reddy, and B. Imam. 1992b. Evaluating land management effects on water quality using multi-objective analysis within a decision support system. 1st International Conference on Groundwater Ecology, Tampa, Florida. 365–373.

Yakowitz, D.S., L.J. Lane, and F. Szidarovszky. 1993. Multi-attribute decision making: dominance with respect to an importance order of the attributes. *Appl. Math. Computat.* 54:167–181.

Chapter 32

Managing Soil Structure, Particularly with Respect to Infiltration, for Long-Term Cropping System Productivity

R. D. CONNOLLY AND D. M. FREEBAIRN

Abstract

The impact of soil structural degradation on the infiltration capacity and productivity of cropping soils in southeast Queensland, Australia was evaluated using the cropping systems model PERFECT-SWIM. The effect of different cropping history and management practices on soil physical characteristics, water balance, and production were simulated. The effectiveness of each management strategy on minimizing runoff and maximizing gross margin was evaluated using multiple objective decision theory. Cropping effected each soil differently, hence the most favorable management strategy varied for each soil. These results highlight the need for physical properties of individual soils to be considered when managing cropping systems on agricultural soils.

Introduction

Declining soil physical conditions under cropping is endemic for many cultivated soils in southern Queensland and northern New South Wales (Bridge et al., 1983;

Dalal and Mayer, 1986; McGarry, 1990). Effects of declining soil physical conditions range from increasing runoff and erosion to severe yield decline. Measured runoff (as a percentage of rainfall) was greater on a light clay soil near Roma (570 mm annual average rainfall) than from catchments on the Darling Downs (black earth soils) with higher rainfall (760 mm average annual rainfall). The relatively high runoff in the more arid area indicated differences in structural stability between the soils. Variation in runoff with changing catchments management and soil condition has been observed in other studies (e.g., Lawrence, 1990). Declining infiltration capacity is a serious problem for many cropped soils in Queensland, limiting water available for crop production and increasing the chances of erosion.

Experimental studies, however, have difficulty characterizing the effects of structural degradation on water infiltration and crop productivity. Direct comparisons of the consequences of declining soil structure at experimental sites are often confounded by climatic variability and site variation. Attempts to extend this shortfall in utility of index based experimentation have centered on the application of models. Technology to predict crop growth and some of the important soil processes controlling crop growth are available, but integrated methods for their application have not been fully developed.

In this paper, a model integrating the effects of soil structural condition on cropping productivity is used to evaluate the relative effectiveness of a range of management strategies on several contrasting soil types.

Methods

The Cropping System Model

The combined SWIM (Soil Water Infiltration and Movement; Ross, 1990) and PERFECT (Productivity, Erosion and Runoff Functions to Evaluate Conservation Techniques; Littleboy et al., 1989) models as described by Connolly (in prep) were used to simulate the effect of changing soil structure on water balance and crop production. PERFECT was used to predict crop growth, evapotranspiration, tillage management, and cover on a daily basis. SWIM was used to predict infiltration, runoff, soil water redistribution, and deep drainage on a short time step (typically 30 min). Models of annual and seasonal changes in soil surface condition were derived empirically and included in the combined model (termed PERFECT-SWIM). The existing surface sealing algorithm in SWIM was replaced with a model which incorporated surface cover and roughness (Connolly, in prep). Effects of rainfall intensity, vegetative cover, and soil roughness on surface seal formation were included in the surface seal model.

Parameters describing surface soil structure changes due to tillage and consolidation by rainfall as well as long-term changes due to crop or pasture phase were derived. Seal initial conductance, surface roughness, and parameters describing runoff varied daily. Seal final conductance and hydraulic conductivity of the subsurface (10 to 20 cm deep) soil layer varied with an annual time step. Details of the models and procedures for estimating parameter values are described in Connolly (in prep).

Climate data, soil water, soil nitrogen, surface soil conditions, and crop agronomy parameters for use in PERFECT-SWIM were derived from data measured at

Table 32.1 Description of Soil Properties for the Three Sites used to Compare Responses to Tillage Management and Crop History

Soil type	Great soil group	Soil taxonomy	Sand (%)	Silt (%)	Clay (%)	Cation exchange capacity (cmol(+)/kg)
Light clay	Affinities toward Solodics and Red Earths	Entic chromustert	48	4	44	34
Texture contrast	Solodic	Udic haplustals	49	16	35	23
Heavy clay	Black Earth	Udic pellerstert	20	16	64	52

the experimental sites. Economics parameters were determined from current operating practices. Simulations for the three sites used rainfall and temperature data from Roma, Queensland.

Experimental Sites

Three soil types were considered, to represent a range of soils in southeast Queensland (Table 32.1). A brief description of each site follows:

- Light clay — cultivated continuously for 26 years for wheat and fodder production, this soil was the most structurally degraded of the three soils. The site is located at the Wallumbilla trial northeast of Roma, Queensland. The soil has a clay loam surface textured surface, but tends toward a brown clay under cultivation, after the surface soil is mixed with a clay subsoil. The soil is moderately hard-setting and exhibits some cracking in a dry condition. Land slope is approximately 3%.
- Texture contrast — this soil was considered the least productive of the three soils, having low water holding capacity and relatively poor infiltration characteristics which are not markedly improved with pasture. The site is located at the Arrowfield trial, northeast of Goondiwindi, Queensland. Land slope is approximately 0.5%. The trial site had been sown to wheat for 5 years.
- Heavy clay — this self-mulching soil has a high water-holding capacity and is considered the most productive of the three soils. The soil represents shallow but well structured soils of the eastern Darling Downs, Queensland. Average land slope is 5%.

Management Strategies

Five management strategies were simulated for each soil type:

- Conventional tillage — the conventional tillage strategy consisted of a wheat monoculture, with a crop grazed 1 in 5 years. Primary tillage was with a chisel implement, followed by scarifier tillage to prepare a seed bed. The site was cultivated 2 to 4 times over the fallow and no fertilizer

was applied. With this management, soil structure declined, exhibited as increased sealing of the surface and reduced hydraulic conductivity of the subsoil, and soil fertility declined. Income from the available land was derived mainly from cropping (4 years in 5 wheat, 1 year in 5 cattle). Australian Standard Wheat prices ($135/t) were assumed to be received for wheat due to low grain protein resulting from declining soil fertility. Grain production costs were assumed to be $80/ha.
- Conventional tillage+N — this strategy was the same as conventional tillage, fertilized to maintain nitrogen levels. Grain production costs were assumed to be $117.5/ha and Prime Hard wheat prices were received for grain ($160/t).
- Rotation — this strategy consisted of gradually phasing in a legume-grass pasture, at the rate of 10% of area per year. After 5 years, half the land was in a pasture phase, and half was cropped with the conventional tillage system described above. From years 6 to 10, land which had been under grass for 5 years was returned to cropping (with a conventional tillage system) and the remaining cropped land established with pasture (20% of area per year). At the end of year 10, the cropped area (50% of total) had all had a history of 5 years pasture, and the remaining 50% of total area was under pasture, waiting to be cropped. Land was continually cycled in this manner, so from year 10 on, all cropping was carried out on "rejuvenated" soil. With the ley pasture rotation option, soil structure was restored during the rotation phase, and soil fertility was maintained. Income was derived from a combination of wheat cropping and cattle. Because of improved soil fertility, wheat yields on rested ground received prime hard grain prices. Gross margin for cattle was assumed to be $59/ha. Wheat production costs were $80/ha.
- Pasture — this strategy simply consisted of continuous pasture with income derived solely from cattle production. Gross margin for cattle production was assumed to be $59/ha.

Multiobjective Analysis Model

Predictions of average annual runoff and gross margin were analyzed using the multiobjective analysis model of Yakowitz et al. (1993). The relative effectiveness of each strategy in maximizing gross margin (most important) and minimizing runoff was determined.

Results and Discussion

Predicted annual average gross margin and runoff for all strategies at the three sites are given in Table 32.2. The ranked order of treatments for each site is given in Table 32.3. The rotation strategy was the most profitable and produced the least runoff of all strategies simulated for the light clay site. Pasture was the most favorable treatment at the texture contrast site because it markedly reduced runoff, even though the gross margin from the rotation strategy was 50% greater than

Table 32.2 Predicted Annual Average Gross Margin and Runoff for the Three Sites

	Light clay		Texture contrast		Heavy clay	
Strategy	Gross margin ($/ha)	Runoff (mm)	Gross margin ($/ha)	Runoff (mm)	Gross margin ($/ha)	Runoff (mm)
Conventional	−2	94.2	32	111.7	44	54.1
Conventional+N	25	91.2	40	107.3	**231**	72.9
Pasture rotation	**124**	27.6	**89**	102.4	202	52.6
Pasture	59	**0.0**	59	**0.0**	59	**0.0**

Table 32.3 Ranked Order of Treatments for the Three Soils, from Best to Worst

Light clay	Texture contrast	Heavy clay
Rotation	Pasture	Rotation
Pasture	Rotation	Pasture
Conventional	Conventional+N	Conventional+N
Conventional+N	Conventional	Conventional

the pasture strategy. Rotation was most favorable at the heavy clay site, again because runoff was considerably reduced.

Pasture at the texture contrast site was the most favorable, with the most profit and least runoff, of the three sites. Rotation at the light clay site was next favorable, followed by rotation at the heavy clay site. The heavy clay site was potentially the most productive site of the three because of its high water-holding capacity, but ranked last as it shed a considerable amount of runoff when cropped. Runoff from the strategies that included cultivation at all sites was high, so pasture was the most appropriate strategy for minimizing runoff while maximizing gross margin.

Conclusions

The physical characteristics of soils at the three sites significantly influenced what was the most appropriate management strategy for minimizing runoff while maximizing cropping system productivity. The two clay soils were most favorably managed with a pasture rotation which improved the infiltration capacity of the soil and reduced runoff. The low productivity, high runoff soil was best managed with continuous pasture. These results highlight how individual soils vary, particularly with respect to soil structural condition and infiltration capacity, and appropriate management strategies should be developed based on the needs of each soil and the consequences of cropping each soil.

Acknowledgments

Thanks to the Grains Research and Development Council (Project DAQ44) for funding this project and the salary of the principal author. Staff of the Queensland Department of Primary Industries are acknowledged for their assistance in collecting and analyzing data.

References

Bridge, B.J., J.J. Mott, and Hartigan. 1983. The formation of degraded areas in the dry savanna woodlands of Northern Australia. *Aust. J. Soil Res.* 21:91–104.

Connolly, R.D. (in prep) Strategies for Improving Infiltration of Rainfall on Structurally Degraded Soils. Thesis submitted for the degree of Doctor of Philosophy. University of Queensland, Brisbane, Australia.

Dalal, R.C. and R.J. Mayer. 1986. Long-term trends in fertility of soils under continuous cultivation and cereal cropping in Southern Queensland. I Overall changes in soil properties and trends in winter cereal yields. *Aust. J. Soil Res.* 24:265–279.

Lawrence, P.A. 1990. The Hydrology of Three Experimental Catchments with Different Land Uses after Clearing. Thesis submitted to fulfill requirements for Master of Philosophy. Griffith University, Brisbane, Australia. 238 pages.

Littleboy, M., D.M. Silburn, D.M. Freebairn, D.R. Woodruff, and G.L. Hammer. 1989. PERFECT: A Computer Simulation Model of Productivity Erosion Runoff Functions to Evaluate Conservation Techniques. Queensland Department of Primary Industries Bulletin QB89005. Brisbane, Australia.

McGarry, D. 1990. Soil structure degradation: extent, nature and significance in S. E. and Central Qld. Pages 25–27 in *Proceedings Soil Compaction Workshop*. QDPI Conference Centre Toowoomba, 15–17 October.

Ross, P.J. 1990. SWIM — A Simulation Model for Soil Water Infiltration and Movement. CSIRO Div. of Soils, Townsville, Queensland, Australia. 59 pages.

Yakowitz, D.S., J.J. Stone, L.J. Lane, M. Hernandanz, P. Heilman, B. Imam, J. Mastertson, and J.A. Abolt. 1993. A Prototype Decision Support System for the USDA Water Quality Initiative. Southwest Watershed Research Centre, USDA-ARS, Tucson, Arizona.

Chapter 33

Protecting Soil and Water Resources Through Multiobjective Decision Making

TONY PRATO AND CHRIS FULCHER

Traditional evaluation of natural and environmental resources utilizes a single objective, reductionist approach. Multiple objective decision making (MODEM) is a powerful and effective holistic approach. Whereas the reductionist approach makes almost exclusive use of the assumptions, prescriptions, and predictions of a single discipline, the MODEM approach integrates concepts and principles from several disciplines.

The optimization framework employed in natural and environmental economics is well suited for MODEM-type evaluations. Integration can be achieved by adding noneconomic objectives as constraints in an optimization problem or by evaluating tradeoffs among competing objectives using the efficiency frontier. The first approach is illustrated by maximizing net returns in an agricultural watershed in Missouri subject to constraints on sediment delivery to and soluble nitrogen concentrations in receiving streams. The second approach is illustrated for a crop farm in northern Missouri.

An interactive, spatial decision support system (ISDSS) is an effective way to make MODEM-type evaluations accessible to nonsophisticated users. Unlike traditional off-line, noninteractive approaches, an ISDSS allows users to evaluate solutions to a MODEM problem based on their own subjective preferences for objectives in an interactive learning and decision-making process. An ISDSS is illustrated which allows users to assess the socioeconomic, environmental, and ecological consequences of alternative land use/management practices and public policies on erosion and nonpoint source pollution in agricultural watersheds.

> In all areas of science, the convergence and integration of information from different points of view, different disciplines, and different approaches are what lead to advances and breakthroughs in understanding.
>
> Gene E. Likens, 1992

Introduction

Soil and water resource degradation from agricultural erosion and nonpoint source pollution have reduced the socioeconomic, environmental and ecological values provided by agroecosystems in North America. In this region, 96% of the soil degradation occurs in agroecosystems dominated by crop and livestock production (World Resources Institute, 1992). Agriculture is a major source of nonpoint source pollution. The U.S. Environmental Protection Agency (EPA) estimates that between 50 and 70% of assessed surface waters are adversely impacted by agricultural nonpoint source pollution (USEPA, 1986).

This paper has four objectives: (1) examine the conceptual basis for MODEM; (2) develop a framework for implementing MODEM which integrates economic, environmental, and ecological objectives; (3) compare noninteractive to interactive approaches to MODEM; and (4) present noninteractive and interactive applications.

Conceptual Basis

Soil and water conservation practices and policies can be evaluated using several approaches which can be arrayed along a spectrum having the *reductionist method* at one end and the *holistic method* at the other. This section examines both approaches and discusses socioeconomic aspects of holistic resource management.

Reductionist and Holistic Methods

The reductionist method is the traditional way of advancing and applying disciplinary knowledge. With the reductionist method, a particular slice of reality is evaluated from a narrow disciplinary perspective. Most scientific inquiry related to soil and water conservation uses a reductionist approach. Reductionism has a long history of use and acceptance in the scientific community. The specialization afforded by reductionist science has advanced the understanding and resolution of a wide range of issues and problems.

The holistic method synthesizes and integrates concepts and information from several disciplines. In this respect, holistic resource management is a systems approach. An holistic approach focuses on the socioeconomic, environmental, and ecological processes that determine the effectiveness and efficiency of soil and water conservation practices and policies. In an holistic approach, the effects of using a conservation practice are examined from a multidisciplinary perspective.

The holistic approach has its share of difficulties. First, it runs counter to the way generations of scientists and practitioners have acquired and applied knowledge. Second, the inherent complexity of the holistic approach requires considerable interaction among the practitioners of several disciplines. Such interaction is difficult and at times frustrating because of differences in theory, methods, and data. Since holistic resource management embraces a multidisciplinary approach and different disciplines emphasize different management objectives, holistic resource management is a MODEM approach.

Socioeconomic Aspects of Holistic Resource Management

Socioeconomic aspects of holistic resource can be illustrated with regard to a pivotal assumption in economics that humans are motivated by selfishness. This assumption underlies the theory of consumer behavior and the theory of the firm which is central to economics. Daly and Cobb criticize the assumption that households maximize utility and firms maximize profit oblivious to social community and biophysical interdependence. They state: "What is neglected is the effect of one person's welfare on that of others through bonds of sympathy and human community, and the physical effects of one person's production and consumption activities on others through bonds of biophysical community" (Daly and Cobb, 1989).

Another socioeconomic consequence of taking an holistic approach to resource and environmental issues is that it raises serious concerns about the macroeconomic objective of pursuing unbridled economic growth. Historically, growth has been the undisputed goal of economic development. While this goal maximizes living standards in terms of material wealth, there is considerable evidence that it contributes to natural resource and environmental degradation which eventually decreases human welfare and the quality of life. In other words, unbridled economic growth is not sustainable.

Another socioeconomic implication of an holistic approach to natural resource and environmental problems is that it requires sociologists and economists to become more familiar with physical and biological principles governing the natural world and to integrate these principles with socioeconomic concepts in addressing resource and environmental problems. In this framework, socioeconomics is viewed not so much as a self-contained body of knowledge, but rather as a set of concepts which in combination with other scientific principles enhances society's understanding of resource and environmental issues.

Integrated Framework for MODEM

A MODEM framework integrates the socioeconomic, environmental, and ecological objectives relevant to the conservation of soil and water resources and the underlying processes which influence the attainment of those objectives. Socioeconomic objectives deal with the social and economic aspects of soil and water resource use. Social objectives address attitudes regarding the acceptability of specific management practices or policies. Economic objectives include the private and social benefits and costs of a management plan or policy, and preferences

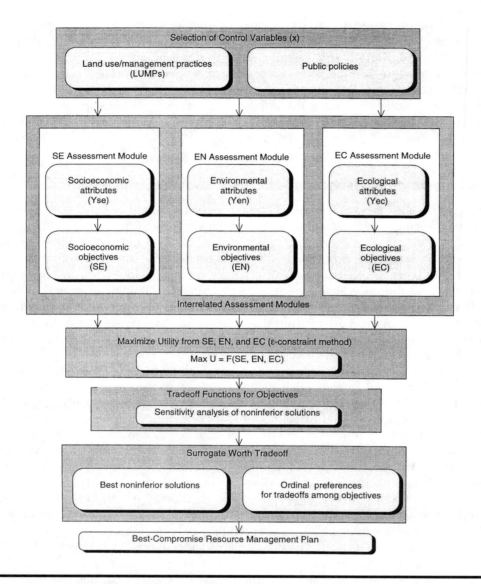

Figure 33.1 Conceptual framework for MODEM.

for the three objectives. Environmental objectives address how practices and policies affect environmental endpoints such as soil erosion and surface and ground water quality. Ecological objectives include the quantity and quality of riparian areas and wetlands and the performance of aquatic and terrestrial ecosystems.

A conceptual framework for MODEM is illustrated in Figure 33.1. The decision maker is assumed to be an individual whose preferences for socioeconomic (SE), environmental (EN), and ecological (EC) objectives are summarized by the following utility function:

$$U = U(SE, EN, EC).$$

U is the level of satisfaction provided by the three objectives. Two features of the objectives are noteworthy. First, they are noncommensurate because they have different metrics; the economic objective is measured in dollars, the environmental objective in mass or concentrations of contaminants, and the ecological objective in species richness and diversity. Second, over some range, the objectives are likely to be competitive with one another. This means that an increase in one objective causes a decrease in another objective.

Each objective has a set of attributes which influences the attainment of that objective. If the SE objective is the viability of farming, then relevant attributes include the mean and variance in net farm returns. If the EN objective is surface water quality in an agricultural watershed, then relevant attributes include mass loading or concentration of nutrients (nitrogen and phosphorus), sediment and chemical oxygen demand in runoff, and stream flow. If the EC objective is the health of aquatic ecosystems, then relevant attributes include species diversity and richness. Attributes related to the same objective can be aggregated. For example, the Index of Biological Integrity developed by Karr et al. (1986) could be used to represent the health of aquatic ecosystems.

Since attainment of an objective depends on the levels of the attributes corresponding to that objective, the utility function can be rewritten as:

$$U = U[SE(\mathbf{y}_{SE}), EN(\mathbf{y}_{EN}), EC(\mathbf{y}_{EC})]$$

where $\mathbf{y}_{SE}$, $\mathbf{y}_{EN}$, and $\mathbf{y}_{EC}$ are vectors of attributes associated with objectives SE, EN, and EC, respectively, and $SE(\mathbf{y}_{SE})$, $EN(\mathbf{y}_{EN})$, and $EC(\mathbf{y}_{EC})$ are utility subfunctions. Maximizing the overall utility received by an individual is tantamount to finding the most preferred set of values for the utility subfunctions, or equivalently, the most preferred set of attributes. This specification of the utility function is common in multiobjective optimization problems (Haimes and Hall, 1977; Changkong and Haimes, 1983; Dinh 1989; Nijkamp and Spronk, 1981; Steuer, 1986; Haimes et al., 1990). The level of each attribute is determined by the selection of one or more control variables which include current and alternative land use/management practices (LUMPs) and public policies for the resource.

Denoting the control variables for a particular farm or watershed by a vector **x** allows the utility function to be rewritten as:

$$U = U[SE(\mathbf{x}), EN(\mathbf{x}), EC(\mathbf{x})]$$

This function states that utility depends on the levels of all control variables.

Although a multiobjective optimization problem does not have a unique solution, it can be used to determine noninferior solutions using the ε-constraint method developed by Haimes et al. (1971) and Cohon and Marks (1973, 1978). Noninferior solutions represent efficient combinations of the objectives. This method maximizes achievement of a primary objective subject to inequality constraints on the remaining objectives. To illustrate this method, let the primary objective be SE. Then, the optimization problem becomes:

$$\begin{array}{c} \text{maximize } SE(\mathbf{x}) \\ \mathbf{x} \end{array}$$

subject to:

$$EN(\mathbf{x}) \leq \varepsilon_{EN}$$
$$EC(\mathbf{x}) \leq \varepsilon_{EC}, \text{ and}$$
$$\mathbf{x} \in \mathbf{X}$$

where ε_{EN} and ε_{EC} are upper limits on attainment levels for the EN and EC objectives, respectively, $\mathbf{x}$ is a set of control variables, and $\mathbf{X}$ is a set of feasible values of $\mathbf{x}$. Any solution to this optimization problem is an acceptable solution to the original constrained utility maximization problem. Noninferior solutions to this optimization problem are determined by solving the problem for different values of ε_{EN} and ε_{EC}. The resulting noninferior solutions are used to derive tradeoff functions for all objectives which are then used to identify the best noninferior solutions for all individuals. Ma (1993) and Xu, Prato and Ma (1995) used the ε-constraint method to generate tradeoff functions for three objectives (maximum net farm returns, minimum soil erosion, and minimum nitrate available for leaching) achieved by alternative farming systems (LUMPs) used on a mid-Missouri farm.

The final element of the MODEM framework involves applying the surrogate worth tradeoff (SWT) method developed by Haimes and Hall (1975). The SWT method combines the best noninferior solutions derived using the ε-constraint method with the individual's ordinal preferences for tradeoffs among objectives to obtain that individual's best-compromise resource management plan for a farm or watershed. Preferences are elicited using the tradeoff information contained in the tradeoff functions. An individual's preferences can be elicited by asking the individual to assign weights to each of the objectives or to respond to a series of questions regarding the satisfaction provided by the objectives.

Applications of MODEM

The MODEM framework can be implemented using noninteractive and interactive approaches. A noninteractive approach involves manually linking the assessment modules in the MODEM model. An interactive approach integrates data, information, and models for the purpose of identifying and evaluating solutions to complex problems involving spatially distributed information.

Noninteractive Applications

The farm-scale application of the noninteractive approach uses a multiobjective mathematical programming model to estimate tradeoff functions for three objectives: maximum net return (NR), minimum soil erosion (SE), and minimum nitrate available for leaching (NL) achieved by the six farming systems described in Table 33.1 (Prato et al., 1994). The farming systems are being evaluated in connection with the Missouri Management Systems Evaluation Area (MMSEA) project located in the Goodwater Creek watershed near Centralia, Missouri.

The economic evaluation indicates that no single farming system achieves all three objectives. FS1 and FS6 have relatively high NRs and high NLs. FS4 has the second highest NR and low NL. FS1 and FS5 are inefficient over the entire range

Table 33.1 Description of MMSEA Farming Systems

Farming system	Crop rotation	Yield (kg/ha)	Tillage system	Nitrogen application rate	Herbicide application rate	Net return ($/ha)
FS1	Corn	1,192	Minimum	High	High	321
	Soybeans	411				
FS2	Sorghum	1,119	Minimum	Medium	Medium	235
	Soybeans	386				
FS3	Corn	976	Minimum	Low	Low	227
	Soybeans	395				
	Wheat	463				
FS4	Corn	993	Ridge	Low	Low	284
	Soybeans	368				
FS5	Corn	955	No-Till	High	High	205
	Soybeans	365				
FS6	Sorghum	1,227	Minimum	High	High	275
	Soybeans	372				

of objectives. FS1 is inefficient because it has a high nitrogen application rate which increases NL. FS5 is inefficient because it has the lowest yields for corn and soybeans, which result in a low NR, and a high nitrogen application rate, which results in a high NL. Results of the tradeoff analysis show that the economic objective (NR) is competitive with the erosion (SE) and water quality (NL) objectives, and the erosion (SE) and water quality (NL) objectives are competitive with one another. A competitive relationship between two objectives means increasing (decreasing) one objective, decreases (increases) the other objective.

The watershed-scale evaluation utilizes a chance-constrained programming (CCP) model to determine how much of the acreage in Goodwater Creek watershed should be planted to each of the six farming systems so as to maximize watershed net returns while achieving specific reductions in sediment yield (SY) and soluble nitrogen concentration in runoff (SN) at the outlet of the watershed (Prato and Wu, 1995). FS1 is chosen as the baseline farming system because it provides the highest net return per hectare of the six farming systems. Farming systems used in the watershed-scale evaluation are the same as those in the farm-scale evaluation (Table 33.1) except for FS6 which is a grass-legume mixture that uses no nitrogen or herbicides and provides a net return of $74/ha.

FS1 has the highest mean and highest standard deviation for NR and SN and FS2 has the highest mean and standard deviation for SY. As expected, FS6 has the lowest mean and standard deviation for NR, SY, and SN. For all net return and water quality reliability levels evaluated with the CCP model, NR decreases at an increasing rate as SN or SY decrease. This indicates tradeoffs between NR and SN, and NR and SY.

Achieving major reductions in SY or SN with high reliability necessitate significant changes in farming systems and/or reductions in planted acreage and watershed NR. A 70% reduction in SY requires idling between 12 and 69% of the cropland acreage and using FS5 on the remaining acreage in the watershed. A 70% reduction in SN requires idling between 23 and 30% of the cropland acreage and using FS2 on the remaining acreage in the watershed. Farming systems that are efficient in reducing SY are not efficient in reducing SN.

Interactive Application

A primary way of implementing an interactive approach is to use an ISDSS. An ISDSS is a knowledge-based computer program designed to assist decision makers in performing their task (Djokic, 1993; Leng, 1991). Unlike traditional off-line, noninteractive approaches, an ISDSS allows the user to evaluate solutions based on his/her own objectives and subjective judgment in an interactive learning and decision-making process (Loucks and Fedra, 1987).

This section describes research in progress to develop a multiobjective watershed policy assessment tool (MOWPAT) which incorporates the elements of an ISDSS. MOWPAT is designed to determine the spatially efficient LUMPs for reducing erosion and nonpoint source pollution in agricultural watersheds based on the MODEM framework illustrated in Figure 1. LUMPs considered by MOWPAT include: crop rotations; tillage practices; conservation practices such as grass waterways and terraces; pollution prevention practices such as timing, rate, and method of application of fertilizers and pesticides; and other landscape elements such as improved vegetative cover in riparian areas and conversion of hydric cropland to riparian wetland.

MOWPAT consists of three assessment modules: socioeconomic, environmental, and ecological. The socioeconomic module evaluates social and economic aspects of a watershed management plan. The environmental module simulates soil and water quality effects of alternative LUMPs. MOWPAT contains two simulation models: AGNPS (agricultural nonpoint source) and SWAT (soil and water assessment tool). The AGNPS model is a distributed parameter model that simulates sediment, runoff, and nutrient transport for alternative LUMPs within agricultural watersheds (Young et al., 1987). It is a storm event-based model that requires dividing the watershed into square, equal-sized cells. The SWAT model simulates the effects of alternative agricultural management practices on erosion, runoff, and water quality in rural basins (Arnold et al., 1994). It is a physically based model that operates on a daily time step. SWAT is capable of simulating results over extended periods of time for the entire basin and for subbasins. MOWPAT uses a GIS (geographic information system) to reduce the time and labor needed to collect, process, and manipulate the input parameters for AGNPS and SWAT. Output from AGNPS and SWAT are used to compare environmental effects of LUMPs relative to a baseline.

The ecological module is used to simulate how in-stream biological characteristics related to fish and invertebrate communities respond to changes in water quantity and quality (mean flow, stability of flow, peak flow, nutrients, dissolved oxygen, sedimentation, and temperature) resulting from different LUMPs and public policies. Fausch et al. (1988) have reviewed models which are suited for such simulations. Except for stream temperatures, inputs to the ecological model are the outputs from the environmental models (AGNPS or SWAT). Ecological performance is evaluated in terms of structural endpoints in the stream, namely species composition for fish and invertebrate communities.

The major value of MOWPAT is that it links changes in LUMPs and public policies to changes in economic (net return), environmental (soil and water) conditions and changes in environmental conditions to changes in the proximate habitat conditions of the stream. In addition, it simulates how changes in habitat conditions are likely to influence the performance of fish and invertebrate communities as indicated by species richness and diversity, and biological integrity. Socioeconomic, environmental, and ecological models are integrated in the IDSS using a GIS.

Conclusion

MODEM provides an holistic framework for evaluating the impacts of alternative land use/management practices and public policies on economic returns, environmental quality, and ecosystem health. The framework integrates the socioeconomic, environmental, and ecological objectives of interest to resource owners, planners and managers, and identifies tradeoffs among competing objectives. Translating this framework into a MODSS allows local decision makers to develop a resource management plan for a farm or watershed that is consistent with their preferences for socioeconomic, environmental, and ecological objectives.

References

Arnold, J.G., J.R. Williams, R. Srinivasan, K.W. King, and R.H. Griggs. 1994. SWAT (Soil and Water Assessment Tool). Reference Manual. Agricultural Research Service, U.S. Department of Agriculture, Temple, Texas.

Changkong, V. and Y.Y. Haimes. 1983. *Multiobjective Decision Making: Theory and Methodology*. Elsevier-North Holland, New York.

Cohon, J.L. 1978. *Multiobjective Programming and Planning*. Academic Press, New York.

Cohon, J.L. and D.H. Marks. 1993. Multiobjective screening models and water resources investment. *Water Resour. Res.* 9:826–836.

Daly, Herman and John B. Cobb, Jr. 1989. *For the Common Good: Redirecting the Economy Toward Community, the Environment, and a Sustainable Future*. Beacon Press, Boston. Page 37.

Dinh, T.L. 1989. *Theory of Vector Optimization*. Springer-Verlag, New York.

Djokic, D. 1993. Towards General Purpose Spatial Decision Support System Using Existing Technologies. Proceedings of the Second International Conference/Workshop on Integrating Geographic Information Systems and Environmental Modeling, September 26–30, Breckenridge, Colorado.

Fausch, K.D., C.L. Hawkes, and M.G. Parsons. 1988. Models that Predict Standing Crop of Stream Fish From Habitat Variables: 1950–85. General Technical Report PNW-213. U.S. Forest Service, Washington, D.C.

Haimes, Y.Y., L. Lasdon, and D. Wismer. 1971. On a bicriteria formulation of the problems of integrated systems identification and system optimization. *IEEE Trans. Syst. Man Cybern.* SMC-1:296–297.

Haimes, Y.Y. and W.A. Hall. 1975. *Multiobjective Optimization in Water Resource Systems: The Surrogate Worth Tradeoff Method*. Elsevier, Amsterdam.

Haimes, Y.Y. and W.A. Hall. 1977. Multiobjective Analysis in the Maumee River Basin: A Case Study on Level-B Planning. Report SED-WRG-77-1. Case Western University, Cleveland, Ohio.

Haimes, Y.Y., K. Tarvainen, T. Shima, and J. Thadathil. 1990. *Hierarchical Multiobjective Analysis of Large-Scale Systems*. Hemisphere, New York.

Karr, J.R., K.D. Fausch, P.L. Andermeier, P.R. Yant, and I.J. Schlosser. 1986. Assessing Biological Integrity in Running Waters: A Method and its Rationale. Illinois Natural History Survey Special Publication 5.

Loucks, D.P. and K. Fedra. 1987. Impact of changing computer technology on hydrologic and water resources modeling. *Rev. Geophys.* 25(2):107–112.

Ma, J.C. 1993. Integrated Economic and Environmental Assessment of Alternative Agricultural Systems. Ph.D. dissertation. Department of Agricultural Economics, University of Missouri-Columbia.

Nijkamp, P. and J. Spronk. 1981. *Multicriteria Analysis: Operational Methods*. Gower, Hampshire, England.

Prato, Tony, Feng Xu, and J.C. Ma. 1994. Estimation of Economic and Environmental Tradeoffs for Alternative Farming Systems. Proceedings of the 4th Annual Water Quality Conference, Missouri Agricultural Experiment Station. Pages 33–47.

Prato, Tony and Shuxiang Wu. 1995. A Stochastic Programming Analysis of Economic Impacts of Improving Water Quality at the Watershed Scale. CARES Report No. 13. University of Missouri-Columbia.

Steuer, R. 1986. *Multiple Criteria Optimization: Theory, Computation and Application*. John Wiley & Sons, New York.

U.S. Environmental Protection Agency. 1986. National Water Quality Inventory. 1986 Report to Congress. EPA-440/4-87-008. Office of Water, Washington, D.C.

World Resources Institute. 1992. *World Resources 1992–1993*. Oxford University Press, New York.

Young, R.A., C.A. Onstad, D.D. Bosch, and W.P. Anderson. 1987. AGNPS, Agricultural Nonpoint Source Pollution Model, A Large Watershed Analysis Tool. Cons. Res. Report 35. Agricultural Research Service, U.S. Department of Agriculture, Washington, D.C.

Xu, Feng and Tony Prato. 1995. Onsite erosion damages in Missouri corn production *J. Soil Water Conserv.* 50:312–316.

Chapter 34

Socioeconomics of Upland Soil Conservation in Indonesia*

Paul C. Huszar

Introduction

Indonesia has been very successful over the past several decades in increasing agricultural production and reducing poverty through its investments in irrigated food crops in the lowlands. These investments, however, have largely bypassed the rainfed agriculture of the uplands where the incidence of poverty among small landholders remains high. Moreover, increasing agricultural conversion of forest land in the uplands and accelerated timber harvesting are increasing soil erosion, which is decreasing soil productivity and exacerbating the problem of poverty.

Indonesia's main program for improving agricultural production and soil conservation is the National Watershed Development Program in Regreening and Reforestation (R&R), which is one of eight Presidential Instruction (INPRES) regional and rural development assistance programs and which channels about US$50 million annually to local governments for watershed activities. Over the last 5 years, the annual budget for this program has increased sixfold. A central component of the program is the construction of bench terraces, though the sustainability of these structures is an increasingly controversial issue (Belsky, 1994; Huszar et al., 1994).

---

* The helpful comments of Sapta P. Ginting and Hadi S. Pasaribu on an earlier version of this paper are gratefully acknowledged, though the opinions expressed in this paper are solely those of the author.

The purpose of this paper is to examine the performance and sustainability of current bench terracing efforts and to suggest approaches for improving this component of upland development efforts. Since the problem of soil erosion is most acute on Java, the focus of this paper is on Java's uplands. Empirical evidence is drawn from the Citanduy Project and the Upland Agricultural and Conservation Project on Java.

Background

Indonesia has an estimated 1995 population of over 201 million spread over nearly 1,000 of its 13,667 islands. While Java constitutes only 7% of Indonesia's total area, it has a population of roughly 130 million (65%), while the other major islands of Kalimantan, Sulawesi, and Sumatra have a population of roughly 68 million (34%). The average population density on Java is over 900 people per square kilometer (km^2) and on the other major islands is approximately 45 people per km^2. The average annual rural household income is about US$450 on Java and about US$600 on Sumatra and Kalimantan (World Resources Institute, 1994; The World Bank, 1990).

Java is approximately 1,000 km (620 mi) long and 160 km (100 mi) wide and has an area of approximately 132,187 km^2 (51,038 mi^2), or roughly the size of the state of New York. A mountain chain dominated by volcanoes runs east and west along the island's spine and is flanked by limestone ridges and lowlands. Roughly one-third of Java's land has slopes ranging from 0 to 8%, one third has slopes from 8 to 30%, and one-third has slopes in excess of 30%. Much of the land is highly fertile due to the relatively recent deposition of volcanic ash (The World Bank, 1990).

Over 12 million of Java's population live in upland areas. Population densities in the uplands vary with soil fertility and, consequently, carrying capacity. Upland areas with volcanic soils have a population density ranging from 550 to 820 persons per km^2 and those with limestone soils have a density ranging from 430 to 570 persons per km^2. With the growth of the uplands' population, agriculture has evolved from shifting cultivation to permanent farms. Average landholdings are approximately 0.4 ha (Barbier, 1990).

Population and development pressures are increasingly felt in the upland areas. Simultaneously, rapid urbanization in the lowlands is reducing and degrading prime agricultural land. An estimated 40,000 ha of cropland are converted annually to other uses and replacing the productive capacity of this land is estimated to cost US$50 to US$100 million/year (The World Bank, 1990). As lowlands are converted to urban uses, increasing pressure is being placed on upland agriculture to replace the lost production. On average, 2.25 ha of uplands are needed to replace each hectare of lost lowland (Huszar, 1991).

Most of Java's watersheds are classified as "critical" in the sense that they are subject to actual or potential degradation due to erosion, though the definition of "critical" is not precise. The total area of Java's watersheds classified by the Indonesian government as "critical" is over 7.8 million ha, of which over 1.9 million ha are located in upland areas and are considered to be the most threatened. The population of the "critical" upland watersheds exceeds 10.7 million, though this figure includes upland cities such as Bandung and Malang.

Table 34.1 Agricultural Land Use, Erosion, and Erosion Costs On Java

	West Java	Central Java	East Java	Total
Agricultural area (ha)	5,010,000	3,487,380	4,326,490	12,823,870
Sawah	1,604,380	1,443,980	1,696,960	4,745,320
Forest	541,270	635,760	1,207,510	2,384,540
Degraded forest	300,920	37,180	61,520	399,620
Tegal	2,563,440	1,335,120	1,360,450	5,259,010
Predicted soil loss (t/year)	404,212,010	180,788,950	112,029,590	697,030,550
Sawah	1,090,780	631,050	582,500	2,304,330
Forest	5,643,630	3,959,540	5,375,750	14,978,920
Degraded forest	30,041,510	1,332,620	2,688,010	34,062,140
Tegal	367,436,090	174,865,740	103,383,330	645,685,160
Predicted soil loss (t/ha/year)	80.7	51.8	25.9	54.4
Sawah	0.7	0.4	0.3	0.5
Forest	10.4	6.2	4.5	6.3
Degraded forest	99.8	35.8	43.7	85.2
Tegal	143.3	131	76	122.8
Estimated annual cost				
(Rp billions/year)	247.6 to 308.4	55.4 to 79.9	238.5 to 267.6	557.9 to 688.4
(US$ millions/year)	152.6 to 189.4	33.5 to 48.4	144.5 to 162.1	340.6 to 406.2

Derived from Magrath and Arens, 1989.

Table 34.1 summarizes agricultural land uses, total amounts of estimated erosion, erosion rates, and the estimated costs of erosion for Java and its three main geographical divisions of West, Central,* and West Java. Sawah, or irrigated rice paddies, is primarily a lowland land use and represents approximately 37% of total agricultural land use on Java. Tegal, or rainfed agriculture, is primarily an uplands land use and represents roughly 41% of total land use. The other major agricultural land use is for forests, where forest and degraded forest land use represents about 22% of agricultural land use on Java.

As shown by Table 34.1, sawah accounts for less than 1% of predicted soil loss, while tegal accounts for nearly 93% and forest land accounts for 7%. Java's predicted soil loss from tegal averages nearly 123 t/ha/year, with West Java having the highest predicted rate of 143 t/ha/year, Central Java has the next highest rate of 131 t/ha/year and East Java the lowest rate of 76 t/ha/year. The 1989 estimated annual cost of soil erosion on Java in terms of lost productivity and off-site costs was Rp57.9 billion (US$340.6 million) to Rp688.4 billion (US$406.2 million).

Regreening Program

The Regreening Program is the government's primary tool for dealing with upper watershed soil erosion. The program is administered by the Ministry of Forestry and is by far the largest upper watershed program on Java, with estimated expenditures of US$166 million during the 1976 to 1986 period. Regreening funds

---

* The province of Yogyakarta has been included in the data for Central Java.

are for privately owned land, as distinct from the Government's Reforestation Program for state forests (World Bank, 1990). The Regreening Program evolved from the FAO-funded Solo Watershed project implemented during the 1972 to 1976 period in Central Java (Belsky, 1994).

The basic approach of the Regreening Program is to establish model farm units and to introduce a package of upland agricultural inputs and conservation practices, emphasizing the construction of bench terraces and the use of new cropping patterns, seed varieties, and inputs of chemical fertilizers and insecticides on land with slopes of up to 50%. For land with slopes of more than 50%, permanent vegetation is established. Subsidies, either in cash or in kind, are provided for the construction of bench terraces and the purchase of inputs. Moreover, the projects transport the improved inputs into the project area. The basic goals of the program are to increase farm production and, therefore, incomes, while reducing soil erosion in densely populated upland areas of Java (Huszar and Cochrane, 1990).

Earlier studies of the program found that incomes were increased and soil erosion reduced. For example, the Citanduy II Project in West Java increased farmer incomes by approximately 25%, while reducing soil erosion by about 40% (Huszar and Cochrane, 1990). More recent studies, however, have found that while incomes are increased and erosion reduced during the implementation of the projects, these effects are not sustainable under the present approach.

Sustainability of Program

The recent study by Huszar et al. (1994) of the Uplands Agriculture and Conservation Project (UACP) has raised serious questions regarding the long-term sustainability of the benefits of the current upland conservation program in Indonesia. The UACP is located within the Jratunseluna Watershed in Central Java and the Brantas Watershed in East Java. UACP's goals of increasing incomes and reducing erosion were similar to those of previous upland conservation projects. These goals were to be achieved by improving farming systems, farm technologies and management (USAID/Indonesia, 1984). Subsidies were used to induce farmers to adopt improved seeds and diversify crops, to employ improved chemical fertilizers and pesticides, and to construct bench terraces and water channels. Funding for the UACP started in 1985 and came from the Government of Indonesia with assistance from the U.S. Agency for International Development (USAID) and The World Bank.

Figure 34.1 shows that project farmers 1 and 2 years after the project ended had higher gross incomes from dryland production than did the control farmers, but by the third and fourth years after the end of the project their incomes had fallen below those of the control farmers.* These results are statistically significant at the 0.05 level. That is, while project farmers seem to initially have higher levels of dryland production, this increase is not maintained over time. Figure 34.1 indicates that after subsidies end, project farmers are not able to sustain the positive changes introduced by the project.

---

* Control farmers have similar characteristics to project farmers, but are outside of the project who were not affected by the project.

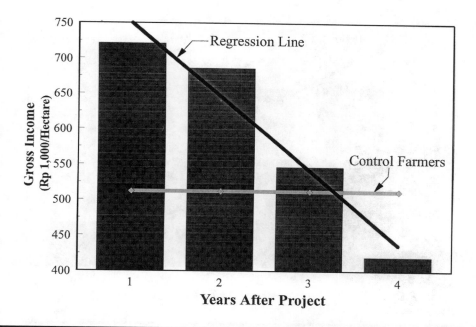

Figure 34.1 Average gross income from dryland.

The same pattern of decline also holds for the net value of dryland production, as shown by Figure 34.2. That is, the surplus of revenues over costs of production initially are higher for project farmers than for farmers without the project, but the net value of production declines rapidly after the project ends, though the differences are only statistically significant at the 0.15 level. Farmers in older project villages tend to earn lower net returns from dryland crops than farmers in more recent project villages.

The regression lines for gross and net incomes explain 95 and 92% of the variation in the observed data, respectively. In both cases, it appears that 4 years after the project ends in a village, farmers who were in the project are worse-off than farmers not affected by the project. One possible explanation for this may be that terracing reduces plantable land area, so that as project farmers return to conventional cropping and cultivation patterns they are at a temporary disadvantage. If this is the case, then it is unlikely that terraces will be maintained and, in fact, they may be intentionally eroded.

While one of the major goals of the UACP was to reduce soil erosion, very little data were collected by the project which measures the effects of the project on soil erosion. This is likely due to the difficulty of obtaining reliable soil erosion measures over time. Data are only available for two relatively primitive measures: a rating score for terraces, which attempts to reflect the effectiveness of the terraces to reduce soil erosion, and an erosion score called Inderosi, which attempts to measure relative changes in erosion (Huszar, et al. 1994). These data were only gathered for a sample of project farmers, so that comparisons are between farmers receiving subsidies and those no longer being subsidized.

The physical quality of terracing was rated by UACP in terms of following the contour, effective planting area, slope of the planting area, main drainage channels, drop structures, drainage ditches, terrace bunds, vegetation on the terrace riser,

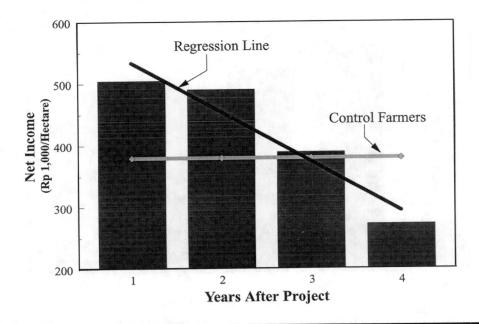

Figure 34.2 Average net income from dryland.

pathways, landslides, gully erosion, and conservation of topsoil. An overall score ranging from 0 to 100 was assigned to farmer terraces in the project, with higher scores reflecting more effective soil conservation.

Figure 34.3 shows the average terrace ratings of former project farmers compared with the average scores for project farmers still receiving subsidies. As shown by Figure 34.3, these scores tend to decline after the project ends, though the regression coefficient is not significant. That is, the quality of terracing appears to decline, indicating that terraces are not being maintained and that these soil conservation measures are not sustainable.

The degree to which terracing reduces soil erosion was also rated by UACP using a rating system developed by UACP called "Inderosi." The Inderosi score is based upon before and after project changes in the vegetative cover during each growing season, the amount and distribution of rainfall, and the quality of terracing. Calculation of the Inderosi score employs a modified version of the Universal Soil Loss Equation (USLE). The Inderosi rating system attempts to provide a relative score for erosion, where higher scores are associated with proportionately greater levels of soil conservation.

As shown by Figure 34.4, these scores are less for former project farmers than for project farmers still receiving subsidies. The regression line indicates some tendency for improvement in the score over time, but the regression coefficient is not significant. That is, soil erosion seems to be reduced less on farms after the project ends, but there is no clear tendency for the reduction in erosion to decline over time.

The evidence suggests that maintenance of terraces declines and erosion increases after the projects end. This evidence agrees with informal observations of earlier upland conservation projects on Java, particularly Citanduy II, and with a study conducted in the Kerinci valley of Sumatra (Belsky, 1994). The study of

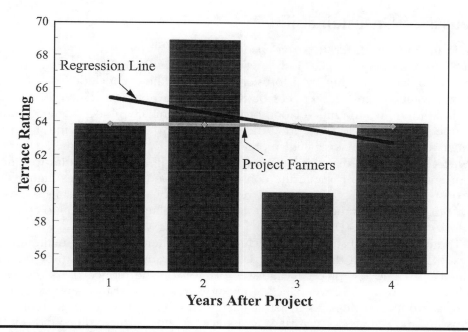

Figure 34.3 Average terrace rating.

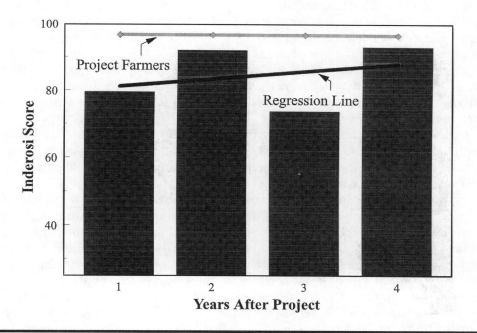

Figure 34.4 Average Inderosi score.

the Regreening program in Sumatra found that terraces were rarely maintained for longer than the 3 years in which the farmer received subsidies from the project. Moreover, erosion may be increased due to poorly maintained terraces which channel runoff more than would be the case for unterraced fields.

Causes of Unsustainability

While upland bench terracing has been promoted by the government since the Dutch colonial period, it has only met with limited sustainable success. An 1889 study of upland terracing by a forester named Berkhout concluded that farmers did not want to construct terraces because they are not accustomed to terracing dryland, it requires too much labor, productivity is initially reduced when top soil is buried by less fertile soil during the construction of the terrace, and it ties the farmer to the land making him more controllable by the government. Farmers are perceived as not maintaining terraces because the terraces require too much labor and farmers do not rate the conservation value of the terraces higher than the lost productivity of the land. A recent study of upland terracing seems to imply that these are valid reasons to this day for the failure of farmers to construct and maintain terraces (Belsky, 1994).

The Belsky (1994) study also concluded that farmers did not maintain terraces because they thought that terraces "desiccated" already poor soils, they reduced planted area and the protective grasses planted on the terrace risers interfered with crop growth. Moreover, since terracing is associated with a shift away from cassava production, females in particular did not want to give up income from preparing and selling cassava cakes and crackers. Finally, the women did not like the additional labor necessary to repair terrace walls and weed (Belsky, 1994).

In order to overcome farmer resistance to terracing, the Regreening program uses subsidies to reduce the initial costs of terracing and the costs of the additional inputs associated with cultivating terraces. The success rate of the program to induce farmers to terrace has been high. Moreover, while the project is in operation, farmers in the area, but not receiving subsidies, also seem anxious to bench terrace and to adopt the introduced farming system (Huszar and Cochrane, 1990). The indication is that contrary to the "conventional wisdom," when conditions are favorable, both subsidized and nonsubsidized farmers find bench terracing profitable.

However, maintenance of the terraces after the subsidies end has been a continual problem. The Belsky (1994) study suggests that this is because bench terraces are the wrong solution, largely because farmers do not like them. We offer an alternative explanation.

Inputs to terraced agriculture are relatively expensive to upland farmers due to poor access to input markets. The farmer must generally travel long distances to the lowlands to purchase the improved seeds, fertilizer, and pesticides needed for the cultivation practices introduced by terracing. The time and effort to obtain inputs represent a significant opportunity cost to the farmer. Without these inputs, however, the productivity of the terraces declines and, therefore, the incentive to maintain the terraces. During the project phase when the terraces are being constructed and subsidies are received, the project provides the needed inputs of seeds, fertilizer, and pesticides locally. But after the project, these inputs are no longer available locally and their real cost increases. Evidence of this effect can be found in the UACP.

Figures 34.5, 34.6, and 34.7 show that the use of the improved seed, fertilizer, and pesticide inputs introduced by the UACP declined after the project ended. While the average value of seed inputs is greater for project farmers than for the control farmers up to 4 years after the project ended, the tendency is one of

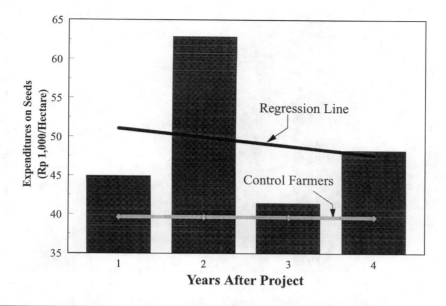

Figure 34.5 Average seed use.

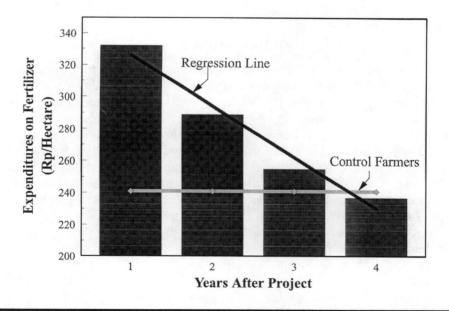

Figure 34.6 Average fertilizer use.

decreased use, as can be seen from Figure 34.5. Fertilizer use of project farmers decreased to approximately the level of the control farmers by the fourth year after the project ended, as shown by Figure 34.6. Figure 34.7 indicates that pesticide use of project farmers fell below the use by control farmers. The more dramatic decline in the use of fertilizers and pesticides may be related to their weight and, therefore, their transportation cost relative to seeds.

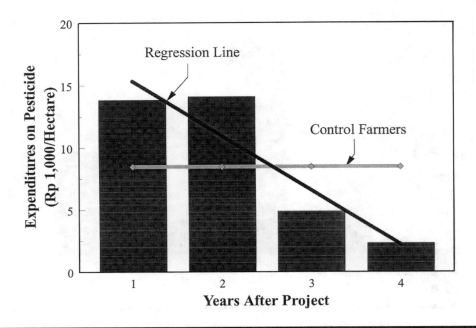

Figure 34.7   Average pesticide use.

Conclusions and Recommendations

The upland soil conservation program in Indonesia seems to be trapped in a cyclical of subsidizing bench terraces and other physical agricultural improvements for short periods after which the improvements rapidly deteriorate for lack of maintenance. The question is why don't the farmers maintain these capital improvements? Rather than being a matter of simple preference, we argue that lack of maintenance is likely due to the higher costs and, therefore, the lower net income associated with the new farming system. If this is the case, then what can the government do to improve the long-term economic feasibility of an upland agriculture that utilizes bench terraces? The upland conservation projects provide technical and financial assistance for the construction of terraces. They also provide the necessary inputs of seeds, fertilizers, and pesticides while the project is being implemented. The problem seems to be that after the project is completed, no provision has been made for the continued local availability of the needed inputs and that the costs associated with obtaining these inputs from distant markets are prohibitive to the upland farmers. Without the improved inputs, the terraces are no longer profitable and, therefore, are allowed to deteriorate.

A simple solution would seem to be for projects to not only provide technical and financial assistance for terracing, but they should also provide for the establishment of local markets for the inputs necessary to making terracing profitable. Besides subsidizing farmers to terrace, projects could subsidize local businessmen to develop enterprises which supply the inputs. This may entail construction of storage facilities, the purchase of vehicles for transporting the inputs from distant distributors, and even the improvement of transportation routes.

Perhaps the most important lesson to be learned is that soil conservation techniques do not exist in a vacuum. The social and economic system within which they operate affect and are affected by these techniques. The sustainability of upland soil conservation projects is not simply a function of the technical characteristics of the techniques introduced, but also depends upon the support of the socioeconomic system within which they operate.

References

Barbier, Edward B. 1990. The farm-level economics of soil conservation: the uplands of Java. *Land Econ.*, 66(2): 199–211.

Belsky, Jill M. 1994. Soil conservation and poverty: lessons from upland Indonesia. *Soc. Nat. Resour.*, 7: 429–443.

Huszar, Paul C. 1991. Recommendations for Economic Impact Analysis of the Upland Agricultural and Soil Conservation Project. Report submitted to Development Alternatives, Inc., Jakarta, Indonesia, January 18.

Huszar, Paul C. and Harold C. Cochrane. 1990. Subsidisation of upland conservation in West Java: the Citanduy II project. *Bull. Indonesian Econ. Stud.*, 26(2): 121–132.

Huszar, Paul C., Hadi S. Pasaribu, and Sapta Putra Ginting. 1994. The sustainability of Indonesia's upland conservation projects. *Bull. Indonesian Econ. Stud.*, 30(1): 105–122.

Magrath, William and Peter Arens. 1989. The Costs of Soil Erosion on Java: A Natural Resource Accounting Approach. Environmental Department Working Paper No. 18. The World Bank, Washington, D.C.

USAID/Indonesia. 1984. Upland Agriculture and Conservation Project. Project Paper, Jakarta.

The World Bank. 1990. Indonesia: Sustainable Development of Forests, Land, and Water. A World Bank Country Study, Washington, D.C.

World Resources Institute. 1994. *World Resources 1994–95*. Oxford University Press, New York.

Chapter 35

Issues and Concerns in Multiple Objective Management of Natural Resources in the Agri-Food Industry

D. PETER STONEHOUSE

Abstract

The considerable contribution by the agri-food industry to both economic development and higher living standards and to natural resource degradation can be attributed in large measure to pervasive technological progress and to the failure of the open market system, respectively. Food has been made abundant and, arguably, artificially cheap (a positive impact or externality). Adverse impacts of soil and water degradation, depletion of wildlife habitat, and reduced biodiversity are felt more off site than on. Such negative externalities devolve more from nonpoint than point source pollution, and frequently leave decision makers ignorant of the damage they are wreaking. Even those who are aware may not be motivated to adopt and use the wide range of conservation measures available because most of those measures are not profitable. The externalities may in any case be safely ignored because they are not traceable to a specific source. The resultant market failure may be the best justification for public intervention.

Public policies and private decisions to conserve soils and wildlife habitat and to abate downstream watercourse pollution need to be developed and applied judiciously and systematically. First, all policy and decision objectives should be clearly and definitively specified, preferably in quantitatively measurable terms, rather than as mere generalized aims. Second, a holistic approach to natural resource conservation and environmental management policy formulation is far preferable to a piecemeal one. Third, natural resource use decisions and resource conservation policy formulation should take place with both broader spatial as well as local and with long-term sustainability as well as short-term objectives in mind. In the temporal dimension, sustainability is necessarily to be evaluated in a long-term context. Shorter-term pollution abatement objectives may not always lead to longer-term sustainability objectives. Fourth, resource conservation and pollution abatement policies should in general not be universally applied. Universal application consistently fails to account for intersite variations in conservation needs, or to interdecision agent variations in conservation efforts expended. Not all conservation efforts lead automatically to reductions in pollution loadings into waterways or to improvements in desirable kinds of wildlife habitat, even though in general resource conservation is to be deemed preferable to degradation. In turn, not all reduced pollution loadings necessarily lead to water quality improvement. Even in the event of demonstrable water quality enhancement, a net benefit to society may not unequivocally follow. This underscores the importance of prespecifying policy objectives. It also implies the need for a targeted approach to resource conservation and environment management. Finally, in the event of conflict among multiple objectives being achieved some mechanism for conflict resolution should be explored.

Background to the Problem

The accomplishments of the agri-food industry during the 20th century throughout the developed world are reflected in the abundance, relative cheapness, generally high quality, and wide variety of foods, frequently offered in highly convenient forms that minimize or even eliminate preparation work. These in turn contribute substantially to improved material living standards and to time available for nonfood production and preparation pursuits. Progress since World War II throughout most of the developing world has been almost as impressive. Spurred by the Green Revolution and other examples of technological progress, starvation and malnutrition are being steadily eradicated.

Not all consequences and implications of this progress in the agri-food industry are positive. Degradation of the natural resource base and environment, depletion of wildlife habitat, ecological imbalances, and a diminution of biological diversity exemplify some of the negative outcomes worldwide (Sfeir-Younis and Dragun, 1993; Baum et al., 1989; Crosson and Brubacker, 1982). In developed countries, technological progress has stimulated a massive restructuring of the agri-food industry, with commensurate rural depopulation and threats to rural community viability (Schertz and Wunderlich, 1982). Such adverse consequences have brought into question the very sustainability of the intensive, highly technical agri-food industry (Paoletti et al., 1993; Edwards et al., 1990). Sustainability is defined here as the ability of a system to meet the current biophysical and socioeconomic needs of human society without detracting from the potential to meet the anticipated needs of future generations. In particular, the heavy dependence of the agri-food industry on nonrenewable fossil fuel energy has been touted as a mark of unsustainability (Kendall and Pimentel, 1994; Giampietro et al., 1992; Stanhill, 1984; Steinhart and Steinhart, 1974).

Problem Statement and Objective

Connections between intensive, high technology production methods and environmental damage and questionable sustainability have been well established (Napier et al., 1994; Sfeir-Younis and Dragun, 1993; Lovejoy and Napier, 1986). Much of the environmental damage has been blamed on farm-related nonpoint source pollution from soil erosion and residues from fertilizers and pesticides (Weersink and Livernois, 1995; Carlson et al., 1993; Crosson and Brubaker, 1982). Difficulties posed by market failure, founded on nonpoint source pollution being more prevalent than point source and on negative externalities being more prominent than on-site costs (Crosson, 1984; Ribaudo, 1986; Clark, 1985), have led to calls for greater public intervention in the agri-food industry in the U.S. (Repetto, 1986; Lovejoy and Napier, 1986; Bromley, 1982), in Canada (Stonehouse and Bohl, 1990; van Vuuren, 1986), and in developing countries (Sfeir-Younis and Dragun, 1993; Stonehouse and Protz, 1992). Further difficulties are then posed by having to decide on the form(s) which such public intervention should best take, and by the potential for conflicts among possible alternative or multiple policy objectives.

The objective of this paper is to raise issues and address concerns associated with the development of a theoretical framework that will assist in the evaluation of multiple objective achievement in the agri-food industry. Emphasis is placed on the development of objective criteria for judging the appropriateness of the form(s) of public policies, and on public and/or private multiple objective conflict resolution.

Overview of Issues in Developing Multiple Objective Management of Natural Resources

The evaluation of multiple objective achievement in the agri-food industry would begin logically with a clear and definitive specification of objectives and an assessment of the potential for compatibility or conflict. Because of the potential for conflict in achieving objectives among different sectors of the economy (e.g., agri-food industry vs. manufacturing industry or private household sector) and because of multiple (human activity) sources of natural resource degradation, arguments in favor of a holistic theoretical framework need to be addressed. This leads logically to a discussion of the spatial dimension, i.e., whether a local geographic region, national, global, or other scale would be most appropriate. Given the developed world's prior experience in multiple objective achievement, there are surely some lessons to be learned by the developing world. This in turn implies a need to incorporate a discussion of time horizon choices. Some arguments should be offered for and against the use of targeted approaches to applications of public policies aimed at achieving specific objectives for and on behalf of the agri-food industry. Finally, because multiple objectives are often found to be potentially conflicting, alternative methods of conflict resolution or circumvention should be examined.

Objectives for Renewable Natural Resources Management

The first and foremost objective of the agri-food industry is to produce food to support the human population from a scarce set of land, air, water, mineral, and energy sources endowed by nature — many would agree that it is to maximize food output, and some that it is to maximize efficient use of resources in the food production process (i.e., to maximize profits). Inherent in this description of the most important objective are the seeds of potential conflict. Physical maximization of food output, as might be espoused by biophysical scientists in the agri-food industry, is typically not commensurate with efficiency maximization of scarce resource use, as usually advocated by economists.

It is the striving to achieve this top priority objective — either physically or economically or perhaps both — that led to so much research, development, and innovation effort throughout the 20th century, but especially since World War II. In turn, this has resulted in a further conflict in objectives. Whether seeking to examine food output or to maximize economic efficiency, research efforts have been narrowly focused. Neither the natural resource degradation/depletion consequences nor the longer-term sustainability implications were factored into research and development programs because of the failure to account for externalities. The nonpoint source nature of much of the agri-food industry's natural resource degradation has allowed its decision agents to ignore these costs to society. Additional incentives to ignore resource degradation and its costs to society stem from the lack of profitability of most conservation measures (Stonehouse, 1995; Stonehouse, 1991).

The scarcity of the natural resource system supporting the agri-food industry contains further potential for objective conflict. The scarce resource base must be shared (a) with competing human activities (e.g., urbanization) and (b) with all other plant and animal species — often collectively referred to as "nature." As natural vegetative cover is cleared for agricultural or other human activity development, the wildlife habitat base is diminished and plant or animal species' survival may be threatened and ecological balance jeopardized (Pimentel et al., 1994; Kendall and Pimentel, 1994). Additional threats to biodiversity and ecological balance emanate from natural resource degradation associated with intensive high technology production methods in the agri-food industry (Kendall and Pimentel, 1994). The same degradation process also imperils the long-term sustainability of the agri-food industry as presently structured and operated.

Specification of Objectives

Given the potential for conflict among multiple objectives, it is argued here that all objectives should be clearly specified prior to the implementation of any policy, program, or planned activity. This procedure would require policy formulators and decision agents to ponder carefully and preplan for the anticipated consequences of their decision activities, and in turn should provide a sense of direction and purpose. Due regard must then be paid to the disciplinary perspective and focus, and to the choice of judgmental criteria. For example, if one has a biological or ecological perspective, one might use "maximization of biodiversity" or "ensuring an optimal ecological balance" as appropriate criteria for judging the effectiveness of one's policies or planned activities. An environmental perspective might instead call for "minimization of environmental damage" or "maximization of renewable natural resource conservation." Agronomists, animal scientists and food aid agencies may seek to "maximize food output," while profit-motivated farmers and food processors may prefer to "obtain the most efficient use of available resources." Traditional economists may also seek to "maximize resource use efficiency," but environmental economists may prefer to "maximize net social benefits" based on the inclusion of all externalities in the objective function.

There are plenty of additional criteria that could — and perhaps should — be considered in evaluating multiple objective achievement. These might include the equitable distribution of net benefits (or costs), administrative ease and resource requirements for program or activity implementation, social and/or political acceptability, potential of the program or policy to mitigate or exacerbate factor price or commodity price distortions, and so on (Bohm and Russell, 1985). In general, the greater the number of objectives, the greater the chances for conflicts and incompatibilities, and the greater the need for multiple criteria for evaluation.

If multiple objectives are specified in definitive, and better yet quantitative terms, rather than general or open-ended terms, then application of judgmental criteria to the evaluation process is made easier (van Vuuren and Stonehouse, 1995). For example, expressing a resource conservation objective in terms of meeting some prespecified maximum allowable limit on eroded soil or sediment loadings into watercourses would make evaluation of policy effectiveness more amenable than stating the objective in more general terms such as "reducing overall soil erosion" or "minimizing sediment loadings into streams." Specifying multiple objectives as definitive targets to be met obviates the trading-off in achieving one objective against another when multiple objectives are specified in general terms. This implies the need to be realistic in setting definitive targets for multiple objectives.

Using a Holistic Approach to Multiple Objective Management of Resources

Degradation of the renewable natural resource base stems not only from human activity in the agri-food industry. Most other human activities can also result in various forms of pollution or depletion of air, water, and land resources. For purposes of mitigating or remedying the degradation, all sources of damage need to be considered (Stonehouse et al., 1995). For example, if downstream watercourse pollution from soil erosion from agri-food industry sources is fully remedied by enjoining all farmers in the requisite watershed to adopt all appropriate soil conservation measures, but manufacturing and construction industry members are not similarly prevailed upon, then pollution from sediment loadings could still occur. Remedying agri-food industry sources of sediment pollution will certainly lessen the overall damage problem, especially if this is the primary source of degradation. Yet overall remediation would require all sources of sediment pollution to be addressed. Remedying agri-food industry sources alone may be a necessary but not a sufficient condition for effecting complete damage relief, or for meeting some minimum resource quality level.

Similarly, the total spectrum of pollutants or resource quality criteria needs to be considered. It is not sufficient, for example in meeting minimum water quality standards for human drinking purposes, that nitrate pollution standards alone be met. Phosphate and other chemical pollution standards must also be met, as must bacteria and other biological pollution limits. One might argue that overall quality of any renewable natural resource is commensurate only with the lowest quality standard of a whole range of physical, chemical, biological, and aesthetic characteristics (van Vuuren et al., 1995).

Since different uses of land, air and water resources are often associated with different quality standards, alternative uses should be considered a relevant part of the mosaic for multiple objective achievement evaluation. For example, heavy metal tolerance levels would be less restrictive for land designated to forestry or recreational uses than they would be for land allocated to animal or human food production. The tolerance levels would be even less restrictive for land earmarked for industrialization.

Given the overlapping nature of renewable natural resource use jurisdictions, and given the composite and complex way in which different resource contaminants from alternative sources are interrelated, a holistic approach to multiple objective management of resources is advocated here. Achievement of the multiple objectives of food production, renewable resource conservation, land and water use allocation, and other agri-food industry policies and planned activities can be more accurately and realistically assessed with a more holistic framework. The level of complexity for policy evaluation frameworks is necessarily raised the more holistic the frameworks are made. This is more than compensated for by the superior evaluations that become feasible.

An example of a holistic theoretical framework for pollution abatement policy evaluation is developed for the case of water resources by Stonehouse et al. (1995), and presented in summary diagrammatic form in Figure 35.1. Effectiveness of abatement policy in this case is predicated upon the use of an economics criterion, namely maximization of net social payoff. Such an evaluation criterion permits full account to be taken of all externalities (public benefits and costs in Figure 35.1), as well as private (on-site) benefits and costs pertaining to water pollution remediation efforts.

Spatial and Temporal Dimensions

In developing an appropriate evaluation procedure for multiple objective achievement, account should be taken of the many alternative spatial scales and time horizons. Degradation of renewable natural resources and the environment recognizes no geographic boundaries, and carries long-term as well as short-term consequences. Clearing the natural vegetative cover from virgin lands for agricultural development purposes usually intensifies soil erosion and watercourse pollution rates and accelerates wildlife habitat loss rates locally. It may also affect climate, biodiversity, and ecological balance more broadly, perhaps even globally. It is likely that both local and broader-scale consequences would be felt well into the future as well as currently.

Practising more intensive techniques in the agri-food industry also is likely to have broad-scale and long-ranging, as well as local and current, consequences.

Issues and Concerns in Multiple Objective Management of Natural Resources 413

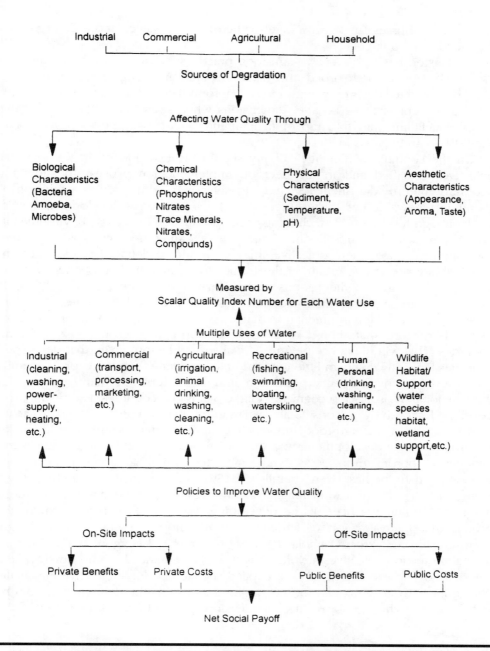

Figure 35.1 A holistic framework for abatement policy evaluation (Stonehouse et al., 1995).

Environmental damage can be greater off-site than on-site, and can be long-lasting in its effects (Ribaudo, 1986; Clark, 1985; Crosson, 1984). A trade-off would appear to exist between causing environmental damage from clearing and developing new lands for agriculture, and increasing environmental damage from intensifying the use of existing resources in the agri-food industry. It is not clear that a reduced intensity in agri-food sector practices, as proposed in the European Union, would lead to a net reduction in environmental damage, because more extensive use of additional resources (in Europe or elsewhere) may be even more damaging to the global environment.

Both extensive and intensive agri-food practices can and do lead to environmental damage in both broad and narrow spatial and temporal contexts. This implies a need for decision agents and policy formulators in the agri-food industry to take account of these broader dimensions when specifying objectives. At the very least, falling into the trap of maintaining a narrow "here-and-now" focus should be carefully avoided.

On the contrary, sufficient flexibility should be incorporated in the multiple objectives specified and in the decision implementation process to allow for necessary interim adjustments. Such adjustments may be needed as a function of changing renewable natural resources conditions, economic development levels, available agri-food industry technologies for production as well as environmental protection, societal attitudes, political acceptability, and other factors. For example, some pollution abatement objectives may be found to detract from longer-term sustainability objectives, requiring the former to be modified subsequently.

There are surely some lessons to be learned from the post-World War II experiences in applying intensive practices in developed countries. The far-reaching nature of environmental damage associated with these intensive techniques is not one that should be exacerbated through extension to developing countries. This will likely not be an easily-absorbed lesson, given that developed countries have so far shown little inclination to moderate their per-capita demands for the sake of environmental protection, and are thereby less than exemplary in their actions. It is also recognized that developing countries lag behind the developed world by a considerable margin in material living standards, that intensification in the agri-food industry has a substantial contribution to make toward economic development, and that environmental protection objectives are generally subordinate to development objectives until a sufficient degree of economic maturity has been established (Stonehouse, 1993b; Rauscher, 1992; Sutton, 1988). These factors mitigate against resisting the temptation to adopt intensive farming practices in the developing world. In fact, recent experience reveals that intensification of agriculture in developing countries has already occurred to some extent (Pagiola, 1995; Sfeir-Younis and Dragun, 1993). Environmental protection objectives nevertheless should be balanced against development and other objectives. There is also an implied need to develop more environmentally benign intensive technologies for the agri-food industry, a burden which should be borne more by the developed than the developing world (Stonehouse, 1994b).

Universal Vs. Targeted Policy Applications

Previous mention has been made of the need for public intervention in response to market failure in the agri-food industry. The question is how best to accomplish this. Arguments are raised here in favor of a targeted approach, as opposed to a universal application of policies and programs on behalf on environmental management in the agri-food industry.

A targeted approach has the advantage of inherent flexibility through offering differential policy and program treatments to different participants in the agri-food industry. This permits account to be taken of (a) differences in physical production capabilities of resources across different sites; (b) differences among agri-food industry decision agents in managerial expertise, especially as this pertains to mitigation or reparation of environmental damage; and (c) differences across sites in environmental sensitivities and conservation needs. Universally-applied policies would, in contrast, treat all sites and all decision agents as having similar needs and capabilities. Yet the intersite differences in natural resource qualities (e.g., topsoil depth, texture, and fertility, topography, etc.) and their proneness to degradation, together with the interagent differences such as attitudes toward resource stewardship, economic ability to undertake environmental protection measures, as well as skills in environmental management, are too great and too important to ignore. Admittedly, universally applied policies are far easier to administer and require far fewer resources for implementation and monitoring. However, through judicious use of farmer classification schemes and with the help of geographic information systems (GIS), satellite photography, and computerized data collection, retrieval, and analysis, targeted approaches should be feasible administratively (Stonehouse, 1996; Stonehouse, 1994a; Stonehouse, 1993a).

It is furthermore important to recognize through a targeted approach to policy application that differences in both conservation needs across sites and conservation efforts across decision agents may have different payoffs (biophysically and/or socioeconomically). This follows from the fact that not all nonpoint source degradation events necessarily result in negative externalities. For instance, not all eroded soil contributes to sediment loadings in watercourses. Much can depend upon distances to nearest watercourses, topsoil characteristics, topography, and other factors. Even where contributions to sediment loadings are made by eroded soil, remediation of land degradation will not always lead to improvement in water quality. This depends to an important extent on uses of water, quality standards associated with those uses, preremediation pollution levels, other sources of degradation, and other pollutants (van Vuuren and Stonehouse, 1995). Furthermore, demonstrated improvements in resource quality from conservation efforts do not automatically translate into net social benefits. That can depend upon correct identification of benefits (van Vuuren et al., 1995), and on overall (private and public) benefits and costs (van Vuuren and Stonehouse, 1995).

Through the adoption of a targeted approach to public intervention, it becomes more feasible to devise environmental management policies and programs capable of attaining prespecified multiple objectives, such as obtaining some measurable amount of positive net social payoff. This furthermore underscores the importance of carefully prespecifying the objectives of environmental management programs. Having biophysical vs. ecological vs. socioeconomic objectives in mind could result in quite different applications and outcomes of the same programs.

Reconciling Conflicting Multiple Objectives

Where multiple objectives encompass biophysical, socioeconomic, political, philosophical, and other perspectives, some conflict in objectives and their simultaneous achievement might be anticipated. In that case, some sort of conflict reconciliation procedure should be developed.

One approach to reconciliation might be to specify one's multiple objectives so as to lessen the potential for conflict. An example is to select a full cost accounting method of achieving economic objectives, so as to narrow the gap between environmental and socioeconomic objectives. Traditional economic objectives focus on maximization of (typically short-run) profits or net present values for the firm or of benefit-cost ratios for society. In the former case, firm objectives rarely include externalities. In the latter case, societal objectives include externalities only sometimes. By inserting all externalities, positive and negative, into economic objectives at either firm or societal (regional, state, national, or international) levels, one is bound to come closer to fulfilling environmental, and perhaps also ecological, objectives at the same time. The temptation to claim complete reconciliation is resisted, even in theory. This is because environmental objectives may call for complete elimination of degradation, or at least as far as is possible, from human activity sources. In contrast, economic objectives may fall short of approving all the expenditures, public and private, that could be required to achieve the environmental objectives. Money is simply not that plentiful. Nor are net social benefits of all remediation or pollution prevention programs necessarily positive.

A second approach would be to follow up on the earlier suggestion that all objectives be specified in quantitative target terms, rather than as open-ended entities. This would require prespecifying set amounts of reductions in soil erosion or sediment loadings into watercourses to be achieved, or measurable improvements in resource quality to be attained, along with targeted levels of net social benefits and social acceptability to be obtained. In turn, quantitatively-based multiple objectives would need to be set at realistic levels, so that reasonably high rates of attainment can be expected on a simultaneous basis. There would always be a risk of trade-off among potentially competing objectives, but this can be lessened through setting quantitatively based objectives with realistic target levels. Prior research into the anticipated outcomes of policy implementation may prove to be the best way of avoiding such trade-offs. Such research should be undertaken using a multidisciplinary team approach, given the direct involvement of many different disciplines with their very different perspectives (Stonehouse, 1996; Stonehouse, 1994b; Stonehouse, 1993a).

A third method would be to adopt a constrained optimization routine. This would entail selecting one of the multiple objectives as the one to be maximized (e.g., sustainability of the agri-food systems or biodiversity) or minimized (e.g., environmental damage), subject to all remaining objectives being achieved in constrained form. Constraints could include variously some maximum acceptable level of societal costs, some minimum levels of social and political acceptability, some reasonably equitable sharing of the cost burdens across society, some maximum acceptable level of distortions to factor and/or commodity markets, some minimum proportion of land in total designated for wildlife habitat use, some minimum levels of food in total and of different types to be produced, and so on. Again, constraint levels would need to be selected with care, so that a reasonable assurance of obtaining a feasible and optimal solution is provided. Given the diverse range of disciplines being drawn upon, a multidisciplinary approach is strongly advocated.

In the event that all above and other possible approaches to reconciliation are held to be infeasible, resort might be had to ranking of one's multiple objectives.

While this could ensure the achievement of the highest priority objectives first, any and all methods of ranking may be regarded as subjective and controversial. Such an approach may therefore not be particularly helpful. Similar arguments would likely be levelled at any system of weighting the multiple objectives.

References

Bohm, P. and C.S. Russell. 1985. Comparative analysis of alternative policy instruments. Pages 395–460 in A.V. Kneese and J.L. Sweeney, Eds. *Handbook of Natural Resource and Energy Economics*. Volume 1. Elsevier Science Publishers, Amsterdam.

Bromley, D.W. 1982. Land and water problems: an institutional perspective. *Am. J. Agric. Econ.* 64(5):834–844.

Carlson, G.A., D. Zilberman, and J.A. Miranowski. 1993. *Agricultural and Environmental Resource Economics*. Oxford University Press, New York.

Clark, E.H. 1985. The off-site costs of soil erosion. *J. Soil Water Conserv.* 40(1):19–22.

Crosson, P. 1984. New perspective on soil conservation policy. *J. Soil Water Conserv.* 39(4):222–225.

Crosson, P. and S. Brubaker. 1982. Resource and Environmental Effects of U.S. Agriculture. Resources for the Future, Washington, D.C.

Edwards, C.A., R. Lal, P. Madden, R.H. Miller, and G. House, Eds. 1990. *Sustainable Agricultural Systems*. St. Lucie Press, Delray Beach, Florida.

Giampietro, M., G. Cerretelli, and D. Pimentel. 1992. Energy analysis of agricultural ecosystem management: human return and sustainability. *Agric. Ecosyst. Environ.* 38:219–244.

Kendall, H.W. and D. Pimentel. 1994. Constraints on the expansion of the global food supply. *Ambio.* 23(3):198–205.

Lovejoy, S.B. and T.L. Napier, Eds. 1986. *Conserving Soil: Insights from Socioeconomic Research*. Soil Conservation Society of America, Ankeny, Iowa.

Napier, T.L., S.M. Camboni, and S.A. El-Swaify, Eds. 1994. *Adopting Conservation on the Farm: An International Perspective on the Socioeconomics of Soil and Water Conservation*. Soil and Water Conservation Society, Ankeny, Iowa.

Pagiola, S. 1995. Environmental and Natural Resource Degradation in Intensive Agriculture in Bangladesh. Paper No.15. The World Bank, Environmental Economics Series, Washington, D.C.

Paoletti, M.G., T.L. Napier, O. Ferro, B. Stinner, and D. Stinner, Eds. 1993. *Socio-Economic and Policy Issues for Sustainable Farming Systems*. Cooperativa Amicizia S.r.l., Padova, Italy.

Pimentel, D., R. Harman, M. Pacenza, J. Pecarsky, and M. Pimentel. 1994. Natural resources and an optimum human population. *Populat. Environ. J. Interdiscip. Stud.* 15(5):347–369.

Rauscher, M. 1992. International economic integration and the environment: the case of Europe. Pages 173–192 in K. Anderson and R. Blackhurst, Eds. *The Greening of World Trade Issues*. Harvester Wheatsheaf, New York.

Repetto, R. 1986. World Enough and Time: Successful Strategies for Resource Management. World Resources Institute, Washington, D.C.

Ribaudo, M.O. 1986. Consideration of off-site impacts in targeting soil conservation programs. *Land Econ.* 62(4):402–411.

Schertz, L.P. and G. Wunderlich. 1982. Structure of farming and landownership in the future: implications for soil conservation. Pages 163–183 in H.G. Halcrow, E.O. Heady, and M.L. Cotner, Eds. *Soil Conservation Policies, Institutions and Incentives*. Soil Conservation Society of America, Ankeny, Iowa.

Sfeir-Younis, A. and A.K. Dragun. 1993. *Land and Soil Management.* Westview Press, Boulder, Colorado and Oxford, England.

Stanhill, G. 1984. Agricultural labour: from energy to sink. Pages 113–130 in G. Stanhill, Ed. *Energy and Agriculture.* Springer-Verlag, Berlin, Germany.

Steinhart, C.E. and J.S. Steinhart. 1974. Energy use in the U.S. food system. *Science.* 184:307–316.

Stonehouse, D.P. 1991. The economics of tillage for large-scale mechanized farms. *Soil Tillage Res.* 20:333–351.

Stonehouse, D.P. 1993). A Targeted Public Intervention Approach to Improved Renewable Resource Stewardship in North American Agriculture. Presentation to Soil and Water Conservation Society 48th annual meeting, Fort Worth, Texas. August, 1993.

Stonehouse, D.P. 1993b. *A Review of GATT and Environmental Issues.* Department of Agricultural Economics and Policy, Wageningen Agricultural University, Wageningen, The Netherlands.

Stonehouse, D.P. 1994a. Canadian experiences with the adoption and use of soil conservation practices. Pages 369–395 in T.L. Napier, S.M. Camboni, and S.A. El-Swaify, Eds. *Adopting Conservation on the Farm: An International Perspective on the Socioeconomics of Soil and Water Conservation.* Soil and Water Conservation Society, Ankeny, Iowa.

Stonehouse, D.P. 1994b. Curriculum development for increasing appreciation of the economic values of the environment. Presentation to the 2nd European Conference on Higher Education in Agriculture. *Managing Change in The Food Chain and Environment.* Gödöllö Agricultural University, Hungary.

Stonehouse, D.P. 1995. Profitability of soil and water conservation in Canada. *J. Soil Water Conserv.* 50(2):215–219.

Stonehouse, D.P. 1996. A targeted policy approach to inducing improved rates of conservation compliance in agriculture. *Can. J. Agric. Econ.* 44(2): in press.

Stonehouse, D.P. and M. Bohl. 1990. Land degradation issues in Canadian agriculture. *Can. Publ. Policy.* 16:418–431.

Stonehouse, D.P. and R. Protz. 1993. Socio-economic perspectives in making conservation practices acceptable. Pages 29–58 in E. Baum, P. Wolff, and M. Zöbisch, Eds. *Topics in Applied Resource Management in the Tropics. Volume 3: Acceptance of Soil and Water Conservation Strategies and Technologies.* Deutsches Institut für Tropische und Subtropische Landwirtschaft, Witzenhausen, Germany.

Stonehouse, D.P., C. Giraldez, and W. van Vuuren. 1995. Holistic Approaches to Natural Resource Management and Environmental Care. Presentation to Soil and Water Conservation Society annual meeting, Des Moines, Iowa. August 1995.

Summers, G.F., Ed. 1983. *Technology and Social Change in Rural Areas.* Westview Press, Boulder, Colorado.

Sutton, J.D., Ed. 1988. *Agricultural Trade and Natural Resources.* Lynne Rienner Publishers, Boulder, Colorado.

van Vuuren, W. 1986. Soil erosion: the case for market intervention. *Can. J. Agric. Econ.* 33(1):41–62.

van Vuuren, W., C. Giraldez, and D.P. Stonehouse. 1995. Importance of benefit identification in evaluating water pollution control programs. *Can. Water Resour. J.* 20(1):1–10.

van Vuuren, W. and D.P. Stonehouse. 1995. Policy instrument effectiveness in water quality control programs in agriculture. Pages 111–138 in A. Weersink and J. Livernois, Eds. *Potential Applications of Economic Instruments to Address Selected Environmental Problems in Canadian Agriculture.* Environment Bureau, Agriculture and Agri-Food Canada, Ottawa, Ontario.

Weersink, A. and J. Livernois, Eds. 1995. *Potential Applications of Economic Instruments to Address Selected Environmental Problems in Canadian Agriculture.* Environment Bureau, Agriculture and Agri-Food Canada, Ottawa, Ontario.

Chapter 36

A Multicriteria Approach for Trade-Off Analysis Between Economic and Environmental Objectives in Rural Planning

G. VAN HUYLENBROECK

Abstract

In this paper a multicriteria methodology is presented for the trade-off analysis between economic and environmental objectives in rural planning. First the discrete version of the methodology is explained and illustrated for a road planning problem. To make the method operational for larger problems, a mixed integer programming formulation of the methodology has been developed. With this formulation also other types of problems can be analyzed such as land use planning problems as is illustrated with a small example.

Introduction

In many of todays' planning or investment problems, the decision problem is focussed on the analysis of conflicts between economic and ecological objectives. In rural planning, decision makers often have to make a trade-off between these rather conflicting objectives due to the scarcity of natural resources such as land, water, etc. Traditional cost-benefit methods fail in these circumstances because it is not easy to express ecological values in terms of money, despite certain attempts in this direction (e.g., the contingent valuation methods or other methods based on the willingness-to-pay principle). Therefore, multicriteria analysis (MCA) (for discrete problems) and multiobjective decision (MOD) (for continuous problems) methods seem to be more appropriate tools to support the decision making in this context. But our experience (i.e., Van Huylenbroeck and Martens, 1992) is that in practice, a straightforward application of MCA or MOD-methods is not evident because the conflicts are so strong that incomparability situations are obtained and decision makers want more information about what they have to give up on the one group of criteria for a performance increase on the other group of criteria, information which remains hidden if alternative solutions are directly compared on all criteria influencing the decision. Another criticism often used against MCA, is that the mechanisms leading to a specific choice are not clear for the decison maker and that certain information is getting lost if opposite criteria are used.

To make the decision procedure more transparant, a three step methodology is proposed (see also Figure 36.1):

1. Separate aggregation of the economic and ecological criteria
2. Visualisation of trade-offs
3. Discrete compromise analysis to support the final choice

In this approach both groups of criteria are first aggregated separately in order to get indicators which can be used in steps 2 and 3 for further analysis. For this first step any multicriteria method can be used, depending on the nature of the data. The result is a trade-off graph identifying the pareto-efficient solutions. The visualization step makes the interpretation of the results for the decision maker more easy and understandable. For the further analysis of the results a discrete compromise programming approach can be followed. To handle more complex problems or continuous problems, an integer programming approach can be used, as will be illustrated later.

For illustration purposes, the proposed methodology will be applied to a typical rural planning problem, the allocation of a new road or railway in a rural area. The approach described in the paper has been followed for the evaluation of the location alternatives for the High Speed Train (HST) through Belgium (Van Huylenbroeck, 1991). However, because this paper emphasizes on the methodology, a more schematic representation of the real-world problem is used for illustration purposes.

The Road Planning Problem

The problem of the optimal location of roads or railways has already been intensively studied in the MCA-literature (Massam, 1982; Siskos and Assimakopoulos, 1989) and can be described as follows: finding the optimal connection path between two points taking into account all relevant criteria. Siskos and Assimakopoulos (1989) are proposing the following evaluation procedure:

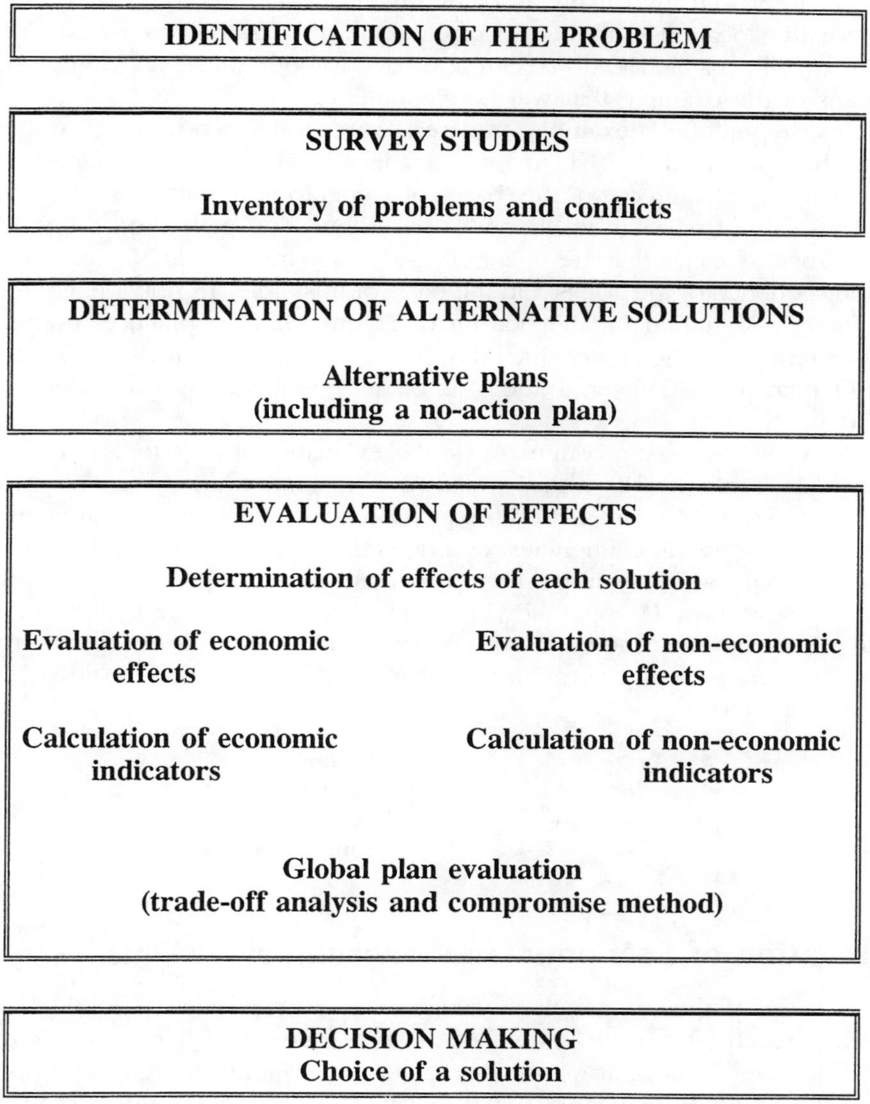

```
┌─────────────────────────────────────────────────┐
│           IDENTIFICATION OF THE PROBLEM         │
└─────────────────────────────────────────────────┘

┌─────────────────────────────────────────────────┐
│                  SURVEY STUDIES                 │
│         Inventory of problems and conflicts     │
└─────────────────────────────────────────────────┘

┌─────────────────────────────────────────────────┐
│        DETERMINATION OF ALTERNATIVE SOLUTIONS   │
│                 Alternative plans               │
│             (including a no-action plan)        │
└─────────────────────────────────────────────────┘

┌─────────────────────────────────────────────────┐
│              EVALUATION OF EFFECTS              │
│        Determination of effects of each solution│
│                                                 │
│  Evaluation of economic    Evaluation of non-economic │
│         effects                    effects      │
│                                                 │
│  Calculation of economic   Calculation of non-economic│
│        indicators                 indicators    │
│                                                 │
│              Global plan evaluation             │
│      (trade-off analysis and compromise method) │
└─────────────────────────────────────────────────┘

┌─────────────────────────────────────────────────┐
│                 DECISION MAKING                 │
│                Choice of a solution             │
└─────────────────────────────────────────────────┘
```

Figure 36.1 Global planning and evaluation framework for LCPs.

1. Division of the area between the two points to be connected into homogeneous zones
2. Definition of the evaluation criteria
3. Determination of the evaluation scores (qualitative or quantitative) for each zone
4. Identification of all possible locations by determining all feasible connection paths
5. Evaluation of each location on the basis of the evaluation scores of the zones that are being crossed
6. Comparison of all possible locations

In this approach, two main problems remain to be solved: how to evaluate the alternative locations on the basis of the zones that are being crossed and how to compare all possible locations. Our approach to both questions will be clarified by means of the example that will be elaborated.

In this (hypothetical) example, the area between the two points A and B that have to be connected, is divided into 14 zones (Figure 36.2). In practice, these zones will be identified on the basis of geographical, natural, or land use characteristics. All possible locations for the connection path can be defined by the sequence of zones that are being crossed. Therefore, Table 36.1 indicates the possible sequence of the zones. On the basis of this table, 18 possible connection paths can be identified (at the bottom of Figure 36.2). In practice, nonfeasible locations (e.g., because of nonsuitability of the land or other technical problems) will be eliminated. However, in the example all possibilities are considered to be feasible.

Besides the succession of the zones, the evaluation scores for each zone are presented in Table 36.1 as well. In total, five evaluation criteria are used: three indicating the (negative) impact on the environment (impact on the fauna, natural resources, and landscape amenities) and two criteria expressing the consequences for the economic activities in the region (agriculture and recreation). As in Siskos and Assimakopoulos (1989) following qualitative scale is used, indicating the negative (for the environmental criteria) or positive (for the economic criteria) degree of impact (negative for the environmental criteria and positive for the economic criteria):

- or +	Low impact
-- or ++	Moderate impact
--- or +++	High impact
---- or ++++	Strong impact

Aggregation of Economic and Environmental Data

On the basis of these data, the evaluation scores for each possible location alternative (each alternative is defined as the sequence of a number of zones) have to be determined. For quantitative criteria this should not give any problem supposed they are additive, as shown in Van Huylenbroeck (1991). Aggregation of qualitative impact scores is more difficult. Siskos and Assimakopoulos (1989) are proposing to use the distribution of the impact scores of the affected zones (the zones that are crossed by a certain alternative). This looks like a good approach with, however, one remark: in the case that the number of affected zones is not equal for all alternatives, this method can give a bias with negative impact scores because a connection path going through a higher number of zones but of lower value, will receive a better classification than an alternative crossing a fewer number of zones, but of higher value. Therefore, their approach is modified in the sense that the number of unaffected zones is taken as evaluation standard. This also allows for a better comparison with the no-action alternative since the environmental value of the area between the two points can be determined as the maximum number of zones in each impact level.

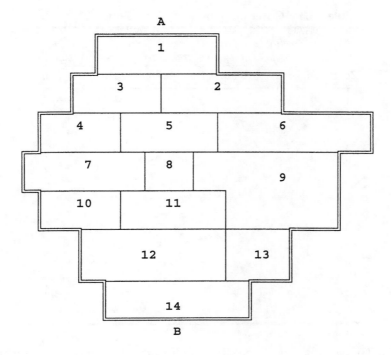

Alternative	Sequence of zones
A1	1-2-6-9-13-14
A2	1-2-6-9-11-12-14
A3	1-2-5-9-13-14
A4	1-2-5-9-11-12-14
A5	1-2-5-8-10-12-14
A6	1-2-5-8-11-12-14
A7	1-2-5-7-10-12-14
A8	1-2-5-7-11-12-14
A9	1-3-5-9-13-14
A10	1-3-5-9-11-12-14
A11	1-3-5-8-10-12-14
A12	1-3-5-8-11-12-14
A13	1-3-5-7-10-12-14
A14	1-3-5-7-11-12-14
A15	1-3-4-8-10-12-14
A16	1-3-4-8-11-12-14
A17	1-3-4-7-10-12-14
A18	1-3-4-7-11-12-14

Figure 36.2 Schematic representation of the hypothetical study area and identification of the possible alternatives.

In Table 36.2 the number of unaffected zones for all impact levels of each criterion is indicated as well as the maximum number of zones in each impact level. For the environmental criteria, the loss of environmental value has to be

Table 36.1 Criterion Scores for Each Zone

Zone	Possible successive zones	Negative impact for			Positive value for	
		Fau	Env	Land	Agri	Rec
1	2,3	--	-	--	++++	+++
2	5,6	---	-	---	+++	++
3	4,5	-	----	----	+++	+
4	7,8	-	---	---	+++	++
5	7,8,9	---	----	--	++	++++
6	9	--	--	---	+++	+
7	10,11	----	----	----	++++	+++
8	10,11	---	---	--	++	++++
9	11,13	--	--	-	++	++
10	12	----	----	--	++	++
11	12	--	--	--	+++	+++
12	14	-	-	--	+	++
13	14	--	----	----	+++	++++
14	-	-	--	-	++	++

minimized. This loss is determined by making the difference between the maximum number of zones in an impact level and the number of unaffected zones in the same impact level. As such, the aggregation problem for the environmental criteria has been translated into a MCA-problem with in this case 18 alternatives (the number of possible locations) and 12 criteria (the number of environmental criteria multiplied with the number of impact scores).

For the economic criteria, the number of affected zones has to be maximized or in other words the difference between the maximum number of zones in each impact level and the number of unaffected zones. In this case a MCA-problem with 18 alternatives and 8 criteria is obtained.

To solve both MCA-problems, the CAM-method (conflict analysis method) described in Van Huylenbroeck and Lippens (1992) and Van Huylenbroeck (1995), which is combining the main features of well-known methods such as ELECTRE, PROMETHEE, ORESTE, and the AHP-method is applied. In this method preference indicators are calculated for each pair of alternatives which take into account both the scores for each criterion as well as the relative importance of the criteria, following the next procedure:

1. For each criterion, the relative difference in impact between two alternatives $\hat{e}_f(a,b)$ is determined by transforming the original data using an appropriate preference function (for the possibilities see Van Huylenbroeck, 1995). In this case a linear preference function is used with absolute preference when all zones of an impact level are unaffected (for negative impact scores) or affected (in the case of positive impact scores). This means that a difference of leaving one more zone unaffected (case 1) or affected (case 2) is translated into a difference in preference of 1 divided by the maximum number of zones for that impact level.

A Multicriteria Approach for Trade-Off Analysis

Table 36.2 Evaluation of the Alternatives: Number of Unaffected Zones for Each Impact Level

Crit.	A1	A2	A3	A4	A5	A6	A7	A8	A9	A10	A11	A12	A13	A14	A15	A16	A17	A18	Max
Fau																			
-	3	2	3	2	2	2	2	2	2	1	1	1	1	1	0	0	0	0	4
--	1	1	2	2	4	3	4	3	2	2	4	3	4	3	4	3	4	3	5
---	2	2	1	1	0	0	1	1	2	2	1	1	2	2	2	2	3	3	3
----	2	2	2	2	1	2	0	1	2	2	1	2	0	1	1	2	0	1	2
Env																			
-	1	0	1	0	0	0	0	0	2	1	1	1	1	1	1	1	1	1	3
--	1	0	2	1	3	2	3	2	2	1	3	2	3	2	3	2	3	2	4
---	2	2	2	2	1	1	2	2	2	2	1	1	2	2	0	0	1	1	2
----	4	5	3	4	3	4	2	3	2	3	2	3	1	2	3	4	2	3	5
Agri																			
+	1	0	1	0	0	0	0	0	1	0	0	0	0	0	0	0	0	0	1
++	3	3	2	2	1	2	2	3	2	2	1	2	2	3	2	3	3	4	5
+++	3	3	4	4	5	4	5	4	4	4	5	4	5	4	4	3	4	3	6
++++	1	1	1	1	1	1	0	0	1	1	1	1	0	0	1	1	0	0	2
Land																			
-	0	0	0	0	1	1	1	1	0	0	1	1	1	1	1	1	1	1	2
--	4	3	3	2	1	1	2	2	3	2	1	1	2	2	2	2	3	3	5
---	2	1	3	2	2	2	2	2	4	3	3	3	3	3	2	2	2	2	4
----	2	3	2	3	3	3	2	2	1	2	2	2	1	1	2	2	1	1	3
Recr																			
+	1	1	2	2	2	2	2	2	1	1	1	1	1	1	1	1	1	1	2
++	3	2	3	2	2	3	2	1	4	3	3	4	3	2	2	3	2	3	6
+++	2	1	2	1	2	1	1	2	2	1	2	1	1	2	2	1	0	0	3
++++	2	3	1	2	1	1	2	2	1	2	1	1	2	2	2	2	3	3	3

2. On the basis of these preference scores, aggregated preference indices are calculated on the basis of the following formula:

$$P(a,b) = \frac{100}{n} \sum_{j=1}^{n} g_j \cdot \hat{e}_j(a,b)$$

with: $\hat{e}_j(a,b)$ = difference in impact score on criterion j for those criteria for which a is better than b, g_j = the relative importance of criterion j, and n = the total number of criteria.

3. These aggregated preference indices can then be compared for each pair of alternatives with certain threshold values in order to determine if there is preference, indifference or incomparability or aggregated into a rank order index as in the Promethee II approach using the following formula (k = the number of alternatives):

$$R(a) = \frac{1}{k-1} \sum_{x=b}^{k} P(a,x) - \frac{1}{k-1} \sum_{x=b}^{k} P(x,a)$$

For the mathematical details and some applications of the first option (pairwise comparison) we refer to Van Huylenbroeck (1991), Van Huylenbroeck and Martens (1992), and Van Huylenbroeck (1995). As the objective in this case is the aggregation of both groups of effects into a common indicator, which can be used in the trade-off analysis, the second option is applied.

In order to give more information to the decision makers the relative position of each location alternative is indicated by extending the set of alternatives (18 in the example) with two hypothetical alternatives: the no-action alternative, leaving all zones unaffected (the ideal situation for the environmental criteria, but anti-ideal for the economic goals) on the one hand and the other extreme, an alternative that is crossing all zones (ideal situation for the economic criteria, but anti-ideal for the environmental objectives) on the other hand.

For the weights it is supposed that all criteria are equally important and that a zone with higher impact level is more important than a zone with a lower impact level. Based on a uniform distribution of the weights it can be proved (Rietveld, 1984, 1989) that the expected average value of the weight-factor is given by:

$$g_j = \sum_{i=r}^{n} \frac{1}{i}$$

with: r = the ranking of criterion j (with r = 1 for the most important criterion and r = n for the least important criterion).

Modifications to this formula make it possible to handle rank orders with ties (several criteria having the same rank) or with a degree of difference (Rietveld and Ouwersloot, 1992). If necessary, the sensitivity of the weights can always be explored by generating a random sample of weights meeting the ordinal conditions or by testing other weight sets.

The results of this exercise are presented in Figure 36.3. In this figure, each location alternative is positioned by its aggregated scores for the environmental criteria on the X-axis and for the economic criteria on the Y-axis. As the objective is to minimize the damage for the environment and to maximize the economic benefits, the line in the graph is connecting the pareto-efficient solutions (in this case solutions A2, A4, A6, A8, and A7). The result of this first step is that the solution-set of possible locations is reduced from 18 to 5.

Identifying the Best Compromise

The best compromise in a decision problem can be defined as the solution closest to the (utopique or infeasible) ideal point (Zeleny, 1973; Yu, 1973; Duckstein and Kempf, 1981). This ideal point is determined as the solution where all objectives achieve their optimum value (Romero and Rehman, 1989). In our example this

A Multicriteria Approach for Trade-Off Analysis

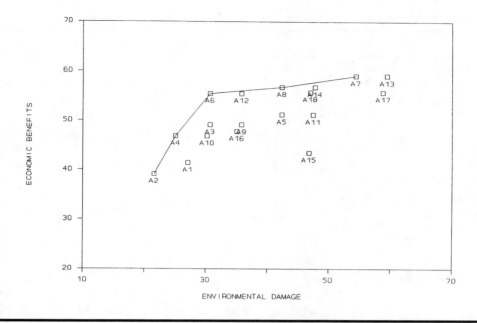

Figure 36.3 Trade-off diagram between the economic and environmental criterion.

means a solution which has a score of 100 for the economic criteria and a score of 0 for the environmental effects.

The best compromise can then be found by minimizing the distance to this ideal point. By using the L_p-metrics, the distance to the ideal point can be determined as follows (with W_j being the weight of criterion j):

$$L_P(W) = \left[\left(\sum_{j=1}^{n} W_j d_j\right)^p\right]^{\frac{1}{p}}$$

with:

$$d_j = \frac{|Z_j^* - Z_j(x)|}{|Z_j^* - Z_{*j}|}$$

In this formula Z_j^* en Z_{*j} are respectively the ideal and anti-ideal solution for criterion j. By changing the W and p parameters different solutions are obtained reflecting different preference structures. In Table 36.3 the results are presented for different weight-sets and different values of p. As can be observed with changing values of these parameters, the best compromise can change. The parameter p act as a weight attached to the magnitude of the deviations, while W_j measures the relative importance given to the main objectives in a decision problem.

Table 36.3 Compromise Solutions Applying the Compromise Method

	A2	A4	A6	A8	A7
Aggregated scores (from Figure 36.2)					
Economic benefits	60.87	53.18	44.54	43.21	40.93
Environmental damage	21.56	25.07	30.67	42.36	54.32
Solutions for W1 = 1 and W2 = 1					
L_1	82.43	78.25	75.21	85.27	95.25
L_2	64.58	58.79	54.08	60.51	68.02
L_3	61.76	54.98	48.94	53.91	61.57
L_∞	60.87	53.18	44.54	43.21	54.32
Solutions for W1 = 3 and W2 = 1					
L_1	204.18	184.62	164.29	171.98	177.11
L_2	183.89	161.50	137.09	136.37	134.27
L_3	182.72	159.75	134.15	131.12	126.24
L_∞	182.62	159.24	133.62	129.63	122.79
Solutions for W1 = 1 and W2 = 3					
L_1	125.55	128.40	136.56	170.28	203.89
L_2	88.82	92.12	102.23	134.21	168.02
L_3	79.16	83.20	95.38	128.71	163.82
L_∞	64.67	75.21	92.02	127.07	162.96
Solutions for W1 = 2 and W2 = 1					
L_1	143.30	131.43	119.75	128.77	136.18
L_2	123.64	109.28	94.21	96.24	98.24
L_3	121.97	106.83	90.28	89.69	89.16
L_∞	121.75	106.36	89.08	86.42	81.86
Solutions for W1 = 1 and W2 = 2					
L_1	103.99	103.32	105.89	127.92	149.57
L_2	74.60	73.09	75.81	95.09	116.10
L_3	67.37	65.15	68.35	88.30	110.55
L_∞	60.87	53.18	61.35	84.71	108.64

W1; W2: Relative weight attached to resp. the economic and environmental objective. L_p: L_p-metric distance with p = power of the function (p = 1,2,3, or ∞). Shaded areas = best comopromise for these weights and L_p-metric.

A Mixed Integer Programming Approach

The above example reveals the conceptual framework that can be used in the case of a small location problem. However, for larger location problems (e.g., in land consolidation planning) the discrete approach is not practical to apply as the number of alternatives can increase rapidly with a higher number of zones (in Siskos and Assimakopoulos, 1989) with 58 zones, the number of feasible alternative tracings went up to 2,703). Therefore, to make the method operational an integer programming formulation of the problem has been developed.

In the integer programming matrix for the routing problem, the objective is to minimize or maximize the number of unaffected zones for a certain objective (minimizing in the case of positive impacts, maximizing in the case of negative impacts). This means that of all objectives (two in the example: the economic and environmental objective) one is chosen as objective function while the other(s) are expressed as constraints of which the boundaries can be changed by parametric programming (for the matrix see Annexe 1).

The zones, through which the road can pass, are represented by binary variables, in this case [X1] to [X14]. If the binary variable has a value of 0 it means that the zone is not in the optimal solution while a value of 1 means that the zone makes part of the optimal route. Besides these variables, auxiliary variables (AUXj) are used to count the number of zones in each impact level (in which AUX represents the name of the criteria and j the impact level, see variables FAU1 to REC4).

The economic benefit coming from agriculture and recreation is maximized by multiplying the auxiliary variables AGR1 till REC4 with their respective weights:

$$\text{MAX } \Sigma \ g_j \cdot \text{AUXj for all AUXj belonging to the objective function}$$

The weights g_j are determined using the g_j-formula discussed earlier.

The activities are subject to three kinds of constraints:

1. Network-constraints in which the conditions about the sequence of zones are stipulated (in the example constraints b1 to b18):

$$\Sigma_i \ [X_i] < [X_k] \text{ with } [X_i] \text{ the subsequent zones of zone } [X_k] \text{ and this for all } k$$

and

$$\Sigma_i \ [X_i] = 1 \text{ for all zones following on the same } [X_k]$$

This last condition implies that only one zone following on $[X_k]$ may be selected;

2. Constraints counting the number of unaffected zones in each impact level: for the negative impact criteria (constraints CFAU1 to CLA4) by expressing that for all $[X_i]$ belonging to a certain impact level j:

$$\Sigma \ [X_i] + \text{AUXj} = \text{FIX}$$

with FIX = the maximum number of zones in that impact level
and for positive impact criteria (CAGR1 to CREC4) that

$$\Sigma \ [X_i] - \text{AUXj} = 0; \text{ and}$$

3. Minimal requirement-constraints for the objectives not in the objective-function (in the example the ENVIR-constraint):

$$\Sigma \ g_j \cdot \text{AUXj} = \tau \text{ for all AUXj belonging to the objective}$$

Hereby τ is the parameter to be fixed in each run of the parametric programming.

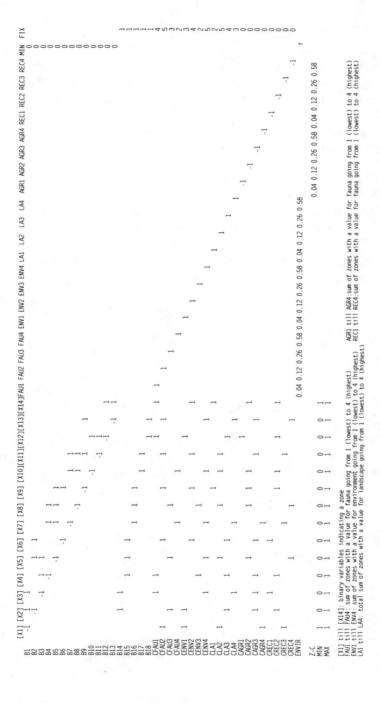

Annexe 1. Matrix for the mixed integer formulation of the road planning application.

By parameterization of the τ-parameter, the pareto-efficient solutions of Figure 36.3 and Table 36.3 are obtained. Once these efficient solutions are determined, the compromise method of par. 4 can be applied. By slight transformation of the matrix, it is even possible to determine directly the best compromise for the different L_p-distances.

Extension

The programming approach makes it not only possible to enlarge the size of the problems, but also to extend the type of planning problems that can be solved. In typical land use planning projects such as land consolidation projects, the method can (for example) be used to determine optimal water management or land use plans. Hence, the problem is the same — determine for each possible site the optimal water level or land use function taking into account the different objectives that have to be fullfilled by that region. This means that on the basis of these objectives for each possible site (e.g., the parcels), one and only one water level or land use type has to be chosen from a list of available options. Extra conditions may be the minimal and maximal areas for each land use type or water level and the exclusion of certain sites for certain functions or levels.

In order to avoid a dispersed solution or isolated sites, some extra conditions can be added (for example) by also counting the values of the surrounding sites. In Annexe 2, the matrix for a small land use planning problem is given: the problem consists of assigning to each of the nine considered parcels one of the two possible functions: agriculture or natural reservate and this based on two evaluation criteria: the appropriateness of each parcel for agricultural and environmental use.

The activities in this case are the possible functions for each parcel. They are represented by the binary variables [Ei] (environmental use) and [Ai] (agricultural use), with i the number of the parcel. Further on auxiliary counting variables (AUXj) with AUX the name of the objective (EN = environmental value of the parcel and AG the value for agricultural purposes) and j the level (from 1 to 4 with 1 the lowest and 4 the highest use value), are used to determine the score of the solution for both objectives. The value of the surrounding parcels is counted by introducing the SE and SA variables, while the EP and AP variables are necessary to calculate the total value of a parcel for resp. the environmental and agrarian function.

Five types of constraints are distinguished (see Annexe 2):

1. A group of constraints indicating that for each parcel only one land use function can be selected (constraints b1 till b9):

    [Ei] + [Ai] = 1 for each parcel i

2. A group of constraints counting the number of zones in each impact level of both criteria (constraints ENV1 till ENV4 for the environmental value and AGR1 till AGR4 for the agricultural value of the parcels). Because the higher the value the better, these constraints can be written as:

432 *Multiple Objective Decision Making for Land, Water, and Environmental Management*

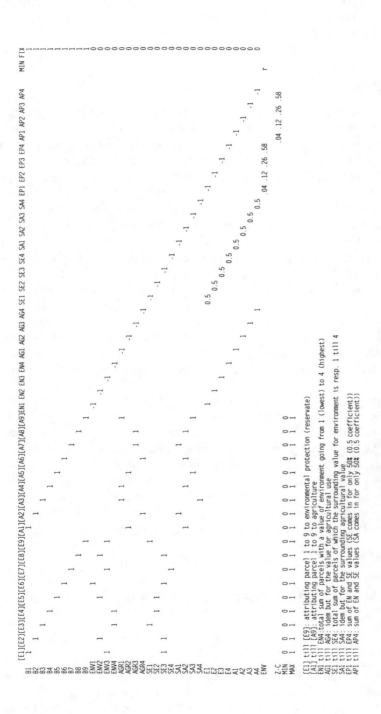

Annexe 2. Matrix for the mixed integer formulation of the land use planning application.

Σ [Ei] − ENj = 0 for all [Ei] belonging to impact level j

Σ [Ai] − AGj = 0 for all [Ai] belonging to impact level j

3. Another group of constraints counting the value of the surrounding zones in each impact level. The surrounding value has been determined as the average impact level of the surrounding parcels. Of course other definitions can be used to determine this value (constraints SE1 till SE4 for the environmental value and SA1 till SA4 for the agricultural value of the surrounding parcels):

Σ [Ei] − SEj = 0 for all [Ei] of which the surrounding value belongs to impact level j

Σ [Ai] − SAj = 0 for all [Ai] of which the surrounding value belongs to impact level j

4. A group of constraints making the addition of the proper value of the sites and the value of the surrounding parcels in each impact level (constraints E1 till E4 for the environmental value and A1 till A4 for the agricultural value). Hereby a different importance can be attributed to both values. In the application the surrounding value (for example) is only counted for 50% (the 0.5 coefficient in the constraints) or:

ENj + 0.5 SEj = EPj for each impact level j

AGj + 0.5 SAj = APj for each impact level j; and

5. The minimal requirements for the objectives not in the objective-function (in the example the ENV-constraint)

Σ g_j . SEj = τ with τ to be parametrized

The objective function consists of maximizing the agricultural function:

MAX Σ g_j . SAj

That the introduction of a value for the surrounding sites is changing the solution is illustrated by Figure 36.4 which indicates that for a value of τ = 1, parcel 3 and 4 are attributed to the natural reservate function if the value of the surrounding sites is not considered, while parcel 1 and 4 are chosen for environmental protection when this is well the case. Parameterization of the minimal environmental value (the τ-parameter) gives other solutions which can further be analyzed and compared using the method explained in Section 4. If, for example, the τ-parameter is increased from 1 to 1.5, parcel 7 is also attributed to environmental protection while the total value for agriculture decreases with 0.28 points. This allows, as in the road planning example to express the trade-offs between both criteria. Combined with a graphical mapping technique, the spatial consequences can be visualized.

Solution without surrounding value

Parcel 1 (2/3)	Parcel 2 (4/2)	**Parcel 3 (1/4)**
Parcel 4 (1/4)	Parcel 5 (2/2)	Parcel 6 (4/1)
Parcel 7 (3/3)	Parcel 8 (3/2)	Parcel 9 (1/3)

Solution with surrounding value

Parcel 1 (2/3)	Parcel 2 (4/2)	Parcel 3 (1/4)
Parcel 4 (1/4)	Parcel 5 (2/2)	Parcel 6 (4/1)
Parcel 7 (3/3)	Parcel 8 (3/2)	Parcel 9 (1/3)

Figure 36.4 Graphical representation of the results for a simple land use planning problem without and with taking into account the value of the surrounding parcels (for a value of the environmental objective = 1). ( ) = the value of the parcel for agricultural use/the value of the parcel for use as reservate; ■ = selected parcels attributed to environmental protection (reservate).

Conclusions

By using the mixed integer programming approach, it becomes possible to handle large planning problems. The size of the matrix will mainly be determined by the number of zones and the network requirements. For road location problems, this part can be large but for other planning problems (e.g., water management planning or allocation of different functions) this network part will be minimal. The counting part only depends on the number of criteria and impact levels distinguished. Finally, it is also possible to work with more than two objectives by extending the number of minimal requirement-constraints.

Another interesting possibility to be further explored is the link or connection of this methodology with graphical mapping techniques like GIS as this will allow the direct graphical analysis of the solutions. In particular for complex allocation problems, as in land consolidation planning, where the function of each land plot has to be decided, this seems a promising research area.

References

Duckstein, L. and J. Kempf. 1981. Multicriteria Q-analysis for plan evaluation. Pages 87–99 in P. Nijkamp and J. Spronk, Eds. *Multiple Criteria Analysis: Operational Methods*. Aldershot, Grower.

Massam, B.H. 1982. The search for the best route: an application of a formal method using multiple criteria. *Sist. Urbani*. 5:183–194.

Rietveld, P. 1984. The use of qualitative information in macro-economic policy analysis. Pages 263–280 in M. Despontin, P. Nijkamp, and J. Spronk, Eds. *Macro-Economic Planning with Conflicting Goals*. Springer-Verlag, Berlin.

Rietveld, P. 1989. Using ordinal information in decision making under uncertainty. *Syst. Anal. Modelling Simulat.* 6:659–672.

Rietveld, P. and H. Ouwersloot. 1992. Ordinal data in multicriteria decesion making, a stochastic dominance approach to siting nuclear power plants. *Eur. J. Operat. Res.* 56:249–262

Romero, C. and T. Rehman. 1989. *Multiple Criteria Analysis for Agricultural Decisions.* Elsevier, Amsterdam. 257 pages.

Siskos, J. and N. Assimakopoulos. 1989. Multicriteria highway planning: a case study. *Math. Comput. Modelling.* 12(10/11):1401–1410.

Van Huylenbroeck, G. 1991. Application of multicriteria analysis in the environmental impact report for the high speed train. *Belg. J. Operat. Res. Stat. Comput. Sci.* 30(4):36–52.

Van Huylenbroeck, G. 1995. The Conflict Analysis method: bridging the gap between Electre, Promethee and Oreste. *Eur. J. Operat. Res.* 82:490–502.

Van Huylenbroeck, G. and G. Lippens. 1992. The Conflict Analysis Method: a decision support method for multicriteria problems. *Tijdschr. Soc. Wet. Landb.* 7(3):222–242.

Van Huylenbroeck, G. and L. Martens. 1992. The average value ranking multi-criteria method for project evaluation in regional planning. *Eur. Rev. Agric. Econ.* 19:237–252.

Yu, P.L. 1973. A class of solutions for group decision problems. *Manage. Sci.* 19:936–946.

Zeleny, M. 1973. Compromise programming. Pages 262–301 in J. Cochrane and M. Zeleny, Eds. *Multiple Criteria Decision Making.* University of South Carolina Press, Columbia.

Chapter 37

Long-Term Explorations of Sustainable Land Use

J.J.E. BESSEMBINDER, M.K. VAN ITTERSUM, R. RABBINGE, AND L.O. FRESCO

Abstract

Long-term land use explorations serve to widen the perspective of decision makers on future land use options. Knowledge on underlying biophysical processes is used to describe alternative production technologies. Current socioeconomic limits to production are excluded or their effects are made explicit. With the Multiple Goal Linear Programming (MGLP) procedure, information on production technologies is confronted with technical constraints and with normative objectives. These objectives are distilled from the various policy views that can be identified in a society. Scenarios can be generated for each of the policy views and as such they make explicit the different dimensions related to the sustainability issue. These scenarios are not of a predictive nature, they only show the scope for policy choices. As such these studies can be a valuable tool in a planning procedure. The methodology is illustrated with examples and some specific features and problems are discussed.

Introduction

Long-term land use planning requires insight in the potential for agricultural production, because changes in current trends are not only likely, they are often also desirable. An exploration of the options for land use under different policy views may help to determine the 'playing field' of policy makers for strategic choices. Mostly there is no direct relation between actual production levels and actual resource use efficiencies on the one hand and biophysical and technical potentials for agricultural production on the other hand. Actual production techniques are influenced strongly by economic considerations, management of the farmer, infrastructure, etc. These factors can change within the time horizon considered in planning studies. Biophysical limits, on the other hand, are more or less stable within this time horizon and they will help to determine the outer boundaries of the 'playing field' for land use planning. This 'playing field' can be determined with a methodology using the MGLP procedure, where biophysical possibilities are confronted with policy views. In this paper the aim and the methodology of these optimization studies are presented and some specific features and problems are discussed.

Main Characteristics of Optimization Studies Exploring Long-Term Land Use Options

The aim of the studies described is exploring future land use options, by confronting the biophysical possibilities and the technical constraints with the normative objectives of stakeholders. This aim has consequences for the methodology used and for the required technical information. It is important to be explicit about the integration level of the study, and about its time horizon (>20 years).

These land use explorations rely on knowledge of underlying biophysical processes, e.g., photosynthesis and effects of growth factors, to quantify new production techniques. Second, production is assumed to take place with the 'best technical means', i.e., available knowledge and available means of production are optimally applied, which precludes any waste or inefficient use of resources. Current economic conditions, not farm infrastructure, present constraints to farming practices (De Koning et al., 1995).

Methodology for Long-Term Land Use Exploration

Figure 37.1 shows a schematic presentation of a land use optimization study. With this figure the different elements of the methodology are explained. Several examples of long-term land use exploration have been published (Veeneklaas et al., 1991; Rabbinge and Van Latesteijn, 1992; Stroosnijder et al., 1994).

The central technique is MGLP. With this optimization, technique information on possible land use and technologies is confronted with technical constraints and a set of objective functions.

Various forms of agricultural production which are related to land use and which are relevant for the region in the time horizon are identified. The soils potentially suitable for these types of land use are identified in a qualitative land evaluation. In a quantitative land evaluation, the potential, water-limited and sometimes nutrient-limited productions are calculated with crop growth simulation models for the suitable soils. These simulation models use knowledge on various growth processes such as photosynthesis to calculate plant growth under different climate and soil conditions. Next, production techniques are defined and their input-output relations are quantified using a target-oriented approach, which means that the inputs to realize a particular output level are quantified, using knowledge on the processes involved. Input-output relations are defined in such

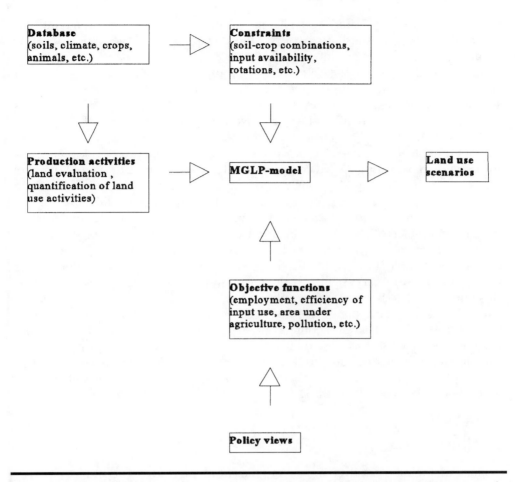

Figure 37.1 Schematic presentation of the methodology for a land use optimization study.

a way that they are valid for many years or cycles of the crop rotation, e.g., no depletion of nutrients is allowed. This implies that no substitution is possible between inputs, like water and nutrients, which are taken up by the plant and which fulfill a specific and essential role. Other inputs (e.g., biocides, labor, mechanization) can replace each other up to a certain degree. It is assumed that inputs are used with the highest technical efficiency, according to the available knowledge and techniques. The choice for the various production techniques depends on the aim of the production activity (i.e., production orientation): high soil productivity, low emissions of nutrients, or biocides per unit product or per unit area, etc. These orientations stand for different conceptions of sustainability. The production orientation determines the combination of substitutable inputs. In a production activity oriented at high soil productivity, control of diseases and pests takes place in such a way that a high productivity is achieved with efficient use of biocides. In an environmentally oriented activity biocides are excluded as much as possible, so that minimum use per unit of acreage is achieved. Lower yields per unit of acreage are accepted in such activities.

Several types of constraints are distinguished. Technical constraints are determined by the biophysical possibilities and can not be changed by man, e.g., available land areas, water available for irrigation. Normative constraints are determined by the desires of man, e.g., the dietary pattern, the export level, or the minimum employment rate.

Various policy views concerning land use can be identified, e.g., views emphasizing self-sufficiency of food, free market, and trade; nature conservation; or environmental issues. These views are operationalized with objective functions, such as maximizing cereal production, maximizing gross revenue of the region, or minimization of the area used for agriculture. Not all policy views can be quantified explicitly in objective functions. For instance, nature development and conservation have strong spatial components, that are hard to catch in an LP model.

Subsequently, land use options can be explored with the MGLP-procedure in several steps. First, the outer boundaries are determined by optimizing each of the objective functions in separate model runs, putting no or only light restrictions on the other objective functions. In this way, the initial freedom of choice (the worst and best value) for each objective is made explicit to stakeholders. In the next step stakeholders have to select the objective with the worst value, which is considered most unacceptable. A tighter bound for that objective is formulated. Subsequently, stakeholders are confronted with the results of a new series of optimization runs and they have to select again an objective with the most unacceptable value. This procedure continues until the stakeholders are satisfied with the compromise between their objectives. This procedure can be repeated with different groups of stakeholders, resulting in different scenarios. The comparison of the different scenarios shows the possibilities to realize objectives and the trade-offs between objectives.

An Illustration

To illustrate the methodology some examples from the study, "Ground for Choices: Four Perspectives for the Rural Areas in the European Community," are presented. A reformulation of the Common Agricultural Policy was needed due to problems related to current land use and agricultural policy in the EC-12. The study was carried out to support reorientation of long-term policy. It shows the importance of objectives for European agricultural policy. The results of this long-term land use exploration served to widen the perspective on future land use and natural resource management and they may help to define strategic policy choices (Rabbinge and Van Latesteijn, 1992).

The time horizon for the study is 25 years. The following forms of land uses are included: cereals, root and tuber crops, pastures, fruit crops, forestry, and nature. A land evaluation was carried out in which characteristics of soils and climate were confronted with requirements for various production technologies of the various crops. For the suitable areas potential and water-limited crop yields were calculated with crop growth simulation models. For perennial crops expert knowledge was used, because no simulation models were available. Table 37.1 shows that in some countries actual wheat yields are close to the water-limited or potential yields, whereas in other countries much more scope for production increase exists. In some countries present socioeconomic conditions and knowledge levels of farmers are favorable and high production levels can be obtained, but in the future these relative advantages over other regions may disappear.

Table 37.1 Actual Wheat Production of the EC-countries, Expressed as Percentage of the Calculated Potential and Water-Limited Production

	% of potential	% of water-limited
Belgium	62	72
Denmark	67	95
France	54	67
Greece	31	38
Ireland	62	69
Italy	29	37
Luxembourg	40	53
Netherlands	74	85
Portugal	12	24
Spain	26	60
United Kingdom	62	77
West Germany	60	67

Reinds and Van Lanen, 1992.

Subsequently, input-output relations on land use were derived for all subregions within the European Community by distinguishing different cropping systems and production orientations. Three production orientations were defined:

1. Yield-Oriented Agriculture (YOA) — aiming at high soil productivity under the given physical conditions
2. Environment-Oriented Agriculture (EOA) — aiming at minimizing negative effects on the environment by restricting the use of biocides per unit area and accepting slight decreases in yields
3. Land use Oriented Agriculture (LOA) — aiming at extensive land use as an alternative for surplus land, accepting low soil productivity (Rabbinge and Van Latesteijn, 1992)

Input levels of nonsubstitutable inputs, water and nitrogen, were determined using the target-oriented approach. With the help of transpiration coefficients of crops, radiation levels, and humidity, water uptake for a particular production level was quantified. Since each of the nutrients has a specific function within the plant, their concentrations are more or less fixed. In this way nutrient uptake for a particular production level can also be calculated (De Koning et al., 1995). Starting point for the quantification of the other substitutable inputs were the common management practices in current high-yielding agriculture (De Koning et al., 1995). For each crop these current practices were modified if more efficient methods of weed, disease, and pest control were judged to be feasible without yield reduction. An example: in maize production in YOA herbicides are assumed to be applied in the rows only while between the rows mechanical control is practiced by hoeing a number of times. This practice allows a reduction in herbicide use of about 40% compared to current agriculture. In EOA herbicide use may still be further reduced, partly by additional mechanical weeding. Nevertheless, according to experts, average yield reductions of about 10% are unavoidable under these practices. Differences in inputs and outputs among regions are based on production orientation, climatic conditions, and soil properties only.

In this study the technical constraints describe the availability of labor, area, water, and manure. The product balances 'regulate' the demand for agricultural products, which corresponds with self-sufficiency or a required export level. The latter constraints are normative constraints, influenced by the dietary pattern included and the policy view.

Four policy views for the EC were distilled from policy documents and discussions: (1) free market and free trade, (2) regional development, (3) nature and landscape, and (4) environmental protection. The aims which play an important role in these views have been translated into eight objective functions:

1. Maximization of soil productivity
2. Maximization of total agricultural employment
3. Maximization of regional agricultural employment
4. Minimization of total pesticide use
5. Minimization of pesticide use per hectare
6. Minimization of total N loss
7. Minimization of N loss per hectare
8. Minimization of total costs for agricultural production

The technical information on production technologies, the constraints, and the objective functions were integrated in the MGLP-model GOAL (General Optimal Allocation of Land use) and land use scenarios were generated for each of the four policy views. The four scenarios show some common results, e.g., all four scenarios show that the objectives can be achieved with less land for agricultural production (Figure 37.2), lower emission of nitrogen, and lower immision of biocides (Figure 37.3), than observed in the current situation. They also show distinct differences between policy views, for instance in the regional development scenario much more land is used with a very different regional land use allocation (Figure 37.2) than in the free market scenario (WRR, 1992). These differences show that there is considerable scope for policy.

Discussion

Data quality is very important in quantitative models such as MGLP-models. Knowledge on biophysical processes in various climates and for different soil types is not always sufficient for quantification of new production activities. Sometimes the knowledge on the processes is available, but data are lacking for proper quantification of processes. For example, for annual crops reliable crop growth simulation models are available, but for perennial crops this is often not the case. Several models exist that describe the fate of pesticides in the soil, however, these models are not validated for, as an example, hot humid climates or for volcanic soils. In most LP problems, input-output coefficients are included which are only estimates and sometimes very rough estimates. It follows that the computed optimal solution is only an approximation. Evidently, *sensitivity analysis* should be part of the study. Given the enormous number of coefficients, sensitivity

Long-Term Explorations of Sustainable Land Use

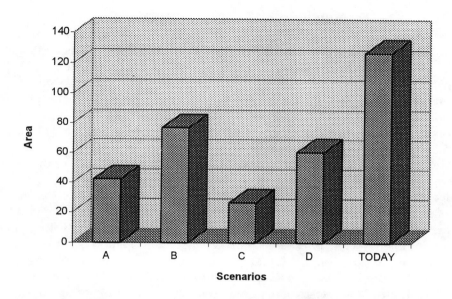

Figure 37.2 Land use in the different scenarios compared with current land use in the EC (in million hectares); A = free market and trade, B = regional development, C = nature and landscape, D = environmental protection. (Rabbinge et al., 1994).

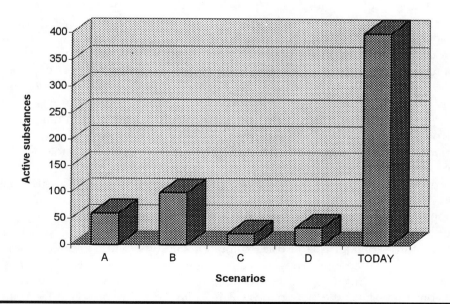

Figure 37.3 Use of crop protection agents in the different scenarios compared with current use in the EC (in million kilograms); A = free market and trade, B = regional development, C = nature and landscape, D = environmental protection. (Rabbinge et al., 1994).

analysis was limited to a few aspects in the studies carried out so far. Most often the effects of uncertainties due to variation in economic variables like prices and costs were analyzed. Uncertainties in technical coefficients of production technologies are hardly ever concerned (Veeneklaas et al., 1991; Scheele, 1992). If the probability distribution of values of the coefficients is known, the effect of uncertainty can be determined using a Monte Carlo approach. If the probability distribution is not known, sensitivity analyses with the most 'optimistic' and most 'pessimistic' values for the technical coefficients may provide an idea of the effect of uncertainty on the final scenarios.

Sustainability and sustainable development have various dimensions: ecological, technical, and socioeconomic dimensions. Operationalization requires normative weighing of more or less conflicting objectives deduced from these dimensions. Attitudes towards sustainability can diverge considerably, depending on the differing weight attached to facts, uncertainties, and risks with respect to the environment (WRR, 1995). Policy makers need information on the possibilities to combine various objectives, and on the trade-off between objectives. The methodology presented in this paper can reveal these possibilities and trade-offs.

In long-term land use explorations, production technologies are designed in such a way that they can be assumed to be valid for many years or cycles of the crop rotation. However, ecological sustainability has a *spatial* and *temporal* aspect (Fresco and Kroonenberg, 1992). By using a GIS in combination with an MGLP-model it is often thought that the spatial aspect of land use is included. This is only partly true, because interactions between different spatial units can not be included in LP models. For instance, the amount of sedimentation in one unit of land depends on the land use in another unit; e.g., if the land use in the mountains is not known beforehand, than a proper description of production technologies in the adjacent plains may not be possible. To model this, nonlinear relations are needed. Many processes are modelled at plot level, and upscaling to higher aggregation levels is done quite often very loosely without taking into account interactions with nearby plots. This might result in overestimation or underestimation of possibilities. Mostly static LP-models are used for land use optimization, implicitly ignoring the temporal aspect of sustainability. Land use is dynamic, although it may seem stable in the short term. Slight fluctuations in nutrient inputs and outputs are no threat to ecological sustainability as long as no continuous nutrient depletion takes place. In static LP-models it is not possible to include these fluctuations between years. However, in multiperiod LP-models, where land use can be adjusted every period, the temporal aspect of ecological sustainability can be taken into account partly.

Optimization studies exploring long-term land use options have no predictive nature, they only present possibilities, without judging the probability of their use in the future. Long-term land use explorations are just one component of a land use planning procedure. They serve to widen the perspective of the decision makers, by showing that the future does not have to be a continuation of the past, but they show the boundaries of the 'playing field', or scope for policy.

References

De Koning, G.H.J., H. Van Keulen, R. Rabbinge, and H. Janssen. 1995. Determination of input and output coefficients of cropping systems in the European Community. *Agric. Syst.* 48:485–502.

Fresco, L.O. and S.B. Kroonenberg. 1992. Time and spatial scales in ecological sustainability. *Land Use Policy.* 155–168.

Rabbinge, R. and H.C. Van Latesteijn. 1992. Long-term options for land use in the European Community. *Agric. Syst.* 40:195–210.

Rabbinge, R., C.A. Van Diepen, J. Dijselbloem, G.H.J. De Koning, H.C. Van Latesteijn, E.Woltjer, and J. Van Zijl. 1994. 'Ground for choices': a scenario study on perspectives for the rural areas in the European Community. Pages 95–138 in L.O. Fresco, L. Stroosnijder, J. Bouma, and H. Van Keulen, Eds. *Future of the Land, Mobilising and Integrating Knowledge for Land Use Options.* Wiley, Chichester.

Reinds, G.J. and H.A.J. Van Lanen. 1992. Crop Production Potential of Rural Areas within the European Communities. II: A Physical Land Evaluation Procedure for Annual Crops and Grass. Working Document 66. Netherlands Scientific Council of Government Policy, The Hague. 82 pages.

Scheele, D. 1992. Formulation and Characteristics of GOAL. Working Document 64. Netherlands Scientific Council for Government Policy, The Hague. 64 pages.

Stroosnijder, L., S. Efdé, R. Van Rheenen, and L. Agustina. 1994. QFSA: a new method for farm level planning. Pages 299–332 in L.O. Fresco, L. Stroosnijder, J. Bouma, and H. Van Keulen, Eds. *Future of the Land, Mobilising and Integrating Knowledge for Land Use Options.* Wiley, Chichester.

Veeneklaas, F.R., S. Cissé, P.A. Gosseye, N. Van Duivenbooden, and H. Van Keulen. 1991. Development Scenarios. Competing for Limited Resources: The Case of the Fifth Region of Mali. Report 4. Wageningen/Mopti: Centre for Agrobiological Research (CABO-DLO)/Etude sur les Systèmes de Production Rurales en 5ème Région (ESPR). 180 pages.

WRR (Netherlands Scientific Council for Government Policy). 1992. Ground for Choices, Four Perspectives for the Rural Areas in the European Community. Reports to the Government nr. 42. Sdu Uitgeverij, The Hague. 144 pages.

WRR (Netherlands Scientific Council for Government Policy). 1995. Sustained Risks: A Lasting Phenomenon. Reports to the Government nr. 44. Sdu Uitgeverij, The Hague. 205 pages.

Chapter 38

A Multicriteria Framework to Identify Land Uses Which Maximize Farm Profitability and Minimize Net Recharge

S.A. Prathapar, W.S. Meyer, E. Alocilja, and J.C. Madden

Abstract

In irrigation areas of southeastern Australia the presence of a shallow watertable exacerbates the damage potential of waterlogging and soil salinization, and threatens the long-term productivity of irrigated agriculture. By limiting farm net recharge, watertable rise and the damage potential of waterlogging and salinization can be minimized. A hierarchical multicriteria framework, *SWAGMAN Options,* is developed to identify profitable nonrice land uses, which complement rice growing and induce discharge from shallow watertables without exceeding critical levels of soil salinity. *SWAGMAN Options* is applied to the Camarooka Project Area in New South Wales, Australia, to study the sensitivity of maximum rice area per farm, rice field water use limit, rainfall conditions, optimal depth, critical salinity, initial piezometric levels, and weights associated with the objective functions on subsurface drainage and farm net returns. Results indicate that complementing rice production with maize and canola, and restricting rice to fields where rice field water use is less than 14 ml/ha, will minimize the watertable rise and improve farm profitability.

Introduction

Rice production in Australia occurs in the irrigated regions of southeastern Australia. The rice industry generates approximately AUD* 500 million per annum in gross revenue, most of it in export earnings. At the crop level rice is grown as a paddy system, where water is ponded for periods of 4 to 6 months. It is inevitable that recharge to groundwater will occur under rice and watertables will rise. When watertables are near the soil surface, salt is deposited within the rootzone and the potential for waterlogging is increased. Shallow watertables also reduce trafficability within farms and damage infrastructure.

Natural discharge of water from watertables is due to plant and soil evaporation, lateral groundwater flow, and groundwater leakage to deeper aquifers. The natural discharge rate is generally less than the rate of recharge, hence it is necessary to adopt practices which minimize the rate of recharge or maximize the rate of discharge.

To control the depth of the watertable, recharge must be removed through the soil surface because the rate of dissipation due to groundwater flows is very small. It could be removed by crops capable of tapping water directly from the watertable by tile drains at or below the optimal depths, or by groundwater pumping at selected sites within the region. The recharging water is of acceptable quality (less than 2 dS/m), and forms a lens on top of relatively saline water (up to 20 dS/m) due to the very slow diffusive and dispersive mechanisms within the soil solution. An ideal mechanism to lower the watertable can remove good quality water only. This would avoid removal of saline water from deeper layers of the unconfined aquifer.

The damage potential of rising watertables within the irrigation areas of southern NSW has been recognized by the state water agencies and the rice industry since the 1940s. The state agencies and the rice industry have negotiated and implemented a series of environmental restrictions which has reduced the rate of watertable rise. Such environmental policies were designed to protect the ecosystem but their impact on farm profitability was not assessed. While it is true that these two major concerns are conflicting at times, the problem has been compounded by addressing them separately. It is essential that irrigated farm profitability and environmental sustainability are considered simultaneously, when agronomic and environmental policies are formulated.

To help achieve this in irrigation areas where rice is predominantly grown, a hierarchical multicriteria framework, *Salt Water And Groundwater MANagement (Swagman) — Options*, is developed. The framework is multidisciplinary in its approach, incorporating information on agronomy, soils, hydrogeology, and economics. It aims to identify optimal land uses at field level which minimize recharge, maintain soil salinity below a critical level, and maximize farm net returns. It is used to evaluate optimal land use practices in the Camarooka Project Area, in a representative subregion of the MIA (Murrumbidgee Irrigation Areas).

---

* Australian Dollars. 1 US$ = AUD 1.39 in July 1995.

Table 38.1 Summary of Routines and Their Solution Sequence within SWAGMAN Options

Submodel	Method	Purpose
RICESELECT	Mixed integer programming	Select fields suitable for rice
WATABRISE	Finite difference groundwater simulation	Determines depth to watertable following rice
REQRECH	Linear programming	Determines subsurface drainage without nonrice land uses
REQIRR	Linear programming	Determines irrigation requirements for land uses while maintaining soil salinity below a critical level
NRETURN	Linear programming	Determines land use which will maximize net returns
REQDISCH	Linear programming	Determines nonrice land use which will maximize discharge
CROPSELECT	Multicriteria optimization	Selects optimal nonrice land use based on weights

Theory and Framework Development

The overall objective of *SWAGMAN Options* is to identify land use, in terms of crop type and acreage for each field within an irrigated subregion, which will maximize net return to the farmer and allow sustainable agricultural production. The environmental issues addressed in the framework are waterlogging and salinization which are preventable by maintaining watertables at or below an optimal depth.

SWAGMAN Options is made up of six optimization routines and a groundwater routine coded in the GAMS language (Brooke et al., 1988). The sequence in which they are solved is given in Table 38.1. Output from a routine is passed on to the subsequent routine as input. Since the optimization routines are solved sequentially there is a possibility of suboptimalities being introduced after the RICESELECT routine. However, since rice is the preferred crop in the region and, other uses are complementary only, the final solution will be a socially acceptable one. There are no possibilities of introducing suboptimalities later during the sequence. A detailed description can be found in Prathapar et al. (1995).

Materials and Methods

Site Description and Land Use Options

The framework was applied to the Camarooka project area in the MIA (Figure 38.1). The project area is 3,750 ha and includes 160 fields owned by 25

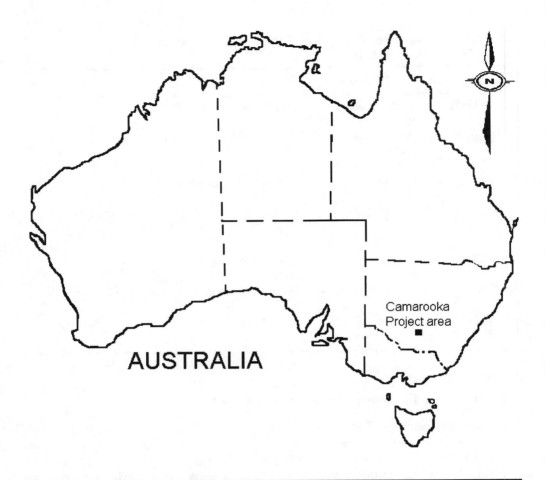

Figure 38.1 Location of the Camarooka Project area.

farmers. Rice is the predominant crop within the project area, but pasture, vegetables, and citrus are grown as well. Descriptions of the soils and hydrogeology of the study area can be found in van Dijk (1961) and Prathapar and Madden (1995). Monthly changes in piezometric levels are monitored by the New South Wales Department of Water Resources, Griffith, by monitoring 138 piezometers. Since rice and citrus are the preferred crops, priority for irrigation water is given to these crops. The land use options (M) available in *SWAGMAN Options* are maize (MAIZE), soybeans (SOYBEAN), sorghum (SORGHUM), sunflower (SUNFLOWER), perennial pasture (PASTURE), lucerne (LUCE), wheat (WHEA), barley (BARL), canola (CANO), and fababeans (FABA). Double cropping options considered are soybeans and wheat (SYWH), soybeans and barley (SYBY), soybean and canola (SYCA), soybeans and fababeans (SYFA), maize and wheat (MZWH), maize and barley (MZBY), maize and canola (MZCA), maize and fababeans (MZFA), sunflowers and wheat (SFWH), sunflowers and barley (SFBY), sunflowers and canola (SFCA), sunflowers and fababeans (SFFA), sorghum and wheat (SGWH), sorghum and barley (SGBY), sorghum and canola (SGCA), and sorghum and fababeans (SGFA). Planting a winter cereal after a rice crop is also an option represented by wheat (RWHEAT), barley (RBARLEY), canola (RCANOLA), and fababeans (RFABA). The annual net return ($/ha), and labor requirements (hr/ha) for the land uses are presented in Prathapar et al. (1995).

Variation in Rainfall and Evapotranspiration

Dry, average, and wet years are categorized by ranking annual rainfall data for the last 114 years and dividing the ranked data into three groups (dry, average, and wet). To account for variability in annual rainfall, the median values of the three groups are used. The median values for dry, average, and wet groups are 283, 383, and 517 mm, respectively. We assumed that 10% of annual rainfall runs off and does not contribute to evapotranspiration or recharge.

The annual reference evapotranspiration (RET) varies from year to year, depending on weather conditions. In order to account for such variability, daily RET values for the last 10 years (1984 to 1993), and annual RET values for the last 30 years (1962 to 1993), are analyzed to calculate RET for dry, average, and wet years. Actual evapotranspiration (AET) for each land use is calculated by adjusting the RET by a crop factor (Prathapar and Madden, 1995).

Sensitivity Analysis

Initially the framework is set up with default options to determine optimal land use practices within the Camarooka Project Area (Run 1). Subsequently, the sensitivity of optimal land uses, subsurface drainage requirement and net returns to maximum rice area permissible for a farm (RICEPCT), rice field water use limit (RWLIM), weather conditions (RAIN), *optimal depth*, critical salinity, initial depth to watertable, and weights on objective functions in CROPSELECT, is determined (Runs 2 to 14). Piezometric levels in September 1991 (31% of the area had a watertable within 1 m of the surface) are considered deep and this figure is used as the initial level. The piezometric levels in September 1988 are considered shallow (90% of the area had a watertable within 1 m of the surface).

Results and Discussions
Run 1

For Run 1, the percentage of area with watertable within 1 m increases to 42% without the introduction of nonrice land uses. When nonrice land uses are introduced, the area with a watertable within 1 m decreases from 42 to 35%. This indicates that, in an average rainfall year, watertable rise due to rice recharge can be controlled by growing complementary crops. Introduction of nonrice land uses also reduces the required subsurface drainage from 426 to 75 ml and increases the regional net return from AUD 430,000 to 1, 323, 900. MAIZE, MZCA, CANOLA, and RCANOLA are the preferred nonrice land uses for the study area.

Run 2

For Run 2, the percentage of the area with a watertable within 1 m decreases to 30% without any nonrice land uses. But the net return from rice decreases from AUD 430,000 to 264,000. This indicates that in an average rainfall year (median annual rainfall 383 mm), reduction in rice area from 30 to 20% results in a marginal reduction in the area with a shallow watertable, but reduces the net returns significantly. When nonrice land uses are introduced, the area with a watertable within 1 m decreases from 30 to 21%. Introduction of nonrice land uses also reduces the required subsurface drainage from 300 to 55 ml and increases the regional net return from AUD 264,000 to 1, 623,000. This is due to higher returns per milliliter of water obtained from some nonrice land uses. MAIZE and MZCA are the preferred nonrice land uses for this run.

Run 3

For Run 3, the percentage of the area with a watertable within 1 m increases to 44% without any nonrice land uses. When nonrice land uses are introduced, the area with watertable within 1 m decreases to 37%. Results from Runs 1, 2, and 3 show that percentage area with watertable within 1 m increases as the allowable rice area increases up to 30% of the farm area. Increases in rice area beyond 30% of the farm did not result in an increase in the percentage of area with a watertable within 1 m. Introduction of nonrice land uses also reduces the required subsurface drainage from 475 to 135 ml and increased the regional net return from AUD 542,000 to 1, 109,000. MAIZE, RCANOLA, and MZCA are the preferred nonrice land uses for this run.

Run 4

Reduction in RWLIM from 15 to 14 ml/ha reduces the area with a watertable within 1 m from 42 to 31%. These percentages are similar to the initial conditions. In an average rainfall year, if RWLIM is restricted to 14 ml/ha, the increase in the area with shallow watertables is minimal. Introduction of nonrice land uses reduces the area with a watertable within 1 m to 23%, reduces the subsurface drainage requirement from 333 to 60 ml, and increases the regional net return from AUD 439,000 to 1, 431, 000. MAIZE, MZCA, CANOLA, and RCANOLA are the preferred nonrice land uses for this run.

Run 5

Increasing RWLIM from 15 to 16 ml/ha increases the area with a watertable within 1 m from 42 to 53%. Introduction of nonrice land uses reduces the area with a watertable within 1 m from 53 to 48%, reduces the subsurface drainage requirement from 490 to 91 ml, and increases the regional net return from AUD 422,000 to 1,231,400. MAIZE, MZCA, CANOLA, and RCANOLA are the preferred nonrice land uses for this run.

Run 6

During a dry rainfall year (median annual rainfall 283 mm), the area with a watertable within 1 m increased to 42%. Introduction of nonrice land uses reduces the subsurface drainage requirement from 425 to 85 ml, and increases the regional net return from AUD 430,000 to 1,186,000. The area with a watertable within 1 m and the drainage requirements are identical to Run 1, though the net return with nonrice land uses is lower. Maybe this is due to a reduction in irrigated crops within the region. MAIZE, MZCA, CANOLA, and RCANOLA are the preferred nonrice land uses for this run.

Run 7

During a wet rainfall year (median annual rainfall 517 mm), the area with watertable within 1 m increases to 42%. Introduction of nonrice land uses reduces the subsurface drainage requirement from 425 to 75 ml, and increases the regional net return from AUD 430,000 to 1,587,000. The area with a watertable within 1 m, and the drainage requirements, are the same as Run 1. However, the net returns with nonrice land uses were higher. This is due to an increase in irrigated crops within the region. MAIZE, MZCA, CANOLA, and RCANOLA are the preferred nonrice land uses for this run.

Runs 8 and 9

As the *optimal depth* increases from 1 m to 1.25 and 1.5 m, so does the area with a watertable within the *optimal depth*. The drainage requirements, without and with nonrice land uses also increases. Increasing the *optimal depth* results in a reduction in net return in the region as a result of a reduction in the area of irrigated crops.

Runs 10 and 11

The framework results are not sensitive to the range of critical salinity values evaluated (1 to 2 dS/m).

Run 12

When the initial watertables are shallow, the area with a watertable within 1 m decreases from 90 to 76%. Introduction of nonrice land uses reduces the area with watertables within 1 m from 76 to 64%, decreases subsurface drainage from 925 to 280 ml and increases the regional net return from AUD 430,000 to 1,300,000.

Run 13 and 14

The framework results are very sensitive to the weights used in the multicriteria optimization module. When w1 was set to one (and w2 = 0), the regional net return is not significantly higher than that obtained in Run 1. But the drainage requirement with nonrice land uses changes from 75 to 170 ml. In contrast, when w1 was set to 0 and w2 to 1, the drainage requirement does not change, but the regional net return decreases from AUD 1,324,000 to 509,000. For Run 14 (w2 = 1.0), BARLEY, RBARLEY, and PAST are chosen.

Conclusions

The current level of rice production (30% of a farm area) as a monoculture increases the area with shallow watertables in an average rainfall year. The increase in area with a shallow watertable resulting from increasing rice area is not linear, and becomes asymptotic after 30% of the farm area is under rice. Introduction of nonrice land uses lowers the watertable and increase the net returns substantially. Growing maize and canola results in higher returns than other nonrice land uses.

The rice field water use limit has a significant impact on the area with a shallow watertable. In an average rainfall year, restriction of the rice field water use limit to 14 ml/ha restricts the watertable rise from rice cropping to the minimum. The introduction of maize and canola increases the net farm returns and reduces the subsurface drainage requirements. Conversely, an increase in rice field water use limit of 16 ml/ha raises the watertable and lowers net returns.

Net returns are lower in dry years due to a reduction in the area under irrigated land uses. The converse is true for wet years. Net returns are not affected when critical soil salinity is increased from 1 to 2 dS/m. An increase in the optimal depth results in reduced net returns and increased drainage volumes.

Increasing the weight for maximizing net returns (w1) from 0.5 to 1.0 does not change the net returns significantly, but increases the drainage requirements. In contrast, increasing the weight for maximizing discharge from 0.5 to 1.0 reduces the net returns significantly, but does not reduce the drainage requirement significantly.

References

Brooke, A., D. Kendrick, and A. Meeraus. 1988. *GAMS. A User Guide*. The Scientific Press, California. Page 289.

Prathapar, S.A. and J.C. Madden. 1995. Determining Optimal Intensity of Irrigation in the Murray Valley. CSIRO Division of Water Resources, Consultancy Report No 95/17.

Prathapar, S.A., W.S. Meyer, E. Alocilija, and J.C. Madden. 1995. SWAGMAN-Options, Final Report to the Rural Industries Research and Development Corporation. CSIRO Division of Water Resources, Consultancy Report No 95/42.

van Dijk. 1961. Soils of the Southern Portion of the Murrumbidgee Irrigation Areas. Soil and Land Use Series No. 40. CSIRO Australia.

Chapter 39

The Use of Crop Models for Land Valuation

F. Doležal, J. Lipavský, J. Křen, P. Novák, J. Šimon, J. Kubát, M. Pýcha, S. Cikánová, P. Klímová, V. Flašarová, and E. Pokorný

Abstract

It is suggested that the system of land valuation used hitherto in the Czech Republic might be improved by simulating, in a set of multiannual runs, the processes taking place on a particular site (crop, soil water, nutrients, etc.). The inputs to the simulation comprise the soil data (known from surveys or interpolated), weather (generated randomly on the basis of the known or interpolated climatic characteristics of the site), and the crop and management data (standardized so as to typify the common agricultural practices). Since the soil and climate are the only variable inputs, the site (i.e., the soil and the climate on a site) can be characterized by the vector of simulation outputs such as the average yields of common crops, their variability in time, the components of water and nutrient balance, the erosion phenomena, etc. A study of feasibility of this approach has started by validating models like EPIC, WOFOST, and the DSSAT crop models, using data from agriculture experiment stations. The second stage assumes that input data will be assembled and simulations carried out for known locations of the former soil exploration pits. Finally, a territory is to be continuously covered so that the quality of each site can be assessed.

Introduction

The existing system of land valuation in the Czech Republic (Group of authors, 1984–1990) is based on two basic data sources:

1. A detailed soil survey (Complex Soil Survey = CSS) carried out during 1960s and 1970s all over the country (Group of authors, 1967)
2. Investigations into economy and performance of large (collective or state-owned) agricultural enterprises carried out during 1970s and 1980s (Group of authors, 1984–1990)

The system is at present composed of:

- A site classification system, capable of ascribing to each spot of agricultural land a five-digit code (cf. Group of authors, 1984–1990, vol. 1) which expresses various aspects of its suitability for agriculture and to some extent also its environmental value. It incorporates the local climate, the main soil unit (the genetic soil type, texture, parent rock, permeability, drainage etc.), the soil profile depth, stoniness, slope, and exposure (N, S, E, W). The basic classification unit, distinguished by its five-digit code, is referred to as a "Valuated Soil-Ecological Unit = VSEU (the abbreviation "BPEJ" is used in Czech).
- The soil maps delineating the approximate spatial extent of patches occupied by individual VSEU at a scale of 1:5,000, gradually revised and digitized.
- A list of the official land prices (per square meter) ascribed to individual VSEU, occasionally renewed (Ministry of Finances, 1990–1994).
- The land cadastre, being rebuilt in order to ascribe to each parcel (of which the owner is recorded by a district cadastre office) a VSEU code.

The aim is that each parcel can be ascribed, at any time, an official price for taxation and land consolidation purposes, by combining, in a district cadastre office's computer database, the parcel identification code and the currently valid official land price list. While the land parcels cannot yet be purchased and sold unrestrictedly (because of the ongoing privatization and property restitution), it is generally accepted that the purchase and selling of the land shall take place at free-market prices.

However, it has been recognized that this system is not detailed and reliable enough, mainly because it had originally been designed for large agricultural enterprises and perhaps also because the genetic soil types, rather than the agronomic soil properties, serve as a basis of the classification. Main drawbacks of the system are as follows:

- A certain VSEU cannot be unambiguously characterized by a certain level of crop yields or a certain mode of environmental functioning.
- Many small pieces of land with properties different from those of the prevailing soil type were considered unimportant and were not mapped at all.

- The actual diffuse boundaries between individual VSEU were delineated in the maps as being sharp, with some degree of arbitrariness.
- The suitability of a land plot for small-scale sustainable agriculture, limited by environmental constraints, may differ from that for an intensive, large-scale agriculture.
- The nonagricultural aspects are not properly allowed for. For example, it is still a matter of dispute to which extent the agricultural land is to be protected against urbanization.

Because of these and other reasons, claims are frequently being raised against the (allegedly) incorrect land classification; the area under dispute has then to be reexamined by soil surveyors and the VSEU maps have to be redrawn, with results that may later appear to be deficient as well. Even worse, economic and environmental assessments of particular landscapes based on the VSEU data cannot be reliable enough simply because the information content of the VSEU classification is not rich enough.

In the meantime, several ways are being investigated how to improve the existing valuation system. It is within this framework that a study is being undertaken by the present authors in order to investigate the use of crop models for a more objective system of land valuation, open to future refinements. Besides the purpose of improving the land valuation system, the crop models combined with pedologic and climatic databases may serve several other purposes, such as the real-time forecasting of agricultural production or assistance in land use planning and nature protection by assessing the feasibility of several alternative agricultural systems in a region. In this sense, the study is genuinely multiobjective. Until now, few authors attempted to apply crop models directly to the issue of land valuation. An approach somewhat similar to ours was adopted in Austria by Stenitzer (1988).

Our basic approach is as follows: with the soil and climate data estimated for each site (using all available data sources) and assuming a standard crop rotation and management (set forth *a priori* so that they are typical for a given region and a given era), a package of selected crop models can be run to simulate a relatively long period (e.g., 50 years) of a steady-state and average-level arable agriculture on this site. The vector of most important simulation outputs, e.g., the means, standard deviations, and other statistics of yields of basic crops, of water and nitrogen balance components and, perhaps, of soil and phosphorus losses by erosion can be used to characterize the value of the site in a complex way. From this vector, the land price as well as many other conclusions may be derived, allowing for the market situation and other circumstances.

The spatial density of land valuation should ideally be such that a land parcel of arbitrary size is adequately assessed. A more realistic way, however, is to proceed in stages:

1. Validating the crop models using selected data from the field experiment stations
2. Creating the corresponding databases and running the simulations for one or several test regions, geographically and agriculturally homogeneous, each of them with area of roughly 300 to 500 km^2), in following steps:

a. Running the simulations for individual sites corresponding to the CSS special pits, of which the average spatial density is about one pit per 30 km^2
   b. The same, for individual sites corresponding to the CSS selected pits (with the spatial density of one pit per 0.7 to 1.8 km^2)
   c. The same, for the sites of the CSS basic pits (one pit per 7 to 18 ha)
   d. The same, for a virtually continuous coverage of the test regions, i.e., a dense discrete coverage allowing for continuous interpolation
3. A whole-country coverage of which the areal density would be gradually increased, in stages and according to practical needs, from that of (a) to that of (c) and (d).

This approach is now being attempted on a small scale in order to check its feasibility. We are yet in the (1)-stage of the above scheme but are preparing further stages, too. A purpose of this paper is to present the approach and to expose it to discussion.

Input Data

Agriculture

Standard crop rotations and management systems have to be set up such that they are typical for the areas to be investigated and stable enough to characterize the agriculture of the preceding period as well as that of the oncoming era. In addition, they should be simulable by the existing crop models. We suggest that a standard crop rotation applicable to most of the arable soils in the Czech Republic, except for the extreme mountainous regions, is the one used within the network of experiment stations of the Research Institute of Crop Production, with the variants allowing for different climatic subregions:

1. Alfalfa (or clover) sown into a cereal crop (spring barley) harvested as fodder
2. Alfalfa (or clover)
3. Winter wheat
4. Maize for grain or silage
5. Winter wheat
6. Spring barley
7. Maize for grain or silage (or sugar beet or potato)
8. Spring barley

The grassland, orchard, and forest soils are outside the scope of our study for the time being.

The doses of fertilizers used in Czech agriculture in the past were fairly high but have been reduced to unsustainably low levels during the postreform period after 1989 (cf. "Green Report", 1994). The standardized doses to be used in this study as inputs to crop models should, however, correspond to the reality expected in future (cf. Burda et al., 1995) and, therefore, should be considerably higher than the present-day averages. In order to keep the simulation strategy manageable,

Table 39.1 Standard Doses of Nitrogen and Phosphorus to be Used as Inputs to the Crop Models for the Purpose of Land Valuation (Crops are Ordered According to the Standard Rotation)

		Average annual doses of mineral fertilizers (kg of element per ha per year)	
Year	Crop	N	P
1	Alfalfa or clover sown into barley for fodder	40	20
2	Alfalfa or clover	0	30
3	Winter wheat	100	40
4,7	Maize for grain or silage	180	50
5	Winter wheat	120	40
6	Spring barley	90	40
7	Sugar beet	150	50
7	Potato	120	50
8	Spring barley	60	40

we exclude from consideration potassium, magnesium, and minor nutrients, assuming that their levels in the soil are always sufficient. Suggested standard doses of nitrogen and phosphorus are presented in Table 39.1. They represent a baseline from which the simulation inputs will deviate in both directions depending upon the actual soil nutrient storage; a feedback of this type will be incorporated in the simulation. For sugar beet and potato, parts of the doses indicated in Table 39.1 may be applied as organic fertilizers.

The liming of acid soils has also been drastically reduced during the last several years. This trend is expected to continue in future but some basic-level liming must be secured in order to keep the topsoil pH at an acceptable level. This can be modelled by adjusting, during a multiyear simulation run, the initial value of pH in the topsoil at the start of each season to a preset constant value. Alternatively, one can execute two parallel simulation runs for each site to demonstrate the effect which a long-lasting refrain from liming may have upon the value of the site. Alkaline and saline soils virtually do not occur in the Czech Republic and, therefore, the harmfully high pH values need not be considered.

Other aspects of agricultural management also have to be standardized, including the planting date, the density of plants, the dates and characteristics of fertilizer applications, and tillage operations. Algorithms have to be elaborated for automated calculation of these parameters depending upon weather, soil, and crop. On the other hand, the submodels of weeds, pests, pesticides, growth stimulators, etc. need not be included in the simulations, because they do not belong to the primary characteristics of the sites.

Soils

The soil properties required as inputs to the crop models have to be derived from the results of the CSS. Since the CSS data had already been used as a basis of the present land valuation system, a natural starting step would be to ascribe certain values of the basic soil properties to each VSEU in a unequivocal way,

Table 39.2 The List of Analyses and Descriptions Made for the Soil Exploration Pits during the CSS

	Category of pits		
Analyses and descriptions	Basic	Selected	Special
Particles < 0.01 mm	+	+	+
Full texture analysis	—	+	+
$CaCO_3$	—	+	+
pH(KCl)	+	+	+
Available P and K	—	+	+
C_{ox}	—	+	+
Cation exchange capacity and exchangeable H^+	—	+	+
$pH(H_2O)$	—	Some	+
Skeleton content	—	Some	+
Water extract	—	Some	+
Bulk density	—	—	+
Empirical hydrophysical characteristics	—	—	+
Retention curve	—	—	Some
Hydraulic conductivity	—	—	Some

Group of authors, 1967; Volume 3.

assuming that a soil exhibits these properties constantly all over the area of occurrence of this unit, with the properties changing abruptly at the boundaries between neighboring units. Attempts at characterizing the main soil units by their physical and chemical properties have already been undertaken (cf. Němeček, 1981; Očadlík and Čečetka, 1989). However, it is intervals of values rather than single values of the soil properties which result from these attempts, and the intervals indicated are in most cases too wide or indications given are only qualitative. It is therefore envisaged that the soil properties have to be ascribed to the sites of individual CSS exploration pits rather than to the soil classification units as such, which means that the primary CSS data (i.e., the analyses from pits and boreholes) have to be used and interpolated. A *basic* open exploration pit was dug during CSS on every 7 to 18 ha and a "*selected*" pit (with more analyses made) on every 70 to 180 ha, while a *special* pit, characterizing a typical pedon, was to represent on average about 3,000 to 4,000 ha of land. An abbreviated list of analyses available for these three categories of pits is given in Table 39.2. The interpolation procedure will be composed of two parts:

1. Estimation of the properties not available for a particular pit from other properties, known for this pit, using a regression derived from the data for pits of higher category
2. Interpolation between the basic pits, using information about relief and other features of the landscape (hydrographic network, artificial drainage, wetland indicators, etc.).

An algorithm which would carry out such interpolations automatically has not yet been devised. A semi-automated procedure is therefore envisaged, using semi-empirical concepts and regressions (cf., e. g., Romanova, 1977; Brubaker et al., 1994).

Weather

The climate at each point of the terrain is to be derived from the existing weather records of the neighboring climatological stations, by inter- and extrapolating them and by using suitable weather generators (like, for example, the WXGEN generator embodied in the EPIC model; cf. Sharpley and Williams, 1990). The climate interpolation has to be based, among other factors, on the latitude, longitude, and altitude of the site, on its slope and exposure, and on its position relative to the regional orography. The parameters subject to interpolation are the basic statistics of the daily weather element values, in particular those needed as inputs into the weather generator. Principally, they have to be regarded as functions of the Julian day or, at least, of the month number.

Irrigation

Under the climatic conditions prevailing in the Czech Republic, the irrigation of agricultural crops is only a supplemental measure, stabilizing crop yields but not always economically warranted. The influence of irrigation upon the value of a site may be expressed, e.g., by providing results of simulation for two parallel runs, one without irrigation (with water supplied by natural rainfall only) and the other with automatic irrigation triggered, e.g., by the values of the available moisture content in the root zone.

Selection of Crop Models

The expression "crop model" means, in the present context, an assembly of algorithms allowing a realistic simulation of the basic processes taking place on (and in) an arable soil, such as weather (if generated stochastically), the crop growth and development, the evapotranspiration, the balance and movement of soil water, the recharge of groundwater or capillary rise from groundwater, the soil anaerobiosis, the frost and snow phenomena, the organic matter and nitrogen cycling, the soil erosion (by both water and wind), and the phosphorus balance. Since the development of a complex crop model is a demanding task not accomplishable within a short period of time, we decided to adopt the existing models as far as possible. Bearing in mind the standard crop rotation listed above and the features of various models known to us, we selected, as a first choice, the following models:

- The crop submodels of the DSSAT package, in particular: CERES-Wheat, CERES-Barley, CERES-Maize, and SUBSTOR (for potato) (cf. IBSNAT Project, 1989; Tsuji et al., 1994) of which a characteristic feature is their explicit capability of a cultivar-specific simulation and of taking into account a heterogeneous soil profile; the crop, water, nitrogen, organic matter and pH submodels of them are of a balanced, medium-level complexity
- The WOFOST universal crop model (Supit et al., 1994), with a more elaborated crop growth and phenology part, a simple soil water submodel (capable, however, of modelling the oxygen stress) and without a dynamic nutrient submodel

- The EPIC model (Sharpley and Williams, 1990), with a medium-complexity crop submodel and a simple soil water submodel, but capable of simulating the surface runoff, the erosion phenomena, and the nitrogen and phosphorus balance

Results and Conclusions

The project is now in its first stage, i.e., the data from several field experiment stations are being used to test the crop models. The test data for three crops — spring barley, winter wheat, and maize — have been collected and are being completed in a form of the DSSAT input files. The DSSAT genetic coefficients have been estimated for some local cultivars of spring barley and winter wheat by combining the information contained in the DSSAT database with the knowledge accumulated in the Agricultural Research Institute Ltd. The CERES-Barley and CERES-Wheat simulation models are currently being validated. The existing soil database is being adapted for a representation and subsequent interpolation within an ARC/INFO geographic information system. The project is going on.

Acknowledgments

This study is being partly financed by the Grant Agency of the Czech Republic as a project no. 502/94/1033. The contribution of several persons to the ideas presented and the data sets used is gratefully acknowledged. These persons are, among others, Eva Fišerová, Jiří Janáček, Ladislav Ludva, Stanislav Mach, Jan Martinec, Miroslav Poruba, Zdeněk Tomiška, Vladislava Vaňková and Anna Žigová.

References

Brubaker, S.C., A.J. Jones, K. Frank, and D.T. Lewis. 1994. Regression models for estimating soil properties by landscape position. *Soil Sci. Soc. Am. J.* 58:1763–1767.

Burda, V., S. Komberec, and J. Lipavský. 1995. Economy of crop production. In Czech. A paper presented at a conference on "Plant Nutrition and Application of Fertilizers", Brno, Czech Republic, 13 September 1995. 10 pages.

FAO/UNESCO. 1988. Soil Map of the World. Revised Legend. World Soil Resources Report 60. FAO, Rome. 119 pages.

"Green Report." 1994. A Report on the State of Czech Agriculture, 1994. In Czech. Ministry of Agriculture of the Czech Republic, Prague. 344 pages.

Group of authors. 1967. Survey of Agricultural Soils of CSSR. In Czech. Volumes 1–3. Ministry of Agriculture and Food, Prague. 246+132+92 pages.

Group of authors. 1984–1990. *Valuation of Czechoslovak Agricultural Aoils and Trends of Their Use*. In Czech. Volumes 1–5. Published by a group of agricultural research institutes, Prague and Bratislava. 132+110+196+86+238 pages.

IBSNAT Project. 1989. *Decision Support System for Agrotechnology Transfer V2.1*. Department of Agronomy and Soil Science, College of Tropical Agricultural and Human Resources, University of Hawaii, Honolulu.

Ministry of Finances of the Czech Republic. 1990–1994. Public notices no. 316/1990, 393/1991 and 178/1994 on prices of structures, land parcels, permanent vegetation, etc. In Czech.

Němeček, J. 1981. Basic diagnostic features and classification of soils of CSR. In Czech. *Studies of the Czechoslovak Academy of Sciences,* No. 8/1981. Academia, Prague. 110 pages.

Očadlík, J. and J. Čecetka. 1989. The Background Data Needed to Set Forth Standards and Legislation Norms in the Field of Land Conservation and Use. In Czech. Final report, project no. P 06-329-813-07, stage 4. Research Institute for Improvement of Agricultural Soils, Prague. 155 pages.

Richardson, C.W. and A.D. Nicks. 1990. *Weather generator description.* Pages 93–104 in A.N. Sharpley and J.R. Williams, Eds. EPIC — Erosion/Productivity Impact Calculator: 1. Model Documentation. U.S. Department of Agriculture Technical Bulletin No. 1768.

Romanova, E. N. 1977. *Microclimatological Variability of the Basic Elements of Climate.* In Russian. Gidrometeoizdat, Leningrad. 300 pages.

Sharpley, A.N. and J.R. Williams, Eds. 1990. EPIC — Erosion/Productivity Impact Calculator: 1. Model Documentation. U.S. Department of Agriculture Technical Bulletin No. 1768. 235 pages.

Stenitzer, E. 1988. SIMWASER. Numerical Model for Simulating Soil Water Regime and Crop Yield on a Site. In German. Report No. 31 of the Federal Institute for Land Reclamation and Soil Water Regime, Petzenkirchen, Austria. 203 pages.

Supit, I., A.A. Hooijer, and C.A. van Diepen. 1994. *System Description of the WOFOST 6.0 Crop Simulation Model Implemented in CGMS. Volume 1. Theory and Algorithms.* Joint Research Centre, European Commission, Catalogue no. CL-NA-15956-EN-C. Office for Official Publications of the European Communities, Luxembourg. 146 pages.

Tsuji, G.Y., G. Uehara, and S. Balas, Eds. 1994. *DSSAT v3.* University of Hawaii, Honolulu.

Chapter 40

An Agroecological Approach to Sustainable Agriculture

ISTIQLAL AMIEN

Abstract

Information on land resources collected by national research centers is not fully utilized in planning agricultural development. To assist policy makers and planners in promoting appropriate agricultural systems an expert system has been developed which assesses the suitability of specific soil and climate conditions to support proper agricultural systems, and selection of crop choices. The input data required are slope, soil texture, acidity, and drainage, which taken together help identify suitable agricultural systems. With additional data on soil moisture and temperature regimes individual crops can be evaluated. Furthermore, if provided with information on water supply, the system can provide recommendations on cropping patterns for annual crops. With additional data on soil mineralogy the system indicates proper soil management practices, i.e., fertilizer management and soil tillage. If some elements of the data are unavailable they may often be inferred from general site/field characteristics. The expert system is designed to be user friendly and can be run using a DOS-based microcomputer with modest memory and disk requirements.

With the agroecological approach, relevant information on soil and climate can be fed into the expert system to delineate areas for conservation and production forests, perennial crop plantation, agroforestry, and annual crop farming. Further crop options for particular agricultural systems, as well as cropping patterns for annual crops, can be generated. The data inputs, and results in the form of digitized maps, are components of a geographical information system (GIS) which facilitates future utilization, updating, and improvement.

Introduction

The results of traditional methods of agricultural research are specific to location, season, cultivar, and management, and may apply only under specific experimental conditions. These methods are unlikely to achieve the ultimate goal of agricultural research: how to prescribe a technology that is appropriate to the land, labor, capital, and management resources of individual farmers (Nix, 1984). The goal is almost impossible to achieve if we do not shift from partial or fragmented research approaches to the holistic approach of integrated, system-based research. A system-based research strategy can be supported by the development of two complementary analytical components: (1) inventories of available resources, usually summarized in database formats and (2) expert systems or crop simulation models that can be used to investigate the potential impact of technologies in terms of both potential production and environmental impact. Simulation models attempt to reproduce mathematically the essential physical, chemical, and biological processes which control plant growth. With simulation models, crop yields can be estimated as a consequence of specific combinations of climate, soil, and management. However, crop simulation models are data intensive and require extensive calibration with many combinations of cultivar and geographical area. Expert systems, on the other hand, seek to capture the best available expert knowledge as descriptive, semi-quantitative or quantitative rules which can be applied in a broad range of circumstances to guide decision making.

Following the early example of the international agricultural research centers in the 1960s, agricultural research in Indonesia is organized on the basis of commodities. The suitability of crops in an area or the transferability of technology between areas can be assessed only when adequate information concerning the requirements of the crop and the nature of the local agroecological conditions are known. Land, soil, hydrology, and, to some extent, climate information have been collected for many parts of the country. However, the use of these data in making decisions in agricultural development is very limited. A proper understanding of the natural resources of any area is imperative when attempting to promote sustainable agricultural development. Delineating agroecological zones on the basis of terrain, soil, hydrology, and climate facilitates crop selection and agrotechnology transfer. With its greater emphasis on the capability of land resources to support specific types of agricultural development, the agroecological approach can improve the efficiency of research and the potential impact of technologies generated by research. In this way, the agroecological approach can help make better use of research resources and reduce research and development time lags.

Climate and Land Resources

Traditionally, land resource surveys and research have placed much greater emphasis on the soil aspects of land, with relatively little regard to climate. Consequently climate resource inventories are seldom associated with soil information. However, climate and soil are closely related. Climate is an important factor in soil formation. Because soil formation is a long-term process, general climatic information on an annual basis is adequate; but to assess whether crops or other agriculture commodities are suitable in a particular climate, more detailed climatic information is required. Monthly data for a period of 20 to 30 years are considered insufficient. In predicting crop performance at the field level, daily weather data are required. Time of planting strongly affects performance and yield of annual crops. The weather variables most affecting crop growth are rainfall, maximum and minimum temperature, and solar radiation. These data can be used in the simulation of dynamic plant growth processes such as photosynthesis and respiration.

Weather and climate information that is limited in quantity, and poor in quality, needs to be enriched to improve its contribution at the detailed planning level. Many soil surveys at the semi-detailed or detail reconnaissance scale provide only general climatic information. With such limited information only crop suitability ratings and no detailed crop performance modelling is possible. If such survey results were supplemented with adequate weather information they could provide better insights into plant management options such as cropping patterns and time of planting, as well as soil management options such as fertilizer inputs or the need for supplementary water application.

Technology recommendations appropriate at the field level can only be made if the decision-making processes of farmers, and the limits imposed on them by the availability of production resources, are well understood. As part of this understanding, the availability of a resource inventory covering both soil and climate is imperative. The Center for Soil and Agroclimate Research, established in 1905, has conducted many soil surveys, but only in 1986 began developing a soil database. The soil information is being linked to a GIS for easy retrieval and updating as more recent data are obtained. At the present moment the database contains only land resources data for Sumatra at a reconnaissance scale. It will be an enormous task to computerize the land resource data obtained from surveys in other areas, in addition to the storage of data from on-going and planned surveys.

Agriculture Systems

Sustainable agriculture can be attained only when proper land utilization is practiced. When land is utilized improperly, productivity rapidly decreases and the ecosystem is jeopardized. Proper land use ensures that people can benefit from nature during their lifetime and also ensures the resources can be used for future generations. Integrated land use management should be directed at an optimal, sustainable use of natural resources. This may include regulation of hydrology, climate stabilization, and preservation of biodiversity such as gene pools, wild life and plant habitats, as well as appropriate research and education. Proper land utilization improves efficiency in the production process because relatively fewer inputs are required to attain the desired output, thus making production more economically profitable.

Forest or permanent tree crops provide natural land cover in the upper watershed. In coastal regions mangrove forests also need to be conserved to protect the coast from being eroded by the sea. In cold or dry areas there are no forests with the variety of trees found in warm humid areas. There is a growing awareness that native vegetation — savanna, shrubs, and trees — needs to be conserved.

Slope is a critical factor because erosion and soil degradation are a real threat to agriculture in hilly regions. Agriculture on steep slopes also limits the use of agricultural machinery and draft animals in soil tillage. Therefore, in such areas, it is usual to grow perennial crops such as forestry, tree plantations, or pasture. Apart from erosion and land degradation problems, energy efficiency in the long run needs to be considered. Labor required to transport agricultural inputs to the farm, and produce out of the farm, will become very expensive. Labor intensive agriculture on sloping lands is unlikely feasible when labor costs are relatively low.

High value horticultural crops such as ornamental crops or vegetables are commonly cultivated on terraced fields in mountainous regions. However, terracing is not feasible on all soils. Soils with loose parent material, like sandstone, are prone to landslide when terraced. Terracing highly weathered acid tropical soils will expose the high aluminum, infertile subsoil and reduce cropping options.

If the soil is suitable, annual crop agriculture is recommended when the slope is about 8% or less. Although the slope does not rule out agriculture on a legal basis, it is not recommended when soils are formed from quartz sand and deep peat, or are high in gravel or stones, making soil tillage difficult. Soil with *'cat clay'* close to the surface can be reclaimed only when reduced conditions can be maintained i.e., when used as paddy fields, or should be left in its natural condition for *'gelam' (Melaleuca leucadendra)* forest. Land with slopes between 8 to 15% is recommended for agroforesty uses in which annual crops are cultivated along with perennial trees. Land with slopes in the range of 15 to 40% are for permanent crops such as perennial tree plantations or forestry.

Sustainable agriculture is determined not only by the condition of the physical environment but also by social and economic aspects, infrastructure, and government policy. The latter becomes more significant with the integration of our economy into international trade and the world economy in general. To be sustainable a technology must not only be technically sound, but it must also be environmentally safe and economically feasible. On other hand, however profitable an agricultural enterprise is, it will not endure if the physical environment is degraded and jeopardized.

Upland agriculture on tropical soils that are generally acidic with low fertility should focus on perennial crops such as tree plantations or pasture rather than on traditional food crops (Amien, 1990). This approach is in line with the need to improve the nutritional value of food to increase protein, vitamin, and mineral intake from diverse agricultural products. By maintaining crop cover, land degradation due to erosion can be reduced. Decomposed plant residues in roots and leaves will improve the soil's physical condition and increase the percolation rate, reduce runoff, and eventually diminish erosion. Organic material also improves soil fertility by direct nutrient supply or by increasing nutrient absorption. Green manure from legume crop residues is reported to detoxify aluminum in acid soils (Hue and Amien, 1989).

Continuous annual food crop cultivation under low input conditions will collapse within a relatively short period, mainly due to weed infestation (Sanchez et al., 1987). This can be seen in the increasing alang-alang *(Imperata cylindrica)* weed infested lands, now abandoned but which were previously cultivated with annual food crops. Continuous monoculture of the same crop does not produce sustained yield because of pathogen buildups (Valverde and Bandy, 1982).

Several ways have been suggested to maintain the productivity of upland farming on acid soils. Von Uexkull (1982, 1984) suggested a low-cost management system based on the establishment of a leguminous cover crop and adoption of a variant of shifting cultivation within the crop cover area. Another way to restore and maintain the soil fertility is alley cropping. This system is based on perennial trees or hedges, using legumes as a source of green manure as well as feed for livestock in farming systems. Trees or hedges are grown along the contours between the annual food crop fields. The deep-rooted, leguminous trees function to recycle the leached nutrients that can not be reached by shallow-rooted food crops as well as preventing erosion. Nevertheless, the long time lapse between planting the trees until they can be pruned for fodder and manure, and the corresponding labor requirements make the approach less attractive. Furthermore, it is still unclear whether the trees or hedges can function well to recycle leached nutrients if they are frequently pruned (Amien, 1990).

Agroforestry is the term given to any type of farming involving trees. Although agroforestry is widely practiced by indigenous people, it is the least studied of all tropical agricultural systems. Research on agroforestry will require an interdisciplinary ecological approach that must include agronomists, anthropologists, geographers, rural sociologists as well as economists. It also implies a perceptual change, for a sustained agriculture yield, and necessitates an integrative approach rather than a substitute orientation (Hecht, 1982). There is still much to be studied on basic agroforestry techniques, like the best plant combinations and their geometric arrangement, which probably will vary for different sets of environments. Interaction between different plant species is usually site-specific, making it difficult to generalize from isolated studies. As technology stands today, it can be regarded as a promising field of research in the humid tropics, but by no means as a system ready to be widely recommended for promoting agricultural development (Alvim, 1982).

Crop Suitability and Cropping Pattern for Annual Crops

Selection of species and cultivars can be assessed based on soil and climatic conditions. The crops suitablity in specific locations is determined by slope, soil texture, and acidity as well as moisture and temperature regimes. Crops that are suitable in a set of environment conditions require less inputs compared to those which are less suitable. Unsuitable crops will yield less, and with lower quality that might be difficult to market and, in long run, the land use may not be economically feasible.

Crop growth is generally constrained by excess or shortage of water and extreme temperatures. Soil constraints are usually easier and less expensive to alleviate. A combination of temperature and moisture regimes stratify the environment into classes in which particular crops can be found to be suitable (Eswaran, 1984).

In annual crop agriculture suitable crops and cropping patterns are determined by water supply as calculated by the number of wet months, i.e., where rainfall exceeds evapotranspiration and other losses. Based on information of moisture and temperature regimes, the selection of perennial crops and hedgerows in alley cropping is also possible.

Zones with sufficient rainfall such as those with *'udic'* moisture regime, can be cultivated with at least three crops annually. Because rice is the main staple food, it is recommended to be cultivated at the beginning of the rainy season. Rice can be intercropped with maize and cassava, followed by other secondary crops. After harvest a more drought-tolerant crop such as cowpea is planted to avoid crop failure in the case of drought. At the end of the dry season cassava will be harvested in order to improve soil aeration as an alternative means of soil tillage for the next crop.

Areas with *'ustic'* moisture regimes, where irrigation is unavailable, can usually be cropped at least twice a year. Rice and maize are recommended early in the rainy season followed by secondary crops, particularly grain legumes. Grain legume crops for soils without acidity problems are normally soybean or mungbean, but peanut or cowpea are grown in acid soils. Peanut requires calcium, particularly during pod filling. Calcium is usually low in acid soils, but because it is needed as a fertilizer rather than a soil conditioner small amounts of either lime or gypsum are adequate.

In semi-aridic moisture regimes rain occurs in only a few months in a year that only one crop is possible. For this area short-duration, drought-tolerant crops such as sorghum or some cultivars of maize are recommended, as also are annual or perennial pigeon pea. The deep rooted pigeon pea can extract water from the relatively moist subsoil during the dry season. Some legume forages such as *'Stylosanthes'* are also known to be drought-tolerant.

Expert Systems

Computer technology has played a large role in agrotechnology transfer and land evaluation. The role of computer technology in database management, simulation, GIS, and expert systems has gained wide acceptance during the last decade (Jones et al., 1986). Expert systems belong to the field of artificial intelligence. The name refers to the use of computers to solve problems in ways that would be natural to humans (Waterman, 1986). An expert system can be considered as a computing system which uses organized knowledge about a specific field to solve a problem. Expert systems can be developed for diagnostics, classification, decision making, tutoring, retrieving information, etc. Although traditional computer programming techniques have been used to solve these types of problems, a major difference exists. In traditional programming, the problem-solving logic and knowledge are integrated together in the program code and become hidden from all but the programmers. In expert systems, the knowledge remains separate and easily accessible to the user, the human expert, and the programmers. This feature of expert systems allows application developers to focus on knowledge presentation (Jones et al., 1986).

The most innovative part of expert systems is the ease with which a variety of information can be represented and used in decision making. This information ranges from quantitative information, including statistical relationships such as regression equations and physical and chemical laws, to less precise general rules of thumb or hunches that have been developed through hard gained experience in the field. The first type of information is referred to as *'algorithmic'* and the second as *'heuristic'*. The latter type of information can now be preserved and utilized in a more systematic way (Yost et al., 1986).

The knowledge base of an expert system can be represented by production rules or frames. Descriptive types of knowledge such as hierarchical classification are best represented by frames, while more *'heuristic'* types are better represented by rules. An expert system operates on a knowledge base of production rules, searching through the rules in an organized way to arrive at a conclusion. In a rule-based expert system there are two primary approaches: data-driven or forward chaining and hypothesis-driven or backward chaining (Waterman, 1986).

Expert systems can also be connected to databases through external programs. This is particularly useful in making decisions or recommendations when only limited data are available from the user. By having access to a database, the expert system can infer site conditions from available data by considering, for example, its geographic location.

Expert Systems for Crop Suitability and Agricultural System Selection in the Tropics

The expert system for crop suitability analysis and production system selection described here has been developed in view of the scarcity of soil and climate data in Indonesia (Amien, 1986). The expert system tries to provide recommendations on appropriate agricultural systems based on land characteristics such as slope, soil texture, acidity, and drainage. Given such factors as steep slope, very coarse soil texture or deep peat, or very low pH as limitations the system will give recommendations on different agricultural systems including paddy rice, annual upland crops, permanent crops, and forestry.

If the crop suitability assessment mode is selected, the system will request additional data on moisture and temperature regimes. Options will be given for a wide range of cereals, root crops, grain legumes, fiber crops, oil crops, beverage crops, vegetables, fruit crops, and cash crops such as sugarcane, tobacco, rubber, and pepper, based on soil and climatic conditions of the land. When the land is just suitable for forestry, ranges of timber species and other forest species are also provided. Crop suitability is generally limited by inadequate or excessive water or extreme temperature.

Furthermore, the system is capable of suggesting cropping patterns when provided with input data on water supply. Although not always accurate, water supply can be inferred from drainage and the number of consecutive months with rainfall more than 100 mm. Recommendations on methods of P and K fertilizer application are given based on soil clay mineralogy. Information on soil mineralogy is also the basis for recommending other soil management options such as organic manuring and appropriate methods of soil tillage. The system also provides cautions on problem soils such as potential acid sulfate soil and soil with alkaline pH. An overview of the expert system is presented in Figure 40.1.

Small scale soil or land unit maps usually have inadequate information to apply traditional land evaluation systems for assessing crop suitability. When the user does not have enough data to feed the system, the system calls upon *'heuristic'* information gained from experience in the field. The system tries to infer most of the input data by asking simple questions on specific soil and land characteristics. Approximate soil acidity is inferred from native vegetation; moisture regime is inferred from soil drainage and consecutive dry months. Poor or ponded

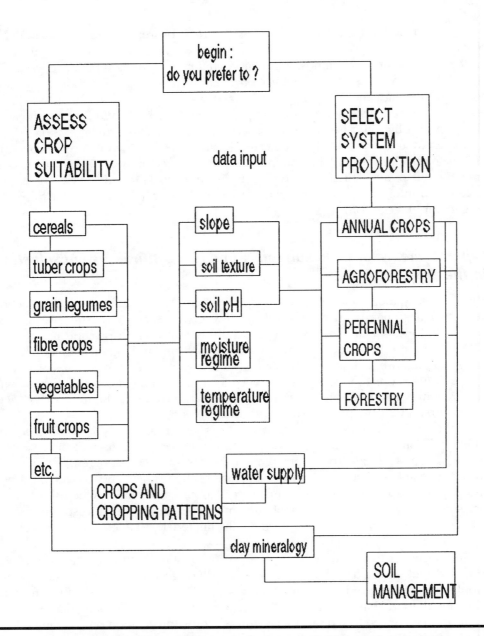

Figure 40.1 Overview of the expert system for crops suitability and agriculture system in the tropics.

drainage indicates that the soil has an *'aquic'* moisture regime. Number of consecutive dry months will determine whether the moisture regime is *'udic'*, *'ustic'* or *'semi-aridic'*. Because the system is designed for tropical regions, the soil temperature regime is inferred from elevation of the land. Elevation less than 750 m above sea level is considered *'isohyperthermic'*, between 750 to 2,000 m is *'isothermic'* and above 2,000 m is *'isomesic'*.

A more complicated inference is involved when determining soil clay mineralogy. This is based on several distinct soil characteristics such as parent material, soil texture, color of the subsoil, and whether or not the soil is cracked in the dry season. Volcanic ash material characterizes noncrystalline amorphous oxides and cracked soil is a sign of smectitic minerals. The color of surface soils is usually darker than subsoil because of higher organic matter content. Therefore, input requested by the system includes the color of subsoil. Oxidic minerals make tile color of subsoil reddish or yellowish. The information on soil clay mineralogy is very valuable in soil management.

Hardware Requirements

Ideally the expert system should have a large, comprehensive knowledge base to provide most detailed information on land use and soil management. However, a larger system requires more sophisticated hardware and development capabilities that is available in many places. Assuming that end users, other than policy makers or planners, are most likely to be extension workers with limited access to computers, the system developed was designed to be compact.

The expert system can be run on any DOS-compatible personal computer with 640 kB RAM (or only with 252 kB RAM if the computer is equipped with a math coprocessor). Because the system does not contain any graphics, no graphic card is required. To make it easy for beginners a batch file has been prepared, so that the system directs the user what to do after turning the computer on.

The system is very user friendly and can become a media for learning soil science, in particular, or agriculture, in general, as well as learning about computer operations. When something is not clear about the question on the screen, an explanation can be obtained by typing the question mark. The rules that require input will be displayed when 'WHY' is typed.

Agroecological Zones and Sustainable Agriculture Development

Sumatra

Sumatra is the second largest island of Indonesia with a land area almost 474,000 km^2 or almost a quarter of the country's total area. Located between 6° North and 6° South and between 95° 1' and 108° 16' East, the island belongs in the equatorial humid tropics. Because of its vicinity to Java, particularly to the capital city Jakarta, Sumatra is the most developed region in agriculture and industry outside Java. Agricultural lands cover almost every part of Sumatra with plantations of oil palm, pepper, coffee, rubber, etc. Favorable climate, especially well-distributed high rainfall, facilitated agricultural development. Agriculture-related industry flourishes from fertilizer plants to paper pulp, and from oil palm and rubber processing to pineapple canning. This is supported by abundant energy resources such as oil, natural gas and coal, as well as hydroelectric power.

Mainly on the basis of relief, and considering that Sumatra has a rather uniformly warm and humid climate throughout the year, Sumatra is divided into six natural regions (Scholz, 1983) which are (1) West Coastal Strip, (2) Bukit Barisan Mountain Zone, (3) Piedmont Zone, (4) Peneplain Zone, (5) Eastern Lowlands, and (6) Western and Eastern Island Chains.

The Western Coastal Strip, Mountain Zone, Piedmont, and the Eastern Lowland as well as the Western Island Chain stretch from the Northern tip of Aceh Province to the south in Lampung Province. The Peneplain occupies areas from the Province of North Sumatra to Lampung. The Eastern Island Chains are mainly situated in Riau and South Sumatra Provinces. The peat and organic soils that occupy about 22% of Sumatra are found in the Western Coastal Strip and Eastern Lowlands. Peat is also found inland on the Peneplain in Jambi and on the mountainous zone of Dairi in North Sumatra. The dominant soils of Sumatra are Ultisols and Inceptisols which together occupy about 47% of the total area. Oxisols are found in about 14% of the total area and Andisols, comprising about 5% of the total area, are found in mountainous zone and the adjacent piedmont. Moisture regimes are either udic or aquic with average rainfall above 2,000 mm annually. The temperature regimes are isohyperthermic in the lowlands and peneplain island chains, and isomesic at high altitudes.

Based on land, soil characteristics, and climatic data, Sumatra is divided into six zones (Table 40.1). As recommended by the expert system zones I, V, and VI must be allocated for forestry to protect natural resources and environment in general, zone II can only be utilized for permanent crops, zone III for agroforestry, and zone IV for annual crop agriculture. Lands are designated for forestry because of very steep slopes or because of poor quality of the soils, which are only fit for conservation forest with native vegetation, or for buffer forest along the coast containing mangroves and other vegetation (such as 'gelam' and 'nipah') on the marine soils. For production forest on the deep peat the possible tree species are 'damar' (*Agathas* sp.),.'ramin' (Gonystyllus bancanus) and 'kruing' (Dipterocarpus sp.).

Selection of crops in permanent crops zone is based on soil and moisture regimes. Because Sumatra is almost uniformly humid, plant performance is limited more by the land elevation. On lands with elevation of up to 750 m above sea level, the crop options are rubber, coconut, oil palm, and fruit trees such as rambutan, durian, jack fruit, banana, and many others. In those areas with elevations of more than 750 m above sea level the crop options are tea, citrus, cinnamon, etc.

The agroforestry zone is also divided into low and high elevations with conventional food crops such as upland rice, maize, peanut, cowpea, and cassava in the warm lowland areas and a wide assortment of vegetables in the cool highlands, cultivated along with tree crops such as 'sengon' (Paraserianthes falcataria) or acacia and other multipurpose trees.

The annual agriculture system is divided into wetland and upland agriculture. The wetland is suitable for rice paddy and in the uplands the warm temperature varieties of cereals, root crops, and grain legumes can be cultivated in multiple cropping systems. Those crops with cool temperature needs are suitable for highland production, e.g., vegetables such as cabbage, carrot, lettuce, beans, etc.

According to the forestry land use consensus, forestry lands that consist of conservation forest, wildlife sanctuary, and production forest in Sumatra are 55,380,000 ha (BPS, 1993). However, the equivalent areas suggested on the basis of agroecological zoning are 14,662,000 ha in mountainous upstream, areas, 1,967,000 ha for buffer forest in the coastal area, and 3,043,000 ha for swamp forest. These suggested values represent only about 41% of the total forest area. This implies that more than 25,000,000 ha of the current forest lands can be converted into perennial tree plantation, agroforestry, or annual crop agriculture.

Table 40.1 Agroecological Zones of Sumatra and Java

Subzone symbol	Elevation (m)	Temperature regimes	Moisture regimes	Physiography	Slope (%)	USDA soil suborder	Drainage
Iax	50–750	Warm	Humid	Volcanic, Hilly	>40	Udults, Udands Ustults, Udalfs Tropepts	Good
Iay*			Somewhat dry				
Ibx	750–2,000	Cool	Humid	Mountain/plateau			
Iby*			Somewhat dry				
IIax	40–750	Warm	Humid	Volcanic, Hilly	16–40	Udults, Udands Udalfs, Tropepts	Good
IIay*			Somewhat dry				
IIaz*			Dry				
IIbx	750–2,000	Cool	Humid	Mountain/Plateau			
			Somewhat dry				
IIIax	50–750	Warm	Humid	Plain, Volcanic, Hilly	8–15	Udults, Udands Udalfs, Tropepts Uderts	Good
IIIay*			Somewhat dry				
IIIaz*			Dry				
IIIbx	750–2,000	Cool	Humid				
IIIby*			Somewhat dry				
IVax1	10–750	Warm	Wet	Alluvial, Marine	<8	Aquents, Aquepts Aquants	Poor
IVby1*			Somewhat dry				
IVbx1	750–2,000	Cool	Humid				
IVax2	10–750	Warm	Humid	Plain Volcanic	<8	Udults, Udands Tropepts, Uderts	Good
IVay2*			Somewhat dry				
IVbx2	750–2,000	Cool	Humid				
V	0–50	Warm	Wet	Marine, alluvial	<3	Aquents, Aquepts	Poor
VI**	0–50	Warm	Wet	Peat dome	<3	Saprist, Hemists	Poor
VII*	0–50	Warm	Humid	Marine	<3	Psamment	Poor

*Found only in Java, **Found only in Sumatra.

On other hand, public and private plantations cover 5,159,891 ha, almost 8 million ha less then the recommended 6,501,000 ha for permanent crops and 6,667,100 ha for agroforestry. Although small-scale plantations and agroforestry have been growing in the last decades, there are still large tracts of land available for tree plantation or annual crop agriculture.

During the population increase of 8.47 million from 1980 to 1991 in Sumatra, agricultural lands increased with 2.68 million ha of perennial crop plantations, 1.35 million ha of annual upland crop agriculture, and 0.49 million ha of rice. Agroecological delineation indicates that millions of hectares of land are still available for agriculture development. However, optimal land utilization through the development of sustainable agriculture in Sumatra is hindered by land classification system where 65% of the lands are within the Department of Forestry land boundry although about 30% of them are not under forest cover (World Bank, 1990).

Java

Java island is among the most densely populated areas in the world. A chain of volcanoes, some still active, like Galunggung in West Java, Merapi in Central Java, and Kelud in East Java have enriched the soils to become fertile agricultural lands. Since ancient times Java has been a center of educational, economic, cultural, and political activities in Indonesia. Until recently, Java has been the main producer of rice and sugar in the country. The fertile plain north of the island along the basins of Citarum, Cimanuk, Jragung, Tuntang, Serang, Lusi, Juana, Bengawan Solo, and Brantas are intensively cultivated throughout the year. To the south, fertile agricultural lands have been developed along the basins of Cimandiri, Citandui, Serayu, Bogowonto, and Progo.

However, because the increasing agrarian population has exceeded the available lands on the plains, people have invaded the hilly and mountainous regions. If farmed without conservation measures this causes erosion and increased flooding. For that reason many reservoirs have been constructed to manage water resources for irrigation and sanitation as well as hydroelectric power. About 60% of the population of Indonesia live in Java and Madura, and almost all of the land is being utilized for agriculture (75%). With a population growth of 1.7% annually, families that depend upon agriculture will increase by about 150,000 annually. This results in the conversion of forest land at the rate of 18,000 ha annually with about 40,000 ha of agricultural lands being used for residential and industrial purposes (Word Bank, 1991).

The well-developed infrastructure and the availability of skilled human resources have made Java the center for industrial expansion that has taken place in many of the fertile agricultural lands. The increasing population of Java is augmented by migration from outer islands attracted by educational, economic, and other social facilities. In particular, population has increased rapidly in urban areas. Three of the biggest urban centers in Indonesia are located on Java. Transmigration and family planning programs are the principal development strategies aimed to overcome these problems.

About two-thirds of Java is hilly and mountain zones, with only one-third being relatively flat. On the sloping lands upland agriculture is the primary undertaking. Upland agriculture differs from lowland rice farming because it is characterized by diverse crops, low productivity and low earnings, less off-farm employment opportunity, and poor infrastructure such as roads and extension services.

Agricultural production plays a greater role in regional development when it generates not only raw materials, but also value-added processed goods from agroindustries. With good planning, such as cultivar standardization within areas of sufficient economic size, future investments in agriculture-based processing industries will offer more income generating and employment opportunities.

Java is divided into six zones, each further divided into six subzones (Table 40.1). In contrast with Sumatra, Java has no significant areas of peat soil, but has a thin strip of sand dunes on the southern coast. This strip, as well as part of the marine soils on the northern coast, must be utilized as buffer forest to protect the coast from erosion. Java and Madura have more diverse temperature regimes with the existence of *ustic* and *semi-aridic* moisture regimes.

As recommended by the expert system, crop options for each zone are more diverse than Sumatra; for example forestry systems can include both teak and mahogany, and perennial tree systems can include *Arabica* coffee in the highlands and *Robusta* coffee in the warm zones, as well as sugarcane, cotton, and tobacco. In areas with a distinct dry season fruit crops also provide greater possibilities, e.g., mango, watermelon, and *'blewah'* (*Cucumis melo*) in the warm areas and apple, grape, lychee (*Litchi sinensis*), and *'longan'* (*Caesalpinia crista*) in the highlands. Annual crops include maize and sorghum for cereals, soybean, peanut, and pigeon pea for grain legumes and *'ganyong'* (*Canna edulis*) and *'garut'* (*Maranta arumdinacea*) for tuber crops. The alternative agricultural systems and crop options for Sumatra and Java are presented in Table 40.2.

Because of the high population density, much of the land on Java that should have been preserved for forestry and conservation purposes has already been utilized for agriculture. The forest area, based on forestry land use consensus in Java and Madura, is 6,026,000 ha (BPS, 1993) and because of increasing agrarian populations, these areas will undoubtedly continue to diminish. The lands currently within the forestry department boundary in Java are about 22% or about 800,000 ha less than the recommended. Further, many of the designated forestry lands are not under forest cover and the geographic position of parts the forest that were inherited from colonial time in down stream area makes it less effective to function hydrologically.

From 1980 to 1991, with an increasing population of 16.31 million in Java, agriculture lands have expanded with 0.04 million ha of tree plantations and 0.19 million ha of annual upland crops agriculture while the rice fields have decreased by 0.08 million ha. High population pressure and industrial development resulted in conversion of agricultural land, mostly irrigated rice fields, into nonagricultural uses. To compensate the losses, 0.49 million ha of annual crop agriculture has expanded to steep lands with slopes over 15%. Although most of this sloping land is terraced, this improper land utilization causes erosion, land degradation, and disrupts the hydrologic function of the watershed resulting in frequent floods and droughts.

Intensive agriculture that has expanded into steeply sloping areas, often caused disasters such as landslides and floods. Agriculture systems on steep land should

Table 40.2 Agricultural Development Alternatives in Sumatra and Java

Subzone symbol	Agriculture system	Crop options	Sumatra km²	%	Java km²	%
Iax	Forestry	Shorea sp., Acacia mangium, Agathis sp., Peronena canescens, teak, mahogany, Dalbergia latifolia, Paraserianthes, Casuarina, Pines	146.220	30.9	34,120	25.7
Iay						
Ibx						
Iby						
Iax	Tree plantation	Coconut. oil palm, rubber and tropical fruits	50.170	10.6	27,107	20.4
IIay		Robusta coffee			700	0.5
IIaz					500	0.4
IIbx		Tea, chinchona, cinnamon, and high land fruits	14.840	3.1	5,170	3.9
IIby		Arabica coffee			1,200	0.9
IIIax	Agroforestry	Legumes trees and shrub rice, peanut, cowpea, cassava, and vegetables	56.110	11.8	9,950	7.5
IIIa		Legume, tree, and shrub maize, peanut, soybean			3,250	2.4
IIIaz		Pigeon pea, sorghum			100	0.1
IIIbx		Legumes tree and shrub high land vegetables	10.610	2.2	3,450	2.6
IVax1	Wetland agriculture	Paddy rice	48.630	10.3	23,380	17.6
IVbx1		High land vegetables	320	0.1	500	0.4
IVax2	Upland agriculture	Rice, maize, peanut, cassava, sweet potato	84.350	17.8	19,460	14.7
IVay2		Pigeon pea, sorghum			100	0.1
IVbx2		High land vegetables	12.260	2.6	500	0.4
V	Forestry	Mangrove	19.670	4.2	3,000	2.3
VI	Forestry	Damar, ramin, keruing sawo kecik	30.430	6.4		
VII	Forestry	Pandanus			100	0.1
Total			473.610	100.0	132,587	100.0

be permanently cropped to protect the soils with the crop's canopy and rooting system. Erosion is caused mainly by the direct impact of rain drops onto the soil surface while the rooting system of trees helps to prevent landslides. By selecting high value crops and by the development of agroindustry, the well being of the people can be improved in the long run.

Sloping land can be protected from erosion if a continuous vegetation cover exists, particularly during the rainy season. Other than perennial trees, many of forage grasses also serve well to prevent erosion because of their rooting systems and dense plant cover at the soil surface. In many studies grass strips have proved to be more effective in preventing erosion than tree crops. Because of their relatively short canopy, grasses do not compete with food crops for sun light in multiple cropping systems. Many of the upland plots in Java have been terraced and the planting of grass strips on the edge of the terrace and terrace raiser serves to strengthen and stabilize them.

The role of forage grasses as feed will become more significant, in an attempt to improve animal nutrition and income opportunities associated with the development of animal husbandry in rural areas. Feed supply from the residue of annual food crops is inadequate. In terrace stabilization programs, forage grass and legumes are cultivated in order to provide better quality feed. The excrement of the livestock can be used as a manure after tapping biogas energy in anaerobic digesters. This also reduces the consumption of fuelwood and that, in turn, will help improve the success of tree planting on hilly lands. Utilization of alternative energy sources such as biogas will become more attractive because of its clean nature, high calorific value, and minimal transportation needs (Amien, 1983).

Database and GIS

Considering the limited quality and quantity of existing data, this first approximation of the agroecological zones of Sumatra, Java, and Madura will be refined as improved data become available. Improvement will be facilitated by storing basic data and analytical results in the database along with digitized maps in a GIS. At a scale of 1:1,000,000 the agrogroecological zones map is appropriate as a basis for national scale agricultural development planning, therefore the database described here has been developed on the more pragmatic basis of a 'minimum database'. The database contains information of terrain, soil, and climate on a mapping unit polygon basis that consists of slope, soil texture, acidity, drainage, elevation and temperature regime, moisture regime and soil suborder based on soil taxonomy. The database also contains administrative locations and the alternative agricultural commodities that are suitable in each area. The agriculture commodities cover an assortment of cereals, root crops, grain legumes, oil, fiber and beverage crops, vegetables, fruit crops, as well as cash crops such as sugarcane, tobacco, rubber, and pepper.

REFERENCES

Alvim, P de T. 1982. A perspective appraisal of perennial crops in the Amazon Basin. Pages 311–328 in S.B Hetch, Ed. Amazonia: Agriculture and Land Use Research. CIAT, Cali, Colombia.

Amien, L.I. 1983. Pupuk organik dan biogas dari limbah pertanian. J. Penelitian dan Pengembangan Pertanian. 2(2):58–63.

Amien, L.I. 1986. Expert system for crops suitability and agriculture systems in the tropics. IARDJ. 8(3&4):72–75.

Amien. L.I. 1990. Utilization of acid tropical soils for sustainable agriculture. IARDJ. 12(2):17–22.
BPS. 1993. Statistik Indonesia. Biro Pusat Statistik, Jakarta.
Eswaran, H. 1984. Uses of soil taxonomy in identifying soil related potentials and constraints for agriculture. In Ecology and Management of Problem Soils in Asia. Food and Fertilizer Technology Center, Taipei, Taiwan.
Hue, N.V. and I. Amien. 1989. Aluminum detoxification with green manure. Comm. Soil Sci. Plant Anal. 20(15&16):1499–1511.
Jones, J.W., H. Beck, R.M. Peart, and P. Jones. 1986. Application of expert system concepts to agrotechnology transfer. Proc. International Soil Sci. Congress. Hamburg, Germany.
Nix, H. 1984. Minimum data sets for agrotechnology transfer. Proc. of International Symposium on Minimum data sets for Agrotechnology Transfer. ICRISAT. Hyderabad, A. P. India.
Sanchez, P.A., J. Benites, and D.E. Bandy. 1987. Low-input system and managed fallows for acid soils in the humid tropics. Pages 353–360 in M. Latham, Ed. *Soil Management under Humid Condition in Asia and Pacific*. IBSRAM Proc. No. 5. Bangkok, Thailand.
Scholz, U. 1983. The Regions of Sumatra and Their Agricultural Production Pattern. SARIF, West Sumatra, Indonesia.
Valverde, C.S. and D.E. Bandy. 1982. Production of annual crops in the Amazon. Pages 243–280 in S.B. Hecht, Ed. *Amazonia: Agriculture and Land Use Research*. CIAT, Cali, Colombia.
Von Uexkull, H.R. 1982. Problem soils for food crops in the humid tropics. Pages 139–153 in Proceedings of the International Symposium–Distribution, Characteristics and Utilization of Problem Soils. Trop. Agric. Res. Centre, Min. Agric. For. Fish., Tsukuba, Japan. Trop. Agric. Res. Ser. No. 15.
Von Uexkull, H.R. 1984. Managing acrisols in the humid tropics. Pages 382–386 in Ecology and Management of Problem Soils in Asia. Food and Fertilizer Technology Center, Taiwan. Book Ser. No. 27.
Waterman, D.A. 1986. A Guide to Expert Systems. Addison Wesley, Reading, Massachusetts.
World Bank. 1990. Indonesia: Sustainable Development of Forests, Land and Water. The World Bank, Washington, D.C.
World Bank. 1991. Yogyakarta Upland Area Development Project (Bangun Desa II). Staff Appraisal Report No. 9113-IND.
Yost, R.S., S. ltoga, Z.C. Li, and P. Kilham. 1986. Soil acidity management with expert system. Proc. IBSRAM Seminar on Soil Acidity, Land Clearing and Vertisols, October 13–20, 1986 Khon Kaen, Thailand.

Chapter 41

Does Existence Value Exist?

JOHN C. BERGSTROM AND STEPHEN D. REILING

Existence value was first introduced as a type of economic value in 1967 by John Krutilla, who argued that an individual may value a resource simply because it exists, independent of present or future use of the resource. Since its introduction, existence value has become an important potential value to include in benefit-cost analyses and natural resource damage assessments. However, no consensus has developed regarding its definition or the motives that lead to existence value. We review the literature to illustrate the various definitions of the concept and motives associated with it. We then offer a new definition that is more restrictive than that used by many other economists. We then argue that noneconomic motives also can lead to a desire to preserve natural resources. Consequently, we conclude that one must examine the motives of individuals when defining and measuring existence value. Maine Agricultural and Forest Experiment Station Publication No. 2033.

Introduction

Since 1960, a number of "new" economic values have been introduced in the economics literature. These include option value, quasi-option value, option price, expected consumer's surplus, existence value, bequest value, preservation value, and intrinsic value. The proliferation of new concepts related to economic value has created confusion among economists and noneconomist alike. Some of the new economic value concepts do not have standardized rigorous definitions, and it seems that some of the concepts may be overlapping. This confusion has led to skepticism among economists and noneconomists, including resource managers, about the validity of the new value concepts and their importance in resource allocation decision processes.

Recent debate has focused on the conceptual definition and empirical measurement of existence value, and the application of existence value measures to benefit-cost analysis and natural resource damage assessment (Kopp, 1992; Rosenthal and Nelson, 1992; Diamond and Hausman, 1994; Hanemann, 1994; Portney, 1994). If existence values do indeed exist, these values may be extremely large in the aggregate. The sheer magnitude of aggregate existence value measures has the potential to dominate the outcomes of benefit-cost analyses of resource projects and natural resource damage assessments filed under CERCLA. Thus, those who stand to gain or lose from the incorporation of existence values into benefit-cost analyses and natural resource damage assessments have a vested interest and motivation to enter into the existence value debate.

The purpose of this paper is to objectively address the question of "Does existence value exist as a conceptually-sound and empirically-measurable economic value concept?" We begin in the next section by reviewing previous theoretical definitions of existence value proposed in the literature. We then propose what we hope is a more standard definition of existence value from both an economic and noneconomic perspective. Implications and concluding comments are presented in the final section.

Previous Existence Value Definitions

John Krutilla is usually credited with introducing the concept of existence value in the economics literature. In his classic 1967 article, "Conservation Reconsidered," Krutilla suggested that people may economically value natural resources simply because they exist. This existence value is different from typical economic values because it is not dependent on current or future use of the resource.

In the almost 30 years since Krutilla's article, a consensus among economists has not emerged on a standard theoretical definition of existence value and techniques for measuring it. An understanding of existence value and its relationship to other economic value concepts may be facilitated by considering the different types of interactions that occur between individuals and natural resources. Table 41.1 represents a classification of activities involving interaction between individuals and natural resources that generate economic value. The activities are classified according to four broad criteria: whether the type of interaction between individuals and natural resources is consumptive or nonconsumptive, and whether the location of the interaction between individuals and natural resources is on-site (in situ) or off-site. Consumptive and nonconsumptive activities are further divided into activities involving direct or indirect interaction between individuals and natural resources.

Consumptive activities involve rival use of a natural resource. Rival use means that use of a natural resource by one individual reduces the quantity or quality of the resource available to other individuals. Consumptive activities involving direct, on-site interaction between individuals and natural resources include hunting and fishing (cell A). Consumptive activities involving direct, off-site interaction include burning firewood or eating berries at home that were collected at some natural area such as a National Forest (cell B).

Consumptive activities involving indirect, on-site interaction include the satisfaction an individual gains from *other* people participating in hunting and fishing

Table 41.1 Classification of Activities which Generate Economic Values

Type of interaction		Location of interaction	
		On-site (in situ)	Off-site
Consumptive	Direct	A Hunting Fishing	B Burning firewood at home Eating berries at home
	Indirect	C Satisfaction gained from hunting and fishing by other individuals in the present Satisfaction gained from hunting and fishing by individuals in the future	D Satisfaction gained from other individuals burning firewood or eating berries at home in the present Satisfaction gained from other individuals burning firewood or eating berries at home in the future
Nonconsumptive	Direct	E Wildlife observation Wildlife photography	F Viewing photos of natural resources at home
	Indirect	G Satisfaction gained from wildlife observation or photography by other individuals in the present Satisfaction gained from wildlife observation or photography by other individuals in the future	H Time devoted to "environmental causes" which benefit a particular natural resource Thinking about natural resource Satisfaction gained from participation in nonconsumptive, off-site activities by other individuals in the present or future

in the present or future (cell C). Consumptive activities involving indirect, off-site interaction include the satisfaction an individual gains from *other* people burning firewood or eating berries at home which are collected at some natural area in the present or future (cell D).

Although the activities falling into cells C and D do not result in direct rival use of natural resources, we classify them as consumptive activities because rival use indirectly occurs on the part of other people in the present or future. In economic terms, this linkage between the individual utilities represents spatial or temporal externality relationships between individuals.

Nonconsumptive activities involve nonrival use of a natural resource. Nonrival use means that use by one individual does not reduce the quantity or quality of the resource available to other individuals. Nonconsumptive activities involving direct, on-site interaction include wildlife observation (e.g., bird watching) and photography (cell E). An example of a nonconsumptive activity involving direct, off-site interaction is viewing a collection of photographs at home that were taken at a natural area such as a National Forest (cell F).

Nonconsumptive activities involving indirect, on-site interaction include the satisfaction an individual receives from participation by *other* individuals in wildlife

Table 41.2 Classification of Economic Values Generated by Activities (from Table 41.1)

Type of interaction		Location of interaction	
		On-site (in situ)	Off-site
Consumptive	Direct	a On-site active use value	b Off-site active use value
	Indirect	c Externality value from on-site active use by other individuals in the present or future	d Externality value from off-site active use value by other individuals in the present or future
Nonconsumptive	Direct	e On-site active use value	f Off-site active use value
	Indirect	g Externality value from nonconsumptive on-site active use by other individuals in the present or future	h Cognitive value associated with time devoted to environmental issues Cognitive value associated with merely thinking about a resource Externality value from nonconsumptive, off-site active or passive use by other individuals in the present or future

observation and photography in the present or future (cell G). Because an individual receives satisfaction through participation by other individuals, cell G activities also represent externality relationships. Nonconsumptive activities involving indirect, off-site interaction include the satisfaction individuals receive from devoting time to "environmental causes" which benefit a particular natural resource (e.g., "save the whales" project) or the satisfaction gained from simply thinking about a natural resource (cell H). Cell H activities also include the satisfaction an individual gains from participation in nonconsumptive, off-site activities by other individuals in the present or future, which represents another type of externality relationship.

Table 41.2 presents a classification of economic values corresponding to the activities listed in Table 41.1. Economic values generated by consumptive activities associated with direct, on-site interaction between individuals and natural resources are termed *on-site active use values* (cell a). Economic values generated by consumptive activities associated with direct, off-site interaction are termed *off-site active use values* (cell b). The values associated with consumptive on-site and off-site *indirect* uses are in the nature of external benefits (cells c and d). External benefits occur because the utility of one individual (in the present) is positively affected by consumption of natural resources *by other individuals* in the present or in the future.

Economic values generated by nonconsumptive activities associated with direct, on-site interaction are termed *on-site active use values* (cell e). Economic values generated by nonconsumptive activities associated with direct, off-site interaction are termed *off-site active use values* (cell f). Economic values generated by nonconsumptive activities associated with indirect, on-site interaction are another type of externality value or external benefit (cell g). Economic values generated by nonconsumptive activities associated with indirect, off-site interaction include *cognitive values* associated with spending time on environmental causes or simply thinking about natural resources, and externality values associated with participation in these type of activities by other individuals (cell h).

Based on this classification of economic values generated by interactions between humans and natural resources, we now turn to existence value definitions. In the economics literature, the term *existence value* has been applied to economic values falling into cells c, d, g, and h in Table 41.2 which are generated by activities falling into cell C, D, G, and H in Table 41.1, respectively. Krutilla (1967) and Krutilla and Fisher (1975), for example, define existence value generally as the value a person places on the mere existence of a natural resource. They identify two major motivations underlying existence values: bequest and sentimental motivations. Bequest motivations give rise to the value an individual places on passing on natural resources for the benefit of future generations. McConnell (1983) identifies bequest value as a type of externality value which could fall into cells c, d, g, or h in Table 41.2. The sentimental motivations discussed by Krutilla (1967) and Krutilla and Fisher (1975) appear to give rise to cognitive values associated with thinking about or reflecting upon the mere existence of a natural resource independent of any present or future active use of the resource by that person or any other person. These cognitive values fall into cell h in Table 41.2.

Randall and Stoll (1983) define existence value generally as the value a person places on simply knowing that a natural resource exists. They identify three types of motivations underlying existence values: interpersonal altruism, intergenerational altruism, and resource altruism. Interpersonal altruism reflects an individual's desire to provide natural resources to other individuals living in the same generation. Intergenerational altruism is an individual's desire to provide natural resources to other individuals living in future generations. Interpersonal and intergenerational motivations give rise to bequest values or externality values which fall into cells c, d, g, or h.

Resource altruism reflects an individual's desire to maintain or protect a natural resource for its own sake or benefit, independent of consumptive or nonconsumptive active use by that individual or any other individual in the present or future. Resource altruism gives rise to cognitive values associated with spending time on environmental causes which benefit a particular natural resource, or simply thinking about a natural resource surviving somewhere in the world. These types of cognitive values fall into cell h.

Madariaga and McConnell (1987) define existence value generally as the value derived from any enjoyment of a natural resource which is not associated with personal, on-site (or in situ) use. This broad definition of existence value potentially captures values in cells b, c, d, f, g, and h. In addition these authors define "pure existence value" as the portion of existence value as broadly defined in the previous sentence which is not connected to personal consumption of some marketed commodity (e.g., travel expenditures). Madariaga and McConnell's (1987) interpretation of "pure existence value" appears to include values in cells c, d, f, g, and h.

In an earlier article, McConnell (1983) also uses the term "pure existence value," and associates it with altruistic motivations and intrinsic motivations. His definition of altruistic motivation coincides with Randall and Stoll's interpersonal and intergenerational altruism which generates externality values in cells c, d, g, and h. His definition of intrinsic motivations appears to coincide with Randall and Stoll's concept of resource altruism which generates cognitive values in cell h.

Brookshire et al. (1983) define existence value broadly as willingness-to-pay (WTP) for the existence or preservation of a natural resource. The thrust of their discussion suggests that they are referring more specifically to values in cells c, d, g, h, and possibly f. Walsh et al. (1984) define existence value generally as an individual's WTP for the knowledge that a natural resource is protected even though no use is contemplated by that individual or any other individual. This definition appears to restrict existence values to cell h.

Brookshire et al. (1986) define existence value as "WTP for attributes of a resource that an individual perceives as inherent in the resource." They also refer to this type of value as a type of intrinsic value. This definition appears consistent with the resource altruism motivation defined by Randall and Stoll (1983) which generate cognitive values in cell h.

Smith (1987) defines existence value generally as the values of a resource not associated with direct, on-site (or in situ) use. For example, it is the value associated with preserving a natural resource when the cost of using the site is so high as to "choke off" direct, on-site use. This definition of existence value appears to include values in all cells in Table 41.2 except for cells a and e. If Smith's (1987) concept of "on-site or *in situ* use" also applies to activities involving on-site collection of resources for later use off-site (e.g., burning firewood collected on-site at home), values in cells b and f would also be excluded as potential types of existence value.

Boyle and Bishop (1987) define existence value broadly as the value an individual places on a natural resource even though the individual is sure he or she will never personally use the resource. These authors identify two primary motivations underlying existence value: bequest motivations and sympathy for the resource itself. Their interpretation of bequest motivations appears consistent with Randall and Stoll's interpersonal and intergenerational altruism which generate externality values in cells c, d, g, and h. The "resource sympathy" motivations appear consistent with Randall and Stoll's concept of resource altruism which generates cognitive values in cell h. Stevens et al. (1991) broadly define existence value as the value derived from knowing that a natural resource exists. They divide existence value into bequest values consistent with externality values in cells c, d, g, and h, and "nonuse values" which appear consistent with cognitive values in cell h.

Bishop and Welsh (1992) define existence value as values for resources that are motivated by individual preferences not associated with personal use of the resource. Personal use of the resource includes any activities involving on-site or *in situ* contact with the resource, personal off-site contact with the resource, or personal off-site consumption of products derived from the resource. Hence, Bishop and Welsh's (1992) definition of existence value appears to include values in cells c, d, g, and h. Silberman et al. (1992) define existence value broadly as the value an individual places on a natural resource when direct use by the individual is constrained to zero. This definition appears to also include values in cells c, d, g, and h.

Larson (1993) defines existence value generally as the "value of a resource to an individual when consumption of complementary goods is held at zero." This broad definition appears to include values in cell f and h, and possibly also in cells c, d, and g. Larson defines "pure existence value" as the value of a resource which is not associated with any changes in observable behavior. This notion of "pure existence value" corresponds most closely to cognitive values associated with merely thinking about a natural resource which fall into cell h.

Towards a More Standard Understanding of Economic and Noneconomic Concepts of Existence Value

As the review of literature in the previous section suggests, there is not a clear consensus among economists as to the economic definition of existence value. In this section we propose a standard economic definition of existence value and distinguish this definition from noneconomic concepts of existence value which have appeared in the literature. Existence value is generally thought of as a type of "nonuse value" (Randall and Stoll, 1983; Smith, 1987). The exact meaning of "nonuse," however, is ambiguous in the literature which is the major contributing factor to the lack of agreement over the exact meaning of "existence value."

"Nonuse" clearly means that direct, on-site interaction between individuals and natural resources does not occur, thereby excluding values in cells a and e (Table 41.2). Although less obvious, it also seems clear that "nonuse" means that direct, off-site interaction between individuals and natural resources does not occur, thereby excluding values in cells b and f. Because externality values in cells c, d, and g involve indirect on-site and off-site use of natural resources, we believe values in these cells also should not qualify as existence values. Thus, we have excluded values in all cells in Table 41.2, except for some of the values in cell h, as candidates for existence value.

Values in cell h are derived from interaction between individuals and natural resources which is purely cognitive in nature (e.g., merely thinking about a natural resource generates economic value). Use by anyone does not occur either directly or indirectly. This type of purely cognitive use, we believe, is the source of true existence value or what has also been called "pure existence value" in the literature (Larsen, 1993; Madariaga and McConnell, 1987; McConnell, 1983).

The title of this paper is "Does existence value exist?" We believe that existence value does exist as a legitimate economic value, but only in terms of "pure existence values" falling into cell h (Table 41.2). Accordingly, we define existence value as follows:

> Existence value is an individual's willingness-to-pay (or willingness-to-accept compensation), measured in terms of a Hicksian compensating or equivalent welfare measure, for a change in a commodity where interaction between the commodity and that individual or any other individual is purely cognitive in nature.

This definition, for example, includes an individual's WTP for devoting time to an environmental "cause" which only provides opportunities for thinking about a natural resource. It would also include an individual's WTP for time spent merely thinking about the natural resource (independent of time devoted to a "cause"), and an individual's WTP for the satisfaction of knowing that other people are enjoying merely thinking about a natural resource.

The definition of existence value proposed above applies specifically to the economic concept of existence value. In economic terms, existence value is properly measured theoretically using a Hicksian compensating or equivalent welfare measure which reflects economic motives and tradeoffs. Hicksian compensating and equivalent welfare measures associated with natural resource changes are measured in terms of willingness-to-pay or willingness-to-accept compensation for the change. Observations or statements of willingness-to-pay or willingness-to-accept compensation for natural resource changes, however, may not always reflect economic motives and tradeoffs. Consideration of the personal motives underlying observations and statements of value associated with natural resource changes is therefore relevant as suggested by a number of previous authors (Madariaga and McConnell, 1987; Edwards, 1986; Sagoff, 1994; Sen, 1979).

In Figure 41.1, motives leading to economic existence value are considered within a broader set of motives. As illustrated in the diagram, motives that give rise to economic existence value are a subset of "social" motives, which in turn, are a subset of "universal" motives.

Consider, for example, the individual who is motivated by a commitment to a particular environmental ethic which is not measurable in terms of a Hicksian compensating or equivalent welfare measure. This "commitment motive" may be reflected, for example, by an individual's monetary contributions to environmental "causes," but where the motive behind these contributions are not consistent with Hicksian welfare measures (Brookshire et al., 1986; Edwards, 1986; Stevens et al., 1991). This represents a noneconomic social motive for the existence of the resource, and would fall in area B in Figure 41.1.

One of the problems facing economists who attempt to measure economic existence value is that motives for the existence of the resource often differ among individuals for the same policy issue. For example, consider two individuals (A and B) who are both willing to pay $100/year to preserve grizzly bear habitat in Wyoming, independent of present or future on-site or off-site use by themselves or others. Individual A is willing to pay $100/year because he or she enjoys thinking about grizzly bears existing in Wyoming. Furthermore, the $100 is the exact amount that makes him or her indifferent between keeping the $100 and not thinking about grizzly bears existing in Wyoming, and paying the $100 and thinking about grizzly bears existing in Wyoming. Thus, the $100 is a Hicksian compensating welfare measure and is a legitimate economic existence value based on the definition presented above. This economic existence value falls in cell h of Table 41.2 and the underlying motive lies in area A in Figure 41.1.

Suppose individual B is willing to pay $100/year because, under a set of social principles held by the individual, he or she feels a social obligation to contribute to the preservation of wildlife habitat. The type, quantity, and location of the wildlife habitat may be of secondary importance to the individual. The decision

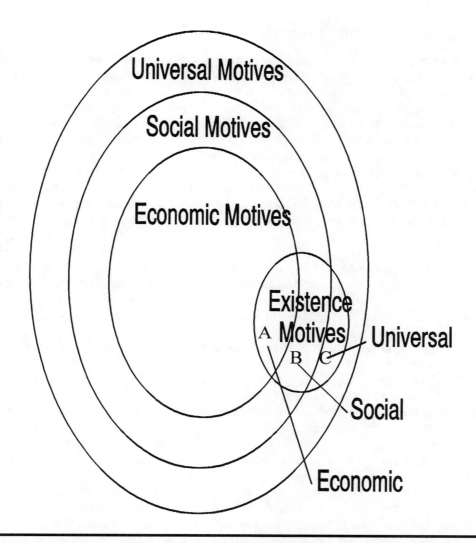

Figure 41.1 Motives for economic existence value within a universal framework of motivations for resource preservation.

to contribute $100 in this case is based on social, noneconomic, motives falling into area B in Figure 41.1. The $100 does not represent the maximum amount of income he or she is willing to pay to make him/her indifferent between the two situations being considered. It therefore does not represent economic existence value as we have defined it.

Philosophers and theologians also suggest that an individual may be motivated to preserve a resource, not out of a social obligation, but because he or she believes resources have a right to exist in and of themselves, independent of individual or social values (Ferre, 1988). These universal motives which fall in area C of Figure 41.1, are nonanthropocentric in nature whereas economic and social motives which fall in areas A and B of Figure 41.1, are anthropocentric.

Summary and Conclusions

In this paper we argue that economic existence values exist but in a narrower sense than is suggested in much of the literature on existence value. For example, we conclude that the bequest and altruistic motives do not produce existence value, as suggested by some previous authors. Instead, we believe that these motives give rise to external values or benefits that stem from interdependent utility functions among individuals in the same or different generations. We conclude that economic existence value is the Hicksian measure of willingness to pay for a change in a resource where the interaction between the resource and the individual is purely cognitive in nature. This occurs when the interaction is nonconsumptive, off site, and indirect. All other forms of activities or uses outlined in Tables 41.1 and 41.2 do not give rise to economic existence value, according to our definition.

Two obvious conclusions stem from our definition of existence value and the motives that give rise to it. First, the magnitude of existence value, as we have defined it, is smaller than it would be under the broader definitions used in the literature. This reduces the likelihood that existence values dominate the other forms of value included in benefit-cost analyses and environmental damage assessments.

Second, the measurement of existence value, as we define it, is perhaps more problematic than when broader definitions are used. The reason for the higher degree of difficulty is that the researcher must assess the motives of the individual to determine whether the stated WTP is consistent with our definition of existence value. As noted above, bequest and altruistic motives for willingness to pay do not qualify for economic existence value under our definition. Likewise, commitment to an "environmental ethic" or nonanthropocentric view of resource value does not yield economic existence value because the underlying motives for the WTP are not economic in nature.

In spite of the inherent difficulties that may exist in measuring the concept, we believe that the definition offered is a more accurate portrayal of economic existence value which eliminates much of the overlap and confusion that currently exists between existence value and other types of value. However, a better understanding of existence value can only be accomplished by examining closely the underlying motives of individuals when attempting to measure the concept. As suggested by previous authors, motives clearly matter when defining and measuring existence value (Madariaga and McConnell, 1987).

We conclude with two suggestions for future research.

First, the narrow definition of existence value offered in this paper needs to be tested using the traditional utility maximizing framework to insure that it is theoretically consistent with underlying consumer theory. Second, the need to assess motives opens the door for greater cooperation in research between economists, psychologists, and philosophers. Economists traditionally have not worried about individuals' motives when measuring economic value. Psychologists and philosophers have more experience in this area and cooperation among the three disciplines should be mutually beneficial.

References

Bishop, Richard C. and Michael P. Welsh. 1992. Existence values in benefit-cost analysis and damage assessment. *Land Econ.* 68(4):405–417.

Boyle, Kevin J. and Richard C. Bishop. 1987. Valuing wildlife in benefit-cost analyses: a case study involving endangered species. *Water Resour. Res.* 23(5):943–950.

Brookshire, David S., Larry S. Eubanks, and Alan Randall. 1983. Estimating option prices and existence values for wildlife resources. *Land Econ.* 59(1):1–15.

Brookshire, David S., Larry S. Eubanks, and Cindy F. Sorg. 1986. Existence values and normative economics: implications for valuing water resources. *Water Resour. Res.* 22(11):1509–1518.

Diamond, Peter A. and Jerry A. Hausman. 1994. Contingent valuation: is some number better than no number? *J. Econ. Perspect.* 8(4):45–64.

Edwards, Steven F. 1986. Ethical preferences and the assessment of existence values: does the neoclassical model fit? *Northeast. J. Agric. Resour. Econ.* 15(2):145–150.

Ferre, Frederick. 1988. *Philosophy of Technology*. Prentice-Hall, Englewood Cliffs, New Jersey.

Hanemann, W. Michael. 1994. Valuing the environment through contingent valuation. *J. Econ. Perspect.* 8(4):19–43.

Kopp, Raymond J. 1992. Why existence value should be used in cost-benefit analysis. *J. Policy Anal. Manage.* 11(1):123–130.

Krutilla, John V. 1967. Conservation reconsidered. *Am. Econ. Rev.* 57:777–786.

Krutilla, John V. and Anthony C. Fisher. 1975. *The Economics of Natural Environments: Studies in the Valuation of Commodity and Amenity Resources*. John Hopkins University Press for Resources for the Future, Baltimore.

Larson, Douglas M. 1993. On measuring existence value. *Land Econ.* 69(4):377–388.

Madariaga, Bruce and Kenneth E. McConnell. 1987. Exploring existence value. *Water Resour. Res.* 23(5):936–942.

McConnell, Kenneth E. 1983. Existence and bequest value. In R.D. Rowe and L.G. Chestnut, Eds. *Managing Air Quality and Scenic Resources at National Parks and Wilderness Areas*. Westview Press, Boulder, Colorado.

Portney, Paul R. 1994. The contingent valuation debate: why economists should care. *J. Econ. Perspect.* 8(4):3–17.

Randall, Alan and John R. Stoll. 1983. Existence value in a total valuation framework. In R.D. Rowe and L.G. Chestnut, Eds. *Managing Air Quality and Scenic Resources at National Parks and Wilderness Areas*. Westview Press, Boulder, Colorado.

Rosenthal, Donald H. and Robert H. Nelson. 1992. Why existence value should not be used in cost-benefit analysis. *J. Policy Anal. Manage.* 11(1):116–122.

Sagoff, M. 1994. Should preferences count? *Land Econ.* 70(2):127–144.

Sen, A. 1979. Personal utilities and public judgments: or what's wrong with welfare economics? *Econ. J.* 89:537–58.

Silberman, Jonathan, Daniel A. Gerlowski, and Nancy A. Williams. 1992. Estimating existence value for users and nonusers of New Jersey beaches. *Land Econ.* 68(2):225–236.

Smith, V. Kerry. 1987. Nonuse values in benefit cost analysis. *South. Econ. J.* 51:19–26.

Stevens, Thomas H., Jaime Echeverria, Ronald J. Glass, Tim Hager, and Thomas A. More. 1991. Measuring the existence value of wildlife: what do CVM estimates really show? *Land Econ.* 67(4):390–400.

Walsh, Richard G., John B. Loomis, and Richard A. Gillman. 1984. Valuing option, existence, and bequest demands for wilderness. *Land Econ.* 60(1):14–28.

Chapter 42

Analyzing Public Inputs to Multiple Objective Decisions Using Conjoint Analysis

DONALD F. DENNIS

Abstract

Resource managers need more than biophysical information and technical expertise to choose among alternate management strategies. Human wants, needs, beliefs, and values also must be considered. Conjoint techniques are well suited for soliciting and analyzing human preferences and values. The goal of this study is to assess and analyze public preferences and acceptable trade-offs for various levels of timber harvesting, wildlife habitats, and recreational opportunities using conjoint techniques on the Green Mountain National Forest.

Introduction

Perceptions and expectations of what forests are, or should be, are changing rapidly as an ever-increasing population demands more of everything from our forests, including preservation. Stepping beyond the clamor of demands, managers seek a holistic means to harmonize a broad array of human wants, needs, beliefs, and values, while preserving the long-run integrity and sustainability of an incredibly complex ecological system.

The management of forest ecosystems requires balancing concerns that range from forest health, biodiversity, and global climatic change to production and sustainability of wood products, wildlife habitats, and opportunities for outdoor recreation. Because needs, wants, perceptions, and values vary widely across society, conflict and expensive litigation frequently arise over the choice of management strategy and the resulting mix of benefits. Conflicts are particularly acute on public lands, where managers often feel trapped in a vice by the competing demands of vocal and unyielding groups of stakeholders.

Solicitation and assessment of public preferences, and understanding the consequences of choices is a crucial component of ecosystem management. Salwasser (1994) stressed the importance of human choice in ecosystem management: "...the success or failure of ecosystem management in protecting environments, revitalizing economies, or restoring healthy communities starts and ends with people and their choices — not with nature preserves, databases, ecological classifications, or any other technological tools that are merely useful means to desired ends." Use of a systematic and rational means to assess preferences and values, understand and analyze choices and the implied trade-offs, and implement these results in decision making may reduce conflict by assuring stakeholders that their preferences were fairly considered. Because these decisions involve a wide range of biophysical, financial, social, and moral variables, they are extremely complex and unyielding to traditional modes of analysis. This study explores the use of conjoint techniques to solicit and analyze public inputs to decisions on the Green Mountain National Forest (GMNF) in Vermont.

The survey has not yet been administered to the public. Focus groups, comprised mostly of research and support staff at the Northeastern Forest Experiment Station and GMNF, were assembled to determine the suitability of the survey for use in soliciting public preferences. The empirical results presented in this manuscript are taken from the focus groups and should not be construed as representing public preferences. The focus of this manuscript is on survey design and analytical techniques, and the empirical data serve to illustrate these techniques.

Conjoint Techniques

Conjoint analysis, a technique for measuring psychological judgments, is frequently used in marketing studies to measure consumer preferences (Green et al., 1988). The objective is to decompose a set of factorially designed stimuli or attributes so that the utility of each attribute can be inferred from the respondent's overall evaluations. In conjoint studies, respondents make choices between alternative products or scenarios displaying varying levels of selected attributes. For example, when studying consumer preferences for station wagon features, respondents might be asked to choose between sample vehicles that vary in cargo space, warranty length, gas mileage, horsepower, and country of manufacture. These data, which outline a respondent's preferences or the trade-offs he or she is willing to make, can be used to solve for the partial utilities for each attribute that are imputed from the overall trade-offs. These partial utilities, or part worths as they are sometimes called, can be combined to estimate relative preferences for any combination of attribute levels. Thus the analyst obtains high leverage between the options actually evaluated by respondents and those that can be evaluated after the analyses.

Conjoint techniques are well suited for soliciting and analyzing the preferences of stakeholders in environmental decisions. These decisions frequently involve trade-offs between costs and benefits that are not efficiently represented in market transactions. For example, Opaluch et al. (1993) described an approach that used paired comparisons to rank potential noxious facility sites in terms of social impacts. Asking respondents to make choices between alternatives mimics the real choices that managers must make and shows stakeholders the consequences of their choices.

There are many ways to design and analyze choice experiments. Respondents may be asked to reveal their preferences by choosing one of two or more options, ranking several options, or assigning numerical ratings to each option. Numerical ratings provide the most information but also place the greatest cognitive demands on respondents. Green (1974), Green and Srinivasan (1978), Louviere and Woodworth (1983), and Louviere (1988) provide information on experimental design in the context of conjoint analysis.

Study Area

The 360,000 acre GMNF comprises approximately 5% of the total land area and half of all public lands in Vermont (USDA, FS 1992). With large blocks of land in remote areas, the GMNF is well suited to provide unique opportunities for backcountry recreation and wildlife habitats in a region characterized by nonindustrial private forests and high population density. Because all public desires for forest-related benefits cannot be met simultaneously, a means for assessing preferences and values must be incorporated in the planning process.

National forest planning occurs on three broad levels: national, regional, and forest. The Forest Plan for the GMNF (USDA, FS 1992) sets goals, objectives, standards, and guidelines that fit within broader direction specified at the national and regional levels. Forest goals are accomplished through management actions that occur on Ranger Districts and subunits of these districts. Although public input and assessment of human values are important throughout the planning process, this study addresses public preferences for actions occurring at the subdistrict level. Techniques and experience developed during this study will be useful in developing broader goals for the next Forest Plan.

The 18,600-acre study area, located on the Manchester Ranger District, is comprised of two adjacent units known as Greendale and Utley. The area contains one developed campground (14 sites) and land classified in the Forest Plan as Management Area's (MA) 2.1, 3.1, 4.1, 6.2, and 8.1. The first three listed MAs include opportunities for roaded natural recreation and emphasize uneven-age silviculture (2.1), even-age silviculture (3.1), and winter deer habitat (4.1). MA 6.2 emphasizes opportunities for semi-primitive recreation while producing high-quality sawtimber by growing trees to an old age. MA 8.1 is the White Rocks National Recreation Area where the emphasis is on protecting wild values. Timber harvesting and roaded recreation are allowed, but restricted. Broad management prescriptions, standards, and guidelines are contained in the Forest Plan. Specific management actions need to be developed for the Greendale and Utley units. The goal of this study is to assess and analyze public preferences and acceptable trade-offs for various levels of timber harvesting, wildlife habitats, and three recreational opportunities: hiking trails, snowmobile access, and all-terrain vehicle (ATV) use.

Timber harvesting on national forests is controversial and the GMNF is no exception. Some publics argue against the environmental disturbances caused by timber harvesting. However, harvesting timber generates revenue for local and regional economies through creation of jobs within the community and cash payments for wood products exported from the area. It is also used as a means to accomplish wildlife habitat and recreation objectives.

The area supports a high species richness with a diverse late-successional community consisting of northern hardwoods, hemlock, and spruce. There are opportunities to enhance habitats for a variety of species through harvesting practices, and creation and maintenance of permanent openings, but it is not clear which species the public would like to see favored. If the public prefers species associated with early-succcession habitats, then management can be tailored to meet these needs. Similarly, management can be structured to favor species associated with mature, contiguous habitats.

Recreational concerns center around opportunities and potential conflicts between motorized and nonmotorized trail uses. The area includes opportunities to feature cultural resources (e.g., old roads, bridges, and farm sites) and to view wildlife (sightings of bear, moose, beaver, and birds are common). There are several opportunities to expand the hiking trail system. Currently 16 mi of travelway are available for snowmobile use, and there is public interest in expanding the available travelway. However, some users are concerned that increased snowmobile access will disrupt other recreational activities and disturb wildlife. ATVs are not currently permitted on travelways, but there is public interest in obtaining ATV access to the area and several potential opportunities have been identified. Those in opposition cite ecological damage and disturbances to wildlife or other recreationists as reasons to deny ATV access.

Survey Design

A conjoint ranking survey was designed to solicit preferences for five forest-related attributes: timber, wildlife habitat, hiking trails, snowmobile use, and ATV access. Three levels covering the range of reasonable alternatives for the Greendale and Utley units were selected for each attribute (Table 42.1). Eighteen alternatives, each depicting a unique bundle of attribute levels, were chosen using an orthogonal design that permits estimation of all main effects over the entire range of possible attribute combinations, with the least number of trials (Addelman, 1962a; 1962b).

The survey was designed to be presented during public involvement meetings conducted by Forest Service personnel. However, first, focus groups were gathered to determine the suitability of the survey instrument. They were given an explanation describing concerns and alternatives for the Greendale/Utley area and an overview of the nature and purpose of the conjoint study. To familiarize respondents with the ranking task, a practice survey was presented. Respondents were asked to imagine that they were considering the purchase of a new vehicle and to rank, in terms of overall preference, nine hypothetical vehicles possessing varying levels of five attributes: gas mileage, horsepower, cargo space, length of

Table 42.1   Choice Attributes and Levels

TIMBER
  1 Do not harvest timber
  2 Harvest timber on 5–10% of the planning area
  3 Harvest timber on 20–25% of the planning area
WILDLIFE
  1 Favor wildlife preferring contiguous unbroken forests
  2 Favor wildlife preferring a mix of young forests and contiguous unbroken forests
  3 Favor wildlife preferring open lands and young forests
HIKING TRAILS
  1 Maintain existing hiking opportunities
  2 Extend the hiking trail system to include 2 additional miles
  3 Extend the hiking trail system to include 6 additional miles
WINTER MOTORIZED
  1 Do not permit snowmobile use
  2 Maintain the existing 16 mi of travelway for snowmobile use
  3 Extend the travelway available for snowmobile use to 23 mi
SUMMER MOTORIZED (e.g., 3- and 4-wheel ATVs, motorized trail bikes)
  1 Do not permit ATVs on travelways
  2 Provide approximately 5 mi of travelway for ATV use
  3 Provide approximately 8 mi of travelway for ATV use

warranty, and country where the vehicle was manufactured. After this task was completed, respondents were provided with a brief verbal explanation of the attributes and levels depicted in Table 42.1 and given the opportunity to ask questions. Additional information, such as expected volume of timber harvests for each level of the timber attribute and lists of species favored for each level of the wildlife attribute, was provided. Forest Service personnel were available to respond to questions concerning any of the attributes or levels. Respondents then proceeded to rank the 18 sample cards, each depicting a unique bundle of forest-related attributes for the Greendale/Utley area. Of the 18 sample cards, 2 are illustrated in Table 42.2. Respondents also completed a series of attitudinal and demographic questions.

Analytical Model

When presented with a set of alternatives, individuals are assumed to make choices that maximize their utility or satisfaction. The utility that the ith individual derives from the choice of the jth alternative (U_{ij}) can be represented as:

$$U_{ij} = \overline{U}_{ij} + e_{ij} = X'_{ij} \beta + e_{ij} \qquad (1)$$

where X_{ij} is a vector of variables representing attributes of the jth alternative to the ith individual, β is a vector of unknown parameters, and e_{ij} is a random disturbance, which may reflect unobserved attributes of the alternatives, random choice behavior, or measurement error (Judge et al., 1985). In the empirical study

Table 42.2 Two Illustrative Sample Cards

Alternative #8	Alternative #14
Harvest timber on 20–25% of the planning area	Harvest timber on 5–10% of the planning area
Favor wildlife preferring a mix of young forests and contiguous unbroken forests	Favor wildlife preferring a mix of young forests and contiguous unbroken forests
Extend the hiking trail system to include 6 additional miles of trails	Extend the hiking trail system to include 6 additional miles of trails
Maintain the existing 16 mi of travelway available for snowmobile use	Do not permit snowmobile use
Do not permit ATVs on travelways	Provide approximately 5 mi of travelway for ATV use
RANK ___	RANK ___

under consideration, a respondent's utility level (U_{ij}) for each of the J alternatives is not observed but an ordinal ranking (Y_{ij}) is. The probability of alternative 1 being ranked above other alternatives is:

$$P_{i1} = \Pr(U_{i1} > U_{i2} \text{ and } U_{i1} > U_{i3} \ldots \text{ and } U_{i1} > U_{iJ})$$
$$= \Pr(e_{i2} - e_{i1} < \overline{U}_{i1} - \overline{U}_{i2} \ldots \text{ and } e_{iJ} - e_{i1} < \overline{U}_{i1} - \overline{U}_{iJ}) \quad (2)$$

Similar expressions hold for each of the remaining alternatives being chosen next in the choice set and the P_{ij}s become well-defined probabilities once a joint density function is chosen for the e_{ij} (Judge et al., 1985).

McKelvey and Zavoina (1975) developed a polychotomous probit model to analyze ordinal level dependent variables. They assume that the e_{ij}s are normally distributed and that the observed variable (Y_{ij}, the ranks for the J alternatives) is related to the unobserved utilities (U_{ij}) in the following way:

$$Y_{ij} = 0 \text{ if } U_{ij} < \mu_1, \; Y_{ij} = 1 \text{ if } \mu_1 < U_{ij} < \mu_2, \ldots Y_{ij} = J \text{ if } U_{ij} > \mu_{J-1}. \quad (3)$$

The μs define the boundaries of the intervals for the unobserved utilities with μ_0 and μ_J assumed to be negative and positive infinity, respectively, and μ_1 normalized to 0. Since the μs are free parameters, there is no significance to the unit distance between the set of observed values of Y; they merely provide the ranking. Estimates are obtained by maximum likelihood and the probabilities entering the log-likelihood function are the probabilities that the observed ranks (Y_{ij}s) fall within the J ranges defined by J+1 μs. The parameters to be estimated are J-2 μs plus the β vector.

The parameters may be used to calculate the estimated probability that a particular alternative will fall within each utility range for individual i.

$$\Pr(Y_{ij} = k) = \Pr(U_{ij} \text{ is in the kth range}) = F(\mu k - X'_{ij}\beta) - F(\mu_{k-1} - X'_{ij}\beta) \quad (4)$$

where k indexes the rankings and $F(\bullet)$ is the cumulative distribution function, assumed normal for the probit specification. Thus, a parameter (β_i) indicates the

Table 42.3 Ordered Probit Results (N = 540)

Variable	Coefficient	t-ratio
Constant	1.4847	7.172
Timber	0.3320*	5.903
Wildlife	0.0557	0.999
Hiking trails	0.0567	1.042
Snowmobile	−0.1089**	−1.935
ATV access	−0.2328*	−4.222

* Significant at the 5% level.
**Significant at the 10% level.

effect of a unit change in an independent variable (X_{ij}) on the estimated probability that a response will fall within each of the ranges. The magnitude of that change will depend on the values for all the estimated parameters and associated variables, as indicated by Equation 4.

Empirical Results of the Focus Groups

An ordered probit model was used to analyze data obtained from 30 respondents participating in focus groups assembled to test survey techniques. Each respondent ranked 18 alternatives providing 540 observed preferences. Selected results of this empirical test are shown in Table 42.3. The dependent variable is the ordinal ranking, of the alternatives which was coded from 0 to 17; higher scores being associated with greater utility. Because these data are used merely for illustrative purposes, detailed analyses and validation of the model will not be undertaken.

General inferences about the preferences of the focus group respondents may be made from the results shown in Table 42.3. The positive sign estimated for the timber attribute indicates that increased levels of timber harvesting will result in a higher estimated probability of a response score falling within the ranges associated with higher utility. Therefore, higher preference scores and greater utility were associated with increased levels of timber harvesting. The opposite was true for snowmobile and ATV access. The coefficients for the wildlife and hiking trail attributes were not significantly different than 0 at the 10% level. Discussions with members of the focus groups revealed that wildlife habitat was important, but many of the respondents addressed their preferences for habitat manipulation by weighting increased timber harvesting heavily in their ranking of the alternatives.

Values for μ_2 to μ_{17} were estimated in addition to the parameters associated with the attributes (β vector) shown in Table 42.3. As discussed in the previous section, the μs define the ranges for the unobserved variable (utility) associated with the observed responses (ordinal ranking). Interpretation of the coefficients for the polychotomous probit model is complicated. The estimated probability that a response will fall within each of the ranges is determined by the estimated coefficients and levels of the associated variables ($X'_{ij}\beta$, Equation 4). Because the probabilities across the ranges must sum to 1, the effect of a unit change in an independent variable is to increase the estimated probability that an alternative will fall within some ranges while decreasing the probability of falling within others. The magnitude of the estimated changes in probabilities depends on the values for all the coefficients as well as the values of the other independent variables at which the change is evaluated. The expected change in the probability of an alternative falling within any of the ranges resulting from a discrete change in an independent variable may be calculated using Equation 4. For example, the estimated probabilities that an alternative providing the middle levels of each attribute will be ranked highest is 0.046 and lowest is 0.045. Increasing the timber attribute by 1 level while holding the other attribute levels constant increases the estimated probability that this alternative will be ranked highest to 0.088, while decreasing the probability of it being ranked lowest to 0.022. Similar calculations can be performed for any changes in individual attributes or combinations of attributes that are of interest to the analyst.

Summary

Resource managers need a means to solicit and analyze human preferences and values. Conjoint techniques are well suited for coping with this task. A conjoint ranking survey was designed for use in soliciting public preferences for various levels of timber harvesting, wildlife habitats, hiking trails, snowmobile use, and ATV access on the GMNF. The survey was tested on focus groups comprised mostly of research and support staff at the Northeastern Forest Experiment Station and GMNF. No significant problems were identified by the focus groups.

An ordered probit model was used to analyze the empirical information collected during the focus group meetings. Although the results do not represent public preferences, they are used to illustrate the analytical capabilities. The results indicate a preference for higher levels of timber harvesting and lower levels of snowmobile and ATV access. Respondents did not show significant preferences for varying levels of wildlife habitats or hiking trails in the ranking survey. Several respondents indicated that wildlife habitat was an important attribute but also recognized that the diversity they desired would be achieved by higher levels of timber harvesting. Public preferences will be solicited during upcoming public involvement meetings. Additional work will include estimation of the effects of socioeconomic variables on preferences.

References

Addelman, S. 1962a. Orthogonal main-effect plans for asymmetrical factorial experiments. *Technometrics*. 4(1):21–46.

Addelman, S. 1962b. Symmetrical and asymmetrical fractional factorial plans. *Technometrics*. 4(1):47–58.

Green, P.E. 1974. On the design of choice experiments involving multifactor alternatives. *J. Consumer Res.* 1:61–68.

Green, P.E. and V. Srinivasan. 1978. Conjoint analysis in consumer research: issues and outlook. *J. Consumer Res.* 5:103–123.

Green, P.E., C.S. Tull, and G. Albaum. 1988. *Research for Marketing Decisions*. 5th ed. Prentice Hall, Englewood Cliffs, New Jersey. 784 pages.

Judge, G.G., W.E. Griffiths, R.C. Hill, H. Lutkepohl, and T.C. Lee. 1985. *The Theory and Practice of Econometrics*. John Wiley & Sons, New York. 1,019 pages.

Louviere, J.J. 1988. Conjoint analysis modelling of stated preferences: a review of theory, methods, recent developments and external validity. *J. Transport Econ. Policy.* 10:93–119.

Louviere, J.J. and G. Woodworth. 1983. Design and analysis of simulated consumer choice or allocation experiments: an approach based on aggregate data. *J. Marketing Res.* 20:350–367.

McKelvey, R.D. and W. Zavoina. 1975. A statistical model for the analysis of ordinal level dependent variables. *J. Math. Sociol.* 4:103–120.

Opaluch, J.J., S.K. Swallow, T. Weaver, C.W. Wesselles, and D. Wichelns. 1993. Evaluating impacts from noxious facilities: including public preferences in current siting mechanisms. *J. Environ. Econ. Manage.* 24:41–59.

Salwasser, H. 1994. Ecosystem management: can it sustain diversity and productivity? *J. For.* 92(8):6–10.

USDA, Forest Service, Eastern Region. 1992. Land and Resources Management Plan: Green Mountain National Forest. Amended. Green Mountain National Forest, Rutland, Vermont.

Chapter 43

Noneconomic Values in Multiple Objective Decision Making

THOMAS A. MORE, JAMES AVERILL, AND RONALD J. GLASS

Introduction: The Role of Value in Resource Management

Throughout the 20th century, scientific management has been one of the most cherished tenets of the various resource management professions. The concept has served us well, halting the rapacious use of U.S. resources that characterized the late 19th century and ushering in a new era of more rational management. Unfortunately, "scientific management" can also create confusion among resource professionals. The phrase itself is an oxymoron — science deals with scientific facts while management deals with values, and, traditionally, values have always been excluded from science. While science deals with what is, management is always for a purpose, to accomplish some human goal or objective, even if that objective is to return an ecosystem to its "natural" state. We contend that the emphasis on scientific management has led to the development of a great deal of scientific fact about natural resources, but that values (with the possible exception of economic values) have been left to shift for themselves, receiving implicit rather than explicit consideration in many management decisions, often as part of the decision's background political context. Therefore, our purpose in this paper is to discuss value theory in the context of resource management, giving particular emphasis to the noneconomic categories of value.

The Concept of Value

Value can be an exceptionally confusing concept because the term is used in several different ways. Some authors use value to describe the functions something serves. For example, Bengston (1993) describes forests as having aesthetic, spiritual, moral, economic, and scientific values. To others, value means worth, usually in a monetary sense — the value of the recreation produced at site x. Still others refer to value as principles — honesty, integrity, kindness, and the like. These differences parallel differences among the various disciplines — economics, psychology, sociology, philosophy — in which the term is used. Each discipline interprets the concept within its own metatheoretical structure thereby heightening differences in meaning. The confusion is further exacerbated by a host of highly related concepts: how does a value differ from a need, a want, a desire, and objective, a goal, or a policy? All these concepts are closely intertwined making it difficult to sort through them. The result is almost certain confusion when one tries to speak of value or values.

In this paper, we adopt the value-as-principle perspective. We define "value" as a criterion by which a state of affairs (an object or a situation) is judged to have or to afford the property x, where x is instantiated by a pro or con predicate (e.g., good or bad, beautiful or ugly) as opposed to a predicate that describes matters of fact (e.g., "green" or "large"). Three things are implicit within this definition. First, there is a valuer — a person or group for whom the state of affairs is good or bad. Second, values are criteria that people use to make decisions; they are the standards by which we judge. And third, values differ from facts. Although each of these could be discussed in more detail, the fact-value relationship is particularly important because of the emphasis the natural resource professions have placed on fact-based problem solutions.

The relationship between facts and values has a venerable history stretching back to the ancient Greeks. In modern (i.e., post Renaissance) philosophy, however, it was David Hume who brought the issue into sharp focus. Hume noticed that his 18th-century contemporaries would describe situations factually (i.e., using "is" statements), and then gradually, almost imperceptibly, slip into using values ("ought" statements). Hume questioned how these "ought" statements arose. He demonstrated that, under standard systems of logic, it is impossible to derive ought statements (values) from is statements (facts). Facts alone never tell us what we ought to do — the ought derives from human values, goals, and objectives. Thus, within the empirical tradition, there is an unbridgeable gulf between facts and values.

Although Hume's point of logic still stands, it set off a controversy that has lasted over 2 centuries. Today, it is virtually impossible to make any statement about the fact/value relationship that is not controversial, with positions ranging from an extreme subjectivism to an extreme objectivism. At the subjective end of the scale are those who insist that all observation is theory laden, so that the very act of observing cannot be divorced from values. This view is popular among multiculturalists because it permits the development of sciences that contrast with traditional western empirical science. For example, there could be an Afro-centrist science or a feminist science; "truth" becomes a matter of where you stand, a position that makes nearly everything a value. In resource management, this is parallelled by the view that culture influences management; i.e., that specific groups such as Native Americans have a different cultural relationship with the land than do other groups and that their management practices flow from this cultural relationship. At the opposite extreme are the realists. While the realists do differentiate between facts and values, they argue that it is possible to make fully objective, factual statements about values. That murder is wrong, for example, does not depend upon our idiosyncratic preferences, but is a factual statement as objectively knowable across all cultures as that water consists of two parts hydrogen and one part oxygen. This position makes nearly everything a fact. Over the past quarter century, the realists have held sway in philosophy (see review by Sayer-McCord, 1988) while the subjectivists have gained in other disciplines.

Our own position is rather traditional and middle-of-the-road. As we see it, facts are objective. That is, they inhere in the object and are considered to be independent of any particular observer. A book for example, can be described objectively with certain facts — length, width, height, number of pages, etc., and these facts remain unchanged no matter who is reading it. But whether or not it is a *good* book depends upon its relationship to you — your interests, goals, objectives, level of reading ability, and so on. That is, the value is unique to the individual and can change across persons. In this sense, values are subjective — of the person or subject — as opposed to objective — of the object.

This position enables us to address the confusion we discussed earlier. In our terms, forests and other natural environments are "states of affairs" that are potentially valu-*able*. That is, forests have attributes (species composition, vastness, beauty, etc.) that can be valued *provided they meet certain criteria*. The criteria (or standards) by which we judge forests (or other states of affairs) are our values; they are the principles that guide us in our relationships with the world. These values help us to *evaluate* — to decide if something is good or bad, right or wrong, beautiful or ugly. Is a particular forest stand beautiful? The answer depends upon the standards we use to evaluate it.

There are, of course, multiple states of affairs in the world with which we must cope, so it is hardly surprising that there are multiple values that we use to evaluate them. One classification of such values might include economic values, moral values, aesthetic values, spiritual values, and rational values. Put simply, economic values are standards for judging goods and services; moral values provide standards for conduct; aesthetic values are standards for beauty; spiritual values are standards for meaning; and rational values are standards for judging truth. Each of these is more complex, of course, and may be subdivided into constituent elements. Moral values, for example, include honesty, altruism, integrity, kindness, etc., each of which may or may not be useful in any specific situation. Moreover, they are not independent of one another, so that we may speak of a person's value system (Rokeach, 1973). In this way, moral values may be suborned for aesthetic ends, or spiritual values may be subject to rational scrutiny, etc. In the remainder of this paper, we will consider each of these values in relation to resource management.

Economic Values

Of the various types of value, economic value is the type most familiar to natural resource managers. Typically, the economic value of a resource refers to its market value. More formally, economic value is the amount that a person is willing and able to sacrifice in order to obtain a good or service. This value is typically expressed monetarily, although it need not be.

How does something come to have worth in an economic sense? The answer lies in a consideration of human goals. We construe human behavior as a set of responses "designed" to achieve some goal or fulfill some function (Averill, 1992; Averill and More, 1993). A person's behavior can be subdivided (at least for theoretical purposes) into three types of systems — biological, psychological, and social — each associated with its own primary goal. The primary goal of biological systems is the sort of inclusive fitness that we associate with long-term health — the optimal functioning of all a person's biological subsystems. Socially, the primary goal is the inclusive fit of the person into society, and, psychologically, the primary goal is the actualization of the self — the realization of one's fullest potential as an individual. These ultimate goals are broken down into subgoals each associated with a subsystem of behavior so that swimming ten laps without stopping, getting a promotion or a new job, and reproducing a Shaker table are all legitimate goals for a person.

Tangible goods and services have worth in an economic sense because they assist a person to fulfill his or her goals. A shirt has economic value because it offers its wearer protection from the elements (reflecting biological goals) and style (a social/psychological goal). A tool has value because it enables a builder to complete a project. Not all goals are equal, of course, and some goods and services can serve multiple goals. Consequently, goods and services differ in their economic value.

Implicit in the above discussion is that economic value is actually a form of social exchange, a medium of negotiated relationships between people or between an individual and a group.

Person x possesses a good or service that is desired by person y. Person y knows what he or she is willing to sacrifice (pay) to obtain it from x, and x knows what he/she is willing to sell it for. There also may be other willing buyers or sellers for the particular good or service, which creates a market. When this market allocates goods and services among willing buyers and sellers under conditions of general equilibrium, the economic value of a particular good or service becomes the market value (or price). The market value of a particular good or service is therefore extremely important information for decision makers.

Unfortunately, not all goods and services are traded on markets. Market allocation can be suboptimal in some cases, and there are some things, e.g., personal liberty or integrity, that simply are not for sale. Many natural resources — wilderness, wildlife, many forms of recreation, water supplies, etc. — lie outside traditional markets and, therefore, are not readily amenable to neoclassical economic analysis. Over the past quarter century economists have devoted a great deal of effort to developing proxy measures of economic value for these resources. Unfortunately, these efforts remain polemic, and at least some economists[*] believe that economic values for such resources are embedded in a broader network of general social values. On that basis, we turn now to the role of noneconomic values in multiple objective decisionmaking.

---

[*] Personal communication. Steven D. Reiling, Department of Resource Economics, University of Maine, April, 1995.

Moral Values

Moral values provide standards for judging conduct. Honesty, fairness, altruism, kindness, justice, and the like form the general substance of this value category. Ordinarily, these values — which constitute the core of ethics — are applied to interpersonal relationships between humans; they are the lubricant for the social world. Over the past half century, however, there have been increasing calls to extend the realm of ethics beyond society into the nonhuman world. While the seeds of this idea are present in the writings of Thoreau and John Muir, they received their most elegant expression in Aldo Leopold's (1953) classic essay "The Land Ethic," in which he called for the extension of rights to the ecological community of plants and animals, just as they had been extended to women and minorities. This call has formed the basis for the emerging discipline of environmental ethics.

Rights and duties are the very essence of human ethics; their extension to the infrahuman world is polemic beyond doubt. Some, following Leopold, would extend rights to rocks and trees (cf. Nash, 1989), while others would be more restrictive. The key issue is to decide what has moral standing (or moral considerability). Most philosophers restrict moral considerability to sentient (feeling) creatures and, therefore, to the animal kingdom. At issue are the capacities of the organism being considered. Partridge (1986), for example, has argued that animals do not have freedom of religion or the right to vote not because we humans are tyrants who refuse to grant them these rights, but because they lack the capacity to worship or to make political choices. Going further, Partridge notes that rocks and trees lack consciousness or the ability to feel or think. If an entity is incapable of caring how it is treated, what sense does it make to attribute any rights to it?

This line of reasoning is also followed by Dawkins (1980, cited in Dennett, 1991). Dawkins argues that it is unlikely that organisms that lack the capacity to remove themselves from danger would evolve the capacity to suffer. There would be no evolutionary point in a tree, for example, having the capacity to suffer in silence. In considering the entire range of organisms that exist on the earth, there needs to be some sort of positive grounds for imputing suffering before we are willing to assign it.

So a conservative treatment would restrict moral standing to members of the animal kingdom. But even with this restriction our attitudes are confused. We willingly kill some animals without a thought, while going to great lengths to protect others. In considering further restrictions of moral standing within the animal kingdom, the two dominant themes are the ability to reason (from Descartes) and the ability to suffer (from Bentham). As Dawkins (1980) points out, these two are not independent. The ability to suffer is a function of the capacity to have articulated, wide-ranging, highly discriminative desires, expectations, and other sophisticated mental states (Dennett, 1991). Of course, these abilities extend well beyond humans. It is possible to cause a great deal of suffering to a Mountain Gorilla, for example, but how about a clam? Is a clam possessed of a sort of consciousness and a set of capacities that can be restricted, or is it more like a mechanical device, with an inability to experience what we might consider genuine suffering?

Further speculation in this area is beyond the scope of this paper. The key point is that with moral values we are considering standards for judging conduct.

We may grant moral standing to some other members of the animal kingdom, but we cannot logically grant it to all. This does not mean that we need not be concerned about other elements in the biosphere, however, but it does make clear that our motive in doing so is self-interest. For example, it may be important to insure stable relationships among soil invertebrates, but the reason for doing so is not that they have moral standing but because it is in our own best interest to see that these relationships are not disrupted. Our understanding of our moral relationship with the earth and its creatures is in rapid flux; we can expect many changes in these sentiments as the world's human population approaches 10 billion (Kennedy, 1993).

Aesthetic Values

Natural environments can produce a beauty that is genuinely awe inspiring. A strong sense of the aesthetic clearly lies behind much of the popular concern about the commercial uses of forests. Unfortunately, we understand very little about natural beauty. How do we come to judge a particular scene beautiful? What criteria do we use to separate the beautiful from the ugly or the simply uninteresting? Typically, the criterion used in most research is preference or liking. But we can like or prefer something for reasons that have nothing to do with beauty. Does liking or preference do justice to the magnificence of an autumn day? Or does it capture Robert Frost's surprising expression of natural beauty in *My November Guest*?

> My Sorrow, when she's here with me,
> Thinks these dark days of autumn rain
> Are beautiful as days can be;
> She loves the bare, the withered tree;
> She walks the sodden pasture lane.

Most of us can empathize with Frost, having had moments when our emotions color our sense of the beautiful, or when something we would ordinarily consider mundane or perhaps even dreary takes on an extraordinary beauty. Are these moments captured by preference or liking?

To consider criteria for the aesthetic requires a historical perspective. Our current sense of beauty — that it is a subjective experience "in the eye of the beholder" — is only about 300 years old. Prior to that (in what we may term the "classical" view or classicism), beauty was very much an attribute of the object. To Plato, for example, beauty was a matter of measure (including anything that can be measured — number, length, swiftness, etc.) and proportion (which concerned the harmonious relations among the measures). These themes, measure and proportion, recur in various guises throughout history, with modern theorists tending to speak of "variety and unity" (Fechner, 1876) or "complexity and order" (Eysenck, 1942). A third category — vitality — was added by Plotinus, the third century neo-Platonist. In this sense, an artificial flower is generally regarded as less beautiful than a natural one, even though it may have better color and definition (Arnheim, 1966). Vitality may be particularly important for understanding the aesthetics of natural environments. Actually being in an environment, and using all five senses to acquire information about it, may provoke a more intense aesthetic response than simply seeing a photograph or painting of it.

The relationship between direct experience and natural aesthetics also raises the issue of context or "meaning." Complexity, harmony, and vitality can be considered to be formal properties of an object, as distinct from its content or meaning. In classicism, beauty is thought to point beyond itself, to some more fundamental underlying truth or ideal. Among contemporary theorists, traces of this theme are found in "contextualism" (as opposed to formalism, which emphasizes the formal aspects of an aesthetic object). Contextualists hold that the total context or setting, including historical, scientific, and other knowledge contributes to the aesthetic appreciation of an object. Put most simply, aesthetic objects have meaning as well as form. Context, particularly in the form of knowledge, also may be a major component of natural beauty.

These four aesthetic criteria — complexity, harmony, vitality, and meaning — form the basis for the classical approach to understanding aesthetic values. The modern approach, which emphasizes subjective experience, was really an outgrowth of the larger transformation that accompanied the Scientific Revolution. Among the changes that occurred in this transformation was a shift from the narrower category of beauty to the broader concept of aesthetics — a term that includes beauty, but might also include elements of the sublime or grotesque. Overturning a rock to expose the scurring creatures underneath, for example, might not be considered beautiful *per se*, but it certainly can be fascinating and, therefore, might well qualify as aesthetic.

The shift to defining aesthetics as subjective experience brought with it a new set of decision criteria: pleasure (enjoyment), absorption (concentration), detachment (a contemplative attitude), and challenge (innovation and mastery). Rather than characteristics of the aesthetic object, these are characteristics of the subjective aesthetic experience.

Although pleasure is generally a prominent feature of aesthetic experience, it is neither a necessary nor a sufficient condition to find something aesthetic. We can experience pleasure in a variety of circumstances that can hardly be considered aesthetic — eating a piece of cake, for example. More importantly, the aesthetic can include both the pleasant and the unpleasant as in tragedy or in Frost's poem.

The second criterion is absorption or concentration. Aesthetic experiences are so powerful that people become absorbed in them, focusing entirely on the experience at hand and losing sight of other, perhaps more trivial concerns. Such concentrations may occur for various reasons, some of which we call "aesthetic." But what is an aesthetic reason? A key line of argument suggests that aesthetic experience involves contemplating an object for its own sake rather than for extrinsic or utilitarian reasons.

The final characteristic of aesthetic experience is challenge or mastery. Aesthetic experiences challenge the mind, leading out of the ordinary and into new territory. They help us to gain increased understanding or new insight into the nature of things, and they help to refine our ability to respond adaptively.

Thus, we have four classical criteria for aesthetic objects paralleled by four modern criteria for aesthetic experience. Note how this division parallels the division between facts and values described above. It is evident that we have a long way to go to understand the aesthetics of natural environments; the simple criterion of preference alone is not likely to take us much further.

Spiritual Values

As biological creatures, people are born with an enormous cognitive capacity that is relatively unstructured. Insects, by contrast, are thought to be born with highly structured minds and behaviors that are largely preprogrammed; ants and worker bees although undoubtedly conscious and possessed of brains do not have to struggle with difficult choices. One result of human unstructured mental capacity is that we all are motivated by a search for meaning — by a desire to interpret the events and circumstances of our lives within a context. Spiritual values provide the standards by which we judge such meaning. As such, they are the overarching set of values within which the other values operate.

Research on spiritual values in relation to natural environments is only just beginning and may take on many different dimensions. Natural environments have always occupied a significant place in organized religions — there have been sacred groves, oracles, and the like. Jesus, Mohammed, and The Buddha all spent time in the wilderness and experienced personal revelation prior to returning to civilization to teach (Harris, 1974). We note, however, that the wilderness they encountered had a vastly different social meaning than the concept does today. It was a hostile environment filled with demons and temptations, rather than the source of spiritual rejuvenation and deep beauty that we know it as today. Natural environments have been the home of sprites, fairies, goblins, ghosts, and the like for many of the world's religions and pantheism. The doctrine of identifying the Deity with the various forces and workings of nature has been a powerful force throughout the world's history. Although officially considered a heresy by the Catholic Church, pantheism continues to exert a strong influence on many. In the U.S., pantheism formed the basis for the doctrines of the Concord transcendentalists, particularly Emerson and Thoreau. More recently, we suspect it continues to influence deep ecologists and other environmentalists.

Throughout the 20th century, psychology has tended to focus on the experiential nature of religious/mystical experience, following William James and his seminal work "The Varieties of Religious Experience." Natural environments and natural beauty are undoubtedly linked to these experiences although the nature of the linkage and the casual mechanisms are unknown. That humans are prone to mystical experiences follows from their unstructured cognitive capacity. These experiences are of two types. If a person is tied to a particular religion, the experience will be interpreted within the doctrine of that religion. If a person is not committed to a religion, they are likely to report a sense of unity with "The One" — a sense of unified "connectedness" with all nature. Such experiences may be contributory to the Deep Ecology movement.

As noted, work on the spiritual values associated with natural resources is only just beginning, and at this point it is unclear how this area will develop. Is it possible to design opportunities for spiritual experiences or to manage for them in some way? Or, given the First Amendment, should public agencies even be concerned with them in any way? What is clear is that spiritual values are immensely important to people and, as such, are powerful determinants of attitudes and behavior. While we may not be in a position to manage for them, neither can we afford to neglect them.

Rational Values

Rational values bring us full circle by providing a link to management decisions. Rational values are the standards we use to judge "truth." It seems a bit odd to consider rationality as a value, but there is a generic quality of "oughtness" to it — rationality is considered good, while irrationality is bad; a rational decision is a good decision, an irrational decision is a bad decision. For this reason, it needs to be considered as a value in multiple objective decision making.

Rationality is both a characteristic of persons, a model of explanation, and a generic value. As a defining characteristic of persons, it is not so much rationality *per se* that matters, but the ability to reason rationally. Some — Aristotle, for example — have considered rationality the single most important characteristic of people. As a model of explanation, rational behavior or decisions are considered self-explanatory; that is, if a decision is considered rational, it ordinarily requires no further explanation. The standards that we use to decide if a decision is rational are encoded in the norms of rationality, i.e., rationality as a value.

There are three conditions required for rational decision making. First, there is a given state of prior information that is (hopefully) logically consistent, empirically testable, and explicitly stated. Second, there must be a predetermined goal or end state — something we are trying to achieve. And third, there must be the ability to select the optimal route to the goal from various alternatives; i.e., the chosen alternative must be more effective than others in achieving the intended goal, or incur less cost (in the broadest possible sense).

If the above conditions are met, then we can apply the rules, or norms, of rationality. These can be divided into logical norms, attitudinal norms, and practical norms. Logical norms for judging the rationality of a decision include objectivity (i.e., universality and impartiality within a domain), internal consistency, and conformity to the rules of inductive and deductive inference (i.e., taking into account base rates, causal relations, etc.). Attitudinal norms are norms of confirmation and/or falsification — a willingness to seek support for the system and extend its domain or, if need be, to test and reject underlying assumptions. Mediation or conciliation is also important — a rational stance strives to balance confirmation and falsification. It is possible, after all, to be unreasonably committed to a particular position or to be unreasonably critical — traps that have often plagued natural resource managers. Finally, there are the practical norms of attainability and suitability; it is irrational to pursue a goal that is unattainable, or to use methods that are inappropriate.

These, then, are the criteria to which we hold management decisions. When decisions meet these criteria, they are considered rational and no further explanation is needed.

Conclusion: Values in Multiple Objective Decision Making

In a densely populated world, natural resource managers will increasingly need to make decisions that incorporate multiple objectives. In this situation, clashes over values will be virtually inevitable. In the past, we have put our faith in facts and the hope that uncovering new facts would resolve problems. Clearly, scientific research and the development of scientific fact will continue to be important.

However, there undoubtedly will be situations in which the disagreement will focus on values and where additional research will be of little help. It is equally clear that we must become more adept at dealing with values.

At the start of this paper we discussed the subjective, idiosyncratic nature if values. To some, this implies that a value judgment is immutable — you like vanilla, we like chocolate — so the discussion ends there, and any further disagreement must be resolved with political clout or judicial action. We disagree. We submit that this can be the case only with the simplest, most basic preferences. In most situations higher-order values are involved and, as our discussion of rational values indicated, it is possible to examine these higher-order values, and the decisions that flow from them, rationally. In fact, the *Summa Theologiae* of St. Thomas Aquinas sets the standard for logical reasoning about values or about any subject, provided, of course, that one accepts the premises.

On this basis, then, rational values — the standards we use for judging truth — are exceptionally important for multiple objective decision making. This is especially the case where multiple public values generate conflicts that must be resolved. The planning and public involvement processes currently used by many state and federal resource management agencies provide an excellent venue to incorporate a plurality of values. In the future we will need to sort through these values without being unreasonably committed to a particular position or unreasonably critical of other positions. The best decisions will be based upon value positions that are clearly rational.

References

Arnheim, R. 1966. *Toward a Psychology of Art*. University of California Press, Berkeley.
Averill, J. 1992. The structural bases of emotional behavior: a metatheoretical analysis. Pages 1–24 in M.S. Clark, Ed. *Review of Personality and Social Psychology*. Volume 13. Sage, Newbury Park, California.
Averill, J. and T. More. 1993. Happiness. Pages 617–629 in M. Lewis and J. Haviland, Eds. *Handbook of Emotions*. Guilford Publications, New York.
Bengston, D. 1993. Changing forest values and ecosystem management. *Soc. Nat. Resour.* 7:515–533.
Dawkins, M. 1980. *Animal Suffering: The Science of Animal Welfare*. Chapman and Hall, London.
Dennett, D.C. 1991. *Consciousness Explained*. Little, Brown, Boston, Toronto, London.
Eysenck, H. 1942. The experimental study of the "good Gestalt" — a new approach. *Psychol. Rev.* 49:344–364.
Fechner, G. 1876. *Vorschule der Aesthetik*. Breitkopf und Hartel, Leipzig, Germany.
Harris, M. 1974. *Cows, Pigs, Wars, and Witches*. Vintage Books, New York. 276 pages.
Kennedy, J. 1993. *Preparing for the Twenty-First Century*. Random House, New York.
Leopold, A. 1953. *Round River*. Oxford University Press, Oxford, England.
Nash, R. 1989. *The Rights of Nature*. The University of Wisconsin Press, Madison. 290 pages.
Partridge. E. 1986. Environmental ethics without philosophy. Pages 136–147 in R. Borden, Ed. *Human Ecology: A Gathering of Perspectives*. Selected papers from the First International Conference of the Society for Human Ecology. The Society for Human Ecology, College Park, Maryland.
Rokeach, M. 1973. *The Nature of Human Values*. Free Press, New York.
Sayer-McCord, G. 1988. *Essays on Moral Realism*. Cornell University Press, Ithaca, New York. 317 pages.

Chapter 44

Quantifying Economic Incentives Needed for Control of Nonpoint Source Pollution in Agriculture

PHILIP HEILMAN, LEONARD J. LANE, AND DIANA S. YAKOWITZ

Multiobjective decision support systems (MODSS) can be a powerful tool to improve natural resource management in agriculture. When the decision is the selection of a land management system from the point of view of all of society, a problem may arise. Some of the objectives will reflect the interests of land managers and others by those affected off-site. An important issue is how to encourage the adoption of improved management systems if they are in society's overall interest, but not in the land manager's interest, as happens with nonpoint source water pollution. Further economic analysis is needed to encourage the adoption of improved management systems, as a complement to MODSS. A farm scale optimization model can be used to estimate the expected cost to a farmer of adopting management systems which will abate the production of agricultural pollutants. An example from the deep loess hills of western Iowa illustrates this approach.

Introduction

A common problem in environmental management is to select a land management system that provides the landowner with an adequate stream of income, maintain long-term productivity, and does not cause negative off-site environmental impacts. MODSS — such as the USDA-ARS Southwest Watershed Research

Center's Multiobjective Decision Support System for Water Quality (Stone et al., 1995) — can be used to rank management systems based on all three primary objectives. However, if an alternative management system is identified to be preferable to the existing system, a pertinent issue is how to encourage the adoption of the preferred management system. If the preferred management system can increase income, sustainability, or reduce the environmental damage directly affecting the landowner, then it is in the landowner's interest to adopt the preferred management system.

On the other hand, the environmental damage may occur off-site and not directly affect the landowner, as happens with externalities such as nonpoint source pollution. In those situations, the landowner does not face the economic incentives needed to encourage the adoption of the preferred management system. This paper describes a method to quantify the cost to a landowner of abating the production of individual pollutants at the farm scale. Management systems highly ranked in a DSS will be implemented only if they are feasible, that is if their resource requirements, such as seasonal labor, do not exceed the quantities available to the farmer.

A method for estimating the economic incentives facing farmers consists of defining a representative farm, consisting of a number of fields, and a set of alternative management systems deemed likely to resolve the major negative off-site effects from each of the fields. Available data relating the effect of potential alternative management systems on the criteria of interest are collected, or if not available, simulated. The current and alternative management systems are then scored in a MODSS. If no alternatives are found which improve upon the current management systems, then no significant improvement is likely and resources can be devoted to other areas where there is a greater potential for improvement. If alternative management systems are identified that are preferred to the current management systems, a farm-scale optimization model is used to analyze the tradeoffs between farm income and the production of individual pollutants.

An understanding of the economic incentives facing farmers to adopt alternative management systems should facilitate the development of more appropriate policies to promote the voluntary adoption of preferred management systems. It may be possible to identify and promote alternative management systems, such as no till tillage, that can improve economic returns to the farmer as well as having positive off-site benefits. The overall benefits of an alternative management system may only be realized at a significant cost to the farmer, for example, by eliminating the use of a particularly cost effective pesticide. Alternative management systems that are ranked highest by MODSS will most likely be adopted if farmers have the economic incentive to adopt those systems. If society's preferred management systems are not the same as the farmer's, economic incentives could be implemented either as charges on management systems or estimated emission levels assuming the "polluter pays" principle is applied.

This paper will step through a multiobjective analysis of a set of alternative management systems for a representative farm in the deep loess hills region of western Iowa. The analysis consists of (1) simulating the effects of the alternative management systems on a number of criteria; (2) using a MODSS to rank the management systems; (3) and using a farm-scale optimization model to estimate abatement cost curves for sediment, nitrogen, and atrazine. Conclusions will be drawn about the need for economic analysis to complement the application of

MODSS, when off-site damage (externalities) exist. Additional details for all stages of the analysis are available in Heilman (1995).

Problem Definition

The Deep Loess Research Station (DLRS) near Treynor, Iowa has been collecting data on the effects of conservation management systems on erosion, runoff, sediment, and water quality from loess soils since 1964. Loess, wind-borne silt, has been deposited on the eastern side of the Missouri River in western Iowa to depths of 5 to 25 m. The soils on the research station, primarily Monona-Ida Series, are moderately permeable with 6 to 18% slopes. The four experimental watersheds are in two sets of pairs located 4 km apart. Observed data on rainfall, storm runoff, baseflow, and sediment yield are available from 1964 to the present, with some nitrogen and phosphorus movement data beginning in 1969 (Saxton et al., 1977; Hjelmfelt, undated).

A farm, representative of the deep loess hills region, consisting of 243 ha in 5 fields was defined. The fields have characteristics of two of the watersheds of the DLRS, Watersheds 1 and 4. Two 30-ha fields are not terraced and are assumed to be exactly like Watershed 1 of the DLRS, which is an unterraced field with a predominant slope of 12%. Three 61-ha fields are terraced and are assumed to be exactly like Watershed 4 of the DLRS. The terraces on Watershed 4 are "double-spaced" or separated by twice the distance in the Soil Conservation Service's specifications for terraces, but as no observed data exist for terraced fields which meet the specifications, Watershed 4 will be considered representative of a terraced field in the deep loess hills region.

A set of alternative management systems was defined that would reduce sediment, nutrient, and pesticide losses when compared to the management system in place on Watershed 1. The management system on Watershed 1 is continuous corn with deep disking, preplant anhydrous ammonia applied at a rate of roughly 168 kg/ha, and atrazine used as an herbicide. The timings and quantities of inputs used with each operation are specified in Heilman (1995), and the rotation, tillage systems, nitrogen application methods and rates, and pesticide subsystems used are described as follows.

All alternatives use a corn-soybean rotation, which is the standard rotation in the area. Two tillage systems will be considered: mulch till, which allows some tillage, as long as there is 30% residue cover at planting (here assumed to be a shallow disking) and no till, which does not allow tillage and has higher residue levels. Three possible nitrogen application methods will be considered: liquid nitrogen and a pre- or postplant application of anhydrous ammonia. It is assumed that liquid nitrogen will be custom applied, so the farmer will have more labor available in May for other activities.

Two different nitrogen application rates are considered, 140 kg/ha (125 lb/acre) and 168 kg/ha (150 lb/acre). Only the corn crop in each rotation receives nitrogen, which is split into two applications. Each corn crop gets 28 kg/ha of nitrogen at planting as starter fertilizer, and the remainder in one of the three methods mentioned above. Preplant nitrogen application ensures that the farmer will not have to devote time to fertilizing between planting and when operations can no longer be performed on a crop. However, the nitrogen is available to be leached below the root zone during the period when large rainfall events are possible and before the crop roots have developed sufficiently to utilize the nitrogen. Postplant application provides nitrogen only later in the season when the growing plants need it, but requires another operation at a time when the farmer would prefer to concentrate on weed control.

Effective weed control often requires a number of different herbicides depending on the timing and type of weed infestations. Only atrazine will be considered as a decision variable, although a number of different herbicides will be simulated. Atrazine is a commonly used herbicide because it is inexpensive and effective. Unfortunately, atrazine is also persistent, particularly once it reaches surface waters. Current measures to control atrazine in Iowa include limits on the amount that can be applied in any given year and a prohibition against using it within 66 ft of waterways.

In total, 24 alternative management systems will be considered on each of the two types of fields. Each management system consists of a combination from each of: two tillage systems, mulch till and no till; three nitrogen application methods, liquid and pre- and postplant anhydrous ammonia; two nitrogen application rates, 140 and 168 kg/ha; and either atrazine or another herbicide.

Simulation Model

Since Watersheds 1 and 4 of the DLRS did not use any of the alternative management systems, a simulation model was used to estimate the effects of the alternative management systems on a number of measures reflecting different objectives. The model was parameterized and run using the Multiple Objective Decision Support System for Water Quality (WQDSS), developed by the Southwest Watershed Research Center of the USDA-ARS. The WQDSS was developed to run under the Unix operating system and implemented using the X Window System and the Motif Libraries* in order to provide a graphical user interface. Components of the WQDSS include databases, input file builders, simulation models, a decision model, and a system driver. The user interface to the system driver and the input file builders which facilitate running the simulation model by using the databases to parameterize the simulation model (Hernandez et al.,1993).

The simulation model is a modification of the Groundwater Loading Effects of Agricultural Management Systems (or GLEAMS) model (Leonard et al., 1987; Davis et al., 1990). Modifications to the model include the addition of a nitrogen leaching component from CREAMS (Knisel, 1980) and the EPIC crop growth component (Williams et al., 1989). A budget generator based on the Cost And Returns Estimator (or CARE) (Midwest Agricultural Associates, 1988), is used to compute the net returns and estimate the amount of time needed for each operation. The simulation model is capable of estimating the sediment yield, nutrient and pesticide loading in runoff and adsorbed to sediment to the edge of the field and the nutrient and pesticide leached below the root zone, as well as net returns for many management systems in rainfed agriculture. A more detailed

---

* Registered trademarks of AT&T, The Massachusetts Institute of Technology, and the Open Software Foundation, respectively. Mention of a tradename does not constitute or imply endorsement by the USDA-ARS.

explanation of the modifications made to the simulation model, hereafter referred to as HGLEAMS, is available in the WQDSS Reference Manual, version 1.1 (Southwest Watershed Research Center, 1994). Erosion (overland detachment) was estimated by the West Pottawattamie County Field Office of the Natural Resources Conservation Service using the Universal Soil Loss Equation (Wischmeier and Smith, 1978).

The model was parameterized using the WQDSS default databases. The most sensitive parameters were modified based on the observed effects of deep disking and ridge till on runoff, sediment yield, baseflow, and corn yield and a method to estimate the runoff curve number based on crop residue presented in Rawls et al. (1980).

Ranking Alternative Management Systems with the WQDSS

The decision theory used in the WQDSS is based on Wymore (1988), Lane et al. (1991), and Yakowitz et al. (1992, 1993a, 1993b). The WQDSS uses score functions for each decision variable and an importance order to calculate an overall score for each management system which is then used to rank the management systems. Ideally, the score functions and importance order would be based on site specific information relating the average annual movement of pollutants from the edge of the field and bottom of the rootzone to off-site damages. For this study, the default scoring functions were used without modification. An importance order from society's point of view was determined by a group of experts familiar with local agriculturally related environmental problems on August 29, 1994.* The importance order determined by the experts is presented in Table 44.1 (both of the atrazine and both nitrogen objectives were given equal importance). The loess soils are so deep that erosion does not significantly reduce yields (Spomer and Alberts, 1984).

Table 44.1 Ranking of Objectives

Rank	Experts' importance order
1	Net returns
2	Atrazine in runoff
	Atrazine in sediment
3	Sediment
4	Nitrogen in runoff
	Nitrate nitrogen in percolation
5	Soil erosion
6	Other pesticides

---

* Marco Buske of the Iowa State University West Pottawattamie County Extension Service; Michael Dea, farmer and chairman of the West Pottawattamie Soil Conservation District; Larry Kramer of the Deep Loess Research Station; Lyle Peterson, Soil Conservation Service; and Roger Webster of Treynor Ag Supply. We appreciate the efforts of all who contributed to this research.

Ranking of the alternative systems is done relative to the "conventional" system that is currently in place. In this example, all management systems were judged relative to a mulch till, preplant anhydrous ammonia, 168 kg/ha N application with atrazine, as that is the management system closest to the one currently in place on Watershed 1. The default scoring functions of the DSS were used along with the importance order listed in Table 44.1, although no other pesticides other than atrazine were considered. On the nonterraced field, no till generally scored higher than mulch till, the 168 kg/ha N scored higher than the 140 kg/ha and the management system without atrazine scored higher than the one with atrazine, while there were no major differences in the scores for the method of nitrogen application (Figure 44.1). The management systems that scored the worst were the combinations of mulch till with atrazine, because of the relatively high amounts of atrazine in runoff. On the terraced field, the results were similar. The management systems with no till tended to score higher than the mulch till systems, the systems with higher levels of nitrogen application scored higher than those with low levels and those which did not use atrazine scored higher than those that did use atrazine (Figure 44.2). The method of nitrogen application did not have much effect on the scores. The management systems using both mulch till and atrazine also had the lowest scores.

Optimization Model

After one or several management systems that score higher than the conventional management system are identified, there is still the problem of inducing the farmer to adopt an improved system. In areas where society would most prefer to see a significant change in management systems, such as situations with large off-site damage from pollution, the farmer probably has little incentive to adopt the management system society would prefer. Efficient economic incentives to control nonpoint source pollution can be fashioned as subsidies/charges on polluting inputs or pollution generated, or as controls on the quantities of polluting inputs or pollution generated (Griffin and Bromley, 1982).

A farm is a system of interconnected activities. Changing one activity on the farm may require other activities to change as well. Because the WQDSS works on a single management unit (at the field scale), and the farm decision-making unit is the whole farm, situations may arise where the management system recommended could be used on an individual field, but not on all similar fields on the farm. For example, labor or machinery availability during a critical period, marketing limits, or government program limits may preclude the adoption of that management system on all of the fields of the farm.

To quantify the cost to the farmer of reducing pollution, a model of how the farmer would react to limits placed on the quantities of pollutants leaving the farm can be estimated using an optimization model. In effect, the optimization model is used to simulate a farmer selecting alternative management systems in order to maximize returns subject to risk aversion and whole farm feasibility. One of the benefits of building a farm scale optimization model is that additional constraints can be imposed on the allowable levels of pollutants leaving the farm's fields. By varying the amounts of the pollutants allowed to leave the fields, an abatement cost curve to the farmer for the pollutant can be estimated. Although

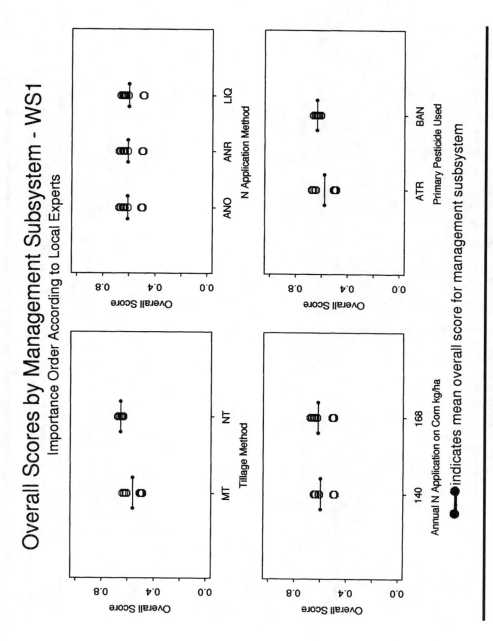

Figure 44.1 Overall score by management subsystem: mulch or no till; postplant, preplant anhydrous, or liquid N application; 140 or 168 kg/ha N application; and atrazine or glyphosate on Watershed 1 (nonterraced).

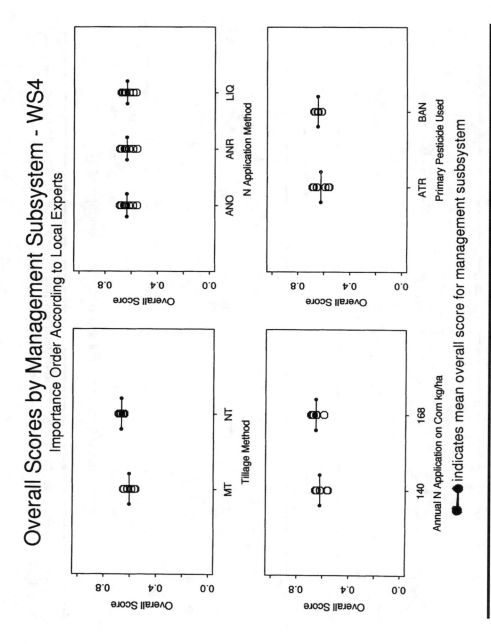

Figure 44.2 Overall score by management subsystem: mulch or no till; postplant, preplant anhydrous, or liquid N application; 140 or 168 kg/ha N application; and atrazine or glyphosate on Watershed 4 (terraced).

the cost of controlling a number of different pollutants could be analyzed, only the three most important will be analyzed here: sediment, nitrogen in percolation, and atrazine in runoff.

The structure of the optimization model used, the generalized mean variance approach, was first proposed by Paris (1979, 1989) to consider aversion to risk in both income and fluctuations in the availability of limiting input supplies. Farmers have to contend with variations in the availability of inputs, for example, when the amount of time available for field work is limited by the weather, or when labor supplies are uncertain. In a mathematical programming framework, such variations correspond to uncertain right-hand sides of constraint equations which ensure that the use of that input is less than or equal to its expected supply.

Using the results of the hydrologic simulation model and the budget generator, a model was built using the GAMS algebraic modeling language (Brooke et al., 1988) that contained constraints limiting the labor to available labor, field area, and time available when the fields are workable. From a base solution, additional constraints were added to limit the quantities of pollutants leaving individual fields to estimate abatement cost curves. Such constraints could be implemented in programs such as the USDA's Conservation Compliance Program which limits eligibility to government programs to those farms following acceptable conservation plans.

As a validation test, the optimization model was used to predict the management systems currently in place. A survey was conducted of farms with predominantly 12% slopes on Ida-Monona soils, in the West Pottawattamie Soil Conservation District to determine what management systems farmers were actually using. The optimization model's predictions were based on farm-size class, the size equipment (the number of rows) that farmers used, and the amount of time farmers reported working in May and June. The predictions made by the optimization model generally matched those reported by farmers, although the model overpredicted the proportion of farms using no till and atrazine (Heilman, 1995).

The farmer was assumed to use 6-row equipment, work an average of 60 hr/week during the months of May and June, and to be moderately risk averse. By changing tillage systems and nitrogen application methods, both labor and field day availability could be modified to maintain feasibility. The base solution without environmental constraints was to use no till, preplant nitrogen at 168 kg/ha and atrazine on almost the whole farm, with only 0.3 ha getting liquid nitrogen. If the farmer wanted to work less, or had 4-row equipment rather than 6-row, liquid nitrogen would have been used on a greater portion of the representative farm.

Sediment Abatement Costs

Although the deep loess hills region is prone to high rates of erosion, most of the soil is redeposited in the grassed waterways and at the base of the slope before leaving the field. For the observed management system on the unterraced Watershed 1, deep disking with continuous corn, the average annual sediment losses were only 14 t/ha from 1973 to 1991 (Deep Loess Research Station, 1992). As would be expected, sediment yields are very low. The most profitable tillage system is no till, which reduces detachment and transport, and in addition, most of the representative farm is terraced (Figure 44.3). Constraining sediment yield to less than 2 t/ha would force the farmer to abandon cultivation on some fields, particularly those that are not terraced.

The implications for conservation agencies are clear: the adoption of no till on similar farms should be encouraged, as both the farmer and offsite water users enjoy benefits. In this case, there is no tradeoff between net returns and sediment yield. Net returns with no till are higher, sediment yield is reduced and no till requires less time than the shallow disking associated with the mulch till system.

Nitrate in Percolation Abatement Costs

Farmers in the area are concerned about nitrogen affecting groundwater quality (Heilman, 1995). The amount of nitrogen leaving the field in surface water or percolation is affected by the natural rate of mineralization of the organic matter in the soils. The Ida Monona soils are high in organic matter, and will lose some nitrogen, no matter what management system is used.

Since the simulation model did not generate percolation at the same rate observed for baseflow at the DLRS, the units in Figure 44.4 are in parts per million concentration, rather than in units of mass. This conservative estimate overstates the concentration of nitrate nitrogen leaving the root zone. A potential problem is that the Maximum Contaminant Level (MCL) for nitrate is set at 10 ppm. The MCL is intended to be used in drinking water on an annual basis and should not be applicable to water leaving the root zone, assuming natural processes continue to reduce nitrate concentrations. The abatement cost curve is generally smooth, as greater proportions of the farm receive both postplant applications of nitrogen and lower rates of nitrogen application, first on the terraced fields and then on the unterraced.

Atrazine in Runoff Abatement Costs

As with most pesticides, it is possible to completely eliminate emissions of atrazine by replacing that pesticide with another, or some other means of reducing damage from the pest. In response to the survey mentioned earlier, the farmers reported that it cost $20/ha for the best replacement for atrazine. Because atrazine is only used every other year on the corn crop, the average annual cost to eliminate atrazine is $10/ha. If atrazine cannot be used at all, it would have a significant impact on farm returns (net of labor and land charges), which were reported to be on the order of $90/ha for 1,400 fields across Iowa in the years 1992 and 1993 (Soil Conservation Service, 1993).

All management systems considered used a corn-soybean rotation, the no till management systems earn more than the mulch till systems while reducing runoff, and the nitrogen application methods and rates had little effect on the quantities of runoff generated. Consequently, the only management choice that could affect the quantity of atrazine leaving the field is the quantity of atrazine applied (Figure 44.5) and the abatement costs are almost linear.

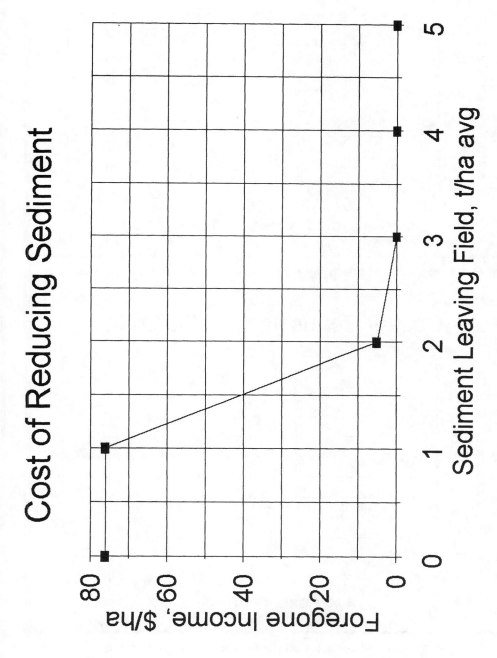

Figure 44.3 Abatement cost curve for sediment leaving the edge of the field.

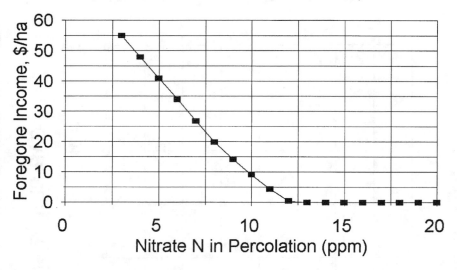

Figure 44.4 Abatement cost curve for nitrate N leaving the bottom of the root zone.

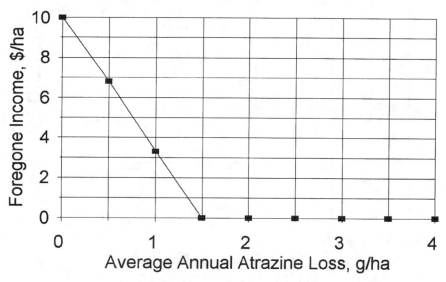

Figure 44.5 Abatement cost curve for atrazine leaving the edge of the field.

Summary and Conclusions

A policy to target farms to encourage the adoption of preferred management systems can be devised, using the information on farms in the deep loess hills region from the simulation model, the decision component of the WQDSS, and the optimization model. Specifically, no till tillage should be promoted to those farms which are not currently using no till, particularly those with small equipment. Management systems with high application rates of nitrogen were highly ranked by the decision component, even though there is a potential problem in exceeding the MCL for nitrate in percolation. Testing of nitrate concentrations under fields cropped in corn and soybean rotations could provide further information about the desirability of restricting nitrogen applications. If farms are using no till, then runoff is reduced, so that even if atrazine is ranked highly as a decision variable, management systems that include atrazine will be highly ranked and also selected in the optimization model. Farms that use mulch till should be the first to reduce atrazine use.

The benefits from using an optimization model as a complement to a MODSS for water quality include the ability to look at farm-scale issues, such as whole farm feasibility and risk aversion, as well as the ability to estimate abatement cost curves by varying constraints on pollutants emitted. However, care must be used in interpreting abatement cost curves. The simulation model estimates the quantities of pollutants leaving the field, rather than the amount of pollution in a given body of water.

Attempts to improve environmental management of nonpoint source pollution by using MODSS may require additional analysis to assess the economic incentives facing the landowners. Voluntary adoption of the preferred management systems will more likely occur if the management systems which lead to both an improved score and increased income for the farmer can be identified and promoted. If the most desirable management systems can only be adopted at some cost to the farmer, the magnitude of the economic incentives needed to make the farmer indifferent between the current system and the preferred system can be estimated and the potential benefits and costs of implementing a system of economic incentives can be examined.

References

Brooke, A., D. Kendrick, and A. Meeraus. 1988. *GAMS: A User's Guide*. The Scientific Press, South San Francisco.

Davis, F.M., R.A. Leonard, and W.G. Knisel. 1990. *GLEAMS User Manual*. Lab Note SEWRL-030190FMD. USDA-ARS and University of Georgia, Tifton.

Deep Loess Research Station. 1992. Treynor, Iowa, unpublished data.

Griffin, R. and D. Bromley. 1982. Agricultural runoff as a nonpoint externality: a theoretical development. *Am. J. Agric. Econ.* 64:548–552.

Heilman, P. 1995. A Decision Support System for Selecting Economic Incentives to Control Nonpoint Source Pollution from Agricultural Lands. Ph.D. dissertation. University of Arizona, Tucson.

Hernandez, M., P. Heilman, L.J. Lane, J.J. Stone, J.A. Abolt, and J.E. Masterson II. 1993. A prototype shell for running field scale natural resource simulation models. In *Application of Advanced Information Technologies: Effective Management of Natural Resources*. ASAE, Proceeding of the 18–19 June 1993 Conference, Spokane, Washington.

Hjemfelt, A. Deep Loess Research Station Summary of Results. Unpublished. USDA-ARS, Columbia, Missouri:

Knisel, W.G., Ed. 1980. CREAMS: A Field-Scale Model for Chemicals, Runoff, and Erosion from Agricultural Management Systems. Conserv. Res. Report No. 26. U.S. Department of Agriculture, Science and Education Administration, Washington, D.C.

Lane, L.J., J.C. Ascough, and T. Hakonson. 1991. Multiobjective Decision Theory — Decision Support Systems with Embedded Simulation Models. Irrigation and Drainage Proceedings 1991, Honolulu, HI, July 22–26.

Leonard, R.A., W.G. Knisel, and D.A. Still. 1987. GLEAMS: groundwater loading effects of agricultural management systems. *Trans. ASAE.* 30(5):1403–1418.

Midwest Agricultural Research Associates. 1988. USER Manual — Cost and Return Estimator. USDA-Soil Conservation Service, Contract no. 54-6526-7-268.

Paris, Q. 1979. Revenue and cost uncertainty, generalized mean-variance and the linear complementarity problem. *Am J. Agric. Econ.* 61:268–275.

Paris, Q. 1989. Revenue and cost uncertainty, generalized mean-variance and the linear complementarity problem: reply. *Am. J. Agric. Econ.* 71:810–812.

Rawls, W.J., C.A. Onstad, and H.H. Richardson. 1980. Residue and tillage effects on SCS runoff curve numbers. Pages 405–425 in W.G. Knisel, Ed. CREAMS: A Field-Scale Model for Chemicals, Runoff, and Erosion from Agricultural Management Systems. Conserv. Res. Report No. 26. U.S. Department of Agriculture, Science and Education Administration, Washington, D.C.

Saxton, K., G. Schuman, and R. Burwell. 1977. Modeling nitrate movement and dissipation in fertilized soils. *Soil Sci. Soc. Am. J.* 41:265–271.

Soil Conservation Service (Iowa State Office). 1993. MAX.

Southwest Watershed Research Center. 1994. A Multiple Objective Decision Support System for the USDA Water Quality Initiative: Reference Manual, Version 1.1. August.

Spomer, R.G. and E.E. Alberts. 1984. Corn yield response to erosion on deep loess soils. Abstract in Proceedings of National Symposium on Erosion and Soil Productivity. ASAE, New Orleans, Dec 10–11. Page 285.

Stone, J.J., L.J. Lane, and D.S. Yakowitz. 1995. A multiobjective decision support system for evaluation of natural resource management systems. Proceedings of Malama 'Aina.

Williams, J.R., C.A. Jones, J.R. Kiniry, and D.A. Spanel. 1989. The EPIC crop growth model. *Trans. Am. Soc. Agric. Eng.* 32:497–577.

Wischmeier, W. and D. Smith. 1978. Predicting Rainfall-Erosion Losses. Agriculture Handbook 537. U.S. Department of Agriculture, Science and Education Administration, Washington, D.C.

Wymore, A.W. 1988. Structuring system design decisions. Pages 704–709 in C. Weimin, Ed. *Systems Science and Engineering.* Proc. Int. Conf. on Systems Science and Engineering (ICSSE '88), Beijing, China. July 25–28.

Yakowitz, D.S., L.J. Lane, J.J. Stone, P. Heilman, and R.K. Reddy. 1992. A Decision Support System for Water Quality Modeling, Water Resources Planning & Management Proceedings of the Water Resources Sessions/Water Forum '92, EE, HY, IR, WR Div/ASCE, Baltimore, Maryland. August 2–6.

Yakowitz, D.S., L.J. Lane, and F. Szidarovsky. 1993a. Multi-attribute decision making: dominance with respect to an importance order of attributes. *Appl. Math. Computat.* 54:167–181.

Yakowitz, D.S., J.J. Stone, L.J. Lane, P. Heilman, J. Masterson, and J. Abolt. 1993b. A decision support system for evaluating the effects of alternative farm management practices on water quality and economics. *Water Sci. Technol.* 28(3–5):47–54.

Chapter 45

CROPS: A Constraint-Satisfaction System for Whole-Farm Planning

NICHOLAS D. STONE, DAVID FAULKNER,
REBECCA K. SCHECKLER, JAMES W. PEASE, AND JOHN ROACH

The Comprehensive Resource Planning System (CROPS) and its planning engine, a constraint-satisfaction-based scheduling system, are presented and described in terms of a general framework for land use planning and decision support. The constraint-satisfaction system was modified to include dynamic constraint relaxation and a simplistic branch-and-bound scheme that allow it to solve the multi-objective farm planning problem. This system has recently been adopted by the Natural Resources Conservation Service (USDA/NRCS) after a 3-year field implementation trial in Harrisonburg, Virginia. The system is scheduled to be incorporated into the NRCS field office computing system over the next 2 years.

Introduction

CROPS is a multiobjective scheduling system that uses heuristics and constraint satisfaction to find acceptable farm-level plans for specific farms (Buick et al., 1992). The program is tailored to individual users (farmers and agricultural agency personnel) and bases its plans on a land and resource inventory of the farm, economic and production targets, and the farmer's personal values. The program allows farmers to focus on the critical factors in their farming operations — the acreage of specific crops they need to produce, their financial needs, the machinery and labor inventory of the enterprise, their environmental priorities — while freeing them from the need to specify which crop/management combination should go into a particular field in a particular season.

The system is built around a constraint-satisfaction (see Nadel, 1989) planning engine that deals with multiple objectives through a simple constraint relaxation scheme. It includes linkages to a geographic information system (GIS) for display of farm maps and spatial calculations and includes several simple evaluator modules to assess yields, environmental hazards, economic return, and other factors.

Functionality and delivery have been major emphases over the last 2 years of this system's development. CROPS has been built using a participatory process and incremental development. End users have been heavily involved in the development process and have been instrumental in generating support for the program's delivery. Our ultimate goal is to deliver a decision tool to the NRCS for use in their field offices across the U.S. to assist in whole-farm and ecosystem level assistance to farmers. Currently, the program is installed for evaluation and further development in the NRCS's Harrisonburg, Virginia, field office and will shortly be installed and tested in field offices in Pennsylvania and Indiana.

This paper describes the overall CROPS design and how the planning engine deals with multiple objectives through constraint satisfaction. We also discuss the status of our delivery project with the NRCS. As a general framework for the discussion, we start with a description of the process of decision support for land use planning (Stone, 1995a). This process is a cycle with the basic steps: inventory, setting goals, generating alternatives, assessment, decision making, implementation, and evaluation (Figure 45.1).

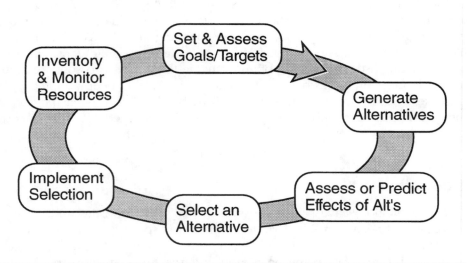

Figure 45.1 Land use planning decision support cycle. Multiobjective decision support systems for land use planning can involve one or more of these steps. The inventory, for example, is often associated with GIS, the assessment with simulation models, and the selection with expert systems or optimization. This cycle can be recursive with decreasing scale: in farm planning, the generation of alternatives requires the previous solution of the field-level planning problem, for example. Redrawn from Stone (1995a).

The cycle begins with a resource inventory, a process in which GIS have been extremely useful (e.g., Laacke, 1995; Williams and Roschke, 1995). Other data are also needed for this inventory in farm planning, including descriptions of farm equipment, labor availability, economics, and descriptions of livestock and other enterprises. The next step is goal setting — this can be as simple as stating that profit should be maximized, but it often involves more complex or potentially conflicting goals such as maintaining or improving the productivity of the land, limiting labor requirements, reducing risks of ground- and surface water contamination, or setting specific cash flow requirements. For land used in different ways by different people (e.g., forests), this aspect of planning can be the most important because it involves identifying and resolving conflicting goals of multiple stakeholders. In farm planning, there is one stakeholder but potentially conflicting objectives: maximize profit, minimize pollution, increase long-term sustainability, etc.

The next step is generation of alternative options. At a field level, this involves identifying appropriate management practices and cropping systems. Such "best management practices" are usually well catalogued. At the farm level, however, the alternatives are farm plans, and their development must be tailored to the individual farm. This farm-level generation of alternatives is the key problem addressed by the CROPS system.

Next, each alternative must be evaluated according to a set of criteria that assess how well each meets the goal(s) of the land owner. This step is generally associated with simulation modeling: one tries to predict the effects of one or more practices over time on the resources in question. Where multiple goals are involved, multiple assessments are needed. Difficulties at this level include the availability of reliable input data for the models and the validity or utility of the models used at the level of resolution needed for farm planning.

Even given perfect knowledge of how each of a set of alternatives will affect a set of resources, farmers must still choose the options they want to implement. This multiobjective decision-making step can be supported by a variety of algorithmic methods, or it can be left to the farmer. In the CROPS system, the three best plans generated are presented to the user for comparison. In principle, CROPS could return any number of plans, but in practice this amount of choice is difficult to deal with.

Implementing a plan will change farm resources, resulting in an updated farm inventory and a renewal of the planning cycle. Reactive planners would automatically modify their plans based on feedback about how the real-world system is tracking the implementation plan, but to date, CROPS is a static planner. If resources, goals, or markets change significantly, the system needs to be re-run to generate a new plan.

CROPS Structure

Based on the framework just described, CROPS is primarily a farm-level alternative generator. However, to produce alternative farm-level plans, it must first determine appropriate sets of field-level operations. Consequently, the program deals with the full cycle of land use planning just described at the field level: inventory of

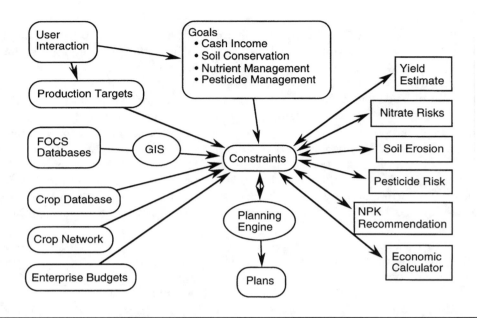

Figure 45.2 The CROPS system broken down by functional component into decision support activities and tools (ovals), evaluator modules (rectangles), and information sources and products (rounded rectangles). Arrows indicate information flow.

resources, setting goals and priorities, generating alternatives, assessing alternatives, and selecting alternatives for each field to become part of a farm-level alternative. Figure 45.2 shows the overall CROPS structure as a combination of assessment tools, input sources, and engines or manipulators.

The inventory is accomplished through a GIS and other input screens. Basic soil, topology, and land use data are accessed through GIS. Other information is gathered to augment these data layers including the presence of field features like gullies or sink-holes, and the ability/willingness of the farmer to use various tillage methods. CROPS also requires information on crop production — appropriate cropping sequences, enterprise budgets for specific crops, erosion factors for crop rotations, and other data to assess, for example, pesticide leaching and runoff risks under specific crop/tillage management scenarios. Most of the information on crops, tillage, and management are stored in a crop network. Nodes in the network are crops, and linkages show transition between crops in time. Most data are associated with the nodes, but some information is tied to the linkage. For example, in a rye-corn rotation, the link describes whether the rye is removed or killed and left in the field in preparation for corn planting.

Because CROPS works both at the farm and the field levels, it needs assessment tools or evaluators at both levels. Field evaluators assess, for example, the extent of pesticide leaching to expect from a specific crop rotation in a specific field. Farm-level evaluators are needed to assess, for instance, average annual net return.

Our research has focused on the use of information returned from evaluators, not on improving the evaluators themselves. There has been considerable work on the latter by others. For example, economic analysis of farm plans has been made much easier in recent years with the development of many computerized budgeting and analysis tools including spreadsheet-style budgeting programs (Gibson, 1988; Stevens and Peggs, 1988; Hoag, 1989), and more sophisticated analysis packages that combine financial analysis with some decision analysis (McGrann et al., 1989; Levins and Rego, 1990; Hilty and Johnson, 1987/88). Likewise there has been considerable work on assessing environmental impacts of agricultural practices using simulation models (e.g., Enfield et al., 1982; Dillaha et al., 1988; Zacharias et al., 1992). In these and other studies, the CREAMS model has been used widely to simulate nutrient leaching and runoff from agricultural systems. Soil loss has been most often simulated by the universal soil loss equation (USDA, 1989) although other models are available (Wischmeier and Smith, 1978). Similarly, the fate of pesticides after their application to agricultural crops and the subsequent environmental risks posed by pesticide applications has been evaluated by the use of simulation models like PRZM (Carsel et al., 1984) and GLEAMS (Leonard et al., 1987), by qualitative methods described in the SCS Field Office Technical Guide (Goss, 1988), by more complex risk assessment procedures (Geter et al., 1992), or by expert systems (Ferris et al., 1992).

Solution Process

The CROPS planner solves the multiobjective problem as a constraint-satisfaction problem (CSP) augmented by heuristics (Stone, 1995b). The CSP is represented in an object-oriented framework (Figure 45.3), consisting of a constraint manager object, objects to represent unary, binary, and n-ary constraints, and node objects to represent variables in the problem. CSP solvers have become popular in the artificial intelligence community for solving various combinatorial optimization problems and complex scheduling problems. The idea is to assign values to a set of variables subject to a set of constraints. The variables are represented as nodes in a constraint graph and the constraints as directed arcs between nodes. If two nodes are linked by an arc, $V_i \rightarrow V_j$, a constraint, $c_{i,j}$, exists in the set of all constraints, C. This means that values $(v_{j,k})$ allowed in V_j are constrained by the value of V_i.

The simplest solution to a CSP is to choose an ordering of the variables, $\{V_1, V_2,...,V_n\}$ with domains $\{\mathbf{d}_1, \mathbf{d}_2,...,\mathbf{d}_n\}$ where $\mathbf{d}_i = \{v_{i,1}, v_{i,2},..., v_{i,m}\}$, and to instantiate the variables in a depth-first manner. After each instantiation, the value, $v_{i,j}$, assigned to V_i is checked against the values of all V_k where $k < i$ and $c_{i,k}$ or $c_{k,i}$ Œ C. If an assignment fails, the next value in the domain is tried. If all values fail for a variable, Vi, the system backtracks to V_{i-1} and instantiates the next value from d_{i-1}. This depth-first tree-search algorithm is combined with arc-consistency checks — elimination of values of variables that are inconsistent with one another based on the constraint network itself. Any value, $v_{i,j}$, for which no consistent value of V_k can be found ($i \neq k$ and $c_{i,k}$ or $c_{k,i}$ C) can be eliminated from d_i.

In CROPS, there is a variable or node in the problem for each of the farm's fields (F) and one or more sets of year nodes (Y_n), each set consisting of one node for each year of the plan, with the number of sets determined by the number of target crops, n, specified by the user. The constraint graph for this problem connects each field to each year-target ($f_i \leftrightarrow y_{j,k}$), every year-target node within a given year ($y_{i,j} \leftrightarrow y_{i,k}$, $j \neq k$), but no connections among nodes of different

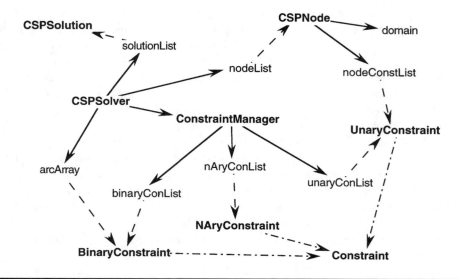

Figure 45.3 Main objects and linkages in the constraint-satisfaction planning engine in CROPS. The bold elements are the objects, standard face names are lists. Solid arrows (→) indicate a "contains" relationship; broken arrows (— →) show a "has elements" relationship; and doubly broken arrows (— - — - →) represent an "a kind of" relationship. The CSPSolver implements the CSP methodology (Nadel, 1989) while the ConstraintManager, the object that contains the constraints, takes care of constraint supervision and relaxation.

years or among fields. Each linkage or arc contains a list of one or more binary constraints. These specify (1) that a crop in a specific year and season in a rotation must match the crop in the field combination for that year and (2) that two target crops cannot conflict within the same year. Unary constraints relate to the user's objectives — each separate objective is encoded as one or more constraints, either on the practices that can be assigned to a field, the set of crops in specific fields that can achieve annual yield, or production and economic targets. Unary constraints are applied before any search of the constraint graph to eliminate domain items.

Consider a 65-ha farm with five fields, with acreages of 5, 10, 12, 17, and 21 ha. The farmer wants to grow about 30 ha of corn and 20 ha of wheat each year, wants to use crop rotation to control corn pests, and wants to minimize soil erosion and application of commercial fertilizer. We will assume that there are many ways to grow corn and wheat in rotations, but that some are more erosive than others, and that they vary in how they utilize legume cover crops, crop residue, and tillage.

This situation is translated in CROPS into several constraints: acreage target constraints limit annual corn acreage to near 30 ha and wheat to 20 ha. There are three ways to combine fields to grow between 28 and 32 ha of a crop, and three other ways to combine fields to grow 18 to 22 ha (Table 45.1). As shown, combos corn-1 and wheat-C, or corn-2 and wheat-A are compatible (can be scheduled in the same year), but other combo pairings are incompatible because they require two crops to be grown in the same field at the same time: corn-2 and wheat-B, corn-3 and wheat-B.

Table 45.1 Field Combinations Totaling Near 30 and 20 ha of Corn (C) and Wheat (W), Respectively, for the Hypothetical Farm Described in the Text

		Corn combos			Wheat combos		
Field	ha	1	2	3	A	B	C
1	5	C	—	—	W	—	—
2	10	C	C	—	—	W	—
3	12	—	—	C	—	W	—
4	17	C	—	C	W	—	—
5	21	—	C	—	—	—	W
Totals	65	32	31	29	22	22	21

Note: A valid "corn combo" is indicated by a column in the table: fields designated with a 'C' are planted in corn, other fields in the same column must not be planted in corn.

This same example would also include several unary constraints. For example: (1) a preference constraint on commercial fertilizer applications that ranks all potential crop rotations in a given field based on estimated applications; (2) a tolerance constraint on soil loss, requiring any crop rotation assigned to a field to result in minimal soil erosion (less than a tolerance level set externally).

Any valid plan must include acceptable rotations for each field that match valid field combinations of crops in each year. This is hardly a trivial exercise. For this farm with just five fields, the search space for 4-year plans is over 2 billion possibilities, assuming approximately ten rotation options.

Dealing with Multiple Objectives: Failure, Relaxation, and Search

Each constraint in CROPS deals with one objective of the user: desired profit, acreage restrictions, soil conservation goals, etc. Evaluating a given option (e.g., a crop rotation in a specific field) against a constraint requires an evaluator. That can be a simple table look up function, a simple rule base, or a model. Simulation models and other numerical or qualitative measures have often been in land use decision support systems (DSS) to assess the success of a given option in achieving one or more objectives. Lane et al. (1991), for example, describe a general approach for using simulation models to evaluate decision variables in natural resource decision support. In general, it is not the evaluation of the achievement of a particular objective that is difficult, it is coming up with an acceptable way to determine the best overall alternative given multiple conflicting objectives that is most difficult and that has spawned the most work. This necessarily involves incorporating value judgments into a quantitative analysis.

The constraint-satisfaction methodology used in CROPS is not an optimizing procedure. It is designed to find feasible solutions in large combinatorial problems. We have augmented the planning engine in two ways to search for nearly optimal solutions: we've added dynamic constraint relaxation, and we've implemented a simplistic branch-and-bound search.

Constraints in the CROPS system are either relaxable or not. Those that are relaxable include, for example, tolerance constraints and range constraints. In the example above, the farmer's goal was 20 ha of wheat planted each year. The system's constraint allowed all combinations that gave 20 ± 2 ha, a 10% error. At a tolerance of zero, there would have been no combinations giving exactly 20 ha of wheat, and the planner would have failed. At that point, this tolerance would have to have been relaxed (e.g., up to 10%) to allow planning to proceed.

Relaxable constraints in CROPS are defined with an "ideal" and a "worst acceptable" case. The planner initially sets all constraints to their ideal setting. For example, no soil loss greater than soil formation rates, low risk of nitrate leaching. These settings almost invariably cause the planner to fail to find any acceptable plan, and at the time of failure, the system calls the object named the "Constraint Manager" (Figure 45.3) asking it to relax a constraint before attempting to plan again.

The Constraint Manager keeps track of each constraint defined in the system, and it has a record of how often a constraint was checked and how often it failed. Each constraint is also linked to one of the user's goals, for which the user has attached an "importance" value. So, for each relaxable constraint in the system following a planning attempt, the Constraint Manager knows which ones have been the most constraining (highest failure rate) and which the user cares the least about maintaining. These values are combined to determine the next constraint to relax.

If all constraints had equal failure rates, the system ought to relax the constraint the user cares the least about. If the user cares about each equally, the system should relax the one that has been most constraining. This is obvious when one considers the extreme case of a constraint failing every time it is checked. Regardless of what the user cares most about, until that constraint is relaxed, the system will go on failing. In practice, CROPS uses both failure rate multiplied by a factor of one minus user-specified importance (defined as a number from zero to one) as the relaxation priority value. This assures that over multiple planning and relaxation cycles that the first solution achieved will be near-optimal considering the user's preferences and the specific resource situation on the farm.

Once one or more solutions have been found, the CROPS planner continues to search for a limited time (modifiable) for solutions that achieve better economic return. During this process, the system does not allow any further constraint relaxation. This process is a kind of hill-climbing search based solely on economics and constrained to keep environmental risks from increasing.

Delivery to NRCS

The ultimate goal of developing the CROPS system was to make it possible for farmers to achieve their planning goals in a sustainable way. It was also apparent that this kind of system would be valuable to farmers wanting to meet the environmental requirements of the 1985 and 1990 farm bills, particularly those provisions that required farm planning.

At the same time, the NRCS (formerly SCS) in Virginia became interested in CROPS as a way to help their field staff incorporate whole-farm planning methodology in the assistance they provide to farmers. Since the 1985 Farm Bill, the

NRCS had taken on a regulatory function, developing and approving farm plans for farms with highly erodible farmland. In July of 1992, Virginia SCS and Virginia Tech began a limited project to adapt the CROPS planning system to serve as a decision aid for SCS field offices. Since that time, a demonstration site has been established in Harrisonburg, Virginia, and there will soon be new sites in Pennsylvania and Indiana.

This implementation and prototyping phase of the system's development has included (1) documenting the farm planning process as it exists in the field offices; (2) modifying the structure of the CROPS system to link to NRCS computer systems, databases, and GIS; and (3) adding components to the system to add functionality at the field office level, beyond the overall goal of generating plans.

NRCS planning is supposed to assess and address problems relating to all farm resources: soil, water, animal, plant, and air (SWAPA) and to all farm enterprises including crops, orchards, livestock, forests, and wildlife. Referring back to Figure 45.1, this requires a considerable inventory, a goals assessment process, a broad set of assessment tools that operate on the resource inventory, and a large set of alternative practices. In short, the scope of NRCS planning is wider than that of CROPS, but the overall is consistent with the general CROPS approach. NRCS planning also is less detailed than that in CROPS. For example, the use of winter cover crops is an NRCS practice whereas CROPS recommends a specific crop for each winter season.

Extending CROPS to include multiple enterprises and all the assessments stipulated by the NRCS planning manual has not been practical. NRCS has not developed or approved assessment tools or objective methodologies to enable all these factors to be included in a planning system like CROPS. For example, practices like manure applications are known to have an effect on air quality, but there is no structured method by which to assess this effect. The delivery of the CROPS system will not, therefore, fully implement the NRCS planning process, but it will provide a methodology to apply the best scientific knowledge available to the crop-based farm planning that NRCS does, and will provide a model for how other considerations can be included in the future.

The economic considerations included in CROPS also had to be modified to meet the needs of NRCS. That agency bases the assessment of a plan on a comparative approach: compare the installed plan to a "benchmark," essentially, the conditions before NRCS involvement. Plans should be evaluated in terms of all the expected *changes* to the farming operation, not just the end result. This comparative approach has proven easy to apply to environmental assessments like soil erosion: CROPS initially compared environmental risks under the planned system to risk assessments based only on the resources (Buick et al., 1992). This change was implemented by keeping a record of the farm's benchmark plan and using the same assessment algorithms to generate the benchmark risks.

Implementing a comparative approach in the economics meant moving to a partial budgeting system in which only the changes to inputs, outputs, costs, and returns were evaluated. This has proven to be simple conceptually, but has required more detail in the representation of costs and inputs (labor, materials, machinery) than in the past. Eventually, this comparative approach will allow NRCS field staff to identify all effects categories which are projected to change significantly as a result of applying a conservation plan, and to describe each change in descriptive and/or quantitative terms. For example, "water infiltration increased *somewhat* in field 5," as a result of the CROPS plan. NRCS now uses a qualitative effects matrix to generate these kinds of assessments; it is a complete mapping of practices (i.e., nutrient management, diversions, grassed waterway) to effects relating to the SWAPA resources.

Conclusions

The constraint-satisfaction planning engine as modified to include dynamic constraint relaxation and limited branch-and-bound search has provided an excellent structure for whole-farm planning. Regulatory requirements on farmers are easily represented as constraints, and the relaxation feature produces a nearly optimal result without specifying any *a priori* ordering of goals or imposing transformations of multiple objectives onto a single scale like utility.

Furthermore, the modular development of CROPS according to the general scheme presented in Figures 45.1 and 45.2 allow the planning system to serve as well as a more general DSS for farm planning in the NRCS field office. Results of the inventory process, for example, include descriptive maps, yield reports, and hazard ratings that have proven to be valuable during the goal setting stage. Likewise, the assessment tools have turned out to be useful in and of themselves during an interactive planning session.

Our partners in NRCS see the use of CROPS in their field office as having tremendous potential. CROPS will:

- Shorten the adoption/diffusion model process time requirements for research findings and other new developments to be extended and become wide spread in use
- Provide a means for promoting sustainable agriculture by containing the latest scientific knowledge base on the topic, all within an automated data accessing system linked to a GIS environment
- Serve to help reconnect economics and ecology through the integration of practical economic considerations with environmental concerns

The CROPS project represents an important experiment in the practical application of state-of-the-art computer technology to modern problems in environmentally sound management of landscapes, watersheds, and ecosystems. It has recently been adopted as part of the national software development effort for NRCS, which means CROPS will be available in NRCS field offices nationwide within this decade. Several challenges remain to be addressed, but this is a major step toward making multiobjective DSS a reality in mainstream U.S. agriculture.

References

Buick, R.D., N.D. Stone, R.K. Scheckler, and J.W. Roach. 1992. CROPS: a whole-farm crop rotation planning system to implement sustainable agriculture. *AI Appl.* 6(3):29–50.

Carsel, R.F., C.N. Smith, L.A. Mulkey, J.D. Dean, and P.P. Jowise. 1984. User's manual for the pesticide root zone model (PRZM): Release I USEPA-600/3-84-109. Athens, Georgia.

Dillaha, T.A., B.B. Ross, S. Mostaghimi, C.D. Heatwole, and V.O. Shanholtz. 1988. Rainfall simulation: a tool for best management practice education. *J. Soil Water Conserv.* 43:288–290.

Enfield, C.G., R.F. Carsel, S.Z. Cohen, T. Phan, and D.M. Walters. 1982. Approximating pollutant transport to groundwater. *Groundwater.* 20:711–722.

Ferris, I.G., T.C. Frecker, B.M. Haigh, and S. Durrant. 1992. Herbicide advisor: a decision support system to optimize atrazine and chlorsulfuron activity and crop safety. *Computers Electronics Agric.* 6(4):295–317.

Geter, W.F., S. Plotkin, and J.K. Bagdon. 1992. National agricultural pesticide risk assessment. Proc. ASAE Summer Meeting, Charlotte, North Carolina.

Gibson, E.L. 1988. Agricultural computer software for bare-bones budgets. *Sun-Diamond Grower.* 7(2):14–15.

Goss, D. 1988. Soil Pesticide Interaction Ratings. Field Office Technical Guide, Section II. USDA/Soil Conservation Service.

Hilty, B.J. and D.M. Johnson. 1987/88. FINPACK, a tool for financial planning. Coop. Extension Service Bulletin. University of Maryland, College Park.

Hoag, D.L. 1989. Budget planner: user-oriented whole-farm budgeting software. *South. J. Agric. Econ.* 21:163–169

Laacke, R.J. 1995. Building a decision support system for ecosystem management: KLEMS experience. *AI Appl.* 9(3):115–127.

Lane, L.J., J. Ascough, and T.E. Hakonson. 1991. Multiobjective decision theory — decision support systems with embedded simulation models. ASCE Irrigation and Drainage Proceedings, Honolulu, Hawaii, July 1991.

Leonard, R.A., W.G. Knisel, and D.A. Still. 1987. GLEAMS: groundwater loading effects of agricultural management systems. *Trans. ASAE.* 30(5):1403–1418.

Levins, R.A. and W.T. Rego. 1990. Agricultural planning expert: a model of farm enterprise selection. *South. J. Agric. Econ.* 22:63–68.

McGrann, J.M., K. Karkosh, and C. Osborne. 1989. Agricultural financial analysis expert system: software description. *Can. J. Agric. Econ.* 37:695–708.

Nadel, B.A. 1989. Constraint satisfaction algorithms. *Computat. Intell.* 5:188–224.

Stevens, M. and A. Peggs. 1988. A computerized farm budgeting program. *J. Agric. West. Aust.* 29:107–110.

Stone, N.D. 1995a. Whole-farm planning for crop/livestock farms: integrating nutrient management. Pages 255–261 in FACTS 95 Conference Proceedings, Cornell University, Morrison Hall, Ithaca, New York.

Stone, N.D. 1995b. Object-oriented constraint-satisfaction planning for whole-farm management. *AI Appl.* 9(1):61–69.

USDA/Soil Conservation Service. 1989. Water Quality Interpretations. National Bulletin No. 430-9-12. USDA-SCS, Washington, D.C.

Williams, S.B. and D.J. Roschke. 1995. Designing configurable decision support software: lessons learned. *AI Appl.* 9(3):103–114.

Wischmeier, W.H. and D.D. Smith. 1978. Predicting Rainfall Erosion Losses — A Guide to Conservation Planning. Agricultural Handbook No. 537. U.S. Department of Agriculture, Washington, D.C.

Zacharias, S., C. Heatwole, T. Dillaha, and S. Mostaghimi. 1992. Evaluation of GLEAMS and PRZM for predicting pesticide leaching under field conditions. ASAE Paper No. 92-2541. ASAE, St. Joseph, Michigan.

APPLICABILITY TO NATIONAL, REGIONAL, AND GLOBAL ISSUES

Chapter 46

Lessons Learned and New Challenges for Integrated Assessment Under the National Environmental Policy Act

S.A. CARNES AND R.M. REED

One of the first government-sponsored demands for integrated assessment to support decision making in the U.S. is embodied in the National Environmental Policy Act of 1969 (NEPA). Over the past 25 years, the Oak Ridge National Laboratory (ORNL) has supported federal agencies in evaluating health and environmental impacts as required by NEPA. Many of ORNL's efforts have focused on complex, programmatic assessments that break new ground and require and integrate expertise from a wide range of technical disciplines. Examples of ORNL projects that illustrate the use of integrated assessment approaches include environmental documentation for: (1) the Department of the Army's Chemical Stockpile Disposal Program, (2) the Federal Energy Regulatory Commission's licensing activities related to the Owens River Basin in eastern California and along a 500-mi reach of the upper Ohio River, and (3) the Nuclear Regulatory Commission's decision regarding restart of the undamaged reactor (Unit 1) at Three Mile Island. Our discussion of these examples illustrates successful integrated assessment approaches and identifies new challenges facing integrated assessment activities.

Introduction

With passage of the NEPA, the U.S. government embarked on the development of an integrated assessment paradigm by requiring federal government agencies to assess the impacts of their actions on the human environment in light of multiple objectives. In the early 1970s, the Department of Energy's national laboratories became involved with preparing NEPA documentation shortly after the Calverts Cliff decision that required independent, third-party preparation of environmental impact statements (EISs). ORNL, along with Argonne and the Pacific Northwest laboratories, were tasked by the Atomic Energy Commission to help prepare EISs on construction of nuclear power plants throughout the country. These projects presented research staff at the laboratories with significant challenges in developing integrated assessments useful to decision makers within well-defined time frames.

Over the past 25 years, ORNL has continued to support federal agencies in evaluating environmental impacts in compliance with NEPA. Many of our efforts have focused on complex, programmatic assessments that break new ground and require expertise from a wide range of technical disciplines. Building on the evolution of NEPA implementation as a response to increasing demands by the public, stakeholders and decision makers to consider and assess the impacts of federal actions in terms of multiple objectives, we describe in this paper several examples of ORNL projects that illustrate the use of integrated assessment approaches either directly related to NEPA compliance or to closely related projects. Our discussion illustrates successful integrated assessment approaches and identifies a number of challenges facing integrated assessment activities, including (1) assessing the distributional effects of federal actions and whether the distribution is proportionate across the demographic landscape (e.g., environmental justice), (2) accounting for impacts of the full life cycle of federal actions, and (3) identifying the appropriate spatial and temporal scales for assessment and satisfying the data and analytical needs associated with those scales in our integrated impact assessments.

Chemical Stockpile Disposal Program NEPA Support

ORNL began its NEPA support for the Department of the Army's Chemical Stockpile Disposal Program in 1984, shortly before Congress directed the Department of Defense to destroy the stockpile of lethal unitary chemical weapons. ORNL's involvement with this program has included preparation of the Draft Programmatic EIS (U.S. Army, 1986) and the Final Programmatic EIS (U.S. Army, 1988), ongoing preparation of site-specific EISs, and ongoing technical support in the development of an integrated emergency preparedness capability surrounding each of the eight stockpile locations in the continental U.S. (located in Maryland, Kentucky, Alabama, Indiana, Arkansas, Colorado, Utah, and Oregon). ORNL also supported development of the environmental assessment associated with movement of the U.S. stockpile located in Germany to Johnston Atoll (U.S. Army, 1990). The following discussion focuses on the integrated assessment approach used in the Final Programmatic EIS (Carnes, 1989). This approach continues to be used in our ongoing site-specific assessments.

Alternatives considered in the Final Programmatic EIS included (1) continued storage (the no-action alternative), (2) on-site disposal (disposal at each of the eight CONUS storage sites), (3) regional disposal at two of the CONUS storage sites, (4) national disposal at one of the CONUS storage sites, and (5) partial relocation/on-site disposal (i.e., relocation of the inventories stored at two of the eight CONUS storage sites with on-site disposal at the remaining storage sites). For each disposal alternative, disposal was defined as reverse assembly of the munition or storage container and incineration of all chemical agent or contaminated solids in functionally specialized incinerators. High temperature incineration was selected as the disposal technology by the Army based on its extensive testing and experience at Rocky Mountain Arsenal in Denver, Colorado, and at the Chemical Agent Munitions Destruction System facility at Tooele Army Depot in Utah, and a recommendation of the National Research Council.

In response to serious concerns expressed by some members of the public, the Final Programmatic EIS focused on the human health risks of alternative disposal strategies, with secondary attention to the environmental risks. In addition to comparing the health and environmental impacts resulting from routine operations (which were found to be relatively minor and could not be used to distinguish the impacts of the different alternative), the analyses used a probabilistic risk assessment (PRA) of the disposal and no-action (continued storage) alternatives to compare the alternatives in terms of health and environmental risks associated with the full suite of accident scenarios identified in the PRA. The source terms resulting from both routine operations and the accidental releases identified in the PRA were used as inputs to environmental pathway models to estimate impacts.

Although the assessment included an aquatic transport model to estimate impacts on surface waters, the inhalation hazard of chemical agents dictated that the bulk of the analysis rely on an atmospheric dispersion model to identify areas at risk. This model, D2PC (Whitacre et al., 1987), assumes a Gaussian distribution of agent in the vertical and cross-wind directions as agent is dispersed downwind and uses agent source strength, release modes (e.g., semi-continuous, instantaneous, and evaporative), and alternative atmospheric or meteorological conditions to estimate the plume parameters for different concentrations of the toxic agent. A computer-based method that integrates demographic, atmospheric, and dose information was then used to estimate the population at risk and fatalities resulting from an accidental release of agent (Hillsman and Coleman, 1989).

The identification of an environmentally preferred alternative, a standard requirement in NEPA documentation intended to integrate and balance the relative importance of all health and environmental impacts (but not all decision variables, such as cost, relationship to other missions, etc.) in the Final Programmatic EIS was based on a multiobjective decision methodology developed at ORNL and reviewed by expert consultants (Carnes, 1989a) (see Figure 46.1). This method, driven largely by the structure of the PRA and estimates of areas and populations at risk, compared alternatives on the basis of human health effects and ecosystem effects with respect to both routine operations and accident scenarios, and the potential effectiveness of emergency planning and preparedness. Four human health risk estimates (i.e., probability of one or more fatalities, maximum number of fatalities, expected fatalities, and person-years at risk) and one ecological risk estimate (i.e., expected plume area) were calculated for the programmatic alternatives

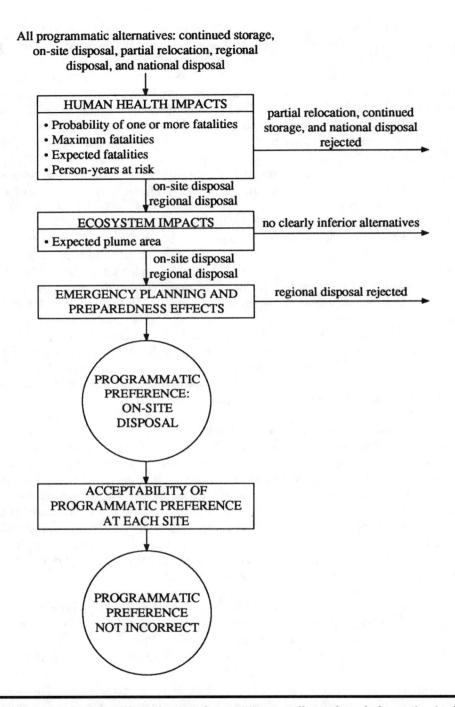

Figure 46.1 Methodology for selecting the environmentally preferred alternative in the Final Programmatic EIS for the Chemical Stockpile Disposal Program.

for the nation as a whole as well as for the diverse impact regions (i.e., those in proximity to the eight storage and proposed disposal locations as well as along cross-country transport corridors). The results of this analysis at the programmatic or national level are presented in Table 46.1.

Table 46.1 Comparison of Risk for Programmatic Alternatives for All Locations (Normalized Values)[a]

Alternatives	Probability of one or more fatalities	Maximum fatalities	Expected fatalities	Person-years at risk	Expected plume area (km^2)
Continued storage	2.4×10^{-3} (7.5)	8.9×10^4 (16.5)	4.5×10^{-1} (478.7)	1.4×10^8 (60.9)	4.4×10^{-2} (95.7)
On-site disposal	3.2×10^{-4} (1.0)	5.4×10^3 (1.0)	9.4×10^{-4} (1.0)	2.3×10^6 (1.0)	4.6×10^{-4} (1.0)
Regional disposal	1.8×10^{-3} (5.6)	4.2×10^4 (7.8)	9.5×10^{-3} (10.0)	5.5×10^6 (2.4)	2.0×10^{-3} (4.3)
National disposal	3.4×10^{-3} (10.6)	4.2×10^4 (7.8)	3.0×10^{-2} (31.9)	5.4×10^6 (2.3)	3.8×10^{-3} (8.3)
Partial relocation	3.7×10^{-3} (11.6)	2.3×10^4 (4.3)	2.5×10^{-2} (26.6)	3.1×10^6 (1.3)	6.6×10^{-3} (14.3)

[a] Normalized values obtained by dividing the smallest value found for each risk measure into the other values for the other alternatives; on any given risk measure, the best alternative has a value of one, and the others are expressed in multiples of one.

Table 2.6.2, U.S. Army, 1988.

Since the size of the total inventory, as well as those at the individual storage sites, was classified for national security reasons at the time the EIS was prepared, results for individual sites and impact regions had to be portrayed in a manner that would not divulge or enable the derivation of classified information. A pictogram approach was developed to communicate differences in impacts among the alternatives for each of the areas potentially impacted by the program. These estimates, along with decision rules regarding uncertainty in the analyses, allowed decision makers and the public to compare alternatives in terms of maximum as well as probability-weighted values.

Federal Energy Regulatory Commission NEPA Support

ORNL has provided NEPA support to the Federal Energy Regulatory Commission (FERC) since the early 1980s. Much of our work has involved assessing the cumulative effects of multiple hydropower development in river basins. We discuss two projects to illustrate different integrated assessment approaches that have proven useful to decision makers.

The licensing of seven small hydroelectric projects proposed for the Owens River Basin in eastern California was evaluated in an EIS published in 1986 (FERC, 1986). The seven projects were to be located on small tributaries of the Owens River in Mono and Inyo counties. The Owens Valley is an arid region that is fed by mountain streams from the eastern slope of the Sierra Nevada to the west and the White mountains to the east. The water resources of the area have been extensively developed to provide water for southern California, as well as to provide electricity generation and irrigation water. This development has resulted in major impacts to streams and stream-dependent resources in the region.

To develop a useful impact assessment for the EIS, an analysis was needed that integrated impacts of multiple projects on important sensitive resources and evaluated impacts from the proposed projects in the context of impacts resulting from other development activities (i.e., past, present, and the foreseeable future). Our approach for the Owens River Basin assessment involved using public, government agency, and environmental group input from the EIS scoping process to identify "target" resources that were of most concern. Once these target resources were identified, the assessment focused on (1) defining attributes for each resource that could be affected by the proposed projects, (2) quantifying potential impacts to the extent possible, (3) assigning value ratings to potential impacts, and (4) displaying the impacts in a matrix format to give decision makers and the public a clear understanding of the comparative impacts of each project and the cumulative impacts of developing all seven projects.

The target resources identified for analysis were resident trout, riparian vegetation, riparian-associated wildlife, aesthetics, recreation, and local economy. Assessment of nontarget resources (e.g., geology, land use, cultural resources) was also evaluated, but the results were not included in the matrix analysis. The impacts on target resources were quantified to the extent feasible using available information on the resources and baseline conditions. Once this step was completed, the FERC staff, using input from developers, government agencies, and the public provided through a series of technical workshops, assigned a rank from 1 to 10 (minimal to severe impacts) to the effects, with a definition given for each rank. The ranking scale was designed so that a value of 6 or greater was judged to be an unacceptable impact to any target resource. The results of this preliminary assessment were presented at a workshop of local resource managers to obtain input on the approach, the assessment, and the values being used. Mitigation was proposed that would lower the ranking, whenever possible, to a value of 5 or less. Table 46.2 shows the results of the analysis, and provides rankings with and without proposed mitigation.

The use of the ranking approach and matrix display proved to be useful in integrating the results of the analysis (McLean et al., 1986). The rankings highlight differences among projects and provides the decision maker with a clear picture of where impacts are likely to occur and an index of the magnitude of the impact. The primary value of the matrix is as a display and communication tool. The defensibility of each rank number resides not in the individual numbers in the matrix, but in the underlying analysis that is presented in the EIS. Cumulative effects for the projects were displayed as the additive ranking values for each resource across the seven proposed projects. These sums provided an index of the combined effects of developing all seven projects and also provided a means of assessing the effectiveness of proposed mitigation measures in reducing adverse impacts (Table 46.2).

A different approach to integrating environmental impacts was used in a second study for FERC that involved licensing actions for 24 proposed hydropower projects at 19 existing dams along a 500-mi reach of the upper Ohio River. We evaluated both cumulative and site-specific impacts that would occur to target and other environmental resources. Target resources that were identified during the scoping process included water quality, fisheries, recreation, wetlands, and river navigation. Four alternatives were analyzed that included different ways of developing the hydropower resources, ranging from production of all the proposed power with little environmental protection to producing 82% of the proposed power while causing no major environmental impacts.

Our approach for this study centered around assessment of interactions among sites and cumulative impacts. We determined that the major environmental change that would occur as a result of installing hydropower at these dams would be redirecting water through the turbines rather than allowing it to spill over the dams. The proposed projects would therefore reduce dissolved oxygen (DO) concentrations below the dams by reducing aeration that occurs when water spilled over the dam or through the gates. The amount of DO provided by the existing dams was quantified using field data collected above and below each dam in the study area. A water quality model was developed to estimate changes in DO concentrations associated with hydropower development at each of the 19 sites. The reduced spillage and altered tailwater flow would initiate a chain of impacts by diverting water through the turbines and subsequently impacting water quality, aquatic biota, and fishery resources.

The analysis found that development of the projects as proposed by the applicants would significantly lower the DO concentrations below the sites and would reduce the growth of many fish species in these areas. Reductions in good fish habitat would adversely affect recreational fishing and possibly wetlands. Cumulative impacts from other activities in the basin were evaluated by including in the DO model the biological oxygen demand (BOD) from major industrial and municipal wastewater treatment plants, as well as BOD loads that simulate

Table 46.2 Relative Ranks for Target Resources in the Owens River Environmental Assessment

Project name	Resident trout		Riparian vegetation		Riparian wildlife		Aesthetics		Recreation		Local economy[a]		Total	
	(w/o[b])	(with[c])	(w/o)	(with)	(w/o)	(with)	(w/o)	(with)	(w/o)	(with)	(w/o)	(with)	(w/o)	(with)
Pine Creek	5	3	9	2	10	2	4	3	10	2	−6	−3	32	9
Horton Creek	10	5	6	1	6	1	7	7	8	4	−2	−2	35	16
Horsetail	9	5	9	2	9	2	7	5	10	4	−2	−2	42	16
Aspen Park	8	3	9	6	10	6	8	7	10	7	−2	−2	43	27
Rancho Riata	7	5	3	2	1	1	3	3	NA	NA	−2	−2	12	9
Big Pine Creek No. 2	10	5	5	2	4	2	8	5	10	4	−4	−3	33	15
Tinemaha and Red Mtn. Ck.	10	5	3	2	5	4	5	5	10	3	−2	−1	31	18
Cumulative effects	59	31	44	17	45	18	42	35	58	24	−20	−15		

[a] A negative rank indicates an increase in personal income of the region.
[b] Without proposed mitigation.
[c] With proposed mitigation.

Tables 5.1-1 and 5.1-2, FERC, 1986.

nonpoint sources of BOD such as runoff, sediment oxygen demand, and decaying algae. The model was calibrated against an existing data set, and impacts were evaluated for several sets of conditions.

Alternatives to the proposed action involved mitigation of the decline in DO concentrations by modifying operations at the proposed projects during certain flow regimes to maintain spill flows that would allow attainment of water quality standards and/or minimization of potential impacts to fisheries. Alternatives also considered not licensing certain projects that had the greatest potential for causing adverse impacts. Figure 46.2 shows the results of the analysis for the four alternatives.

The Ohio River assessment is a good example of a study where initial scoping of the problem allows a critical variable to be defined that serves as a driver for the assessment of other impacts and allows integration of the impact analysis for closely related resources. Identifying DO as the major variable that was affected by altering the path of river flow at navigation dams provided a means for predicting changes to other resources that were dependent on changes in DO. Being able to model the effects of multiple projects on DO also provided a means for determining the interrelationships among projects. Such an approach is not always possible and can be overly simplistic, but when possible, it serves to integrate a number of impact drivers and provides a framework for understanding the interactions among various resources.

Three Mile Island Assessment

Integrating knowledge and understanding across the entire environmental domain (i.e., physical, life, and social scientific expertise) is the ideal. It is often the case, however, that knowledge and understanding within a single element of the environmental domain is sufficiently problematic that intradisciplinary integration (i.e., across the physical sciences, the life sciences, or the social sciences) is needed before a more holistic interdisciplinary integration can occur.

In response to a decision by the U.S. Court of Appeals for the District of Columbia in 1982 requiring an assessment, under the purview of NEPA, of the psychological and community well-being impacts of restarting the undamaged reactor (Unit 1) at Three Mile Island (TMI), the Nuclear Regulatory Commission (NRC) engaged support from ORNL to prepare such an assessment (Sorensen et al., 1983; Sorensen et al., 1987). Building on the extensive literature on human and organizational response to natural and technological hazards and disasters, this assessment integrated views, concepts, and theories from several social science disciplines. As outlined in Figure 46.3, the assessment utilized an analytical, cause and effect, framework based on prior investigations into how people and groups perceive and adjust to hazards in the course of their normal lives (e.g., see Burton et al., 1978), how individuals and societies respond to the threat of impending disasters and warnings (e.g., see Perry et al., 1981), and individual and organizational behavior in and about specific disasters (e.g., see Drabek and Key, 1984).

Using multiple research methods and approaches (including an extensive literature review, focus group meetings with proponents and opponents of restart,

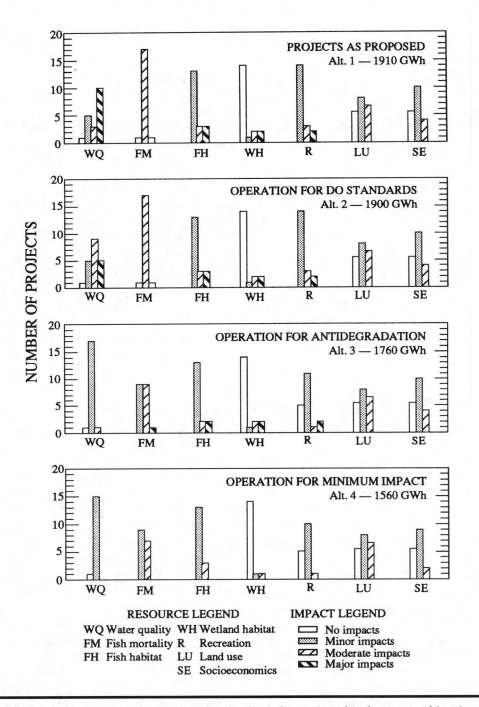

Figure 46.2 Summary of the impacts for the four alternatives for the FERC Ohio River EIS (FERC, 1988).

a review of analogous situations, and community and functional social group profiling), this assessment examined individual psychological impacts, manifestations of stress, and group-level and community-level impacts. Table 46.3 summarizes the major types of impacts revealed by the various approaches.

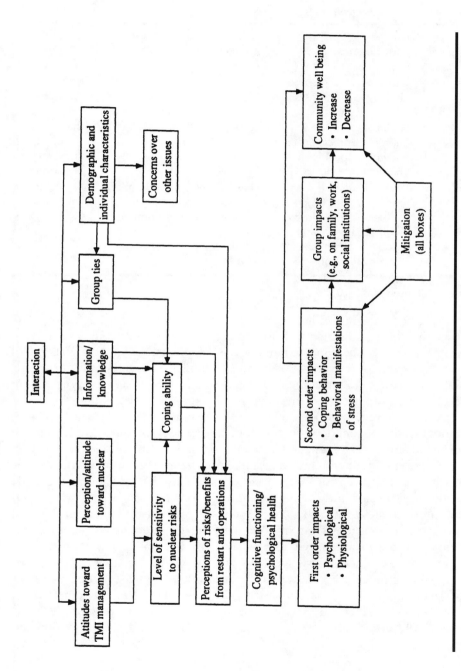

Figure 46.3 A causal model of TMI-restart psychological and community well-being impacts (adapted from Sorensen et al., 1987).

Table 46.3 TMI-Restart Psychological and Community Well-Being Impacts, by Method

Impact category	Method/approach		
	Focus groups	Profiling	Reviews
Individual psychological impacts	Fear, anxiety; hopelessness	Distrust; fear of health effects; anxiety; worry	Concern; some anxiety
Manifestations of stress	Disease; sleeplessness; lethargy; loss of job	Long-term health problems	Minor somatic effects
Group-level impacts	Social stigma; loss of home; conflict; protest; neighborhood loss; harassment; prosperity	Protest; violence; hostility	Intergroup conflict; loss of family cohesion
Community impacts	Higher taxes; outmigration; property values decline; loss of well-being; economic growth; stable taxes; lower utility bills	Higher taxes; outmigration; property values decline; increased development; lower taxes; property values increase; lower utility bills	No measurable impacts

Adapted from Sorensen et al., 1987.

By comparing the empirically derived impacts identified in the focus-group discussions and the profiling with those from reviews of analogous situations, it was possible to establish some general levels of confidence in the validity of various impact types. Moreover, the assessment identified a number of intervening contextual and perceptual factors that might shape the magnitude of restart impacts: (1) level of openness in decision making, (2) types and levels of information people receive about restart, (3) operating experience after restart occurs, and (4) the extent to which efforts are made to mitigate impacts. In assessing the distribution of impacts among populations in the vicinity of TMI, we found differential vulnerability to restart impacts among the functional social groupings in the TMI area and increased vulnerability for families in the early stages of their life cycles and for persons closest to the facility.

Although we believe that this assessment of the psychological and social impacts of restart constitutes an important effort to integrate social scientific understanding of the problem, we also believe that integrating the results of this integrated social scientific analysis with the other relevant domains (i.e., with the life and physical sciences) is the approach that should be taken when assessing hazardous technologies. This is particularly the case given the Supreme Court's reading of NEPA — that social (and psychological) impacts are cognizable under NEPA if and only if they are related to changes in the biophysical environment.

New Challenges for Integrated Assessment in NEPA

Although new challenges facing NEPA assessment derive from a variety of public, stakeholder, and decision-maker concerns, in the final analysis we need to do a better job of relating the cause and effect relationships in the natural and human environments and of integrating our knowledge and understanding. This requirement necessarily involves a better understanding of natural and human systems and their linkages. In particular, and as demonstrated from the examples discussed above and in other ORNL health and environmental assessments, we see three fundamental challenges in the context of integrated assessment for NEPA analyses — assessing the distributional effects of federal actions and whether the distribution is equitable across the demographic landscape (e.g., environmental justice), accounting for impacts of the full life cycle of federal actions, and identifying the appropriate spatial and temporal scales for assessment and satisfying the data and analytical needs associated with those scales in our integrated impact assessments.

It is no longer sufficient, since promulgation of Executive Order 12898 (Clinton, 1994), to project impacts to undifferentiated human populations. With the executive order and the environmental justice movement that preceded and now accompanies it, we must now be able to project impacts to minority and low-income populations within regions of impact for federal actions that trigger NEPA assessments (Vogt et al., 1995; Carnes and Wolfe, 1995). Equally important our assessments of the potential for environmental justice impacts must develop or use methodologies that are responsive to the concerns and perceptions of those populations; the executive order suggests that this latter objective is to be achieved by involving potentially affected minority and low-income populations in the development of impact assessment research designs, methodology development, and data collection and analysis.

Although the research community has some experience with considering the full life cycle of problem or policy domains (e.g., the nuclear fuel cycle), and the environmental assessment community also has some relevant experience (e.g., cumulative impact assessments, programmatic and generic assessments), it is frequently the case that the full life cycle of proposed federal actions is not adequately considered in NEPA documents. The challenge is to evaluate the full life cycle in a meaningful way when information is limited, inadequate, and/or unavailable. Addressing the full life cycle is increasingly important for decision makers, but needs to be done using bounding analyses (Saylor and McCold, 1994) and recognizing the uncertainties associated with those parts of the life cycle that lack clear definition.

Finally, environmental professionals and other scientists increasingly face challenges of identifying and assessing issues at appropriate spatial and temporal scales. In terms of spatial analyses, we deal with problems ranging from site- and regional-specific issues of ecosystem management to issues of national energy policy and global climate change; in terms of temporal scales, we are asked to predict immediate fatalities resulting from acute exposures to toxic materials, latent morbidities resulting from chronic exposures to toxic materials, and health and environmental threats to future generations resulting from loss of containment of toxic materials hundreds or thousands of years into the future. These different kinds of problems clearly call for different levels of spatial and temporal resolution.

These and other challenges lead us to believe that approaches that are strictly bottom-up or top-down are either misrepresentations of reality or of limited relevance to decision makers. Use of advanced computer techniques, geographical information systems, remote-sensing data, and data from regional, national, and international databases to address these and other issues becomes an ever more important aspect of providing decision makers with appropriate data, information, and analyses in time frames relevant for decision making (i.e., before all uncertainties can be identified, much less quantified). Knowing and understanding the linkages between the natural and human environments across space and time, and being able to demonstrate those linkages effectively and efficiently to decision makers and the publics, are fundamental challenges for all scientists.

References

Burton, I., R. Kates, and G. White. 1978. *The Environment as Hazard*. Oxford University Press, New York.

Carnes, S.A., Guest Ed. 1989. Integrated environmental assessment for a major project. *Environ. Profess.* 11(4):277–446.

Carnes, S.A. 1989. Decision making for the CSDP health and environmental assessment. *Environ. Profess.* 11(4):422–433.

Carnes, S.A. and A.K. Wolfe. 1995. Research and institutional dimensions of environmental justice: implications for NEPA documentation. Pages 286–291 in *Environmental Challenges: The Next 20 Years, NAEP 20th Annual Conference Proceedings*. National Association of Environmental Professionals, Washington, D.C.

Clinton, William J. 1994. Federal Actions to Address Environmental Justice in Minority Populations and Low-Income Populations. Executive Order 12898. The Executive Office, Washington, D.C. February 11, 1994.

Drabek, T. and W. Key. 1984. *Conquering Disaster: Family Recovery and Long-Term Consequences*. Irvington, New York.

Federal Energy Regulatory Commission (FERC). 1988. Final Environmental Impact Statement, Hydroelectric Development in the Upper Ohio River Basin. FERC/FEIS-0051. Office of Hydropower Licensing, Washington, D.C.

Federal Energy Regulatory Commission (FERC). 1986. *Final Environmental Impact Statement, Owens River Basin, Seven Hydroelectric Projects, California*. FERC/EIS-0041. Office of Hydropower Licensing, Washington, D.C.

Hillsman, E.L. and P.R. Coleman. 1989. Integrating demographic, atmospheric, and dose information to estimate effects from accidental release of chemical agents. *Environ. Profess.* 11(4):354–366.

McLean, R.B., D.M. Evans, G.F. Cada, C.H. Petrich, and R.M. Reed. 1985. Determination of the effects of multiple small-scale hydroelectric projects on resources in a California river basin: a matrix approach. Pages 1736–1745 in *Waterpower '85, Proceedings of an International Conference on Hydropower*. ASCE, Las Vegas, Nevada. Sept. 25–27.

Perry, R., M. Lindell, and M. Greene. 1981. *Evacuation Planning in Emergency Management*. Heath and Company, Lexington, Massachusetts.

Saylor, R.E. and L.N. McCold. 1994. Bounding analyses in NEPA documents: when are they appropriate? *Environ. Profess.* 16:285–291.

Sorensen, J.H., J.E. Soderstrom. E.D. Copenhaver, S.A. Carnes, and R. Bolin. 1987. *Impacts of Hazardous Technology: The Psycho-Social Effects of Restarting TMI-1*. State University of New York Press, Albany.

Sorensen, J.H., J.E. Soderstrom, R. Bolin, E.D. Copenhaver, and S.A. Carnes. 1983. Restarting TMI Unit One: Social and Psychological Impacts. ORNL-5891. Oak Ridge National Laboratory, Oak Ridge, Tennessee.

U.S. Army. 1986. Chemical Stockpile Disposal Program Draft Programmatic Environmental Impact Statement. Program Manager for Chemical Demilitarization, Aberdeen Proving Ground, Maryland.

U.S. Army. 1988. Chemical Stockpile Disposal Program Final Programmatic Environmental Impact Statement. Program Executive Office — Program Manager for Chemical Demilitarization, Aberdeen Proving Ground, Maryland.

U.S. Army. 1990. Global Commons Environmental Assessment and Finding of No Significant Impact (FONSI) for the German Retrograde Program. Office of the Assistant Secretary of the Army, Chemical Demilitarization Agency, Aberdeen, Maryland.

Vogt, B.M., J.H. Sorensen, and H. Hardee. 1995. Environmental Assessment and Social Justice. ORNL/TM-12919. Oak Ridge National Laboratory, Oak Ridge, Tennessee.

Chapter 47

Sustainable Agriculture and the MCDM Paradigm: The Development of Compromise Programming Models with Special Reference to Small-Scale Farmers in Chile's VIth Region

CLAUS KÖBRICH AND TAHIR REHMAN

Abstract

A typical view of sustainable agriculture is that it "… should involve the successful management of resources for agriculture to satisfy changing human needs, while maintaining or enhancing the quality of the environment and conserving the natural resources." Further such systems have to be economically viable, environmentally sound, and socially acceptable. Many different approaches have been proposed to evaluate these systems including: measures related to factor and total productivity; economic, environmental and social indicators; and single- and multiple-objective mathematical (mostly linear) programming models for analyzing economic and environmental issues. The main features of the compromise programming models proposed for Chilean agriculture are related to the objectives being considered and the scale that is represented. They consider both public and private objectives as well as the criteria of economic viability, ecological soundness, and social acceptability by *maximization of farm profit, minimization of risk, minimization of soil erosion*, and *minimization of differences in income*. The models developed represent the Chilean agriculture at two levels. The first, or lower level, is related to the decision centers that is a set of farming systems (FSs). Survey data on farm operations, household aspects, land use, agricultural practices, and livestock are analyzed with multivariate techniques, to construct groups of homogenous FSs. A representative farm for each group is selected for direct survey and the data collected are then used to construct individual FS models. The aggregation to the second or upper level provides the representation of a microregion. Each model is intended to evaluate a specific government development policy for the microregion under study. The compromise solutions generated should allow the policy-maker to examine the effects of each policy on the overall sustainability of the agricultural systems and their development.

Introduction

The sustainability of agricultural systems around the world is threatened by accelerated soil erosion, chemical pollution, salinity, water logging, organic waste materials, loss of genetic diversity, and habitat changes (Soule et al., 1990; Tivy, 1990). However, which of these threats is relevant to an agricultural system depends on local or regional characteristics. The study of sustainability, therefore, requires the identification of the relevant issues for a specific situation. The considerable body of literature on sustainability deals mostly with the conceptualization of the problems involved. Only a few contributions have operationalized these concepts, possibly due to the lack of a unique and widely acceptable definition of sustainability itself. To address these problems, a research project is currently underway at the University of Reading in association with the University of Chile to examine the most relevant concepts of sustainability appropriate to the development of agriculture in Chile and to develop a framework for the evaluation of sustainability. Based on it, Multiple Criteria Decision Making (MCDM) Models will be constructed to evaluate and select appropriate agricultural development policies in the light of relevant sustainability criteria. This paper presents that MCDM approach.

First a suitable definition of agricultural sustainability is given and then its basic relationship with FSs and agricultural policies is explored. A description of the study area and the procedure followed to identify and define the relevant farming systems is then given. Next the proposed MCDM models are specified emphasising the two main features: (1) the four objectives and (2) their simultaneous inclusion in the construction of a microregional model. Finally, some concluding remarks on the progress of the work are presented.

Agricultural Sustainability

At least three broad approaches are discernible from the literature on the definitions of sustainability and sustainable agriculture. The first, the North-American and European perspective, views sustainable agriculture as a production system with 'no-use' or 'low-use' of external inputs, leading to organic farming and low input-sustainable agriculture. The second approach, the standpoint from low-income countries (LICs), looks for practices which reduce the environmental impact of agriculture without necessarily reducing the use of external inputs. The advocacy of sustainability must recognize the need for enhancement of productivity, resulting from the application of scientific knowledge, technology, and good practice (Ruttan, 1990). In LICs, sustainable agriculture cannot be equated with subsistence farming or consistently low yields (Tandon, 1990). The Food and Agriculture Organization (FAO) takes a similar view by saying that sustainable agriculture "... should involve the successful management of resources for agriculture to satisfy changing human needs, while maintaining or enhancing the quality of the environment and conserving the natural resources" (FAO, 1989). A third but narrower approach asserts that agricultural sustainability is the ability to maintain productivity, that is resilience, whether of a field or a farm or a nation, in the face of stresses or shocks (Conway and Barbier, 1990).

For this paper, however, the FAO definition is appropriate because it has wider implications. First, it recognizes that the farmer is the principal agent in the local environment, who manages the resources to achieve his objectives or goals, within the environmental restrictions; second, there is no bias against the use of external inputs or high input technology, as long as a system remains sustainable; third, the needs are not restricted to food or other agricultural output, but include the social and cultural aspects. Finally, the current generations have a moral obligation to recognize the inviolable rights of future generations to use natural resources and inherit a benign environment, and that such rights are not tradable (Spash, 1993). Thus, no agricultural system is inherently either sustainable or unsustainable.

Most of the approaches to develop appropriate technologies for LICs emphasize a systems framework of analysis (Altieri, 1989) with its five constituent elements: the resources and the external environment consisting of physical, biological, economic, and social components; the enterprises or elements; the manager (decision maker); and his objectives (Churchman, 1968). So a FS has to be viewed as an arrangement of enterprises managed in response to an environment and in accordance with the manager's objectives and resources. Given this perspective, two questions remain. Who is the manager and what are his objectives? The objectives being pursued by a manager are a blend of his own objectives and his household's aspirations. Farmers' objectives have been classified in four groups (Reijntjes et al., 1992). The first two groups (productivity and stability) focus on productive sustainability, the third (continuity) on environmental sustainability, and the last one (identity) on social sustainability. These objectives can be and usually are conflicting. Notwithstanding this knowledge of a systems objectives provides the performance measure for the whole system, and allows us to understand why it behaves as it does. Inevitability then, sustainable FSs have to find an optimal balance between these objectives instead of striving for a maximum level of anyone of them individually. The idea of sustainability is then definable and measurable in terms of these objectives.

The concepts on sustainability must answer questions such as 'is a system sustainable?,' 'is this sustainability changing over time?,' 'which system is more sustainable?,' etc. To answer these questions one has to construct a particular definition of sustainability appropriate to field, regional, or national levels. There are many indicators for measuring economic performance of a FS such as profit, gross margin, net farm income, etc.; but from both economic and environmental points of view, the relevant question is, is there a single state variable that can measure sustainability, or should one use a set of relevant control variables (Harrington, 1992)?

A state variable measures directly the condition of a particular component or parameter of a system. As sustainability involves economic, environmental, and social components, it is necessary for the state variable to reflect them all. This poses two further problems. First, how are individual components measured, considering that the determinants of each one are multiple, differ between FSs, and change over time. Second, how can the individual components be combined into a single measure for sustainability; that is, which function gives the best fit between sustainability, its components, and their interactions. Hitherto, these problems are unresolved; therefore an estimate of sustainability can be made through the use of a set of control variables, such as total factor productivity (TFP), that is the relation between total value of all outputs and total value of all inputs (Lynam and Herdt, 1989); total productivity (Harrington et al., 1994), which includes off-farm and environmental costs and benefits; and intertemporal and interspatial TFP including the unpriced contributions of both natural resources and production flows (Ehui and Spencer, 1993). A problem with productivity measures is that sustainability is evaluated ex-post. Other measures of sustainability include indicators of environmental quality and ecological soundness, productivity, and socioeconomic level (Neher, 1992); agroenvironmental indicators (Parris, 1994); approximated sustainability index (Gutierrez-Espeleta, 1993); and environmental sustainability index (Sands and Podmore, 1994). For most of these indicators, however, the trade-offs between determinants cannot be considered and the aggregation of values can conceal specific problems.

The design of policies and the development of technologies for improving the sustainability of agriculture have to consider a region or microregion as the unit of analysis and the FS as the decision-making point. This requires first defining the environmental problem and then identifying the sources of that problem. For the present a microregion is a geographic planning area which has similar agroecological features, similar water availability, a given pattern of FS, and, a recognizable unit of socioeconomic integration (INDAP, 1993).

Within such a microregion, a great variety of farming systems coexist, and this diversity has to be recognized when designing policies. Because it is not feasible to consider all FSs individually and simultaneously, they have to be grouped. Multivariate techniques like factor and cluster analyses can be used to identify and define typical and representative FS. Once the relevant FSs have been identified, the actual and feasible production activities can be defined, along with their relationship to output (or income) and risk, and their environmental impact. Modelling such FSs and aggregating them over the microregion allows the prediction of the FS response to given development policies.

Since one of the critical issues related to sustainability is the presence of multiple and often conflicting objectives, multicriteria methods provide a convenient tool of analysis. Optimization models are generally well-suited for environmental and economic research, because many activities and restrictions can be examined simultaneously; an explicit and efficient optimum seeking procedure is involved; results from changing variables (parameters) can be calculated; and new production techniques can be incorporated (Wossink et al., 1992). Finally, another reason in favor of using MCDM techniques for analyzing sustainability is their ability to consider the whole FS instead of just parts of it, which is frequently the case when solo indicators are used.

An MCDM Model for Examining Agricultural Sustainability

The theoretical underpinnings of this model are the threefold structure of sustainability and the bilevel of the decision problem. Achieving sustainability satisfies the household's and society's demand for agricultural products, preserves or enhances the environment and the natural resource base, and is socially acceptable and equitable. The bilevel structure implies that the needs of two decision levels have to be satisfied. The lower level is made up by hundreds of farmers who decide how to use the natural resources being managed by them. The farmers' main objective is of a productive and social nature — to survive and to grow — leaving the environmental concern as conditioned by the achievement of the previous objectives. The higher level is given by the rest of society, represented through governments and policy makers, whose influence on the farmers' decisions is through norms and regulations, incentives or subsidies, public campaigns to achieve moral persuasion, or investment in research and extension (Zekri and Romero, 1991) which stimulate some types of activities or restrict the farmers' scope of action.

Definition and Identification of Farming Systems

Chile is divided into 12 administrative regions and a metropolitan area. Each region is subdivided in a variable number of provinces and districts ('comuna'). The microregion under study is constituted by the districts of Litueche, Marchigüe, and Pumanque (34°00′–34°45 South and 71′30°–72′00° West). This microregion, with an area of 1,754 km², covers a major part of the VIth region's coastal mountains, especially its eastern or interior declivity. Its climate is marine mediterranean with rain during the cold season, a dry period during the hot season and a subtropical temperature pattern. Winter is mild and there is a frost-free period of more than 4.5 months. The average maximum temperature is between 10 and 20°C during the cold months and over 21°C during the hot months. The rainfall varies between 450 and 900 mm/year, and the dry season is over 5 months long (INIA, 1989). The agricultural capability of the area without irrigation is mainly for winter cereals, winter legumes, and pastures. With irrigation, it is possible to grow corn, potatoes, and orchards for apples, citrics, etc.

A survey was undertaken during 1994 to collect data from 67 farms in the area about the manager, available labor, available land, land use, and livestock. This data set was used to identify and define the FSs to be studied further. Prior to defining various FSs missing, nonrelevant, nondiscriminating, and correlated variables were discarded, in order to reduce the problem's dimensionality. From the initial set of 39 variables, only 11 remained, which were used to construct factors for cluster analysis. The idea behind factor analysis is to explain the observed variation in the variables through a smaller number of new variables

```
              C   C   C   C           C   C   P   P   P   P
              L   L   L   L   P   P   L   L   1   1   1   1
              1   2   3   4   7   5   5   6   8   2   7   9

        (N)  (23)(9)(18)(2) (1) (1) (2) (7) (1) (1) (1) (1)
          1 +XXXXXXXXXXXXXXXXXXXXXXXXXXXXXXXXXXXXXXXXXXXXXXXXX
            |XXXXXXXXXXXXXXXXXXXXXXXXXXXXXXXXXXXXXXXXXXXXX    .
    N     2 +XXXXXXXXXXXXXXXXXXXXXXXXXXXXXXXXXXXXXXXXXXXXX    .
    u       |XXXXXXXXXXXXXXXXXXXXXXXXXXXXXXXXXXXXX XXXXX      .
    m     3 +XXXXXXXXXXXXXXXXXXXXXXXXXXXXXXXXXXXXX XXXXX      .
    b       |XXXXXXXXXXXXXXXXXXXXXXXXXXXXX XXXXX   XXXXX      .
    e     4 +XXXXXXXXXXXXXXXXXXXXXXXXXXXXX XXXXX   XXXXX      .
    r       |XXXXXXXXXXXXXXXXXXXXXXXXXXXXX XXXXX    .    .    .
          5 +XXXXXXXXXXXXXXXXXXXXXXXXXXXXX XXXXX    .    .    .
    o       |XXXXXXXXX   XXXXXXXXXXXXXXXXX XXXXX    .    .    .
    f     6 +XXXXXXXXX   XXXXXXXXXXXXXXX   XXXXX    .    .    .
            |XXXXXXXXX   XXXXXXXXXXXXXXX     .      .    .    .
    C     7 +XXXXXXXXX   XXXXXXXXXXXXXXX     .      .    .    .
    l       |XXXXXXXXX   XXXXXXXXX           .      .    .    .
    u     8 +XXXXXXXXX   XXXXXXXXX           .      .    .    .
    s       |.   XXXXX   XXXXXXXXX           .      .    .    .
    t     9 +.   XXXXX   XXXXXXXXX           .      .    .    .
    e       |.   XXXXX   XXXXX     .         .      .    .    .
    r    10 +.   XXXXX   XXXXX     .         .      .    .    .
            |.    .      XXXXX     .         .      .    .    .
         11 +.    .      XXXXX     .         .      .    .    .
            |.    .       .        .         .      .    .    .
         12 +.    .       .        .         .      .    .    .
```

Figure 47.1 Clustering history for the last 12 clusters. (N = number of farmers; Cl^n = cluster number 'n'; and P^n = observation 'n' of Pumanque District).

(factors). Principal components analysis was used to create the factors (SAS, 1985). Seven factors were retained explaining 85.4% of the total observed variation and at least 70.0% of the variation in selected variables. These seven factors were then used in cluster formation. Ward's minimum variance criterion was used as the clustering method. Figure 47.1 shows the dendrogram with the clustering history of the last 12 clusters.

Based on this analysis the following groups were established (Figure 47.1):

- Group I   = cluster 1
- Group II   = clusters 2 and 3
- Group III = cluster 4, and observations P7 and P5
- Group IV   = cluster 5
- Group V   = cluster 6

Four observations (P18, P12, P17, and P19) remained unclassified, because they were merged with other groups or clusters in the later stages of the clustering process. Two main groups can be recognized (I and II). The main differentiating features of each group are

- Group I   = mostly without partner or other family members working on farm, small number of cows, small farm size
- Group II   = with partner on-farm but without other family members working on it, low number of cows, small farm size

Table 47.1 Cross Tabulation of the Farms According to Group and Productive Orientation

Productive orientation	Group				
	I	II	III	IV	V
Wheat-sheep	1	4			
Wheat-legume-sheep	7	11	1	1	1
Wheat-maize-sheep		1			1
Wheat-pasture-cattle	11			5	
Wheat-pasture-cattle-legume-maize		4	1	1	
Wheat-vineyard-sheep	4	7	2		
TOTAL	23	27	4	7	2

- Group III = with partner and others working on-farm, large farm, with small areas of artificial pastures, large numbers of cattle, goats and sheep
- Group IV = with other family members working on-farm, a considerable part of the land may be taken for sharecropping, medium sized farm
- Group V = farmer works alone on-farm and a considerable part of the farm is given out for sharecropping

The main conclusion is that labor variables are very important for this classification, and all groups (except I from V) can be differentiated according to these.

These farms were also classified according to their 'productive orientations' (POs), which corresponds to the main cropping or livestock activities present on the farm. A cross tabulation of the groups with the POs allows further classification of these farms, with greater emphasis on current activities (Table 47.1). Each group-PO pair was then defined as a FS.

Next, for every FS with four or more observations the distance between each farm and its cluster mean vector or centroid, defined as the average value for all clustering variables of each FS, was computed. The total distance was defined as the sum of the squared standardized difference between every variable for the observation and its group average. This distance was minimized to select the most representative farm for each FS. These were then subject of further detailed survey, to determine the coefficients to be used in the individual FS models.

The MCDM Model

To outline the models, it is necessary to make certain assumptions. First, the higher decision level (say Ministry of Agriculture) is interested in reducing the problem of soil erosion, but to achieve that the lower level objectives and goals of farmers have to be reconciled and harmonized with the aim of reducing soil erosion. Second, at the lower decision level, that is the individual farmer, the objectives of gross margin maximization and risk minimization are pursued. The models have a bilevel structure; the lower level is a set of FS models (FSMs), and the second level is a microregional model (MRM), which is derived by aggregating the FSMs. However, some additional restrictions have to be added to limit off-farm labor use. The MRM has the additional objective of minimizing income differences across farm types by not allowing these differences to go beyond their current levels.

The first objective is to maximize gross margin (GM), that is total output less variable costs. According to economic theory, the leading objective of a firm is maximization of utility. GM is used as a surrogate measure of that utility. This objective is of a private nature and represents the criterion of economic viability.

The second objective is to minimize risk. Every farm plan is in effect a set of states or outcomes, with associated probabilities of occurrence or nonoccurrence and consequences for each pair of action and state (Selley, 1984). Risk minimization is included in the model because any measure towards its reduction benefits the farmer (Anderson and Dillon, 1992) and is modelled using the target-MOTAD (minimization of total absolute deviations) method which is easily incorporated into MCDM models (Tauer, 1983). The difference between the value of a farm plan and a given target level (e.g., a 'safety-first' level of income) under different states of nature is minimized.

Minimizing soil erosion is the third objective. Soil erosion is a serious problem in the study area. Erosion reduces land productivity and the resulting sedimentation is one of the major forms of downstream water pollution (Tivy, 1990). Erosion reduction therefore represents the criterion of environmental soundness and is of interest to both farmer and public. Erosion is estimated using the Universal Soil Loss Equation or USLE (Wischmeier and Smith, 1978):

$$E = R * K * L * S * C * P$$

where: E = predicted soil loss, R = rainfall and runoff factor, K = soil erodibility factor, L = slope length factor, S = slope gradient and steepness factor, C = soil cover and management factor, and P = erosion control practice factor.

R and K are common for the microregion; L and S are specific to each farm, and C and P are specific to each crop and management system. Thus, the model has to allow for the inclusion of fields with different slopes and specific defined activities for each soil type.

The fourth and final objective is to generate a fair distribution of economic growth by minimizing income differences. Perhaps the most frequently used indicator of income distribution is the Gini coefficient. The idea of this coefficient is to group the population according to some measure of income and then compare the average income of lower income groups with the average income of the whole population (Cowell, 1977; Sugden, 1981). To incorporate this concept into the MCDM model the following approach is proposed, using GM as an indicator of income. Let

$$GM_i \leq GM_j \quad \text{for all } i < j$$

where GM_i is the average GM of group i. An improvement in the income distribution is achieved when two conditions are met: (1) the difference in the income for two groups is reduced and (2) the income of any group is not reduced, relative to its original income level, namely

$$GM_{0i} - GM_{0j} \geq GM_{1i} - GM_{1j} \quad \text{for all } i < j; \text{ and}$$

$$GM_{0i} \leq GM_{1i} \quad \text{for all } i$$

GM_{0i} represents the average gross margin of group i on the starting or original position (0) and GM_{1i} the GM of group i after the FS has been improved (1).

The following are constraints of the FS models.

1. Available land. Two distinctions are made: between irrigated and dry land and between flat, hilly, and mountainous.
2. Available labor. Days of labor available to the household after adjusting for off-farm use, and any hired labor during a particular month.
3. Available working capital. Determined by expenses and incomes. It is assumed that negative cash flow cannot occur in any season.
4. Available forage. Must be sufficient to feed the current livestock in all seasons. If needed, feed transfers between seasons are possible and forage can be purchased.
5. Agricultural restrictions. They specify maxima and/or minima on certain activities including the balance required between production and use of certain outputs.
6. Household restrictions. This set of restrictions is necessary to satisfy a household's specific consumption or production patterns. These restrictions include minimum area for wheat and vegetables to satisfy the food consumption, or minimum levels of seasonal farm gross margin to cover fixed costs or off-farm expenditures.

The Farming System Mathematical Programming Model

Maximize gross margin
$$Max \; Z_1 = \sum_{j=1}^{m} GM_j x_j$$

Minimize economic risk
$$Min \; Z_2 = \sum_{r=1}^{s} n_r$$

Minimize soil erosion
$$Min \; Z_3 = \sum_{j=1}^{m} e_j x_j$$

Subject to:

Available land for each land type (i)
$$\sum_{j=1}^{m} lu_{ij} x_j \leq la_i \quad (i = 1,\ldots,q)$$

Farm labor for each season (k)
$$\sum_{j=1}^{m} l_{jk} x_j + hl_k - ol_k \leq al_k \quad (k = 1,\ldots,n)$$

Off-farm labor for each season (k)
$$ol_k \leq dl_k \quad (k = 1,\ldots,n)$$

Working capital for the first season
$$ik + mi_1 + bi_1 - me_1 - be_1 - wk_1 \geq 0$$

Working capital for the next seasons (k)	$wk_{k-1} + mi_k + hi_k - me_k - he_k - wk_k \geq 0$ $(k = 2,...,n-1)$
Working capital for the last season	$wk_{n-1} - ik \geq 0$
Forage balance for each season (k)	$\sum_{j=1}^{m} f_{jk} x_j \geq 0$ $\quad (k = 1,...,n)$
Agricultural rotations	$y_j x_i - y_i x_j \leq 0$
Household consumption	$x_j \geq \min_j$
Risk vectors for each year (r)	$\sum_{j=1}^{m} gm_{rj} + n_r - p_r \geq t$ $\quad (r = 1,...,s)$

where: GM_j = gross margin for activity $j = 1,...,m$, x_j = value of activity j, e_j = expected erosion for activity j, lu_{ij} = land type i used in activity j (vector of 0 and 1), la_i = available land of type $i = 1,...,q$, l_{jk} = labor used in activity j in season k, hl_k = hired labor during season k, ol_k = off-farm labor used during season k, al_k = available labor during season k, dl_k = maximum off-farm demand for labor during season k, ik = capital initially available, mi_k = monetary income during season k, hi_k = household's off-farm income during season k, me_k = monetary expenses during season k, he_k = household expenses during season k, wk_k = available working capital in season k, f_{jk} = forage production or consumption for activity j in season k, y_j = years of crop j in rotation, $\min_j$ = minimum level of activity j to satisfy household's consumption, gm_{rj} = GM of activity j on each of the $r = 1,...,s$ seasons, n_r = negative deviation of expected GM from target for year r, p_r = positive deviation of expected GM from target for year r, and t = target value for the total GM.

The Microregional Mathematical Programming Model

Maximize farm profit	$Max\, Y_1 = \sum_{l=1}^{u} w_{1l} Z_{1l}$
Minimize economic risk	$Max\, Y_2 = \sum_{l=1}^{u} w_{2l} Z_{2l}$
Minimize soil erosion	$Max\, Y_3 = \sum_{l=1}^{u} w_{3l} Z_{3l}$
Minimize differences in farm gross margin	$Min\, Y_4 = \sum_{l=1}^{u-1} w_{4l} d_l$

Subject to:
One set of FS model restrictions for each FS

Expected income vectors	$Z_{1l} - Z_{1(l+1)} - d_l \geq 0$	for all $l = 1,...,u-1$
Minimum expected GM	$Z_{1l} \geq aGM_i$	for all l FS

where: Z_{1l} = expected GM (i.e., objective 1) in FS l, Z_{2l} = economic risk (i.e., objective 2) in FS l, Z_{3l} = expected soil erosion (i.e., objective 3) in FS l, d_l = difference in expected GM between two subsequent FSs sorted by actual GM, w_{hl} = weight of each of the l FSs in the objective function h, and aGM_i = observed GM of FS i.

Compromise programming (CP) is one of the MCDM methods and is based on vector optimization techniques. CP exploits the concepts of Pareto efficient solutions and the *ideal point*. The *ideal point* refers to that 'hypothetical solution' which corresponds to the optimum value for each of the objectives being considered in the analysis. Since in reality achieving the respective optimum values for all objective functions individually and simultaneously is impossible, CP proceeds on the basis that "alternatives that are closer to the ideal are preferred to those that are farther away. To be as close as possible to the perceived ideal is the rationale of human choice" (Zeleny, 1982). To get the most desirable solution requires determining the *ideal* and to measure the distance between it and any feasible alternative. First, CP establishes an ideal solution, i.e., the vector obtained through optimization of the individual objectives, and then the distance between it and any efficient solution is computed. Although an infinite number of metrics do exist, CP can only calculate two of them; L_1 represents the longest geometric distance and L_∞ minimizes the maximum of the individual deviations (Romero et al., 1987). The solutions obtained with these metrics characterizes the bounds of the compromise set (Yu, 1973), that is any desirable efficient solution in terms of a different metric will lie between these two points.

Concluding Remarks

Any approach towards the measurement of sustainability of farming systems has to start with a clear definition of this concept and its determinants. Four major determinants are recognized: (1) economic viability, (2) environmental impact, (3) equity, and (4) acceptability. Evaluation of these four components is a problem. The proposed approach calls for the recognition of a reduced number of indicators which are measurable and can be included into the MCDM models. These indicators, which represent public and private objectives, are basic to the definition of the conceptual MCDM model being applied to the farming systems. These FS were generated through typification and classification, allowing to construct the models for each of the FS, and through their aggregation for the whole microregion.

The next stage of this project involves the transformation of the prototype models into operational ones, which can be used to evaluate the set of feasible policies. This is an essentially interdisciplinary process because of the variety of thematic areas considered. The iterative process of validation and calibration also highlights the heuristic nature of the proposed method to evaluate the impact of alternative policies on the sustainability of the farming systems.

Acknowledgment

The authors are grateful to Fundación FIA of the Chilean Ministry of Agriculture for the financial support given to this project.

References

Altieri, M. 1989. Agroecology: a new research and development paradigm for world agriculture. *Agric. Ecosyst. Environ.* 27(1):37–46.

Anderson, J.R. and J.L. Dillon. 1992. Risk Analysis in Dryland Farming Systems. Farm Systems Management, Number 2. Food and Agriculture Organization, Rome.

Churchman, C.W. 1968. *The Systems Approach*. Dell Publishing, New York.

Conway, G. and E.B. Barbier. 1990. *After the Green Revolution. Sustainable Agriculture for Development*. Earthscan Publications, London.

Cowell, F.A. 1977. Charting inequality I. Pages 17–39 in *Measuring Inequality*. Philip Allen, Oxford, UK.

Ehui, S.K. and D.S.C. Spencer. 1993. Measuring the sustainability and economic viability of tropical farming systems: a model from sub-Saharan Africa. *Agric. Econ.* 9(4):279–296.

FAO, Eds. 1989. Sustainable Agricultural Production: Implications for International Research. FAO Research and Technology Paper. TAC Secretariat and CGIAR. Food and Agriculture Organization, Rome.

Gutierrez-Espeleta, E.E. 1993. Indicadores de Sostenibilidad: Instrumentos para la Evaluación de las Políticas Nacionales. Conferencia en el 50avo Aniversario de la Facultad de Ciencias Económicas. Universidad de Costa Rica. San José, Costa Rica. 19 de Noviembre, 1993.

Harrington, L., P. Jones, and M. Winograd. 1994. Operacionalización del Concepto de Sostenibilidad: un Método Basado en la Productividad Total. Sexto Encuentro de RIMISP. Caminas, Brasil. 11–14 April, 1994.

Harrington, L.W. 1992. Measuring sustainability: issues and alternatives. Pages 3–16 in W. Hiemstra, C. Reijntjes, and E. van der Werf, Eds. *Let Farmers Judge: Experiences in Assessing the Sustainability of Agriculture*. Intermediate Technology Publications, UK.

INDAP. 1993. Guía de microregionalización. Instituto Nacional de Desarrollo Agropecuario. Santiago, Chile.

INIA. 1989. Mapa agroclimático de Chile. INIA, Santiago, Chile.

Lynam, J.K. and R.W. Herdt. 1989. Sense and sustainability: sustainability as an objective in international agricultural research. *Agric. Econ.* 3:381–398.

Neher, D. 1992. Ecological sustainability in agricultural systems: definition and measurement. In R.K. Olson, Ed. *Integrating Sustainable Agriculture, Ecology and Environmental Planning*. Food Product Press, London.

Parris, K. 1994. Developing a Set of Indicators for Use in Agricultural Policy Analysis. Paper presented to the Agricultural Economics Society of Ireland. Dublin, Ireland. 19 September 1994.

Reijntjes, C., B. Haverkort, and A. Waters-Bayer. 1992. Sustainability and farmers: making decisions at the farm level. Pages 24–34 in C. Reijntjes, B. Haverkort, and A. Waters-Bayer, Eds. *Farming for the Future: An Introduction to Low-External-Input and Sustainable Agriculture*. MacMillan Press Ltd., London.

Romero, C., F. Amador, and A. Barco. 1987. Multiple objectives in agricultural planning: a compromise programming application. *Am. J. Agric. Econ.* 69(1):78–86.

Ruttan, V. 1990. Sustainability is not enough. Pages 400–404 in C.K. Eicher and J.M. Staatz, Eds. *Agricultural Development in the Third World*. Johns Hopkins University Press, Baltimore, Maryland.

Sands, G.R. and T.H. Podmore. 1994. Development of an environmental sustainability index for irrigated agricultural systems. Electronic Conference on Sustainability Indicators. January-April 1994.

SAS. 1985. *SAS User's Guide: Statistics. Version 5 Edition*. Statistical Analysis System Inc., Cary, USA.

Selley, R. 1984. Decision rules in risk analysis. Pages 53–67 in P.J. Barry, Ed. *Risk Management in Agriculture*. Iowa State University Press, Ames.

Soule, J., D. Carré, and W. Jackson. 1990. Ecological impact of modern agriculture. Pages 165–188 in *Agroecology*. C.R. Carroll, J.H. Vandermeer, and P. Rosset, Eds. McGraw Hill, New York.

Spash, C.L. 1993. Economics, ethics, and long-term environmental damages. *Environ. Ethics.* 15:117–132.

Sugden, R. 1981. Interpersonal comparisons. Pages 50–66 in *The Political Economy of Public Choice: An Introduction to Welfare Economics*. Martin Robertson, Oxford, UK.

Tandon, H.L.S. 1990. Fertilizers and sustainable agriculture. Pages 15–27 in V. Kumar, G.C. Shrotriya, and S.V. Kaore, Eds. *Nutrient Management and Supply System for Sustaining Agriculture in the 1990s*. IFCO, New Delhi, India.

Tauer, L.M. 1983. Target MOTAD. *Am. J. Agric. Econ.* 65(3):606–610.

Tivy, J. 1990. Agriculture and the environment. Pages 243–260 in *Agricultural Ecology*. Longman Scientific and Technological, England.

Wischmeier, W.J. and D.D. Smith. 1978. Predicting Rainfall Erosion Loss — A Guide to Conservation Planning. Agricultural Handbook N. 537. U.S. Department of Agriculture, Washington, D.C. 58 pages.

Wossink, G.A.A., T.J. de Koeijer, and J.A. Renkema. 1992. Environmental-economic policy assessment: a farm economic approach. *Agric. Syst.* 39(4):421–438.

Yu, P.L. 1973. A class of solutions for group decision problems. *Manage. Sci.* 19:936–946.

Zekri, S. and C. Romero. 1991. Influencia de las preferencias del centro decisior y de los incentivos económicos en la reducción de la contaminación por sales. *Invest. Agrar. Econ.* 6(2):223–239.

Zeleny, M. 1982. Displaced ideal: an operational model. Pages 152–183 in *Multiple Criteria Decision Making*. McGraw-Hill, New York.

Chapter 48

Use of a DSS for Evaluating Land Management System Effects on Tepetate Lands in Central Mexico

Mariano Hernandez, Philip Heilman, Leonard J. Lane, Jose L. Oropeza-Mota, and Hector M. Arias-Rojo

Abstract

Increases in population and subsequent increases in demand for agricultural production have resulted in agricultural systems that cause a severe decline in productivity of the soil resource in some areas of Mexico. The central region of Mexico contains areas in which the topsoil has been severely eroded. The exposed subsoil in these areas is characterized by bare and hard surfaces locally named Tepetates. Tepetates are volcanic soils that consist of a duripan exposed through erosion of the overlying soil. In completely exposed soils, the average annual rate of erosion is approximately 6 ton/ha which contributes to the detriment of water quality and loss of storage capacity in reservoirs in the area. Reclamation of tepetate lands for agriculture has been established as one alternative to help meet the demand for food production in that region of Mexico. The U.S. Department of Agriculture — Agricultural Research Service (USDA-ARS) in Tucson has developed a prototype decision support system (DSS) with a multiobjective framework. As a case study, the prototype DSS is applied in Texcoco, Mexico to evaluate crop productivity of maize in tepetate lands using straight row farming, contour row farming, narrow-base terraces, and bench terraces as management systems. The decision variables selected to evaluate the management systems are crop yield, total cost of terrace construction, sediment yield, and runoff. The weather generator CLIGEN is used to reproduce a 20-year record of daily precipitation and the GLEAMS (Groundwater Loading Effects of Agricultural Management Systems) model to estimate sediment yield, crop yield, and runoff for the same time record. The application of the DSS in selecting conservation management systems in tepetate lands in Mexico provided an improved basis for decision making and revealed problems which will probably be common to many applications of multiobjective decision support technology in developing countries.

Introduction

Mexico is one of the countries most heavily affected by soil erosion. Soil erosion processes are the major cause of nonpoint pollution, and the effects of excessive sediment loading on receiving waters include the deterioration or destruction of aquatic habitat, loss of storage capacity in reservoirs, and accumulation of sediments that inhibit normal biological life. In addition, agricultural productivity in Mexico may be severely affected if soil erosion processes are not adequately controlled; Mexico's capacity to become food self-sufficient is at risk. Oropeza-Mota (1995) points out that soil degradation in Mexico is caused by several factors: (1) an increase in population has caused modification of land use to meet the demand for food production; (2) lack of research, promotion, and publication of simple and income-producing soil conservation management systems in rural environments; and (3) exploitation of natural resources at rates greater than self-recovery capacity.

In the Central Highlands of Mexico, many slopes are covered by duripans which locally are named Tepetates. The word tepetate comes from the Nahuatl language and means hard stratum. Ancient Mexicans used the word tepetate to characterize areas in which topsoil has been severely eroded and unconsolidated weathered rock exposed as a result of deforestation. Tepetate lands are composed of hardened volcanic materials. Their hardness and almost complete lack of nitrogen, phosphorous, and organic matter render them unproductive. Nevertheless, ancient Mexican civilizations cropped tepetate lands using bench terraces on sloping lands to reduce runoff and increase infiltration.

Today, the same technique is used to reclaim heavily eroded soils using heavy machinery. The main crops grown in this region are maize, beans, and squash. Soil conservation management systems are not practiced by farmers, thus soil productivity has diminished with time. Arias-Rojo (1992) reported that the average annual rate of erosion is approximately 6 ton/ha in bare runoff experimental plots, and tepetate lands are the main source of sediment carried by rivers nearby which contributes to the detriment of water quality and loss of storage capacity in reservoirs of the area. According to Zebrowski (1992), the total area covered by tepetate lands in Mexico is unknown. However, 27%, approximately, 30,700 km^2 of the Mexican volcanic axis is characterized by having tepetate lands. In some states of Mexico, for example the state of Tlaxcala, 54% of the area is covered by tepetate lands.

The Federal Government of Mexico established Proyecto Lago de Texcoco in 1973 to help solve some of the problems due to deforestation and changes in land use, and as one alternative to help meet the demand for food production in that region of Mexico. The program involved land reclamation and reforestation. Many benefits resulted from the program, such as incorporating tepetate lands to forest and agricultural activities, reducing soil loss rates, aquifer recharge, flood control, and a general improvement of the environment (Llerena-Villalpando and Sanchez-Bernal, 1992). Within the reclamation program, several studies have been carried out to determine the best management system in tepetate lands. Among management systems studied are contour row across the main slope, and narrow-base and bench terraces (Arias-Rojo, 1992; Pimentel-Bribiesca, 1992).

Purpose and Scope

The purpose of this paper is to provide a basis for identifying the best conservation management system alternatives in tepetate lands in Central Mexico that maximize crop productivity, enhance water quality, and minimize the rate of soil erosion at a minimum cost by applying the USDA-ARS Multiple Objective Decision Support System (Southwest Watershed Research Center, 1994). In addition, research data will be identified that is required for future applications of the DSS in Mexico.

Overview of the Prototype DSS

The USDA-ARS in Tucson, Arizona has developed a prototype DSS with a multiobjective framework. Multiobjective decision theory is one method of evaluating alternative management systems. The methodology involves ranking in order of importance or utility the objectives for different scenarios. The USDA-ARS DSS is a computer-based system which incorporates a decision model based on multiobjective decision making, a default database, a hydrologic/erosion/pesticide/nutrient/economic simulation model, and an interface shell. The DSS system runs on a workstation platform using the UNIX operating system under the X Window graphical interface environment. A full description of the DSS is given by Yakowitz et al. (1992 a, b) and Stone et al. (1995) and so only a brief description follows.

Decision Model

The decision model combines the use of scoring functions developed in Wymore (1988) with a modification of the decision tools described in Yakowitz et al. (1992a). Scoring functions are a means of scaling between 0 and 1 decision variables (i.e., runoff, sediment, nitrogen concentration, economic returns) which have different units and magnitudes. Four generic shapes of the scoring functions (Wymore, 1988) are shown in Figure 48.1. The decision variables are parameterized by either measured data or output from a modification of the GLEAMS (Davis et al., 1990) model discussed below. To make a decision with this methodology, several alternative management systems (i.e., no-till, ridge till) are selected to be compared against a conventional management system (i.e., continuous corn). The average annual, maximum, and minimum values of each decision variable of the conventional system are used to construct a scoring function chosen from the

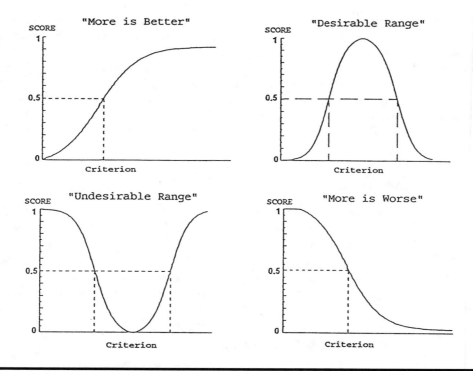

Figure 48.1 Generic scoring function types.

four generic scoring function types shown in Figure 48.1. The score for each individual decision variable for the conventional system is defined to be 0.5 and the combined score of all the decision variables is thus equal to 0.5. The average annual value of a decision variable for an alternative management system (i.e., contour row farming) is scored using the scoring function of that decision variable derived from the conventional system. For each alternative management system, the scores for all the decision variables are combined and the management system is ranked according to the combined score. The management system with the highest combined score is considered to be the best management system among those being evaluated.

Simulation Model

The primary purpose of the simulation model is to quantify the decision variables when values are not available through a database or expert opinion. The GLEAMS model was chosen for the prototype because it simulates many of the processes necessary for evaluating the effects of management systems on water quality. It is important to point out that the decision model described above is not dependent on whether field data or simulation data are used or in the particular model selected. GLEAMS has three major components: hydrology, erosion/sediment yield, and pesticides. The hydrology component uses daily climate data to compute the water balance in the root zone. The erosion component computes estimates of rill and interrill erosion on overland flow areas as well as channel erosion for each storm event. Sediment enrichment is computed for use in the estimation of adsorbed pesticide transport. The pesticide chemistry component simulates the

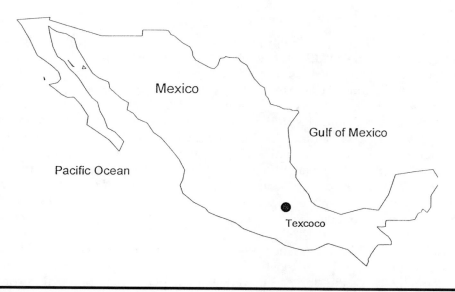

Figure 48.2 Location of experimental site.

movement of pesticides over the surface and through the root zone. A more detailed explanation of the simulation model is available in the WQDSS Reference Manual, version 1.1, (Southwest Watershed Research Station, 1994).

Application of the DSS in Central Mexico

Description of the Site

The study area is located approximately 5 km east of the City of Texcoco and 50 km northeast of Mexico City and is operated as a research facility by Colegio de Postgraduados (Figure 48.2). Mean annual precipitation on the area is about 640 mm with 84% occurring during the summer thunderstorm season of May to October. Mean maximum and minimum temperatures are 27 and 4°C, respectively. According to the FAO-UNESCO, the soil classification is Andisols and Inceptisols. The soils are not well drained and are composed of hardened volcanic materials. Soil may have fragments of duripan in some horizons within a depth of 50 cm. The pan is mostly impenetrable by plant roots. Soil texture is sandy loam with 66% sand, 22% clay, and 12% silt by weight.

Management Alternatives

Soil conservation management systems are not usually practiced by farmers, thus soil productivity has been diminished with time. As stated before, the government has established the program for reclamation of tepetate lands for agriculture as one alternative to help meet the demand for food production in this region. Within this program, farmers are encouraged to use contour row farming, narrow-base and bench terraces, as opposed to straight row farming which is the conventional management system in the region.

Experimental Design

In 1976 and 1977, Colegio de Postgraduados conducted experiments on ten runoff plots (Trueba-Carranza, 1978). For the purpose of this study, only six plots were considered. Plot treatments consisted of contour row farming and two types of terraces, bench and narrow-base. There were two replica plots for each treatment (Figure 48.3). Each plot was 95 by 74 m in size with the long axis perpendicular to the slope. A concrete channel perpendicular to the main slope was constructed for each plot to collect runoff and a flow meter was placed at the end of each channel to measure runoff from each plot. Maize (*Zea mays*) was the main crop grown in all plots. The planting day was July 5 of 1976 and the vegetative cycle of the maize was 120 days. The application rate of nitrogen was 137 kg/ha at the planting day in the form of ammonium nitrate. Crop yield production for each treatment, total sediment yield for the 2 years, and total costs of construction are listed in Table 48.1.

Later, a second experiment (1990 to 1993) was carried out by Colegio de Postgraduados to evaluate straight row farming parallel to the slope and contour row farming across the slope. The average annual crop yield production was 1.83 ton/ha for straight row farming parallel to the slope and 2.37 ton/ha for contour row farming across the slope. The average annual sediment yield for straight row farming was 1.0 ton/ha and 0.493 ton/ha for contour row farming.

Selection of Decision Variables

Four decision criteria were selected to evaluate agricultural management systems in tepetate lands in Central Mexico. The decision criteria were selected to reflect the farmers' and government's interests to increase crop productivity and reduce runoff and sediment yield at a minimum cost. The four criteria are

1. Crop yield
2. Total cost of terrace construction
3. Sediment yield
4. Runoff

Only 2 years (1976 and 1977) of data were available for the terrace management systems (bench and narrow-base) and 4 years (1990 to 1993) of straight row farming parallel to the slope (conventional) and 6 discontinuous years (1976 to 1977 and 1990 to 1993) of contour row farming across the slope. Thus, the GLEAMS model was used to estimate mean annual values for runoff, sediment yield, and crop yield for each of the management systems for a 20-year period. Climatological data were available at a nearby station for the period 1958 to 1984. Based on this record, two conditional probabilities were calculated: the probability of a wet day following a dry day, and the probability of a dry day following a wet day. The weather generator CLIGEN (Nicks and Lane, 1989) was used to estimate a 20-year time series of daily rainfall based on the calculated conditional probabilities, and mean monthly precipitation, monthly standard deviation, and coefficient of skewness of daily precipitation, monthly maximum and minimum temperatures, and solar radiation.

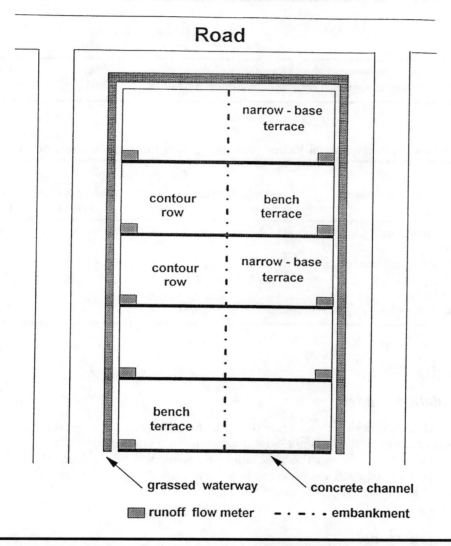

Figure 48.3 Experimental site and distribution of management systems at Lomas de San Juan, Chapingo, Mexico.

Table 48.1 Maize Yield Production, Sediment Yield, and Total Costs of Construction

Treatment	Maize yield		Total sediment yield		Costs of construction
	1976 (ton/ha)	1977 (ton/ha)	1976 (ton/ha)	1977 (ton/ha)	(pesos)
Contour row farming	2.90	2.12	0.356	0.433	75
Narrow-base terrace	2.72	1.90	0.292	0.339	2,195
Bench terrace	2.13	2.34	0.058	0.070	4,517

Trueba-Carranza, 1978.

Table 48.2 Results of the Calibration using the Straight Row Management System.

	Decision variables		
	Runoff (mm)	Sediment yield (ton/ha)	Crop yield (ton/ha)
Observed average annual values (4 years)	155	1.00	1.83
Average annual values based on a 20-year record	139	0.83	1.74

Table 48.3 Average Annual Values over the 20-Year Runs of the Simulations

Agricultural management systems	Decision variables			
	Crop yield (ton/ha)	Terrace total cost (pesos)	Sediment yield (ton/ha)	Runoff (mm)
Straight row parallel to the slope (conventional)	1.74	0.00	0.83	139.46
Contour row (alternative 1)	2.18	75.00	0.38	111.05
Narrow-base terrace (alternative 2)	2.23	2,195.00	0.30	59.54
Bench terrace (alternative 3)	2.20	4,517.00	0.080	18.33

Results

Simulation Model

The GLEAMS model was calibrated using the observed average annual values corresponding to the straight row management system parallel to the slope (conventional). Results of the calibration are shown in Table 48.2. The average annual values for each decision criterion determined by the model GLEAMS are presented in Table 48.3.

Decision Model

To obtain the scores, scoring functions were designed for each of the criteria. For example, for crop yield (ton/ha) a "more is better" scoring function (see Figure 48.1) was selected. This generic function is then customized in the DSS by including a lower threshold at 0.0 ton/ha, the minimum annual yield predicted by the conventional system simulation run and an upper threshold at 3.96 ton/ha, the maximum annual yield predicted by the conventional system. The average annual crop yield (1.74 ton/ha) from the conventional system determines the baseline value, for which this scoring function is 0.5. The slope of the function at the baseline value is a function of the threshold and baseline values. The baseline values can be determined using a standard or conventional system, federal regulation, or by expert opinion. All of the other alternative management systems are scored relative to the conventional management system for each criterion. A management system which performs better than the conventional system with

Table 48.4 Default Importance Decision Variable Order

Agricultural management systems	Decision variable order			
	Crop yield (ton/ha)	Sediment yield (ton/ha)	Runoff (mm)	Terrace total cost (pesos)
Straight row parallel to the slope (conventional)	0.500	0.500	0.500	0.500
Contour row (alternative 1)	0.858	1.000	0.854	0.485
Narrow-base terrace (alternative 2)	0.884	1.000	0.995	0.139
Bench terrace (alternative 3)	0.870	1.000	1.000	0.006

regard to a specific criterion will score >0.5 for that criterion and one that performs worse will score <0.5. The default importance ordering of the criteria ranks highest that criterion which has the potential for the greatest change in score when a small change in the criteria near the conventional system is observed (Yakowitz et al., 1992b). Ranking the decision criteria by the normalized value of the slopes of the scoring functions at the baseline values resulted in a default importance order listed in Table 48.4.

Based on the established importance order of the decision criteria, best and worst composite scores are determined by the DSS for each of the alternatives by solving the linear programs given by Yakowitz et al. (1992b). The solutions to these linear programs are the most optimistic and pessimistic composite scores (weighted averages) consistent with the importance order given above. Figure 48.4 shows the range of composite scores from best to worst for each alternative for the default importance decision variable order.

Based on the default importance order and composite scores, all the alternatives score better than the conventional. The best alternative is the contour row since its composite average score is the highest among the other two alternatives. This scenario could reflect the government's perspective if it wants to encourage farmers to maximize crop productivity and reduce soil erosion through an established National Soil Conservation Program that may fully subsidize the cost of construction and maintenance of terraces.

The DSS allows the user to redefine an importance order for the decision variables. This is a realistic feature designed to accommodate a particular preference associated with a different user's perspective of the decision criteria. For example, a farmer with little capital may wish to place a higher importance on the terrace cost than other decision variables. A new importance decision variable order was selected in which the terrace cost and crop yield are the most important decision variables (terrace cost > crop yield > sediment yield > runoff). Figure 48.5 shows the range of composite scores from best to worst for each alternative for the new importance decision variable order. In this scenario, the average composite scores of both terrace management systems were lower than the conventional, but the contour row farming scored better than the conventional.

A farmer with more capital may wish to change the decision variable order to place a higher importance order on crop yield, terrace cost, and sediment yield, therefore the new decision order is as follows: crop yield > terrace cost > sediment yield > runoff. Figure 48.6 shows the range of composite scores from best to

580 *Multiple Objective Decision Making for Land, Water, and Environmental Management*

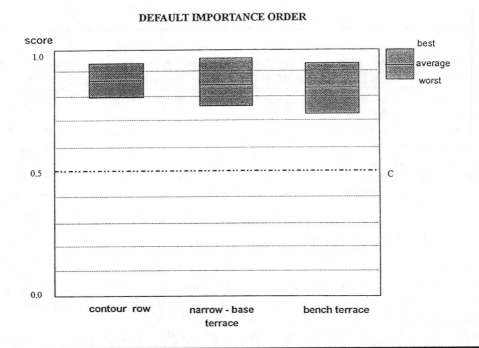

Figure 48.4 Range of scores for the four management systems and default importance order.

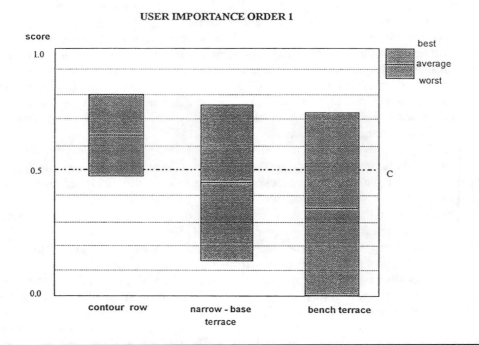

Figure 48.5 Range of scores for the four management systems and user importance order 1.

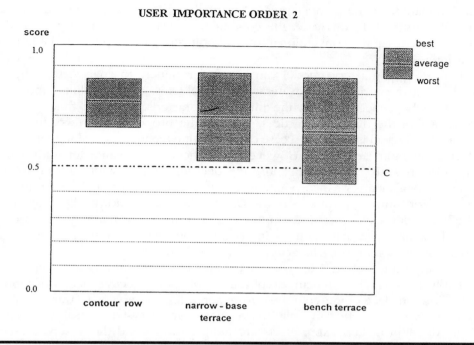

Figure 48.6 Range of scores for the four management systems and user importance order 2.

worst for each alternative for the new importance decision variable order. Notice that by shifting the order of importance of crop yield, the average composite scores of both terrace management systems performed better than the conventional.

Conclusions

The application of the DSS in selecting conservation management systems in tepetate lands in Mexico provided an improved basis for decision making. Of the three importance decision variable order sets: (1) default set, (2) terrace cost most important decision variable set, and (3) crop yield most important decision variable set, contour row farming scored higher than the conventional and terrace management systems. This result suggests that by implementing contour row farming in the region, maize yield production may not be as high as crop yields from terrace systems, however, the small difference in production may be offset by taking into account the low cost of contour row farming construction compare to the cost of terracing. Based on the information available, we were not able to assess offsite effects on water quality of surface and groundwater, recharge of local aquifers, and other conflicting societal issues. Under other scenarios, terraces might score better than the conventional and contour row farming. One potential application of the DSS is in planning the promotion of conservation technologies. Contour row farming does well in all importance orders constructed, and is the only system which does better than the conventional from the capital-constrained farmers' point of view. If an alternative were clearly better than all others when an importance order which reflects all of society is used, but that alternative is not dominant from the farmers' point of view, then a subsidy to adopt that system might be justified. In this case, the contour row system should be promoted: it scores higher than the conventional from all points of view and it is in the farmers' own interest.

Before any simulation model can be used routinely to analyze the physical processes and economics of water quality problems, a tremendous amount of data must be collected about the physical characteristics of the fields and the economics of the different practices. Without the proper parameters, the use of a simulation model is a potential case of information technology making misleading information more accessible. However, with the proper parameter information, simulation models can help quantify the decision variables when data are not available as a means of incorporating the "best science" in the decision-making process. For future applications of the DSS in Mexico, decision makers, scientists, local and federal agencies, and farmers will need to collaborate together in order to build a database that includes information on climate, watershed geometry, management system, soil characteristics, and economics.

The application of the DSS to tepetate lands has revealed problems which will probably be common to many applications of multiobjective decision support technology in developing countries. In general, less observed data are available which describe the effects of alternative management systems on the objectives of the decision maker. Extra efforts are required to extend the observed record to fill data gaps, that will require the use of probably simulation models. When these less-certain decision criteria are used to make a decision, in the absence of confidence limits, the results, particularly between closely ranked alternatives, should be interpreted jointly by scientists, farmers, and decision makers.

References

Arias-Rojo, H.M. 1992. Tepetate reclamation: an alternative for agriculture, livestock, and forestry production. Pages 309–317 in *Suelos Volcanicos Endurecidos*. Primer Simposio Internacional, Mexico, October 20–26, 1991. TERRA Volume 10.

Davis, F.M., R.A. Leonard, and W.G. Knisel. 1990. GLEAMS User Manual. Lab Note SEWRL-030190FMD. USDA-ARS and University of Georgia, Tifton. 38 pages.

Llerena-Villalpando, F.A. and B. Sanchez-Bernal. 1992. Tepetate reclamation in the east part of the Mexico Valley. Pages 302–308 in *Suelos Volcanicos Endurecidos*. Primer Simposio Internacional, Mexico, October 20–26, 1991. TERRA Volume 10.

Nicks, A.D. and L.J. Lane. 1989. Weather generator. Pages 2.1–2.19 in L.J. Lane and M.A. Nearing, Eds. USDA-Water Erosion Prediction Project: Hillslope Profile Model Documentation. NSERL Report No. 2. USDA-ARS National Soil Erosion Research Lab. W. Lafayette, Indiana.

Oropeza-Mota, J.L. 1995. Mexico, uno de los paises mas afectados por la erosion del suelo. In La erosion, nuestro mayor problema ambiental, *La Jornada Ecologica*, Año 3, numero 33, jueves 5 de enero de 1995.

Pimentel-Bribiesca, L. 1992. Reclaming tepetate badlands in Mexico. Pages 293–301 in Suelos Volcanicos Endurecidos. Primer Simposio Internacional, Mexico, October 20–26, 1991. TERRA Volume 10.

Southwest Watershed Research Center. 1994. A Multiple Objective Decision Support System for the USDA Water Quality Initiative: Reference Manual, Version 1.1. August.

Stone, J.J., D.S. Yakowitz, and L.J. Lane. 1995. A multi-objective decision support system for evaluation of natural resource management systems. In First International Conference on Multiple-Objective Decision Support Systems for Land, Water and Environmental Management: Concepts, Approaches, and Applications, July 23–28, 1995, Honolulu, Hawaii.

Trueba-Carranza, A. 1978. Evaluacion de la Eficiencia de cuatro practicas mecanicas para reducir las perdidas de suelo y nutrimentos por erosion hidrica en terrenos agricolas de temporal. M.S. thesis. Colegio de Postgraduados, Chapingo, Mexico.

Wymore, W.A. 1988. Structuring system design decisions. Pages 704–709 in *Proceedings of International Conference on Systems Science and Engineering 88*. International Academic Publishers, Pergamon Press.

Yakowitz, D.S., L.J. Lane, J.J. Stone, P. Heilman, and R.K. Reddy. 1992a. A decision support system for water quality modeling. Pages 188–193 in Proceedings of the Water Resources Planning and Management Conference of the Water Resources Sessions/Water Forum '92, Baltimore, Maryland. August 2–6, 1992.

Yakowitz, D.S., L.J. Lane, and F. Szidarovszky. 1992b. Multi-attribute decision making: dominance with respect to an importance order of the attributes. *Appl. Math. Computat.* 54:167–181.

Zebrowski, C. 1992. Indurated volcanic soils in Latin America. Pages 15–23 in Suelos Volcanicos Endurecidos. Primer Simposio Internacional, Mexico, October 20–26, 1991. TERRA Volume 10.

Chapter 49

Climate Change Effects on Agricultural Productivity in the Midwestern Great Lakes Region

B.Z. Littlefield, J.L. Ehman, W. Fan, J.J. Johnston, B. Offerle, and J.C. Randolph

Abstract

The purpose of this paper is to report preliminary results of a study that examines trends since the 1940s and models climate change effects on agriculture production in the midwestern Great Lakes region, which is composed of Illinois, Indiana, Michigan, Ohio, and Wisconsin. Changes in crop productivity for corn and soybeans were modeled using both process models and statistical models. The process model, CENTURY 4.0, a soil organic matter model utilizes climate, soil parameters, and management practices as input data for the modeling process. The statistical model uses climate variables to estimate parameters and forecast future crop yields. Output data from the CENTURY 4.0 model was produced through 2060; forecasts from the statistical model were generated through 2017. While the modeling process using CENTURY 4.0 was applied to all counties in the five-state region, the statistical model was applied to selected counties in the region for intermodel comparison.

Effects of Carbon Dioxide and Climate Change on Crop Potential

Crop yields have steadily increased during the last 56 years, a result of improved technology in the form of certified seed, agrichemicals, and equipment applied in concert with scientific farm management practices. Between-year variance in corn and soybeans yields has increased as average crop yields have increased during the past 56 years. County average for corn production has varied 60 bushels or more per acre from one year to the next for counties within the Great Lakes region. This represents a shift in revenues of $100 or more per acre after accounting for price shifts in the commodities market. The increase in average yields over the study period can be attributed to technological change in agriculture. Differing climate conditions, particularly in growing season temperature and precipitation when combined with technological advances, appear responsible for the increased year-to-year variance in crop yield. Figure 49.1 illustrates the increase in annual yield and the between-year variance in corn production for Boone County, Indiana from 1937 to 1992. (Boone County was selected as representative of rural agricultural counties within the study region. Boone County is primarily an agrarian county in central Indiana, that produced results consistant with those generated by other counties in the study region.) Figure 49.2 displays soybean yield for Boone County during the same time period; this also shows both the increase in average annual yield and an increase in between-year variance of annual yield. Figure 49.3 displays mean July temperature and Figure 49.4 displays total July precipitation. July is the critical month for corn pollination and ear development (Ritchie and Hanway, 1984), and pod development on soybeans (Ritchie et al., 1985). This is not to say that temperature or precipitation in June or August are not important; extreme variations from the normal weather patterns for any period of time during the growing season can have a detrimental affect on that season's yield. Given this pattern of crop yield behavior, what can be expected from crop yields in the next quarter century?

Current and Future Weather Patterns

Impacts of increased levels of atmospheric carbon dioxide (CO_2) on temperature and precipitation have been predicted by general circulation models (GCMs). While not without uncertainties, most GCMs predict that doubled effective CO_2 concentrations will significantly change global temperature and precipitation regimes, including adding variation and volatility to weather patterns. (Smith and Tirpak, 1990). Effective CO_2 concentrations aggregate the effects of all greenhouse gasses and report them in CO_2 units for comparative purposes. Doubled CO_2 may result in global temperatures rising as much as 5°C and precipitation increasing more than 10% (Hansen, 1989; Schneider, 1989; Smith and Tirpak, 1990). If these predictions are true, it is important to understand how crop growth may be affected by the direct and indirect effects of increasing atmospheric CO_2.

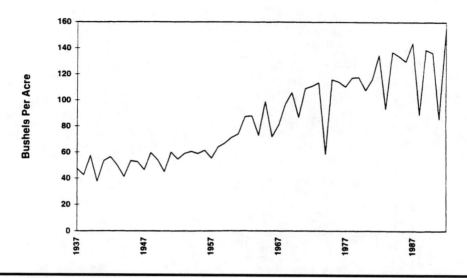

Figure 49.1 Boone County corn production, 1937 to 1992.

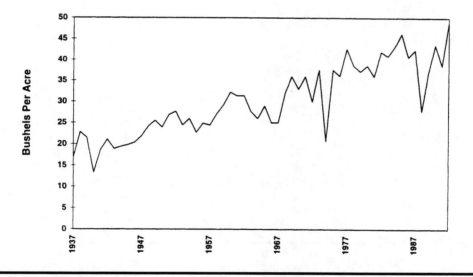

Figure 49.2 Boone County soybean production, 1937 to 1992.

Expected Implications to Agriculture

Agricultural potential will be affected by increased CO_2 concentrations through indirect effects of changes in land suitability and direct physiological effects on crops (Parry et al., 1990). Land suitability changes include alterations in climatic patterns that may result in spatial shifts of agricultural potential. Direct physiological effects of increased atmospheric CO_2 include changing photosynthetic, respiration, and transpiration processes. In turn, these direct effects of increased

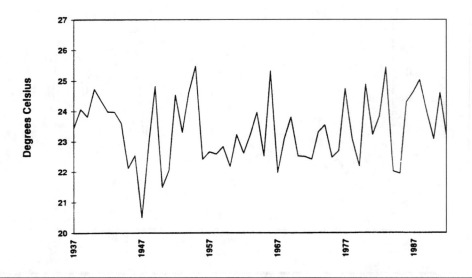

Figure 49.3 Boone County mean July temperature, 1937 to 1992.

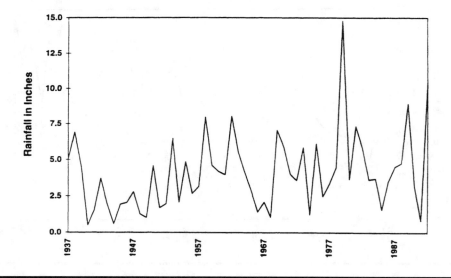

Figure 49.4 Boone County July precipitation, 1937 to 1992.

atmospheric CO_2 may be impacted by indirect effects such as temperature increases, or longer crop growing seasons. Growing season length is an important component in crop and cultivar selection as it determines when a crop will reach maturity. Increases in temperature also result in increased evapotranspiration requiring increased water inputs to sustain plant growth (Allen and Gichuki, 1989).

Numerous studies have investigated the direct effects of increased atmospheric CO_2 on plant growth and plant physiological processes (Easterling et al., 1991; Peart et al., 1989; Ritchie et al., 1989; Rosenzweig, 1989). The focus of the above research has primarily been on changing photosynthetic and transpiration rates in response to increased concentrations of atmospheric CO_2. There is general agreement that increased levels of CO_2 will increase short-term productivity, especially for crops which use the C_3 photosynthetic pathway. Subsequent decreased transpiration per unit leaf area results in increased areal water-use effect (WUE) in C_3 species. However, expected greater total leaf area could lead to elevated total canopy respiration and raise water demand, increasing the need for irrigation. Tissue adaptation rates to increased levels of CO_2 are as yet unknown. With the relative uncertainties of predicted direct effects due to interactive processes, this study focuses on the indirect effects of increased atmospheric CO_2 on plant growth particularly changes in temperature and precipitation within the midwestern Great Lakes region.

Selection of Region for Study

In the Great Lakes region, which includes the midwestern Great Lakes states (Illinois, Indiana, Michigan, Ohio, and Wisconsin) and Iowa, Minnesota, and Missouri, agriculture is the single largest land use, with 52% of the land devoted to crops and pastures (Smith and Tirpak, 1990). In the early 1980s, the Great Lakes States produced 26% of the total U.S. agriculture products worth $36 billion (Federal Reserve Bank of Chicago, 1985). In 1993, about 59% of all U.S. cash income from corn and 40% of the cash income from soybeans came from this region.

Several studies of climate change effects on agriculture have focused on the mid-latitudes, and many researchers believe that temperature and precipitation will change more dramatically in mid- and high-latitudes than in tropical regions (Parry, 1990). Jones and Kiniry (1986) have suggested that temperature increases alone could reduce crop yields everywhere in the Great Lakes region except the northernmost latitudes, especially for corn. As climate conditions change, more northerly regions of North America may have increased agricultural potential. While several studies have modeled crop response, few studies have considered available soil nutrients as a factor that could impact crop response to climate change.

Selection of Crops to Study

The crop with the most acres planted for each of the five states of the midwestern Great Lakes region is corn, generating an average of 39% of the total U.S. corn production from 1975 to 1991. Corn production more than quadrupled soybeans production in total yield in the five-state region from 1975 to 1991. As a farm product, corn can be sold on the market or fed to the producer's livestock. Corn can also be cut for silage before reaching maturity and drying, to be stored and fed to livestock, especially cattle on dairy farms. Acreage harvested for silage average 12 to 20% of corn acres planted for Michigan and Wisconsin, 5 to 10% for Ohio, and 1 to 3% for Illinois and Indiana (NASS, 1994). Silage is not a market commodity; the number of acres planted for silage most likely will be affected by changes in economic markets for livestock, milk prices, and the need for silage rather than changes in crop yield.

Soybean acres account for the second largest agricultural land use in four of the five states in the study region. (Soybeans are fourth in Wisconsin behind corn, oats, and potatoes.) The five-state region accounted for 34.71% of the soybeans produced in the U.S. from 1975 to 1991 (NASS, 1994). Soybeans, unlike corn which may be fed to the producers livestock, are always produced as a cash crop. Of the soybeans produced in the U.S., 30% are exported (NASS, 1994). Of those not exported, 70% have the oil extracted and are then ground for meal to be used to supplement livestock feed (NASS, 1994).

Both corn and soybeans are used for more purposes than feedstocks and table oils (Harpstead, 1992). Corn is also processed for use in sweeteners, fuels, pharmaceuticals, and plastics. Soybeans are used in newspaper ink, adhesives, engine oil additives, paint, and varnishes. While the U.S. share of the world soybean market has declined during the past 15 years, the total world market has continued to grow, indicating that demand will exist for these crops into the next century (Vitosh and Diers, 1992).

Models Applied

This study involves the comparison of crop production under a scenario of doubled atmospheric CO_2 using change in temperature and precipitation values generated by the Oregon State University (OSU) GCM. The OSU GCM was selected for its less extreme climate forecasts (Randolph and Lee, 1993). A linear ramp was used to apply the temperature and precipitation changes that are predicted to occur from 1980 to 2060. Outputs generated by the CENTURY 4.0 model and forecasts generated by a statistical model were compared visually as the CENTURY 4.0 output was in the form of difference in production over a modeled baseline production. The statistical outputs were forecast as average bushels per acre on a county-wide basis.

Counties were selected as the unit of evaluation for the study, because most historic crop data are collected and maintained with the county as the minimum data unit. Most of these data are collected and compiled by each state's Agricultural Statistical Service with assistance from county agricultural extension agents. CENTURY 4.0 was applied to model all counties within the study region. The statistical model was applied to six counties within Indiana that were predominately agrarian and located within different climate divisions. The counties selected were Adams, Boone, Gibson, Parke, Porter, and Scott.

Process Model

CENTURY 4.0 is a soil organic matter model developed by researchers with the USDA-ARS Great Plains Systems Research Unit at Colorado State University. "The primary purposes of the model are to provide a tool for ecosystem analysis, to test the consistency of data and to evaluate the effects of changes in management and climate on ecosystems" (Metherell et al., 1993). CENTURY 4.0 simulates carbon, nitrogen, phosphorus, and sulfur cycling within the model ecosystem over multiple-year time periods. The program incorporates climate, soil properties, and management practices into the analysis, and generates values for net primary production, carbon, nitrogen, phosphorus, and sulfur for the area being modeled.

Several model runs were performed, using both baseline climate conditions and doubled CO_2 conditions. CENTURY is able to model both the direct and indirect effects of doubled atmospheric CO_2. Only the indirect effects were included for these preliminary model runs. Two major groups of model runs were made: those to simulate a baseline climate scenario and those to simulate a changed climate scenario. For the baseline runs, current temperature and precipitation patterns throughout the region were modeled as being representative of historic ones. The OSU-based changed climate runs utilized a linear ramping of temperature and precipitation in each county from current conditions to predicted conditions in 2060.

To examine only the physiological effects of changing climate on agricultural production in the region, management practices were held constant in both the baseline and changed climate scenarios. The management practices which are available in CENTURY include planting and harvesting dates, crop rotation patterns, frequency and magnitude of fertilizer applications, and tillage and irrigation events. While it is recognized that these management practices might indeed change in light of altered climatic conditions, the potential impacts of these activities were not included for model runs at this small spatial scale. Future runs will allow for changes in these practices in response to changed climatic conditions.

Output from these model runs included net primary productivity (NPP) for the crop growing season. To provide for a comparison with the multivariate model predictions, NPP results from changed climate runs were converted to crop yield in bushels/acre using a corn harvest index of 0.8 and a soybeans harvest index of 0.35. These harvest indices provided the best agreement when matching CENTURY-modeled NPP with historic yield data.

Statistical Model — Multivariate Model

May, June, July, August, and September temperature and precipitation data for the time period of 1937 to 1992 were used as the independent variables to fit multivariate models for the dependent variables of corn and soybeans. These variables were considered to provide the "most appropriate model" for representing crop-weather relationships by Huff and Neill (Garcia et al., 1987). One corn yield study estimated coefficients for temperature variables producing a positive coefficient for June, and negative coefficients for July and August. Precipitation variables were estimated to have negative coefficients for May rainfall, and positive coefficients for July rainfall (Garcia et al., 1987).

Only 14 data values were missing out of the 3,360 data values for the 56 years of data in the six county data sets. An ARIMA model was estimated for each missing data value, and the missing value was forecast from the Box-Jenkins parameter estimates. The statistical software Regression Analysis of Time Series (RATS) was used for multivariate analysis.

High potential yields associated with corn production have led to the crop being planted on marginally suitable soils, and under less than optimal climate conditions. The ability to change soil quality characteristics in statistical modeling is limited by available data and multivariate techniques. During the 25-year forecast, soil conditions were kept constant. Annual variations in temperature and precipitation within the growing season are expected to account for the variance in annual yield, after removal of the upward trend associated with technological improvement. Temperature and precipitation patterns for May through August were analyzed to determine if the climate variables met the condition of stationarity; a state of stochastic equilibrium (Mills, 1990). It was determined that all temperature and precipitation patterns met the condition of stationarity, and therefore were not cointegrated with the yield data. This was tested by analyzing the autocorrelation functions (ACF) and partial autocorrelation functions (PACF) to determine that they were displaying "white noise," thus the data was thought to be in stochastic equilibrium. Testing for unit roots in the climate data was inappropriate as the data did not display a trend or indicate any distinct pattern within the raw data or the ACFs and PACFs.

Developing Model Structures and Estimating Parameters

The multivariate model was developed for Boone County and tested by withholding 5 years of crop yield data from the initial model run for model validation. Two models were designed to compared the effects of climate variables and lagged yield variables on crop yield. The crop yield data was transformed using logarithms to remove the volatility of the data due to the variance in annual yields. When a single difference was applied to the crop yield data, the appropriate model for both crops was determined to be a single step moving average model or MA(1) model. Since an infinite autoregressive (AR) model is the inverse of an MA(1) model, lagged values of the dependent variable were included in the models. Two models were developed for each crop, the difference between the models was the number of lagged variables included in each model. The full corn model had five lagged variables of the dependent corn yield variable. The full soybean model had three lagged variables of the dependent soybean yield variable. The restricted model for both crops had two lagged variables of the dependent corn yield variable. In addition to the lagged variables, linear terms for May precipitation, June temperature, August temperature, and linear and quadratic terms for July temperature and July precipitation were included in the model. The quadratic terms are included to capture and parameterize the negative effects of excessive temperature and precipitation.

The corn models developed were as follows:

$$Y = a + b*lagY\{1\} + b*lagY\{2\} + b*lagY\{3\} + b*lagY\{4\} + b*lagY\{5\} + b*MP + b*JNT + b*JLT + b*JLTSQ + b*JLP + b*JLPSQ + b*AT + e$$
$$Y = a + b*lagY\{1\} + b*lagY\{2\} + b*MP + b*JNT + b*JLT + b*JLTSQ + b*JLP + b*JLPSQ + b*AT + e$$

The soybean models were:

$$Y = a + b*lagY\{1\} + b*lagY\{2\} + b*lagY\{3\} + b*MP + b*JNT + b*JLT + b*JLTSQ + b*JLP + b*JLPSQ + b*AT + e$$
$$Y = a + b*lagY\{1\} + b*lagY\{2\} + b*MP + b*JNT + b*JLT + b*JLTSQ + b*JLP + b*JLPSQ + b*AT + e$$

where: Y = the annual corn or soybean yield, lagY{year lagged} = prior year yields, MP = May precipitation, JNT = June temperature, JLT = July temperature, JLTSQ = July temperature squared, JLP = July precipitation, JLPSQ = July precipitation squared, and AP = August precipitation.

The reduced corn model was developed after the five lag model had been estimated, when the results of the model parameterization revealed that lag 3 and lag 4 were not statistically significant at the 95% confidence level. To model the system parsimoniously, the reduced parameter model was developed to be tested against the five lag model for statistical significance and forecasting ability. The multivariate procedure in RATS was applied to the data set first to estimate the parameters for the five lag model, then for the two lag model.

Lag 4 and lag 5 were not statistically significant in the soybean model. The full model for soybeans included three lag variables of the yield variable plus the climate variables listed above. The reduced model included two lags and the climate variables listed above.

Other climate variables considered for the corn and soybean models included May temperature, June and August precipitation, and September temperature and precipitation. The variables were rejected for the following reasons: May temperature is not a significant variable for midwestern corn growth as most corn is planted mid to late in the month and generally does not emerge until the last week of May. Soybeans are planted at the end of the May or in early June. June and August temperature were included in earlier modeling applications, and rejected as failing to add statistically significant substance to the model. June temperature is important to crop germination. August temperature is important for corn ear development and soybean pod fill, but is secondary to August precipitation in effect on crop development and yield. September temperature is not an issue, as the crops have completed their growth, and are drying in the field. September precipitation does not affect crop growth. However, it can be a factor in slowing harvest, and increasing the need for grain drying after harvesting. (Ritchie and Hanway, 1984; Ritchie et al., 1985; Schweitzer, 1994)

Interpretation of Model Results

Both statistical models produced statistically significant parameter estimates for the lagged values and July precipitation. In the five lag model, lagged variables one and five, and July precipitation variables were significant at the 95% confidence level. The positive coefficient estimate for July precipitation indicated that an increase in yield is associated with increased precipitation. The quadratic term for July precipitation was significant at the 90% confidence level. This indicated that extreme levels of precipitation will result in reduced yields.

The two lag model produced similar results as the five lag model. Lagged variable one and the July precipitation were significant at the 95% confidence level. The other variables were not statistically significant. The parameter estimate for July precipitation was almost the same for both the five lag and two lag models (0.02), other parameter estimates varied. June, July, and August precipitation all had positive parameter estimates; May precipitation was negative, as were the parameter estimates for the lag variables. Both five lag and two lag models had

Table 49.1 R² Values for Counties Modeled with Multivariate Process

County	Corn	Soybeans
Adams	0.875	0.696
Boone	0.868	0.712
Gibson	0.813	0.841
Parke	0.832	0.777
Porter	0.822	0.642
Scott	0.895	0.839

F statistics that were significant at the 99% confidence level. The Durbin-Watson statistic was not tested here, as it is not valid when using lagged variables, since occurrence lags bias analysis of this statistic toward a failure to reject the null hypothesis.

Testing for Serial Correlation

An M-test was applied to the residuals of the two lag and five lag models for both corn and soybeans to determine if serial correlation existed within the residuals. None of the parameter estimates for the residuals in the M-test model were significant at the 90% confidence level. The residuals from the M-test were viewed as "white noise," although a sine wave pattern is displayed in the residuals. The residuals from the two lag model M-test displayed more "noise" than the residuals from the five lag M-test, but there is less of a pattern with these residuals.

Forecasting Future Yields

The parameter estimates of the five lag multivariate procedure were utilized to forecast new corn and soybean yield values for the next 25 years. After determining that the variables selected for the multivariate model produced reasonable results for Boone County, parameter estimates were generated for an additional five Indiana counties to forecast corn and soybean production for the time period of 1993 to 2017. Table 49.1 displays the R^2 values for the corn and soybean regression equations for each county.

Forecast estimates for Adams County corn are displayed in Figure 49.5. Figure 49.6 displays Adams County soybeans forecasts. Both forecasts are displayed along with the actual yield values for the 1937 to 1992. Figure 49.7 displays Parke county corn forecasts and Figure 49.8 displays Parke county soybean forecasts. All of the forecasts initially increase then decline to levels of production near the current mean, supporting the significance placed on lagged variables by both the corn and soybean models.

The between-year variance discussed at the beginning of this paper is not addressed in applying either model. However, the variance that may occur with

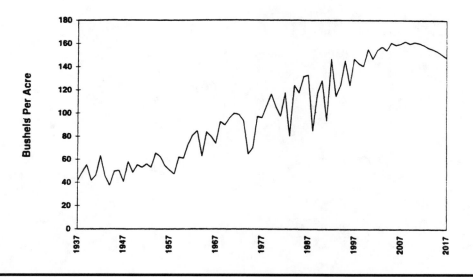

Figure 49.5 Adams County corn production, 1937 to 2017.

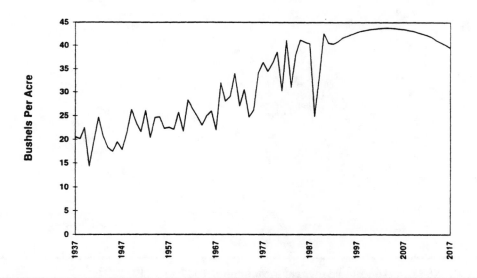

Figure 49.6 Adams County soybean production, 1937 to 2017.

climate change is an issue to be faced by producers. Introduction of variance by way of stochastic disturbance in the climate conditions for the multivariate models was inappropriate as the CENTURY model utilized in this study did not include variance in climate conditions. To place an economic perspective on the variance in yield, each bushel of corn adds about $2.00 in revenues, and each bushel of soybeans adds $5.50 in revenues. This means the difference in the variance of annual crop yield can be greater than $100/acre.

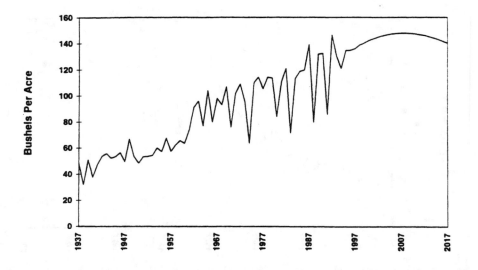

Figure 49.7 Parke County corn production, 1937 to 2017.

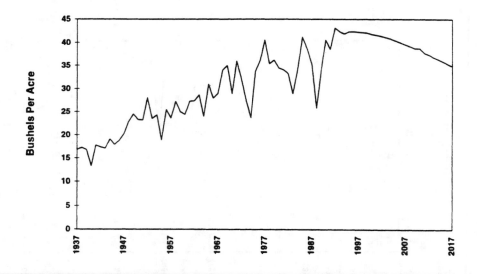

Figure 49.8 Parke County soybean production, 1937 to 2017.

Comparison of Results

An initial comparison of the CENTURY results with the multivariate forecasts indicates that initial multivariate yield forecasts for both crops are higher than the CENTURY forecasts for the 1993. The multivariate forecast for corn yield increases at a higher rate than the CENTURY forecast for corn yield, but begins to decline at 2010. Multivariate corn yield forecasts remained higher than the CENTURY forecasts for the 25-year forecast period. Multivariate forecasts for soybean yield also began higher than the CENTURY soybean yield predictions, but began to decline by 2005, and dropped below the CENTURY forecasts by the end of the forecast period.

CENTURY has the ability to introduce stochastic climate disturbances during the model run. Early attempts to incorporate stochastic disturbances with changed climate have met with mixed results. Further effort is needed in this area to incorporate stochastic disturbances into the modeling process. If the predicted climatic changes result in increasing variance in annual temperature and precipitation, modeling techniques must incorporate this factor into estimates of agricultural productivity. At the same time, improved cultivars that are more tolerant of varied climate conditions must be developed.

When the final results of this CENTURY modeling effort are complete, the output data will be used as input data into modeling programs such as Purdue Crop/Livestock Linear Program to investigate long-run strategic changes and the economic and policy decisions associated with those changes.

Conclusion

The multivariate model allows forecasts to be generated when data is limited. The initial forecasts, especially for corn, are in agreement with the results of the CENTURY model. The advantage of using a process model is the ability to develop relationships between input variables and modify process rates under changing conditions. CENTURY, being a soil organic matter model, has the ability to model crop growth as an ecosystem. Not being limited to plant growth and productivity, CENTURY models the building blocks of crops. This interaction with the soil and the atmosphere, when incorporating the direct and indirect effects of doubled CO^2 and stochastic disturbances, has the potential to produce highly accurate estimates of crop yields.

Acknowledgment

Research support provided by the National Institute for Global Environmental Change — Midwest Regional Center.

References

Allen, R.G. and F.N. Gichuki. 1989. Effects of projected CO_2-induced climate changes on irrigation water requirements in the Great Plains states. In J.B. Smith and D.A. Tirpak, Eds. The Potential Effects of Global Climate Change on the United States. U.S. Environmental Protection Agency, Washington, D.C.

Easterling, W.E. II, M.S. McKenny, N.J. Rosenburg, and K.M. Lemon. 1991. Report IIB: a farm-level simulation of the effects of climatic change on crop production in the MINK region. In Processes for Identifying Regional Influences of and Responses to Increasing Atmospheric CO2 and Climate Change. U.S. Department of Energy, Washington, D.C.

Federal Reserve Bank of Chicago and the Great Lakes Commission. 1985. *The Great Lakes Economy: A Resource and Industry Profile of the Great Lakes States*. Harbor House Publishers, Boyne City, Michigan.

Garcia, P., S. E. Offutt, M. Pinar, and S. A. Changon. 1987. Corn yield behavior: effects of technological advance and weather conditions. *J. Climate Appl. Meteorol.* 26:1092–1102.

Hansen, J.E. 1989. The greenhouse effect: impacts on current global temperature and regional heatwaves. In D.E. Abrahamson, Ed. *The Challenge of Global Warming*. Island Press, Washington, D.C.

Harpstead, D.D. 1992. Corn. In *Corn and Soybeans: Status and Potential of Michigan Agriculture-Phase II. Special Report 51*. Michigan State University, Agricultural Experiment Station, East Lansing.

Jones, C.A. and J.R. Kiniry. 1986. *CERES-Maize: A Simulation of Maize Growth and Development*. Texas A&M University Press, College Station.

Metherell, A.K., L.A. Harding, V.C. Cole, and W.J. Parton. 1993. CENTURY Soil Organic Matter Model Environment Technical Documentation. Agroecosystem Version 4.0. Great Plains System Research Unit Technical Report No. 4.

Mills, T.C. 1990. *Time Series Techniques for Economists*. Cambridge University Press, Cambridge, England.

National Agricultural Statistical Service. 1994. U.S. Agricultural Statistics. U.S. Department of Agriculture, Washington, D.C.

Peart, R.M., J.W. Jones, R.B. Curry, K. Boote, and L.H. Allen, Jr. 1989. Impact of climate change on crop yield in the southeastern U.S.A.: a simulation study. In J.B. Smith and D.A. Tirpak, Eds. The Potential Effects of Global Climate Change on the United States. U.S. Environmental Protection Agency, Washington, D.C.

Perry, M.L., J.H. Porter, and T.R. Carter. 1990. Agriculture: climate change and its implications. *TREE*. 5:318–322.

Randolph, J.C. and J.K. Lee. 1993. Climate change effects on forests of the Eastern United States. Pages 483–491 in Proceedings from 7th Annual Symposium on Geographic Information Systems in Forestry, Environment and Natural Resources Management. Volume 1. Vancouver, British Columbia. February 15–18, 1993.

Ritchie, J.T., B.D. Baer, and T.Y. Chou. 1989. Effect of global climate change on agriculture Great Lakes Region. In J.B. Smith and D.A. Tirpak, Eds. The Potential Effects of Global Climate Change on the United States. U.S. Environmental Protection Agency, Washington, D.C.

Ritchie, S.W. and J.J. Hanway. 1984. *How a Corn Plant Develops. Special Report No. 48*. J.C. Clayton, Ed. Cooperative Extension Service, Iowa State University of Science and Technology, Ames.

Ritchie, S.W., J.J. Hanway, H.E. Thompson, and G.O. Benson. 1985. *How a Soybean Plant Develops. Special Report No. 43*. J.C. Clayton, Ed. Cooperative Extension Service, Iowa State University of Science and Technology, Ames.

Rosenzweig, C. 1989. Potential effects of climate change on agriculture production in the Great Plains: a simulation study. In J.B. Smith and D.A. Tirpak, Eds. *The Potential Effects of Global Climate Change on the United States*. U.S. Environmental Protection Agency, Washington, D.C.

Schneider, S.H. 1989. The greenhouse effect: science and policy. *Science*. 243:771–781.

Schweitzer, L.E. 1994. *Soybean Growth & Development — Environmental Effects*. Department of Agronomy, Purdue University, West Lafayette, Indiana.

Smith, J.B. and D.A. Tirpak, Eds. 1989. The Potential Effects of Global Climate Change on the United States. U.S. Environmental Protection Agency, Washington, D.C.

Vitosh, M.L. and B.W. Diers. 1992. Soybeans. In *Corn and Soybeans: Status and Potential of Michigan Agriculture-Phase II. Special Report 51*. Michigan State University, Agricultural Experiment Station, East Lansing.

Chapter 50

Decision Support Systems for Irrigated Zones: An Integrated Approach to Land Use Planning and Management in Southern Europe

L. MIRA DA SILVA, J. PARK, AND P.A. PINTO

Abstract

Alentejo is an administrative region in the South of Portugal covering 46% (1,840,000 ha) of the total agricultural area of the country. The characteristic Mediterranean climate, with hot and dry summers, provides considerable potential for irrigated crop production. However, the average annual irrigated area during the last 5 years was less than 70,000 ha. The Alqueva project is a large irrigation scheme in Alentejo that received final approval in 1993. The main dam will have a total water holding capacity of 4,150 hm^3 and will be used for irrigation, electricity production, and urban and industrial supply. Despite the apparent suitability of the region for irrigated agriculture there are still many differing opinions about the implementation of the Alqueva project. Critics refer to steep slopes and unsuitable soils, uncertainty in water availability, and poor water quality. Further questions about the economic rationale and social implications of the project reinforce the need for an assessment framework capable of synthesizing information from a range of data sources.

This paper provides a brief analysis of the existing irrigation schemes in Alentejo and a critique of the Alqueva project reviewing its history and some of the conflicting opinions. A dynamic and integrated decision support system (DSS) for irrigated land use planning and management is then proposed and discussed. The system uses existing models to quantify biophysical variables such as crop yields, crop water requirements, and soil erosion and combines these variables with socioeconomic data in the evaluation of different policy and management strategies. It is concluded that the integration of biophysical and socioeconomic data, facilitated by qualitative and quantitative models, is necessary if the highest potential use of irrigation water is to be achieved while at the same time considering appropriate cropping sequences and environmental sensitivity. Land use planning will only be improved if relevant data can be analyzed and presented in a manner that is able to inform decision-making processes.

Introduction

The transition from rainfed to irrigated agriculture is a complicated process that often leads to considerable changes in crop production systems, land use and farm ownership, size, and structure. The implementation of large irrigation schemes in southern Portugal, some of them currently characterized by poor performance due to low uptake of irrigation, is a good example of the consequences of underestimating these changes and misunderstanding the system into which water is introduced. Furthermore, it suggests that the methodologies applied in land use planning and management in some projects have failed because they were unable to anticipate the impact of introducing irrigation.

These examples in Alentejo clearly highlight the consequences of adopting a technical perspective without understanding the criteria upon which farming decisions are made, the behavior resulting from those decisions and the impact of that behavior upon the natural resource base. Where the social context of physical change is not considered, a range of unforeseen consequences may arise, which will invariably result in a mismatch between the objectives of that intervention and the eventual outcome. Decisions are influenced by a complex set of factors which do not differentiate between physical and social processes (Lovejoy and Napier, 1988; Redclift, 1994). To improve the understanding of this complexity there is a need to move away from clearly defined disciplinary boundaries towards an integrated and interdisciplinary method. While the insights obtained from traditional science about natural processes are of great importance, to be relevant to decision making they must be related to the socioeconomic and cultural environment in which they occur (Lemon and Park, 1993).

Agricultural systems research is attempting to overcome the "perspective issue" by focusing on the integration of biophysical and socioeconomic subsystems (Stomph et al., 1994; Park and Seaton, 1995) and there is increasing support at a funding level for the integration of the natural and social sciences (Newby, 1992). In some cases social enquiry can actually succeed where traditional scientific methods have proved inappropriate in providing information about changes in natural phenomena (Lemon et al., 1994). It is crucial therefore that decision support models incorporate those elements that form the basis for decision making and are not only restricted to, or even based upon, those factors that can be easily handled by "traditional" scientific approaches.

In parallel to the development of new integrative and interdisciplinary methodologies, there is a need to make the results of such research available at the decision-making level. Quantitative models, together with new software packages such as "user friendly" database management systems (DBMS) and geographic information systems (GIS), are creating new possibilities for the implementation of powerful analytical and communicative tools.

This paper proposes the development of a DSS, based on an integrated and interdisciplinary approach, to be used in irrigated land use planning and management. The associated research project comprises several sequential activities that will be presented and discussed in the following sections. To summarize:

1. Assessment of the large irrigation schemes in Alentejo
2. Description and analysis of the Alqueva project
3. Identification of developments required in land use planning and management in irrigated areas
4. Selection and assessment of an irrigation scheme to be used as a case study
5. Design and implementation of a DSS for irrigated land use planning and management

Assessment of the Large Irrigation Schemes in Alentejo

Since the 1930s irrigation has been considered a priority in several regional development plans in the Alentejo region of southern Portugal. Without water, the usual hot and dry summers do not provide many alternatives for crop production but with irrigation there is a great potential for agriculture. In 1937 the Portuguese government passed a law (Decreto-Lei 1949, 15/02/1937) supporting the implementation of large subsidized irrigation schemes in Alentejo. The first of these schemes, Vale do Sado, started to irrigate in 1948 and the second, Campilhas, in 1954. More than 10 years passed until the publication of a regional irrigation plan proposing the implementation of several other projects (Alentejo Irrigation Plan, MOP-DGSH, 1965). Apart from Vale do Sado and Campilhas (Alto Sado was later integrated in this project), already in use at that time, this plan included all the actual large irrigation schemes in Alentejo (Table 50.1).

In spite of the apparent suitability for irrigated agriculture in Alentejo there are still conflicting opinions as to the real usefulness of certain irrigation schemes. Skeptics usually refer to the performance of the schemes where the actual area of land that is currently irrigated is considerably less than the projected potential area (Table 50.2). In Mira and Odivelas, for instance, the actual irrigated area only exceeded half of the potential area in recent years. Despite the general increase in utilization rates (actual irrigated area/potential irrigated area), there is still an average of more than 18,000 ha every year (about 40% of the total potential irrigated area) in which the irrigation infrastructures are not used.

Water availability is often the main reason for low utilization rates. For instance, in Campilhas e Alto Sado, in 1983, the water volume in the dam at the beginning of the irrigation period was 7% of the net dam volume (Table 50.3) and only 17%

Table 50.1 Main Characteristics of the Large Irrigation Schemes in Alentejo

Irrigation scheme	Dam	Begin of exploration	Basin area (km²)	Lagoon area (km²)	Net volume (hm³)	Potent. irr. area (ha)
Campilhas/Alto Sado	Campilhas	1954	109	3.33	26.2	5963
	Mt. da Rocha	1972	246	11.00	97.8	
	Fonte Serne	1979	30	1.05	3.7	
Divor	Divor	1965	43	2.65	11.9	488
Caia	Caia	1968	571	19.70	192.3	7237
Roxo	Roxo	1968	351	13.78	89.5	5040
Mira	Santa Clara	1970	520	19.86	240.3	12000
Odivelas	Odivelas	1974	430	9.73	70.0	6381
	Alvito	1979	212	14.75	130.0	
Vale do Sado	P. do Altar	1948	743	7.98	94.0	6171
	Vale do Gaio	1948	509	5.50	63.0	
Vigia	Vigia	1985	125	2.88	15.9	1505

Note: Three "large" schemes are not considered in this and the following tables: Lucefecit, Monte Gato e Miguéis and Vale do Sorraia. The former ones because of their size. The potential irrigated area is only 228 ha in Lucefecit and 134 ha in Monte Gato e Miguéis. The latter due to its location. More than 80% of the dominated area in Vale do Sorraia is in Ribatejo, the region above Alentejo.

Table 50.2 Potential Irrigated Areas and Utilization Rates in the Main Large Irrigation Schemes in Alentejo

Irrigation scheme		Camp./ A. Sado	Divor	Caia	Roxo	Mira	Odivelas	Vale do Sado	Vigia	Total
Potent. irrig. area (ha)	81/84	5,963	488	7,237	5,040	12,000	6,381	6,171		43,280
	85/91	5,963	488	7,237	5,040	12,000	6,381	6,171	1,505	44,785
Utilizat. rate (actual irrig. area/ potent. irrig. area)	1981	37%	0%	51%	3%	35%	37%	54%		37%
	1982	37%	26%	53%	21%	40%	22%	85%		43%
	1983	17%	1%	57%	10%	47%	12%	62%		37%
	1984	52%	48%	49%	31%	45%	30%	88%		49%
	1985	57%	65%	52%	29%	44%	33%	90%	12%	49%
	1986	53%	66%	50%	29%	43%	31%	93%	18%	49%
	1987	55%	50%	51%	32%	34%	34%	92%	30%	47%
	1988	64%	66%	64%	39%	38%	35%	91%	51%	53%
	1989	69%	44%	64%	43%	43%	34%	89%	52%	55%
	1990	80%	85%	65%	54%	41%	41%	96%	59%	60%
	1991	68%	85%	61%	49%	46%	44%	97%	59%	59%

Adapted from Daehnhardt, 1993.

of the potential area was irrigated (Table 50.2). In 1981 and 1982 and from 1984 to 1989 this improved but water was still limiting. In 1990 there was enough water to irrigate all of the potential area and the utilization rate was 80%. However, part

Table 50.3 Net Dam Volumes, Potential Irrigated Areas, Capacity for Interannual Regularization (Net Dam Volume/Potential Irrigated Area) and Water Availability at the Beginning of the Irrigation Period in the Main Large Irrigation Schemes in Alentejo

Irrigation scheme		Camp./ A. Sado	Divor	Caia	Roxo	Mira	Odivelas	Vale do Sado	Vigia
(1) Net volume (hm³)		127.7	11.9	192.3	89.5	240.3	200.0	157.0	15.9
(2) Pot. irr. area (ha)		5,963	488	7,237	5,040	12,000	6,381	6,171	1,505
(1)/(2) (m³/ha)		21,415	24,385	26,572	17,758	20,025	31,343	25,442	10,565
Water availability (net water volume at the beginning of the irrigation period/net dam volume)	1981	42%	30%	59%	0%	61%	23%	13%	
	1982	29%	43%	62%	13%	54%	15%	71%	
	1983	7%	13%	33%	1%	26%	3%	14%	
	1984	39%	47%	68%	22%	48%	19%	96%	
	1985	61%	99%	95%	37%	73%	54%	99%	
	1986	50%	72%	89%	33%	71%	46%	97%	100%
	1987	37%	52%	79%	27%	69%	47%	96%	100%
	1988	64%	72%	84%	48%	92%	56%	94%	100%
	1989	46%	45%	55%	32%	97%	36%	51%	87%
	1990	99%	100%	93%	100%	92%	84%	99%	100%
	1991	88%	100%	94%	79%	72%	87%	98%	84%

Adapted from Daehnhardt, 1993.

of the potential area was still not irrigated, despite the availability of water. It is possible to conclude that the lack of water is not the only reason for nonadoption of irrigation.

In fact many of the schemes can be considered to be oversized with respect to irrigation water requirements. Excluding Roxo and Vigia, the relationship between the net dam volumes and the potential irrigated areas is always higher than 20,000 m³/ha (Table 50.3: (1)/(2)). If the dams were completely replenished every year the water availability would exceed irrigation requirements. Oversizing the dams allows the regulation of interannual water supplies, permitting irrigation when the water flowing to the dams is not enough to replenish them. Caia and Mira are successful examples of this strategy. However, some of the other schemes do not have sufficient water in dry years. In Alvito (Odivelas), for example, the total net water volume of the dam (130.0 hm³) is more than three times the estimated annual water flow (40.0 hm³). Thus, without an external water source it takes on average more than 3 years, without consuming water, to fill the dam. Similar conclusions can be reached about Roxo, Monte da Rocha (Campilhas e Alto Sado), Divor, and Vigia.

There are other reasons that explain the low uptake of irrigation. In all the schemes mentioned, excluding Vigia, water is distributed in open channels by gravity and sprinkler irrigation can require costly electrification and pumping equipment. In these cases, some farmers prefer to continue to grow rainfed crops. There are also examples of an inadequate initial selection of the irrigated areas in which delimitation was mainly based on topographic data (Ramos et al., 1992). Indeed, a common argument to justify the low uptake is that some soils and slopes are not suitable for irrigation. Several problems are related to deficient planning and management and a consequent absence of actions to increase the irrigation uptake. A governmental agency report states that schemes were planned and implemented without considering the regional economic conditions or farmers' problems, and that this can explain the lack of success of the associated agricultural policies (MAI-CCRA, 1982). Road and electricity infrastructure, soil drainage, farmer training, and applied research are some of the core issues that can determine the success of a project. Other common constraints are related to credit rates, marketing, extension and technical advice, farm size and structure, weak farmers' associations, and the instability of agricultural policies.

Thus, the implementation of irrigation infrastructure in a region does not necessarily mean that all the farmers are prepared to use water in a rational way. Only an adequate planning and management strategy for irrigated areas can help to increase the uptake of irrigation and improve the utilization of the infrastructure. This has been achieved in some situations but the reasons, like the causes for the low performance of the less successful projects, are often unclear. Thus, valuable information can be gathered by assessing these schemes, together with the exploration of the reasons for inefficient implementation.

Description and Analysis of the Alqueva Project

The Alentejo Irrigation Plan was presented by the Portuguese government in 1957. It was the first exhaustive and detailed proposal to irrigate Alentejo. The plan included several dams and considered different water uses, namely irrigation, electricity production, and urban and industrial supply. The main infrastructure was a dam on the Guadiana river near the Alqueva village. The Alqueva dam, as it was called, was to be integrated with other existing dams in Alentejo as well as with a number of new structures. Over the last 40 years the project has been discussed and assessed in detail but the main dam has still to be constructed. It was only in 1993 that a detailed study led to the decision to implement the project. The Alqueva dam will have a total water holding capacity of 4,000 hm^3 (net volume — 3,150 hm^3) and should be able to irrigate 110,000 ha, produce electricity, and supply water to overcome part of urban and industrial regional requirements. Some of the characteristics of the dam are presented in Table 50.4. The new irrigated areas are still dependent upon a recent environmental impact assessment (SEIA, 1994).

Table 50.4 Main Characteristics of the Alqueva Dam

Total basin area	55,000 km^2
Total lagoon area	250 km^2
Total volume	4,150 hm^3
Net volume	3,150 hm^3
Maximum water level	152 m
Minimum water level	135 m

The colorful history and long-term nature of the project has meant that it has been the subject of considerable debate, a *resume* of which is presented here.

Project Dimensions

The large size of the dam and project have been criticized and, as an alternative, the implementation of smaller irrigation projects has been suggested. Some people believe that Alqueva cannot meet regional needs because it is confined to a specific area, whereas smaller dams are less costly, avoid water pumping costs, and could be better directed to suitable areas (Feio, 1988; Leitão, 1988). The opposing view is that the project has a number of purposes and is a strategic water reserve that cannot be substituted. However, the most convincing argument is that Alqueva and the smaller dams do not exclude each other but are, indeed, complementary. Alqueva has the potential to regulate the characteristic interannual rainfall variations and the project will be integrated with other dams that cannot assure water availability in the dryer years (Ferreira, 1989). Moreover, the project can be considered versatile because it is constituted by several independent irrigation schemes that can be implemented over the years or modified as convenient in the future (Costa, 1989; DR-CCRA, 1993).

Water Availability

Skeptics state that the water flow in the Guadiana river will not be enough to fulfill the estimated requirements of irrigation, electricity production, and urban and industrial supply (Feio, 1988; Leitão, 1988, 1989). Leitão (1988) claims that the Guadiana's basin in Portuguese territory is not able to provide enough water and that the flow from the Spanish basin will tend to be reduced because there are more than 400,000 ha in that country with potential for irrigation using water from the river. In addition, the water flow from soils in the Guadiana's basin will tend to diminish due to the increasing concern over erosion and the consequent afforestation of high slope areas. Despite these opinions, in a recent assessment on the availability of water, which considers the estimated difference between water flow and consumption in a future stable scenario, the predicted average annual water flow in the Guadiana river in Alqueva is 2,475 hm^3 (COBA, 1995). This is enough water to irrigate, produce electricity, and provide for urban and industrial supply.

Water Quality

A related problem is that the quality of water originating in Spain cannot be guaranteed. Leitão (1989) claims that water is severely polluted and saline and that quality will tend to decrease over the years. However, a recent assessment concludes that water quality may increase with the construction of new water treatment stations in Spain and because of the increasing concern over the utilization of chemicals in agriculture (DR-CCRA, 1993). The most recent and detailed environmental impact assessment (SEIA, 1994) states that salinity problems are not to be expected beyond minor problems with soil permeability. One of the conclusions of the assessment is that the water can be classified as good for irrigation and if adequate crop production, drainage and irrigation techniques are used then serious problems should not arise.

Water Cost

Water cost will certainly be a major concern, because water will have to be pumped from the Alqueva dam to the irrigated areas, some of them located at much higher altitudes. This may prevent the cultivation of water demanding crops, such as rice, and reduce crop selection alternatives. Good management of irrigation projects, a strict control of water use and the utilization of adequate irrigation practices by farmers, in order to increase water efficiency, may overcome this problem.

Soil Suitability and Slopes

This discussion is highly speculative because the irrigated areas are not yet completely defined. There are claims that the soils are not suitable and slopes too great (Feio, 1988), but they usually refer to the poor performance of existing irrigation schemes to support them and are not based on local assessments. Only after the final decision about the areas to irrigate will it be possible to assess the soils and slopes and evaluate their suitability for irrigation.

Economic and Social Implications

Agricultural development, via the introduction of 110,000 ha of irrigated land, is considered as the main aim of the Alqueva project. Due to the climate in Alentejo, especially the seasonal distribution of rain through the year (i.e., wet winters and autumns, irregular springs, and very dry summers), it is impossible to achieve consistently high yields in rainfed crops. Other returns, such as electricity production and urban and industrial water supply, need to be considered. Some of the social and cultural benefits are difficult to quantify in economic terms. However, farmers' profits have been decreasing and the population in rural areas is diminishing. The project may open new avenues for regional development.

General Environmental Impact

Environmental impact was considered in all the recent assessments of the Alqueva project (GCA-EDP, 1987; CCE-DGPR, 1992; SEIA, 1994). The most important issues raised in these reports are related to water quality, land degradation, and natural habitats. Importance is attached to the control of new residential, industrial, and tourism zones; roads and electricity power lines; the monitoring of agricultural pollution sources; the prevention of soil erosion; and the preservation of the regional fauna and flora diversity. The consequences of flooding are not excessive because the majority of the dam area (about 90%) is characterized by high slopes and poor soils, presently under extensive crop production systems producing low economic returns (Figueira, 1988). Moreover, there are some benefits related to the regularization of the river water flow, the improvement of the local climate (relative humidity increase and temperature decrease in summer), and the development of new leisure areas.

Despite the wealth of debate, the recent confirmation of the start of the project means that decision-making tools are required to aid the planning of new irrigated areas and that research needs to be undertaken to encourage the efficient utilization of water. The low levels of uptake in some previous schemes suggest that land use planning and management can be improved.

Identification of Developments Required in Land Use Planning and Management in Irrigated Areas

In a report produced by a government agency about the implementation of one of the most recent irrigation schemes in Alentejo (Vigia), the need for the improvement of planning methodologies was clearly affirmed in one of the introductory statements (MAI-CCRA, 1982):

> We know that the method used in the past in the development of irrigation schemes in Alentejo is not the best. It is also known that this must be modified but the problem exists and the advances have been slow. ... It seems that the usual methods will be applied in Vigia, farmers will be abandoned to their luck and the project will be, like some of the others, a heavy investment without returns.

Ten years after irrigation started in Vigia the highest uptake of irrigation was still only 60% of the potential and the reasons for this low performance are unclear. Similar situations exist in some of the other large schemes. There is a difference between the suggested potential and the actual irrigated land and this, at least partially, is due to an incomplete understanding of the complexity of physical, economic, social, and cultural systems into which irrigation schemes were introduced. This opinion is shared by the agencies responsible for the development of the projects, as shown by the example of Vigia. Usually, they suggest a lack of understanding with respect to farmers needs and a greater need for planning, management, research, extension services and farmer training.

If the methods used in planning and management in the Alqueva project are similar to those applied in the former irrigation schemes then similar issues of irrigation uptake may be experienced. Research is clearly required in order to develop these methods for new irrigated areas. The assessment of the existing irrigation schemes is also essential to avoid some of the errors of the past. There is still time to rethink the planning and management methods to apply in Alqueva in order to assure a high level of irrigation uptake. The success of the project may depend upon an integrated approach to assess alternative and more effective land uses for the new irrigated areas.

Here we propose the development of a DSS to be applied in land use planning and management in irrigated areas. A major part of the suggested methodology is based on the fundamental principles introduced by the FAO (Food and Agricultural Organization of the United Nations) for agroecological assessment, land evaluation, and planning in irrigation (FAO, 1976; Doorenbos and Pruitt, 1977; Doorenbos and Kassam, 1979; FAO, 1978–1981; FAO, 1985; Fresco et al., 1990; FAO, 1993). Similarly, former attempts to use models in interdisciplinary and multiple objective land use planning and management have been referenced. The CropSyst model (Stockle et al., 1994), the IBSNAT project and the DSSAT system (Uehara and Tsuji, 1993; Jones, 1993), the EPIC model (Williams et al., 1987; Jones et al., 1991), the WOFOST model (van Diepen et al., 1989), the CRIES system (Schultink, 1986), and the CREAMS model (Knisel, 1980; Zhu et al., 1993) are valuable examples. However, these tools are usually concerned with the biophysical environment and not easy to adapt for irrigation planning and management in Alentejo. A simple, flexible, and updatable tool is required, which is able to integrate information from a diverse range of sources, to be used iteratively by decision makers.

Selection and Assessment of an Irrigation Scheme to be Used as a Case Study

An existing irrigation scheme in Alentejo has been identified to help isolate relevant biophysical and socioeconomic factors affecting land use and to assess their effects and interactions on the irrigated cropping systems. If the DSS is to be applied in irrigation projects in the South of Portugal, it has to be developed and validated under representative conditions. Vigia was selected to be used as the case study because it has many similar attributes to Alqueva. In spite of the size differences between the schemes, the most important biophysical and socioeconomic properties are comparable.

The Vigia scheme is the most recent large irrigation project in Alentejo. The dam and other infrastructures were constructed between 1976 and 1985, when irrigation started. The scheme was managed by a government agency until 1991, when control was transferred to an association of farmers. The potential irrigated area is 1,505 ha and water is distributed under pressure allowing sprinkler irrigation to be utilized without any additional water pumping costs. This is an important characteristic when considering the choice of Vigia as a case study, as water will be under pressure in Alqueva.

The Vigia case study has been used to qualitatively identify and establish the relations among the variables that characterize the irrigated cropping systems. The first step was to establish the processes, agencies, and interactions that characterize the system under study. A set of semi-structured interviews with individuals involved in activities related to agricultural production in the area (mainly farmers, extensionists and decision makers in governmental agencies) was undertaken with this purpose. This has provided the basis for a second set of more structured interviews with farmers designed to obtain quantitative data relating to processes and relationships that emerged from the first phase. These structured questionnaires will focus on the past use of land and water, on local field attributes, and on structural and regional socioeconomic characteristics. A broad characterization of farmers (labor and technology investments, level of education), their perceived hazards (uncertainty of price support, virus and disease), main crop production systems (time schedules, inputs and outputs) and irrigation techniques will also be obtained. These data complement those already collected from regular sources, such as weather and soil data, and provide a basis to support model implementation.

Design and Implementation of a DSS for Irrigated Land Use Planning and Management

This section introduces the DSS. The tool is in the design phase and only a conceptual scheme (Figure 50.1) and a brief description of its main characteristics will be presented. A set of databases covering information from different sources is being developed. The characterization of land units includes soil characteristics for the most representative soil types, a series of 30 years of monthly weather data, farms and farmers attributes, and field data. Land units can be defined in this context as "homogeneous fields," which are areas of land that can be considered uniform and characterized by the same soil, weather, farm, farmer, and field data. Structural and socioeconomic regional characteristics are described in a separate database, because the information is shared by all the land units. Another database is used to specify the requirements of the main actual and potential crop growth systems. These result from the combination of crops with management conditions, which are mainly defined by the level of inputs, including products (seed/plant, fertilizers, agrochemicals), labor (specialized and nonspecialized workers), and machines (tractors, reaping-machines, ploughs). Crop growth systems should be initially set but later identified or modified by the user.

Modelling can be summarized in three steps:

1. Crop growth requirements are matched with the structural and socioeconomic regional characteristics and with the data that characterize the land units (weather/soil/field/farm and farmer) in order to select the possible crop growth systems for each land unit.
2. Crop yields, crop water needs, and soil erosion are estimated. Crop yields and crop water needs are then transferred to a crop budgets database. A net crop income is calculated for each crop growth system/land unit combination. The crop budgets, defined for each crop growth system according to their management characteristics, are descriptions of all the monetary inputs and outputs. The database is supported by the crop growth system requirements database and by an input/output prices database, where the costs of labor, machines, and products are defined.
3. The third and last step is crop allocation or the final selection of a rotation for each land unit. The model takes all possible crop growth systems in a land unit and selects the rotation that maximizes the average net income, considering a range of alternatives identified by the user (crop rotation alternatives database). Some constraints defined by the user, such as the maximum levels of erosion, the total water availability, and the minimum absolute net crop income can also affect the selection.

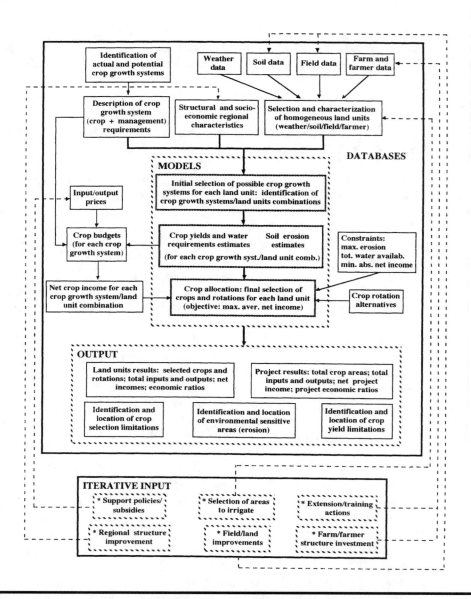

Figure 50.1 Diagrammatic representation of the decision support tool.

The output includes the selected rotation and crops, their corresponding inputs, outputs, and net incomes and some economic ratios, such as the net income per unit of applied water or required labor unit. The identification of variables affecting crop selection and the location of potential erosion areas are examples of more specific results. The output will be displayed using a GIS. This facilitates spatial analysis and overlays to identify more subtle features that otherwise could be lost in the complexity of such a volume of data. If land use planning and management is to be improved then the output must be presented in a clear and versatile way, in order to suit different users together with their decision-making needs.

The system is being designed in such a way that it should be flexible enough to allow the decision maker to explore the use of support policies and subsidies (such as input and output prices, crop subsidies, or set-aside characteristics), regional structure improvements (such as new roads or local markets), farm or farmer investments (such as new machines or the increase of store capacity), field or land improvements (such as lime applications, field drainage, or soil erosion protection), and extension or training actions (leading to new crop technologies or different irrigation methodologies). In addition, new land units can be included or the existing ones can be excluded or modified. Thus, the DSS can be used to assess and select irrigated areas, to identify appropriate cropping growth systems, to locate environmentally sensitive areas, or in more broader socioeconomic analysis.

The final stage of this project will be the application to the Alqueva project. The DSS will be tested for a range of policy, structural, cultural, and management scenarios, taking into account the potential uptake by farmers and a broad range of socioeconomic and biophysical variables, in order to estimate the best areas to supply water to with respect to productive and sustainable agriculture.

Discussion

The first two stages of the approach developed in this paper, the assessment of the large irrigation schemes in Alentejo and the analysis of the Alqueva project, had three main purposes:

1. To identify the main issues associated with the introduction of irrigation into a region
2. To highlight the need for improved land use planning and management in irrigated areas
3. To aid the selection of an irrigation scheme to be used as a case study

This initial phase of the research has thus provided a substantive understanding of the system upon which to base the decision support tool. In particular it has highlighted the need to collect and use data from both the natural and social sciences if the main system components are to be represented. In previous schemes it appears that assumptions have been made or key variables have been ignored when irrigation projects have been planned and introduced. This has led to problems of water shortage at one extreme and low levels of water use at the other.

At present this research is exploring methods for handling large amounts of data, but in particular how more traditional physical data on soil types and climate can be integrated with the data output from social enquiry and how this can be incorporated into a decision support tool. A questionnaire that is currently being distributed in the irrigation schemes in Alentejo will provide a considerable amount of relevant data. There is a clear need to improve our understanding about how social structures affect and are affected by the natural system and how these interactions can be represented in a manner which is responsive to policy and management instruments that promote changes on the natural, social, structural, or cultural configurations. This suggests that there is a requirement for individual researchers who have a broad traditional science background but who are capable of exploring and utilizing data and techniques associated with social enquiry.

An important feature of this decision support tool is its iterative and updatable nature, as irrigation in the Alqueva project will develop over more than 20 years. In such a period of time, it is extremely difficult to estimate changes in the socioeconomic, political, cultural, or even in the biophysical characteristics of the system under evaluation. For example, in a period of 20 years the crops cultivated in the region could be considerably different. If the input data can be modified, for instance the crop specifications or the socioeconomic regional characteristics, the DSS might be used to assess different policy, structural, cultural, or management scenarios and will remain effective over several years. The research is therefore considering methods for building flexibility into the decision support tool.

A large part of this project will continue to involve the exploitation of methods for handling large volumes of data from different sources. The increased access to powerful computers and the release of new software that facilitates the development of models and the analysis and presentation of data allows the construction of complicated but "user friendly" tools. From a decision-making point of view there are advantages in constructing simple (or transparent) models which can be used to explore the implications of change, whether these be focused at the farmer or policy level. However, the development of complex quantitative models and the presentation of results using sophisticated software packages can mean that weaknesses in input data and model construction are difficult for the user (the decision maker) to isolate and question. In many instances it is the questioning of relationships and assumptions which form the basis of a model that allows the decision maker to gain a better understanding of the perspective from which the system has been analyzed and modeled and to gauge the importance of the tool in the overall decision-making process. Thus, the proposed DSS should be viewed as an exploratory tool to help land use planning and management in irrigated areas rather than the definitive method of selecting new areas to irrigate in the Alqueva project or find solutions for the low uptake of irrigation in the existing schemes.

Acknowledgment

This work has been partially supported by the PRAXIS XXI program which is a research program cofinanced by the European Commission and the Portuguese government.

References

CCE-DGPR (Comissão das Comunidades Europeias — Direcção-Geral das Políticas Regionais). 1992. Empreendimento de fins múltiplos de Alqueva. Estudo de avaliação global. Lisboa.

COBA (Consultores de Engenharia e Ambiente). 1995. Recursos hídricos do rio Guadiana e sua utilização. Instituto da Água. Lisboa.

Costa, M.R.A. 1989. Guadiana: tarefa de gerações sem alternativa. *Rev. Ciênc. Agrár.* 12(1):33–41.

Daehnhardt, E. 1993. Perímetros de rega em exploração. Actualização de algumas características e elementos estatísticos até ao ano de 1991. Direcção-Geral de Hidráulica e Engenharia Agrícola. Lisboa.

Doorenbos, J. and A.H. Kassam. 1979. Yield Response to Water. FAO Irrigation and Drainage Paper 33. Food and Agriculture Organization, Rome

Doorenbos, J. and W.O. Pruitt. 1977. Guidelines for Predicting Crop Water Requirements. FAO Irrigation and Drainage Paper 24. Food and Agriculture Organization, Rome.

DR-CCRA (Dossier Regional — Comissão de Coordenação da Região Alentejo). 1993. Empreendimento do Alqueva. Évora.

EDP (Electricidade de Portugal). 1988. Projecto-base de Alqueva. Lisboa.

FAO (Food and Agriculture Organization of the United Nations). 1976. A Framework for Land Evaluation. FAO Soils Bulletin 32. FAO, Rome.

FAO (Food and Agriculture Organization of the United Nations). 1978–1981. Report on the Agro-Ecological Zones Project. Volume 1, Methodology and Results for Africa; Volume 2, Results for Southwest Asia; Volume 3, Methodology and Results for South and Central America; Volume 4, Results for Southeast Asia. FAO World Soil Resources Report 48. Volumes 1–4. FAO, Rome.

FAO (Food and Agriculture Organization of the United Nations). 1985. Guidelines: Land Evaluation for Irrigated Agriculture. FAO Soils Bulletin 55. FAO, Rome.

FAO (Food and Agriculture Organization of the United Nations). 1993. Agro-Ecological Assessments for National Planning: the Example of Kenya. FAO Soils Bulletin 67. FAO, Rome.

Feio, M. 1988. Alqueva. Um regadio com más condições. *Vida Rural.* 11:54–64.

Ferreira, J.F.F. 1989. A problemática actual do projecto de Alqueva — alternativas. *Rev. Ciênc. Agrár.* 12(1):5–15.

Figueira, J.P.R. 1988. Avaliação dos impactes na agricultura. *Rev. Ciênc. Agrár.* 11(4):123–133.

Fresco L., H. Huizing, H. van Keulen, H. Luning, and R. Schipper. 1990. Land Evaluation and Farming Systems Analysis for Land Use Planning. FAO Guidelines. Food and Agriculture Organization, Rome.

GCA-EDP (Gabinete Coordebador de Alqueva — Electricidade de Portugal). 1987. Estudos de impacte ambiental do empreendimento do Alqueva. Lisboa.

Jones, C.A., P.T. Dyke, J.R. Williams, V.W. Kiniry, V.W. Benson, and R.H. Griggs. 1991. EPIC: an operational model for evaluation of agricultural sustainability. *Agric. Syst.* 37:341–350.

Jones, J.W. 1993. Decision support systems for agricultural development. In F.W.T. Penning the Vries et al., Eds. *Systems Approaches for Agricultural Development.* Kluwer Academic Publishers, The Netherlands.

Knisel, W.M. 1980. CREAMS: A Field Scale Model for Chemicals, Runoff, and Erosion from Agricultural Management Systems. Conservation Research Report No. 27. U.S. Department of Agriculture, Washington, D.C.

Leitão, J.D. 1988. Alqueva envolvida por sérias razões negativas provenientes de escassez de água com origem em Espanha. *Rev. Ciênc. Agrár.* 11(4):13–18.

Leitão, J.D. 1989. Regar o Alentejo a partir dos recursos próprios sem o Alqueva. *Rev. Ciênc. Agrár.* 12(1):43–49.

Lemon, M. and J. Park. 1993. Elicitation of farming agendas in a complex environment. *J. Rural Stud.* 9(4):405–410.

Lemon, M., R. Seaton, and J. Park. 1994. Social enquiry and the measurement of natural phenomena: the degradation of irrigation water in the Argolid Plain, Greece. *Int. J. Sustain. Dev. World Ecol.* 1:206–220.

Lovejoy, S. and T. Napier, Eds. 1988. Conserving Soil: Insights from Socioeconomic Research. Soil Conservation Society of America, Iowa.

MAI-CCRA (Ministério da Administração Interna — Comissão de Coordenação da Região do Alentejo). 1982. Aproveitamento hidroagrícola da Vigia. Breves considerações. Évora.

MOP-DGSH (Ministério das Obras Públicas — Direccão-Geral dos Serviços Hidráulicos). 1965. Plano de Valorização do Alentejo; Rega de 170,000 hectares. Lisboa.

Newby, H. 1992. "Join forces in modern marriage". Higher Times Educational Supplement, 17/01/1992. pp20.

Park, J. and R. Seaton. 1995. Integrative research and sustainable agriculture. Agric. Syst. in press.

Ramos, J.B., J. Bragança, and R.L. Faria. 1992. Caracterização dos aproveitamentos hidro-agrícolas do Alentejo e Algarve. Direcção-Geral de Hidráulica e Engenharia Agrícola, Lisboa.

Redclift, M. 1994. Reflections on the sustainable development debate. *Int. J. Sustain. Dev. World Ecol.* 1:3–21.

Schultink, G. 1987. The CRIES resource information system: computer-aided spatial analysis of resource development potential and development policy alternatives. In K. Beek, P.A. Burrough, and D.E. McCormack, Eds. *Quantified Land Evaluation Procedures*. International Institute for Aerospace Survey and Earth Science (ITC), Enschede, The Netherlands.

SEIA (Sociedade de Engenharia e Inovação Ambiental). 1994. Estudo integrado de impacte ambiental do empreendimento do Alqueva. Lisboa.

Stockle, C.O., S.A. Martin, and G.S. Campbell. 1994. CropSyst, a cropping systems simulation model: water/nitrogen budgets and crop yield. *Agric. Syst.* 46:335–359.

Stomph, T.J., L.O. Fresco, and H. van Keulen. 1994. Land use system evaluation: concepts and methodology. *Agric. Syst.* 44:243–255.

Uehara, G. and G.Y. Tsuji. 1993. The IBSNAT project. In F.W.T. Penning the Vries et al., Eds. *Systems Approaches for Agricultural Development*. Kluwer Academic Publishers, The Netherlands.

van Diepen, C.A., J. Wolf, H. van Keulen, and C. Rappoldt. 1989. WOFOST, a simulation model of crop production. *Soil Use Manage.* 5:16–24.

Williams, J.R., C.A. Jones, and P.T. Dyke. 1987. The Erosion Productivity Impact Calculator. Draft USDA Agricultural Research Service, Economics Research Service and Soil Conservation Practice, Temple.

Zhu, M., D.B. Taylor, and C.S. Subhash. 1993. A multiple-objective dynamic programming model for evaluation of agricultural management systems in Richmond County, Virginia. *Agric. Syst.* 42:127–152.

Chapter 51

Management of Natural and Renewable Resources on Watershed Basis in India

RAM BABU AND B. L. DHYANI

Abstract

Diminishing land-man ratio, subsistance agricultural systems, and unscientific management of renewable and nonrenewable resources are responsible for land degradation in India thereby creating a critical poverty erosion cycle. Continuous decline in productivity and increased unemployment are the direct impact, while high runoff rates, increased sediment yield, and mass wasting are indirect consequences of land degradation. Integrated watershed management (WSM) approaches — adopted to manage natural and renewable resources under Operational Research Projects, River Valley Projects, and Flood Prone Area Projects in India during the 1980s — have produced convincing results.

Implementation of WSM programs enhanced the annual productivity of arable and nonarable lands from 1.2 to 21.6 q/ha and 5.5 to 243 q/ha, respectively. Runoff volume reduced by 2 to 42% and soil loss by 10 to 80%. Additional employment from 14 to 40 man-days/ha was created through WSM activities. WSM approaches were found to be economically viable with benefit:cost ratio ranging from 1.10 to 4.50 and internal rate of return from 17 to 66%. The success and sustenance of the program was high in those watersheds where people's participation was maximum.

Introduction

Among the major natural resources available in the country, the most vital is the land which is comprised of soil, water, and vegetation. These elements are the basic ingredients of a primary production system. In India, about three-quarter's of the population still lives in rural areas and depends on the primary production system to meet their basic requirements. Diminishing land-man and land-livestock ratio, subsistance agricultural systems, unplanned cash cropping, improper water management, and unscientific management of renewable resources put the once "golden bird" (India) into a critical poverty erosion cycle (Singh and Dhyani, 1992). Of the 329 m ha geographical area of India, 173 m ha is in various forms of degradation and losing 7.84 m ha land in terms of productive base annually out of which 3.8 m ha is irreversibly lost (Das, 1994). Declining productivity, undernurishment, and underemployment are other direct consequences of our multiple interventions on the natural ecosystem. High volume of runoff, soil loss, sedimentation rates, and increasing quantum of loss through natural calamities (floods, drought, mass wasting, etc.). are the indirect effects of irrational utilization of natural resources.

Indian Scenario

Research was intensified by establishing Regional Soil Conservation Research, Demonstration and Training Centres during the very first Five Year Plan (1951 to 1955) to conserve natural resources. These technology packages were demonstrated on farmer's fields as well as on common lands under various soil conservation programs. Up to 1993 to 1994, India had spent Rs. 35915 m (US$ 1,200 m) to treat 37.34 m ha (22% of its problem area). Until the early 1980s, the WSM Program had not performed up to expectation due to the fact that these programs were single targeted, top-down administered, and least coordinated (Dhyani and Singh, 1991). During the 1970s, the Central Soil and Water Conservation Research and Training Institute, Dehradun demonstrated an integrated WSM approach in three model watersheds located in different agroclimatic regions of the country. The success of these WSM programs in achieving the conflicting goals of production, protection, employment generation, and improvement in the health of natural and cultural resources have opened new vistas of development (Ram Babu et al., 1994; Dhyani et al., 1994; Dhyani et al., 1993, Agnihorti et al., 1989). Consequently, WSM become synonymous with balanced development in rural India. Since then, various rural development and soil conservation programs have been in progress on watershed basis whether it is a River Valley Project, Flood Prone River Project, Drought Prone Area Project, or other WSM Programs. During the VIIIth Plan, soil conservation and WSM programs are being implemented through 5,000 watersheds. In this paper the achievements of some Operational Research Projects, River Valley Projects, and Flood Prone River Projects in accomplishing the conflicting goals of the farmers and the nation as a whole are presented.

Watersheds Description

The watersheds selected for the study are 16 Operational Research Projects (ORP), 8 River Valley Projects (RVP), and 6 Flood Prone River Projects (FPR) representing 10 agroecological zones and 13 States of the country (Figure 51.1). These watersheds cover major land degradation problem areas of the country. In each program

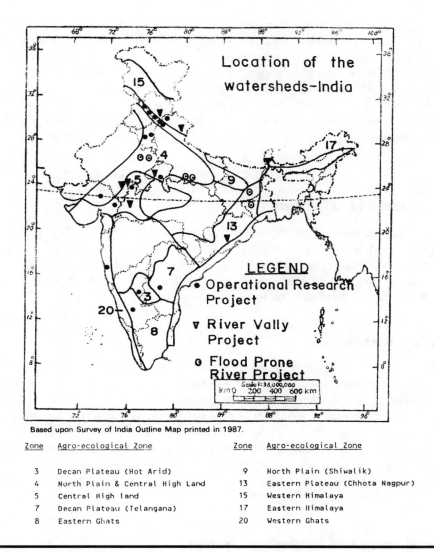

Figure 51.1 Location of watersheds — India.

a model agency was identified and made responsible for overall planning, coordination, monitoring, etc. The implementation was mostly done by State Line Departments with technical support from the Indian Council of Agricultural Research and State Agricultural Universities.

Resources and Problems of Watersheds

The watersheds under study covered arid, semi-arid, subtropical to humid regions with varied annual rainfall (525 to 3,000 mm) and elevation ranging from 120 to 3,000 m above sea level. The size of the watersheds varied from 90 ha (Nada) to 0.31 m ha (Ram ganga). Further, arable land occupied major portions of most of the watersheds except in Himalayan region. The nine soil groups represented in the study are alluvial, black cotton, red, lateritic, red yellow, silty loam, red brown, loam, and black red. Denudation and mass wasting are the major problem in the Himalayan region, while the Shiwalik region is dominated by problems of denudation, flash floods, high sedimentation rates, and droughts. In the Northern plain and Central highlands, sheet, rill, gully and ravines are the major forms of land degradation. Sheet, rill, and drought are the major problems for efficient utilization of land in Malwa region, while the Chhotanagpur plateau is affected by flood, rill, and ravines. Eastern Himalaya and part of Orissa suffer from problems of shifting cultivation (Table 51.1).

Watershed Management

It was envisaged that in all the watersheds, a comprehensive WSM plan would be prepared on the basis of results of a benchmark survey on physiography, hydrology, soil, land capability, vegetation, irrigability and socioeconomic aspects in consultation with the farmers of the region. The plan would include following types of soil and water conservation measures.

Foundation Structures

This relates to the construction of various mechanical measures to reduce the velocity of runoff water, arrest transported material within the structure, and ensure safe disposal of excess water. The measure also helps in storing runoff water for multiple uses. Under mechanical measure small dams, tanks, spillways, gully plugs, check dams, silt detention basins, trenching, embankment, terracing, levelling, bunding and dug out pond were included. Water distribution systems were also included in the watersheds wherever desirable, keeping in view the technical feasibility and need of the people. About 60 to 80% of the total expenditure of watershed was utilized by this sector.

Super Structure of Production System

Efficient use of conserved resources were made by putting the land under most suitable productive purposes. For extension of improved farm technology, a large number of demonstrations were conducted on farmers' fields. The basic aim of these were to demonstrate the efficiency and efficacy of available technologies in enhancing production potentiality of land with improvement in soil health and environment on a sustained basis. Works were implemented mostly by involving local people with the objective to make it a people's program.

Watershed Responses to Multiobjectives

Production with protection (eco-friendly) on a sustained basis and generating gainful employment within the watershed are the multiple objectives of the WSM programs. Responses of selected watersheds on these aspects are presented.

Table 51.1 Basic Resources and Major Problems of Watersheds

Watershed	Rainfall (mm)	Area (ha) Arable	Area (ha) Nonarable	Major soil group	Program	Ref.
WESTERN HIMALAYA						
Fakot	1,900	80	290	RY	ORP	Dhyani et al., 1993
Ram Ganga	1,800	93,180	217,420	RY	RVP	ASC, 1991
Sutlej	1,800	6,374	56,185	SL	RVP	ASC, 1991
EASTERN HIMALAYA						
Upper Gail Khola	2,313	442	1,024	A	RVP	IN-RIMT, 1994c
NORTHERN PLAIN (SHIWALIK)						
Sukhomajri	1,120	50	85	A	ORP	Anonymous, 1988
Nada	1,116	35	55	A	ORP	Anonymous, 1988
Bunga	1,116	336	127	A	ORP	Agnihotri et al., 1989
Maili	1,136	8,560		A	Kandi area	Singh et al., 1991
Chohal	1,136	2,225		A	Devp. Prog.	Singh et al., 1991
NORTHERN PLAIN and CENTRAL HIGHLAND						
Bajar Ganiyar	640	820	270	A	ORP	Sharma, 1994
Siha	640	424	236	A	ORP	Sharma, 1994
Khar khan Kal.	525	850	850	A	FPR	IN-RIMT, 1994d
Kishangarh	565	575	1,725	A	FPR	IN-RIMT, 1994d
Matatila (U.P.)	787	15,063	7,968	RB	RVP	AFC, 1988
Tejpura	940	526	250	LO	ORP	Hazra et al., 1987
GYIJ (Gomti)	835	1,254	96	A	FPR	IN-RIMT. 1994d
GK 3 a(Gomti)	1,074	1,589	386	A	FPR	IN-RIMT, 1994d
Navamota	819	218	95	R	ORP	Singh et al., 1994
CENTRAL HIGHLAND (MALWA)						
Rebari	960	370	200	SL	ORP	Singh et al., 1994
Chhajawa	874	453	—	B	ORP	Prasad et al., 1994
Mandavarsa	825	3,210	620	BR	RVP	AFC, 1992a
EASTERN PLATEAU (CHHOTANAGPUR)						
Upper Jayantia	1,268	663	285	A	FPR	IN-RIMT, 1994b
Taldengra	1,380	1,220	1,002	La	FPR	AFC, 1991
Machkund sileru (ORP)	760	29,181	107,819	La	RVP	ASC, 1987
DECEAN PLATEAU						
Joladarasi	528	509	61	B	ORP	Rao et al., 1994
Chinnatekur	654	815	306	BR	ORP	Rao et al., 1994
G.R. Halli	601	151	169	R	ORP	Adhikari et al., 1991
WESTERN GHATS						
Khumbhave	3,000	121	89	LA	ORP	Talashilkar, 1990

A-Alluvial, B-Black cotton, R-Red, La-Lateritic, RY-Red Yellow, SL-Silty loam, RB-Red brown, Lo-Loam, BR-Black red.

Productivity and Production

Proper management of incident precipitation and runoff through land management/configuration and vegetation was given top priority in WSM keeping farmers preferences in view. Harvested water in various water harvesting structures and recycling through efficient water distribution systems increased the irrigated area by 40 to 333%. Better *in situ* moisture conservation in combination with suitable crops, high yielding varieties, and other improved agrotechniques increased the cropping intensity in the watersheds by 10 to 110%. With the result, the productivity of arable land enhanced by 1.2 to 21.6 q/ha (10 q = 1 tonne) depending upon the severity of degradation, agroclimatic conditions, and nature of crops raised in the watershed. In nonarable lands (land capability class V to VII) were used for plantation of multipurpose trees and grasses. Consequently, the productivity of nonarable land increased by 5.54 to 243 q/ha (Table 51.2). Substitution of low yielding animals (local cows, sheep, and goats) with high yielding animal breeds and availability of good quality fodder increased the milk production by 40 to 350 thousand lit/annum in different watersheds.

Protection

Watershed approaches facilitated monitoring the hydrological responses and sediment yield from the watersheds. The protective benefits from watersheds are reduction in runoff volume, peak discharge rates, sediment yield, and an increase in lean period flow with time and recharge of groundwater. Data collected from different watersheds indicated that WSM programs are successful in achieving their objectives (Figure 51.2). The reduction in runoff ranged from 2 to 42%, soil loss from 10 to 80%, and peak discharge from 20 to 40% in Gomti and Upper Gail Khola watersheds, respectively. It helped to reduce siltation in ponds, reservoirs, and moderating flood peaks. Protective measures also provided opportunities for groundwater recharge. This resulted an increase in groundwater table (0.8 to 2.0 m), and volume of lean period flow and minimized the stream widening and other associated downstream environmental degradation problems. Thus, the ill effects of drought were moderated to a great extent. Further, it also helped in changing the Indian agriculture from the stage of risk/uncertainity to the stage of certainity where farmer can develop his farm plan for prosperity to himself and to the nation.

Sustainability of the Program

Employment Generation

WSM programs may yield productive and protective benefits in perpetuity if and only if the components of the programs are economically sound, provide gainful employment, and become integral parts of the farming system. Execution of mechanical measures, i.e., foundation structures, generated enough casual (short period) employment opportunities. Enhanced productive potential owing to a change in land and animal husbandary practices from extensive to intensive and

Table 51.2 Impact on Irrigation, Cropping Intensity, and Productivity

Watershed	Irrigated area (ha) Pre	Irrigated area (ha) Post	Increase in cropping intensity (%)	Increase in productivity of land (q/ha) Arable	Increase in productivity of land (q/ha) Nonarable
WESTERN HIMALAYA					
Fakot	11.2	24.0	56	10.5	130
EASTERN HIMALAYA					
Upper Gail Khola	20.0	47.0	14.0	5.0	1.0
NORTHERN PLAIN (SHIWALIK)					
Sukhomajri	—	29.0	82	14.5	74
Nada	—	31.5	78	21.6	42
Bunga	—	243.0	110	21.0	36
NORTHERN PLAIN and CENTRAL HIGHLAND					
Bajar Ganiyar	125.0	395.0	21	8.7	40
Siha	115.0	302.0	23	10.4	46
Kharkalan	230.0	257.6	17	2.5	Increased
Kishangarh	220.0	233.2	10	1.2	Increased
Matatila (U.P.)	187.5	374.5	41	2.0	N.A.
Tejpura	20.0	510.0	97	20.0	243
GYIJ (Gomti)	234.0	712.0	68	12.4	N.A.
GK 3a (Gomti)	365.0	671.0	41	8.6	N.A.
Navamota	17.7	35.0	26	9.0	10.0
CENTRAL HIGHLAND (MALWA)					
Rebari	9.0	55.3	20	9.0	5.5
Chhajawa	32.5	260.3	26	5.6	60
Mandavarsa	8.5	17.5	31	3.0	N.A.
EASTERN PLATEAU (CHHOTANAGPUR)					
Upper Jayantia	14.0	35.0	12	2.4	N.A.
Taldengra	N.A.	109*.0	31	9.4	N.A.
Machkund-Sileru	—	—	21	5.0	N.A.
DECEAN PLATEAU					
Joladarasi	N.A.	3.0	18	13.0	33.0
Chinnatekur	217.0	354.0	25	11.0	130.0
G.R. Halli	12.0	52.0	12	N.A.	N.A.
WESTERN GHATS					
Khumbhave	N.A.	N.A.	40	4.4	35

traditional to improved, generated regular employment opportunities. Data presented in Table 51.3 clearly indicate that rational utilization of natural resources may generate employment of 0.01 to 25.4 m man-days, depending of size of the watershed. On an average, various WSM activities generated gainful employment at the rate of 14 to 40 man-days/ha/year which may be useful in sustaining a family for 2 to 6 months in a year.

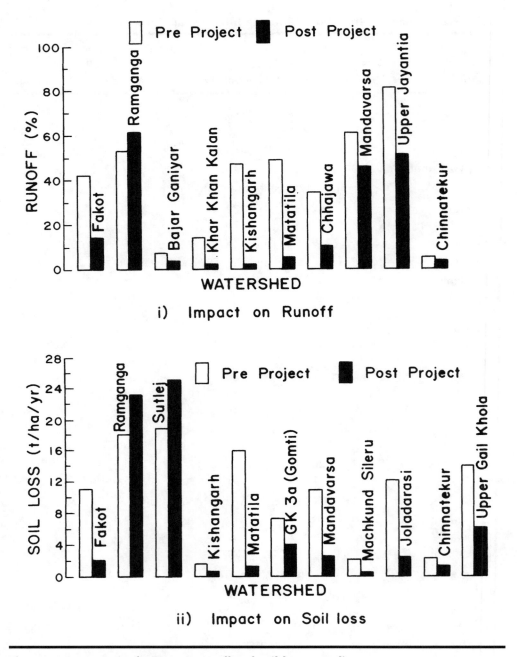

Figure 51.2 Impact of WSM on runoff and soil loss — India.

Economic Viability

Economics of investment in WSM for the management of natural resources was judged by various workers and agencies/organizations in India. Generally, budgeting techniques were employed. For economic evaluation the project life varied from 10 years (Tejpura) to 50 years (Maili and Chohal). The range of discount

Table 51.3   Employment Generated through WSM Programs

Watersheds	Employment generated		Region
	Casual (man-days/ha)	Regular (man-days/ha/year)	
Fakot	203	32	Western Himalaya
Ramganga	65	10	Western Himalaya
Sutlej	429	6	Western Himalaya
Matatila	326	10	Western Himalaya
GYIJ (Gomti)	130	5	Northern plain and central highland
GK 3a (Gomti)	506	12	Northern plain and central highland
Navamota	240	11	Northern plain and central highland
Rebari	88	6	Northern plain and central highland
Mandavarsa	46	2	Central highland (Malwa)
Taldengra	196	23	Central highland (Malwa)
Machkund Sileru	87	5	Eastern plateau (Chhotanagpur)
Chinnatekur	268	45	Decean plateau

rate for obtaining present value of cost and benefit streams were from 10 to 15% as they are the rate of interest charged by nationalized banks for works of such nature. Results presented in Table 51.4 amply demonstrated the economic viability (BCR ranged from 1.10 to 7.06) and financial feasibility (IRR ranged from 16.8 to 66.5%) of such projects on private viewpoint.

Temporal and Intergeneration Equity

Proper arrangement for temporal equity, i.e., equidistribution of benefits among all the watershed farmers from Common Property Resources (CPRs) was given due consideration. It was done through establishment of Watershed Resource Management Society. All the affected farmers of the watershed were admitted as a member of the Society. They elected their representatives/office bearers for the Society. Representatives from technocrats, administrator, and developmental organizations working in the project were also nominated as member of the Society. These Societies were free to manage/create CPRs through their own norms within the broad guidelines of the Society Act.

It was observed that these societies performed well in those watersheds where newly created CPRs generated substantially higher benefits to the individual than cost of their participation in the program in perpetuity, e.g., Sukhomajri, Nada, Bunga, Fakot, and Upper Gail Khola watersheds. This helped in development of many other CPRs in the watershed and improved the economy and environmental wealth (Table 51.5). On the contrary, the Societies could not function properly in those watersheds where CPRs were not selected properly in the first stage or the arrangement for distribution of benefits was improper and/or large differences existed in the political or social affluence within the members of the society. Anyone of these may have led to even the destruction of the CPRs themselves, e.g., Behdala(Una), G.R. Halli, Mata tila, Tal dengra etc.

Table 51.4 Economic Evaluation of WSM Program

Watershed	Project life (years)	Discount rate (%)	BCR	NPV (m Rs)	IRR (%)
WESTERN HIMALAYA					
Fakot	25	10	1.92	0.5	24
Ram ganga	45	10	5.16*	1,201.5	—
Sutlej	45	10	7.06*	4,203.3	—
NORTHERN PLAIN (SHIWALIK)					
Sukhomajri	25	12	2.06	—	19
Nada	30	15	1.07	—	—
Bunga	30	12	2.05	—	—
Mali	50	15	1.10	2.4	—
Chohal	50	15	1.12	1.6	16.8
NORTHERN PLAIN and CENTRAL HIGHLAND					
Bajar-Ganiyar	20	15	1.58	—	17.0
Khar Kalan	15	15	6.07	10.9	—
Kishangarh	15	15	2.35	5.8	—
Matatila (U.P.)	12 (Agric.)	12	3.80	9.3	41
	20 (For.)	12	4.50	—	—
Tejpura	10	10	3.42	—	—
GYIJ	21	15	3.94	12.5	50.0
GK 3a	21	15	1.97	3.7	25.8
Navamota	30	12	2.00	0.8	—
CENTRAL HIGHLAND (MALWA)					
Rebari	20	12	2.65	0.9	37.5
Chhajawa	20	10	2.24	131.7	—
Mandavarsa	20	15	1.97	1,264.9	66.5
EASTERN PLATEAU (CHHOTANAGPUR)					
Upper Jayantia	15	15	1.28	0.3	214
Taldengra	25 (Agric.)	15	1.62	15.6	41
	25 (For.)	15	5.23	16.9	50
Machkund sileru	25	10	4.39*	—	—
DECEAN PLATEAU					
Joladarashi	15	15	1.45	1.7	—
Chinnatekur	15	15	1.81	18.5	—
GRHalli	15	15	1.48	0.9	—
WESTERN GHATS					
Khumbhave	20	15	2.1	—	—

* Include protective benefits.

Intergeneration equity was ensured by reduction in soil loss, improvement in soil health, good vegetation cover, increase in the volume of lean period flow in the streams, etc. Presently, most of these programs have become the part of farming system and hence expected to yield these benefits in perpetuity. They

Table 51.5 Effectiveness of Watershed Resource Management Societies

Watershed	Effectiveness	Remark
Fakot	Satisfactory	Five informal groups for each sector were formed; four are working satisfactorily, one (sprinkler irrigation group) defunct due to property dispute
Sukhomajri	Efficient	Society is economically sound; contributing in development of new CPRs like village road, light, school, common sitting place, etc.
Nada	Satisfactory	Maintaining old created CPRs (water bodies and afforested area)
Bunga	Efficient	Society is economically sound; contributing for development of new CPRs e.g., school, veterinary hospital, village street, common place for various functions
Maili	Unsatisfactory	Major CPRs created during project (water harvesting structures) did not fulfill the aspirations
Behdala	Unsatisfactory	Cost of participation was higher than benefits; large socioeconomic and political affluence differences in the society members
Upper Gail Khola	Satisfactory	Effectively managing existing CPRs, however not creating new CPRs due to lack of resources in the society
Mandavarsa	Satisfactory	Informal groups were formed by farmers themselves; maintaining existing CPRs

will improve the microclimate and continue to improve the health of natural resources which can be used by future generation. Thus WSM programs help to maintain intergenerational equity in the use of natural resources.

Acknowledgment

The authors are thankful to the Director of the Institute for guidance in preparation of draft. Miss Nisha Verma (for typing) and Mr. Nirmal Kumar and Mr. Roopak Tandon (for computer processing work) also deserve thanks. We are also thankful to Mr. M.P. Juyal and Mr. Deepak Kaul for cartographic works.

References

Adhikari, R.N., S. Chittaranjan, M.S. Rama Mohan Rao, and B.S. Thippannavar. 1991. Operational Research Project, G.R.Halli, Karnataka. Annual Report. Central Soil and Water Conservation Research and Training Institute, Dehra Dun.

AFC. 1992. Report on Evaluation Study of Soil Conservation in the River Valley Project of Chambal Catchment. Agricultural Finance Corporation Limited, Bombay.

AFC. 1991a. Evaluation Study in the Catchment of Flood Prone River Project of Rupnarayan, West Bengal. Agricultural Finance Corporation Limited, Bombay. 180 pages.

AFC. 1991b. Evaluation Study in the Catchment of Flood Prone River Project of Ajoy. Agricultural Finance Corporation Limited, Bombay. 204 pages.

AFC. 1988. Evaluation Study of Soil Conservation in the River Valley Project of Matatila, Nizam Sagar and Ukai: Summary Report. Agricultural Finance Corporation Limited, Bombay. 58 pages.

Agnihotri, Y., S.P. Mittal, and S.L. Arya. 1989. An economic perspective of watershed management project in Shiwalik foot hill village. *Ind. J. Soil Conserv.* 17(2):1–8.

Anonymous. 1988. *Operational Research Project on Integrated Watershed Management*, Progress Report. Central Soil and Water Conservation Research and Training Institute, Dehradun.

ASC. 1991. Soil Conservation Scheme, River Valley Projects, Ram Ganga (U.P.) and Sutlej (H.P.). Administrative Staff College of India, Hyderabad.

ASC. 1987. River Valley Projects, Evaluation of Centrally Sponsored Soil Conservation Schemes: A Case Study in Machkund Sileru in Andhra Pradesh and Orissa, Pochampad in Maharashtra. Administrative Staff College of India, Hyderabad. 174 pages.

Das, D.C. 1994. Soil and water conservation for achieving goal of Panchayat Raj. Souvenier. *Nat. Conf. Soil and Water Conservation for Sustainable Production and Panchayat Raj*. Soil Conservation Society of India, New Delhi. Pages 9–18.

Dhyani, B.L., Ram Babu, Sewa Ram, V.S. Katiyar, Y.K. Arora, G.P. Juyal, and M.K. Vishwanatham. 1993. Economic analysis of watershed management programme in outer Himalaya — a case study of Operational Research Project, Fakot. *Ind. J. Agric. Econ.* 48(2):237–245.

Dhyani, B.L., Ram Babu, M.C. Agarwal, and Nirmal Kumar. 1994. Impact of watershed management technology on farm income. *Ind. J. Soil Conserv.* 22(3):71–77.

Dhyani, B.L. and G. Singh (1991). Socio economic aspects of environmental security and development in Himalayas. Pages 155–162 in Pradeep Monga and P. Venkataraman, Eds. *Energy Environment and Sustainable Development in Himalayas*. Indus Publishing, New Delhi.

Hazra, C.R., D.P. Singh, and S.P. Singh. 1987). Soil and water conservation for efficient crop production on watershed basis at Tejpura, Jhansi. *J. Soil Water Conserv. India*. 31(3&4):229–239.

IN-RIMT. 1994a. Evaluation Study of Mg5f and mb2p Watersheds, Sahibi-FPR Catchment Rajasthan. Indian Resource Information and Management Technologies, Hyderabad. 127 pages.

IN-RIMT. 1994b. Evaluation Study of Af1g and Ag 3g Watersheds of Ajoy FPR Catchment, Bihar. Indian Resource Information and Management Technologies, Hyderabad. 55 pages.

IN-RIMT. 1994c. Evaluation Study of Tc2m and TK3g Watersheds, Teesta RVP Catchment, West Bengal. Indian Resource Information and Management Technologies, Hyderabad. 59 pages.

IN-RIMT. 1994d. Evaluation Study of GK3a and Gyij Watersheds, Gomti FPR Catchment. Indian Resource Information and Management Technologies, Hyderabad. 122 pages.

Prasad, S.N. 1994. Operational Research Project, Chhajawa (personal communication).

Ram Babu, B.L. Dhyani, and M.C. Agarwal. 1994. Economic evaluation of soil and water conservation programmes. *Ind. J. Soil Conserv.* 22(1&2):279–289.

Rao, M.S. Rama Mohan. 1994. Operational Research Projects on Watershed Management in Semi Arid Region of India (personal communication).

Sharma, A.K. 1994. Operational Research Project on Watershed Management in Aravali Hills of Haryana (personal communication).

Singh, G. and B.L. Dhyani. 1992. Programmes and barriers in land protection — an Indian experience. Pages 322–332 in *Proceedings of the 7th ISCO Conference, Sydney: People Protecting Their Land*. Volume 1. International Soil Conservation Organisation, Sydney.

Singh, H.B. 1994. Operational Research Projects of Watershed Management in Gujarat (personal communication).

Singh, Karam, H.S. Sandhu, Nirmal Singh, and Balbir Kumar. 1991. Kandi watershed and area development project: benefit cost analysis of investments in two watersheds. *Ind. J. Agric. Econ.* 46:132–141.

Talashilkar, S.C. 1990. Success story of Kumbhave watershed management project. *Ind. J. Soil Conserv.* 19(3):23–30.

Chapter 52

Water Management Scenario Simulation for Decision Support in Multiobjective Planning

SAFWAT ABDEL-DAYEM, SHADEN ABDEL-GAWAD, AND KHALED ABU-ZEID

Abstract

A mathematical model has been developed for the *SI*mulation of *Wa*ter management in the *A*rab *R*epublic of *E*gypt (SIWARE). The model predicts the effect of changes in irrigation water management, hydraulic and hydrologic conditions, cropping pattern, and crop characteristics on the quantity and salinity of drainage water. The model also calculates the effect on other decision parameters such as crop yield as a function of evapotranspiration and soil salinity. It is a tool for supporting water management decisions aiming at multiple objectives such as improving water use efficiency, saving on water supply quantities, reclaiming land for agriculture, use of other conventional and nonconventional water resources, and providing more farmer freedom in crop selection.

This paper presents how a simulation model such as SIWARE can be used for simulating different water management scenarios for the sake of selecting the most favorable option. Planning of water resources on a regional scale involves multiple objectives that need to be achieved by the existing water supply system. Two water management scenarios are evaluated as examples of steps in the decision-making process. The first scenario deals with the reduction of total water supply to the Middle Nile Delta with the objective of saving water or coping with a shortage in water supply. The second scenario deals with the shift to a so called "free cropping pattern" policy with the objective of providing more freedom to the farmer in selecting crops that he wishes to cultivate depending on his vision of the local and international market prices. The impact of these scenarios on crop evapotranspiration and soil salinity are evaluated.

Introduction

The decision makers and planners in Egypt have to face the increasing demands of different users, through the conjunctive use of the fixed share of Nile water supply (55.5 billion m^3), available agricultural drainage water of adequate quality, and groundwater abstracted from the Nile system aquifers (about 4.2 billion m^3) that originates from Nile water through deep percolation. Drainage water from the Nile Delta flows through an intensive network of drains to the Mediterranean sea and Northern Lakes. Under the decreasing per capita share of fresh water, reuse of drainage water is inevitable. At the present time, about 4.0 billion m^3 of agricultural drainage water is reused for irrigation. It is planned to increase this quantity up to 7.0 billion m^3 by the year 2000.

One of the fundamental limitations in the planning process of water resources in Egypt lies in the uncertainty of future quantity and salinity of reused drainage water for irrigation as a result of changes in water management and cropping pattern. A proper procedure for predicting changes in such complex situations is to formulate all relevant physical and functional relationships and combine them in a simulation model. To this effect the regional simulation model **SIWARE** has been developed (Abdel Gawad et al., 1991).

A monitoring system of drainage water in the Nile Delta (Figure 52.1) provides a continuous flow of the data necessary to evaluate the current situation (Abdel-Dayem, 1994) and help in predicting future changes. The SIWARE model together with a database of historical drainage water quantity and quality serve as powerful tools of a decision support system (DSS) for water resources planning and management in the Nile Delta (DRI, 1995).

The Siware Model

The **SIWARE** model package includes a number of physically based submodels, where each submodel has a specific function in the simulation of a regional water management system (Figure 52.2). The submodels included in SIWARE are WDUTY, DESIGN, WATDIS, and REUSE. They have the following functions.

WDUTY is a submodel developed for the determination of the local crop water requirements on the scale of a **SIWARE** calculation unit, which has an average acreage of around 6,500 ha. These water requirements are also referred to as farmers' demand. The required input data include parameters such as local climate, hydrological conditions, initial soil moisture conditions, leaching requirements, irrigation practices, etc. The output is generated on a 10-daily basis per calculation unit.

The DESIGN submodel has been implemented to synthesize an irrigation canal system based on the design parameters obtained from the Ministry of Public Works and Water Resources (MPWWR). Since all irrigation command canals, together

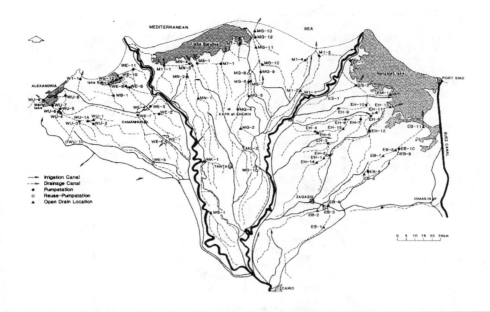

Figure 52.1 The monitoring program of drainage water in the Nile Delta.

with all control structures, should be included in the system definition, and available data sets are not complete and subject to changes, the inclusion of such a model in the **SIWARE** package appeared mandatory. **DESIGN** computes (1) the water allocation to the main command canal intakes from the river Nile and (2) the target water levels at control structures.

The **WATDIS** submodel is designed for calculating the irrigation water distribution within the system. It effectively computes the amount of irrigation water lifted by the farmers from the lowest order irrigation canals irrigation uptake, as well as the spillway losses and conveyance losses from these canals. Moreover, **WATDIS** also provides the quantities lost from the command canals at the tail-ends. Typically, sink and source terms, such as municipal and industrial abstraction and reuse of drainage water, are included in the calculations. The output is generated on a 10-daily basis, although the actual time step may vary between 1 and 3 hr to account for the farmers' practices.

The **REUSE** submodel contains two levels. The first one is regional keeps track of the following:

1. Organization of input and output data for the module **FAIDS** (*F*ield level *A*gricultural *I*rrigation and *D*rainage *S*imulation module) (local level), where the on-farm water management is simulated
2. Distribution of irrigation water supplied to the calculation units for the different field crops
3. Simulation of crop rotation, i.e., crop grown in the next season following harvesting;
4. Simulation of the unofficial reuse of drainage water (farmers directly abstracting drainage water from drains to compensate for their unmet freshwater irrigation needs)

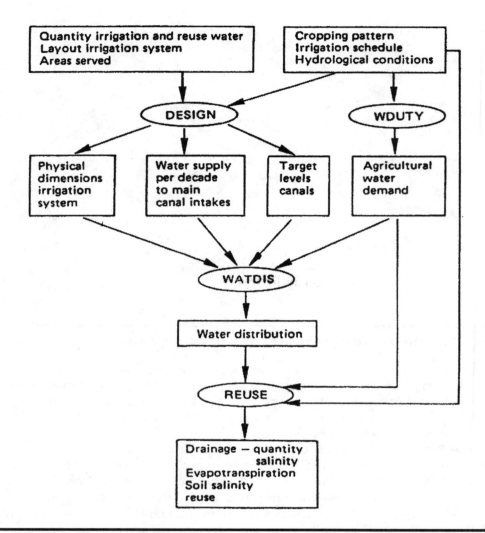

Figure 52.2 Schematization of the SIWARE model, its submodels, input, and output.

5. Simulation of the irrigation water salinity after mixing with drainage water at reuse pump stations
6. Simulation of the changes in the salinity of the irrigation water due to official and/or unofficial reuse of drainage water
7. Calculation of conveyance time lags in the drainage canal system
8. Preparation and presentation of simulated results and outputs

The second level covered by the **REUSE** submodel is the local level at the calculation unit scale. At this level all parameters are considered uniformly distributed. The relevant processes are simulated by the on-farm water management model **FAIDS** (*F*ield level *A*gricultural *I*rrigation and *D*rainage *S*imulation module). This model, which forms the central core of the SIWARE package, computes the following variables describing water and salt cycles in each cropped field:

1. Irrigation and field water losses (on-farm losses)
2. Actual evapotranspiration
3. Leaching of salts from the crop root zone
4. Movements of salts in the soil
5. Leakage (percolation) to and seepage from the groundwater aquifer
6. Fast subsurface drainage through soil cracks
7. Subsurface drainage to open or underground drainage systems

Potentials of the Siware Model Package

The power of SIWARE as a tool for planning and decision making may be illustrated by some multiobjective applications. The following two water management applications are typical examples of what SIWARE can do.

Reduction in the Total Water Supply

The SIWARE model package has been used to test the possibilities of reducing the Nile water supply to the Middle Delta in order to save fresh water that can be diverted to new reclamation project areas in other parts of the Delta. The objective of this water management option is to reduce the water supply as much as possible while avoiding adverse side effects. The adverse effects from reduced water supply may be increased soil salinity, reduced crop yield, seawater intrusion, and increased pressure for nonagricultural water uses.

The Middle Delta includes about 1.5 million acres of cultivated land. The reference supply rate of Nile water to this part of the Delta amounts to 11.10×10^9 m^3 of water assuming normal conditions. This quantity is gradually reduced in this scenario in five steps, each of 5% of the total supply, while the water supply for nonagricultural uses is maintained without reduction. In conjunction with the supply reduction, a water distribution based on actual (local) crop water requirements has been tested in order to better match water supply with demand, resulting in less operational losses. The present water distribution is based on average crop water requirements (regional average) over the whole Nile Delta.

The effects of reduced water supply have been evaluated in terms of crop evapotranspiration and soil salinity for the short and long terms. Increased soil salinity or reduced water supply have major long-term effects on evapotranspiration. One evaluation criterion is the location and size of the area where a reduction in the evapotranspiration is more than 20% and where the soil salinity exceeds 3.0 mmho/cm. Both values are seen as threshold values at which adverse effects on crop production may appear. These criteria indicate the extent to which total water supply can be safely curtailed without violating the sustainability of agriculture.

Model calculations show that up to a 10% reduction of Nile water supply, using the present average crop water requirement, will result in only a slight loss of about 1 to 2% in the average relative evapotranspiration. The area with insufficient evapotranspiration (i.e., less than 80% of the optimum value) will probably increase from 9% under the reference situation to some 12 to 15% of the area. Curtailing the supply within these limits is apparently a rather safe

Table 52.1 Different Components of the Agricultural Water Supply System for the Reference Situation and 4 Simulated Reductions

	Reference	−10%	−15%	−20%	−25%
Allocation based on average crop water requirements × 10^6 m³/year					
Nile water supply	11,113	10,002	9,447	8,892	8,337
Official reuse	582	460	397	349	306
Total supply	11,695	10,462	9,844	9,241	8,643
System losses	2,798	2,062	1,742	1,468	1,234
Irrigation uptake	9,195	8,701	8,404	8,092	7,713
Unofficial reuse	1,010	962	934	904	870
Groundwater use	424	424	424	424	424
Total crop supply	10,629	10,087	9,762	9,420	9,007
Allocation based on local crop water requirements × 10^6 m³/year					
Nile water supply	11,113	10,002	9,447	8,892	8,337
Official reuse	582	430	375	329	288
Total supply	11,695	10,432	9,822	9,221	8,625
System losses	2,263	1,669	1,445	1,250	1,095
Irrigation uptake	9,724	9,034	8,649	8,239	7,808
Unofficial reuse	1,108	1,052	1,015	974	930
Groundwater use	424	424	424	424	424
Total crop supply	11,256	10,510	10,088	9,637	9,162

practice. Nile water supply shortages are compensated (partly) by some 14 to 26% savings on lower system losses (at tail ends of irrigation system).

In Figure 52.3 the various components of the water balance are presented graphically, where their values during the reference simulation conditions have been set to 100%. The irrigation water uptake from the irrigation canals is roughly represented by the difference between the total supply, including official reuse, and the irrigation system losses. It appears that although the official reuse falls more than proportionally when compared to the reduction in Nile water supply, the irrigation water uptake is affected less severely (Table 52.1). Meanwhile, the reduction in unofficial reuse does not result in big changes.

According to the simulation, a reduction of 10% of the Nile water supply causes a reduction of only 4% in irrigation water uptake by farmers, a reduction of 2% of the crop evapotranspiration and a 5% increase in areas where a reduction of 20% of the crop evapotranspiration will be expected. An additional reduction totaling 25%, it turns out, is predicted to reduce irrigation water supply only 16%, reflecting an increased water use efficiency. However, reducing irrigation water supply has a severe adverse effect, because the area where the reduction in crop evapotranspiration is more than 20% increases from 9% of the total cultivated area in the reference situation, to 31%.

Better results, however, could be obtained when the water distribution is based on actual crop water requirements. Using spatially varying values for the crop

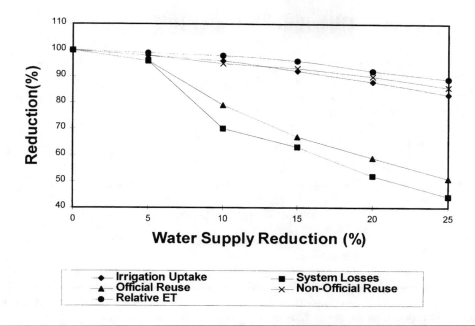

Figure 52.3 Relative components of the irrigation water balance as a function of various supply reductions using regional average crop water requirements.

water requirements would lead to an average aggregated relative evapotranspiration of 91% (in the case of 10% supply reduction), which is the same value as that for the reference simulation (the case with no supply reduction). Moreover, the areas with lower relative evapotranspiration than 80%, reduced from a 9% of the total area for the reference case, to a mere 6% for the 10% reduction in water supply case. Prospects for the reuse of drainage water become less promising as a result of deterioration through increased salinity, however evapotranspiration requirements of the crops can be supplied by additional irrigation water resulting from some 40% savings on lower irrigation system losses. The simulation results suggested that better water management can be achieved when the water allocation to the main canals and the water distribution is based on local crop water requirements rather than on a regional average. Figure 52.4 shows that without any supply reduction, the crop evapotranspiration increases 3% compared to the reference situation. A better match of the irrigation distribution with local crop water requirements also results in significantly lower system losses (spillway and tail end losses) and in a higher official and unofficial reuse of drainage water. On 2% of the area, the crop evapotranspiration is less than 80% of the optimum, illustrating a significantly improved water distribution. A supply reduction of 10% together with water distribution based on actual crop water requirements performs even better than the reference situation without any reduction in supply. The crop evapotranspiration is almost similar to the reference situation, while only in 6% of the area, the crop evapotranspiration is less than 80% of the optimum conditions. Supply reductions of more than 10% showed better conditions in the case using spatially varying crop water requirements than that using average crop water requirements.

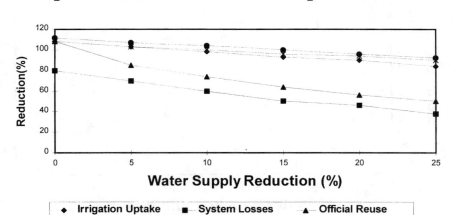

Figure 52.4 Relative components of the irrigation water balance as a function of various simulated reductions using spatially varying crop water requirements.

Unexpected Changes in Cropping Patterns

The overall irrigation efficiency in Egypt, according to the definition given by Thompson (1988) is classified as high. This is attributed to strict planning and operational procedures based on water distributions that are proportional to demand and partial control of cropping pattern. Also the reuse of agricultural drainage water contributes to water use efficiency. However, it has been decided to liberate the cropping pattern as a part of recent economic reform. This new policy, however, leaves the planners uncertain about the quantity of water required to satisfy the agricultural demands. The complexity of this situation might be clear when it is known that water supply is controlled by the High Aswan Dam (HAD) about 1,000 km south of Cairo. Water released from the reservoir of HAD reaches Cairo in about 12 days. Water is released from HAD according to the water requirements of a given crop pattern. The actual crop pattern may deviate under a more liberal crop pattern policy.

The effects of such unexpected deviation of crop patterns can be evaluated with the assistance of the SIWARE model package in terms of (1) crop evapotranspiration (as an indicator for crop yield) and (2) drainage water availability for reuse.

The SIWARE model has been applied to the Middle Delta to evaluate the effects of the deviation in cropping pattern on the system efficiency. For the analysis, it was assumed that the Nile water supply to the Middle Delta should match the total demands on the basis of a representative cropping pattern. It was decided to adhere to standard management procedures of regulating internal irrigation water distribution followed by the Ministry of Public Works and Water Resources.

The results were evaluated for an increase of approximately 10 and 20% in the area cropped with high water consuming summer and winter crops with a reduced

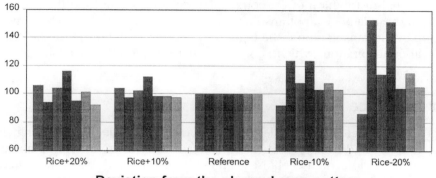

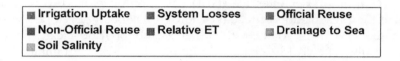

Figure 52.5 Relative components of the irrigation water balance as a function of deviations from the planned cropping pattern.

area of crops with low water consumption. A similar evaluation was performed with 10 and 20% decrease in the area cropped with high consuming crops. The cropping pattern and water management of 1987 has been used as a reference.

The simulation showed that growing 20% more of high water consuming crops increased the total crop water requirements with 2,700 million m³/year, which cannot be compensated by the Nile water supply. The irrigation system will be operated under stress conditions resulting in a less water losses of about 16% as shown in Figure 52.5. As a result of the increase in irrigation uptake, the irrigation system efficiency will increase up to 80% with an increase of 4% compared to the reference conditions. The marginal increase in official reuse by 3%, and the sharply increased nonofficial reuse with 171 million m³/year (16% increase), added to the 441 million m³/year of freshwater irrigation abstracted by farmers could not fully compensate for the increased demand. As a consequence, the average crop evapotranspiration is 86% of the optimum, with a decrease of 5% compared to the reference situation. The spatial distribution of the evapotranspiration is rather poor: it's less than 80% of the optimum evapotranspiration on 18% of the area, and less than 90% of the optimum on 55% of the area.

Increasing the area cultivated by crops with a low water consumption by 20% more than is expected, resulted in a demand which was 1,700 million m³/year less than necessary for the average crop pattern. This resulted in a very low efficiency of the irrigation system, amounting to 65%, with a decrease of 11% compared to the reference situation. The irrigation system losses increased dramatically with almost 1,400 million m³/year (51% increase), caused by reduced irrigation water abstraction by farmers with about 1,300 million m³/year or 15%. The official reuse of drainage water increases by 14% as a result of an increased drainage water availability, but the unofficial reuse of drainage water drops about 195 million m³/year or 18% due to the reduced demand.

The spatial distribution of evapotranspiration is excellent for a threshold value of 80%, and a relative evapotranspiration of 90% was realized in 88% of the area. The crop evapotranspiration is the highest for all options with 95% of the optimum crop evapotranspiration, which is 3% more than in the reference situation.

User Interface

The SIWARE model is a versatile tool for planning and decision making. It has been decided to add this model to the other models used by the Planning and Management Sectors of the Ministry of Public Works and Water Resources. In order to facilitate the job of the end users, the SIWARE package has been ported to the PC-DOS/WINDOWS environment as well as the UNIX environment. Furthermore, a Graphical User Interface (GUI) has been developed to facilitate use and enhance the presentation of results. The GUI has been developed in consultation with the end users to meet their needs and provide clear and easy to understand results. The GUI is interactive and enables users with minimal training to handle the complicated model.

Conclusions

Water management simulation models can be used as decision support tools. The SIWARE model has been developed to simulate the water management in the Nile Delta. Water management scenarios with multiple objectives demonstrate the power of such tools for short and long-term water resources planning.

The model showed that a reduction of the Nile water supply of 10% (1.1×10^9 m^3 annually) is possible without serious adverse consequences for the total agricultural production. Within certain limits, the local adverse effects of reducing the Nile water supply can be neutralized by a water distribution which is based on actual (local) instead of regional average crop water requirements. Water distribution based on local crop water requirements will lead to significant reduction in system losses, and reduction in the area where significant crop yield reductions would occur with reduced water supply. The SIWARE package has a capability for the calculation of local (spatially varying) crop water requirements according to different soil and climate zones.

SIWARE also suggest that free cropping pattern may endanger the efficiency of the planning process due to unexpected deviation of the area cropped with high water-consuming crops. Shortage or surplus water supply may occur resulting either in reduction in crop yields or loss of the valuable fresh water.

References

Abdel-Dayem, S. 1994. Potential of Drainage Water for Agricultural Reuse in the Nile Delta. 8th IWRA World Congress on Water Resources: Satisfying Future National and Global Water Demands, Cairo, Egypt.

Abdel Gawad, S.T., M.A. Abdel Khalek, D. Boels, D.E. El Quosy, C.W.J. Roest, P.E. Rijtema, and M.F.R. Smit. 1991. Analysis of Water Management in the Eastern Nile Delta. Reuse of Drainage Water Project Report 30. Drainage Research Institute, Kanater, Cairo, Egypt and The Winand Staring Centre, Wageningen, The Netherlands.

DRI. 1995. Reuse of Drainage Water in the Nile Delta; Monitoring, Modelling and Analysis. Reuse Report 50. Reuse of Drainage Water Project, DRI/DLO Winand Staring Center, Cairo, Egypt.

Thompson, S.A. 1988. Patterns and trends in irrigation efficiency. *Water Resour. Bull.* 24(1):57–64.

Chapter 53

Use of Goal Programming in Aquaculture Policy Decision Making in Southern Thailand

KEVIN A. PARTON AND AYUT NISSAPA

Development planning of the Outer Lake of the Songkhla Lake of Southern Thailand has multiple decision-maker and multiple criteria characteristics. While aquaculture is becoming a significant economic activity, it competes with capture fisheries for the use of the lake and it produces environmental degradation in the form of residues in the water and in sediment. An attempt is made in the present paper to model the trade-off between the various economic and environmental goals when observed from the perspective of policy planners representing different organizations. Weighted goal programming is the central method of the analysis, which incorporates nonlinear production and residue functions. While the environmental goal turns out to be an important determinant of the optimal solutions, further expansion of aquaculture in particular locations would still be possible without serious degradation.

Introduction

Development planning of the Outer Lake of the Songkhla Lake of Southern Thailand is influenced by various interest groups, and each interest group has multiple objectives. In general even for a single individual, these objectives are conflicting (e.g., an economic objective of increasing regional income vs. an environmental objective of improving environmental quality). When a number of organizations and their representatives become involved in regional decision making, there is a need to integrate objectives that are conflicting both for individuals and between them. Therefore, developing a multiple objective analysis (MOA) that incorporates the preferences of different individuals can play a crucial role in assisting the policy decision makers.

While cage aquaculture is becoming a significant economic activity, it competes with capture fisheries for the use of the lake and it produces environmental degradation in the form of residues in water and in sediment. An attempt is made in the present paper to model the trade-off between the various economic and environmental goals when observed from the perspective of regional policy planners representing different organizations.

There are several MOA techniques (Cohon and Marks, 1975; Loucks, 1975; Nijkamp, 1977). Goal programming (GP) is perhaps the oldest approach within the field of MOA (Romero, 1986). It is simple and bears close resemblance to linear programming (LP) which has had numerous applications. The variants of GP, including lexicographic goal programming (LGP), weighted goal programming (WGP), integer goal programming, and stochastic goal programming, are now well established and have direct relevance to model building in planning processes (Charnes and Cooper, 1977; Contini, 1968; Romero and Rehman, 1985). In this paper the WGP is used to establish plans for aquaculture development taking into account the conflicting objectives of the two communities who benefit most from using the Outer Lake.

Figure 53.1 shows the Outer Lake of the Songkhla Lake and the areas where representative aquaculture farms were selected for the present study. The Outer Lake is unique in terms of agroecological characteristics and resource endowments. It is situated in the southeastern part of the Songkhla Lake connecting with the Pacific Ocean by a single outlet. The lake is relatively shallow though the water inside is well mixed. The lake is also connected with the other parts of the Songkhla Lake by a single canal. The impacts of seawater and freshwater are observed in salinity levels that vary according to season. In the rainy reason, salinity levels may drop to 1 to 2 ppm and increase to 33 to 35 ppm in the dry season. The other main water quality determining factors are dissolved oxygen, primary productivity and water flow, and factors related to aquaculture such as feed, stocking density, and effort expended in cleaning cages. Their interaction results in farm residues such as ammonia in water.

This paper is organized into six sections. Following this introduction, the second and third sections include formulation of the problem and explanation of the GP method, respectively. The fourth section presents a regional model with linkages between aquaculture producers and other sections of the community. The objectives or goals of regional decision makers are also included. The fifth and sixth sections contain analytical results and conclusions for policy planning, respectively.

The Problem

There are currently two species, seabass *(Lates calcarifer)* and tiger prawns *(Penaeus monodon)*, cultured in cages in the Outer Lake. Seabass was introduced into the Outer Lake more than 2 decades ago to supplement the decreasing catch

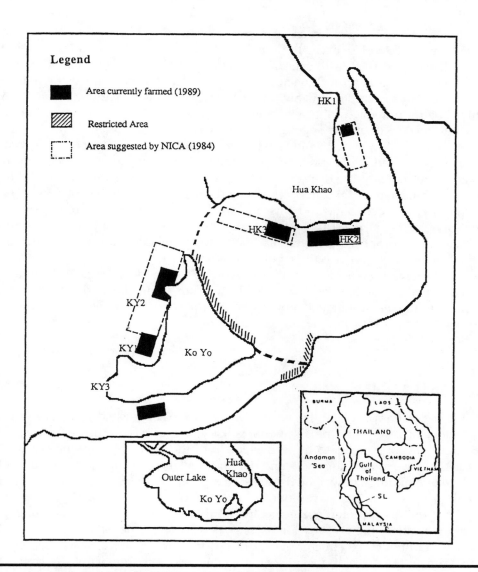

Figure 53.1 Location of study sites at Ko Yo and Hua Khao.

from the sea. Tiger prawns in cages, on the other hand, were introduced in the late 1980s because of the high export demand for this species. The Thai Department of Fisheries has attempted to develop aquaculture production through development of culturing technologies, new species, and new techniques of production. This development is aimed at increasing regional income and making full utilization of the natural resources. However, concerns have been expressed in recent years about the impact of aquaculture production on water and sediment quality; competition for water space between aquaculture and other users; and effects of aquaculture structures and its activities on the natural environment as manifest in water regime, salinity level, diversity of flora and fauna, and aesthetic values.

From 1974 to 1986, the number of marine prawn farms in Thailand increased from 1,518 to 5,534, while their area and average yield increased from 12,091 to

45,368 ha and 147 to 395 kg/ha/year, respectively (Thai Department of Fisheries, 1995). In the case of seabass, the number of farms increased from 70 in 1982 to more than 250 in 1988, with similar productivity improvements to marine prawns. As shown in Figure 53.1, aquaculture in Ko Yo and Hua Khao has expanded outside the area suggested by the National Institute of Coastal Aquaculture (NICA). This expansion involves not only more farms and more cages per farm, but also the introduction of a new species, tiger prawns.

Such expansion of aquaculture without proper direction is expected to cause an increase in the above problems to communities in the region, including the aquaculture industry itself. In the present study, an attempt is made to take two groups of the Outer Lake into consideration to form a regional plan with two main objectives, economic and environmental. These two groups are seabass and tiger prawn farmers, and capture fishermen. Some level of compromise is expected to exist in the proposed regional plan because the model is formulated such that overall regional objectives are simultaneously incorporated.

Specifically, the objectives of the study were to introduce MOA/WGP to assist decision making at the regional level by (1) analyzing the trade-offs between economic and environmental objectives and between aquaculture and capture fisheries, and (2) showing the consequences of developing the aquaculture industry in various ways.

Regional Model

(GP is an extension of LP. However, GP can be distinguished from LP by four key elements: (1) conceptualization of objectives as goals, (2) assignment of priorities and/or weights to the achievement of the goals, (3) presence of deviational variables to measure over- and underachievement from target or goal levels, and (4) minimization of weighted-sums of the deviational variables to find solutions that best satisfy the goals (van der Zel and Walker, 1988; North, 1990).

GP requires a preference structure in the objective function defining the decision maker's preferences for each goal relative to the others. In the WGP model, the preference structure is expressed in terms of cardinal weights which approximate the decision maker's value judgments of the merits of the goals. All goals in the WGP are considered simultaneously at the same priority level provided all goals are commensurable. If the goals are not commensurable, as in the present study, a percentage conversion factor may be applied relating to the percentage deviation from the goal target.

There are several methods of specifying the weights that most resemble the decision maker's judgment. Each of the methods has its own advantages and limitations (Barnett et al., 1982; Dyer et al., 1979; Ignizio, 1976; Ijiri, 1965; McMillan, 1975). The method employed in the current study was to work interactively with the decision makers. The specification of targets should be made by the concerned decision makers whose preference structures are elicited. In those cases where it is not possible to specify the decision maker's targets, maximum or minimum attainable values are used to ensure that the solution obtained by GP analysis is nondominated. Some targets are prespecified, for example, the standard values of environmental quality parameters. These values are set by the relevant institutions to maintain acceptable levels of environmental quality with respect to those parameters.

Apart from the goal constraints, resource constraints are also included in the WGP model. The resource constraints are to be satisfied before the goal constraints are considered.

Sensitivity analysis can be carried out in the case of GP by varying parameters such as preference structure or weight, technical coefficients and targets or resource levels. The parameterization of the preference structure and targets makes GP in general a promising and flexible decision model. By variation of these parameters, a transformation curve (TC) or feasibility frontier may be obtained. Illustration of the use of the TC can be found in Ignizio (1976), Romero and Rehman (1985), and Cohon and Marks (1973).

The regional GP model of the current study is based on an aggregation of six representative aquaculture farms and a single capture fisheries model. There are two seabass farms and one tiger prawn farm for each of the two locations, Ko Yo and Hua Khao, and one capture fisheries model for the whole Outer Lake.

The constraint structure of each aquaculture model is shown in Table 53.1. It consists of three main components of the farm, viz.: (1) the production response functions, (2) residue functions, and (3) farm linear constraints. The production response function and residue functions are Cobb-Douglas in form, acknowledging that the generation of outputs from farms have nonlinear relationships with inputs and biological and environmental conditions of the water. The explanatory variables in these nonlinear functions may exhibit either generation or destruction mechanisms depending on the nature of the relationship. The third component is the farm linear constraints. These constraints include several farming activities such as labor activities, residue elimination, working capital, water supply and its contributions to water quality inside the cages, and lower and upper limits of input use, production, and residues. The constraints also include transferring and balancing activities to complete the formulation of the farm planning model.

There are real-world mechanisms that are modelled to link these constraint components together in the aquaculture planning sub-models. The first is the removal of residues in water and sediment. Labor for cage repair is an example of this which is performed and controlled by the farmers. The generation of residues, however, is influenced not only by the farm inputs but also the biological and environmental parameters of the water. These parameters, in turn, are influenced by quality of water flow in the cages and the transmission factors of the cages. The water flows bring chlorophyll-a, dissolved oxygen, and salinity into the cages. They also remove residues from the cages. It is assumed that together natural water flows and labor for cage repair are able to remove residues down to the levels found at the control sites outside the cages.

The production response and residue functions were estimated by Nissapa (1995), and the farm constraints were established from survey work conducted by NICA and the authors of the current paper.

As mentioned above, there are six aquaculture planning models. Each model represents an average farmer whose farming techniques are regarded as representative of all farms in that vicinity. The seabass farmers in Ko Yo and Hua Khao have different production and residue functions, while the tiger prawn farmers in both areas are assumed to share the same production and residue functions,

Table 53.1 Constraint Structure of the Aquaculture Models

Variables constraints	Variables	Deviational variables						RHS
	X_j	d_1^-	d_1^+	d_2^-	d_2^+	d_3^-	d_3^+	
Nonlinear constraints								
1. Production function								
$Y = f(FEED, FING, WV, LCR, FAE)$								
2. Residue function								
$NH3W = f(FEED, FING, CHA, SY1, SY2)$								
$TNW = f(FEED, FING, SAL, SY1, SY2)$								
$TPS = f(FEED, CHA, SY1, SY2)$								
$SULS = f(FEED, FING, DO, SY1, SY2)$								
3. Farm linear constraints								
– Water supply								Amount of flow per crop
– Lower and upper limits								
– Residues elimination								$\leq$ Environmental standards
– Labor activities								
– Working capital								$\geq$ Carrying capacity of cages
– Transfer and balancing								Farm resource levels
4. Goal constraints								
G1: Net revenue (baht)		+1	−1					$= b_1$
G2: Ammonia level (ppm)				+1	−1			$= b_2$
G3: Labor for cage repair (person days)						+1	−1	$= b_3$
5. Objective function minimize		$\dfrac{w_1 \times 100}{b_1}$			$\dfrac{w_2 \times 100}{b_2}$		$\dfrac{w_3 \times 100}{b_3}$	$= Z$

reflecting the more homogeneous type of technology in prawn production. However, there are some differences in terms of activities and constraints added to match the individual circumstances of the farmers.

The numbers of farms in each homogeneous group are used as population weights. Using number of farms as weights is considered to be an appropriate aggregation procedure because the representative farms are approximately the mean representations for their group (Hazell and Norton, 1986). This aggregation, however, depends heavily on assumptions of technological homogeneity, pecuniary proportionality, and institutional proportionality described by Day (1963). Although the models established come close to fulfilling these criteria, aggregation bias is inevitable and it is not possible to eliminate it completely.

The first step in establishing the objective function of the combined model was to select policy planners from relevant institutions. The first institution was NICA, where basic research on aquaculture, and water and sediment quality of the Songkhla Lake had been done. At present, development of aquaculture technology, new species, and appropriate techniques of aquaculture production are being studied in this institution. The second institution was Prince of Songkla University (PSU), where environmental monotoring of the Songkhla Lake has been conducted since the 1970s. This institution is technically linked with the Office of the National Environment Board in Bangkok. Some more general studies of Songkhla Lake also have been completed. The third institution was the Songkhla Department of Fisheries (SDF) whose duties are to implement regulations for capture fisheries and to control development of aquaculture. It plays an important role in providing extension services to the farmers and acts as a linkage between the above-mentioned institutions.

One policy planner from each of these three institutions (PP1NICA, PP2PSU, and PP3SDF) was interviewed in order to find his weighting of goals with respect to aquaculture development and its impacts on the capture fisheries and the environment.

The full set of regional objectives observed were: Goal 1 regional income generated from aquaculture (income-aquaculture), Goal 2 regional income generated from capture fisheries (income-fisheries), Goal 3 regional employment generated from aquaculture (employment-aquaculture), and Goal 4 ammonia levels in the Outer Lake (ammonia-aquaculture). Hence, there are three economic-oriented goals and one environmental-oriented goal. In the objective function, the deviational variables corresponding to these goals are minimized in percentage terms as:

$$\text{Minimize } Z = \frac{W_1 \times 100}{b_1} d_1^- + \frac{W_2 \times 100}{b_2} d_2^- + \frac{W_3 \times 100}{b_3} d_3^- + \frac{W_4 \times 100}{b_4} d_4^+ \quad (1)$$

where b_1, b_2, b_3, and b_4 are targets for goals 1, 2, 3, and 4, respectively. The underachievement variables of the economic-oriented goals d_1^-, d_2^-, and d_3^- are minimized because achievement or over-achievement of these goals is desirable. The overachievement variable of the environmental-oriented goal (d_4^+) is minimized because achievement or underachievement of this goal is desirable. The policy planners were interviewed to find their weights or relative importance of the specified goals. A normalized rank reciprocal technique proposed by Bakus et al. (1982) and Bain (1987) was used to quantify the weights. The policy planners were expected to be able to rank the goals according to their importance to their institutions.

The goal constraints consist of goals, their corresponding deviational variables, and targets. The target levels for the three decision makers were developed by an informal Delphi-type process. The agreement reached at the conclusion of the process was that realistic target levels were: 30 million baht for income-aquaculture, 20 million baht for income-capture fisheries, 3,000 jobs in aquaculture, and 0.06 ppm for ammonia residue. The four goals in the model are

Goal 1: Income-aquaculture (million baht)

$$\sum_i p_{yi} n_i Y_i + d_1^- - d_1^+ = 30 \qquad (2)$$

where $Y_i = f_i(X_i)$ are output from the production functions with explanatory variables $X_j = 1,2,\ldots n$, p_{yi} is price of seabass and tiger prawn, n_i is number of farmers sharing the i-th production function, $d_1^- - d_1^+$ are under- and overachievement variables, respectively, and i is representative farm, $i = 1,2,\ldots,6$.

Goal 2: Income-capture fisheries (million baht)

$$p_{cf} N(CF) + d_2^- - d_2^+ = 20 \qquad (3)$$

where $CF = f(EFFORT, \Sigma Y_i, \Sigma NH3W_i)$ is output from a catch-effort model, p_{cf} is the average price of fish and prawns caught from the lake, N is total number of fishermen, $EFFORT$ is total amount of capture effort in person days per year, ΣY_i is total aquaculture production, $\Sigma NH3W_i$ is total ammonia generated from aquaculture farms, and $d_2^- - d_2^+$ are under- and overachievement variables, respectively.

Goal 3: Employment-aquaculture (jobs)

$$\sum_i n_i e_i Y_i + d_3^- - d_3^- - d_3^+ = 3,000 \qquad (4)$$

where n_i is population weight or the total number of aquaculture farms which the production function represents, e_i is the employment coefficient representing number of jobs per kg of production, and $d_3^- - d_3^+$ are under- and overachievement variables, respectively.

Goal 4: Ammonia-aquaculture (ppm)

$$\sum_i n_i (NH3W_i) - f(WATER) + d_4^- - d_4^+ = 0.06 \qquad (5)$$

where $WATER$ is total amount of water passing through the Outer Lake system, and $d_4^- - d_4^+$ are under- and overachievement variables, respectively.

The constraints are those related to the representative farm models, the capture fisheries model, and other regional constraints. The resource constraints of the farm models are labor, working capital, and water volume. The other farm constraints are those of a technical and institutional nature which represent the aquaculture and environmental conditions, and transfer and balancing constraints that provide linkages among the included constraints. Included in these constraints are the Cobb-Douglas production and residue functions, and the catch-effort model. The coefficients in the constraint sets include technical and employment coefficients, prices, and population weights.

The calculation of population weights presented some difficulties because farmers do not have a uniform culturing period. The best procedure to find the population weights is to equalize all production activities to a 1-year timeframe. The 2-year culturing period of seabass is converted to a 1-year equivalent assuming that one-half of the output is produced. In the case of tiger prawns, production of two crops per year is assumed so double weight is attached to the per crop production level.

The resource levels at the regional-planning level are regional carrying capacities based on dissolved oxygen, maximum number of fishing days per year for the region, and maximum water surface and water flows. The maximum number of fishing days per year for the region is calculated by considering number of days that the fishermen are committed to social and religious functions, days reserved for recreation and days on average with heavy rainstorms. The water surface and water flows per year are calculated using flow rate, distance of water flow, and cross-section of water body.

The GAMS/MINOS software package was used in the study. This package is capable of handling nonlinear programming problems, and hence was appropriate for solving this problem involving nonlinear production and residue functions embedded in the constraint set.

Results

The results of the Outer Lake planning model based on the WGP technique are presented in Table 53.2. The three policy planners are different in terms of weights assigned to the goals. The PP1NICA is more concerned about the environmental situation of the lake, so the ammonia goal is given a high weight followed by employment-aquaculture, income-fisheries, and income-aquaculture. The optimal solution for this planner involves expansion of aquaculture only into areas where there is little risk of not achieving the environmental target. A large expansion of seabass would be permitted at KY1 and a small increase at KY2. This decision maker would prefer little expansion of seabass at Hua Khao and, given the polluting effects of tiger prawns, would like considerable reduction in their level of production. This conservative plan satisfies the ammonia level, income-fisheries, and income-aquaculture goals.

The PP2PSU and PP3SDF optima are the same as each other. They give more emphasis to the employment-aquaculture goal, and the results reflect this and lead to a higher level of production and income-aquaculture. It is observed that if high weight is given to the employment-aquaculture goal, the solution appears

Table 53.2 WGP Results of the Outer Lake Planning Model

Variable and goal	Actual 1989 level	Policy planner	
		PP1NICA	PP2PSU and PP3SDF
Aquaculture production			
1.1 Seabass at KY1 (kg/year)	37,660	165,230	204,000
1.2 Seabass at KY2 (kg/year)	45,523	52,683	52,683
1.3 Tiger prawn at KY3 (kg/year)	30,013	14,820	60,192
1.4 Tiger prawn at HK1 (kg/year)	6,591	2,967	18,728
1.5 Seabass at HK2 (kg/year)	65,170	66,848	252,640
1.6 Seabass at HK3 (kg/year)	36,596	36,973	36,973
Capture fisheries			
2.1 Effort (person days/year)	345,268	365,750	365,750
2.2 Total catch (kg/year)	910,000	691,000	691,000
2.3 Catch-effort ratio	2.64	1.89	1.89
Goal attainment			
3.1 Income-aquaculture (baht)	20×10^6	30×10^6 (satisfied)	61×10^6 (satisfied)
3.2 Income-fisheries (baht)	18×10^6	20×10^6 (satisfied)	20×10^6 (satisfied)
3.3 Employment-aquaculture (job)	254	345 (unsatisfied)	518 (unsatisfied)
3.4 Ammonia level (ppm)	0.06	0.06 (satisfied)	0.08 (unsatisfied)

to be in favor of income-aquaculture since these two goals are complementary. The solution for these two planners involves the same small increase in production of seabass at KY2 and HK3 as for PP1NICA, and also a rapid increase in production at KY1. In addition, policy planners PP2PSU and PP3SDF prefer a considerable increase in seabass production at HK2 and also an increase in output of tiger prawns in both locations.

This less conservative plan clearly results from the lower relative weight attached to the ammonia goal. This is a significant aspect of the results because the solutions obtained are highly sensitive to a small change in the environmental goal. However, it should be noted that the levels of ammonia in the less conservative plan are not too different from those of the alternative, and are still considered to pose little environmental threat. This implies that, as long as an appropriate plan is implemented, it would be possible to expand aquaculture production considerably without environmental loss. The locations for this expansion are shown in Figure 53.2.

Various sensitivity analyses were performed The analysis related to the target of income-aquaculture for PP1NICA are presented in Table 53.3. When the level of the income-aquaculture target is parameterized up to the level of 50 million baht/year, the PP1NICA optimum suggests a gradual increase in seabass production in KY1 and HK2. The KY2 and HK3 farmers have reached their maximum production capacity so there is no change in the level of production. The production of tiger prawns in both Ko Yo and Hua Khao areas is suggested to increase. If the target of income-aquaculture is parameterized up to 70 and 100 million baht, the maximum level of income-aquaculture achieved is 61 million baht. In order to obtain an additional income of 11 million baht, the seabass farmers at HK2 and tiger prawn farmers at both Ko Yo and Hua Khao must increase their production level. It is also observed that the employment-aquaculture

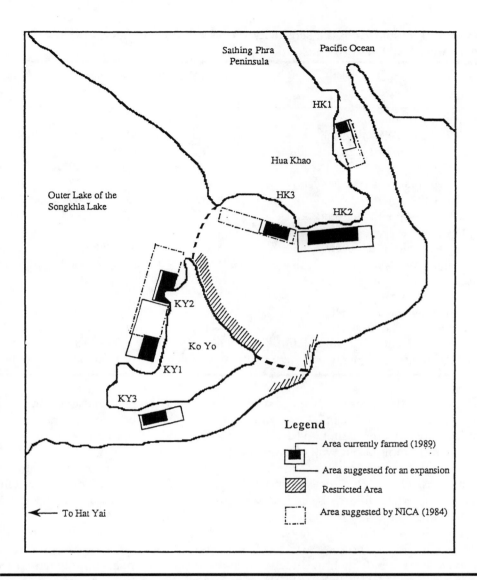

Figure 53.2 Areas suggested for aquaculture compared with current zoning.

levels increase following the increase in aquaculture production while the amount of catch per effort of capture fishermen remains unchanged. As would be expected, as more emphasis is applied to the income-aquaculture goal the ammonia target receives less relative weight, and the environment receives more pressure.

Conclusion

Planning at the regional level involves multiple objectives to satisfy a number of communities in the region, and several decision makers to form and implement the plan. A multiple objective framework via WGP is used to include these multiple objective characteristics in a regional-planning model. There are four objectives

Table 53.3 Results from the Sensitivity Analysis of Income-Aquaculture Goal Target for PP1NICA

Variable and goal	Level of income-aquaculture goal			
	30×10^6	50×10^6	70×10^6	100×10^6
Aquaculture production				
1.1 Seabass at KY1 (kg/year)	165,230	204,000	204,000	204,000
1.2 Seabass at KY2 (kg/year)	52,683	52,683	52,683	52,683
1.3 Tiger prawn at KY3 (kg/year)	14,820	17,547	60,192	60,192
1.4 Tiger prawn at HK1 (kg/year)	2,967	10,092	18,728	18,728
1.5 Seabass at HK2 (kg/year)	66,848	236,640	252,640	252,640
1.6 Seabass at HK3 (kg/year)	36,973	36,973	36,973	36,973
Capture fisheries				
2.1 Effort (person days/year)	365,750	365,750	365,750	365,750
2.2 Total catch (kg/year)	691,000	691,000	691,000	691,000
2.3 Catch-effort ratio	1.89	1.89	1.89	1.89
Goal attainment				
3.1 Income-aquaculture (baht)	30×10^6	50×10^6	61×10^6	61×10^6
3.2 Income-fisheries (baht)	20×10^6	20×10^6	20×10^6	20×10^6
3.3 Employment-aquaculture (job)	345	470	518	518
3.4 Ammonia level (ppm)	0.06	0.06	0.08	0.08

specified in the present study expressing socioeconomic and environmental requirements of the plan. These objectives relate to income generated from aquaculture, income generated from capture fisheries, employment from aquaculture, and ammonia in water generated from aquaculture.

The above objectives were evaluated by three policy planners from different institutions which are directly responsible for aquaculture development. Weights were assigned to these objectives according to their elicited personal view of relative importance.

The results of the WGP model reveal that the policy planners have different ideal solutions for the expansion of aquaculture. The degree of expansion and its location depend on how much emphasis the decision makers attach to the environmental goal. Despite this, there would still seem to be scope for considerable expansion of aquaculture in appropriate locations of the Outer Lake of the Songhkla Lake without compromising the quality of the water and imposing external costs on capture fisheries.

The use of WGP in the present study has provided a means of finding the number, types, and location of farms that both produce economic benefit and have limited environmental cost to the region. However, such plans are not perfect. The particular limitations of these plans at the regional level need, of course, to be recognized and continuous improvements must be made as they are used to aid those who make policies. Nevertheless, as a first step in an iterative procedure between analyst and policy maker the plans are considered to be a useful contribution. Moreover, an important contribution of the analysis is in the communication between policy planners. In this context, the discipline of constructing the model forces the decision makers to understand clearly their own objectives and to become more aware of the position of other planners.

References

Bain, M.B. 1987. Structured decision making in fisheries management: trout fishing regulations on the Au Sable River, Michigan. *North Am. J. Fish. Manage.* 7:475–481.

Bakus, G.J., W.G. Stillwell, S.M. Latter, and M.C. Wallerstein. 1982. Decision making: with applications for environmental management. *Environ. Manage.* 6:493–504.

Barnett, D., B. Blake, and B.A. McCarl. 1982. Goal programming via multi-dimensional scaling applied to Senegalese subsistence farms. *Am. J. Agric. Econ.* 64(4):720–727.

Charnes, A. and W.W. Cooper. 1977. Goal programming and multiple objective optimizations. *Eur. J. Operat. Res.* 1:39–54.

Cohon, J.L. and F.H. Marks. 1973. Multiobjective screening models and water resource investment. *Water Resour. Res.* 9(4):826–836.

Cohon, J.L. and F.H. Marks. 1975. A review and evaluation of multiple objective programming techniques. *Water Resour. Res.* 11(2):208–220.

Contini, B. 1968. A stochastic approach to goal programming. *Operat. Res.* 3:576–586.

Day, R.H. 1963. On aggregation linear programming models of production. *J. Farm Econ.* 45(4):797–81.

Dyer, A.A., J.G. Hof, J.W. Kelly, S.A. Crim, and G.S. Alward. 1979. Implications of goal programming in forest resource allocation. *For. Sci.* 25(4):535–543.

Hazell, P.B.R. and R.D. Norton. 1986. *Mathematical Programming for Economic Analysis in Agriculture*. Macmillan, New York.

Ignizio, J.P. 1976. *Goal Programming and Extension*. Lexington Books, Massachusetts.

Ijiri, Y. 1965. Management goals and accounting for control. In H. Theil, Ed. *Studies in Mathematical and Management Economics*. Volume 3. North-Holland, Amsterdam.

Lee, S.M. 1972. *Goal Programming for Decision Analysis*. Auerbach, Philadelphia.

Loucks, D.P. 1975. Planning for multiple goals. In C. Blitzer, P. Clark, and L. Taylor, Eds. *Economy Wide Models and Development Planning*. Oxford University Press, New York.

McMillan, C. 1975. *Mathematical Programming*. 2nd ed., John Wiley & Sons, New York.

Nijkamp, P. 1977. *Theory and Application of Environmental Economics*. North-Holland, Amsterdam.

Nissapa, A. 1995. Planning for Aquaculture Development in the Outer Lake of the Songkhla Lake of Southern Thailand. Unpublished Ph.D. thesis. Department of Agricultural and Resource Economics, University of New England, Armidale, Australia.

North, R.M. 1990. Application of MOA/GP to Water Project Reallocation. Paper presented for Center for Water Policy Research, University of New England, Armidale, Australia.

Romero, C. 1986. A survey of generalized goal programming (1970–1982). *Eur. J. Operat. Res.* 25(2):183–191.

Romero, C. and T. Rehman. 1985. Goal programming and multiple criteria decision-making in farm planning: some extensions. *J. Agric. Econ.* 36(2):171–185.

Thai Department of Fisheries. 1995. Statistics of Marine Fish Farms. Report No 9/1995. Department of Fisheries, Ministry of Agriculture and Cooperatives, Bangkok.

van der Zel, D.W. and B.H. Walker. 1988. Mountain catchment management with goal programming. *J. Environ. Manage.* 27(1):25–51.

Chapter 54

A Case Study in the Use of an Expert System as a Multiobjective Decision Support System (MODSS) — Boobera Lagoon Environmental Management Plan

P.L. MATTHEW AND B.A. PEASLEY

Abstract

Boobera Lagoon in northwestern New South Wales (NSW) is of environmental and cultural significance to the Australian community. Overuse of the area by the agricultural sector is causing significant degradation of the ecosystem. This degradation is also destroying the lagoon's native cultural value. An environmental management plan based on land resource assessments was devised to control this degradation. An expert system, LANDCAP, was used as decision support for the assessment of land degradation potential, agricultural production potential, and the general sustainable management practices. LANDCAP formalized and made more consistent the expert judgments of the land resource assessors. The natural language format and the transparency of the system processes and outputs aided the resolution of disputes over the land assessments and facilitated the community acceptance of the expert judgments. LANDCAP provided a significant advancement in the public relations and extension effect of the development and acceptance of the environmental management plan for the Boobera Lagoon.

Introduction

Agricultural land capability assessment forms one of the foundations of rural land use planning in NSW. In the past, assessments on land degradation hazards and agricultural potential of land were made as expert judgments. Although the expert judgments may have been made on incomplete, imprecise, and sometimes incorrect data sets they have stood the test of time and have enabled ecologically sustainable management systems to operate for many years. Nevertheless, there have been many bitter disputes over the assessment of the land resources. In response to the perceived subjectivity of land assessment, a prototype expert system (LANDCAP) was developed and validated for northwestern NSW (Matthew et al., 1992) . LANDCAP was to overcome the inherent difficulties of land assessment, such as subjective judgments and the interpretation of large data sets. This system had its first trial in the development of an environmental management plan (EMP) for the Boobera Lagoon. The use of LANDCAP assisted the conservation officers to rebut many of the arguments and objections that were raised by those opposing the conservation of the lagoon and reallocation of land uses in the foreshore areas.

Perspective to the Land-use Conflict

Boobera Lagoon is a unique aquatic habitat located in the Macintyre River catchment of northwestern NSW (Figure 54.1). The waterway complex is 10.5 km long with an average width of 150 m. Permanent water bodies, such as Boobera Lagoon, are rare in the semi-arid environment of the Murray/Darling Basin. As such, it is considered by the Australian community to be of State and National significance in terms of its environmental importance.

To the local community Boobera is more important than just its environmental value. This importance hinges on six major issues, namely:

1. Aboriginal (native) cultural and religious rites
2. Supply of irrigation waters
3. Stock water for grazing lands
4. Irrigated and dryland crop production
5. Grazing land
6. Public uses for recreation such as water-skiing, camping, fishing, etc.

The community pursuit of these competing uses without consideration of the environmental concerns has driven the ecosystem to displaying many of the symptoms of stress, and possible near collapse. The control of the degradation of Boobera Lagoon hinges on the resolution of conflicts (between competing groups) for the area's land and water resources.

A Case Study in the Use of an Expert System as a MODSS

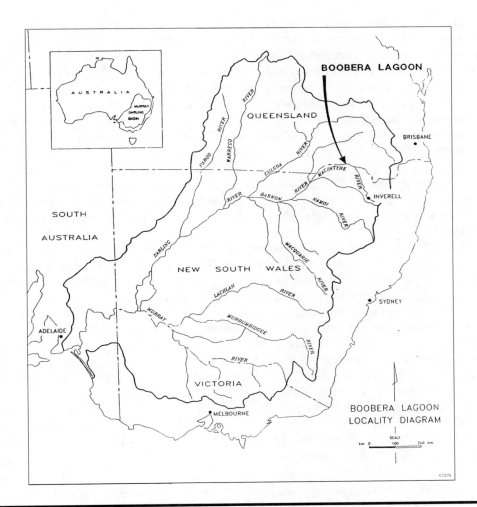

Figure 54.1   Location of Boobera Lagoon.

The economic development of the local community is dependent on the agricultural production from the area. Irrigators draw water from the lagoon for distribution on the adjacent lands. Within the 15,758 ha catchment of Boobera Lagoon, land uses have changed from no flood irrigation in 1962, to 42% of the total area in November 1994 (Peasley, 1993). For instance, Boobera Lagoon supplies the water to irrigate 130 ha of cotton which uses approximately 950 to 1,050 Ml per season. Graziers also demand access to the lagoon for stock water and the foreshores for grazing of cattle and sheep. Some 150 head of cattle and 5,000 sheep use the lagoon and its foreshores.

The high water use of the local irrigators causes rapid draw down of the lagoon water levels which kills the near foreshore water plants and reeds (Eigeland, 1993). In general, the intensity of agricultural activities has led to increased soil erosion and ruin of aboriginal cultural values, e.g., exposure of burial grounds and destruction of scar and canoe trees.

The agricultural sector is the main culprit in the destruction of area. This destruction is the source of the conflict that is evident between the different groups that want to use the Boobera Lagoon and its foreshores. These groups, their special interests, and environmental problems come in three forms.

First, the aboriginal community believes that Boobera Lagoon is central to their religious beliefs and lifestyle. They draw strength from the traditions that believe the lagoon is the resting place of a giant spirit being, which is the local equivalent of the "Rainbow Serpent." It is estimated that the area surrounding Boobera Lagoon contains up to 1,000 scar and canoe trees and several million stone artefacts. The anthropological significance of this site is such that it is considered by many to be the most important aboriginal cultural site in the upper part of the Murray/Darling Basin. The beliefs and traditions relating to Boobera Lagoon are part of contemporary Aboriginal law (Thompson, 1993). The area is currently under a native land claim.

Second, the environmental protection groups maintain that the site's ecosystem value demands that the area be preserved for wildlife. The lagoon and its foreshores are seriously degraded. The loss of riparian vegetation is causing a loss of habitats, excessive erosion, and sedimentation of the lagoon

Boobera Lagoon is the local refuge area for native fish stocks. These stocks then replenish the surrounding river systems after drought. These fish are a recreational resource for anglers and a food supply for water birds. The risk to this part of the ecosystem is the total removal of the once prevalent reed beds along the lagoon edges (Patten, 1993), and turbid waters and lagoon contamination with pesticides from the surrounding agricultural areas.

On land the lack of native vegetation has resulted in a serious decline in the number and diversity of the wildlife. This habitat depletion is primarily due to human activity in the foreshore areas of the lagoon. In a recent study some of the wildlife depletion indicators were

1. Lack of bird species (e.g., Black Swan and Reed Warbler) that feed on water plants
2. Low to nonexistent small mammals (e.g., Common Dunnart, Bush Rat) and grassland birds (e.g., quail, Richard's Pipit)
3. Small numbers of reptiles (15 out of 45 possible species)
4. No raptorial birds (hawks, owls, kestrels)

Currently, various degradation problems at Boobera Lagoon prevent it from realizing its previous role as an important natural resource of the Murray/Darling Basin (Hawes, 1993).

Finally, the public require access to the lagoon for water sports and passive recreation activities. Boobera Lagoon Reserve acts as a regional focus for recreational and social outings. An estimated 7,500 to 10,000 people visit the area per annum (Fish, 1993). Recreational use also degrades the ecosystem resulting from vehicular access to the lagoon foreshores, boat access to the water and the inadvertent and inevitable wake wash when power boats are active on the lagoon.

Resolution of the conflicts are difficult. However, environmental protection and the conservation of cultural heritage can be compatible in their general goals and final outcomes. Public recreation requirements may be managed in a way that is sympathetic to other interest groups. The major conflict is between agricultural uses and all the other more passive competing land and water users. The environmental management plan for the site must address this major conflict to ensure that the goals and needs of the different interest groups are accommodated as much as possible.

Development of the Management Plan Using a MODSS

The planning process followed a systems approach (McLoughlin, 1969) with community perceptions of planning problems, goals, and objectives to guide the management plan. A detailed environmental audit was performed to characterize the site and provide possible management options. The audit consisted of two parts. First, an evaluation of the current state of the natural and cultural assets of the site, and second, an evaluation of the agricultural potential and the 'land degradation hazards' was undertaken. The second part of the evaluation used a MODSS (LANDCAP) to analyze and interpret the large amount of land resource data.

LANDCAP — An Expert System Structure as a MODSS

The expert system (LANDCAP) was used to quantify and make consistent the expert judgments used in the environmental audit of Boobera Lagoon and its environs. The expert system had three functions:

1. An assessment of the erosion hazard and land degradation potential of the site
2. An assessment of the plant production potential of the site
3. A final assessment of land capability with its consequent land use management packages

The general structure of the expert system (Figure 54.2) followed a nested programming approach. There were several large rule bases which contained a number of other smaller sets of rules. The facts from the first rule base were stored in MS-DOS sequential files which were accessed at the beginning of the next rule base. The rule bases were formatted to be a set of inferences and conclusions that are consistent with the objectives of land degradation assessment, agricultural production assessment, and consequent sustainable management. Within the rule bases the structure of the expert system employed a hierarchy of knowledge where the inferences or conclusions from one set of rules were used as facts in a subsequent set. The search of the rule lists used the backward chaining method. The structure and the search method resulted in a build up of conclusions about the land use potentials of each examined site.

The knowledge base made assessments from raw site data that a planner normally collected during the initial stages of a land capability study (the soil data card). The format of the input data was consistent with conventional nomenclature used by soil conservation staff. For example, the soil horizons were presented as "A" and "B," or texture classes used terms such as "loam." The general terminology used, made the questions posed by the expert system easier for soil conservation staff to understand. The form of the rules were a simple "IF — THEN" logic

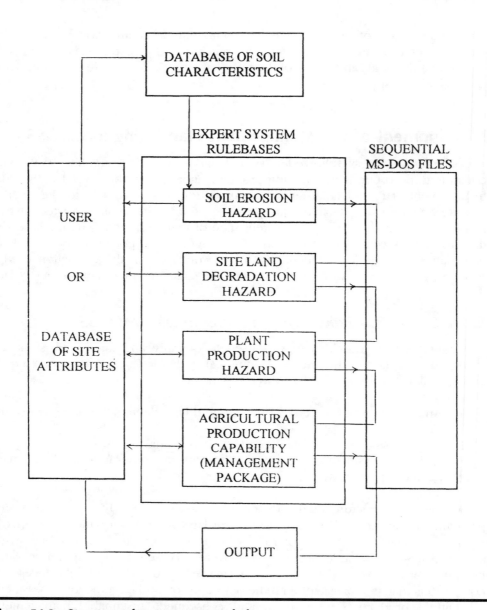

Figure 54.2 Structure of expert system rule bases.

statement. Consequently, each unique set of attributes lead to a specific conclusion, for instance:

RULE A8   Compared with   RULE A9

IF   b_texture = clay loam or   IF   b_texture = clay loam or
        b_texture = silty clay loam or   b_texture = silty clay loam or
        b_texture = fine sandy clay loam &   b_texture = fine sandy clay loam &
        b_fabric = earthy or   b_fabric = earthy or
        b_fabric = rough ped and   b_fabric = rough ped &
        b_color = grey to black   b_color = yellow to mottled
THEN b_drainage = good   THEN b_drainage = seasonally waterlogged

The expert system had the facility to use soil type instead of individual attributes for each horizon to develop the inferences and reach conclusions. The nomenclature describing the soil types was based on the Australian great soil groups (Stace et al., 1968). A database of soil types contained the soil attributes as facts that circumvented a significant portion of the rule bases. The database was in the DBASE 3+ format and was accessed by forward chaining from the first rule base in the inference sequence.

Facts describing textures, soil moisture holding ability, runoff potential, erodibility assessment, etc., (where appropriate) were included in the database. The database contained cruder measures of the soil site attributes than full data entry. Consequently, the great variability of the soil attributes within the great soil groups meant that the judgments were more general with greater expected error. This factor plus the large information gaps in the database precluded the use of the database as part of the land assessment exercise for the Boobera Lagoon.

Application to the Plan

The use of the expert system in the development of the environmental audit and land management plan followed a parallel process of MODSS and human expert judgment.

To achieve the three functions of the expert system, the foreshore and catchment area around the lagoon was divided into a series of mapping areas. Each of these areas had a full description of the land resource information, such as a detailed characterization of the soil, terrain assessment, climate, and the like (Banks, 1993). The expert system then assessed the coded information.

Data collection requirements for the land resource management experts were limited to more general descriptions of the land resources in the first instance. Then they were exposed to the more detailed expert system data. The experts assessed the land and derived their own set of recommendations.

The integration of the parallel assessments followed three processes. First, similar assessments were used as cross-checks to ensure consistency and correctness of the interpretations. At times the difficulty was in deciding whether the human judgment was the check or whether it was the expert system. Second, when significant differences occurred, the data set was reassessed in detail, by the human experts working through the expert system's method. The logic of this approach was to ensure that the expert had made the decisions based on a full data set and to identify any error in the rule base of the expert system. Third, the expert system was used by human experts when they could not make up their minds. During times of doubt about the interpretation of the land unit data, the expert system proved to be extremely useful. The human expert using the expert system was forced to look at the full data set and in following the expert system logic, to reassess their own cognitive process of evaluating the land information. Anecdotal evidence suggests that experience with the expert system equips the experts to be better at their land resource assessments than they were before.

We Came, We Saw, We Conquered
(MODSS Applied to Land Management Outcomes)

The philosophy behind the use of LANDCAP as a MODSS was that it was a decision support to the human experts, not a decision substitution. As such, conflicts were resolved and support given to compatible land uses between different interest groups by the conservation experts with the assistance of the expert system.

Arable Land Assessments

The agricultural sector was provided with a series of management prescriptions to enable the land user to have a sustainable productive system that controls land degradation to an acceptable level. Lands incapable of supporting arable agricultural activities were identified, such as the irrigation lands watered from the lagoon, and some dryland cropping areas. The limits on the lagoon as a source of irrigation water decreased the rapid and unnatural draw-down of water which exposes the fragile shoreline protective vegetation to rapid drying and depletion.

The identification of these lands caused significant disputes between land users (and their agricultural experts) and the assessments of the conservation officers (and their expert system). One particular area of land was identified by agricultural sector as prime arable land which the conservation experts and LANDCAP identified as poor quality grazing land. The essential difference between the two assessments was the frame of reference of the judgments. The agricultural sector looked at existing use with a short-term planning time horizon (less than 5 years). The agriculturalist's approach looked at production requirements and devised a management package to achieve it. The conservation sector and LANDCAP had a long-term planning time horizon (a sustainable farming system of greater than 50 years). The conservationists looked at the limitations of the land resources and managing within those limitations to achieve the available production.

The factors that identified the land as grazing were

1. A highly saline layer at approximately 1 m (in the natural and undisturbed state) which with the passage of time and present irrigation practices and technologies would eventually encroach on the root zone.
2. The cost of proper agricultural drainage would be in excess of $10,000/ha and uneconomic if a sustainable irrigation management plan was implemented.
3. A dispersive and hard setting soil surface which impedes infiltration and seedling establishment.
4. A generally high exchangeable sodium percentage and dispersion hazard which predisposes the land unit to erosion.
5. The foreshore location of the site makes it a significant hazard to any water body.

Although the experts used a similar data set, it was interpreted in a different manner. The agricultural sector with its different frame of reference could mount a powerful case based on existing determinations, models of management, and short-term history of the land unit under study. The long-term judgments of the conservation experts had no such support. The expert system as a MODSS supported the expert judgments and made the logic transparent to nonexperts.

Grazing and Stock Water Control

From the grazing land perspective, the expert system was used to define characteristics of the lagoon and its foreshores that made it incompatible with present high levels of grazing and unrestricted stock access for water. This information was used to control stock numbers and to identify areas that may be used as restricted livestock watering points. As an example, low slope ramps into the lagoon were identified by the grazing sector as the most appropriate zones to allow stock access for watering. The experts and LANDCAP identified these areas as requiring stock exclusion or very limited stocking rates. The land resource assessment showed that these ramps were water courses for the recharge of the lagoon and the soils were highly erodible. These stream bank areas that were so sensitive as to exclude livestock and have pumps, pipes, and tanks installed for trough-based watering more distant from the lagoon.

Modern and Ancient Australian Cultural Needs

Areas identified as suffering from unacceptable levels of land degradation hazard were flagged as requiring special livestock, recreational, and vehicular management plans for their development and use. For instance, sensitive shore areas were identified and water-skiing on the adjacent lagoon waters restricted. Boat ramps, camping grounds, parking and access tracks will be kept away from the more sensitive areas.

The cultural heritage of the aboriginal site is to be protected by further management activities in addition to those proposed under LANDCAP. The agricultural users were the major mode of site destruction and the control of their activities will complement conservation efforts by the Aboriginal people.

Perspiration vs. Inspiration (Mathematical Models vs. Expert Systems)

The question that must be raised is "why not use a normal mathematical model instead of an expert system to assess the salinization, plant production, and erosion problems?" The argument against the model approach was fourfold.

First, the universal model to describe the operations of the world in this environment does not exist. The power of a model is directly proportional to the extent by which it addresses the specific planning problems faced by the community. This means that if there is a universal model there is a need to modify it (maybe extensively) to answer the questions that are posed. The proper modification procedures would have taken significant time and expense which was not available.

Second, the data shortages would make the validation of any model for the specific problems of the site suspect, and open to being discredited. There is no reason to embark on a modelling exercise when the mathematical equations do not have the reliability of formalized expert judgment.

Third, the real issue in the land assessment was not "when the degradation would occur" but rather "if it would occur?" If the model merely confirms something which is obvious, it is a waste of valuable time. A rapid expert judgment would satisfy the requirements of this question.

Finally, the output of a complex mathematical model and its operation is incomprehensible to the majority of the community. The EMP evolved into a conflict resolution exercise. This is a public relations approach to the environmental management issue. The arguments and decision making of any assessments must be fully understood if the cooperation of the interested parties may be achieved. A model could not provide the simple language description of an expert system such as LANDCAP. LANDCAP used simple English, in common usage, to be the symbol representations of the land attributes. The members of the public could read and interpret the operation and output of LANDCAP directly without assistance.

All that Glitters is not Gold (Risks in using MODSS)

The application of LANDCAP to the environmental plan has some inherent risks that it shares with other MODSS. The most disturbing problem is that the users may actually believe and trust the system. This is not only true for the lay persons in land assessment, but there is a risk that the experts may rely on the computer representation instead of their own judgment. This scenario is not the intent in the design of LANDCAP.

It is inappropriate to blindly trust the system for several reasons. The rigid data requirements may restrict the assessment to only those issues in the system, i.e., the data boundary is described by the expert system. This rigidity depletes the greatest strength of heuristic judgment, its ability to learn from new situations and synthesize new knowledge. In addition, data gaps may lead to interpretations of the expert judgments that may not be valid, e.g., nondescriptive nature of the planning time horizon used by the expert system; or specific criteria for the monitoring of the remaining activities in the area are not identified. However, on balance, when using expert judgment, it is better to have an expert system description of the expertise than not to have it.

Conclusion

The expert system aided the decision making of the different government instrumentalities, e.g., Conservation and Land Management over cropping land management, stocking rates, and watering methods; Water Resources Commission in substantiating irrigation license conditions and license relocations in line with agricultural land capability; and the determination of likely compensation payable in the possible resumptions of Crown land for special purposes. The expert system gives a set of defined standards to assess development proposals that may be put forward in the future.

The use of the expert system (LANDCAP) has been a significant advance in the assessment of land use options and development of sustainable land management packages for these areas of New South Wales. The consistency and transparency of the logic provided by this MODSS has the advantages of:

1. Formalizing the judgments and enhancing the validity of experts
2. Defining the logic of decisions about land quality to form a basis for communiy assessment of the expert judgments
3. Providing defined guidelines for land quality assessment to form a basis for the development of land management plans
4. Assisting in the resolution of disputes between competitive land users and experts
5. Providing a set of standards to assist the administrative arm of agencies to implement responsible land management policy and plans

The determination of the area's land capabilities under a LANDCAP MODSS considered the land degradation issues that are both catchment wide and transcend administrative boundaries. The provision of nonbiased, scientific, and national land assessments to determine levels of land management is essential to sustain the intrinsic values of Boobera Lagoon.

Epilogue

The future of Boobera Lagoon as a valuable community resource will depend on the adoption of a more cooperative management approach. Landholders and other land users attended a series of meetings and workshops where the results of the EMP were discussed and evaluated. Several plans of management were developed by the groups. Each plan reflected the aims and objectives of the groups and were framed within the land assessment guidelines determined by LANDCAP. The final plan reflected the land degradation and production potentials identified by the land assessment. In spite of the general agreement within the community one section of the agricultural land users refused to cooperate and insulted and disrupted the other community groups. The political action of this group in the NSW government led to the rejection of the plan (a recent change in the party politics of the ruling sector of NSW government may address this problem). The protection of the environmental values and the cultural heritage of the Boobera Lagoon now rests with the Australian Federal Government. It is hoped that the political leaning and courage of the federal government will override the self-interested and short-sightedness of a small part of the community.

In this light a newly formed Boobera Landcare group, which comprises representatives of all user groups, focussing on making physical changes to the immediate lagoon precinct, will be the precursor to community cooperation. If such an approach is not further pursued and developed, the cultural, natural, recreational, educational, and farming resources of the area are in jeopardy and eventually everyone will lose.

References

Banks, R. 1993. Soils. Pages 131–166 in Anon. 1993. Boobera Lagoon Environmental Audit. Department of Conservation and Land Management, NSW.

Eigeland, N. 1993. Water quality, hydrology, groundwater. Pages 119–130 in Anon, 1993. Boobera Lagoon Environmental Audit. Department of Conservation and Land Management, NSW.

Fish, R. 1993. Recreational Study. Pages 283–361 in Anon. 1993. Boobera Lagoon Environmental Audit. Department of Conservation and Land Management, NSW.

Hawes, W. 1993. Significant Findings — Flora and Fauna. Pages 6–7 in Anon. 1993. Boobera Lagoon Environmental Audit. Department of Conservation and Land Management, NSW.

McLoughlin, J.B. 1969. *Urban and Regional Planning: A Systems Approach*. Faber and Faber, London.

Matthew, P.L; B.A. Peasley, and J.A. Duggin. 1992. An expert system for agricultural land capability assessment. Pages 732–737 in *Proceedings of the 7th ISCO Conference Sydney*.

Patten, J. M. 1993. Fish. Pages 75–86 in Anon. 1993. Boobera Lagoon Environmental Audit. Department of Conservation and Land Management, NSW.

Peasley, B. A. 1993. Landuse and soil erosion changes. Pages 168–176 in Anon. 1993. Boobera Lagoon Environmental Audit. Department of Conservation and Land Management, NSW.

Stace, H.C.T., G.D. Hubble, R. Brewer, K.H. Northcote, J.R. Sleeman, M.J. Mulcahy, and E.G. Hallsworth. 1968. *A Handbook of Australian Soils*. Rellim Technical Publications S.A.

Thompson, P. 1993. Aboriginal cultural heritage. Pages 206–281 in Anon. 1993. Boobera Lagoon Environmental Audit. Department of Conservation and Land Management, NSW.

Chapter 55

Multiple Objective Decision Support Systems Used in Management of Temperate Forest Ecosystems in Southeast Australia

JOHN ROSS AND IAN HANNAM

Introduction

The total area of forest and woodland in New South Wales (NSW), Australia is approximately 14 M ha of which 5.1 M ha occurs on private property (Commonwealth of Australia, 1992). This paper discusses the management of the 2.5 M ha of temperate forest and woodland on private property in southeast Australia which occurs on **Protected Land** (see Figure 55.1). Protected Land is ecologically sensitive areas of the NSW landscape, mapped under the New South Wales Soil Conservation Act 1938 (**SCA**) (Abraham et al., 1992). Although there is other NSW legislation with provisions to control the destruction of trees on private land, Protected Land covers the largest area of forest ecosystem, and importantly, all major forest and woodland types found in southeast Australia are represented on Protected Land. For this reason, it has a significant conservation role, particularly in management of biological diversity. Various decision support systems (DSS) have been developed by the Department of Land and Water Conservation to manage the legal, ecological, and social information required to make good decisions about the conservation of forest on Protected Land (Ross, 1994).

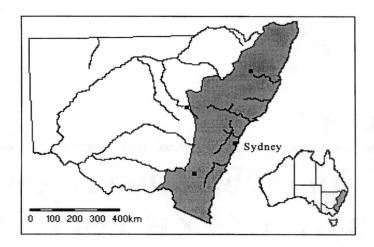

Figure 55.1 Areas covered by protected land maps.

The environmental law which applies to decision making on Protected Land is complex, necessitating development of sophisticated expert systems to manage field collected data and manage interactive statutory decision-making procedures. The SCA manages the soil environment, but tree destruction proposals are also bound by the environmental assessment responsibilities of the Environmental Planning and Assessment Act 1979 (**EPAA**) and Environmental Planning and Assessment Regulation 1994 (**EPAR**) in relation to all aspects of the environment, and to the Endangered Fauna (Interim Protection) Act 1991 (**EF(IP)A**), in relation to endangered fauna habitat conservation. The responsibilities of these Acts as they apply to Protected Land is referred to as the *environmental assessment regime* (Hannam, 1995c).

The first part of this paper discusses the ecological criteria, environmental standards, and technical standards on which the administrative procedures are dependent, and these are not always clear or apparent in the legislation. Guidelines for the development of these criteria and standards are found in a range of sources including the above legislation, precedent from relevant court cases, the wider body of natural resources environmental law, and the scientific literature. The second part of the paper explains the specifics of three multiple objective decision support systems (MODSS) currently being used to manage forest ecosystems on the Protected Land, the reason for their development, and how they were developed. The integration with the overall legal decision-making responsibilities for managing forest ecosystems is also discussed.

Temperate Forest Systems of Southeast Australia

The native forest and woodland in southeast Australia includes tropical, subtropical, and temperate rainforest; mangrove and swamp forest, wet eucalypt forest, ash forest, coastal eucalypt forest, dry forest and woodland; river red gum forest; native pine forest and woodland (Commonwealth of Australia, 1992). This representation means that the Protected Land *environmental assessment regime* requires

DSS to accommodate the vegetative and geomorphic variability of these forest ecosystems. Types of disturbances which DSS must accommodate include:

- Selective logging of rainforest, wet eucalypt forest, coastal eucalypt forest, and river redgum forest
- Integrated logging of wet and dry eucalypt forest
- Regrowth clearing of wet and dry eucalypt forest
- Logging and clearing of native pine forest
- Clearing of woodland

Concern over the accelerated loss of ecosystem functioning and biological diversity has emerged as an objective in forest management and virtually all NSW forest (Commonwealth of Australia, 1992), has been disturbed by selective removal of trees, fire, climate, and human interference, resulting in a constantly changing pattern of the size, shape, and arrangement of the ecosystems (McNeeley, 1994; Commonwealth of Australia, 1995). Conservation efforts need to give greater attention to ecosystem characteristics and Protected Land legislation can be used to ensure that forests on private land are used in a more ecologically sustainable manner. The legal, policy, and administrative procedures for environmental assessment and decision making are reasonably clear, but cumbersome to implement. Owners or occupiers of Protected Land require a legal authority under the **SCA** to injure or destroy trees, and this process, aided by comprehensive MODSS, can be used to achieve good conservation standards and sustainable forest use.

The Environmental Law and Policy Relating to Private Property Forest Management in Southeast Australia

The environmental law relating to forest management in southeast Australia falls into two categories of law: (1) the legislation applying to State Forest (public) land and (2) the legislation applying to non-State forest land, or private land and Crown Land. The *environmental law regime* associated with the second category, which includes Protected Land, activates an environmental assessment process (Hannam, 1994) which brings to light many ecological characteristics of forest on private property (Hannam, 1995a,c), including a poor understanding of the sociological characteristics of private property forestry, a poor policy basis, and only a very limited amount of research information (Tracey, 1991). MODSS help to close these information gaps.

Soil Conservation Legislation and Forest Conservation

The SCA was created in 1938 with an express concern to control soil erosion and to mitigate the effects of soil erosion and siltation. The Protected Land provisions were included by amendment in 1972 specifically to control the destruction or injury of "trees" on critical parts of the landscape. Although Protected Land actually covers forest and woodland areas, it was never envisaged in 1972 that these provisions would be implemented for broader forest conservation purposes. With the creation of the EPAA in 1979, the Commissioner of Soil Conservation became bound to this Act's 'environmental assessment' requirements on Protected Land (Hannam, 1994) before he can give permission to injure or destroy trees. The legislation's ecosystem philosophy presents a facility for the environmental assessment and decision-making process to focus on forest as a community (Commonwealth of Australia, 1992) as well the soil conservation matters. Although, the SCA does not have any express environmental assessment or environmental management requirements for forest land use, the general legal, policy, and land management considerations for forests which society has established (Commonwealth of Australia, 1992) are applied in conjunction with the Commissioner's general duty to the environment, to set up the assessment rules for soil and forest protection and management.

Environmentally acceptable standards for forest management are developed from the general *conditions* in the Act (ss21D(3)(a)-(i) and (3A)(a)-(f)). These provisions are reworked into the land management guidelines for forest activities (e.g., Department of Conservation and Land Management, 1993b) and the technical relationships for the DSS. Significant court rulings (*Bailey v Forestry Commission of New South Wales (1989) 67 LGRA 200-218*) have also influenced the preparation of DSS so that they fully satisfy the legal environmental assessment requirements.

The soil conservation provisions cover a wide range of aspects of the environment and can be used to manage ecosystem composition, structure and function, particularly ecosystem diversity, complexity, and stability (see Noss, 1990; Franklin, 1993). The conditions, which underlie technical components of the DSS are categorised into three groups:

- **Conditions which control the amount of forest disturbed** (these restrict the amount of landscape disturbed, and contribute to conservation of forest ecological integrity (see Zavala and Oria, 1995, particularly fauna habitat management).
- **Forest debris management** (includes burning, which is critical for soil erosion control, survival of fauna occupying the forest floor, and understorey and ground condition).
- **Soil erosion and land degradation prevention** (these prevent soil erosion from forest operations and clearing activities including the effect of road and access track construction on slope stability and catchment hydrology. They are also used to develop operational guidelines (Department of Conservation and Land Management, 1993a, b)).
- **Other environmental conditions** (these are used to manage aesthetic, recreational, scientific, environmental quality, anthropological, endangered species of fauna and flora, and cumulative adverse effects on the environment).

Management of the Environment in General

The EPAA establishes the regime for comprehensive environmental assessment of temperate forest on Protected Land, which, in conjunction with the above SCA conditions, present the parameters of the DSS. This *environmental law regime* also requires decision makers to consider environmental impact of any forest disturbance proposal by examining and taking into account 'to the fullest extent possible all matters affecting or likely to affect the environment by reason of that activity' (EPAA, s111(1); Hannam, 1994). The factors set out in the legislation are general in nature and need to be interpreted in the context of forest land management (EPAR, ss82(1)(b)-(2)(a)-(o)). To properly execute the environmental assessment responsibility within the parameters of the law, creditable DSS are developed following an interpretation of three statutory procedures.

- **Factors to be considered** — Court cases, including the *Bailey* (1989) case, have determined the factors which must be investigated to determine the likely significant affect of a forest disturbance proposal on the environment. These include flora and fauna protection, hydrologic management, soil management, and measures to prevent degradation of the vegetation environment. In the case of Protected Land, the DSS are the main tool used to assess these factors so that the ecological integrity of the forest ecosystem can be determined. In particular, the DSS predict if there will be a "transformation" of the area (EPAR cl82(2)(a)). The application of the general meaning of transform (*The Macquarie Dictionary*, 1985) to an area of forest would be to change it's form; change to something of different form; or to change it's appearance, condition, nature or character, especially completely or extensively. The environmental process must be based on scientifically valid techniques, able to interpret the impact of a proposal on the ecology of the forest, and define sustainable approaches to forest use.
- **Environmental assessment procedure** — The diversity of selective logging, integrated logging, clearing for grazing, and regrowth clearing activities, necessitates flexibility in decision support procedure. They often include a number of forest types, in various structural stages, and condition, where the motivation for disturbance includes commercial values, or the value that may accrue from removing an area of forest (i.e., to establish pasture for grazing). The environmental assessment process around which the DSS are established distinguishes these characteristics so that, at an appropriate stage, the right type of prescriptions can be devised to manage disturbance within ecologically sound guidelines and standards. Responsibility clearly rests with a decision maker to develop a creditable methodology, process, and standards to carry out the environmental assessment function and to support the final decision. A general guideline has been established (Hannam, 1995b) to determine 'impact on the forest environment,' the concepts upon which this process is based (i.e., EPAA s111(1)) consider the *'likely impact on the environment'* from logging or clearing activities. Should there be a significant environmental impact then an environmental impact statement must be prepared before a final decision is made.
- **Meeting the requirements of the threshold test** — Another significant feature of the *environmental law regime* is the statutory test to determine *significant effect [of a proposal] on the environment* of a forest area. This question has been argued in various cases (e.g., *Jurasius v Forestry Commission of New South Wales 1990, Randwick Municipal Council v Crawley (1986) 60 LGRA 277, and Bailey V Forestry Commission of New South Wales (1989)*), so suitable precedent exists to draw on for useful guidelines for

MODSS development. In determining the "significance" of an integrated logging proposal, in *Jurasius*, Hemmings J at 94–95 said:

> 'In my opinion, by its very nature the integrated logging activity, whether on a local or regional viewpoint, has inevitably a significant effect of converting the environment from that of an old growth forest to that of a different and regenerated forest. The forest must be fragmented and flora is likely to be reduced in species and diversity. The full extent of the likely impact on flora in this environment is difficult to assess because there has been no comprehensive survey or research on impacts on non-commercial species.'

Importantly, he went on to outline *a number of specific forest disturbance activities (e.g., roading) and resultant ecosystem changes and disturbances* (e.g., disturbance of, and changes to, fauna habitat and water quality), which are likely to significantly effect the environment. These are a good representation of the type of disturbances that can take place in forestry operation, but there are also other geomorphic and vegetation circumstances where a change, can result from logging and clearing activities (see McNeely, 1994, Hannam, 1995a, Zavala and Oria, 1995).

- Forest ecosystems and communities — One of the important features of the legislative regime is its requirement for the 'environment' of the forest to be considered in a 'community' or 'ecosystem context (EPAR cl82(2)(b),(c)). This is critical from an ecological integrity perspective as well as for the development of creditable DSS as the scientific methodology which provides information on the type of biological processes operative, and which characterizes the particular area of forest, must also be able to predict the changes (i.e., the "significant effects"), that may occur from the proposed forest disturbance activities. The ecosystem concept becomes the logical framework in which to "consider" (EPAA s111(1)) Protected Land forest areas because it focuses on the biological components and characteristics of the natural environment, and it is also an effective framework into which the proposals of humans can be considered. Some of the important components and characteristics which determine ecosystem integrity (see Willison et al., 1992) include the size and location of the community, the representativeness of the community on the local topography, the number of species present and species diversity, structural diversity (including presence of old growth), and evidence of effects of modifying influences (human induced or natural affects).

Forest and Woodland Fauna

The introduction of the EF(IP)A 1991 brought with it a very comprehensive set of factors to be assessed by a determining authority in respect of the environment of the habitat of endangered fauna (Hannam, 1994). This Act provides the best opportunity in NSW to fulfill the objective of conserving habitat of forest fauna, which, in many circumstances, has reached critical levels (see Lunney, 1991). This legislation introduced a new regime for the protection of endangered fauna and was primarily based on many of the arguments in the landmark case Corkill v Forestry Commission of New South Wales and Ors (1991) 73 LGRA 126.

In the absence of guidelines for the interpretation and implementation of the EF(IP)A, an extensive interpretation was made of the 'seven tests of significance' in the Act (EPAA s4A), and a methodology was prepared to assess the environmental effects of logging and clearing on endangered forest fauna on Protected Land (Smith and van der Lee, 1992, Hannam, 1994). This procedure has been adapted to decision support technology which embodies a specific environmental assessment process, which, in turn, suggests ecologically sustainable methods for logging and clearing activities. The Guidelines (Smith and van der Lee, 1992) include specific strategies to: use macrohabitat as a basis for predicting the occurrence of species with unknown or poorly known microhabitats; assume that species occur in all areas of potentially suitable microhabitat; and conservatively estimate average densities of species. The four basic steps to implement the legislation (and upon which the DSS is based) include:

1. Identification and mapping of endangered fauna habitats (or environments) on the Protected Land
2. Identification of endangered species likely to occur in each of the habitat types present
3. Assessment of the likely significance of clearing and logging impacts on endangered fauna habitats present
4. Setting prescriptions for mitigation of logging impacts on endangered forest fauna and their habitats

Environmental Assessment Procedures on Protected Land

Role of Delegated Officers

A system exists which allows the Commissioner of Soil Conservation to delegate to other officers his powers to carry out the environmental assessment regime and make determinations on applications to log or clear trees on Protected Land (s30A SCA, Hannam, 1994). These officers possess varying qualifications, skills, and experience. Most officers have tertiary qualifications in one or more areas of natural resource management and are employed either as full time Protected Land officers or have Protected Land duties as a component of their overall duties which include a range of soil conservation considerations. Other officers, Field Service Managers, have backgrounds in field supervision of earthmoving machinery (a major component associated with both agricultural and forestry soil conservation activities) but do not necessarily have any formal qualifications in natural resource management. The common area of speciality of both tertiary qualified and nontertiary qualified officers however, is in soil conservation. Notwithstanding the differences in qualifications and skills, all delegated officers are required to follow the same procedures in assessing the environmental impact of logging and clearing activities and to make interpretations on how the environmental legislative regime will apply to the activity in question (Department of Land and Water Conservation, 1995; Hannam, 1995b). Regardless of the particulars of the activity, the environmental law regime has to be considered in the same manner to ensure that decisions consider the sustainable use of forests (see Cameron and Aboucher, 1991; Recher, 1992).

Inconsistencies

The varying experience of field officers, large turnover of applications, and particular areas of speciality of full-time Protected Land officers has allowed them to develop routines and individual approaches involving all of the components associated with assessing logging and clearing applications. Variations between the biogeographic regions in which the officers work causes inconsistencies in field interpretations and decisions. The lack of practical guidelines for addressing the environmental legislation involved has a multiplying effect on these inconsistencies and differences in standards and rigor with which applications are examined.

The growth in scrutiny by environmental groups of activities which involve the cutting down of trees in forests in southeast Australia has necessitated stringent assessment and compliance procedures (Environs, 1994). There have been successful court cases brought against determining authorities (e.g., Bailey, 1989; Van son v Forestry Commission of New South Wales (1995), Supreme Court of New South Wales, Equity Division, 3485/93, Cohen J) for not properly carrying out their determining authority roles in relation to approval of activities which may have a significant effect on the environment. Inconsistencies in assessment approaches and rigor increases the chance of inadequate or inappropriate environmental controls and conditions (SCA, s21D(3)-(3A)) being applied to logging and clearing activities, thus increasing the likelihood of degradation of the ecosystems associated with those forests. These inconsistencies also leave the Commissioner of Soil Conservation open to prosecution for not properly carrying out his roles under the various legislation (i.e., s124 EPAA). It is for these reasons that MODSS are being developed to assist in the ecological sustainable management of forest ecosystems in southeast Australia (Commonwealth of Australia, 1991; Recher, 1992).

MODSS Currently In Use

The need to develop DSS to increase the level of efficiency and consistency in decision making for logging and clearing activities undertaken on Protected Land has been discussed by Ross (1992; 1993; 1994), and Hannam (1995c). There are at present two computer DSS being utilized by field officers to address specific land management objectives on Protected Land. These are LOGSPERT© and HABASYS©. Both of these DSS were developed to improve the understanding by delegated officers of legislative requirements associated with Protected Land activities, and to provide them with a vehicle for satisfying those requirements in the processing of logging and clearing applications. In achieving this, both systems have had to possess four basic characteristics to ensure their widespread use by delegated officers.

- Practical — Both DSS have to be able to fulfill a useful role in the routine of delegated officers in processing applications to log or clear. The DSS have to be able to process field collected and operational information and provide the officers with consistent and appropriate ecological land management recommendations in less time than it would take them to arrive at a decision without the assistance of a DSS (i.e., by processing the field and operational information manually).
- Simple — Because both DSS are computer-based and the computer skills of delegated officers varies considerably, the input and processing functions of the DSS has to be easy to understand and simple to use. This reduced the possibility of mistakes or misinterpretations in the processing of field and operational information (which may lead to inappropriate decisions being made and open to legal challenge), a common occurrence from manual processing.
- Legal — Both DSS have to aim to present land management recommendations which are consistent with specific legal obligations (SCA, s21(3)-(3A); EPAR, cl82(2); EPAA, s4A) . The interpretations must accommodate a precautionary approach to the legal rules so that the Commissioner's obligations to the environmental law regime are satisfied (Cameron and Aboucher, 1991).
- Informative — Adequate supplementary information, in the form of documentation and help screens, has to be included so that delegated officers are aware of the basis for the land management recommendations which the DSS proffered, and so that they can comprehend difficult ecological concepts and technical relationships between field data and management prescriptions. This characteristic is particularly important given the disparities in experience between delegated officers.

LOGSPERT©

There has been increased pressure by the community through environmental groups and regulatory authorities (e.g., Environment Protection Authority) to ensure the environmental sustainability of logging operations in forests of southeast Australia (Watson, 1990; Rowland, 1992). This has increased the amount of time spent by field officers assessing Protected Land under application for logging because the increase in standards requires a more sophisticated environmental assessment process. The need for LOGSPERT© became apparent due to the large number of applications received by the Department of Land and Water Conservation and the need to maintain consistency and continuity in processing. LOGSPERT© forms part of a process to implement the soil erosion and sediment control principles contained in the Soil Erosion Mitigation Guidelines for Logging (SEMGL) (Department of Conservation and Land Management, 1993b). The SEMGL allow forest planners to develop site-specific conditions suitable to protect the physical environment of forest areas (see Atkinson et al., 1992).

- Implementation — LOGSPERT© was developed to accommodate the soil conservation principles in the SEMGL and to determine the instances under which each principle would be relevant in a field situation. A set of legally enforceable conditions, derived from the biophysical principles in the SEMGL, can be imposed on logging approvals. Criteria for each condition was developed through interviewing experienced logging supervising officers. They were asked to identify circumstances under which they would apply each condition (developed in consultation with those same officers) to a logging Authority. This criteria was based on the combination of the physical attributes of the site and operational details of the logging application. For example:

SEMGL principle:	"In areas of high erosion hazard, the grade of snig tracks should not exceed 25°."
Authority Condition:	"The grade of snig tracks shall not exceed 25°."
Criteria for Condition:	▪ maximum slope on site to be logged >25°
	▪ logging contractor intends to construct snig tracks
	▪ erosion hazard of site is high (based on Universal Soil Loss Equation adapted for southeast Australian forests (Department of Conservation and Land Management, 1993b))

In addition to the general conditions derived from the SEMGL a number of special conditions have been developed for specific combinations of land attributes and operational details which were not contained in the SEMGL. Criteria for when each of those conditions would be appropriate for inclusion on an approval was also produced.

The structure of the LOGSPERT© computer DSS (developed in the Level 5 Object© expert system shell) is a series of input screens where the ecological and operational data are entered. Screens are logically ordered and help screens (Figure 1) attached to each input screen give information on how to respond to each input screen. These provide conceptual and practical information associated with each input (e.g., functions of filter strips, method for assessing soil erodibility [Murphy, 1984]). LOGSPERT© identifies the critical biophysical relationships between the data which require special attention (i.e., environmental protection) in the form of legal conditions of approval. The rule base consists primarily of the "criteria" (i.e., legal and technical) developed for each condition, and where the combination of field and operational information collected in the field meets the criteria, then the condition is recommended.

HABASYS©

Introduction of the EF(IP)A placed specific obligations on the Commissioner to consider the likely impact of logging and clearing activities on the habitat of endangered fauna (Hannam 1994; 1995c). Although an obligation to consider the effect of logging and clearing activities on endangered fauna already existed in the EPAA it was too general and could be satisfied without detailed assessment of macro- and microhabitat characteristics. The seven ecological questions contained in the EF(IP)A, which must be addressed in relation to each activity are outside the area of expertise of most delegated officers and they are unable to confidently approve any logging or clearing activities because they cannot adequately address the requirements of the legislation. The development of a uniformly applied manual assessment system (Smith and van der Lee, 1992) overcame this dilemma (i.e., *The Guidelines*).

HABASYS© is based on *The Guidelines* (Smith and van der Lee, 1992) and established a procedure easily followed by delegated officers with little or no habitat assessment skills. The process examines location, macrohabitat, and microhabitat features of the site under application and helps determine the types of habitat present, with specific reference to habitats of endangered forest fauna and the maintenance of ecological sustainability (see McKinnell et al., 1991; Recher, 1992). The main assumption of *The Guidelines* (Smith and van der Lee, 1992) is that, if all of a species' essential habitat requirements are present on the site under application, and the site is within the estimated distribution range of the endangered species, then the species is assumed to be present (Smith and van der Lee, 1992). If the species habitat is present then mitigation prescriptions should be imposed to protect those habitats from destruction during the logging or clearing operation (Lunney, 1991).

Although the process set out in *The Guidelines* (Smith and van der Lee, 1992) for determining the presence of endangered fauna habitat is simple and can easily be followed by delegated officers, it is time consuming if carried out manually because of the necessity of checking distribution maps and performing a number of calculations. For officers who only process a few applications per year the process can be laborious because they do not have the opportunity to develop routines and develop familiarity with the process. In effect, the process contained in *The Guidelines* (Smith and van der Lee, 1992) has to be relearned each time such officers process an application to log or clear.

Full time officers who process applications regularly are very familiar with the processes in *The Guidelines* (Smith and van der Lee, 1992) and are able to complete the process a lot quicker. Nevertheless, the time spent determining the likely effect of the proposed logging or clearing activity on the habitat of endangered fauna can constitute up to 50% of the total application processing time, particularly if there are a number of macrohabitats present on the site under application. To facilitate the endangered fauna habitat assessment component of the overall application processing time, and to reduce the chance of human error in processing the field data collected for the assessment, HABASYS© was developed.

Like LOGSPERT©, HABASYS© was also developed in the Level 5 Object© shell and its structure is similar to LOGSPERT© in that a succession of screens containing questions relating to location, macrohabitat, and microhabitat details are presented to the user. These are answered after field data is collected from one or more transects in the area under application. The four basic characteristics of DSS are contained in HABASYS© in similar forms as in LOGSPERT©.

Total Environmental Assessment System (TEAS)

While the above two DSS have fulfilled a useful role in the administration of Protected Land, they are separate components of the overall environmental assessment procedure and decision-making process. The next logical step in MODSS development for Protected Land is the "one-stop" system which addresses all of the requirements of the *environmental legislation regime* in management of forest ecosystems, encompassing the assessment and decision-making principles of LOGSPERT© and HABASYS©.

The need for such a DSS becomes more apparent due to the lack of guidelines which interpret the legislation for practical implementation by field officers carrying out the Commissioner's legislative responsibilities. The case law concerning Protected Land indicates that decision makers have a responsibility to develop environmental assessment procedures which allow them to be consistent in their decision making (i.e., interpretations). A failure to do this can lead to litigation (Gibson, 1993). The differences in interpretations of the *environmental legislation regime* as it relates to logging and clearing on Protected Land present a potential credibility problem if ecologically based, consistent, and objective assessment and decision-making processes are not developed (Ross, 1994).

A need was seen to develop a DSS which aims to promote confidence, not only in the processes used to assess the land and satisfy the legislation, but which benefits the ecological integrity of the forest ecosystems. A third, more comprehensive, DSS is currently being developed to address these needs. The Total Environmental Assessment System (TEAS) will more satisfactorily meet the Commissioner's obligations under the full range of environmental legislation

Methodology

The assessment procedure for Protected Land is not simply inventory, but has a legislative nucleus. It requires legal decisions to be made in direct response to the physical and operational data collected. To adequately satisfy the environmental legislation regime a process-oriented approach is needed which identifies critical interactions between the activity and area of land on which the activity is planned to take place. The methodology used to develop the assessment procedure used in TEAS is described below.

Four major steps are involved in developing TEAS.

i) *Identify relevant provisions in the legislation, and interpret what the legislation requires the Commissioner to do in order to fulfill his legal responsibilities.*

> The *environmental legislation regime* discussed above provides a comprehensive list of *factors for consideration* (i.e., SCA, s21(3)-(3A); EPAR, cl82(2); EPAA, s4A) which must be considered by the Commissioner before approval can be issued to carry out a logging or clearing activity. Although a standard set of guidelines has been prepared to assist decision makers in following a legally acceptable procedure (Department of Planning, 1995), they are considered too general for assessment of forest ecosystems. Under the circumstances these factors for consideration provide an appropriate focus for the development of TEAS.
>
> In some cases the factors for consideration can be easily interpreted and the data needed to determine the effect of the proposal on the particular factor is well-defined (e.g., impact on soil stability). However, in many of the factors for consideration (EPAR, s82(2)), there may be a range of interpretative techniques to assess environmental impact. Perceptions of which information needs to be collected, analyzed, and interpreted may vary significantly between groups and individuals. Analysis of the case law (e.g., Bailey 1989; Jurasius, 1990)

indicates that there can be major discrepancies in interpretation of ecological concepts on the part of decision makers. However, the absence of legal guidelines for logging and clearing operations on Protected Land necessitates that the Commissioner make an interpretation as to what each of the factors for consideration requires him to do to fulfill his legal obligations (see Hannam, 1995b, c).

ii) *Interpret which data needs to be collected to properly address each of the factors for consideration.*

After interpretation of each of the factors for considerations has been made, specific data sets to properly address the requirements inherent in each factor need to be identified. This is done for the range of logging (e.g., selective, native plantation, exotic plantation, integrated) and clearing (e.g., pasture establishment, banana plantations, powerlines, roads) activities likely to be encountered on Protected Land. The data requirements are ordered into a practical sequence for collection in the field. The data set is realistic in what it requires the delegated officer to collect, because the main purpose of the Protected Land assessment is to determine whether or not the logging or clearing activity is likely to have a significant effect on the environment within the meaning discussed above in Jurasius (1990).

Classes of data

There are three classes of data which need to be collected to properly address the factors for consideration presented in the legislation.

1. Primary data — In some instances a single item of data may be able to lead directly to decisions being made as to the likely impact of an activity on the environment. This may be data in a measurable form (e.g., slope gradient, rainfall), or by presence/absence. For instance, the legislation requires protection of habitat of endangered species. If the site in question contains the colonial roost site of an endangered bird, then any tree clearing activity is likely to have a significant effect on that site. The range of possible options for any proposed activity thus decreases as the effect becomes more significant (Jurasius, 1990) and mitigation prescriptions are less likely to be successful. Several options in this particular case are now possible.

   a. The proponent may still wish to go ahead with the logging or clearing proposal, in which case the Commissioner is obligated to ask the proponent to prepare an environmental impact statement (EIS) before he can further consider the application.
   b. The proponent may abandon the logging or clearing proposal, in which case there is no need for further action.
   c. The proponent may modify the logging or clearing proposal so that the area of the colonial roost site is eliminated from the proposal.

   Primary data triggers immediate options for the proponent (similar to those above), because the decisions associated with the data do not vary when combined with other items of data.

2. Secondary data — In other cases a combination of data may be required to determine certain impacts of activities on the environment. The individual components in themselves may not lead directly to identification of an impact, but in conjunction with other variables may become significant (e.g., see Rab, 1992). For instance, the legislation requires protection of the environment from pollution. If a logging operation were to take place on a hill slope adjacent to a river then, in order to comply with the legislation, the determining authority would need to ensure that the activity did not cause increased levels of sediment to enter the river. To be able to assess the likelihood of sedimentation the determining authority would need to assess (among other things) the erosion hazard (Department of Conservation and Land Management, 1993b) of the site to be logged. The components of that index (slope gradient, slope length, soil erodibility, groundcover, erosivity, and management practices) are all data which may not in themselves highlight the need for any specific management options, but when combined, may indicate a high or extreme erosion hazard (Department of Conservation and Land Management, 1993b) which may cause the Commissioner to determine that the proposal would have a significant effect on the environment (Jurasius, 1990) and offer options to the proponent similar to those of primary data set out above. Some data may also be both primary data and secondary data, and it may assist in addressing more than one of the factors for consideration from the legislation, either as an individual item of data or in combination with other items of secondary data.
3. Operational data — Similarly, operational data may work individually or in combination with other data in providing information with which to assess the factors for consideration of the legislation. This data is provided primarily by the applicant and includes matters such as the intensity of operation, the intention of constructing roads, snig tracks, log dumps, and machinery to be used. Each of these may assist the Commissioner to determine the likely effect of the activity on each one of the factors for consideration.

iii) Identify the critical relationships and thresholds of the data and what the likely effect of the activity will be on those relationships and thresholds

The critical relationships between the operational, climatic, geological, hydrological, vegetation, soils, and social/anthropological elements is determined in response to the data collected. A "hazard identification" approach which aims to identify the critical combinations of data is more applicable in the management of Protected Land than classification/suitability systems (e.g., FAO, 1976; Emery, 1986). "Hazard identification" considers the whole area under application but filters out only those areas which may pose a threat to the forest environment (e.g., soil conservation, biodiversity, anthropological significance, etc.), as interpreted from the factors for consideration. Where there are no significant data combinations then there is no need to consider the area further for any special management prescriptions. This increases the time efficiency of inspecting officers because only the combinations of data which are potential hazards are identified and addressed in TEAS.

For items of primary data thresholds are often readily determinable, especially for data which is collected on a presence/absence basis (e.g., level of existing erosion, anthropological site, colonial roost sites etc.). Other items of data such as slope (van de Graaff, 1990), site drainage (Crouch et. al, 1991), rainfall zones (Edwards, 1979) and macrohabitat (Smith and van der Lee, 1992) can be readily classified and limitations for certain land uses identified. In many cases though, the significance of both primary and secondary data does not become apparent until it is considered in conjunction with the operational data, and an overall determination of the effect of the proposal cannot be made until this is done. The next phase of TEAS provides best management options for the critical combinations of data.

iv) Determine how the interaction between the critical relationships and the activity should be managed.

In the assessment process, wherever critical relationships of data are identified a decision must be made within the environmental legislation regime on how to best manage that combination. If any of the thresholds identified in the previous step are reached then the outcome of the decision may take several forms. The aim of TEAS is to provide a guide to the Commissioner as to which of these forms is the most appropriate for each combination. This step identifies:

1. The circumstances where an EIS would be required (e.g., proposals to log old growth forests, proposals to clear large areas of forest, proposals which pose competition to existing land uses in area)
2. When additional information from the proponent (e.g., order of works, timing of operations), or more detailed assessment from the delegated officer (e.g., laboratory analysis of soils, geotechnical advice sought), is required on certain aspects
3. The legal conditions of authority which would be suitable to manage various aspects of the proposal (on the same principles as LOGSPERT©)
4. Advice on operational considerations which may not warrant a legal condition, but which may guide the field supervisor to a particular principle (e.g., which habitat trees may be most suitable for retention, the most suitable part of the landscape to locate log dumps, appropriate methods of protecting archaeologically significant sites).

Conclusion

LOGSPERT© and HABASYS© demonstrate the ability to develop DSS which satisfy multiple objectives. Both DSS assist the Commissioner to follow a reasonable assessment and decision-making process, consistent with the ecological principles of certain factors for consideration, and to fulfill his obligations under the environmental legislation regime. In addition to the fulfillment of legal obligations, LOGSPERT© and HABASYS© offer field officers a practical tool of data collection, assessment, and interpretation which provides consistency, objectivity, and accountability in decision making for forests in southeast Australia (Ross, 1994).

TEAS is an extension of the principles on which LOGSPERT© and HABASYS© are based and provides a comprehensive assessment and decision-making system, addressing the full range of factors for consideration under the environmental legislation regime. TEAS provides a practical process of data collection, inference, and administration for field officers to manage logging and clearing operations in a range of temperate forest landscapes. The data collection process is logically ordered and only data which is relevant to the activity/landscape relationship, and which is readily collectable by persons of varying experience and qualifications is required.

Management options provided through TEAS to confront each of the "hazards" identified through the data collection and assessment process are also based on both legal and practical requirements. The options provide advice as to when legal solutions are required to manage combinations of physical and operational data (e.g., when an EIS may be required; control of activities through legal conditions), and when practical solutions can be reached without the need for legal parameters (e.g., need for additional information, provision of educational material to logging contractors and farmers).

Legal and practical objectives of all three DSS are inseparable. Logging and clearing activities deal with multiobjective decision-making perspectives. Although aimed at satisfying important rules of environmental law, the three DSS have to consider both the ecological and human elements associated with the sustainable management of temperate forests. Developing a system aimed at satisfying only the ecological aims of the legislation without considering the human variables (e.g., diversity in qualifications and experience of field officers, motivational and economic considerations of the proponent) would be inadequate (Zavala and Oria, 1995).

The three DSS make significant contributions to the ecologically sustainable management of Australia's southeast temperate forests. In doing so, they contribute to the conservation of one of Australia's most valuable resources. Other areas of natural resource management may also benefit from the legal and practical premises on which the three DSS are based including management of water resources, pastoral and arable land uses, and management of recreation areas.

References

Abraham, N.A., I.D. Hannam, and P. Spiers. 1992. Protected land and its role in land management. Pages 179–186 in *Proceedings of 7th International Soil Conservation Organisation Conference,* Sydney Australia, 27–30 September 1992.

Atkinson, G., R.A. Attwood, J.J. Kingman, and R. Saul. 1992. Soil Conservation Issues, Compartments 168–170, Oakes State Forest. Report prepared for the Forestry Commission of New South Wales by Department of Conservation and Land Management.

Cameron, J. and J. Aboucher 1991. The Precautionary Principle: A Fundamental Principle of Law and Policy for the Protection of the Global Environment. *Boston Coll. Int. Compar. Law Rev.* 19(1):1–27.

Commonwealth of Australia. 1992. A Survey of Australia's Forest Resource. Resource Assessment Commission, Australian Government Publishing Service, Ch 3, Table 3.1.1.

Commonwealth of Australia. 1995. Native Vegetation Clearance, Habitat Loss and Biodiversity Decline. Department of The Environment, Sport and Territories, Biodiversity Series, Paper No 6 Biodiversity Unit.

Crouch, R.J., K.C. Reynolds, and R.W. Hicks. 1991. Soils and their use for earthworks. In P.E.V. Charman and B.W. Murphy, Eds. *Soils— Their Properties and Management.* Sydney University Press.

Department of Planning. 1995. Is an EIS Required? Best Practice Guidelines for Part 5 of the Environmental Planning and Assessment Act 1979. New South Wales Government.

Department of Conservation and Land Management. 1993a. Guidelines for Mitigation of Erosion and Land Degradation for Permanent Clearing on Steep Protected Land.

Department of Conservation and Land Management. 1993b. Soil Erosion Mitigation Guidelines for Logging.

Department of Land and Water Conservation. 1995. Draft Policy. Management of Forest on Protected Land in New South Wales.

Edwards, K. 1979. Rainfall in New South Wales — With Special Reference to Soil Conservation. Technical Handbook No. 3. Soil Conservation Service of New South Wales, Sydney.

Emery, K.A. 1986. Rural land capability mapping. Scale 1:100 000, Soil Conservation Service of New South Wales.

Environs. 1994. The North Coast Magazine. Logging in Private Forests. October/November:19.

FAO. 1976. A Framework for Land Evaluation. Soils Bulletin No. 32. Food and Agriculture Organization, Rome.

Franklin, J.F. 1993. Preserving biodiversity: species, ecosystems or landscapes? Ecol. Appl. 3(2):202–205.

Gibson, R.A. 1993. Environmental assessment design: lessons from the Canadian experience. *Environ. Profess.* 15:12–24.

Hannam, I.D. 1994. Implementation and Enforcement of the Protected Land Provisions of the New South Wales Soil Conservation Act 1938. *The Australian Centre for Environmental Law Conference,* University of Adelaide, Adelaide 7–8 May 1994.

Hannam, I.D. 1995a. Soil conservation in steep forested catchments in New South Wales. In G.S. Brierley and F. Nagel, Eds. *Geomorphology and River Health in New South Wales.* Proceedings of a Conference held at Macquarie University October 7, 1994. Graduate School of the Environment, Macquarie University, Working Paper 9501.

Hannam, I.D. 1995b. *Guidelines for Deciding When an EIS is Required in Relation to the Administration of Protected Land Under the Soil Conservation Act 1938.* Department of Conservation and Land Management.

Hannam, I.D. 1995c. Environmental law and private property forest management in New South Wales. In Proceedings of 2nd Public Interest: Defending the Environmental Law Conference. Australian Centre for Environmental Law, University of Adelaide, Adelaide Australia, 20–21 May 1995.

Hansard. 1972.

Lunney, D. 1991. *Conservation of Australia's Forest Fauna.* The Royal Zoological Society of New South Wales.

McKinnell, F.H., E.R. Hopkins, and J.E. Fox, Eds. 1991. *Forest Management in Australia.* Surrey Beatty and Sons, Chipping Norton, NSW.

McNeely, J. 1994. Lessons from the past: forests and biodiversity. *Biodivers. Conserv.* 3:3–20.

Murphy, B.W. 1984. A Scheme for the Field Assessment of Soil Erodibility for Water Erosion. Technical Paper 19/84. Wellington Research Centre, Soil Conservation Service of New South Wales, Sydney.

Noss, R.F. 1990. Indicators for monitoring biodiversity: a hierarchical approach. *Conserv. Biol.* 4:355–364.

Office of Technology Assessment. 1987. Technologies to Maintain Biological Diversity. U.S. Government Printing Office, Washington, D.C.

Rab, M.A. 1992. Impact of Timber Harvesting on Soil Disturbance and Compaction with Reference to Residual Log Harvesting in East Gippsland, Victoria — A Review. Department of Conservation and Environment, VSP Technical Report No 13.

Recher, H.F. 1992. Paradigm and paradox: sustainable forest management. In M. Rowland. Ed. *Sustainable Forest Management*. Proceedings of Seminar, University of Newcastle Australia. Sponsored by Institute of Foresters and Board of Environmental Studies.

Ross, J.D. 1992. An expert system for the administration of protected land in New South Wales. Pages 756–761 in *People Protecting Their Land — Proceedings of 7th International Soil Conservation Organisation Conference,* Sydney Australia, 27–30 September 1992.

Ross, J.D. 1993. An expert system for soil erosion mitigation in logging operations on steep land. *AI Appl.* 7(4):69–70.

Ross, J.D. 1994. Forestry Decision Making — Consistency, Objectivity and Accountability through Expert Systems. *Decision Support 2001— Proceedings of 17th Annual Geographic Information Seminar and Resource Technology '94 Symposium, Toronto Canada 12–16 September, 1994.*

Rowland, M. 1992. *Sustainable Forest Management*. Proceedings of Seminar, University of Newcastle Australia. Sponsored by Institute of Foresters and Board of Environmental Studies.

Smith, A.P. and G. van der Lee. 1992. Guidelines for Assessing the Significance of Logging and Clearing on Endangered Forest Fauna on Protected Land in New South Wales. NSW Department of Conservation and Land Management.

The Macquarie Dictionary. 1985. Revised Edition. Macquarie Library.

Tracey, J. 1991. Insecurity in the Logging Industry. *Proceedings of the Forest Industries Inaugural Logging Conference*. Launceston, 7–11 November 1991.

van de Graaff, R.H.M. 1988. Land evaluation. Pages 258–281 in R.H. Gunn, R.E. Beattie, R.E. Reid, and R.H.M. van de Graaff, Eds. *Australian Soil and Land Survey Handbook*. Inkata Press, Melbourne.

Watson, I. 1990. *Fighting Over the Forests*. Allen and Unwin, Sydney.

Willison, J.H.M. et al. Eds. 1992. *Science and the Management of Protected Areas*. Developments in Landscape Management and Urban Planning. Elsevier, Amsterdam.

Zavala, M.A. and J.A. Oria. 1995. Preserving bological diversity in managed forests: a meeting point for ecology and forestry. *Landsc. Urban Plan.* 31:363–378.

Chapter 56

Seasonal No-Tillage Ridge Cropping System: A Multiple Objective Tillage System for Hilly Land Management in South China

XIANWAN ZHANG, SHI CHEN, TONGYANG LI, AND MENGBO LI

The large area of hilly land in the southern regions of the Yangtze river is an important agricultural region in China. In this area, the land has been overexploited for many years due to heavy population with high demand for food supply. Soil erosion, soil fertility degeneration, and land productivity decrease are severe problems in this region. **Seasonal No-Tillage Ridge Cropping System (SNTRCS)**, a multiple objective tillage system for hilly land management, was developed to address these problems in the Sichuan Basin of south China. SNTRCS consists mainly of four components: (1) developing a 30- to 50-cm living ploughlayer to ameliorate subsoil (15 to 60 cm) structure and nutrition conditions by ridging soils, applying organic compost, and periodically alternating ridges with furrows; (2) building up a framework structure in fields for storing water and controlling soil erosion; (3) applying compound storied cropping patterns to increase total crop yields; (4) establishing agroforestry and water supporting systems to improve the environment and sustainability of resources. In 10-year experiments, SNTRCS successfully integrated the development and conservation of land resources into one tillage system and achieved an ecological balance between high production and long-term sustainable agriculture while reducing environmental impact. This system was implemented on 338,000 ha of the target areas.

Physical Characteristics and Constraints of Hilly Land in South China

The land in south of the Yangtze river is predominantly hilly terrain with a subtropical humid monsoon climate. The total land area is approximately 1.8 million km^2 which occupies 18.8% of total land area in China. More than half of the total population of the country lives in this area. Highly weathered red earth soils dominate the area and other major soils include the yellow earth soils, purple soils, and paddy soils. In this area, the annual mean temperature is 14.7 to 22.6°C, ≥10°C accumulated temperature 5,000 to 7,000°C, sunshine duration 1,500 to 2,000 hr/year, frostless season 250 to 300 days/year, and annual mean precipitation 1,000 to 1,400 mm (maximum 2,074 mm); 50 to 80% of the annual precipitation falls between May and September and a maximum episode ranges from 100 to 200 mm.

During the summer, the air current in the area is controlled by the subtropical high pressure of the Pacific Ocean which provides the area with plenty of moisture. With plentiful rainfall and warm temperatures in the same season, the south hilly land becomes an important agricultural region with a large population. In 1989, the area produced 55% of total food crops, 40% of total cotton, 60% of total vegetable oil, and over 90% of total rice in China (Zhong and Xie, 1989). However, soil erosion and a declining soil fertility have become severe barriers for food production in this region. According to the figures from the 1980s (Shi, 1983), soil erosion affected 11 provinces in south China covering 28% of the total land area in the region and 200,000 km^2 area had severe soil losses. The average erosion modulus of 81 counties in Sichuan province reached 4,886 mg/km^2; 20% of the farm lands had a soil depth less than 20 cm.

Seasonal No-Tillage Ridge Cropping System

In order to utilize the rich resources, overcome the physical constraints, and provide a solution for the problems described above in the south hilly land, we developed a multiple objective tillage system in the early 1980s. The system was designed to develop soil resources, reduce soil erosion, and increase the efficiency of utilizing water and soils. We hoped that the system would create a sustainable model for utilizing land resources, which would integrate the development and conservation of land resources and improve the environment of agriculture.

As a soil conservation tillage technique, no-tillage has been implemented in the U.S., for many years. Although the no-tillage technique successfully prevented soil losses from run-off and wind in an economic way, it might also concentrate soil nutrients from organic matter in top layers and prompt breeding of pests (Cai et al., 1993; Buhler, 1992). In the south hilly land, a majority of precipitation is concentrated in the summer causing soil erosion. When the first heavy rain falls in early summer, the most intensive soil losses always occur on sloped lands where the ploughed soils are scarcely covered by seedling crops (Figure 56.1). Modifying no-tillage techniques (Mc Isaac et al., 1986; Neibling et al., 1984; Pimental et al., 1989), we conducted the ridge-furrow tillage with remaining stubble and mulch plants, no-tillage only on ridges in summer. This approach effectively controlled soil losses in the south hilly land. Based on this practice, the SNTRCS

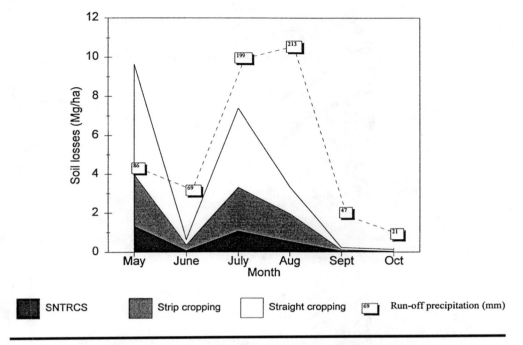

Figure 56.1 Soil losses under different cropping systems (1985 to 1989).

was developed (Zhang et al., 1990). It is conducted as follows: (1) apply about 15,000 kg/ha fabric organic compost to the fields, (2) ridge soils to 30 cm high and 100 cm wide with the same width furrows along the contour of the slope and plow furrows deeply to 20 cm, (3) leave stubble and mulching plants on the ridges with no-tillage in early summer and less-tillage in autumn, (4) dispose compound storied plants between the ridges and furrows, (5) alternate the ridges with furrows and apply fabric organic compost again in 3 to 5 years. In addition to the above procedures, SNTRCS also employs agricultural support systems with the supply of organic materials and irrigation to ensure its sustainability.

The original SNTRCS experiment was conducted at Yanting Purple Soil Agroecological Experiment Station, the Ecological Network of the Chinese Academy of Sciences. The experiment site is situated at N 30°58'–30°30' and E 105°12'–105°42' in the north central portion of Sichuan Basin. The experiment site was set on a typical purple orthent soil with a 4° sloped land. The strip tillage with local conventional cropping systems was used as a control. The extended experiments were conducted in four kinds of purple orthents on 10° of sloped lands. All experiments were conducted continuously for 10 years from 1985 to 1994 and the results are presented below.

A 30- to 50-cm Living Plough-layer To Enrich Cropping

A main barrier to agricultural production on sloped land in the south hilly area of China was the thin soil layers resulting from soil erosion. Under the local traditional cropping systems, the average living plough-layer has 15 cm in the area. Therefore, one of the major goals of SNTRCS was to increase the soil

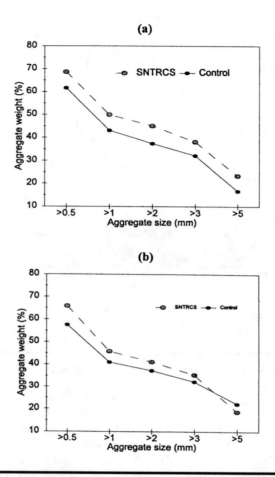

Figure 56.2 SNTRCS increased soil aggregation in 0 to 35 cm depth (1988); (a) water stable aggregate distribution in 0 to 15 cm, (b) water stable aggregate distribution in 15 to 35 cm.

thickness of living plough-layers. With the ridge cropping, 35-cm soil living-layers with well-flocculated organo-mineral complex structures were developed. After alternating furrows with ridges, up to 50-cm soil living-layers were developed on the ridges while the soil layers in furrows reached 20 cm. By several alternations of the ridges and furrows, the living plough-layers in an entire field reached 30 to 50 cm. The soils within living-plough layers have good physical, chemical, and biological properties.

Soil Structure Properties

SNTRCS significantly ameliorated soil structure, especially improving the soil structure below 15 cm. Compared to the controls, SNTRCS increased soil aggregation including water-stable structure aggregates in top soils (0 to 15 cm) and subsurface soils (15 to 35 cm) (Figure 56.2). Soil microaggregates analysis (Table 56.1) showed that >0.01 mm microaggregates in 0 to 60 cm soil layers

Table 56.1 SNTRCS Ameliorated Soil Aggregation

	Soil depth (cm)	The ratio (%) of microaggregates in different sizes (mm)							
		1.0–0.25 (mm)	0.25–0.05 (mm)	0.05–0.01 (mm)	0.01–0.005 (mm)	0.005–0.001 (mm)	<0.001 (mm)	<0.01 (mm)	>0.01 (mm)
SNTRCS	0–15	5.2	30.0	37.4	10.1	13.3	4.1	27.5	72.5
	15–35	7.2	31.4	35.5	9.3	12.3	4.3	25.9	74.1
	35–60	5.2	33.3	35.4	11.2	9.2	5.7	26.1	73.9
Control	0–15	4.7	29.8	37.7	11.6	12.9	3.4	27.8	72.2
	15–35	6.7	28.0	36.6	11.8	11.8	5.1	28.7	71.3
	35–60	6.2	28.1	36.8	11.4	13.2	4.4	28.9	71.1

Table 56.2 SNTRCS Increased Soil Organo-Inorganic Complex

	Soil depth (cm)	Organic C (%)	C heavy fraction (%)	Complex quantity (%)	Complex degree (%)
SNTRCS	0–15	0.65	0.60	0.58	88.9
	15–35	0.58	0.53	0.51	88.1
	35–60	0.44	0.42	0.41	92.5
Control	0–15	0.89	0.68	0.65	73.4
	15–35	0.43	0.38	0.36	84.5
	35–60	0.40	0.28	0.27	89.8

under SNTRCS were significantly increased, but the <0.01 mm microaggregates decreased correspondingly. It is possible that the small microaggregate are stacked into larger ones under SNTRCS.

The characters of organo-inorganic complex (Table 56.2) also indicated the SNTRCS increased carbon in the heavy fraction by 45%, complex quantity by 45%, complex degree by 3.6% within 15 to 60 cm soil layers compared to the controls. According to successive measurements of soil bulk density within ridges for 10 years (Figure 56.3), the soil bulk density within subsurface layers (15 to 35 cm) decreased and has remained less than 1.3 g/cm^3 for 10 years. This improvement was also found in the subsoils (35 to 50 cm) where bulk density dropped below 1.4 g/cm^3 after alternating ridges with furrows. The results showed that the soil could maintain good structure steadily under SNTRCS management in the south hilly land up to 5 to 10 years (Khakural et al., 1992).

Soil Porosity

An analysis of micromorphology on soils under SNTRCS management showed that the aeration porosity of 1 to 0.2 mm was increased from 7.11 to 21.2% in surface layers (0 to 15 cm) and 11.8 to 16.2% in subsurface layers (15 to 35 cm) compared to the controls.

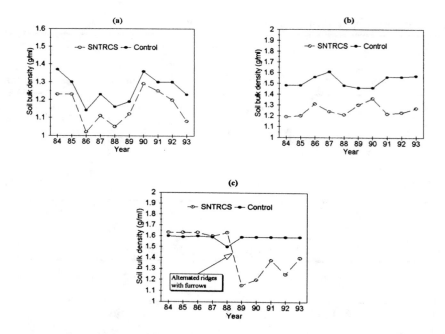

Figure 56.3 SNTRCS decreased soil bulk density in 0 to 50 cm depth (1984 to 1993); (a) 0 to 15 cm, (b) 15 to 35 cm, (c) 35 to 50 cm.

Soil Nutrients and Biological Activities

According to measurements of soil nitrogen, organic matter, soil enzyme activities, and microbes (Table 56.3), SNTRCS significantly increased soil N nutrition and biological activities in the subsurface layer (15 to 35 cm) and the subsoil layer (35 to 50 cm), but little differences in the top layers (0 to 15 cm). Total N, organic matter, total microbes, aerobic fiber-decomposing bacteria, urease activity, alkaline phosphatase, invertase were increased by 20 to 80% or more in the subsurface layer and subsoil layer. This improvement of the nutrition status in the subsoils is significant for crop growth. For example, our ^{15}N urea experiment (1993) suggested that the uptake rates of N by wheat were 11.1% when the fertilizer was applied in surface layer (2 to 3 cm), 12.5% when applied in sublayer (20 to 25 cm), 15.6% when applied in subsoil layer (40 to 45 cm), and corresponding ratio of wheat yields is 78.2:93.4:100.

Crop Root Development

Most farming fields under the management of traditional local cropping systems in the south hilly land have a surface plough layer less than 15 cm. In the control plots with these cropping systems, wheat roots were found within the top 11 cm (Figure 56.4). However, wheat roots were distributed within 20 cm depth under SNTRCS and spread to 60 cm depth after ridges were alternated with furrows. The results suggest that SNTRCS ameliorated the soil structure improved the nutrition state below surface layers deep to 50 cm, which significantly increased root development in subsurface soil layers.

Table 56.3 SNTRCS Improved Soil Nutrients, Microbes, and Enzyme Activity

	Soil depth (cm)	Organic matter (g/kg)	Total N (g/kg)	Total microbes (10,000/g)	Aerobic fiber decomposing bacteria (10,000/g)	Urease activity (mg NH2-N/ g.24hr)	Alkaline phosphatase (mg Phenol/ g.24hr)	Invertase (mg glucose/ g.24hr)
SNTRCS	0–15	10.2	0.82	28.9	0.38	0.77	1.56	17.95
	15–35	8.4	0.70	107.2	0.28	0.84	1.10	13.13
	35–60	6.0	0.65	15.7	0.07	0.70	0.75	10.23
Control	0–15	10.4	0.83	27.7	0.04	0.82	1.67	18.80
	15–35	6.0	0.58	17.0	0.22	0.53	0.89	7.75
	35–60	4.5	0.43	11.5	0.05	0.01	0.30	1.75

Note: Microbes data measured in 1993, enzyme activity data measured in 1989, and others measured average of 1985–1993.

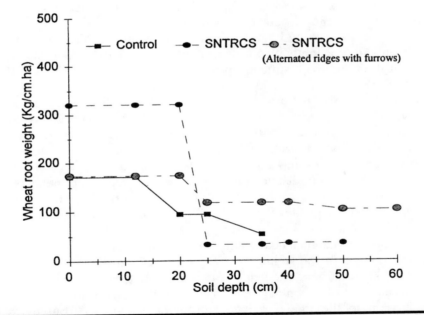

Figure 56.4 SNTRCS improved wheat root development (1989). The wheat root weights are average values in different depths: control l 0 to 12, 12 to 25, 25 to 36 cm; SNTRCS in 0 to 20, 20 to 35, 35 to 50 cm; SNTRCS (alternating ridges with furrows) in 0 to 20, 20 to 40, 40 to 60 cm.

Framework Structure in Fields Reducing Water and Soil Losses

As mentioned above, a major erosion risk in the south hilly land is intensive rainfall during summer causing severe soil losses on sloped lands. In order to reduce run-off, SNTRCS built up a framework-like structure in fields for storing water and controlling soil erosion in the following way.

Table 56.4 The Contributions of Selected SNTRCS Components
to Reducing Soil Losses (1985–1989)

SNTRCS components	Ridges	Furrows	Small-blocks	No-tillage	Mulching	Total
Retained soils, mg/ha (%)	5.8	9.3	1.7	1.8	2.1	20.7
	28.2	44.9	8.3	8.5	10.1	100
Retained run-off, m3/ha (%)	376.7		39.5	45.8	51.5	513.5
	73.4		7.7	8.9	10.0	100

1. Small soil blocks 10-cm high and 10-cm wide were piled at 5 to 7 m intervals in the furrows to form a framework structure in fields. This approach built up hundreds of small reservoirs in the furrows formed by ridges and soil blocks. This field structure not only effectively prevented soil losses during heavy rainfall, but also stored this water in the field. According to 5-year measurements (1985 to 1989) in the purple orthent soil field on a 4° sloped land, the average soil loss in the fields with framework structure was 5.22 mg/ha, while 19.16 mg/ha in the control fields. The annual average run-off in the fields with framework structure was 616.4 m^3/ha and 1,431 m^3/ha in the control fields. The other experiments in four kinds of purple orthents soils on 10° sloped lands showed that the framework structure reduced soil losses by 83.6% with the losses of 4.05 mg/ha on average. In addition, run-off was reduced by 37.5% with the run-off of 818.7 m^3/ha on average. The run-off retained by the framework infiltrated into the soil profile as "reservoirs" in the fields. According to 4-year (1986 to 1989) measurements (intervals every 2 weeks), the framework structure of SNTRCS increased soil water content 0.6% on average in ridges and 1.1% in furrows within 0 to 50 cm soil layers.
2. We estimated the relative contributions of selected components of SNTRCS to reducing soil erosion (Table 56.4). Although ridge-furrow-block system accounted for greater than 80% of reductions in soil losses, no-tillage in early summer also played an important role in reducing soil losses caused by the first summer heavy rain after shifting crops.
3. In addition to the framework structure built in the fields, back ditches, side ditches around the fields, water pits, and sediment collection pits were installed as protecting systems. These installations further reduced run-off erosion to the minimum and stored the sediments from fields. These sediments were returned back to the fields during slack farming season. Using ^{137}Cs technique to measure soil erosion under different cropping systems, Quine et al. (1992) found that these protective installations reduced soil erosion by 1 to 6 mg/ha/year while 10 mg/ha soils were lost per year in the fields without these installations.

Compound Storied Cropping in the Ridge-Furrow

The spatial differences between ridges and furrows of SNTRCS created various microecological environments. The sunlight competition among crops has abated between ridges and furrows. Various crops can be planted to utilize slack ecological

Table 56.5 Annual Total Production and Profits of Different Cropping Systems and Patterns (1988–1989)

	Patterns	Total yield (mg/ha)	Total profit (*Yuan/ha)	Net profit (Yuan/ha)
SNTRCS	Ridges: wheat-sweet potato Furrows: rape-corn	8.45	8,375	7,776
SNTRCS	Ridges: barley-peanut Furrows: green bean-early corn-autumn corn	9.27	11,048	10,514
SNTRCS	Ridges: barley-sweet potato-corn Furrows: green bean-early corn-soybean	12.80	12,828	12,245
Control	Wheat-early corn-sweet potato	8.76	8,007	7,587

* Yuan is China dollar.

niches in the surface and subsurface of the plots. The ridge, the heart of the fields, has a 30 to 50 cm fertile living soil layer that is well-drained and aerified with relative larger range of diurnal temperature. It is adapted to low water requirement and shorter stalk crops such as peanut, sweet potato, pepper. However, ridges are also suitable for corn and cotton that require higher soil fertility. The fertilized furrow with high moisture content is adapted to taller crops which need more water such as corn and rape vegetable. During early summer and late autumn, the slack ecological niches in the ridge-furrow can be planted with some crops such as green manure, sweet potato, and vegetables to utilize sunlight, heat, and water. The summer no-tillage saved labors to seed advanced and covered barer soil surfaces early. This cropping pattern can increase crop production as well as reduce soil erosion.

In the experiment comparing different cropping systems in 1988 to 1989, we found that production and profit increased significantly with increasing utilization of microecological environments in the ridge-furrow system (Table 56.5). For example, compared to the conventional intercropping system of wheat-corn-sweet potato, the SNTRCS cropping system of barley-sweet potato-corn on ridges and green broadbean-early corn-soybean in furrows increased total yield by 46.1%, total profit by 60.2%, and total net profit by 61.4% (Table 56.5). A long-term experiment comparing SNTRCS and local strip cropping systems was conducted in 44 blocks during 1984 to 1993. The results (Figure 56.5) showed that SNTRCS maintained good yields and increased production by 15.0% over the control on average. After alternating ridges with furrows, the yield reached 13.3 mg/ha, a 26.5% increase over the control. In the demonstration area of Wuan township, Shehong county of Sichuan province, farmers planted spring corn using plastic films to cover seedling plots in the furrows, and planted wheat, then early summer corn on ridges with intercropping sweet potato. The annual productions were recorded in 1991 as wheat (4.5 mg/ha), spring corn (4.4 mg/ha), early summer corn (4.5 mg/ha), sweet potato (7.4 mg/ha), and total (20.8 mg/ha).

Agricultural Resource Support Systems

In an agroecosystem, all individual components are mutually dependant. Tillage is a node in the network of the agroecosystem and requires support from other

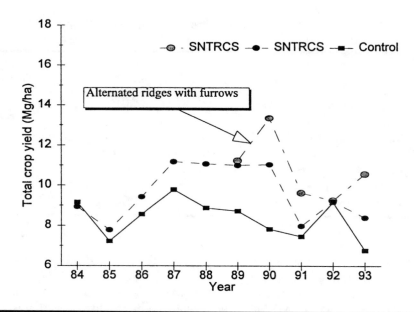

Figure 56.5 SNTRCS has increased crop total production for many years.

subsystems. Although SNTRCS has capabilities of fertilizing subsurface soil layers and storing rainfall as soil water supply, it is necessary to establish resource support systems such as organic compost resources and irrigation systems in the agroecosystem.

Alder and Cypress Agroforestry System

Agroforestry has been generally accepted as an important approach for sustainable agriculture since the 1970s. It not only provides various products to meet both ecological and economic needs, but also reduces soil losses from farmland with biologic hedges (Napoleon, 1987). Adapted to the subtropical climate and lime purple soils, an agroforestry system of farm land inlaid in the mixed alder and cypress was established to support SNTRCS. The mixed forest was located on the tops of low mountains and hills, and the upland fields lay on terraces and paddy soils in valleys. The ratio of forest to farmland was 2:1. First, compared to the control with no-forest reserve area, the agroforestry system decreased the annual mean temperature by 0.47°C and increased the relative humidity by 2.24%. Correspondingly, the annual absolute lowest temperature rose 0.95°C and the annual absolute highest temperature dropped 1.66°C. Second, the agroforestry system reduced soil losses in the entire system. According to our measurements, the annual run-off modulus was 3.49 l/s.km^2 in the forested area and 7.64 l/s.km^2 in the control area. The captured run-off gradually infiltrated into the soils to increase water storage in the agroforestry system. The annual soil erosion in the forest area was 25.3 to 182.4 mg/km^2 while the nonforest area was 179.6 to 1,034.3 mg/km^2. Third, the forest provided the farm fields with deciduous leaves at a rate about 5 mg/ha/year.

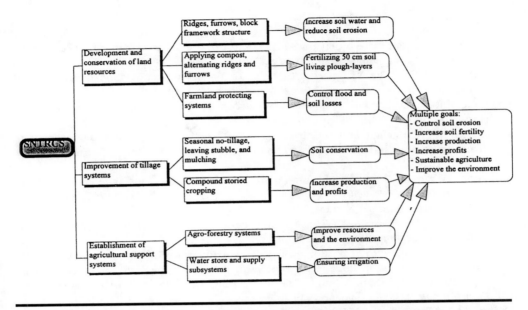

Figure 56.6 SNTRCS serves as a multiple objective tillage system for hilly land management in south China.

Water Storage and Supply Systems

The small agroecosystem that the SNTRCS experiment was set in has a total area of 17.95 km². In this area, a total of 594,000 m³ of surface water was stored by small dams, dikes, weirs, and winter waterlogged fields. It irrigated 47% of the catchment area. Besides surface water, there existed 260,000 m³ of groundwater that could be used as irrigation. These waters would satisfy approximate 73% of plant water requirement in the area. If considering the forestry hydrologic effect that could reduce 79,000 m³ water requirement, the constraint of water shortage in the area was greatly released (Zhuang et al., 1985). With the backing of water and organic fertilizer from these supporting systems, SNTRCS created and has been maintaining many small sustainable agroecosystems in the south hilly land since the 1980s.

Conclusion

SNTRCS has integrated increase of production, maintenance of soil fertility, preservation of soil water, reduction of soil erosion, and improvement of agricultural environment into one system. In the 338,000 ha of implemented areas of hilly land in Sichuan Basin, SNTRCS has increased food crop production by 2,723,000 mg, rape 10,000 mg, cotton 1,000 mg, watermelon 4,300 mg, and green manure 13,000 mg. The total net profit was increased by RMB (China dollar) 160 million yuan. This system is summarized as a multiple objective tillage system for hilly land management in south China (Figure 56.6).

References

Buhler, D.D. 1992. Population dynamics and control of annual weeds in corn (*Zea mays*) as influenced by tillage system. *Weed Sci.* 40:241–248.

Cai, Dian Xong, Xiao Bin Wang, and Xu Ke Gao. 1993. The examination of sustainable conservation tillage systems. *China Progr. Soil Sci.* 21(1):1–7.

Khakural, B.R., G.D. Lemme, and T.E. Schumacher. 1992. Effects of tillage system and landscape on soil. *Soil Tillage Res.* 25:43–52.

Mc Isaac, C.F., J.K. Mitchell, and M.C. Hirschi. 1986. The effect of contour and conservation tillage on runoff and soil loss from the Tama silt loam soil. *Am. Soc. Agric. Eng.* 86:2600–2617.

Napoleon, T. Vergara. 1987. Agroforestry: a sustainable land use for fragile ecosystems in the humid tropics. Pages 7–18 in H.L. Gholz, Ed. *Agroforestry: Realities, Possibilities, and Potential.* Martinus Nizhoff Publishers.

Neibling, W.H., O.R. Stein, and T.E. Logan. 1984. Soil loss from new and no-tillage ridges. *Am. Soc. Agric. Eng.* 84:2577–2557.

Pimental, D., T.W. Colliney, I.W. Buttler, D.J. Reinemann, and K.B. Beckman. 1989. Low-input sustainable agriculture using ecological management practices. Pages 3–24 in M.G. Paolett, B.R. Stinner, and G.G. Lovenzoni, Eds. *Agriculture Ecosystems and Environment.* Elsevier Science Publisher, Amsterdam.

Quine, T.A., D.E. Walling, Xin Bao Zhang, and Y. Wang. 1992. Investigation of soil erosion on terraced fields near Yanting, Sichuan province, China, using Caesium-137. Pages 155–168 in D.E. Walling, T.R. Davies, and B. Hasholt, Eds. *Erosion, Debris Flows and Environment in Mountain Regions.* IAHS Press.

Shi, D.M. 1983. Soil erosion and conservation in red earth region. Pages 237–253 in Q.K. Li, Ed. *The Red Earths in China.* China Science Press, Beijing.

Zhang, Xian Wan, Shi Chen, and Tong Yang Li. 1990. Treatises of studies on seasonal no-tillage ridge cropping system. *China Bull. Soil Agrochem.* 5(1&2).

Zhong, Gong Fu and Yue He Xie. 1989. The views on the development and utilization of south hilly land in China. Pages 3–7 in Natural Resource Society of China, Ed. *The Studies on the Integrative Development and Utilization of South Hilly Area of China.* China Science Press, Beijing.

Zhuang, Ti Ren, Z.G. Sheng, and W.Q. Jiang. 1985. Priminary studies on combined utilization of water from four different sources in middle Sichuan Basin. *China J. Water Soil Conserv.* 34:29–33.

Chapter 57

The Use of Multiobjective Decision Making for Resolution of Resource Use and Environmental Management Conflicts at a Catchment Scale

ROGER SHAW, JOHN DOHERTY, LINDSAY BREBBER, LEX COGLE, AND ROB LAIT

Abstract

The increasing role by community groups in catchment scale decision making on sustainable resource use and management requires much more comprehensive tools for managing ecosystems on a temporal and spatial scale. Integrated Catchment Management (ICM) is being strongly promoted in Australia as a strategic framework where individuals, groups, and government agencies with a vested interest in catchment outcomes can make group decisions on regional and strategic development and management strategies for sustainable resource use.

For these decisions to be effective on a catchment scale, there is a need for multiobjective decision-making tools to optimize as transparently as possible the diverse needs and expectations of stakeholders. In such a participatory decision-making environment, the challenge is to convey data and simulation model outcomes in a useful manner for evaluation of "what if" scenarios for resource use and management strategies. Economic and social factors together with simulation modelling are important components of multiobjective decision making.

The conceptual linkage of models and data into decision support systems (DSS) involves the integration of point scale simulation models to predict impacts on a catchment scale. This process involves space and time scale factors of several orders of magnitude with consequent magnification of errors. A time and space matrix of hydrologic processes is presented and criteria to assist in the choice of appropriate models to suit client group and stakeholder needs is discussed.

Simulation of complex processes is a compromise between complex multiparameter models that cannot be easily parameterized or calibrated, and simpler lumped parameter empirical models. Aspects of these approaches are discussed in relation to model prediction uncertainties and management of risk and the impact on catchment scale predictions. A brief synopsis of current DSS in use in Australia for natural resource management issues is outlined.

A discussion of appropriate decision support strategies for catchment scale issues is presented using a case study irrigation area in north Queensland. Water table levels under part of the irrigation area are currently rising at 0.5 m/year. Expansion of irrigation and increased infrastructure development is proposed. A range of viable water table control options have been proposed by the stakeholders. These need to be assessed within the context of economically and ecologically sustainable development where the cost benefits to all stakeholders can be quantitatively and openly assessed.

Introduction

The strong emphasis on environmental sustainability for agricultural, urban, and industrial issues has resulted in the need to evaluate the consequences of proposed developments and current practices. In Australia, the agreement by all States and the Commonwealth to the principles of Ecologically Sustainable Development (ESD) has put increasing pressures on agriculture. The proposed Natural Resource Management legislation for Queensland will incorporate acceptable minimum codes of practice for agriculture. Codes need to be developed that are soundly based and that adequately assess the expected impacts in the short and long term.

The multidisciplinary nature of ESD and the concept that management of individual properties can no longer be considered in isolation but is an integral part of a catchment scale system, means that decision making on the most appropriate management practices has become much more involved.

Natural resource management is a complex equilibrium. Ecological sustainability and agriculture comprise a fine balance between competing objectives. A change in one management option can change other parameters. For example, the shift to minimum till and stubble retention in cropping systems has dramatically reduced runoff and erosion from agricultural fields. Reduction in runoff has resulted in increased drainage below the root zone with consequent rising groundwater levels in sensitive landscapes, movement of agrochemicals to groundwater, and the mobilization of salts in the unsaturated zone due to increased water movement through the soil profile.

Since catchment scale issues involve all the catchment stakeholders comprising the community, producers, industries, urban community, and government agencies, a decision system whereby resource management issues can be openly examined and evaluated is required. In particular, the downstream or the catchment-wide consequences can become an important consideration of how site management should be undertaken to meet a range of sometimes competing objectives.

This paper outlines the background factors determining the style of multiobjective decision support systems (MODSS) in Australia to resolve resource use and management issues. The choice of an appropriate level of technical input into the decision-making process is identified and some current decision support approaches being used in Australia are reviewed briefly. The Cattle Creek catchment of the Mareeba-Dimbulah Irrigation Area of North Queensland is used as a case study of the multiobjective decisions that are required for sustainable management in this catchment.

Background

Several government and community initiatives that have been taken in Australia over the last 10 years have been significant in determining the direction for development of DSS for natural resource management.

Community Involvement in Decision Making

In the 1980s, the conservation movement expressed considerable concern about the degradation of the environment by agriculture, urban developments, and various land and water development projects. Farmer and business groups opposed these claims because the implications would have a major impact on their productivity. Then in 1988, the peak farmer's group in Australia, the National Farmer's Federation, and the peak conservation group, Australian Conservation Foundation, together with the Federal Minister for Primary Industries agreed to work together on a concept called Landcare. The Landcare movement began in Australia with government support in 1989 (McDonald and Hundloe, 1993). The 1990s have been called the 'Decade of Landcare' (Commonwealth of Australia, 1991) Landcare has been quoted by several authors as being "one of the most significant developments in land conservation in Australia" (Curtis et al., 1994).

Landcare is a voluntary movement between landholders, government agencies, conservation groups, businesses, and others who meet together in local areas to develop sustainable and productive practices for land management. In just 6 years, there have been over 2,200 Landcare groups formed in rural Australia which reach over 30% of farmers. One of the national goals of Landcare is: "All Australians working together in partnership for sustainable land use." People are coming together to discuss degradation issues of common concern and to develop strategies to reverse the degradation. Responsibilities are spelled out for Commonwealth and State governments, and landcare groups as a partnership approach (Commonwealth of Australia, 1991).

This cooperative focus has been expanded to land, water, and vegetation management through the government-sponsored movement called ICM. The ICM movement comprises local stakeholder groups forming committees to develop strategies for implementation of sustainable catchment management practices. It is an outcome of government policies which strongly promote regionalization and participatory democracy in local decision making. One of the principles of ICM is: "In a democratic society, sound land and water management is best achieved through the informed action of individual users and managers of these resources" (Queensland Government, 1991).

Communities of Common Concern have been established in the Murray Darling Basin as a strategy to overcome severe resource degradation and to develop sustainable resource use. The Murray-Darling Basin, which represents one-seventh of the area of Australia and produces about one-third of the economic output of rural industries, faces serious threats to sustainability from irrigation, nutrients, and salinity. One of the group responsibilities is "to prepare action plans for their localities consistent with regional objectives." One of the government responsibilities is "to provide technical, financial, social, policy and legislative support" (Murray-Darling Basin Ministerial Council, 1990).

These strategies reflect the principle that complex intractable problems are best researched together with people, rather than by researchers for people, or, on the behavior and changes required of the people that need to implement the recommendations of planners and resource management policy makers (Bawden and Packham, 1991). This approach requires an openness and ability to communicate complex issues in readily understandable forms.

Thus with this framework of strong community and stakeholder participation in natural resource management, and the range of different objectives of the various stakeholders, there is a major requirement for sound technical data and MODSS to allow appropriate and sustainable resource management options to be selected. These community and stakeholder groups require the presentation of technical and socioeconomic issues in a readily understood manner to facilitate making decisions on complex issues.

Ecologically Sustainable Development

The Australian Government and all state governments are signatories to the concept of ESD (Commonwealth of Australia, 1992). This requires strategies in place by all Australian governments to address the recommended policies. Proposals under the ESD agreement need to demonstrate that they have a net beneficial effect economically and environmentally. This can only be done using technically validated examples and simulation models of the effects under different resource management options with adequately quantified socioeconomic inputs and nonmarket valuations of natural resources.

Additionally, Environment Protection Legislation requires proposed developments to submit an Environmental Management Plan. The plan is an agreement between the proponent, the responsible government agency, and the community. The plan comprises a statement of a policy on environmental sustainability for the various elements of importance, performance criteria on which compliance with that policy can be assessed, monitoring procedures from which performance can be assessed, and reporting and corrective actions that will be implemented should the specified criteria not be met.

A more technically sound decision process is required to assess the likely effectiveness of these performance management options, their acceptability to the regulatory authorities, and the corrective actions that may be required as the number and complexity of interdisciplinary issues increase. For this aspect to be achieved, simulation models and objective decision making is needed.

The formulation of new Natural Resource Management legislation in Queensland will integrate a range of existing acts into a new policy and legislation on the sustainable use and management of natural resources (Department of Primary Industries, 1994). Two of nine principles in the formulation of this legislation are

1. The holders and users of natural resources are primarily responsible for the resources held under their care.
2. Natural resource management is best achieved through effective processes of cooperation and coordination within government, between government and the community, and within the community.

For an increased number of stakeholders with different and sometimes competing objectives, it can be extremely difficult to reach a near consensus view on the most appropriate strategy for management of the resources. Also in decision making at this level, the technical understanding by stakeholders is very diverse and whether soundly based decisions can be made without technical input and a formal and objective decision process is open to question (Foran and Stafford Smith, 1990).

In summary, a major requirement for sustainable natural resource management on a catchment scale is sound technical and decision-making support for community-based groups in a participatory decision-making context. A similar approach is required for project-based developments to ensure that proposed actions will be sustainable. These issues are setting the background for the development of MODSS in Australia.

Concept

Consideration of processes at a catchment scale, with an increased number of stakeholders involved, and more complex natural processes to be assessed, means decision making has jumped an order of magnitude in complexity.

DSS can be defined (after Thompson et al., 1992) as the integration of expert knowledge, management models, and timely information to assist in making day to day operational and long-range strategic decisions. Key concepts are the ability to evaluate "what if" questions and to predict the effects of decisions. An integrated approach to DSS is illustrated in Figure 57.1 where data, geographical information systems (GIS), models, information, and DSS are linked on a multidisciplinary basis.

DSS have most frequently been used to assist in making decisions on complex environmental problems, poorly formed problems, or where the choices are numerous. Janssen (1992) considers DSS are necessary where the complexity of the system, the time scale of the processes, and the diversity of the effects are beyond the imagination of most people. DSS can assist in evaluating the untried options for situations where traditional practices are more likely to be adopted than innovative practices or new managements with greater environmental sustainability, because of a lack of awareness of alternatives,

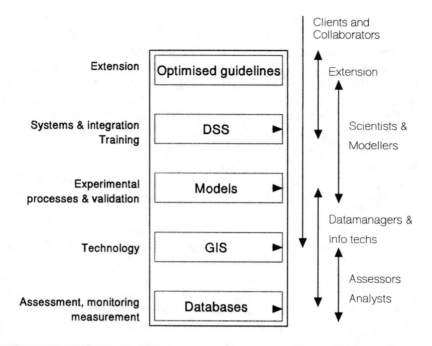

Figure 57.1 A schematic framework linking data through GIS and models to a DSS illustrating the involvement of different stakeholders and different disciplines.

Expert systems alone are less useful where an economic analysis is part of the decision making (Berry et al., 1992). Economic analyses require quantitative results from data or simulation model outputs. Van Diepen et al. (1991) discuss a variation of the linear programming model and decision theory, called multiple-goal programming, that has been applied in the case of conflicting objectives in regional land use planning. This approach is broadly similar to that discussed later of Lane et al. (1994).

Some principles of multiobjective decision making that need to be adopted in the context of community groups are as follows:

- Decisions need to be made by the stakeholders and not the DSS developers or the technical people.
- A transparent and rigorous system with decision-making variables tied to scientifically sound system responses. This means that the consequences of actions are based on measured or predicted responses so that effective and balanced decisions can be made.
- Catchment or ecosystem scale is required.
- Some changes in catchment response to resource management are very slow (for example, salinity), so that good predictions of small changes over long time periods are needed (not a simple task).
- A multidisciplinary approach is essential so that consequences across a very wide range of issues can be considered.

- Risks need to be quantified as far as possible so that the impacts of a proposed management change can realistically be evaluated. One aspect illustrating this point is the widespread promotion, a few years ago, of tree planting to combat salinity problems. The costs of tree planting were relatively high compared to the potential for reclamation or control of the salinity problem. There are many good reasons for planting trees but salinity control is not the best reason. This was confirmed by Doherty and Stallman (1992) who used groundwater modelling and simple salt balance recharge assessment to evaluate the impact and financial gain of a range of dryland salinity management strategies including planting of trees.
- Models are needed to adequately integrate climate variability (particularly rainfall) and spatial variability. Models are only tools and while the literature is full of detailed specific models which assist in understanding, the outcome needs to be assessed in conjunction with other variables and criteria within a DSS. All options need to be considered by stakeholders.
- DSS with different stakeholders need to have transparent and open systems where the stakeholders can follow the logic and be able to challenge the assumptions and measurements as well as evaluate the sensitivity of the decisions to the potential risk in either the model predictions or the climate and/or economics of the proposals. This is particularly important where stakeholders may have conflicting objectives.
- The presentation method needs to be as simple as possible with the complexity and functionality present but not intrusive into the decision process.

The Resource Assessment Commission (1992) evaluated the use of multicriteria analysis for natural resource management in Australia. They considered that multicriteria analysis provided a structured yet flexible approach and was particularly useful for complex natural resource management issues. It was considered an additional advantage where an understanding of the divergent social values will assist in arriving at a sound decision. A weighting scheme for the comparison of not directly comparable values was considered a difficulty with the approach. In reality, this difficulty is also a strength where different criteria can be compared.

There is an increasing trend towards defining an appropriate valuation of environmental resources (Commonwealth of Australia, 1995). Future strategies will require the full market costs of the use, maintenance, and replacement of environmental resources. Such a move would better reflect the costs of sustainable development.

Optimizing the Form of a MODSS

There are several dilemmas to be resolved in the technical and biophysical aspects of a MODSS design as a result of the background given above. They can be identified as (1) space-time scale integration, (2) model complexity-outcome accuracy, and (3) awareness — technical detail-group decision making.

Space-Time Scale Integration

In some scientific disciplines there is great difficulty in moving from point scale understanding (coming from a reductionist background) to a catchment scale understanding of processes. Spatial variability in the parameters needed to drive hydrologic models is enormous and the ability to scale up to catchments is in doubt for some parameters (Shaw, 1994; Hatton et al., 1994; Shaw et al., 1995). Catchment scale integrated measurements, e.g., streamflow and mass loads of nutrients exported, are achievable but very difficult to disaggregate to specific regions within catchments or to the processes operating.

The emergence and increasing availability of larger and cheaper computing capability, databases, and GIS has promoted a systems approach to natural resource management issues. A systems approach endeavors to encompass many of the processes determined by reductionist science into spatial systems models. While a systems approach to some parts of agriculture has been around for some time (Squires, 1991), catchment scale implications are relatively new for agriculture. Catchment scale considerations are better developed for groundwater hydrology and ecosystem sciences where larger spatial entities are necessary to allow the modelling to be carried out.

GIS are unsuitable as a shell for hydrologic models as pixels and polygons need to interact with each other. Dynamic and routing models can be developed to interact with GIS but have to write new data to each polygon during a separate modelling phase (Shaw, 1994). Johnson (1995) considered GIS needs a stronger analysis and modelling capability if it is to be used in a general way for spatial data. The important distinction is between spatial description of a multitude of layers and process understanding on a spatial scale. The latter is important as well in catchment management DSS.

Bouma (1986) identified the difficulties of applying different levels of simulation model complexity in determining soil water availability on a spatial scale. He identified that one end of a spectrum represented local experience which had wide spatial applicability but limited degree of detail while at the other end, deterministic models had a large degree of detail but very limited spatial applicability. Expert judgment and simple models were in between the above options. Thus a compromise is required between spatial applicability and the degree of detail in the models required. Simple models would appear to be a useful compromise.

Often in evaluating sustainable agriculture and environmental risks, the consequences are required over long time periods. There is an enormous scale and time problem when using deterministic classical soil physical approaches to water movement based on the concepts of short time and point measurement application of the Richard's equation. This is illustrated in Figure 57.2 from Shaw et al. (1995) where the time and space changes from point scale and hourly-based to century-based soil physics and catchment scale processes is in the order of millions of scale factors. This can be likened to observation and modelling of wave action at the beach by either being in the wave or observing the waves from an aircraft several kilometers above the waves. It is the latter approach that is required in DSS in the first instance to give big picture, predictive outcomes of the consequences of developments on environmental issues. For high risk situations, currently developed point and deterministic mechanistic models can be used for additional quantification where required.

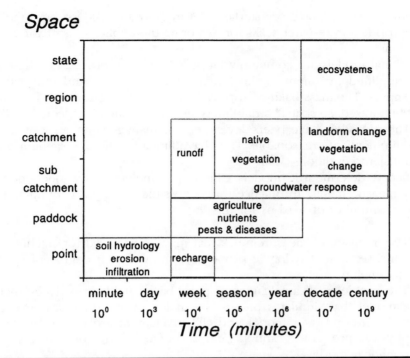

Figure 57.2 Time and space relationships for various processes which impact on agriculture and the sustainability of natural resources (from Shaw et al., 1995).

Hatton et al. (1994) were critical of building complex hydrological models for landscapes because of the degree of parameterization required and the enormous difficulties and uncertainties in spatial scaling which is probably a nonlinear function with distance and landscape features.

Model Complexity-Outcome Accuracy

The outcome required of models for DSS are predictions of the consequences of various management options under the variability of climatic and management practices. The model has to be valid in the field. Conceptually, it must be reasonably well understood by all stakeholders rather than be perceived as a black box, and it must have a practical use in environmental and resource use planning activities.

Addiscott and Wagenet (1985) reviewed the different approaches to modelling, with particular emphasis on water movement and solute leaching. They classified the main types of models as deterministic and stochastic. Their concepts as well as the applicability of the modelling approaches in a DSS context on a catchment scale are discussed below.

Deterministic models, where each process is considered to give a uniquely defined outcome, can be considered as mechanistic or functional. Mechanistic models are based on rate processes. These models incorporate the most fundamental of known processes. Restrictive boundary conditions and high data requirements have limited their use to research applications. Generally they have been used with results of laboratory studies. Bevan (1989) considers these models are most appropriate in research tasks for exploring the implications of various model assumptions.

Functional models are commonly based on capacity processes. These models give a simplified treatment of the fundamental processes and thus have more modest inputs. The mass balance approaches fit into this category. They are mass- and not flux-based, so spatial variability is less of an issue (Addiscott and Wagenet, 1985). These models have been developed for management purposes since the data requirements are reasonable. They have limitations in determining processes under transient conditions.

In theory, deterministic models can be parameterized by independent field property measurements. In practice this is possible to only a very limited extent. There are a number of reasons for this.

- The intensity of measurements that may be required for catchment properties (e.g., soil hydraulic properties) show a high degree of spatial heterogeneity.
- The problems of amalgamating point-based field property estimates into a model that simulates process on at least a paddock scale, or over an entire groundwater model grid cell or element are very large. "Average" physical properties cannot yield "average" model outcomes for highly nonlinear models.
- Other difficulties arise from the scale-dependence of certain parameters, the cost of data-gathering, imperfect descriptions of physical and chemical processes, etc.

Stochastic models are useful where the outcomes are uncertain and the prediction of the statistical limits of the system response are required. They are less useful in DSS unless a prior sampling and analysis has been undertaken.

Complex Models with Numerous Parameters

As attempts are made to conjunctively simulate more interrelated and multidisciplinary environmental processes at the catchment scale, large, complex, physically based models are being developed. This new generation of simulation models is replacing older types of models which, while based on gross conceptual simplifications of reality, or empirical relationships without a strong physical basis, had the powerful advantage that they were easily applied on a catchment scale and could be calibrated against catchment-scale data. The newer generation of models which simulate environmental processes in greater detail allow a user to test major changes in catchment management strategies by altering the relevant, identifiable, model inputs.

However, the use of large, physically based models in a DSS framework is fraught with danger. This danger is often not apparent to those who apply such models to particular catchments, and is rarely, if ever, apparent to nonspecialist catchment stakeholders. Invariably, even physically based models need to be calibrated against catchment-scale data. However, the calibration process is much more difficult than that of the simpler models which they replace. The large number of model parameters results in extremely high parameter correlation levels and, consequently, a high degree of uncertainty in estimates for these parameters achieved through the calibration process. These uncertainties may or may not diminish model predictive accuracy. In any case, it is fair to say without any exaggeration that the issue is not sufficiently recognized by most modellers.

Many of these models are not able to be calibrated, or have a time and space framework inconsistent with the available data or the space and time frame of the results required. Konikow and Bredenhoeft (1992) reviewed the difficulties of general model validation and concluded that models cannot be validated, they can only be invalidated. Most models are calibrated manually; computer-based nonlinear estimation methods are ignored because the methods are not well understood, and suitable model-independent parameter estimation software has not been available until recently. The recent advent of suitable software, e.g., PEST (Watermark Computing, 1994), increases the awareness of the nonuniqueness and correlation problems of model calibration. With manual calibration methods, no indication of parameter uncertainty levels (a quantity forthcoming from computer-based nonlinear estimation methods) and hence of model predictive uncertainty can be made. The "sensitivity analysis" undertaken as an adjunct to many manual calibration exercises ignores the possibility that an apparently sensitive (and hence "well determined") parameter may be strongly correlated to another parameter such that joint variation of both parameters produces no net change to model outcomes.

So the question must be asked as to whether the current range of physically based models forms a suitable basis for decision support. Their educative role cannot be questioned. But where they must be calibrated before they can simulate processes within a specific catchment, in spite of the fact that they are not designed to be uniquely calibratable, their use as a decision-making tool must be questioned. Perhaps, with the expanding use of DSS, a new suite of models should be built. The design criterion that models replicate reality as closely as possible should be rejected in favor of the notion that they replicate reality only to a degree that is consistent with an ability to be able to be calibrated at a catchment scale. Few such models exist at present.

Physically Based Empirical Models

Physically based empirical models have a number of advantages in the context of catchment scale DSS. There are arguments presented in the literature supporting the use of more empirical, soil property based, physically sound models for landscape hydrological processes. For example, Van Genuchten and Leij (1992) evaluated the possible use of soil particle size data to estimate soil hydraulic properties and behavior. They concluded with some recommendations that included directions for future research and development. One recommendations was

> The applicability of the Richard's equation for water flow in soils in situations with macropores is uncertain. The current relationships between the water retention and hydraulic conductivity near saturation are too complicated for the simple mathematical functions in current use. More sophisticated mathematical modelling is necessary.

Bouma (1989) evaluated the dilemma of requiring descriptive soil survey information to be useful for quantitative evaluation of land and land use impacts. He considered that current models cannot use the data usually collected. Bouma criticized the solution to this problem by modellers which involved the incorporation of matching factors without scientific basis. He considered the use of pedotransfer functions as viable approaches. Pedotransfer functions are relationships between variables of interest and either continuous functions such as clay content, or discrete functions such as soil type. Continuous functions with properties such as clay content are possible surrogates.

Because of the complexities of quantitative data and models, some authors (e.g., Moffatt (1990)) suggest a greater reliance on causal mechanisms and processes than on quantitative relationships for complex environmental problems.

Models need to be balanced with expert judgment. As Yaalon (1994) discussed, most processes in the soil-water movement area have been developed and understood in the laboratory rather than the landscape dimension. An interpretation and understanding of the processes must accompany or proceed modelling. This is particularly so where ecosystems are considered as all the components cannot be described, let alone modelled, so that in depth precise modelling of one part of the system may not be the most appropriate use of resources in the planning and evaluation phase.

Novotny and Chesters (1981) graphically related the potential errors with a change of scale for a range of water quality models. Many of the criticisms of scale issues discussed earlier in this section have been based on hydrologic models, which are orders of magnitude better than the predictive capability for other water quality parameters such as sediment concentration, phosphorus, nitrogen, and organic chemical concentrations. Thus the appropriate choice of measurement variable is an important part of the process of choosing relevant simulation models and outcomes for DSS on a catchment scale for use in group decision making.

Ongoing monitoring, using appropriate environmental sustainability indicators, can track the outcome of a decision where considerable uncertainty exists which should be backed up by a reevaluation of the predictions made. A toolkit comprising a range of available models of varying complexity is required in decision making, depending on the importance of the consequences of the decision.

Awareness — Technical Detail — Group Decision Making

The involvement of a broad range of stakeholders in decision making on natural resource management issues requires at least a minimum level of awareness of the issues for informed debate. As awareness increases, a greater depth of understanding of the implications of various management options is required. Time, and openness by all stakeholders, is necessary for this to occur. There is commonly a lot of general information available but the usefulness of this information is often inadequate except for rudimentary decision making. Whittaker (1993) described the relationship between the quantity of data and the relative usefulness and value of data. Figure 57.3 is drawn from the concepts of Whittaker (1993). The figure illustrates the basic scientific method in seeking the interpretation of a large number of observations. Information is only of use if it is organized and relevant to the issues at hand.

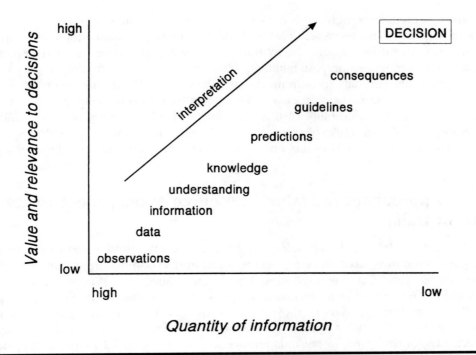

Figure 57.3 Relationship between information quantity and information value in relation to decision making. Figure has been derived from Whittaker (1993).

Multistakeholder decision making imposes other significant complexities. Group decision making in the presence of multiple conflicting objectives is complex and difficult (Lewis and Butler, 1993) as optimization of individual judgments and preferences may not align leaving consensus, negotiation, and compromise methods as an important decision process. In this case, Lewis and Butler cite Beck and Lin (1983) who proposed two techniques for group decision making: (1) to maximize agreement and (2) to minimize regret about a proposed decision as an additional means of reaching a decision. Minimizing regret procedures were particularly appropriate for large-scale ranking problems.

Group decision-making processes identify a multitude of options that may or may not be appropriate for further consideration. This is because the frame of reference or the scope and understanding of the systems approach on a spatial scale may be limited. Two methodologies are appropriate:

1. Consensus on the best-bet options. The difficulty in this approach is that hard decisions are very difficult to make as the group will lean towards the easy and often trivial options that don't upset the current practices. For example, the planting of trees to control shallow water tables is a very commonly promoted option. Alternatively, the group will tend to satisfy the most vocally or articulately presented case.
2. Technically sound predictions of consequences. Where a technically sound prediction of the consequences of certain management options can be made, the decision is more likely to be based on the future outcomes and consequences than on the immediate impacts.

Thus a scaled approach in complexity is required. Where the issues can be identified in yes/no terms, simple 'back of the envelope' calculations and expert opinion may suffice. Where the interactions between processes or managements are more complex, more quantitative, broader scale, and complex models with associated greater data requirements are necessary. There has been a major emphasis on deterministic process modelling of more recent times which has led to an exaggerated credibility of models compared to more empirical or intuitive techniques. Johnson (1995) expressed it as "intuitive techniques for exploring data in a spatial context have often been ignored in favour of elegant formal modelling."

DSS Approaches to Natural Resource Management Issues in Australia

Published guidelines, handbooks, expert systems, DSS, and spreadsheet based 'what if' scenarios have all been used to assist in decision making for many years. For simple issues, a one-page guideline is quite adequate. Where the decision maker has to optimize on essentially single issues with limited interactions, simple DSS are appropriate. Where the issues are difficult to conceive, expert systems models and DSS come into play. Where there are a number of stakeholders whose decisions for their own interest also impact on other stakeholders, the appropriate decisions are increasingly more difficult to make and MODSS are required.

For scenarios where the interaction of rainfall, cropping practices, and nutrition interrelate, resulting in crop yields that may be economically marginal, modelling approaches are needed to integrate the variables. A classic case of this approach is water balance crop yield models, e.g., PERFECT (Littleboy et al., 1989).

Cropping systems modelling has been a very common form of decision-making tool for paddock scale productivity. Carberry et al. (1991) have linked cropping systems models to economics for the semi-arid tropics. The advantage of this modelling approach is that climatic variability over time can be used to develop probabilities of economic returns. Also climate can overshadow many management practices and is responsible for the high variability in runoff and hence erosion which are issues for catchment scale land sustainability.

There are no MODSS approaches commonly available for natural resource management issues in Australia. A wide range of agricultural productivity DSS are available which have not been discussed here. The following are DSS that provide aspects of natural resource management issues. These are, briefly:

- PRIME — **P**lanning; **R**esearch; **I**mplementation; **M**onitoring and **E**valuation is a procedure for developing ICM Plans by stakeholder groups developed by Syme et al. (1994). It is a staged decision framework. In the *planning phase*, the problem is defined, available knowledge is collated, priorities formulated, objectives negotiated, gaps identified, the basic plan devised, and resources required and criteria for evaluation and monitoring determined. In the *research phase*, feasible solutions from the literature and elsewhere are identified, barriers to adoption identified and collaborative applied research programs developed. In the *implementation phase*, the implementation strategy is derived and resources determined and allocated for priority activities. Similar activities are carried out for the monitoring and evaluation phase. The process is followed by planning review phase where coordination is a priority issue. This process is currently being used in a series of large catchments in Australia where integrated solutions to catchment scale dryland and water salinity problems are being sought. PRIME is a decision support process rather than a software-based DSS.

- ALES — **A**utomated **L**and **E**valuation **S**ystem developed by Rossiter (1990) which is an expert system framework for land evaluation. Johnson et al. (1994) integrated economics and biophysical data with ALES. The process uses diagnostic attributes and decision trees to relate land characteristics to land use requirements based on expert knowledge.
- LUPIS — **L**and **U**se **P**lanning and **I**nformation **S**ystem has been developed to solve land allocation problems, in particular where more than one land use is proposed within a planning area. It takes into account the site-specific land attributes of the area (Ive et al., 1989; Ive, 1992). It uses decision tree and/or attribute rating and weighting methods to solve site selection for land use. Alternative uses for a site can only be selected if the consequences of a land use can be maintained within an acceptable agreed range with respect to preset criteria of acceptable change.
- ASSESS — **A** **S**ystem for **SE**lecting **S**uitable **S**ites was developed as a rating procedure to select radioactive sites in Australia (Berry, 1994). It comprises a series of GIS layer coverages, called themes, and a flexible weighting and optimization procedure to determine the desired outcomes against a set of defined criteria. Development is continuing.
- AEAM — The original **A**daptive **E**nvironmental **A**ssessment and **M**anagement concepts from Canada as reported by Grayson and Doolan (1995) have been used as a catchment scale in some areas of Australia. The AEAM process is aimed at providing links between communities with a problem and the technical resources available. Major benefits have been identified as the creation of a common understanding and ownership among stakeholders, as well as the development of the computer simulation model using the best available information (Grayson and Doolan, 1995). Disadvantages are the need for skilled modellers in structured workshops, and validation of the models is not generally done except as a qualitative comparison with informed technical opinion. A more realistic belief in the limitations of models follow from the process (Grayson and Doolan, 1995). The ability to interactively progress the planning phase overcomes some of the community skepticism and impatience if long time periods are spent in planning (Syme et al., 1994).
- CMSS — **C**atchment **M**anagement **S**upport **S**ystem is a catchment scale model for estimating nutrient loads in runoff, particularly of phosphorus and nitrogen. It is based on data for nutrient generation rates and areas to calculate loads (Davis et al., 1991). The DSS can consider policy issues for the determination of what level of development with consequent nutrient loadings should occur within catchments. There is an associated compendium of nutrient data on generation rates from across Australia that can be used to determine the suitability of areas for particular developments.

All of these DSS programs with the exception of PRIME and ASSESS run on PC computers. ASSESS is being further developed to run on a PC. There are major advantages of PC-based systems where the very strong community involvement in catchment based decisions on natural resource management systems are required.

Decision Optimization Process

Evaluation of options on the basis of "what if" scenarios is required with stakeholder negotiations leading to accepted decisions. Algorithms and processes for decision optimization are presented in the literature (for example, Bana e Costa, 1990; Janssen, 1992).

Lane et al. (1994), in a collation of several papers from his group, developed a prototype MODSS for the U.S. water quality initiative. Their system considered some innovative approaches to the ranking of alternative management options and assessment of the important multiobjective decision variables chosen by the stakeholders (Yakowitz et al., 1992, 1993).

Two of these approaches are particularly applicable to the Australian scene:

1. A normalized scoring approach which allows the nonrelated variables in a multiobjective system to be compared. A dimensionless scoring of 0 to 1 with conventional practice at 0.5 is done. Each decision variable is parameterized by using data, expert knowledge, or simulation models and a preselected response function (Yakowitz et al., 1992)
2. A priority order of the decision variables is chosen to reduce the problems of how to weight the different decision variables with each other and the difficulty of deciding the actual weighting. The software then computes the best, worst, and average scores for that variable by maximizing and minimizing the decision functions. This is done within the constraints of the chosen priority order of the variables together with the slope of the scoring function about the point 0.5 which represents the conventional treatment. Different priority orders for the decision variables can then be selected and compared from which promising options can be evaluated further (Yakowitz et al., 1993). This is a more simplified approach suitable to environmental management issues.

Situation Statement for Cattle Creek Catchment

The Tinaroo Dam was constructed in 1959 to supply water to the developing Mareeba-Dimbulah Irrigation Area, North Queensland. Tobacco was a significant crop on the lighter soils. In the Cattle Creek subcatchment of some 18,000 ha, tobacco was grown on the lighter soils. Rice expanded onto the heavier clay soils with irrigation. With the decline of the tobacco market and the impending deregulation of the tobacco industry, other horticultural crops such as mangoes and coffee are being grown in the upland areas. With the collapse of the rice industry in North Queensland in 1990, sugar cane has become a significant crop in the area being some 150,000 tonnes in 1993 and projected to be 300,000 tonnes in 1994. There is the potential for an expansion of irrigated sugar in the catchment where some 2,500 ha of potentially irrigable uncleared land is available.

Groundwater monitoring in the catchment since the mid-1980s has indicated a significant rise in groundwater levels of around 0.5 m/year with water levels under cane in some areas currently about 5 m below ground (Lait, 1992). The salinity (electrical conductivity) of the groundwater is 28 dS/m with very high sodicity in the region where water levels are rising fastest. Other parts of the catchment are showing hydrologic stress with seepage areas, permanent streamflow, and the development of saline discharge areas. If the hydrologic imbalance is not reduced and stabilized, or reversed, irrigated agriculture will be severely restrained. The catchment falls within the Mitchell River catchment, a priority ICM catchment in Queensland. A local Cattle Creek Groundwater Management Group comprising landholder, industry, and government has been formed to address the concerns for the future of the catchment.

The varied stakeholders in the catchment include the irrigators on different parts of the catchment, industry groups, Department of Primary Industries Water Resources supplying the water, other government agencies, local shire council, and downstream tobacco irrigators in the Dimbulah irrigation area who require good quality irrigation water for tobacco. The local community and ICM groups are very interested in sustainable development. A proposal is under consideration for a new sugar mill in the area with likely demand for an expanded irrigated sugar industry close by. The sustainability of the irrigation enterprise on a catchment basis is the major concern.

Catchment Management Options

A list of options has been proposed by different stakeholders as solutions to the problems. These include some proposed by The Cattle Creek Groundwater Management Strategy Committee (1993), Department of Primary Industries, and others. Some of the proposed solutions will require restrictions on some stakeholders for the benefit of others in the catchment. Those who are affected by the restrictions will never have a problem on their property due to the upslope location of their property in the catchment.

The only rational way to resolve these issues is to use a MODSS to examine the "what if" scenarios according to the needs of the various stakeholders. To do this, a groundwater-plant water balance-irrigation demand model of the catchment is being constructed so that the consequences of each strategy can be quantitatively assessed. This will then be linked to a MODSS where the objectives and outcomes for each stakeholder can be assessed objectively and transparently.

A project funded and supported by various government and industry organizations and funding agencies and in collaborative development with the U.S. Department of Agriculture, Agricultural Research Service is underway to model the hydrology of the catchment, link it to a MODSS and, in conjunction with the stakeholders, evaluate the options available from which the question of who pays, and how the costs and long-term productivity benefits can be constructively and openly evaluated. Resource economics and the nonmarket value of natural resources are being included in the evaluations.

There is a long history worldwide where the consequences of resource management changes are not adequately assessed before developments begin. This is partly because of the complexity of the issues and the need for a true multidisciplinary approach. Because the time frame for developments is often short and stakeholders need to see action on the ground quickly, approaches like this are an important development in the new paradigm of finding technically sound management options for ecologically sustainable development of catchments.

References

Addiscott, T.M. and R.J. Wagenet. 1985. Concepts of solute leaching in soils: a review of modeling approaches. *J. Soil Sci.* 36:411–424.

Bana e Costa, C.A., Ed. 1990. *Readings in Multiple Criteria Decision Aid.* Springer-Verlag, Berlin.

Bawden, R.J. and R.G. Packham. 1991. Improving agriculture through systematic agriculture research. Pages 261–271 in V. Squires and P. Tow, Eds. *Dryland Farming A Systems Approach: An Analysis of Dryland Agriculture in Australia.* Sydney University Press, Sydney.

Beck, M.P. and B W. Lin. 1983. Some heuristics for the consensus ranking problem. *Computers Operat. Res.* 10:1–7.

Berry, G. 1994. A Radioactive Waste Repository for Australia: Site Selection Ssudy — Phase 2. A discussion paper. National Resource Information Centre, Canberra.

Berry, J.S., W.P. Kemp, and J.A. Onsager. 1992. HOPPER decision support system: rangeland grasshopper management for the 1990s. Pages 610–615 in D.G. Watson, F.S. Zazveta, and A.B. Bottcher, Eds. *Computers in Agricultural Extension Programs.* ASAE Publication 1-92. ASAE, Michigan.

Bevan, K. 1989. Changing ideas in hydrology — the case of physically-based models. *J. Hydrol.* 105:157–172.

Bouma, J. 1986. Using soil survey information to characterise the soil water state. *J. Soil Sci.* 37:1–7.

Bouma, J. 1989. Using soil survey data for quantitative land evaluation. *Adv. Soil Sci.* 9:177–213.

Carberry, P.S., A.L. Cogle, and R.L. McCown. 1991. An Assessment of Cropping Potential for Land Marginal to the Atherton Tableland of North Queensland. Rural Industries Research and Development Final Report. Project No. CSC-16A.

Commonwealth of Australia. 1991. Decade of Landcare Plan. Commonwealth of Australia, Canberra.

Commonwealth of Australia. 1992. National Strategy for Ecologically Sustainable Development. Commonwealth of Australia, Australian Government Printing Service, Canberra.

Commonwealth of Australia. 1995. Techniques to Value Environmental Resources: An Introductory Handbook. Australian Government Publishing Service, Canberra

Curtis, A., J. Daly, D. Keane, and T. DeLacy. 1994. Landcare in Queensland: Getting the Job Done. Report No. 20. Johnstone Centre of Parks, Recreation and Heritage, Charles Sturt University, Albury.

Davis, J.R., P.M. Nanninga, J. Biggins, and P. Laut. 1991. Prototype decision support system for analyzing impact of catchment policies. *J. Water Resour. Planning Manage.* 117:399–414.

Department of Primary Industries. 1994. The Sustainable Management of Queensland's Natural Resources, Policies and Strategies: A Discussion Paper. Queensland Department of Primary Industries, Brisbane.

Doherty, J.D. and A. Stallman. 1992. Land Management Options for a Selected Catchment on the Darling Downs. Queensland Department of Primary Industries Project Report QO92010.

Foran, B.D. and D.M. Stafford Smith. 1990. A comparison of development options on a northern Australian beef property. *Agric. Syst.* 34:77–102.

Grayson, R.B. and J.M. Doolan. 1995. Adaptive Environmental Assessment and Management (AEAM) and Integrated Catchment Management. Land and Water Resources Research and Development Corporation, Occasional Paper No. 1/95.

Hatton, T., W. Dawes, and J. Walker. 1994. Predicting the impact of land use change on salinity — the role of spatial systems. Pages 556–568 in Proceedings, Resource Technology 94. Melbourne.

Ive, J.R. 1992. LUPIS: Computer assistance for land use allocation. *Resource Technology 1992 Taipei: Information Technology for Environmental Management*. November 1992, Taipei, Taiwan ROC.

Ive, J.R., K.D. Cocks, and C.A. Parvey. 1989. Using LUPIS land management package to select and schedule multisite operations. *J. Environ. Manage.* 29:31–45.

Janssen, R. 1992. *Multiobjective Decision Support for Environmental Management*. Kluwer Academic Publishers, Dordrecht.

Johnston, A.K.L. 1995. The use of models to analyse performance of sugarcane production systems: use in spatial analysis. Pages 103–116 in M.J. Robinson, Ed. *Proceedings, Research and Modelling Approaches to Examine Sugarcane Production Opportunities and Constraints*. University of Queensland, St Lucia, November 1994.

Johnston, A.K.L., R.A. Cramb, and J.R. McAlpine. 1994. Integration of biophysical and economic data using an expert system: results from a case study in northern Australia. *Soil Use Manage.* 10:181–188.

Konikow, L.F. and J.D. Bredehoeft. 1992. Groundwater models cannot be validated. *Adv. Water Resour.* 15:75–83.

Lait, R.W. 1992. Groundwater Occurrence and Management in the Cattle Creek Catchment, Mareeba-Dimbulah Irrigation Area, North Queensland. QDPI unpublished report.

Lane, L.J., D.S. Yakowitz, J.J. Stone, M. Hernandez, P. Hellman, B. Imain, J. Masterson, and J. Abolt. 1994. A multi objective decision support system for the USDA Water Quality Initiative. Report, South-West Watershed Research Center, USDA-ARS, Tucson, Arizona, July 1994.

Lewis, H.S. and T.W. Butler. 1993. An interactive framework for multiperson, multiobjective decisions. *Decision Sci. J.* 24:1–22.

Littleboy, M., D.M. Silburn, D.M. Freebairn, D.R. Woodruff, and G.L. Hammer. 1989. PERFECT — A Computer Simulation Model of Production Erosion Runoff Functions to Evaluate Conservation Techniques. Queensland Department of Primary Industries Publication, Training Series QE89010.

McDonald, G.T. and T.J. Hundloe. 1993. Policies for a sustainable future. Pages 345–383 in G. McTanish and W.C. Boughton, Eds. Land Degradation Processes in Australia.

Moffatt, I. 1990. The potentialities and problems associated with applying information technology to environmental management. *J. Environ. Manage.* 30:209–220.

Murray-Darling Basin Ministerial Council. 1990. Natural Resources Management Strategy: Murray-Darling Basin. Murray Darling Basin Commission, Canberra.

Novonty, V. and G. Chester. 1981. *Handbook of Nonpoint Pollution: Sources and Management*. Van Nostrand Reinhold, New York.

Queensland Government. 1991. Integrated Catchment Management: A Strategy for Achieving the Sustainable and Balanced Use of Land, Water and Related Biological Resources. Queensland Department of Primary Industries, Brisbane.

Resource Assessment Commission. 1992. Multicriteria Analysis as a Resource Assessment Tool. Resource Assessment Commission, Research Paper No. 6.

Rossiter, D.G. 1990. ALES: a framework for land evaluation using a microcomputer. *Soil Use Manage.* 6:7–20.

Shaw, R.J. 1994. GIS-Hydrologic Modelling-DSS; Report on a Technical Workshop. Occasional Paper No 11/94. Land and Water Resources Research and Development Corporation, Canberra 1994.

Shaw, R.J., A.L. Cogle, R. Lait, J.D. Doherty, J.J. Stone, and L.J. Brebber. 1995. The role of decision support systems for ecologically sustainable productivity in the sugar industry. Pages 117–124 in M.J. Robinson, Ed. *Proceedings, Research and Modelling Approaches to Examine Sugarcane Production Opportunities and Constraints.* University of Queensland, St Lucia, November 1994.

Squires, V.R. 1991. A systems approach to agriculture. Pages 3–15 in V.R. Squires and P. Tow, Eds. *Dryland Farming: A Systems Approach.* Sydney University Press, Sydney.

Syme, G.J., J.E. Butterworth, and Nancarrow. 1994. National Whole Catchment Management: A Review and Analysis of Processes. Land and Water Research and Development Corporation, Occasional Paper No. 01/94. Canberra.

The Cattle Creek Groundwater Management Strategy Committee. 1993. Managing Groundwater and Avoiding Salinity in the Cattle Creek Catchment — An Action Plan. Publication of the Cattle Creek Groundwater Management Strategy Committee.

Thompson, I.L., J.R. Barrett, and Jones. 1992. Decision Support Systems for U.S. Agriculture — A Status Report. Pages 670–675 in D.G. Watson, F.S. Zazveta, and A.B. Bottcher, Eds. *Computers in Agricultural Extension Programs.* ASAE Publication 1-92. ASAE, Michigan.

Van Diepen, C.A., H. Van Kessler, J. Wolf, and J.A.A. Berkhout. 1991. Land evaluation: from intuition to quantification. *Adv. Soil Sci.* 15:139–204.

van Genuchten, M.Th. and F.J. Leij. 1992. On estimating the hydraulic properties of unsaturated soils. Pages 1–14 in M.Th. van Genuchten, F.J. Leij, and L.J. Lund, Eds. *Proceedings of the International Workshop on Indirect Methods for Estimating the Hydraulic Properties of Unsaturated Soils.* University of California, Riverside. October 1989.

Watermark Computing. 1994. PEST Model Independent Parameter Estimation User Manual. Watermark Computing, Brisbane.

Whittaker, A.D. 1993. Decision support systems and expert systems for range science. Pages 69–81 in J.W. Stuth and B.G. Lyons, Eds. *Decision Support Systems for the Management of Grazing Lands: Emerging Issues.* The Parthenon Publishing Group.

Yaalon, D.H. 1994. On models, modelling and process understanding. *Soil Sci. Soc. Am. J.* 58:1276.

Yakowitz, D.S., L.J. Lane, J.J. Stone, P. Heilman, R.K. Reddy, and B. Imam. 1992. Evaluating land management effects on water quality using multiobjective analysis within a decision support system. Pages 365–374 in First International Conference on Ground Water Ecology. U.S. Environmental Protection Agency, American Water Resources Association, Tampa, April 1992.

Yakowitz, D.S., L.J. Lane, and F. Szidarovszky. 1993. Multiattribute decision making: dominance with respect to an importance order of the attributes. *Appl. Math. Computat.* 54:167–181.

SUMMARY AND CONCLUSIONS VI

Chapter 58

A Synthesis of the State-of-the-Art on Multiple Objective Decision Making for Managing Land, Water, and the Environment

A.C. JONES, S.A. EL-SWAIFY, R. GRAHAM, D.P. STONEHOUSE, AND I. WHITEHOUSE

Observations and Conclusions

A number of general conclusions were formulated from the material presented at the "First International Conference on Multiple Objective Decision Support Systems (MODSS) for Land, Water, and Environmental Management" in Honolulu, Hawaii (see the Introduction, Chapter 1 in this volume). Most of the material presented at the conference is published in the preceding chapters within the sections which represent the corresponding themes of the conference. In this concluding chapter we present a synthesis of the thematic concepts, emerging challenges, and recommendations which we feel reflect the collective thoughts and ideas shared by participants during formal presentations as well as informal discussions. We recognize that not all participants may agree with our synthesis; the following synopsis is the translation of what the authors heard and understood in an effort to present the "big" picture of MODSS for land, water, and environmental management.

Relevance of and Rationale for MODSS

People intuitively exercise multiple objective decision making, the selection of choices, and arriving at compromises for conducting many aspects of their lives, e.g., when planning tasty and nutritious meals or making investments for secure capital and rapid growth. In land use, research and technological accomplishments and innovations have focused largely on, and succeeded in, assuring impressive increases in global agricultural productivity. However, authors of Chapters 1 and 3 make the case that we can remain eternally optimistic, as these successes appear to be nearing their limits. A variety of reasons are given, most of which are driven by the scarcity of productive land resources and the continuing expansion of earth populations and their needs at alarming rates.

"Sustainable development," "conservation-effectiveness," and "land husbandry" are among the relatively recent concepts or paradigm shifts that emerged in response to these concerns. All aim to integrate "conventional, production-driven" objectives with "protective, environmentally sound" objectives of natural resources use and management. Realizing such ideals, however, is often hampered by conflicts among these competing objectives. In Chapter 3, the author argues that conflicts arise primarily because of multiple client needs, diverse expectations, and the diverse impacts of human actions within ecosystems. Resolving such conflicts requires a careful holistic articulation of land use objectives, recognizing potential problems, defining the "multiple-criteria" which must be addressed to evaluate the ecosystem's responses to selected management alternatives, selecting the appropriate databases and/or models for quantifying system responses and tradeoffs between competing objectives, expert setting of a framework and threshold values for judging changes in ecosystem attributes, and formulating recommendations for optimal actions. Different stakeholder needs and preferences must be addressed not with single options but by a spectrum of choices; identifying optimal choices underpin the need for multiple objective decision making.

The chapters in Sections II and IV demonstrate that significant progress has recently been made in understanding and quantifying the influence of physical, biological, management, and socioeconomic-policy factors on the productivity, quality, and sustainability of diverse agroecosystems. Section III shows that natural resource models, multiple resource databases, and computer-based expert systems allow developing powerful tools for quantitative, *ex ante* evaluation of land uses and specific management practices. When these tools are combined into decision support systems (DSS), they allow formulating recommendations for addressing complex situations such as those encountered in "multiple objective decision making." Section V illustrates the utility of these concepts in addressing selected local, national, regional, and international issues.

Addressing MODSS User Needs

To be effective in meeting the needs of concerned stakeholders, developers of DSS in general, and MODSS in particular, as well as their institutions, must be mindful of the following:

1. Building a DSS that users need and can use: the MODSS must function in a working environment and direct users of the tool must be identified early in the development stage. There are several tiers of users including model developers, scenario constructors, domain experts, model testers, and the conflict resolution specialist or decision maker. Users should be informed of the data, model, and DSS limitations and assumptions should be made explicit so that the model is more believable.
2. Involving the users: users should be part of the MODSS development process. This helps ensure that relevant issues are addressed and are easily understood. Also, users are more likely to use the MODSS if they feel a sense of ownership. Involvement of users in DSS development also creates a unique learning environment that will result in a more meaningful DSS. In industry, this is called "total engineering." Although direct involvement of the users is best; indirect involvement via surveys, workshops, and public hearings can also be beneficial.
3. Overcoming institutional and disciplinary barriers: communications across disciplines, ethnic groups, cultures, and socioeconomic groups is often difficult. When human energy is expended in breaking down real or perceived barriers, the result is often not a product, but instead a process. Disciplinary, vague, and complex jargon also detracts from effective communication, even with intelligent, but nontechnical, users.
4. Enhancing the reward system: bureaucracies tend to reward outputs not outcomes. Reward systems rarely recognize researchers or DSS developers for effective communications with and involving the users. This lack of recognition is then a limiting factor to user involvement.
5. Choosing an appropriate approach: if decision makers are many, knowledge is largely insufficient, and the problem is value-laden, then the process of making decisions needs to be an iterative rather than a linear one. Multiple objective problems are quite difficult to sort out and solve. This is especially true when at least some aspect of the problem is value-laden and many decision makers are involved. MODSS can facilitate the decision-making process by iteratively helping decision makers come to a consensus decision to which all choose be committed.

The importance of addressing user needs as part of the overall success of a MODSS is illustrated by presented case studies from several countries (Section II). Investigations on biomass energy in the U.S. during the past 15 years suggest that spatial scale of the issues is critical to developing a meaningful and useful MODSS. At different spatial scales, different economic and environmental concerns need to be modeled. Also, the users of MODSS change with scale. Policy makers are more likely to be interested in aggregated national scale models whereas landowners are more likely to be interested in detailed farm and watershed scale models. In New Zealand, the identification and introduction of efficient and sustainable land management practices has used a community-based approach to seek solutions within the Land-Care program (Allen et al., Chapter 4). This effort is facilitated by a comprehensive knowledge-based DSS. Resource issues as well as multiple value-laden social perspectives influence the acceptability of specific management alternatives. The "process" approach allows for the required information flows to be rethought and reformulated as society, the economy, and the environment coevolve. Landowner needs, community needs, as well as scientific knowledge are embedded in the DSS. Sharing an understanding of how different groups see the world and what they do in it creates a learning environment that can promote constructive and voluntary behavioral change. This iterative process focuses on outcomes and has largely diminished institutional and disciplinary barriers as everyone's input has value and new information can be sought to fill identified "knowledge gaps."

Methodologies, Components, and Integration

In general, natural resource models, multiple resource databases, and computer-based expert systems allow developing powerful multiple objective decision aids for quantitatively evaluating alternative land uses and management. The following observations are based on the material presented and demonstrated at the conference, most of which comprise Section III of this volume:

1. MODSS appear to be developing from a truly multidisciplinary team perspective. Teams include social, physical, and biological scientists, engineers, model developers, DSS specialists, stakeholders, and the ultimate user or client. Such joint involvement strengthens communication and provides the team with a better understanding and appreciation of each other's perspective and priorities.
2. There is an abundance of MODSS computer programs available and being developed. Data needed to drive MODSS appear to be increasing in quantity and quality. In addition, multiple objectives are incorporated to a greater extent than before to address the multiple priorities of stakeholders. Nevertheless, important issues are being raised that impact further development and refinement of MODSS. Data are often gathered from many different research sources. As a result, databases for MODSS may be incomplete, inconsistent, and inaccessible at the micro or macro levels. Data quality and availability, model accessibility and level of sophistication, and multiplicity of MODSS and stakeholder objectives are among the priority topics being addressed by those currently involved with MODSS development.
3. Complexities of MODSS and its associated processes are being recognized. As complex problems are better defined and delineated, the interrelatedness and interdependency of knowledge impacting decision making become more apparent.
4. MODSS tools may be advancing more rapidly than MODSS methodologies. While MODSS is complex, it may be that the tools being developed are too complex for the end user. The tools are not always accessible, responsive, nor user-friendly. Alternative MODSS and tools of varying complexity are needed to provide choices that are suitable for different clientele. Therefore, more emphasis is needed to ensure that the validity, appropriateness, and complexity of evolving MODSS are such that they attract the interest of decision makers for use in making sound resource management decision.

5. Not all objectives of the stakeholders are being articulated for incorporation in available MODSS. Many people in various society sectors are stakeholders in the outcomes of a land use decision. General concerns of society and associated externalities must be considered as well.
6. MODSS developers need to be more holistic in their approaches. Current interests and documented history tell us about a select group of land use and management alternatives. However, other (innovative) options, that are comprehensive in scope and cognizant of spatial and temporal contexts, should also be part of the complement of alternatives using MODSS as a vehicle for change.

The importance of using a relevant spatial and temporal scale in dealing with sustainability issues was highlighted in Chapter 2, Section I of this volume. Section III provides a variety of approaches to developing and using MODSS. Examples include work in Egypt and the U.S. In Chapter 20, Rafea illustrated the use of integrated DSS consisting of three components, namely crop management expert systems, databases, and multimedia capabilities. While the purpose of the first two is obvious, the latter maximizes user-friendliness by acquiring and displaying major symptoms of disorders that may inflict growing crops. This substitution for narrative descriptions also minimizes the errors in making diagnostic characterization of the crop's disorders. In the U.S., other approaches to natural resource management resulted in MODSS that include such features and components as a Windows user interface, a simulation model, a decision model, simple databases, and a report generator (e.g., Imam et al., Chapter 19). Selection of a best management practice must satisfy the multiple constraints of environment, technology, and society. A multistep approach is used to order decision variables and rank management alternatives using utility functions. Applications of this approach include the effects of dryland agricultural management systems on water quality, evaluation of irrigation management systems on water quality, and evaluation of trench cap design systems for mixed waste burial sites. Several alternative water quality simulation models are incorporated into these DSS approaches.

Socio-Cultural-Economic, Policy and Sustainability Issues

Decision making is a human-based sociological process. Quantitative and qualitative parameters as well as values and other intangibles impact this process. Thus, sociology, economics, policy, culture and concerns with risk, uncertainty, and sustainability must be incorporated into the design of DSS, when and where appropriate. Several factors that determine success in dealing with these issues include accessibility to DSS, dealing with DSS in a human context, effective communications, and the nature of leadership. The material in Section IV also point to the following conclusions:

1. Differences in culture and standard of living among countries will determine our ability, at the global level, to be effective stewards of land, water, and air resources. Individual countries generally survive within a limited geographical niche on the Earth. Differences in geographic distribution, associated natural resource base, and socioeconomic-policy factors produce

substantial differences in access to wealth. Accessibility to wealth can markedly influence decision-making processes, independent of cultural differences and value systems. It is likely that wealth, to a large extent, will strongly influence what we choose to do as stewards of natural and renewable resources, both on a short- and long-term basis. Monetary wealth gives countries ready access to pertinent knowledge, facilitates the acquisition of quantitative information and the development of tools such as DSS, and allows enhancing their population's intellectual capacity.

2. Models are not the real world — only our best interpretation of the way we understand the real world and our best translation of it. Most DSS emphasize the incorporation of quantitative information which then provide different outcomes that reflect risk or uncertainty. Even if a quantitative measure of risk is not directly incorporated in a model or DSS, decision makers can use it to manage risk by simply deriving multiple outcomes and interpreting the associated consequences as part of the decision-making process.

3. DSS are intended to provide information that will aid in decision making; decision making itself is a subsequent, very value laden process. A DSS, especially one that is computer-based, does not, and should not be used as a tool to provide the one perfect solution to any given problem. Rather, it should be used as a tool to help support and incorporate as much information as possible that is likely to impact a decision. This represents a unique opportunity for scientists to be an integral part of shaping not only a decision-making process but also long-term policy.

4. Contrasts between short- and long-term considerations also pose challenges to MODSS modelers. Compared to most physical and biological processes, social, economic, and political conditions can change quite rapidly over time. Thus, to be effective, models and DSS should describe current conditions and also allow projecting future conditions. Producing dynamic outcomes broadens the views of decision makers.

5. DSS facilitates the decision-making process by helping people see the world in a different light and providing them with options that are different from those that might otherwise be chosen. A beneficial MODSS will inspire decision makers to view the problem under consideration in a way that is different from their past views of the same problem. Thus, not only can DSS allow improved decision-making processes and ultimately exercising an optimum decision, but also can be effective tools for promoting behavioral change.

6. In addition to the external information incorporated into the DSS, each decision maker brings a unique set of valuable internal information to the process. This internal information may be gained from experiences, intuition, values, age, culture, and other sources. The synergy between both external and internal information can help decision makers to better formulate instructional problems, acquire deep insight into the issues at hand, and ask the right questions. The level of decision making is also an important consideration in the development and use of DSS. Strategic decision making requires a more political set of information inputs than does operational decision making.

7. DSS modelers are human. If modelers do not consider all of the factual information available or misrepresent that information, a bias will result that will likely influence the alternative solutions presented by the DSS. Thus, it is important for them to be introspective and comply with the professional ethics that guide the work of credible scientists and engineers. In addition, any interpretation of data brings with it some value judgment so scientific objectivity is essential. Scientific credibility and ethics will be among the priority issues by which external groups and individuals judge the scientific community and our ability to provide unbiased information for their decision-making process. Furthermore, it is unlikely that even the best DSS will capture the entire range of decisions or alternatives that may be made. Decision makers should be encouraged to be flexible in their thinking and look beyond outcomes addressed strictly by the DSS.
8. Communications and working relationships between stakeholders, scientists/engineers, and decision makers must be strong. Stakeholders and decision makers need to be involved in the process from problem identification and information compilation to synthesis, DSS design, and testing. A historical perspective of DSS and decision making indicates that, in too many instances, the lack of stakeholder and decision-maker input in the DSS development stages produces irrelevant, misleading, or unrealistic products from an economic, social, political, and scientific point of view.
9. DSS may or may not incorporate values in a quantitative or qualitative manner. This is essential as "judgments" of alternative outcomes must be exercised by the decision maker. The incorporation of values into a DSS may or may not be appropriate depending upon the problem under consideration. In some situations we will have value-based information to which we can assign quantitative or qualitative measures. In other situations, values are simply not a part of the DSS but can be incorporated into the decision-making process through different pathways. Alternatively, whatever we choose to represent or not represent in the DSS also displays the values of those having input.
10. The research community has been labeled by some as "a land of excellence and knowledge surrounded by a sea of noncommunication." Many of today's scientists and engineers were educated in a university setting where the importance of communication skills, sociological considerations, and economics may have been diminished. As future generations of scientists and engineers are educated, much more emphasis needs to be placed on communication effectiveness and a broad base of knowledge. This is especially true for those involved with DSS. Within the DSS context, communication must be improved with stakeholders and decision makers, as well as among the scientists and engineers themselves. This First International MODSS Conference is an excellent example of communications on multidisciplinary issues that involved both the quantitative "hard" sciences and the more qualitative "soft" sciences.
11. Achieving sustainability will reflect the role of land managers and decision makers as leaders of their community. Production, economics, environmental concerns, and societal desires must be in balance and be allowed to evolve over time. What is considered good and correct, i.e., sustainable, today may change tomorrow because of our changing values, knowledge

base, or socioeconomic-policy factors. Thus, sustainability is not static but, rather, a dynamic and broad goal that we move towards. It can be reached by using a variety of practices in the environmental or agricultural arenas. The complexity of sustainability issues is often misunderstood or ignored. Furthermore, some multiple objective problems and solutions may have unstated or undeclared objectives.

Studies from India and other countries reported in Section IV provide unique examples of the sociocultural, economic, policy, and sustainability issues surrounding decision making. High erosion rates, runoff, sedimentation and mass wasting are among the indirect consequences of misuse of natural and renewable resources in India (Ram Babu and Dhyani, Chapter 51). Past efforts to mitigate problems have been unsuccessful because problems were evaluated singularly and in isolation, planned and administered in a top-down manner, and carried out with little coordination. Integrated watershed management approaches have been used since the mid-1980s and have proved successful in accomplishing diverse competitive and conflicting goals for farmers and the nation. Based on 25 watershed management projects it has been shown that participatory, bottom up, and client first approaches yielded desirable results. Specific vegetative plantings reduced runoff and soil erosion thus moderating drought and flood hazards. Persistent adoption of conservation practices were found to be useful only if the rural masses developed a steward ethic. Thus, the multiobjectives of farmers and the nation can be achieved with available technologies on a sustained basis if the planning and implementation processes involve local people on a watershed basis. Several authors discuss the challenge of dealing with nonmarket values which do not lend themselves to a quantitative measurement but should be of primary consideration in natural resource management and policy formulation. Approaches such as conjoint analysis and gaming theory are being tested and appear to be promising for addressing these issues.

Global and Regional Issues

The truly international participation in this MODSS conference provided many opportunities for exposing many contrasting views in dealing with natural resource use and environmental issues. Among the issues raised are the following:

1. Sustainability may be the luxury of the rich. A subsistence standard of living is dominated by the day to day concerns of producing enough food for survival. Sustainability, by its very definition, considers production in relation to other priorities for profitability, environmental quality, and social justice. Without adequate food supplies, these other priorities do not even enter into the production equation. Wealth allows a people to purchase goods and services from other sectors so as to lessen the environmental impact of the immediate surroundings. Concurrently, a wealthy nation can accept less production and even omit some types of production to meet ecological standards or in the name of social justice. These arguments notwithstanding, we believe that sustainability is a necessity of the poor,

including those who have little or no purchasing power. Their very survival depends upon the ability to utilize the resource base in a way that sustains productivity. At the national level, a short-term or narrow focus of natural resource management will not suffice as ecosystem productivity and quality are so strongly entwined that, given enough time, the decline of one is a serious threat to the other.

2. Global environmental awareness will continue to increase. Environmental treaties may result in imposing trade barriers that require continual efforts to improve agricultural, forest, and other land uses so that they are more ecologically sound and environmentally safe. This will provide many opportunities for MODSS tools to help evaluate current and future land use and management practices from an environmental perspective. Consumers are beginning to demand safe food. Market research of Asian consumers shows that they want nutritious food that is pesticide free and produced in an environmentally sound manner. Global environmental conventions such as Agenda 21 and conferences on greenhouse gases and disposal of toxic waste will continue to address international environmental issues. Where MODSS have an opportunity to address such international problems, even more emphasis will be placed on communications among stakeholders, model developers, and decision makers to ensure that all pertinent information is incorporated into the decision aid and that it is correctly interpreted according to value systems of the interested parties. An International Environmental Court of Law could very well be established, thus providing additional opportunities and challenges for DSS on a global scale.

3. Knowledge will be increasingly globalized. Meetings such as this one, computer technology, and electronic communications all serve to break down traditional barriers to the exchange of knowledge. As knowledge becomes more portable, more people will be able to make better decisions. Also, there is a likelihood that people trying to solve similar problems may converge in their approach to problem solving; thus, using the best approaches from all different sectors of the planet.

4. Restoration, rehabilitation, and recycling of natural and renewable resources will be growth industries. DSS will clearly have an important role here to provide acceptable options based on well-defined sustainability criteria.

5. Many DDS support the current power structures and decision-making processes. The power of science, the power of researchers, the power of big business and the power of bureaucrats is supported and reinforced by the now-evolving DSS. In a few instances MODSS may support farmers but these tools do not appear to be supporting nongovernment organizations, indigenous people, or other organizations within our society.

6. Concurrently, all too often, scientists may choose not to be involved in shaping policy or may be excluded from the power structure in terms of making meaningful contributions to decision making on environmental and sustainability issues. If more such involvement takes place, MODSS will then serve as mechanisms by which scientists attempt to gain some power and be able to influence decisions and shape policy.

7. DDS are largely designed, developed, and tested by people who are rather well educated, have an above average standard of living, and not necessarily representative of mainstream stakeholders. On the other hand, a review of the global literature clearly portrays that women are the major decision makers at the farm level. Very few decision support tools attempt to address the issues raised in behalf of different genders or indigenous people.
8. Many evolving DSS reinforce individual rights rather than collective rights. To date, there has been an emphasis to make individual decisions based exclusively on farm level information with little regard for the rights or needs of the larger community. Many DSS actually reinforce litigation rather than conflict resolution. Now, more emphasis is being placed on problem solving at the watershed level. However, the watershed approach tends to group individual farms within geographic boundaries rather than to consider the watershed as a community of individual farms. Maximization of watershed goals is often emphasized rather than community acceptability. Collective decision making, collective involvement, and collective rights constitute the model of how indigenous people successfully solve environmental problems.
9. Most DSS are used for tactical or operational decision making rather than strategic decision making. Strategic thinking and decision making has been out of favor for several decades. Most existing DSS and decision makers focus on "how to get the job done" instead of "which job is to be done." Fortunately, the visioning process appears to be returning and will likely mark the end of one millennium and the beginning of another in terms of natural resource and environmental management.

These issues will be interpreted differently by diverse cultures because of their differential access to wealth, education, political land use policies, and global knowledge of resource management. As shown in Section V, the U.S. and Egypt portray very different paradigms for contrast. On the Hawaiian Island of Oahu water may be abundant on the whole, but is not well distributed geographically. This was highlighted during the MODSS Conference's field trip, mostly in the context of issues arising from rapidly changing land use in Hawaii as a whole, and on Oahu in particular. Issues include water rights, geographic and use-based water allocation, impacts on the quantity and quality of groundwater recharge and sustainable water yield from aquifers, and alternative patterns of tourist industry and residential development. Among the socioeconomic factors affecting the resolution of these issues are the impact of preserving open space on tourism and the need to replace diminishing plantation agriculture with diversified agriculture. Addressing these issues and factors represents a classic need for applying MODSS. Interactive multicriterion optimization allows the identification of most desirable land use options.

In contrast, water in Egypt is a very scarce commodity, so good irrigation water management and maximum possible recycling are essential to the survival of agriculture. A MODSS has been developed to predict the effects of changes in irrigation management, hydraulic conditions, cropping patterns and crop characteristics on the quantity and salinity of drainage water (Abdel-Dayem et al., Chapter 52). This model is a tool to address multiple objectives such as improving water use efficiency, reducing water use, reclaiming land for agriculture, use of additional or nonconventional water sources, and providing more farmers freedom in crop selection. When water quantity and quality is in short supply, this approach provides assistance in strategic decision making at the policy level and operational decision making at the watershed or farm level.

Additional Recommendations

A number of recommendations are incorporated or inferred in the preceding sections and those need not be reiterated here. In addition, we believe that all those who are involved with the development, implementation, and application of MODSS have many challenges ahead. MODSS, as presented at this conference, is a relatively new but extremely pertinent contribution to decision making. Some of the important additional "take home" messages from the authors are listed below. Hopefully, the reader will also find these points enriched with "food for thought" and ideas for guiding further development of MODSS in support of sound management of natural resources and enhancement of environmental quality:

1. Consider and prioritize the global issues where DSS can help people make better decisions.
2. Always remember the people side of decision making.
3. Upscale yourself and your organization with the skills needed to work with public multistakeholder groups.
4. Use a steering group of users to guide your DSS development and or research.
5. If a donor, provide funds stipulating direct involvement of the users so they can participate in setting research priorities and assure obtaining relevant research information.
6. Maintain a thirst for learning as well as a thirst for knowledge.

In addition, we wish to emphasize that this book is based on what was THE FIRST international MODSS conference dedicated to land, water, and environmental management. We, therefore, advocate continued efforts to hold follow-up conferences periodically in order to communicate further state-of-the-art developments, identify global issues of mutual concern, seek collaborative problem solving opportunities, promote MODSS transferability and adoption, and deploy appropriate efforts towards shaping sound environmental policy at the national, regional, and global levels.

INDEX

INDEX

Index

A

Acidity Decision Support System (ADSS), 177–184
Action-specific risks, comparative analysis of risk alternatives, 143–151
Adaptive Environmental Assessment and Management (AEAM), 711
AEAM (Adaptive Environmental Assessment and Management), 711
Aesthetic values, 508–509
AGFADOPT, 167–174
Aggregation rule for ranking alternatives, 205–214
AGNPS, 392
Agribusiness, 16, 407–417
   holistic approach, 411–412, 413
   issues in multiple objective management, 409
   objectives for renewable natural resource management, 410
   problem statement and objective, 409
   reconciling conflicting objectives, 415–417
   spatial and temporal dimensions, 412–414
   specification of objectives, 410–411
   sustainability issues, 408–409
   universal versus targeted policy applications, 414–415
Agriculture, 4, 6
   adoption of conservation practices, Scioto River study, 337–346
   AGFADOPT, 167–174
   in Australia
      Boobera Lagoon environmental management plan, 655–665
      maximizing profitability, 447–454
   case studies and examples, 17–19
   in Chile, 557–567
      MCDM model, 561–567
      sustainability definitions and issues, 558–561
   in China, no-tillage cropping system, 685–695
   constraints, resource-based, 11–13
   crop models for valuation of land, 455–462
   CROPS, 527–536
   database and multimedia integration of expert systems, 233–239
      databases, 235–236
      expert systems, 234–235
      multimedia, 237–239
   Egypt
      decision making models, 61–72
      SIWARE, 629–638
   Great Lakes region productivity, climate change effects, 585–597
   Indonesia
      agroecological approach, 465–479
      upland soil conservation, 395–405
   interactive, spatial decision support system, 385–393
   livestock ration formulation, 291–297
      case study, 296–297
      database, 295–296
      model base, 292–295
      user interface, 296
   long-term land use options, 440–442, 443
   MADM, fuzzy logic applications, 313–321
   in Mexico, 571–582
      application of DSS, 575–578
      environmental issues, 572–573
      prototype DSS, 573–575
      results, 578–581
   multiattribute processes
      decision making, 269–279
      ranking alternatives, 209–210
   nutrient management, 177–184
   optional averaging schemes, 217–231
      case study, 224–230
      stochastic decision criteria values, 221–224
      USDA-ARS WQDSS, 218–221
   PARI DSS, 299–310
   pesticides, *see* Pesticide Economic and Environmental Tradeoffs (PEET)
   pollution control, quantification of incentives, 513–525
   in Portugal, irrigated zone DSS, 599–612
   production versus environmental goals
      convergence of, 13–15
      divergence of, 16–17
   short-term and single-objective decision making in Egypt, consequences of, 61–72
   SMART PITCHFORK, 189–194
   SOILCROP, 349–361
   soil structure management, 379–384
   steep land cropping, 93–103
      analysis, 97–101, 102, 103

approach, 95–96, 97
    decision criteria, 94
    decision making, 101, 103
    objectives, 94–95
  WATERSHEDDS, 241–248
Agroforestry, 167–174
  ADSS, 182
  characteristics of knowledge, 178–181
  difficulties in developing decision aids, 180–181
  FACS, 183
  IITA, 183
  in Indonesia, 469
  N manager for Windows, 183
  PDSS, 183
  propa, 182–183
  survey of system models, 181–182
  uncertainty, 183
  United Nations University Expert System, 234
Ahupua'a concept, 17–18
ALES (Automated Land Evaluation System), 711
Alternative analysis
  agricultural pollutant control, incentive analysis, 516–518, 519, 520
  CROPS, 529
  forested ecosystem, 143–151
  INFORMS-R8 rulebase technology, 253
  ranking of
    MADM problems with attribute value discrimination, 317–318
    multiattribute tools for, 205–214
    optional averaging schemes, 217–231
  tepetate land management DSS, 579
Amish farmers, 339–330, 334
Application developers, DSS, 327
Aquaculture in Thailand, 641–652
  expansion of industry, 642–644
  regional model, 644–649
  results, 649–652
ARIMA, 591
Artificial intelligence, distributed, 301–303
Asia, 6
ASSESS (A System for Selecting Suitable Sites), 711
A System for Selecting Suitable Sites (ASSESS), 711
Attributes
  data maintenance, 331
  layered technology data model, 330–331
  value discrimination, MADM problems, 317
Australia
  Boobera Lagoon environmental management plan, 655–665
  catchment management, 697–714
  temperate forest management, 667–682
    ecosystems and types of disturbances, 668–669
    environmental assessment procedures on protected land, 673–678
    law and policy relating to private property, 669–673
    TEAS development, 678–681
Automated Land Evaluation System (ALES), 711

B

Beef cattle ration formulation, 291–297
Behavior
  adoption of conservation practices, Scioto River study, 337–346
  incentives, *see* Incentives
  upland soil conservation in Indonesia, 404
Best management practices, WATERSHEDDS, 241–248
Biological data, PEET data, 90
Biological diversity, 12
Blackboard model
  cooperative expert systems, 181
  PARI DSS, 304–308, 309
Boobera Lagoon environmental management plan, 655–665
Border disputes, Flathead River, 35–44
British Columbia, Flathead River dispute, 35–44
Budgeting tools, 332
Business functions, DSS, 333

C

Canada
  Flathead River dispute, 35–44
  PARI DSS, 299–310
Carbon dioxide, climate change modeling, 585–597
Catchment management, 697–714
  community involvement, 699–700
  concept, 701–703
  decision optimization, 712
  DSS approaches to resource management, 710–712
  optimizing MODSS form, 703–710
    awareness, technical detail, group decision making, 708–710
    model complexity and outcome accuracy, 705–708
  options, 713–714
  situation statement for Cattle Creek Catchment, 712–713
  sustainable development, 700–701
Catchment Management Support System (CMSS), 711
Cattle ration formulation, 291–297
Century 4.0, 585–597
CERCLA, 121
CERES models, 281–288
  comparison of observed versus modeled yields, 283, 284
  crop models for valuation of land, 461
  field experiment, 282–283
  simulation experiment, 283
CHAMBER, 235
Chemical Runoff and Erosion from Agricultural Management Systems (CREAMS), 153–164, 218–221
Chemical-specific risks, comparative analysis of risk alternatives, 143–151
Chemical Stockpile Disposal Program, 542–546
Chemical transport properties, PEET data, 90

Index

Chile, agriculture in, 557–567
China, no-tillage cropping system, 685–695
  agricultural resource support system, 693–695
  compound storied cropping in ridge-furrow, 692–693
  constraints and characteristics of hilly land, 686
  framework structure, 691–692
  plough layer, 687–691
  system, 686–687
C-Language Integration Production System (CLIPS), 251, 255, 262, 264, 265, 266, 267
Cleanup risks, see Remediation
Climate
  agricultural issues, 11
  agriculture expert systems, 234
  CERES models, 283, 286
  Chilean farming regions, 561
  CLIGEN, 576
  crop models for valuation of land, 461
  rangeland P-DSS, 365–375
  stochastic decision processes, optional averaging schemes, 221–230
Climate change effects on Great Lakes region agricultural productivity, 585–597
CLIPS (C-Language Integration Production System), 251, 255, 262, 264, 265, 266, 267
CMSS (Catchment Management Support System), 711
Co-learning, 51–58
  integrated system for knowledge management, 53–57
    accessing relevant knowledge, 54
    community dialogue, 55–56
    monitoring and adaptive management, 56–57
    scoping goals and objectives, 53, 54
  methodological challenges, 52–53
COMAX, 234
Community-based research, see Co-learning
Comparative analysis, cleanup alternatives, 143–151
Comprehensive Resource Planning System (CROPS), 527–536
  features of, 527–529
  implementation and extension of, 534–536
  solution process, 531–534
  structure of, 529–531
Compromise, PARI DSS reusable agents, 308
Compromise method
  agriculture in Chile, 557–567
  trade-off analysis, 426–427, 428
Conan, 30
Conflict, 6
Conflict analysis methods, trade-off analysis, 424
Conflict resolution, 23–45
  agribusiness, 409, 410, 415–417
  agriculture, PARI DSS, 299–310
  catchment scale, 697–714
  Flathead River dispute, 35–44
  GMCR II, 30–44
    existing systems for multiple participant decision making, 30–31
    system features, 31–35
  goals and methods, decision-making, 23

graph model, 25–29
  modeling, 26–27
  stability analysis, 27–29
  trade-off analysis, 419–434
Conjoint analysis, public inputs, 493–500
Conservation planning, 7
Constrained optimization routine, agribusiness modeling, 416
Constraints, weighted goal programming, 645, 646–649
Constraint-satisfaction-based scheduling, CROPS, 527–536
Cooperating expert systems, PARI DSS, 299–310
Cooperative decision making, see also specific MODS approaches
  agroforestry models, 167–174
  co-learning, 51–58
Cost
  agricultural pollutant control, analysis of incentives, 521–524
  landfill cover design, 155
  PEET data, 91
  ration formulation for cattle, 292
  remedial action alternative selection criteria, 149
  tepetate land management DSS, 577, 578
  upland soil conservation in Indonesia, 404
  weed control, PEET, 76
CREAMS (Chemicals Runoff, and Erosion form Agricultural Management Systems), 153–164
CROPMAN, 309–310
Crop models for valuation of land, 455–462
Cropping
  AGFADOPT, 167–174
  database and multimedia integration of expert systems, 233–239
  Egyptian strategies, 105–120, see also Egypt, cropping strategies in
  in Indonesia, 469–470
  MODEM applications, 390–393
  no-tillage system in China, 685–695
  rice production in Australia, 447–454
  SIWARE model, 636–638
  SMART PITCHFORK, 189–194
  SOILCROP, 349–361
  soil structure management, 379–384
  steep land, 93–103
  tropical systems, 471–472
CROPS (Comprehensive Resource Planning System), 527–536
Culture, see also Values
  agricultural issues, 12
  Boobera Lagoon in Aboriginal tradition, 657–658, 663
  Mennonite-Amish farmers, 339–330, 334

D

Dams
  Alqueva, 604–607, 611
  Aswan, 18
Darby Creek, Ohio, 337–346

Database
  agricultural expert system integration with, 233–239
  agroecology of Indonesia, 479
  layered technologies, 329
  rulebase toolkit requirements, 258–260
  SOILCROP application to FESLM-based DSS, 357–360
Data maintenance, layered technologies, 331
Data management tools, 330
Data Quality Objectives, ecological risk assessment, 121–132
DBL-CROP, 234
Decision criteria, MADM model issues, 317
Decision makers, DSS user management, 328
DecisionMaker system, 30
Decision model, rangeland P-DSS, 366–367, 371–375
Decision support system elements, 323–334
  challenges, 325–328
  layered technologies, 328–333
  nature of decision problem, 324–325
Deep Loess Research Station (DLRS), 515
Deforestation, 7, 12, 13
Desertification, 61–72
Deterministic approach to expert systems, 182
Disimprovements, graph model, 28
Distributed artificial intelligence, PARI DSS, 301–303
Domain specialists, DSS, 327–328
DSSAT, crop models for valuation of land, 462
DSS INTERACT, 33
Dynamic risk assessment models, 133

E

Ecological modules, MODEM, 387–390, 392
Ecological risk assessment, *see* Risk assessment
Economic development, sustainability of intensive agriculture, 414
Economic impact criteria, cropping pattern strategy evaluation, 111
Economic policy, *see* Policy
Economics, *see also* Socioeconomic issues
  adoption of conservation practices, Scioto River study, 337–346
  agribusiness, 16, 407–417
  agriculture
    Chilean farming study objectives, 563, 564, 565–567
    steep land cropping calculations, 95–96
    SWAGMAN Options, 447–454
  aquaculture model goal constraints, 648
  existence value, 481–490
  incentives, *see* Incentives
  Indian subcontinent watershed management, 622–623, 624
  MADM problem, 316, 321
  trade-off analysis, 419–434
  valuation of land, crop models for, 455–462
Ecosystem health concept, agricultural policy, 17
Ecosystem management, Northeast Decision Model, 197–203

Ecosystem unit, 4
Egypt, 18–1
  cropping strategies in, 105–120
    accumulated weights of experts' impact weights, 109
    experts' revised ranks for environmental criteria, 116–117
    experts' weights in different types of impacts, 112, 113
    impact criteria, 110–111
    multicriteria evaluation expert system, 107–117
    normalized experts' strategy impact weights, 109
    normalized experts' weights, 112
    two strategies evaluation by more than one expert, 114–115
  cropping strategies in multicriteria evaluation, 117–120
  short-term and single-objective decision making, consequences of, 61–72
    bias and unfair competition, 68–69
    data quality, 66–67
    degradation of land, 63–71
    investment in conservation, rich and poor farmers, 71–72
    lack of options and alternatives, 70–71
    land, water, people, food, 62–63
    limited definitions and assessments, 64–65
    specialist versus multidisciplinary approach, 67
  SIWARE, 106, 629–638
ELECTRE, 424
Elevation
  Indonesia, upland soil conservation, 395–405
  as unmanageable factor, 178
EMYCIN, 254, 255, 256
Entities
  data maintenance, 331
  INFORMS-R8, 260, 261
  layered technology data model, 330
EPIC, crop models for valuation of land, 462
EPINFORM, 234
Equilibrium states, graph model, 27
Eritrea, 18–19
Erosion, 7
  AGFADOPT, 167–174
  Chilean farming study objectives, 563, 564, 565
  crop models for valuation of land, 462
  fuzzy logic approaches, 319–320
  in Indonesia, 468, 478, 479
  Indonesian uplands, 396, 397
  landfill cover design, 153–164
  Nile, 18
  rangeland P-DSS, 372
  SOILCROP indicators, 353, 354
  steep land cropping, 93–103
Evaluation
  MADM, 269–279
  SOILCROP application to FESLM-based DSS, 357–360
Evaluators, DSS user management, 328
Existence value, 481–490

previous definitions, 482–487
standardization of definition, 487–489
Expert system
  Boobera Lagoon environmental management plan, 655–665
  shells, 332
External models, DSS, 332

F

FAIDS, 631, 632
Failure, CROPS constraints, 533, 534
Fate and transport models, 133
Federal Energy Regulatory Commission, 546–549, 550
Federal Facilities Agreement (FFA), 121
Fertility Advice and Consulting System (FACS), 177–184
Fertilizer/soil nutrients
  adoption of conservation practices, Scioto River study, 337–346
  agribusiness practices, 408
  CERES models, 281–288
  MODEM applications, 391
  pollution control, quantification of incentives, 513–525
  SMART PITCHFORK, 189–194
  upland soil conservation in Indonesia, 403
  WATERSHEDDS, 241–248
FESLM (Framework for Evaluation of Soil and Land Management), 351–352
Financing agencies, international, 68–69
Flathead River dispute, 35–44
Food resources, 6, 7
Foresight, graph model, 27–28
Forest management, 11
  in Australia, 667–682
    ecosystems and types of disturbances, 668–669
    environmental assessment procedures on protected land, 673–678
    law and policy relating to private property, 669–673
    TEAS development, 678–681
  conjoint analysis of public inputs, 493–500
  conversion to agricultural land in Indonesian uplands, 395
  ecological risk assessment, 121–132
    assessment endpoints, 127, 129
    conceptual framework, 122–125
    data gaps, 128–129
    lessons, 132
    measurement endpoints, 128, 129
    Oak Ridge Reservation features, 122
    problem statements and decisions, 127
    process, 126–132
    review of existing framework, 125–126
    specific tasks, 129–132
  in Indonesia, 467, 474
  INFORMS-R8 reasoning toolkit, 251–268
  Northeast Decision Model, 197–203
Fractal process, agricultural systems as, 178

Framework for Evaluation of Soil and Land Management (FESLM), 351–352, 357–360
Fuzzy logic, principles, 318–320
Fuzzy MADM model, 313–321
Fuzzy sets, 143–151, 318

G

Game theory, 24
General metarationality, 28, 29
General nurse approach to cooperative expert systems, 181–182
Geographic information systems, 328
  agroecology of Indonesia, 479
  crop models for valuation of land, 462
  in cropping pattern strategy evaluation, 107, 117–120
  CROPS, 528, 529
  layered technologies, 329–330
  MODEM applications, 393
  SOILCROP, 349
Gini coefficient, 564
GLEAMS
  rangeland P-DSS, 363–376
  tepetate land management DSS, 573–575, 576, 578
GLIGEN, 576
Global Assessment of Land Degradation (GLASOD), 12
Global control expert, PARI DSS, 304–305, 308, 309, 310
Global issues
  applicability to, 6
  biological diversity, 12
Global knowledge base, PARI DSS, 304–305
Goal programming
  agriculture modeling, 291–292
  aquaculture modeling, 642, 644–645
Goals, user-selected, 180
GPFARM DSS evaluation, 269–272
GRAIN MARKETING ADVISOR, 234
Graph model, DSS, 25–29
  existing systems for multiple participant decision making, 30–31
  GMCR II
    Flathead River Dispute, 30–44
    system features, 31–35
  modeling, 26–27
  stability analysis, 27–29
Grasslands, 11
Great Lakes region agricultural productivity, 585–597
  carbon dioxide and climate change on crop potential, 586–589
  crops to study, selection of, 589–590
  models applied, 590–595, 596
  region for study, selection of, 589
  results, 596–597
Great Plains Framework for Agricultural Resource Management (GPFARM), 269–279
Green Mountain National Forest, conjoint analysis of public inputs, 493–500

Groundwater Loading Effects from Agricultural Management Systems (GLEAMS), 218–221

H

HABASYS, 676–677, 677, 681–682
Habitat mapping, Oak Ridge Reservation ecological risk assessment, 129, 130, 131
Hawaii, agricultural issues, 17–18
Hazard assessment, PEET, 86–89
Hazardous contaminants
    pesticides, *see* Pesticide Economic and Environmental Tradeoffs (PEET)
    U.S. Federal Facilities
        cleanup risk management, 133–140
        Oak Ridge Reservation ecological risk assessment, 121–132
Hazardous pollutants, National Environmental Policy Act, 541–554
Hazard rating, INFORMS-R8, 254, 255
HELP (Hydrologic Evaluation of Landfill Performance), 153–164
Herbicides, *see* Pesticide Economic and Environmental Tradeoffs (PEET)
    long-term land use options, 441
    pollution control, quantification of incentives, 513–525
Heterogeneous reusable agents, PARI DSS, 299–310
Holistic methods
    agribusiness, 411–412, 413
    comparison with reductionist methods, 386–387
Hydrologic Evaluation of Landfill Performance (HELP), 153–164

I

Images
    agricultural multimedia tools, 237, 238, 239
    remote sensor, habitat mapping and analysis, 130, 131
Incentives
    adoption of conservation practices, Scioto River study, 337–346
    agribusiness practices, 410
    for pollution control, quantification of, 513–525
India, watershed resource management, 615–625
    description, 616–617
    management, 618
    multiobjectives, responses to, 618, 620, 621
    resources and problems, 617–618, 619
    sustainability of program, 620–625
Indonesia
    agroecological approach to sustainable agriculture, 465–479
        agricultural systems, 467–469
        climate and land resources, 466–467
        crops, cropping practices, 469–470
        ecological zones, 473–479
        expert systems, 470–473
    upland soil conservation, 395–405
        causes of unsustainability, 402–404

        population pressures and resource loss, 396–397
        Regreening Program, 397–398
        Regreening Program sustainability, 398–402
INFER, 234
Information transfer, 52
INFORMS-R8, 251–268
    design of toolkit, 260–262
        metadata and other data tables, 262–263
        rule types, 260–261
        user interface, 263
    project background, 253–258
    rulebase toolkit requirements, 258–260
    toolkit role in rulebase development, 262–267
        eliciting knowledge, 263
        encoding rulebases, 264–267
        rule construction, 264
Institutional factors, *see* Policy
Integer programming, trade-off analysis, 428–434
Integrated catchment management (ICM), 697–714
Integrated decision making, fuzzy MADM model, 313–321
Integrated systems for knowledge management (ISKM), 53–57
Integrity, data maintenance, 331
INTERACT, 30
Interactive, spatial decision support system (ISDSS), 385–393
Interdisciplinary approach, agribusiness policy, 416
Interdisciplinary Relevancy factors, agroforestry models, 182–183
International Agricultural Research Institutes (IARIs), 4
International Joint Commission (IJC), Flathead River dispute, 35–44
International Monetary Fund, 68–69
Irrigation
    in Australia, 447–454, 655–665
    computer model (SIWARE), 106
    crop models for valuation of land, 461
    degradation of resources, 12
    expert systems, 235–236
    long-term land use options, 440
    PEET data, 90
    in Portugal, 599–612
    weed control, PEET, 76
Island ecosystems, 6
    Indonesia, *see* Indonesia
    small, tropical, volcanic, 17–18

J

Java, 476–479

K

Knowledge
    agricultural, characteristics of, 178–180
    DSS structuring, 326
    integrated systems for management (IKSM), 53–57

PARI DSS, 304–305
SOILCROP, 350–351
Knowledge of preferences, graph model, 28

L

Land Analysis and Decision Support (LANDS), 324, 334
LANDCAP, 655–665
Landcare, 699
Landfill cover design evaluation, 153–164
Land managers, ISKM framework, 56–57
LANDS (Land Analysis and Decision Support), 324, 334
Land use, 7
   agriculture expert systems, 234–235
   short-term and single-objective decision making effects, 61–72
   steep land cropping, 93–103
Land Use Planning and Information System (LUPIS), 711
Land valuation, crop model use for, 455–462
Layered technologies, 328–333
Legal issues, 94, *see also* Policy
Legislation
   in Australia, 667, 668
   in Egypt, 71–72
   land use policy, 94
   in U.S.
      CERCLA, 121
      National Environmental Policy Act, 541–554
Lexicographic goal programming, 642, 644
Limit-move stability, 28
Linear programming, SWAGMAN Options, 449
Linguistic variables, fuzzy logic, 318–319
Livestock, 7
   Boobera Lagoon conflicts, 652, 663
   MADM problem, 316
   rangeland P-DSS, 363–376
   ration formulation, 291–297
LOGSPERT, 675–676, 677, 681–682
Long-term planning
   agribusiness issues, 412–414
   explorations of land use options, 437–444
LUMPS, 392
LUPIS (Land Use Planning and Information System), 711

M

MADM (multiattribute decision making), 269–279
Management, *see* Co-learning
Manure application
   adoption of conservation practices, Scioto River study, 343–344
   SMART PITCHFORK, 189–194
Marginal areas
   AGFADOPT, 167–174
   agricultural expansion, 11
Mathematical models, expert systems versus, 663–664

Mathematical optimizer, 330
Mechanized agriculture, 16–17
Mennonite farmers, 339–330, 334
Metagame Analysis, 30
Metarationality, 28, 29
Methodologies, 6
Mexico, tepetate land management, 571–582
   application of DSS, 575–578
   environmental issues, 572–573
   prototype DSS, 573–575
   results, 578–581
Mining, Flathead River dispute, 35–44
Mixed data, MADM model issues, 317
Mixed integer methods
   SWAGMAN Options, 449
   trade-off analysis, 428–434
Monitors, DSS user management, 328
Moral values, 507–508
MOWPAT, 392–393
Multiagent systems, PARI DSS, 301–303
Multiattribute decision making (MADM), 269–279
   case study, 275–278
   concepts, 273–274
   evaluators, 271–272
   format, 272–273
   fuzzy logic applications in, 313–321
   key issues in MADM, 316–318
   reasons for evaluation of DSS, 270–271
Multiattribute tool for ranking alternatives, 205–214
   aggregation tool, 208
   decision, 209
   farm management example, 209–210
   hierarchy of criteria, 212
   indicator values for large heterogeneous sites, 212–213, 214
   multiple states of nature, 210–212
   ranking procedure, 207
Multicriteria optimization, SWAGMAN Options, 449
Multimedia, agricultural expert system integration with, 233–239
Multiobjective decision theory, rangeland P-DSS, 366–367
Multiple criteria decision making (MCDM) techniques, 24
   GMCR II, 30–44
   graph model, 25–29
Multiple Goal Linear Programming (MGLP), long-term land use options, 437–444
Multiple objective decision making (MODEM)
   agriculture, 385–393
      applications, 390–393
      integrated framework, 387–390
      reductionist versus holistic methods, 386–387
Multiple resource prescriptions, Northeast Decision Model, 197–203
Multivariate analysis tools, RATS, 591–597

N

Nash stability, 27–28, 29
National Environmental Policy Act (NEPA), 541–554

Chemical Stockpile Disposal Program, 542–546
Federal Energy Regulatory Commission support, 546–549, 550
new challenges for integrated assessment, 552–554
Three Mile Island assessment, 549–552
Natural resource inventory, CROPS, 529
Natural resource management, 4
Negotiation, PARI DSS, 308–309
NERISK, 235
New South Wales, temperate forest, 667–682
New Zealand, community-based approaches to management, 57–58
Nile River
 agriculture, 13, 18–19
 SIWARE, 629–638
Nitrogen, *see* Fertilizer/soil nutrients
Noncommensurate data, MADM model issues, 317
Nonmyopic stability, 28
Nonpoint source pollutants, WATERSHEDDS, 241–248
Northeast Decision Model (NED), 197–203
Nurse approach to cooperative expert systems, 181–182
Nutrients, soil, *see* Fertilizer/soil nutrients
Nutrition, livestock ration formulation, 291–297

O

Oak Ridge facility
 forested ecosystem ecological risk assessment, 121–132
 NEPA support, 542
Objectives, MODEM framework, 387–390
Object-oriented data models, 331
Optimization, 332
 agricultural pollutant control, analysis of incentives, 518, 521
 long-term land use options, 438–444
Orchard, AGFADOPT, 167–174
ORESTE, 424

P

Parkland Agriculture Research Initiative (PARI) DSS, 299–310
 conflict management technique applications, 304–309
 cooperating experts paradigm, 303–304
 distributed AI, 301–303
 scenario, 309–310
PARMS, 309
Participatory research, *see* Co-learning
Payoff function, graph model, 26
PERFECT-SWIM, 379–384
Pesticide Economic and Environmental Tradeoffs (PEET): developer's perspective, 83–91
 implementation, 90–91
 system design, 84–90
  display of results, 90
  economic impact calculation, 85–86
  groundwater hazard, 86–90
  user inputs, 84–85
Pesticide Economic and Environmental Tradeoffs (PEET): users' perspective, 75–80IBRD
 ease of use, 80
 execution time, 79–80
 help, 80
 setup menu, 79
Pesticides
 agricultural expert system, 234, 235
 MODEM applications, 391
 pollution control, quantification of incentives, 513–525
 steep land cropping calculations, 95–96
 upland soil conservation in Indonesia, 403, 404
Pest management
 hazard rating values in forest management, 254, 255
 in Indonesia, 468
Phosphorus, *see* Fertilizer/soil nutrients
Phosphorus Decision Support System (PDSS), 177–184
Planning, 7
 aquaculture, 641–652
 CROPS, 527–536
 short-term and single-objective decision making in Egypt, consequences of, 61–72
 trade-off analysis, 419–434
Planning; Research; Implementation; Monitoring and Evaluation (PRIME), 710–711
Plantations
 AGFADOPT, 167–174
 Indonesian, 467, 473, 474
PLANTER, 234
Plant pathology
 expert systems, 235
 images, 237, 238, 239
 in Indonesia, 468
Policy, 6
 agriculture
  divergence of production and environmental goals, 16–17
  industrial-scale/agribusiness, 407–417
  resource degradation, 12
 Australian, temperate forest and woodland, 667–682
 community-based research approach, *see* Co-learning
 conjoint analysis of public inputs, 493–500
 long-term explorations of sustainable land use, 437–444
 National Environmental Policy Act, 541–554
 short-term and single-objective decision making in Egypt, consequences of, 61–72
  bias and unfair competition, 68–69
  data quality, 66–67
  degradation of land, 63–71
  investment in conservation, rich and poor farmers, 71–72
  lack of options and alternatives, 70–71
  land, water, people, food, 62–63

Index 741

limited definitions and assessments, 64–65
  specialist versus multidisciplinary approach, 67
Political impact criteria, cropping pattern strategy evaluation, 111
Pollution, 4, *see also* Agriculture
  agribusiness as source, 409, 410, 411, 412
  in Indonesia, 468
  MODEM applications, 392
  National Environmental Policy Act, 541–554
  WATERSHEDDS, 241–248
Population
  agricultural issues, 10
  Indonesia, 396
  in Mexico, 572
  resource pressures in Egypt, 62–63
Portugal, irrigation in, 599–612
  assessment, 601–604
  case study, selection of area, 608–609
  description and analysis, 604–607
  DSS design and implementation, 609–611
  identification of developments required in planning and development in irrigated areas, 607–608
Preferences, gaph model, 28
Preference structure, goal programming, 644
PRIME (Planning; Research; Implementation; Monitoring and Evaluation), 710–711
Principal components analysis, Chilean farming study, 562
Process models, Great Lakes region agricultural productivity, 585, 590–591
Production functions, aquaculture model constraint structure, 646
Productivity
  agricultural land use practices, 14
  Great Lakes region agriculture, 585–597
  Indian subcontinent watersheds, 620, 621
  indicators and objectives of soil and land management, 351
  preservation of, 11
  rangeland, 363–376
  tepetate land management DSS, 577
Professor Papaya, 182
PROMETHEE, 424

Q

Query rules, INFORMS-R8, 260, 261

R

Rangeland management, prototype DSS evaluation, 363–376
  DSS components, 366–369
  evaluation with decision model, 371–375
  evaluation without decision model, 369–370
  watershed description and data collected, 365–366, 366
Ranking of alternatives, multiattribute tools for, 205–214
Rational values, 511

Ration formulation for livestock, 291–297
RATS (Regression Analysis of Time Series), 591–597
RCRA cap, 154–155, 156
Reachability matrix, 26
Reductionist methods, comparison with holistic methods, 386–387
Regional issues
  applicability to, 6
  community-based research approach, *see* Co-learning
Regression Analysis of Time Series (RATS), 591–597
Relaxation
  CROPS constraints, 533, 534
  PARI DSS, 308–309
Relevancy factors, agroforestry models, 182–183
Reliability of MODSS, 664
Remediation, 7
  comparative analysis of alternatives, 143–151
    application to hypothetical marshland, 147–151
    level 1 analysis, 145–146
    level 2 analysis, 146–147
  U.S. Federal Facilities
    cleanup risk management, 133–140
    forested ecosystem ecological risk assessment, 121–132
Remote sensor images, habitat mapping and analysis, Oak Ridge Reservation, 130, 131
Resource-based constraints, agriculture, 11–13
Reusable agents, PARI DSS, 299–310
REUSE, 631–632
Revised Universal Soil Loss Equation (RUSLE), 95, 96
Risk assessment, 6
  forested ecosystem, 121–132
  U.S. Federal Facilities
    cleanup risk management, 133–140
    large forested ecosystems, 121–132
River systems, *see also* Nile River; Watersheds
  Flathead River dispute, 35–44
  Scioto River, Ohio, 337–346
Road planning problem, trade-off analysis, 420–422
Rulebase technology, INFORMS-R8, 251–268
Rural planning, trade-off analysis, 419–434

S

Salinity/salinization, 13, 106
Scheduling tool, 332
Science, ISKM framework, 57
Scioto River Watershed, 337–346
  descriptive findings, 340–343
  multivariate findings, 343–245
  study region, 338–340
Scoping process, participatory management process, 53, 54
Seasonal No-Tillage Ridge Cropping System (SNTRCS), 685–695
Security
  DSS elements, 326
  indicators and objectives of soil and land management, 351

Sedimentation
  abatement costs, analysis of incentives, 521–522
  Nile, 18
Sensitivity analysis
  goal programming, 645
  long-term land use options, 442, 444
  SWAGMAN Options, 451
Sequential stability (modeling concept), 28, 29, 30
Silviculture
  agroforestry, 167–174
  INFORMS-R8 reasoning toolkit, 251–268
  Northeast Decision Model, 197–203
Simple deterministic approach to expert systems, 182
Simulation of Water Management in Arab Republic of Egypt (SIWARE), 106, 629–638
SIWARE
  applications, 633–638
  submodels in, 630–633
Slope
  in Indonesia, 468, 477, 479
  no-tillage cropping system in China, 685–695
  as unmanageable factor, 178
Small island ecosystems, *see* Island ecosystems
SMART PITCHFORK, 189–194
Social impacts criteria, cropping pattern strategy evaluation, 111
Social policy, *see* Policy
Socioeconomic issues, *see also* Economics; Values
  agricultural systems study, 600–601
  irrigation projects, 606
  MODEM applications, 392
  MODEM framework, 387–390
  upland soil conservation in Indonesia, 395–405
Soil, *see also* Agriculture; Erosion
  crop models for valuation of land, 459–460
  data for PEET implementation, 90
  nutrient management, 177–184
  PARI DSS, 299–310
  resource-based constraints, 11–13
  structure management, 379–384
SOILCROP, 349–361
  application to FESLM-based DSS, 357–360
  evaluation framework, 351–352
  farmer knowledge and indicators, 350–351
  prototype expert system, 351–357
  symbolic classes, 353–357
Solution concepts, graph model, 27–28, 29
SPANNS, 30
Spatial decision making system, interactive (IDSS), 385–393
Spatial rules, INFORMS-R8, 260–261
Spatial tools, layered technologies, 329–330
Spiritual values, 510
Stable states (modeling concept), graph model, 27
STARRT, 309
State Elimination (modeling concept), 33, 35
States (modeling concept), graph model, 26
Statistical models, Great Lakes region agricultural productivity, 585, 591–594
Steep land, 7

cropping
  analysis, 97–101, 102, 103
  approach, 95–96, 97
  decision criteria, 94
  decision making, 101, 103
  no-tillage cropping system in China, 685–695
  objectives, 94–95
  slope as unmanageable factor, 178
Stochastic decision criteria, climate process effects on farm management alternatives, 221–224
Strategic disimprovements, graph model, 28
Sumatra, 473–476, 478
Sustainability, *see also specific issues and approaches*
  agricultural land use practices, 14
  community-based research approach, *see* Co-learning
  long-term land use options, 442, 444
Sustainable development, 6
SWAGMAN Options, 447–454
SWAT, 392
SWIM, PERFECT-SWIM, 379–384
Symbolic classes, SOILCROP, 353–357
Symmetric metarationality, 28, 29
Syntax checking, rule construction, 265
System effectiveness measures, 269–279
System integration with databases and multimedia, 233–239

T

Task manager, DSS, 333
Terracing
  in Indonesia, 468
  upland soil conservation in Indonesia, 395–405
Thailand, aquaculture in, 641–652
Three Mile Island, 549–552
Tillage
  adoption of conservation practices, Scioto River study, 343–344, 345
  in Indonesia, 468
  long-term land use options, 441
  MODEM applications, 391
  PEET data, 90
  steep land cropping, 93–103
  weed control, PEET, 76
Total environmental assessment system (TEAS), 677–678
Trade-off analysis, 419–434
  aggregation of economic and environmental data, 422–426
  extension of mixed integer approach, 431–433, 434
  identification of best compromise, 426–427, 428
  mixed integer programming approach, 428–431
  road planning problem, 420–422
Transfer of technology (TOT) model, 52
Transitivity, graph model, 28
Transportation, trade-off analysis, 420–422
Transport models, 133
Transport properties, PEET data, 90

Index

Tree crops
  AGFADOPT, 167–174
  Indonesian, 467
Treynor, Iowa case study, 224–230
Tropical rainforests, 12
Two strategies system, 107, 108, 114–115

U

Uncertainty estimation, agroforestry models, 182, 183
Unilateral improvements, graph model, 27
Unilateral moves, graph model, 28–29
United Nations University Agroforestry Expert System, 234
United States, *see also specific regions*
  Flathead River dispute, 35–44
  Northeast Decision Model, 197–203
United States Agency for International Development (USAID), Structural Adjustment policies, effects in Egypt, 68–69
United States Department of Agriculture, 571
  Agricultural Research Service WQDSS, 218–221
  CROPS, 527–536
  expert systems, 234–235
  multiattribute decision making, 269–279
  rangeland prototype DSS, 363–376
United States Federal Facilities
  cleanup risk management, 133–140
  Federal Facilities Agreement (FFA), 121
  landfill cover design evaluation, 153–164
  National Environmental Policy Act, 541–554
  Oak Ridge forested ecosystem ecological risk assessment approach, 121–132
United States Forestry Service INFORMS-R8 reasoning toolkit, 251–268, 255
Unmanageable factor identification, 178
Upland regions of Indonesia, 395–405, 468
User, DSS structuring, 326, 327–328
User interface
  DSS, 333
  SIWARE, 638
User-selected goals, agricultural systems, 180

V

Valuation of land, crop model use for, 455–462
Values
  Boobera Lagoon in Aboriginal tradition, 657–658, 663
  conjoint analysis of public inputs, 493–500
  existence value, 481–490
  noneconomic, 503–512
    aesthetic, 508–509
    concept of value, 504–505
    economic value and, 505–506
    moral, 507–508
    rational, 511
    role in decision making, 511–512
    role of value in resource management, 503
    spiritual, 510

W

WATDIS, 631
Water, 6, *see also specific MODS approaches*
Water management scenario simulation, 629–638
Water Quality Decision Support Service (WQDSS), 218–221
Water Quality DSS, quantification of incentives, 516–525
Water Resources Management Decision Support System (WRMDSS), 106, 107
WATERSHEDDS, 241–248
  development and objectives of, 242–243
  structure and components, 243–247
    bibliographic links, 246
    database, BMP, 246
    education and information sections, 245–246
    instructions for use, 247–248
    modeling tool, 246
    pollutant budget spreadsheet, 247
    user interface, 243–244
Watersheds
  in Egypt, 13, 18–19, 629–638
    degradation assessment, 15
    land-use planning, 14
    large river basins, 13, 18–19
  in India, resource management, 615–625
  Scioto River, 337–346
Water use, short-term and single-objective decision making effects, 61–72
WDUTY, 630
Weather, *see* Climate
Weed control
  community-based approaches to management, 57–58
  pesticides, *see* Pesticide Economic and Environmental Tradeoffs (PEET)
Weeds
  in Indonesia, 468
  loss calculations, 85–90
Weighted goal programming, aquaculture modeling, 642, 644–652
Weighting, *see also specific DSS approaches*
  goal programming, 644–649
  SOILCROP, 352, 353,3 54
Wetlands trading system, Scioto River study, 344, 345–346
WOFOST, crop models for valuation of land, 461

Y

Yellow page approach to cooperative expert system, 181
Yield data
  nutrient management, 177–184
  PEET data, 90, 91
  SMART PITCHFORK, 192
  steep land cropping calculations, 95–96